Lecture Notes in Electrical Engineering

Volume 778

The book series *Lecture Notes in Electrical Engineering* (LNEE) publishes the latest developments in Electrical Engineering - quickly, informally and in high quality. While original research reported in proceedings and monographs has traditionally formed the core of LNEE, we also encourage authors to submit books devoted to supporting student education and professional training in the various fields and applications areas of electrical engineering. The series cover classical and emerging topics concerning:

- Communication Engineering, Information Theory and Networks
- Electronics Engineering and Microelectronics
- Signal, Image and Speech Processing
- Wireless and Mobile Communication
- Circuits and Systems
- Energy Systems, Power Electronics and Electrical Machines
- Electro-optical Engineering
- Instrumentation Engineering
- Avionics Engineering
- Control Systems
- Internet-of-Things and Cybersecurity
- Biomedical Devices, MEMS and NEMS

For general information about this book series, comments or suggestions, please contact leontina.dicecco@springer.com.

To submit a proposal or request further information, please contact the Publishing Editor in your country:

China

Jasmine Dou, Editor (jasmine.dou@springer.com)

India, Japan, Rest of Asia

Swati Meherishi, Editorial Director (Swati.Meherishi@springer.com)

Southeast Asia, Australia, New Zealand

Ramesh Nath Premnath, Editor (ramesh.premnath@springernature.com)

USA, Canada:

Michael Luby, Senior Editor (michael.luby@springer.com)

All other Countries:

Leontina Di Cecco, Senior Editor (leontina.dicecco@springer.com)

**** This series is indexed by EI Compendex and Scopus databases. ****

More information about this series at http://www.springer.com/series/7818

Ankur Choudhary · Arun Prakash Agrawal ·
Rajasvaran Logeswaran · Bhuvan Unhelkar
Editors

Applications of Artificial Intelligence and Machine Learning

Select Proceedings of ICAAAIML 2020

Editors
Ankur Choudhary
Department of Computer Science
and Engineering
Sharda University
Greater Noida, Uttar Pradesh, India

Arun Prakash Agrawal
Department of Computer Science
and Engineering
Sharda University
Greater Noida, Uttar Pradesh, India

Rajasvaran Logeswaran
Asia Pacific Centre for Analytics (APCA),
Asia Pacific University of Technology
and Innovation (APU)
Kuala Lumpur
Malaysia

Bhuvan Unhelkar
Information Technology
University of South Florida Sarasota–
Manatee Campus
Sarasota, FL, USA

ISSN 1876-1100 ISSN 1876-1119 (electronic)
Lecture Notes in Electrical Engineering
ISBN 978-981-16-3069-9 ISBN 978-981-16-3067-5 (eBook)
https://doi.org/10.1007/978-981-16-3067-5

This Springer imprint is published by the registered company Springer Nature Singapore Pte Ltd.
The registered company address is: 152 Beach Road, #21-01/04 Gateway East, Singapore 189721, Singapore

ICAAAIML-2020

General Chair

Prof. Parma Nand, Dean, School of Engineering & Technology, Sharda University, Greater Noida, India
Prof. Rajasvaran Logeswaran, Head, Asia Pacific Centre for Analytics (APCA), Asia Pacific University of Technology and Innovation (APU), Kuala Lumpur, Malaysia
Prof. Bhuvan Unhelkar, Professor, University of South Florida, USA

General Co-Chair

Prof. Nitin Rakesh, Head of Department, Department of CSE, Sharda University, Greater Noida, India

Convener & Conference Chair

Prof. Ankur Choudhary, School of Engineering & Technology, Sharda University, Greater Noida, India
Prof. Arun Prakash Agrawal, School of Engineering & Technology, Sharda University, Greater Noida, India

Organizing Chairs

Prof. Rani Astya, SET, Sharda University, Greater Noida, India
Prof. Gaurav Raj, SET, Sharda University, Greater Noida, India
Prof. Abhishek Singh Verma, SET, Sharda University, Greater Noida, India

Publicity Chairs

Prof. Vishal Jain, SET, Sharda University, Greater Noida, India
Prof. Ranjeet Rout, National Institute of Technology, Srinagar, India

Preface

Artificial intelligence (AI) has become a buzzword in the last two decades. The capability of digital computers or computer-controlled robots to perform activities, viz. reasoning, finding importance, summing up, and gaining knowledge, from experience is called artificial intelligence. Systems that are blessed with intellectual processes—earlier considered a characteristic of humans only—are called artificially intelligent systems. Unlike passive machines, artificially intelligent algorithms make decisions by frequently utilizing real-time data. AI systems join the data from a wide range of sources—sensors, remote inputs, or digital information—analyze the data instantly, derive insights from this data, and follow it up. AI is an exciting horizon in the world of computer science, empowering technology to advance in ways never before possible.

Machine learning (ML), deep learning, and neural networks are the three fundamental concepts of AI. While AI and machine learning are sometimes considered interchangeable terms, AI covers a broad domain with the rest of the terms as a subset of it. Machine learning is a part of AI that enables machines to learn without explicitly programming them to perform a task.

Keeping in view the importance of AI and ML in today's era, ICAAAIML'2020 provided a forum for researchers, engineers, and practitioners from computer science, data analytics, medical informatics, biomedical engineering, healthcare engineering, and other engineering disciplines to share and exchange their knowledge and progresses of current research issues, technologies, and recent advances and applications of AI in various domains.

This proceedings volume includes 54 articles selected from those presented at ICAAAIML'2020, addressing a wide spectrum of important issues such as artificial intelligence and its applications in smart education, big data and data mining, challenges of smart cities future research directions, image/video processing, machine learning applications in smart healthcare, manufacturing, security and privacy challenges and data analytics, soft computing and smart infrastructure and resource development and management using artificial intelligence and machine learning. The editors would like to thank all the authors for their excellent work and the reviewers from all over the world for their valuable critiques and commitment to help the authors in improving the quality of their manuscripts. Special thanks are extended to Springer for publishing this proceedings volume and to our families for their support.

Greater Noida, India — Ankur Choudhary
Greater Noida, India — Arun Prakash Agrawal
Kuala Lumpur, Malaysia — Rajasvaran Logeswaran
Sarasota, USA — Bhuvan Unhelkar

Contents

About the Editors

Ankur Choudhary is a Professor of the Department of Computer Science and Engineering at Sharda University, India. He completed his Ph.D. from Gautam Buddha University (GBU), India. His areas of research are nature-inspired optimization, artificial intelligence, software engineering, medical image processing, and digital watermarking. Dr. Choudhary has over 15 years of academic and research experience. He has published several research papers in conferences and journals. He is associated with various International Journals as a reviewer and editorial board member.

Arun Prakash Agrawal is currently a Professor with the Department of Computer Science and Engineering at Sharda University, India. He obtained his Masters and Ph.D. in Computer Science and Engineering from Guru Gobind Singh Indraprastha University, New Delhi, India. He has several research papers to his credit in refereed journals and conferences of international repute. He has also taught short-term courses at the Swinburne University of Technology, Melbourne, Australia, and Amity Global Business School, Singapore. His research interests include machine learning, software testing, artificial intelligence, and soft computing. He is an active professional member of the IEEE and ACM.

Rajasvaran Logeswaran is Head of the Asia Pacific Centre for Analytics and full Professor of Computing and Engineering at Asia Pacific University of Technology and Innovation (APU), Malaysia. He got his B.E. (Hons) Computing at Imperial College London, the United Kingdom. He completed his Masters and Ph.D. from Multimedia University, Malaysia and post-doctoral research in Korea. His research interest areas are image processing, data compression, neural networks, and data science. Dr. Logeswaran has over 150 publications in books, peer-reviewed journals, and international conference proceedings to his credit. Prof. Logeswaran has been a recipient of several scholarships and awards. He is a Senior Member of the IEEE and Chair of the award-winning IEEE Signal Processing Society Malaysia Chapter.

Bhuvan Unhelkar (B.E., MDBA, MSc, Ph.D.) is an accomplished IT professional and Professor of IT at the University of South Florida at their Sarasota-Manatee campus. He is also Founding Consultant at MethodScience and PlatiFi, with mastery in business analysis and requirements modeling, software engineering, big data strategies, agile processes, mobile business, and green IT. He has a Doctorate in the area of "Object Orientation" from the University of Technology, Sydney, in 1997. His areas of expertise include big data strategies, agile processes, business analysis and requirements modeling, corporate agile development, and quality assurance and testing. His industry experience includes banking, finance, insurance, government, and telecommunications where he develops and applies industry-specific process maps, business transformation approaches, capability enhancement, and quality strategies. Dr. Unhelkar has authored numerous executive reports, journal articles, and 20 books with internationally reputed publishers.

Artificial Intelligence and Its Applications in Smart Education

Building a Language Data Set in Telugu Using Machine Learning Techniques to Address Suicidal Ideation and Behaviors in Adolescents

K. Soumya and Vijay Kumar Garg

Abstract Taking one's own life is a tragic reaction to stressful situations in life. There is a noticeable increase in the ratio of number of suicides every year in Telangana [1]. Most of them are adolescents and youngsters and others too. So there is an urging need of research to be done on suicidal ideation and preventive methods to support mental health professionals and psychotherapists. So this paper aims in developing technological solutions to the problem. Suicides can be prevented if we could identify the mental health conditions of a person with ideations and predict the severity in earlier [2]. So in this paper, we applied machine learning algorithms to categorize persons with suicidal ideations from the data that is maintained or recorded during visit of an adolescent with a mental health professional in textual form of questionnaires. The data is recorded in native Telugu language during the session, as most of cases are from illiterates [1, 3]. So in order to classify the patient test data with more accuracy, there is a need of language corpus in Telugu with ideations. So this paper would give a great insight into creation of suicidal language or ideation corpora in native language Telugu.

Keywords Machine learning · Classification algorithms · Telugu data set · Telugu corpus · Suicidal ideation · Mental illness

1 Introduction

Suicide is one of the major and leading cause of deaths happening in the Telangana [1]. And, it is mostly the poor and less-literate people who committed suicide, according to NCRB data show [3]. Reasons for the suicide may be different from case to case,

K. Soumya (✉) · V. K. Garg
Computer Science and Engineering Department, Lovely Professional University, Phagwara, Punjab, India

V. K. Garg
e-mail: vijay.garg@lpu.cop.in

K. Soumya
VBIT, Hyderabad, India

A. Choudhary et al. (eds.), *Applications of Artificial Intelligence and Machine Learning*, Lecture Notes in Electrical Engineering 778,
https://doi.org/10.1007/978-981-16-3067-5_1

but it could be reduced with proper and prior prediction in most cases. Suicidal ideation is perceived as proneness of a person to end one's own life. There is a lot of study going on the unnatural deaths happening due to suicides in India.

Thoughts that may lead to suicide include for example:

- Looking dull and depressed most of the time
- Frequently expressing the feeling like killing one's self, or willing to die, or hatred toward his/her own birth in simple situations
- Feeling hopeless every time in simple circumstances
- Change in the normal life style
- Doing unhealthy things
- Having frequent mood swings
- Getting the means to kill one's life like pills.

2 Suicidal Data Set Creation in Telugu

First step of any psychotherapy is knowing the reasons for suicidal ideation of a person, and it is conducted in native language. In order to do further analysis on documented data or gathered data by applying advanced ML techniques, there must be a specific language data set consisting of suicidal ideation and symptom related sentences and words in Telugu.

But even today, there are less resources available in Telugu in terms of tools, data set or linguistic corpus, so that any advanced techniques could be applied on it [4]. So in order to predict the severity of ideation among adolescents using ML techniques or any other, data set must be created.

3 Technique Used for Data Gathering to Create Required Data Set

Because of the scarcity of data availability in Telugu, we used web scraping in order to create it [5].

Web scraping is also termed as web data extraction, web harvesting or screen scraping too, in which large amount of data posted in Internet web is downloaded and can be saved into the local drive and later can be used for different purposes (Fig. 1).

There are various software's and tools to perform web scraping, but each and individual tool is application specific. So we used Scrapestorm tool here for our task of text scraping.

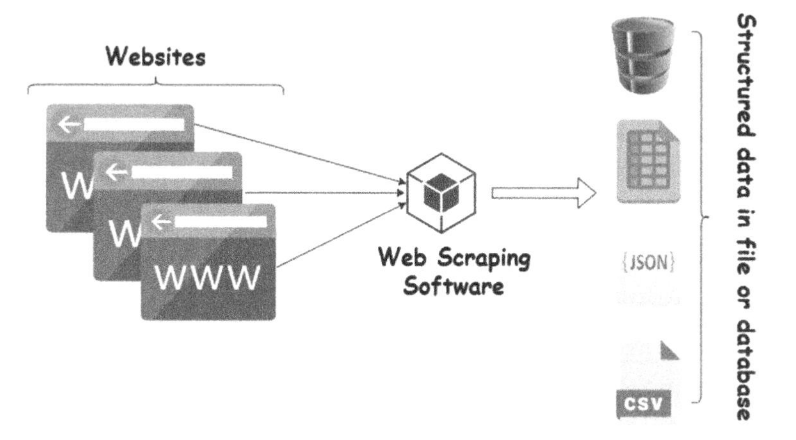

Fig. 1 Scraping data from website

4 Data Collection

Here, we would like to present about resources from where raw data were accumulated and used in data set creation. There are only few resources available in Telugu language which consists of complete word set of the language. So it is must to depend on other sources and create the required data set. And that also may not cover all the word sets. As social media and Internet is the major resource for all kinds of Telugu text for various stories [6]. Collected data from the above said sources them and filtered the required sentences out of it. And extracted sentences were preprocessed to remove headings, sub-headings, extra symbols, any non-Telugu characters, etc. This process can be depicted as follows (Fig. 2).

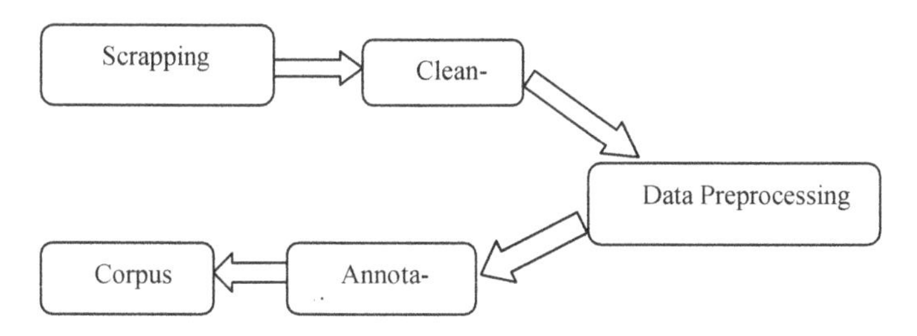

Fig. 2 Process of creating corpus

4.1 Scrapping

As there are only few websites that are created specially to publish in Telugu text form, scrapping is done on few Telugu stories websites, and te.wikipedia.org/wiki/ఆత్మహత్యand news websites too. Like kathalu.wordpress.com, sukatha.com, eenadu and sakshi.net [7].

4.2 Cleaning

200 articles were collected from news website, and almost 13 stories were collected which were based on ideation. And later extracted data was preprocessed for noise reduction, eliminating extra sentences which are not relevant to data set, and removing headings, etc. Then as part of preprocessing, required featured sentences about suicidal ideation are populated.

4.3 Annotation

Annotation is a requisite step of data preprocessing, where text in the sentences is tagged with its description too. Descriptions are based on the requirement. There are various types of annotation. Like

- Semantic annotation
- Image and video annotation
- Text categorization and content categorization
- Entity annotation
- Entity linking
- Phrase linking.

Text categorization and entity annotations are used in building suicidal ideation data set here. Where in text categorization, sentences filtered are considered and assigned with predefined categories. Categories were predefined very carefully that would describe the symptoms of suicidal ideation. Then finally in entity annotation, sentences are labeled with above said categories. As an outcome of sentences is tagged with ideation polarities, few polarities are shown in Table 1.

For example, the following sentence should be tagged as suicidal ideation as it clearly indicates or expresses the ideation.

Telugu: "నా వాళ్ళు నన్ను వదిలి వెళ్ళిపోయారు ఇక నాకు బ్రతకాలని లేదు".

English: They left me. I don't want to live anymore.

Table 1 Words and categories that describe suicidal ideation

ఆత్మాహుతి	ఆత్మహత్య	నన్ను చంపుకున్నా
నాజీవితాన్ని ముగించండి	నా సూసైడ్ నోట్	శాశ్వత నిద్ర
నాకు బ్రతకాలని లేదు	నాకు జీవించే హక్కు లేదు	ఒంటరిగా చావడం
నా మణికట్టు కోసి	జీవితం మీద విసుగు	చావడానికి సిద్ధం
ఉరి వేసుకోవడం	బలవన్మరణాం	మనస్థాపాం

4.4 Data Set Statistics

From the selected 200 articles and 13 stories and 40 sentences that describe depressive state of mind and suicidal ideation in Telugu from wiki. From the above count of articles from various sources, a total of 4800 sentences were filtered and tested for polarity and annotation. The process is explained above, where still some sentences were filtered and finally could find 4200 sentences for data set.

After the annotation reliability of the annotations is also measured using Cohen kappa coefficient, which is a statistic used to measure the effectiveness of annotation by categories called inter-rater reliability and is calculated using the following formulae [8, 9].

$$k = \frac{p_0 - p_e}{1 - p_e} \tag{1}$$

where p_0 is relative observed agreement, p_e is agreement by chance. Normal accepted range of k value is 0.6–0.8. So as result of calculation obtained 0.62.

5 Methodology

Here, the procedure or steps involved in creating a corpus which could then be used for (1) training a model and (2) to create a doc2vec to be used in machine translation model.

So, the steps involved are

Step 1: Acquire the raw data

Step 2: Annotate the corpus obtained by applying web scraping on various resources (as mentioned above) with defined polarities.

Step 3: Training raw data to doc2vec model that results in a sentence vector.

Step 4: Perform binary classification using various ML classifiers.

Step 5: Validate the classifier with more iterations on data.

Step 6: Choose the classifier with best performance and end (Fig. 3).

Now let us look into the steps briefly.

- Acquire raw data—the raw data in Telugu was obtained from "Indian Languages Corpora Initiative (ILCI)", that consisted of more than 6laks raw sentences in Telugu.
- Annotations of raw data is done as described in the previous sections.
- Training raw data to doc2vec model is done with the help of Gensim tool in Python for topic modeling that gives sentence vector as an outcome [10].
- Applied various binary classifiers of ML like, logistic regression, SVM, decision trees, and random forest by using scikit-learn tool kit to appropriately label the data.

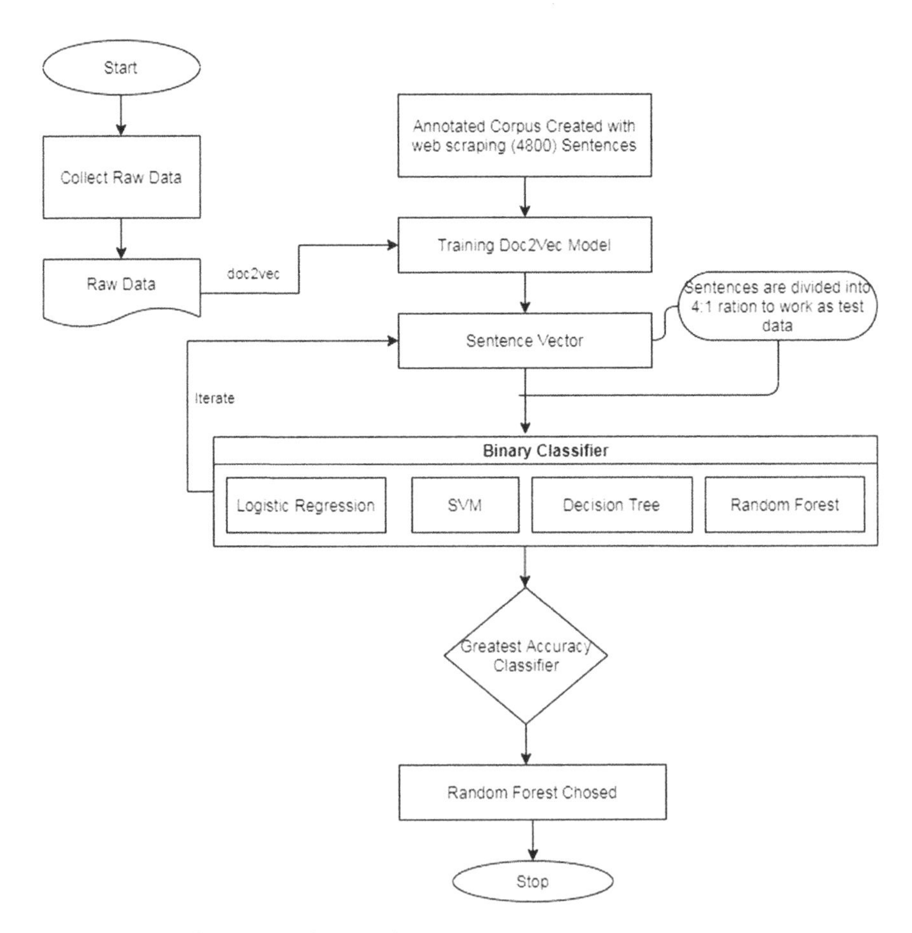

Fig. 3 Flowchart for complete framework

- Validated the accuracy of labeled data by more iterations done on test data selected randomly [11] from the corpus with all binary classifiers.
- As random forest has performed well in all iterations, it is chosen for identifying a test data to classify as with ideation or without ideation.

6 Experiments and Results

As shown in the above flowchart, we performed experiments on corpus created by applying various machine learning classifiers to find out which classifier could label it perfectly for all test data in all iterations [12, 13].

6.1 Accuracy by ML Classifiers

Various machine learning classifiers are applied on the corpus sentences by splitting them into test set and training data sets. And in multiple iterations, they performed with following accuracies. So the algorithm with highest accuracy is chosen (Fig. 4; Table 2).

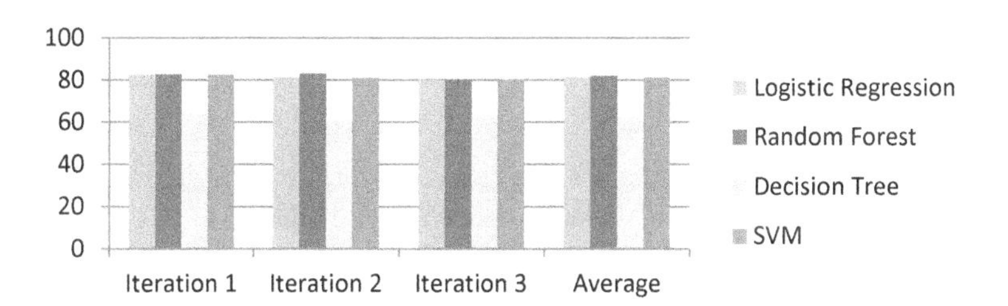

Fig. 4 Performance of various ML classifiers

Table 2 Iterations and the result of accuracy by ML classifiers

	Iteration 1	Iteration 2	Iteration 3	Average
Logistic regression	82.71	81.46	80.25	81.47
Random forest	82.82	83.15	80.19	82.05
Decision tree	64.23	61.03	62.96	62.74
SVM	82.60	81.15	80.31	81.35

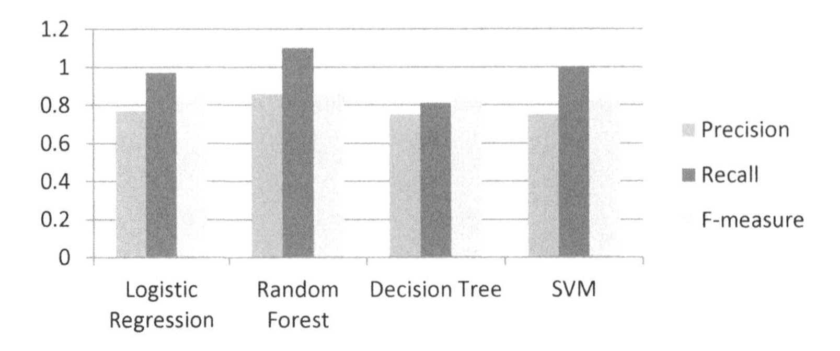

Fig. 5 Precision, Recall, F-measure for binary classifiers

Table 3 Observations of measures

	Precision	Recall	F-measure
Logistic regression	0.77	0.97	0.85
Random forest	0.86	1.1	0.86
Decision tree	0.75	0.81	0.83
SVM	0.75	1.0	0.83

6.2 *Binary Classification Measures*

Precision, Recall and F-measures of binary classification are used in experiments.
Where

Precision and Recall are numerical evaluations of algorithms that gives a single real value as an outcome to indicate the performance (Fig. 5; Table 3).

Precision formulae is:

$$\text{Precision} = \frac{\text{True Positives}}{(\text{True Positives} + \text{False Positives})} \tag{2}$$

Recall formulae is:

$$\text{Recall} = \frac{\text{True Positives}}{(\text{True Positives} + \text{False Negatives})} \tag{3}$$

F-measure formulae is:

$$f1 = \frac{(2 * \text{Precision} * \text{Recall})}{(\text{Precision}/\text{Recall})} \tag{4}$$

7 Conclusion

In this paper, we described the process of corpora creation in Telugu, all the steps involved in collecting raw data, processing it to filter and annotate, finally creating the corpora. And in this paper, we found that random forest performs well on the corpus created to properly classify between ideation and non-ideation sentences.

8 Future Work

There is a lot requirement of a bilingual corpora for Telugu—English or English—Telugu with more number of sentences that would help in further research work on most used native language of Telangana. Many other techniques can be applied on corpus to validate the corpus. The validation using nearest neighborhood approach, sentence similarity and neural machine translation techniques can be applied.

References

1. Nilesh V (2019) Telangana has third-highest suicide rate in India. NCRB. https://www.newindianexpress.com/states/telangana/2019/nov/11/telangana-has-third-highest-suicide-rate-in-india-ncrb-2060087.html
2. Choudhary N, Singh R, Bindlish I, Shrivastava M (2018a) Emotions are universal: learning sentiment based representations of resource-poor languages using siamese networks. arXiv preprint arXiv:1804.00805
3. Rohit PS, State records highest suicide rate in country. https://www.thehindu.com/news/national/telangana/state-records-highest-suicide-rate-in-country/article8433720.ece#comments_14219168
4. Naidu R, Bharti SK, Babu KS, Mohapatra RK (2017) Sentiment analysis using Telugu SentiWordNet. In: 2017 international conference on wireless communications, signal processing and networking (WiSPNET), Chennai, pp 666–670. https://doi.org/10.1109/wispnet.2017.8299844
5. Magdum D, Dubey MS, Patil T, Shah R, Belhe S, Kulkarni M (2015) Methodology for designing and creating Hindi speech corpus. In: 2015 international conference on signal processing and communication engineering systems, Guntur, pp 336–339. https://doi.org/10.1109/spaces.2015.7058279
6. Gangula RR, Mamidi R (2018) Resource creation towards automated sentiment analysis in Telugu (a low resource language) and integrating multiple domain sources to enhance sentiment prediction. In: Conference: language resources and evaluation conference, At Miyazaki (Japan)
7. Srirangam V, Abhinav A, Singh V, Shrivastava M (2019) Corpus creation and analysis for named entity recognition in Telugu-English code-mixed social media data. In: Proceedings of the 57th Annual Meeting of the Association for Computational Linguistics: Student Research Workshop. https://doi.org/10.18653/v1/p19-2025
8. Abdelali A, Guzman F, Sajjad H, Vogel S (2014) The AMARA corpus: Building parallel language resources for the educational domain. In: Proceedings of the ninth international conference on language resources and evaluation (LREC'14). European Language Resources Association (ELRA), Reykjavik, Iceland, pp 1856–1862

9. Lu X (2017) Automated measurement of syntactic complexity in corpus-based L2 writing research and implications for writing assessment. SAGE J. https://doi.org/10.1177/0265532217710675
10. Choi Y, Wiebe J (2014) Effectwordnet: sense-level lexicon acquisition for opinion inference. In: Proceedings of the 2014 conference on empirical methods in natural language processing (EMNLP), pp 1181–1191
11. Wołk K, Marasek K (2014) A sentence meaning based alignment method for parallel text corpora preparation. Adv Intell Syst Comput 275:107–114. arXiv:1509.09090
12. Chen T, Guestrin C (2016) Xgboost: a scalable tree boosting system. In: Proceedings of the 22Nd ACM SIGKDD international conference on knowledge discovery and data mining. ACM, pp 785–794
13. Aguilar WG, Alulema D, Limaico A, Sandoval D (2017) Development and verification of a verbal corpus based on natural language for Ecuadorian dialect. In: IEEE 11th International Conference on Semantic Computing (ICSC), San Diego, CA, 2017, pp 515–519. https://doi.org/10.1109/icsc.2017.82
14. Choudhary N, Singh R, Bindlish I, Shrivastava M (2018b) Sentiment analysis of code-mixed languages leveraging resource rich languages. arXiv preprint arXiv:1804.00806

Feature Selection and Performance Comparison of Various Machine Learning Classifiers for Analyzing Students' Performance Using Rapid Miner

Vikas Rattan, Varun Malik, Ruchi Mittal, Jaiteg Singh, and Pawan Kumar Chand

Abstract Information technology revolution and affordable cost of storage devices and Internet usage tariff have made it easy for educational bodies to collect data of every stake holder involved in. This collected data has many hidden facts, and, if extracted, it can give new insights to every concerned contributor. The educational bodies can use educational data mining to examine and predict the performance of students which helps them to take remedial action for weaker students. In education data mining, classification is the most popular technique. In this paper, emphasis is on predicting students' performance using various machine learning classifiers and the comparative analysis of performance of learning classifiers on an educational dataset.

Keywords Data mining · Student performance prediction · Feature selection random tree · KNN · Random forest · Naïve Bayes

V. Rattan · V. Malik · R. Mittal (✉) · J. Singh
Chitkara University Institute of Engineering and Technology, Chitkara University, Rajpura, Punjab, India
e-mail: ruchi.mittal@chitkara.edu.in

V. Rattan
e-mail: vikas.rattan@chitkara.edu.in

V. Malik
e-mail: varun.malik@chitkara.edu.in

J. Singh
e-mail: jaiteg.singh@chitkara.edu.in

P. K. Chand
Chitkara Business School, Chitkara University, Rajpura, Punjab, India
e-mail: pawan.chand@chitkara.edu.in

A. Choudhary et al. (eds.), *Applications of Artificial Intelligence and Machine Learning*, Lecture Notes in Electrical Engineering 778,
https://doi.org/10.1007/978-981-16-3067-5_2

1 Introduction

Data mining (DM) is one of the consistently developing and very much perceived field of computer science. DM is a process of discovering knowledge in the form of useful and unknown patterns from databases [1]. DM has been applied in an extraordinary number of fields, including retail deals, bioinformatics, banking, etc. Information technology (IT) rapid evolution and ease of data collection means made almost every organization relying on IT [2]. There has been expanding enthusiasm for the utilization of data mining to examine scientific questions within educational settings is known as educational data mining (EDM). Lately, analysts from different areas have begun to investigate the utilization of data mining to encourage research in the area of education for the assistance to all the partners and particularly the students [3]; along these lines, EDM is an undeniably rising field of study that utilizes data mining, artificial intelligence (AI) and statistical methods to investigate and break down the data and information gathered for educational research [4]. The most progressive educational institutes of the present era regularly use data mining models to inspect the data gathered and to extricate information and knowledge to keep themselves abreast and reactive to current dynamics [5]. The fundamental focus is to comprehend the area of EDM from alternate points of view and afterward to create strategies to comprehend the students and their learning ecosystem [6]. With the expanded utilization of educational ERP software, and other allied learning tools like MOODLE, Blackboard, educational bodies are assembling the enormous amount of data adding to massive data stores [7]. This data can be considered as the most significant resource and can be utilized to get some significant bytes of knowledge to support all the stake holders and the overall education system [8].

Forecasting the performance of students is one of the most popular applications of EDM and prominent practitioners in the area of such data analysis treat this process as a combination of multiple disciplines such as statistics, text and emotion analysis, descriptive and predictive tasks, etc. [9, 10] with an objective to comprehend students' learning behavior and anticipate their knowledge ingestion [11]. However, anticipating student performance is not simple, and many attributes are there to influence students' performance. These variables may include learner family background, past academic performance, and interaction between student and educator [12]. This paper aims to analyze the student performance dataset taken from UCI repository [13, 14] for predicting student's performance using various machine learning classifiers such as random tree, random forest, Naïve Bayes, and KNN and for comparing the predictive accuracy of various classifiers using the educational version of Rapid Miner 9.7.

A brief introduction of various classifiers is given below:

Random Forest: This learning classifier follows the ensemble approach and can tackle problems of both classification. It is a multitude of decision trees where each of the trees is formed from different samples and subsets of the training data with the same parameters [15]. Information gain ratio criteria are chosen for the best splitting attribute. Here, the outcome of the random forest is decided by confidence voting.

Random Tree: It works on the functionality of the CART decision tree with just one difference that for each split only a random subset of attributes is available. Here also, we chose information gain ratio criteria for best splitting attribute minimum leaf size and minimum size for splitting are kept 4 and 2, respectively.

Naïve Bayes: This algorithm uses probability theory particularly Bayes theorem for classification problems. It is relatively very quick to classify objects based on certain features in the large dataset [16] using Eq. (1).

$$P(TV|F1,\ F2\ldots Fn) = \frac{P(F1|TV)P(F2|TV)\ldots P(Fn|TV)P(TV)}{P(F1)P(F2)\ldots P(Fn)} \tag{1}$$

where TV is a target variable whose label will be predicted based on $F1$, $F2$, ..., Fn.

K-Nearest Neighbor: K-nearest neighbor algorithm works on the principle of "similar things exist nearby." KNN derives the proximity of data points with other points by calculating the distances between data points. It comes under the category of lazy learning algorithm because it does not learn anything from training data. Rather at the time of evaluation, data object class is determined by calculating k closet neighbors [17]. In this study, we used Euclidian distance and K is put to 10.

2 Literature Review

Mudasir Ashrafa investigates and concludes that the performance of various classifiers can be based on the use of various filtering approaches, SMOTE, under-sampling technique, or ensemble methods have a considerable impact on the prediction accuracy of learning classifiers [18]. Preet Kamal et.al developed a prediction ensemble model using NB, DT, and SVM. They found that consumption of alcohol, tobacco, and mishap in the past year affects the performance of student [19].

Pornthep Rojanavasu applied Apriori algorithm and DT for the extraction of important facts about admission planning and predict whether the student will get the job in the IT sector or non IT sector [20]. Bindhia K. Francis et al. predict the student performance using various demographic attributes, behavioral features, and academic-related features using SVM, NB, DT, neural network (NN), and observed significant relationship between academic behavior and performance [21].

Authors estimated the ability of student's homework problems solving in the MOOC platform using item response theory (IRT) and TrueSkill. Data mining models were developed and compared on the basis of IRT and TrueSkill. Both IRT and TrueSkill-based data mining models showed a comparable predictive power for large datasets [22]. Libor Juhanak analyzed students' behavior in online learning activities and detected specific patterns of interaction in learning management software (LMS) using process mining. During analysis of event log of LMS, it was found that the standard quiz taking behavior of students was different from non-standard quiz taking behavior of students [23].

David Azcona et al. developed a predictive model using used k-nearest neighbor (KNN), random forest, logistic regression, linear SVM, Gaussian SVM to automatically detect students "at risk" in computer programming subjects based on their weekly assignment and to simultaneously support adaptive feedback [24]. Kabra R R and Bichkar R S predicted unsuccessful first-year engineering students with the 90.7% accuracy [25].

Raheela Asif et al. predict the performance of undergrad students using a decision tree. Predictions were made only based on their marks in the first four semesters and found from the decision tree that few of the subjects of the first and second semesters were the best indicators of poor and bad performance in the degree program [26]. Anne-Sophie Hoffait et al. developed models using random forest, logistic regression, and artificial neural networks algorithms to identify the difficulties faced by fresher students in their first academic year in university and thus to identify the students, who were at high risk of failure. Model formed using random forest was able to identify 21.2% of the students facing a high risk of failure [27].

Ali Daud et al. predicted the academic performance of the students based on family income, family expenditure, family assets, and student personal information. It was discovered that the rise in the expenditures of family shrinks the opportunities for a student to grow up and excel in their studies and model using SVM was declared best for the given attributes [28].

3 Research Methodology

The dataset for this study has been derived from UCI repository [13] and was understood using metadata and visualization. Initially, this dataset was having 131 instances and 22 attributes and the description of the attributes is shown in Fig. 1.

Now, during the pre-processing and transformation stages, marital status attribute is dropped because every instance is having the same status; secondly, end semester performance (ESP) is chosen as a target variable; and thirdly, among the remaining 20 attributes, 8 best attributes (ARR, AS, ATD, IAP, ME, SH, TNP, TWP) have been chosen based on information gain ratio for developing model as shown in Fig. 2.

In this experiment, to find out the set of relevant attributes for analysis, first of all, data is imported in the local repository of Rapid Miner in the repository panel. After importing the data, feature weights are calculated for relevant features using various available operators like weight by information gain ratio, weight by information gain, weight by chi-square, weight by RELIEF, and optimized selection (evolutionary) as shown in Fig. 2.

After data cleaning and transformation, various machine learning classifiers are used on the final dataset such as random forest, random tree, k-nearest neighbor and Naïve Bayes to classify the dataset and the approach used for classification is kept to be cross-validation using the education version on Rapid miner 9.7 tool. We have

Attributes	Description	Possible values
Gender	Student's sex	{M, F}
CST	Student's category	{G, MOBC, OBC, SC, ST}
TNP	10th class percentage	{Best, Very good, Good, Pass, Fail} If percentage >= 80 then Best If percentage >= 60 but less than 80 then Very good If percentage >= 45 but less than 60 then Good If percentage >= 30 but less than 45 then Pass Percentage < 30 then Fail
TWP	12th class percentage	{Best, Very good, Good, Pass, Fail} Same as TNP
IAP	Internal assessment percentage	{Best, Very good, Good, Pass, Fail} Same as TNP
ESP	End semester percentage	{Best, Very good, Good, Pass, Fail} Same as TNP
ARR	Compartment status	{Yes, No}
MS	Marital status	{Married, Unmarried}
LS	Location of living	{Rural, Urban}
AS	Admission seat	{Free, Paid}
FMI	Family income	{Very high, High, Above medium, Medium, Low} If EMI >= 30000 then very high If EMI >= 20000 but less than 30000 then High If EMI >= 10000 but less than 20000 then Above medium If EMI >= 5000 but less than 10000 then Medium If EMI is less than 5000 then Low
FS	Family size	{Large, Average, Small} If FS > 12 then Large If FS >= 6 but less than 12 then Average If FS < 6 then Small
FQ	Father's qualification	{IL, UM, 10, 12, Degree, PG} IL = Illiterate UM = Under class X
MQ	Mother's qualification	{IL, UM, 10, 12, Degree, PG} IL = Illiterate UM = Under class X
FO	Father's occupation	{Service, Business, Retired, Farmer, Others}
MO	Mother's occupation	{Service, Business, Retired, Farmer, Others}
NF	Number of friends	{Large, Average, Small} If NF > 12 then Large If NF >= 6 but less than 12 then Average If NF < 6 then Small
SH	Study hours	{Good, Average, Poor} >= 6 hours Good >= 4 hours Average < 2 hours Poor
SS	Category of school attended at 10th class	{Govt., Private}
ME	Medium of instruction in 10th class	{Eng, Asm, Hin, Ben}
TT	Home to college travel time	{Large, Average, Small} >= 2 hours Large >= 1 hours Average < 1 hour Small
ATTD	Class attendance percentage	{Good, Average, Poor} If percentage >= 80 then Good If percentage >= 60 but less than 80 then Average If percentage < 60 then Poor

Fig. 1 Description of student performance dataset [14]

emphasized on supervised learning technique for analysis and have chosen to run the experiment using various classifiers such as KNN, random tree, random forest, and Naïve Bayes as they are the most widely used approaches, and moreover, the attributes in our dataset are nominal and polynomial in nature.

4 Experiment and Results

In the experiment, various machine learning classifiers are trained and tested using 10/15/20 folds cross-validation in rapid miner using random tree operator, KNN operator, random forest operator, and Naïve Bayes operator along with cross-validation operator. ESP is a polynomial target variable having possible values (pass, good, very good, best), given dataset is having 27 instances with value "pass," 54 instances with

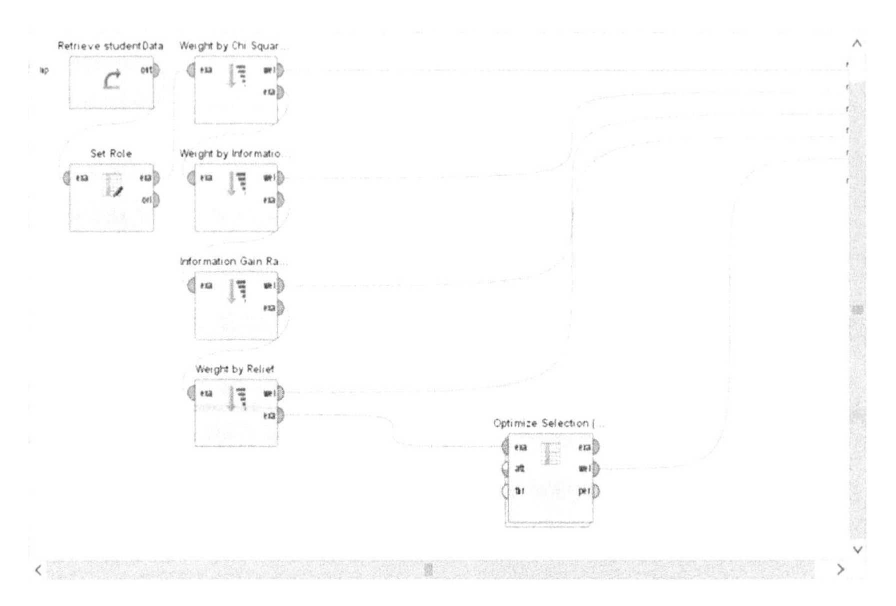

Fig. 2 Demonstration of feature selection on students' dataset

value "good," 42 instances with value "very good," and 8 instances with the value "best." Therefore, stratified sampling is used to preserve the imbalanced class distribution in each fold. Cross-validation is the best technique to test the effectiveness of the model [29]. Results drawn from various machine learning classifiers in the form of correctly, incorrectly classified instances, and their respective percentages are shown in Table 1.

It is observed that all four classifiers as shown in Table 1 have shown the accuracies for correctly classified instances in the range of 55–67%. The results have shown that the Naïve Bayes algorithm is resulting in maximum accuracy of 66.41% with 15-fold cross-validation and is showing less accuracy with 20%. This is due to the problem of imbalanced distribution of the class variable (ESP) in the overall dataset.

The results have shown that Naïve Bayes algorithm is classifying 87 instances out of a total of 131 instances as correctly classified with 15-fold cross-validation as shown in Fig. 3, whereas KNN algorithm is classifying 85 instances as correctly classified with 15-fold cross-validation as shown in Fig. 4. However, random tree and random forest classification algorithms have given the highest accuracy of 61.83% as per Fig. 5 and 64.12% as shown in Fig. 6, respectively, with 20 folds cross-validation approach.

Table 1 Classification results based on Naïve Bayes, KNN, random tree and random tree

	Naïve Bayes			KNN			Random tree			Random forest		
Folds	10	15	20	10	15	20	10	15	20	10	15	20
Correctly classified instances	85	87	85	81	85	79	80	73	81	83	83	84
Incorrectly classified instances	46	44	46	50	46	52	51	58	50	48	48	47
Correctly classified %	64.89	66.41	64.89	61.83	64.89	60.30	61.06	55.72	61.83	63.35	63.35	64.12
Incorrectly classified %	35.11	33.59	35.11	38.17	35.11	39.7	38.94	44.28	38.17	36.65	36.65	35.88

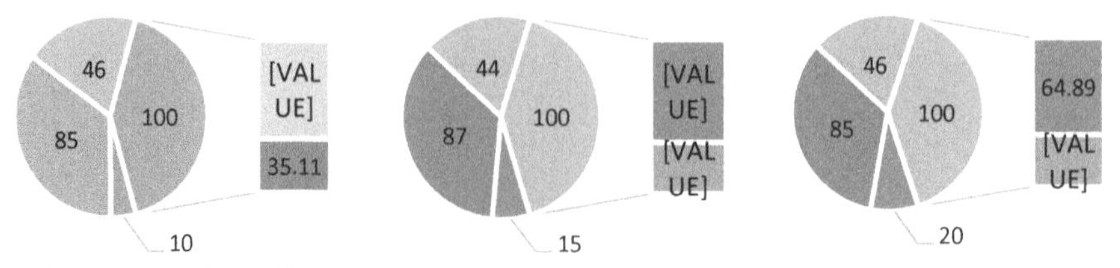

Fig. 3 Classification based on Naïve Bayes algorithm using 10, 15 and 20 folds cross-validation, respectively

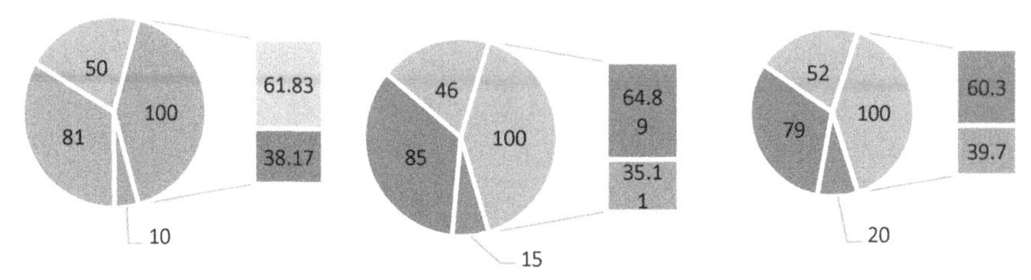

Fig. 4 Classification based on KNN algorithm using 10, 15 and 20 folds cross-validation, respectively

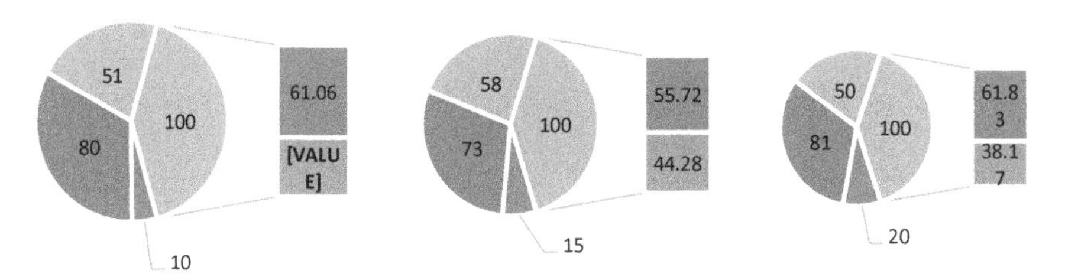

Fig. 5 Classification based on random tree algorithm using 10, 15 and 20 folds cross-validation, respectively

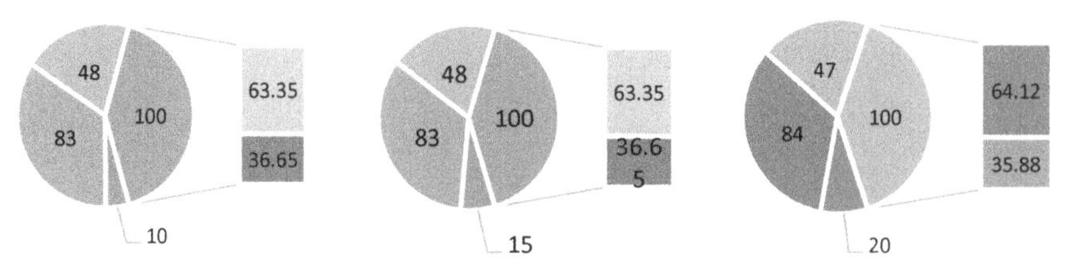

Fig. 6 Classification based on random forest algorithm using 10, 15 and 20 folds cross-validation, respectively

5 Conclusion

The paper is intended to analyze the accuracy of different classifiers on the students' academic performance dataset. This dataset involves the analysis of all the categorical attributes and 131 instances of students. Attribute representing the marital status of students is dropped during data pre-processing. To enhance the results, eight most relevant attributes are picked. In this experiment, four machine learning classifiers namely naïve Bayes, KNN, random tree, and random forest are used, using Rapid Miner 9.7, for conducting the comparative analyses of their respective predictive accuracy. Results have displayed that the predictive accuracy of correctly classified instances in case of naïve Bayes is highest with 15-fold cross-validation while random tree and random forest algorithms are depicting good predictive accuracy of correctly classified instances with 20-fold cross-validation. This may be due to the imbalanced distribution of the class variable in the entire dataset and can be considered as the limitation of this study. In the future scope of the study, we wish to work on the student's dataset by applying Synthetic Minority Oversampling Technique (SMOTE) to balance the distribution of target class variable and compare the results with/without SMOTE.

References

1. Klösgen W, Zytkow JM (eds) (2002) Handbook of data mining and knowledge discovery. Oxford University Press, Inc., Oxford, pp 10–21
2. Mittal R, Rattan V (2019) Evaluating rule based machine learning classifiers for customer spending in a shopping mall. J Adv Res Dyn Control Syst 11(08):716–719
3. Carver CA, Howard RA, Lane WD (1999) Enhancing student learning through hypermedia courseware and incorporation of student learning styles. IEEE Trans Educ 42(1):33–38
4. Barnes T, Desmarais M, Romero C, Ventura S (2009) Educational data mining 2009. In: Proceedings of the 2nd international conference on educational data mining, pp 1–3
5. Saa AA, Al-Emran M, Shaalan K (2019) Factors affecting students' performance in higher education: a systematic review of predictive data mining techniques. Technol Knowl Learn 24(4):567–598
6. Peterson PL, Baker E, McGaw B (2010) International encyclopedia of education. Elsevier Ltd.
7. Salloum SA, Shaalan K (2018, September) Factors affecting students' acceptance of e-learning system in higher education using UTAUT and structural equation modeling approaches. In: International conference on advanced intelligent systems and informatics, pp 469–480. Springer, Cham
8. Mostow J, Beck J (2006) Some useful tactics to modify, map and mine data from intelligent tutors. Nat Lang Eng 12(2):195–208
9. Romero C, Ventura S (2007) Educational data mining: a survey from 1995 to 2005. Expert Syst Appl 33(1):135–146
10. Romero C, López MI, Luna JM, Ventura S (2013) Predicting students' final performance from participation in on-line discussion forums. Comput Educ 68:458–472
11. Romero C, Ventura S (2010) Educational data mining: a review of the state of the art. IEEE Trans Syst Man Cybern Part C (Appl Rev) 40(6):601–618
12. Araque F, Roldán C, Salguero A (2009) Factors influencing university drop out rates. Comput Educ 53(3):563–574

13. UCI Machine Learning Repository: Student Academics Performance Data Set. Available: https://archive.ics.uci.edu/ml/datasets/Student+Academics+Performance. Accessed 28 Sept 2020
14. Hussain S, Dahan NA, Ba-Alwib FM, Ribata N (2018) Educational data mining and analysis of students' academic performance using WEKA. Indonesian J Electr Eng Comput Sci 9(2):447–459
15. Breiman L (2001) Random forests. Mach Learn 45(1):5–32
16. Agarwal S (2013, December) Data mining: data mining concepts and techniques. In: 2013 international conference on machine intelligence and research advancement, pp 203–207. IEEE
17. Saini I, Singh D, Khosla A (2013) QRS detection using K-Nearest Neighbor algorithm (KNN) and evaluation on standard ECG databases. J Adv Res 4(4):331–344
18. Ashraf M, Zaman M, Ahmed M (2020) An intelligent prediction system for educational data mining based on ensemble and filtering approaches. Procedia Comput Sci 167:1471–1483
19. Kamal P, Ahuja S (2019) An ensemble-based model for prediction of academic performance of students in undergrad professional course. J Eng Des Technol 17(4):769–781
20. Rojanavasu P (2019) Educational data analytics using association rule mining and classification. In: 2019 joint international conference on digital arts, media and technology with ECTI northern section conference on electrical, electronics, computer and telecommunications engineering (ECTI DAMT-NCON), pp 142–145. IEEE
21. Francis BK, Babu SS (2019) Predicting academic performance of students using a hybrid data mining approach. J Med Syst 43(6):162
22. Lee Y (2019) Estimating student ability and problem difficulty using item response theory (IRT) and TrueSkill. Inform Discov Deliv 47(2):67–75
23. Juhaňák L, Zounek J, Rohlíková L (2019) Using process mining to analyze students' quiz-taking behavior patterns in a learning management system. Comput Hum Behav 92:496–506
24. Azcona D, Hsiao IH, Smeaton AF (2019) Detecting students-at-risk in computer programming classes with learning analytics from students' digital footprints. User Model User-Adap Inter 29(4):759–788
25. Kabra RR, Bichkar RS (2011) Performance prediction of engineering students using decision trees. Int J Comput Appl 36(11):8–12
26. Asif R, Merceron A, Ali SA, Haider NG (2017) Analyzing undergraduate students' performance using educational data mining. Comput Educ 113:177–194
27. Hoffait AS, Schyns M (2017) Early detection of university students with potential difficulties. Decis Support Syst 101:1–11
28. Daud A, Aljohani NR, Abbasi RA, Lytras MD, Abbas F, Alowibdi JS (2017, April) Predicting student performance using advanced learning analytics. In: Proceedings of the 26th international conference on world wide web companion, pp 415–421
29. Diamantidis NA, Karlis D, Giakoumakis EA (2000) Unsupervised stratification of cross-validation for accuracy estimation. Artif Intell 116(1–2):1–16

Internet of Things (IoT) Based Automated Light Intensity Model Using NodeMcu ESP 8266 Microcontroller

Shyla and Vishal Bhatnagar

Abstract Internet of Things (IoT) is a ubiquitous technology for connecting anything from anywhere impacting the life drastically by expanding its reach in economical, commercial and social areas. In this paper authors used NodeMcu ESP 8266 microcontroller for modelling automated system by embedding wifi moduled. The light intensity is continuously captured and is transmitted over cloud network. The captured data is then used for analysing voltage fluctuations using python. The automated system works according to the intensity of light, if the light intensity falling on light dependent resistor (Ldr) is low then light emitting diode will switch on and if the light intensity falling on light dependent resistor (Ldr) is high then the light emitting diode will remain off making system energy efficient which works automatically according to the light intensity.

Keywords Light dependent resistor · Light emitting diode · Resistor · Microcontroller

1 Introduction

Internet of Things (IoT) is used to connect anything with anyone from anywhere. It is the aggregation of multiple devices which provides the ability to control all the connected devices from anywhere. The interconnected devices make a smart city with smart homes, smart buildings, smart equipment and smart electronic devices using IoT. It is the combination of various connected objects, appliances, services and humans by exchange of information among several applications to achieve the objectives. Jian et al. [1], found that the emergence of IoT devices transforms the living pattern drastically by introducing smart devices to perform daily chores. IoT includes

Shyla (✉) · V. Bhatnagar
Ambedkar Institute of Advanced Communication and Technologies, New Delhi, India

Shyla
GGSIPU, New Delhi, India

A. Choudhary et al. (eds.), *Applications of Artificial Intelligence and Machine Learning*, Lecture Notes in Electrical Engineering 778,
https://doi.org/10.1007/978-981-16-3067-5_3

smart cities, smart organisations, smart homes, smart transportations and smart equipment all around. IoT is defined as the internet of everything where multiple devices interact with each other by creating a global network.

The major propaganda of IoT devices is the ability to manage flow of information in the network with minimum human intervention. Zhang [2], found that the things in IoT refers to any person with heart monitoring system, any animal with biochip, any device including sensors and any other automated devices that can be associated with internet protocol address to build a network for communication. The organisations and industries are inclining towards IoT to expand their service's distribution by understanding the requirements of end users. IoT system includes smart automated devices that include processors, hardware devices, embedded systems and sensors to collect data. Sahtyawan and Wicaksono [3], found that the data captured by these sensors is transmitted using IoT gateways to cloud and then analysed locally by using data analysis tools. The complete procedure of information exchange occurs without interference of humans. The data collection and analysing procedure include machine learning and artificial intelligence to make devices smarter and generate outcomes on the basis of learning.

Organizations are benefitted from IoT in different ways that can be industry-specific and can be used across several industries. Hussain et al. [4], found that IoT can benefit businesses by monitoring the tasks, improving customer services by saving time and money, improving employees' work performance, creating adaptive models, increasing revenue and generating profits. IoT has vast range of applications that can be used in automations, telecommunications and energy saving models which includes smart appliances, smart electronic devices, wearable sensor equipment, patients monitoring devices, smart farming, smart city and smart buildings to connect anything from anywhere.

1.1 Problem Statement

The problem statement concisely defines the major issue of wasting electric energy by ignoring regular chores. The careless attitudes of people using electricity for outdoor lighting in day time cause light pollution. The artificial brightness of the night sky also causes energy wastage. To save energy consumption authors model a smart electronic equipment where light switch on and off automatically depending on the light intensity for smart cities with minimum human interaction.

1.2 Objectives

- The objective is to design a model to minimise energy consumption.
- To design the automated system by using sensors and analysing the captured data using python.

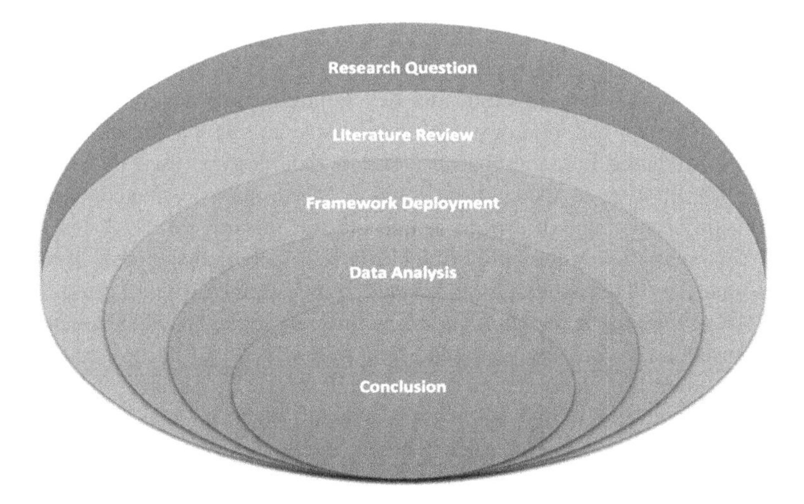

Fig. 1 Research methodology [5]

The paper is divided into various sections as: Sect. 1 shows Introduction, Sect. 2 as Research Methodology, Sect. 3 as Literature survey, Sect. 4 as Performance Analysis and Sect. 5 as Conclusion and Future Scope.

2 Research Methodology

Research methodology defines the procedure to evaluate the overall study. It includes discrete structure which explains how to conduct research in systematic manner to obtain the results.

Figure 1 shows the methodology adapted by authors to deploy the findings. The research questions define the need for the required analysis on the basis of literature review. The literature review represents the related work in the area of IOT and microcontrollers where research framework is designed and deployed for achieving the objectives. In this paper authors used IOT devices to generate data and transmit the data using IoT hub in the network and the accumulated data is then analysed using anaconda python data analysis tool.

3 Literature Survey

IoT integrates multiple smart devices to transmit information through each other in the network with minimum human intervention. In the area of computing, IoT is fastest growing technology. Ngu et al. [6], found that the security methods are to be improved to secure IoT systems. The machine learning and deep learning techniques are used for enhancing the security measures and facilitation of secure communication among devices. The authors presented an analytical survey on machine learning techniques and deep learning methods for developing the security methods. Alrawi et al. [7], proposed a methodology that researchers can take on for enhancing the security for IoT-based home appliances. The methodology is used to understand occurrence of different attacks and stakeholders. The authors evaluated 45 devices to identify different research aspects which are left unexplored. Seok [8], found that IoT can be applied in the field of blockchain for data management, data reliability and integrity. This approach can be used in industrial internet of things (IIoT) network which includes actuators, sensors and program logics. The blockchain technique is based on hashed cryptographic functions which can be implemented in IIoT. The authors used hashed functions for lightweight blockchains in IIoT and proposed architecture for hash-based cryptographic functions that can alter hash algorithms according to network traffic. The future of IoT will impact the life drastically by expanding its reach in economical, commercial and social areas. Traditionally cryptographic approaches are used in making IoT systems more secure but this is not enough for newly emerging attacks. Hussain et al. [4], authors systematically observed the several attack patterns and security requirements and then emphasised on machine learning and deep learning techniques for resolving different security hinderance in IoT. Dwivedi et al. [9], found that healthcare area is now adopting IoT technology. The IoT-based wearables which enclose sensors and embedded systems to combat health issues which are becoming the most important segment of human lives. The authors try to understand the need of patient monitoring equipment, healthcare devices and blockchains with IoT for information security. The cryptographic functions and approaches are used for making system more secure and vulnerable.

Sahtyawan and Wicaksono [3], authors designed a light monitoring device using relay module, NodeMcu ESP 8266 microcontroller with Wifi module. The device is used to control led lights in real-time from anywhere from handphone using Blynk application. The authors modelled 4 switches to on and off Led and is connected by wifi. The devices can be controlled by using common blynk user IDs from anywhere.

4 Performance Analysis

The framework illustrates the theoretical structure to systematically deploy research plan in a systematical approach. To achieve the defined objectives authors used IoT devices, IoT hub and Data analyses. The Ldr is used for resisting light intensity depending on the amount of light falls on Ldr. The Ldr is connected to IoT hub which act as gateway for collecting light voltage fluctuations and the collected data is analysed using python (Fig. 2).

- **IoT Devices**—The devices are hardware systems that include sensors, antenna and embedded system to transfer captured data from one device to another. Zhang

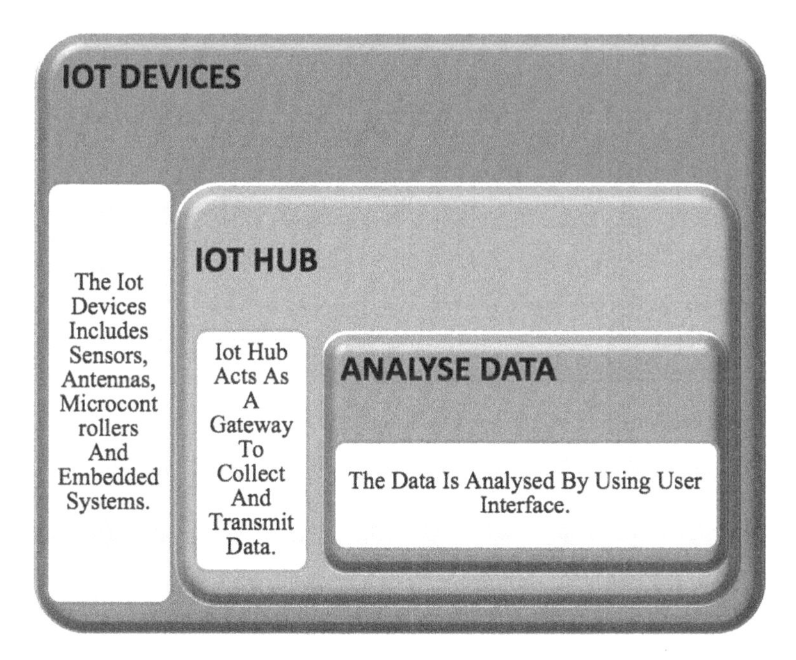

Fig. 2 Research framework [2, 5]

[2], found that the devices include wireless systems, actuators, mobile devices and medical devices.

- **IoT Hub**—IoT Hubs are used for bi-directional communication among devices. The information is transmitted from IoT devices to cloud and cloud to IoT devices. The information is secured in the backend of any cloud by using gateways and hubs as the central system.
- **Data Analysis**—The information that is accumulated from IoT devices is analysed by using data analysis tools and machine learning techniques.

The automated system is modelled by authors to conserve energy which includes the usage of hardware and software systems. The hardware includes Node Mcu, light emitting diode (Led), Resistor, Ldr, connecting wires and bread board. The software include Audrino, ThingSpeak and Python. The major component of model is light dependent resistor that changes its resistance according to the light intensity, if the light of high voltage falls on Ldr the led will switch off and if the light intensity that falls on Ldr is low then led will remain on. The data is captured on ThingSpeak cloud from the Ldr sensor and is analysed by using python.

The Ldr is used for resisting light intensity depending on the amount of light falls on it.

$$\mathrm{RL} = \frac{500}{\mathrm{LUX}} \tag{1}$$

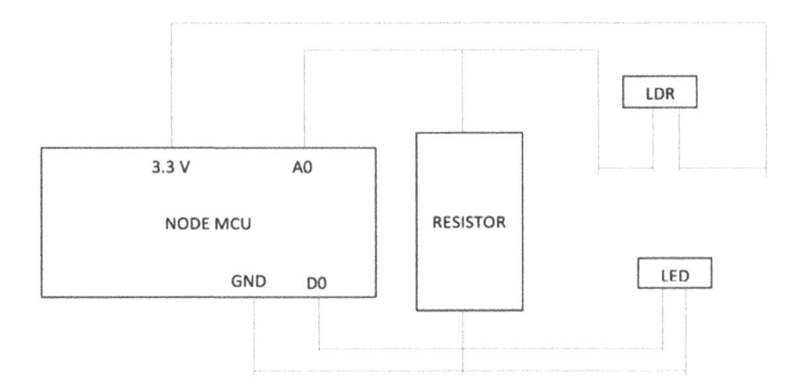

Fig. 3 Circuit diagram [3]

Here RL is Resistance and LUX is Light Intensity. The Ldr is getting 5 V supply using ESP8266 and is connected to 3 K resistors. The obtained voltage output is,

$$V = 5 \times \frac{\mathrm{RL}}{\mathrm{RL} + 4} \tag{2}$$

$$\mathrm{LUX} = \frac{\frac{2500}{V} - 500}{4} \tag{3}$$

The circuit for the system is used for creating the model. The Node Mcu ESP 8266 microcontroller is used which is connected with resistor, Ldr and Led. The A0 pin works as input and D0 as output.

Figure 3 shows the circuit diagram for the energy-efficient model. The circuit diagram is used for connecting the hardware on the bread board to create the automated system.

4.1 Analysis Outcomes

The different voltage levels data is captured by Ldr and is transmitted to ThingSpeak cloud. The authors collected the 135 voltage readings to capture the voltage fluctuations. The Ldr voltage data is analysed by using python.

Figure 4 shows the low voltage readings which is captured by the system. In the presence of low light the value of voltage is varied from 25 to 40 V depending on the area where system is placed.

Figure 5 shows that the Led switches on automatically in the absence of light or in presence of low light. The system acts as smart sensored device which switches on and off automatically in presence or absence of light.

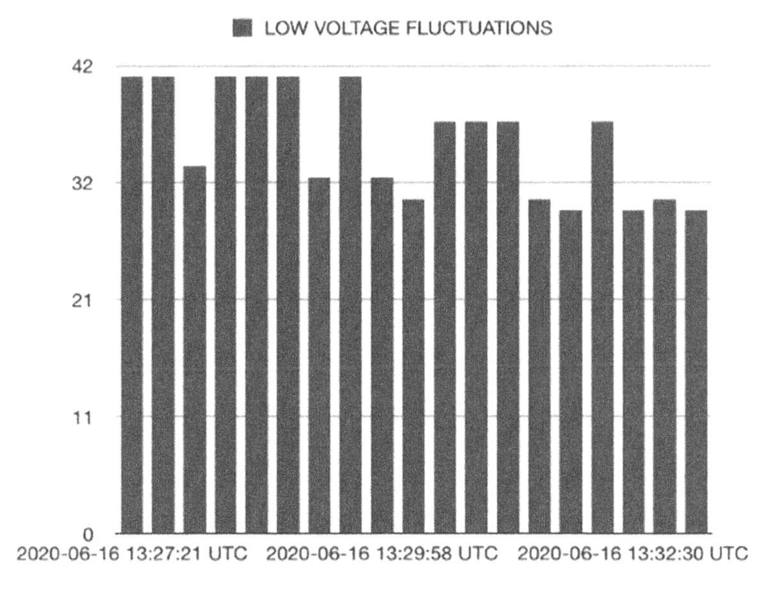

Fig. 4 Low voltage fluctuations

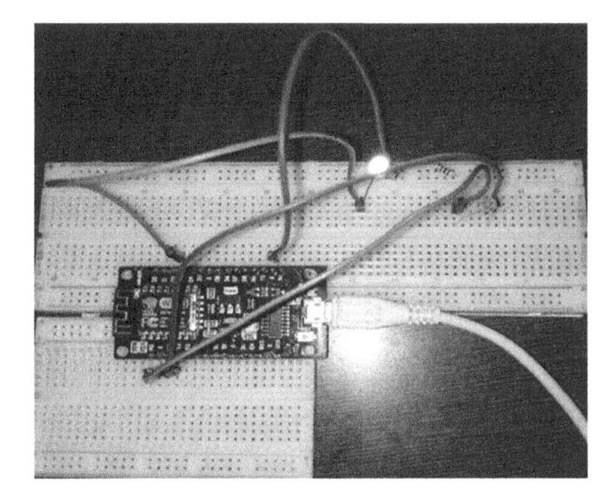

Fig. 5 LED on during low voltage

Figured 6 shows the high voltage readings which is captured by the system. In the presence of high light emitting device due to which the value of voltage is varied within 250–1024 V depending on the exposure of light.

Figure 7 shows that the LED switches off automatically in the presence of high light intensity. The system acts as smart sensored device which switches on and off automatically in presence or absence of light.

Figure 8 shows the dip and hike of voltage values observed during experimental setup. The Led will switch on or off automatically by considering the readings. In

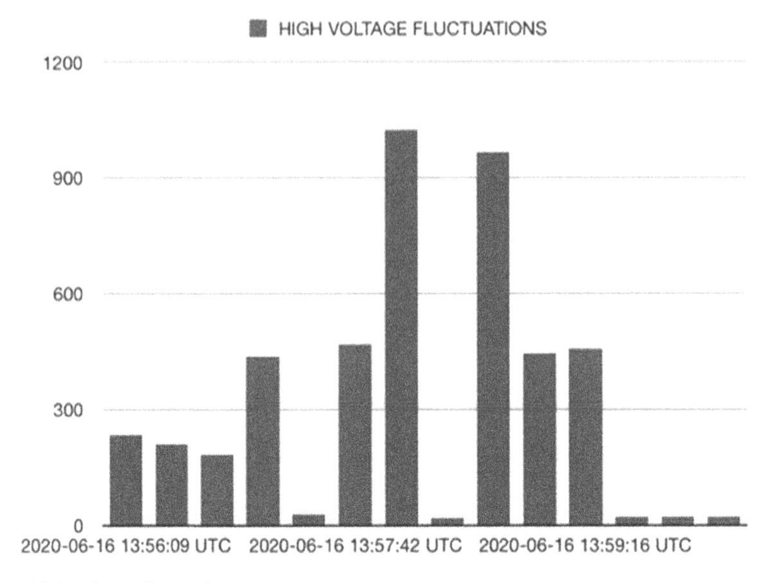

Fig. 6 High voltage fluctuations

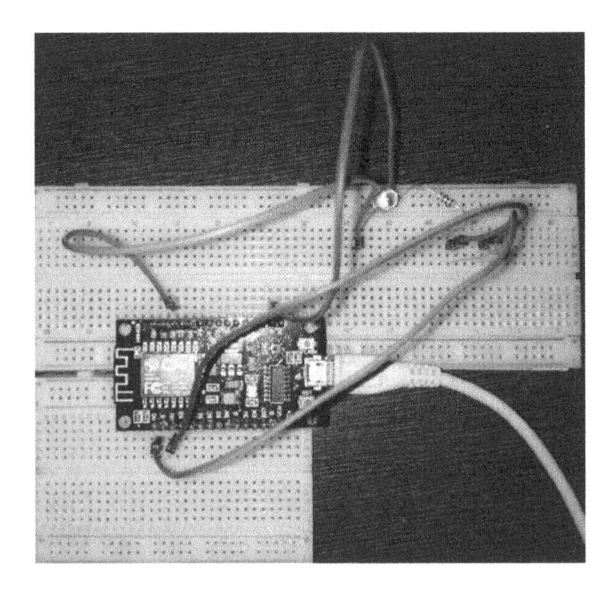

Fig. 7 LED off during high voltage

this authors maintains the voltage level of 150 below which the Led switches on and above that threshold value the Led switches off.

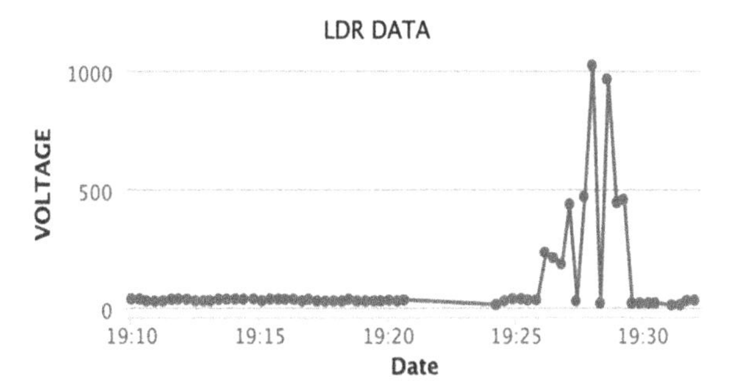

Fig. 8 Voltage fluctuation

5 Conclusion and Future Scope

The IoT is the most crucial technology for establishing the interconnected global network of multiple devices for communication with each other. The major aspect of IoT devices is the ability to manage flow of information with minimum human intervention in the network. In this paper, the vision is to analyse a IoT system with sensors and actuation functions which is blended with each other to generate information. The NodeMcu ESP 8266 microcontroller is used for modelling automated system which embeds wifi module. The Ldr continuously captures the light intensity and is transmitted over the cloud network using ThigSpeak. The automated system works according to the intensity of light, if the voltage levels are low the Led will remain on and if the voltage levels become high then Led will switch off.

- To create user-oriented software that is easy to deploy for things to human collaboration.
- To introduce nanotechnology devices which support cognitive processing.
- To create self-repairing network which resolves any faults using self-learning.

The model plays an important role for making an energy-efficient system which works automatically according to the light intensity. The IoT is the crucial technology for development of future smart housing and smart city.

References

1. Jian A, Xiao LG, Xin H (2012) Study on the architecture and key technologies for internet of things. Adv Biomed Eng 11:329–335
2. Zhang Y (2011) Technology framework of the internet of things and its application. In: Electrical and Control Engineering (ICECE), pp 4109–4112
3. Sahtyawan R, Wicaksono AI (2020) Application for control of distance lights using microcontroller Nodemcu Esp 8266 based on internet of things (IoT). Compiler 9(1):43–50
4. Hussain F, Hussain R, Hassan AS, Hossain E (2020) Machine learning in IoT security: current solutions and future challenges. Commun Surv Tutor 22:298–444

5. Bhatnagar V, Ranjan J, Singh R (2011) Real time analysis on finding the significance of DM on CRM of service sector organizations: an Indian perspective. Int J Electron CRM (IJECRM) 5(2):171–201
6. Ngu AH, Gutierrez M, Metsis V, Nepal S, Sheng QZ (2016) IoT middleware: a survey on issues and enabling technologies. Internet Things J 4(1):1–20
7. Alrawi O, Lever C, Antonakakis M, Monrose F (2019) Sok: security evaluation of home-based iot deployments. In: Symposium on security and privacy (SP), pp 1362–1380
8. Seok B, Park J, Park JH (2019) A lightweight hash-based blockchain architecture for industrial IoT. Appl Sci 9(18):37–40
9. Dwivedi AD, Srivastava G, Dhar S, Singh R (2019) A decentralized privacy-preserving healthcare blockchain for IoT. Sensors 19(2):320–326
10. Chen S, Xu H, Liu D, Hu B, Wang H (2014) A vision of IoT: applications, challenges, and opportunities with china perspective. Internet Things J 1(4):349–359
11. He C, Xin C (2020) Application and strategy of internet of things (IOT) in college education informationization. In: Data processing techniques and applications for cyber-physical systems, pp 1089–1097
12. Praba TS, Laxmi SP, Sethukarasi T, Harshitha RD, Venkatesh V (2020) Green IoT (G-IoT): an insight on green computing for greening the future. In: Advances in greener energy technologies, pp 579–600
13. Cao Y, Li W, Zhang J (2011) Real-time traffic information collecting and monitoring system based on the internet of things. Pervasive Comput Appl (ICPCA) 45–49

Handwritten Mathematical Symbols Classification Using WEKA

Sakshi, Shivani Gautam, Chetan Sharma, and Vinay Kukreja

Abstract Machine learning tools have been extensively used for the prediction and classification of mathematical symbols, formulas, and expressions. Although the recognition and classification in handwritten text and scripts have reached a point of commensurate maturity, yet the recognition work related to mathematical symbols and expressions has remained a stimulating and challenging task throughout. So, in this work, we have used Weka, a machine learning tool, for the classification of handwritten mathematical symbols. The current literature witnesses a limited amount of research works for classification for handwritten mathematical text using this tool. We have endeavored to explore the potential classification rate of handwritten symbols while analyzing the performance by comparing the results obtained by several clustering, classification, regression, and other machine learning algorithms. The comparative analysis of 15 such algorithms has been performed, and the dataset used for the experiment incorporates selective handwritten math symbols. The experimental results output accuracy of 72.9215% using the Decision Table algorithm.

Keywords Handwritten · Symbols · Characters · Machine learning · Decision table · J48 · Bayes algorithm · Classification · Weka

Sakshi · V. Kukreja
Institute of Engineering and Technology, Chitkara University, Punjab, India
e-mail: sakshi@chitkara.edu.in

V. Kukreja
e-mail: vinay.kukreja@chitkara.edu.in

S. Gautam
School of Computer Applications, Chitkara University, Himachal Pradesh, India
e-mail: shivani.gautam@chitkarauniversity.edu.in

C. Sharma (✉)
Chitkara University, Himachal Pradesh, India
e-mail: chetan.sharma@chitkarauniversity.edu.in

A. Choudhary et al. (eds.), *Applications of Artificial Intelligence and Machine Learning*, Lecture Notes in Electrical Engineering 778,
https://doi.org/10.1007/978-981-16-3067-5_4

1 Introduction

The emerging trend noticed in the continuous evolution of the tremendous scientific community has witnessed the generation and usage of a massive amount of scientific documents. It is a remarkable and prodigious achievement, as the scientific and research zones are consistently and constantly generating supplementary knowledge, which is represented in a different form. And this knowledge and developed information need to be represented in the form of scientific and mathematical notations. Nowadays, it is crucial to have this information digitalized, to make feasible searching, and to access a set of relevant documents. Being an important part of the scientific and engineering literature, the classification and recognition of the mathematical expressions have become an exciting and stimulating research area of the pattern recognition with unlimited real-world implications [1]. However, it is for analyzing or accessing research works or be it for information retrieval for other scientific and educational purposes. Also, the recognition of images in the field of artificial intelligence is prevalent among researchers for many years, as this can be used in various areas. Machine learning tools have gained commensurable popularity in the last decade. Thus, today individuals or organizations are using the machine-dependent application in every sector to perform their tasks. Thus, the interest of this research work is inclined towards implementing and exploring the potential of machine learning tools like WEKA. Moreover, recognizing and classifying a handwritten math symbol is an arduous classification problem, requiring real-time identification for all the symbols containing an input as well as the complex 2D relationships between symbols and subexpressions [2]. So, the researchers need to access math information recurrently while working with computational systems. Mathematical character or symbols recognition is yet challenging, and the emerging field of research [3].

In the educational area, the recognition of math symbols and characters is incorporated with a marking system to evaluate the marks or scores of mathematical questions and exercises automatically. However, learning notations like LATEX and MathML, or using graphic editors is an essential requirement for introducing math notation into an electronic device. The process of recognition of math symbols and characters aims at building recognition systems and models that can automatically understand mathematical content provided by humans in the form of printed or handwritten characters or symbols. Optical character recognition mainly focused on the machine printed output, where the number of font styles can be used to write any character or symbol. In that inconsistency between the character font, style and attributes are small, whereas when a character is written impersonal, their variability is relatively high. This variability and distortion make recognition very challenging [4]. The writing style of any individual varies from person to person, and this variety of writing style of community creates distortion and variation in the dataset [5]. Identification of character or symbol from where refined features can be extracted from the data is one of the primary task to make recognition rate more accurate, and to locate such region from the data; various sampling techniques are used in the field of pattern recognition [6]. So, it becomes more critical to extract stable and reliable features to enhance system performance. In future, character and symbol recognition in the field of mathematics might serve as a foundation stone to start the paperless strategy by digitizing and processing the saved hard copy documents. Handwritten data is vague by nature as they don't contain the sharp and perfectly straight lines so, the goal of recognition is to extract the essential information from any raw image data [7]. Tokas and Bhadu [8] Illustrate structural, statistical, and global transformation classification methods of feature extraction techniques. The analytical approach is used to select the data, and it uses the information related to the statistical distribution of pixels in the image. Neves et al. [9] Conduct study on the NIST SD19 dataset using SVM based offline handwritten digit recognition system and concluded that SVM outperforms in their experiment. Perwej and Chaturvedi [10] Convert the handwritten dataset into electronic data and used the NN approach to make machines capable of recognizing the dataset.

2 Literature Review and Related Work

Study is conducted on single character recognition of math symbols by the use of Support vector machines. They focus on improving the classification of InftyReader, which is optical character recognition (OCR) software used to recognize text, scientific figures, and math symbols. SVM is used to improve the classification of InftyReader. InftyReader confuses in the classification of pairs letter, so the author firstly compares the performance of SVM kernels and features of pair letters. Then they illustrate the multi-class classification with SVM, utilizing the ranking of alternatives within InftyReader's confusion clusters. The proposed technique decreases its misrecognition rate by 41% [11].

Author considered machine printed and handwritten document images from three Indic scripts (Bangla, Devnagri, and Roman) for their study. They applied the OCR technique on printed and handwritten document images. The author has taken 277 document images from both the methods in three mentioned scripts. They used a Multilayer perceptron classifier with 5-fold cross-validation, an average accuracy rate of 98.75% for Bangla, 100% for Devnagari, and 100% for Roman scripts are obtained. When they combined all three scripts, the average accuracy rate of 98.9% is obtained [12].

Author proposed solution to locate the mathematical formula in any PDF document using machine learning and heuristic rule methods. They recommended four new features in their study for preprocessing and post-processing techniques. LibSVM-R-D, LibSVM-R, LibSVM-P, Logistic regression, MLP, J48, Random Forest, BayesNet, PART, Bagging-RF, AdaBoost-RF learning algorithms are taken by the author to experiment on Ground-truth dataset which is now publically available. The author concludes that they increased overall accuracy through the proposed system by 11.52 and 10.65% compared to the previous studies [13].

Study was conducted on Devnagri script, which is widely used in many languages. The author has taken 60 handwritten Devanagari symbols from different writers; out of 60 characters, 50 are letters, and 10 are digits. 60 sample of each character has been taken so in total; 3600 samples are taken for feature extraction. The author performed classification through Multilayer perceptron, K-Nearest Neighbour, Naive Bayes classifier, and Classification tree on the selected dataset in WEKA. They compare the performance of the chosen classifier and found that a multilayer perceptron performs well among all classifiers and achieves a 98.9% accuracy rate from the multilayer perceptron classifier [14].

Study is done to discover the software tool which is capable of identifying the character or digits. The different writing skills of an individual make this field challenging. The author used the MNIST image dataset to perform classification. Various steps are followed to achieve rating on the considered dataset, and firstly dataset is chosen, then the input image is converted to a grayscale image. Finally, all images are converted to binary format. WEKA tool is selected to experiment, and as WEKA accepts CSV and ARFF format, so all processed images are converted to a comma-separated file for training and testing purposes. They used Random forest, Decision tree, and Hoeffding Trees machine learning algorithms to perform classification on the selected dataset. On the base of different parameters, the author results in that the Hoeffding tree is the best classification technique for considered datasets out of all classification techniques [15].

Author conducted this study to recognize the handwritten digit, which is a significant problem in the field of pattern recognition. Researchers are working in this field to develop an efficient algorithm for identifying the handwritten numbers as input from the user through digital devices. They used a collection of 3689 digit datasets, which is making available by the Austrian Research Institute for Artificial Intelligence, Austria. Out of 3689, 1893 samples are taken to train the system, and 1796 samples are taken for testing on the train system. J48, Multilayer Perceptron, Random forest, SVM, Bayes classifiers, Random Tree machine learning algorithms were considered by the author to conduct their study to recognize the digits using WEKA. The author found a 90.37% accuracy rate in Multilayer Perceptron, out of all considered machine learning algorithms. So, it results in that Multilayer Perceptron algorithms perform significantly well to recognize the digits [4].

Author illustrates the approach for offline recognition of handwritten mathematical symbols. The study included symbol recognition for over more than 300 classes. The objective of designing the classifier to recognize these 300 symbols. Firstly they describe the issues related to segmentation using SLIC and study experiment results shows different accuracy rate for different algorithms. They achieved 84% for the kNN classifier, 57% for HOG, 53% for LBP. The author modified 87 classes using the LeNet and gained a 90% accuracy rate. SqueezeNet is used to pre-trained the 101 classes and result in a 90% accuracy rate [16].

Study is conducted on handwritten offline Urdu character recognition using different machine learning techniques. They created their dataset with 9600 instances from the various native writers. The author used edge histogram descriptor, ColourLayout, and Binary Pyramid to extract the feature from the considered dataset.

They applied different machine learning algorithms like MLP, SVM, SMO, and simple logistic using the WEKA 3.8 tool an achieved 98.60% of accuracy through SVM [17].

3 Proposed Work

- **Data Collection**: In any recognition system, the first step is to collect the data, and data can be obtained in any form. Data sets are created by taking handwritten documents from different users. Further, these documents are scanned through digital types of equipment and develop a scanned image for extracting feature purposes.
- **Registration**: Images collected are mostly in the RGB scale, so after receiving the data, these images have to be converted into a grayscale format using proper threshold values to avoid the loss of information and after that these images are converted to binary format so that feature extraction can be done efficiently.
- **Preprocessing**: Preprocessing is an essential factor to be considered when we take any data for recognition as we know any data which is raw by nature contain some noise factor, vague and inconsistent data that is required to remove to achieve better performance. Preprocessing is also used to enhance the signal of the binary image data. Images are preprocessed in different matrix values like 5×7, 14×10, 32×32, and many more to get better recognition.
- **Feature Extraction**: In feature extraction processed image is represented in feature vector and the main goal through this is to extract a set of feature which helps the system to maximize the recognition rate. In handwritten documents, this is very challenging to get a useful feature due to the high degree of variability. It can be solved by dividing the processed image is into the $N \times M$ zone to obtain local characteristics rather than global characteristics. The feature can be extracted through statistical, structural, and global transformations methods.
- **Classification**: Once the features are extracted from the images, then we can classify them through different classification techniques like Hoeffding tree, random tree, Bayes Classifier, J48, Neural Networks, Random Forest, Support Vector Machines (SVM), etc. No classification method is considered to be the best classification method as the use of classifiers depends upon factors like training dataset, test dataset, number of features, and many more (Fig. 1).

4 Implementation

Dataset to conduct this study is downloaded from https://www.kaggle.com/guru001/hasyv2, which is publically available for the experiment. Dataset contain 369 symbols with 168,236 png images for symbols, each 32px $\times$ 32px from the different writers. We had taken 35 mathematical symbols shown in Fig. 2 out of 369 symbols with 3650

Fig. 1 Stages in character recognition system [14]

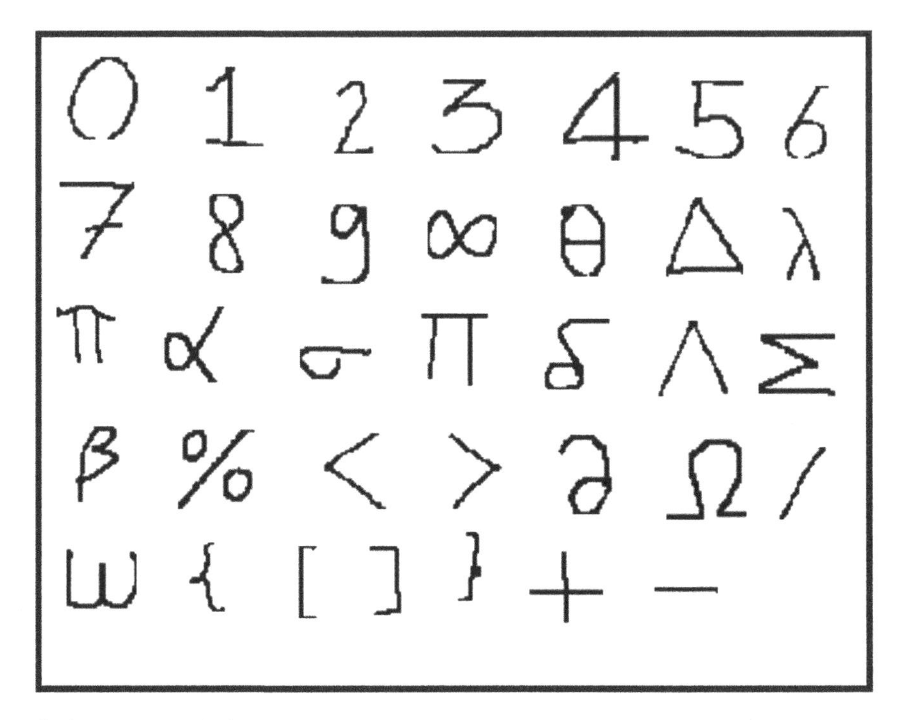

Fig. 2 Dataset handwritten mathematical symbols

png images each of 32px × 32px to perform our experiment on different classification methods. Firstly, in WEKA, we preprocessed dataset images included in the .csv file using a simple color histogram filter and change other attributes to nominal values for classification.

WEKA 3.6 machine learning tool written in Java and developed at the University of Waikato is used to conduct our experiment as this tool provides us with different classifiers to examine the performance. WEKA is used to evaluate different data mining tasks like pre-processing, classification, regression and many more. WEKA accepts .csv and .arff file format and the chosen dataset have already created the required data in the mentioned format. To determine and inspect the performance of different classification methods mentioned in Tables 1 and 2 comparison has

Table 1 Experiment results based on correctly classified instances and model time

Algorithms	Correctly classified instances (%)	Incorrectly classified instances (%)	Time taken to build the model (s)	Kappa statistic
J48	72.36	27.64	1	0.454
Hoeffding tree	67.45	32.55	6.71	0
Decision stump	67.45	32.55	0.16	0
Random tree	72.68	27.32	0.84	0.461
REPTree	67.45	32.55	0.09	0
Bayesnet	72.76	27.24	0.44	0.46
Naïve Bayes	72.68	27.32	0.09	0.462
Multinomial Naive Bayes	67.45	32.55	0.01	0
Decision table	72.93	27.07	15.05	0.456
Jrip	68.90	31.10	8.39	0.083
One R	6.69	93.31	0.08	0.0009
PART	71.80	28.20	3.08	0.444
Zero R	67.45	32.55	0	0
Input map classifier	67.45	32.55	0.01	0
Kstar	72.76	27.24	18.93	0.449

been performed. The algorithms named Naïve Bayes, Naïve Bayes Mutlinomial are Bayesian classifiers which belongs to the family of simple "probabilistic classifiers" based on Bayes' theorem and Decision Stump, Hoeffding Tree, Hoeffding Option Tree, Hoeffding Adaptive Tree are the Decision tree classifiers, which is popular supervised machine learning classification algorithm. Authors use the same method or procedure as per the WEKA tool suggestions and dataset is considered as instances and features in the data. To understand the experiment results we divided the results into two subparts for easier analysis and evaluation. First part of the results are shown in Table 1 which contain the correctly, incorrectly classified instances, time taken to build model, and kappa statistic. In the second part Table 2 contain the different errors during the simulation in WEKA. We run the different classifiers on the considered dataset in WEKA, and their results are shown in Tables 1 and 2.

5 Conclusion and Future Scope

In this paper, we have demonstrated several machine learning-based algorithms for identifying the classification rate of the handwritten mathematical symbols for determining and comparing the classification rate of these considered algorithms. More

Table 2 Experiment results based on different errors

Algorithms	Mean absolute error	Root mean squared error	Relative absolute error (%)	Root relative squared error (%)
J48	0.01	0.05	0.60	0.85
Hoeffding tree	0.01	0.06	1.00	1.00
Decision stump	0.01	0.06	0.91	0.98
Random tree	0.01	0.05	0.61	0.85
REPTree	0.01	0.06	0.96	1.00
Bayesnet	0.01	0.05	0.72	0.84
Naïve Bayes	0.01	0.05	0.69	0.84
Multinomial naive bayes	0.01	0.06	1.00	1.00
Decision Table	0.01	0.07	1.24	1.02
Jrip	0.01	0.06	0.92	0.98
One R	0.01	0.12	1.63	1.86
PART	0.01	0.06	0.60	0.88
Zero R	0.01	0.06	1.00	1.00
Input map classifier	0.01	0.06	1.00	1.00
Kstar	0.01	0.05	0.63	0.81

importantly, it becomes crucial to analyze the performance based on several metrics and compare the results meticulously. This paper compared 15 classifiers used for the recognition of different handwritten mathematical symbols. All considered algorithms performed well on the considered dataset except the one R algorithm. The accuracy rate for each algorithm is mentioned in Table 1, and we conclude that the Decision table presents exceptionally well with an accuracy rate of 72.9251% out of all algorithms. We propose to extend this work in the future by using different preprocessing methods and considering an extended and modified dataset with other exclusive handwritten mathematical symbols and diverse machine learning and deep learning algorithms.

References

1. Afshan N, Afshar Alam M, Ali Mehdi S (2017) An analysis of mathematical expression recognition techniques. Int J Adv Res Comput Sci 8(5):2021–2026
2. MacLean S, Labahn G (2015) A Bayesian model for recognizing handwritten mathematical expressions. Pattern Recognit 48(8):2433–2445
3. Liu C-L, Nakashima K, Sako H, Fujisawa H (2003) Handwritten digit recognition: benchmarking of state-of-the-art techniques. Pattern Recognit 36(10):2271–2285

4. Shamim SM, Miah MBA, Angona Sarker MR, Al Jobair A (2018) Handwritten digit recognition using machine learning algorithms. Glob J Comput Sci Technol 18(1)
5. Plamondon R, Srihari SN (2000) Online and off-line handwriting recognition: a comprehensive survey. IEEE Trans Pattern Anal Mach Intell 22(1):63–84
6. Das N, Sarkar R, Basu S, Kundu M, Nasipuri M, Basu DK (2012) A genetic algorithm based region sampling for selection of local features in handwritten digit recognition application. Appl Soft Comput 12(5):1592–1606
7. AlKhateeb JH, Pauplin O, Ren J, Jiang J (2011) Performance of hidden Markov model and dynamic Bayesian network classifiers on handwritten Arabic word recognition. knowl Syst 24(5):680–688
8. Tokas R, Bhadu A (2012) A comparative analysis of feature extraction techniques for handwritten character recognition. Int J Adv Technol Eng Res 2(4):215–219
9. Neves RFP, Lopes Filho ANG, Mello CAB, Zanchettin C (2011) A SVM based off-line handwritten digit recognizer. In: IEEE international conference on systems, man, and cybernetics, pp 510–515
10. Perwej Y, Chaturvedi A (2012) Machine recognition of hand written characters using neural networks. arXiv Prepr.arXiv1205.3964
11. Malon C, Uchida S, Suzuki M (2008) Mathematical symbol recognition with support vector machines. Pattern Recognit Lett 29(9):1326–1332
12. Obaidullah SM, Das N, Roy K (2014) An approach to distinguish machine printed and handwritten text from document images for indic script. Int J Appl Eng Res 9(20):4670–4675
13. Lin X, Gao L, Tang Z, Baker J, Sorge V (2014) Mathematical formula identification and performance evaluation in PDF documents. Int J Doc Anal Recognit 17(3):239–255
14. Shelke SV, Chandwadkar DM (2016) Handwritten Devnagri character recognition. GRD J Glob Res Dev J Eng 1(5):67–70
15. Lavanya K, Bajaj S, Tank P, Jain S (2017) Handwritten digit recognition using hoeffding tree, decision tree and random forests—a comparative approach. In: International conference on computational intelligence in data science (ICCIDS), pp 1–6
16. Nazemi A, Tavakolian N, Fitzpatrick D, Suen CY et al (2019) Offline handwritten mathematical symbol recognition utilising deep learning. arXiv Prepr.arXiv1910.07395
17. Jameel M, Shuja M, Mittal S (2020) Improved handwritten offline urdu characters recognition system using machine learning techniques. Int J Adv Sci Technol 29(6):4865–4978

Enhancing Sociocultural Learning Using Hyperlocal Experience

Smriti Rai and A. Suhas

Abstract In today's scenario of a global pandemic and stay-at-home orders for 1.5 billion people worldwide, the future will be hyperlocal. As per predictions by experts on how different the world will be in five years, it is believed that a change model established in mutual benefit evolving in collective action will be led by hyperlocality. It has gradually led to a revival of a type of human interaction that had been driven to near extinction by technological and social shifts—hyperlocal collaboration. Hyperlocal is information oriented across a community with its principal focus targeted toward the interests of the people in that community. It refers to all businesses in your vicinity, the neighboring general store, market, mall, restaurant, and other products and service providers. Hyperlocal platforms resolve the challenge of equaling immediate demand with the nearest available supply in the most optimized manner. Thus, it can be said that hyperlocal content has two main aspects: geography and time. These two aspects when combined help us identify types of hyperlocal content. Nowadays, the hyperlocal content includes GPS enabled mobile applications which further accentuate the geographic and time aspects. Geolocation services have evolved as one of the major sectors in our country and have transformed our lifestyle, health, and relationships. The legacy offline systems have been re-engineered with the help of technology to create a fresh and unforgettable experience. Hyperlocal has become the hottest business trend for many e-commerce firms setting the pathway for concepts like hyperlocal delivery, hyperlocal marketing, and hyperlocal forecasting with a sole objective of giving the user an enhanced hyperlocal experience. This paper presents a sociocultural learning model which is enhanced by providing a hyperlocal experience.

Keywords Hyperlocal experiences · Hyperlocal content · Sociocultural learning · Hyperlocal forecasting

---

S. Rai (✉)
Nitte Meenakshi Institute of Technology, Bengaluru, Karnataka, India
e-mail: smriti.rai@nmit.ac.in

A. Suhas
Infosys Ltd., Mysore, Karnataka, India

A. Choudhary et al. (eds.), *Applications of Artificial Intelligence and Machine Learning*, Lecture Notes in Electrical Engineering 778,
https://doi.org/10.1007/978-981-16-3067-5_5

1 Introduction

The sociocultural learning theory revolves around the idea that a learner's environment plays a crucial role in his/her learning development. Professionals consider sociocultural perspectives, and the role culture, interaction, and collaboration have on quality learning. Human development and learning originate in social, historical, and cultural interactions [1].

The learning community is on the brink of a new era of online learning. Online learning has been endorsed as being more convenient and economical as compared to the traditional learning environments. It also provides more opportunities for more learners to continue their education. Online learning has been defined as any class that offers at least part of its curriculum in the online course delivery mode, or as a transmission of information, and/or communication via the Internet without instructors and students being connected at the same [2]. Today, however, online learning is defined more clearly as any class that offers its entire curriculum in the online course delivery mode, thereby allowing students to participate regardless of geographic location, independent of time and place [3]. In other words, online learning has advanced to the point where learners need not meet face-to-face to complete a course. Research has proved that social presence influences results along with the learners and the instructor's satisfaction with a course. In online learning environments, social presence plays a vital role and has direct relation to students' learning experiences and satisfaction with the instructor.

Dropout rate is one measure of the effectiveness of an online program. Program quality can be determined, in part, by calculating student completion rates [4], and these rates tend to be lower for online classes [5]. This indicates that online courses may be less appropriate for certain students than the more traditional face-to-face type of training. The factors that influence a learner to leave a course are issues of isolation, disconnectedness, i.e., lacking one-to-one interaction with the instructors, schedule conflicts, learning environment was too de-personalized, etc.

This paper proposes an online learning model which attempts to overcome the challenges of learner retention to an online learning course/program.

2 Literature Survey

Online learning is education that is structured to be held over the Internet. It is referred to as e-learning among many other terms. It could be considered as one of the types of distance education/learning which acts as an umbrella terminology for any learning activity that occurs across distances and not in a traditional class environment. Online learning enrollments have continued to rapidly increase among the students and those at organizations who are looking at enhancing their skills quickly.

Given the number of smartphone users that exist in the world, it would be easy to assume that e-learning is easily available to everyone. However, the rate at which technological capabilities are growing at, it is still evident that computer literacy is still far away from justifying e-learning as a solution to all in the society. In India, the average computer illiteracy stands at around 25%, which is about a quarter of its population still having roadblocks in having access to e-learning. This scenario is similar in every other developing country with high economic plans toward education. E-learning boasts of a lot of features that make it the most sought after means of learning. These include convenience, flexible learning hours, efficient access to course materials, real-word skills and lifelong learning methods, global connectivity, and lastly financial benefits. In an ever-changing world with timeless opportunities for new technological possibilities, the scope and reach that e-learning has to offer to learners and educators is extremely widespread. Yet, these models have a lot of drawbacks which the proposed system is likely to help overcome, opening much more options in the ever-growing industry.

E-learning ends up taking away much more time to that of offline sessions. Online sessions are mainly textual based. Any efforts to communicate with an educator or learner happens via messages, comments, or post replies. Many a times, learners are not able to convey what they intend to resolve via text and vice versa for educators. Reading resource materials takes more time than listening to live lectures. It takes more effort to understand concepts online. The courses and platforms are structured in such a way that they allow learners to be more self-paced at finishing the curriculum. This requires immense sense of managing time and resources, which is not the case when it comes to a model that involves a mix of both online and offline sessions. These models also have a very rigid structure that allows either a learner to swim across easily provided he/she is an active learner or ends up sinking in chunks of pending lectures. Platforms have a target to meet bundle multiple courses together in order to lure learners of the best deals they can get. Once enrolled, there is really none to keep a track of your progress or make you aware of the consequences of not finishing the course in the stipulated time. This gives learners a sense of freedom, which is something that learners are not capable of handling wisely.

Communication is a key skill that is necessary irrespective of the mode of learning. The current model often neglects this aspect since there is little or no interaction among peers. In the long run, this only leads to learners being theoretically sound but cannot seem to excel at applying them practically. Lastly, online learning is best suited for a limited set of genres only. Many vocational and non-professional courses need a certain degree of hands-on sessions and other practical experience to help understand and learn better. Even though simulations and other means of instruction cater to these needs, the systems are not yet at a point where this method is completely sufficient.

A general online learning course/platform evaluation survey was taken up to better understand the users' perspective of current learning models, their interests of learning, and other demographic details. The chart in Fig. 1 shows the various interests of learning of users who would potentially like to enroll or already taken courses online. Users from the age group of 18–24 and 35–44 years were identified to

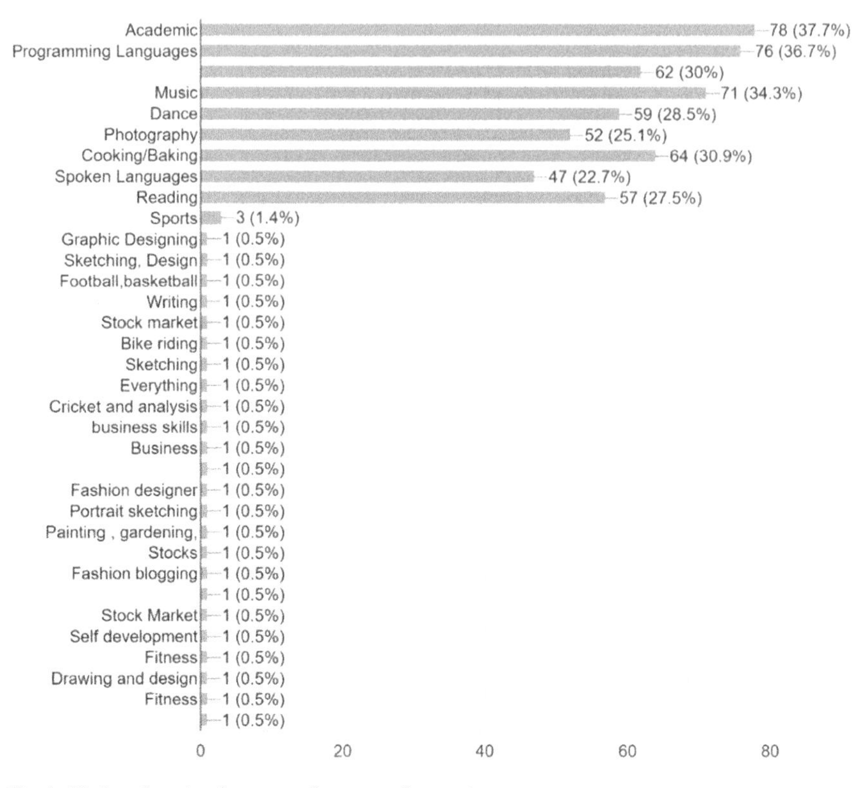

Fig. 1 Various learning interests of users prefer to take up

have their interests in academic, professional courses, and other vocational training such as music, dance, photography, and stock market analysis. Users between the age group of 25–34 preferred more career-oriented learning programs, whereas those between 45 years and above-preferred leisure activities and other community-based programs.

Figure 2 is a chart that shows the overall satisfaction that users have with the online learning platforms and courses that are currently available. About 52.7% of the users who have taken up courses online believed the current model is a mere 7 or below on a 10-point grade scale.

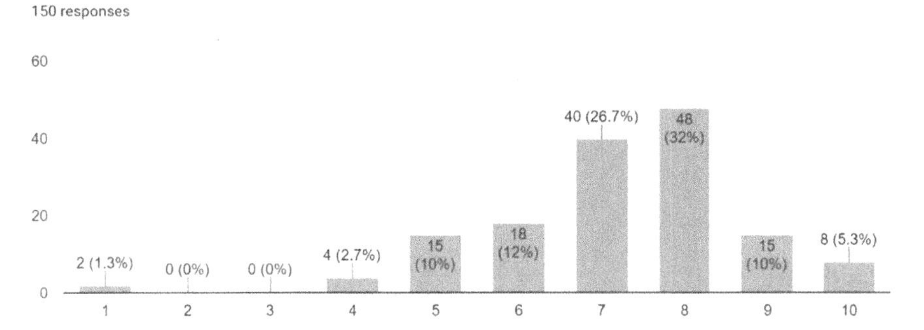

Fig. 2 Overall satisfaction with online learning platforms/courses

3 Working of the System

3.1 Overview

The proposed system for sociocultural learning using hyperlocal experience provides users with a focused and comfortable technological product that allows the access to an in-person model of learning by helping them find educators, artists, or courses on the basis of social validation. It is a one-stop application for all learning needs—academic oriented, vocational, or hobby based. This product proposes to help you find the most appropriate courses or educators depending on your choice of interests. It presents a map-based interface that connects learners to their educators depending on their geographic location—a Swiggy for e-learning, to be precise.

Smartphones have become universal, not only in developed countries but also in the developing ones. According to the "Emerging Markets" report, nearly a quarter of Indian consumers are using smartphones as their primary devices. Indian consumers' content preferences are social media (70%), education (50%), travel (49%), and health (40%). Thus, education seems to be one of the most in demand type of content on smartphones. Every smartphone has global positioning system (GPS) module inbuilt in them which when connected to wireless Internet can make instant location tracking possible. A subset of users in a specific location can be easily tracked by using a GPS enabled mobile device.

3.2 Hyperlocal Forecasting

The hyperlocal experience aspect is dependent on a process known as hyperlocal forecasting. Hyperlocal forecasting is an artificial intelligence-based solution that takes into account each demand tag of a user and their location to provide accurate forecasts on the most basic level. One of the features of hyperlocal forecasting is that it allows the forecasts made to be extremely accurate. Instead of having a general demand–supply ration between the learners and educators, hyperlocal forecasting allows the systems to easily predict if a particular community with a dense population of teenagers is more likely to be interested in picking vocational courses over academic courses. This allows the local service providers to cater to the needs of such communities than to be targeting a geographically infeasible region.

3.3 *Architecture*

The product offers users with a simple Web application that allows users to enter personal details, choice of interests/preferences of learning and other demographic data to help understand the user better. Once the users' data is collected, the application presents him/her with a map-based UI that shows various hotspots, most happening or live feeds of results based on the interests of the user. The user has to simply hover around the interface and tap on his/her liking. The default filters are customized to only offer the solution and services that are geographically closer to the users' location. A hyperlocal engine is the powerhouse to the model being offered. As shown in Fig. 3, the hyperlocal engine is an integration of machine learning algorithms' and map APIs that help in forecasting accurate results. This involves supervised learning, unsupervised learning, and reinforcement learning.

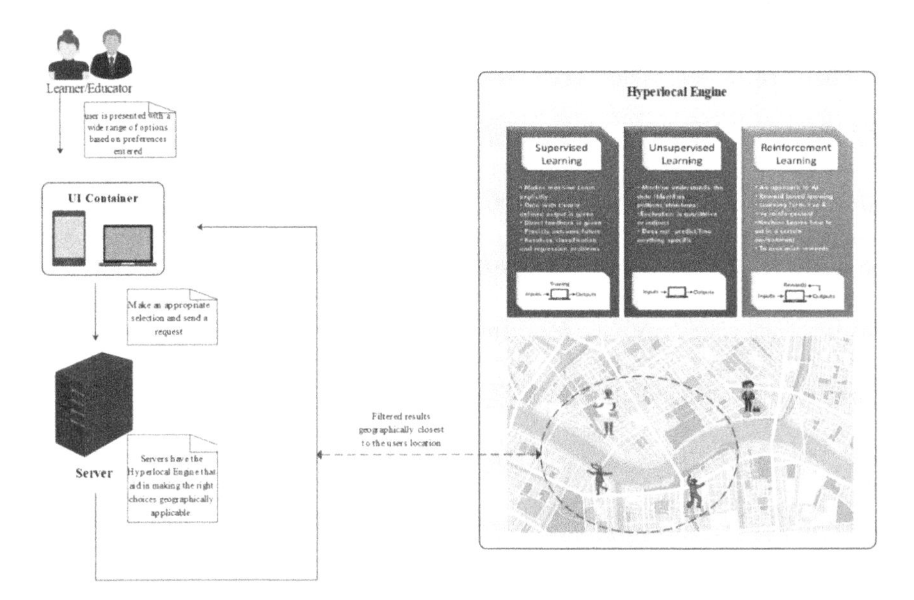

Fig. 3 Sociocultural learning using hyperlocal experience—architecture

3.4 Methodology

Supervised learning helps the machine learn explicitly. The output of this learning model is data that is clearly defined with direct user feedback taken into consideration. It helps resolve problems due to regression and classification. With data collected over a period, supervised learning helps predict outcomes in the future. Unsupervised learning is an indirect and qualitative-based learning model which allows the machine to understand data by identifying patterns and structures. It is incapable of making predictions. Reinforcement learning is an artificial intelligence-based learning process that mainly works on a reward-based approach. The machine learns decision making and is capable of acting accordingly in simulated environments. Being a reward-based model, the outcomes of this model are more dynamic compared to the other two learning methods. Location-based services are class of computing services that use the consumer's location data in order to personalize and enable features and services. Services include those for mapping and tracking, emergency response and disaster management, and location-based advertising. Of these services, location-based advertising, also known as hyperlocal marketing, accounts for the largest market share [6].

Figure 4 shows a modular view of how hyperlocal experience is used to generate tailored results. User requests from the application are directed to a hyperlocal engine that is deployed on the remote servers. These user demand requests are run through various hyperlocal forecasting and machine learning algorithms. The data that is processed consists of various filters, log files, and other system properties to be compared to historical data to provide accurate forecast results. The options presented to the user post forecasting is said to be the hyperlocal tailored results.

Figure 5 shows how a hyperlocal experience can be conceptually applied to enhance learning. The whole idea of going hyperlocal is to immensely benefit the local communities and building responsible peer-to-peer oriented learning environments that help eradicate social differences in being able to get access to learning.

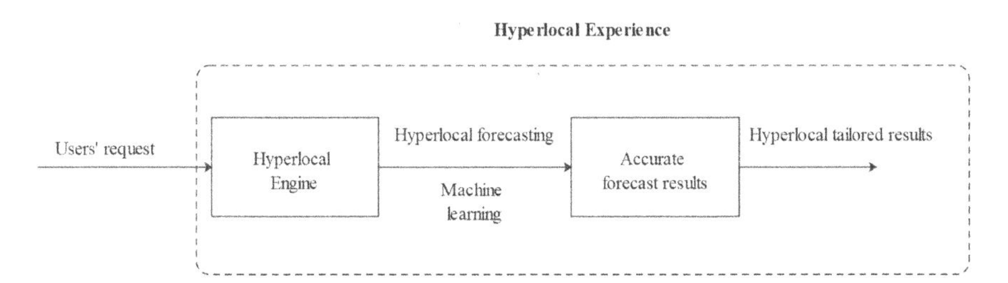

Fig. 4 Generating tailored results for user requests using hyperlocal experience

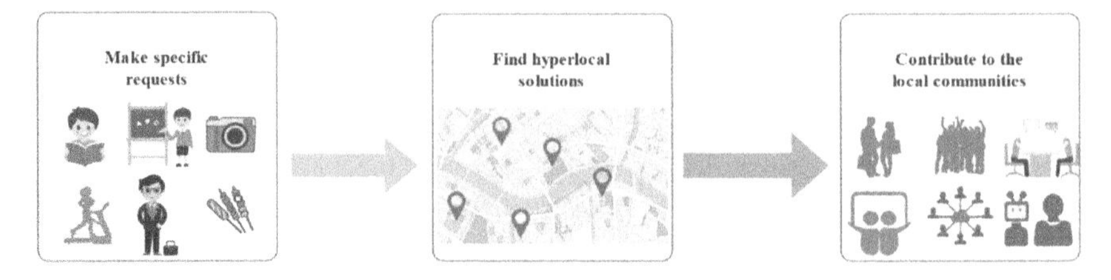

Fig. 5 Conceptual hyperlocal experience applied to learning

3.5 Proposed Solution

The solution provides the users with an interface that presents them with results from the hyperlocal area catering to their specific learning needs. The basic interface collects data like name, age, gender, and e-mail from the user on his/her first visit. It also collects the user's learning interests and preferences concentrating mainly on areas like educational, hobbies like music, dance, photography, or any other vocational areas like cooking, baking, art, etc. The users will also have the option to mention whether they would be interested in taking up the role of an instructor or a learner in the mentioned areas of interest. On every subsequent visit to the application, the user is presented a map-based interface showing hotspots/most happening/closest solutions based on choice of interests. Users will be able to scroll through the application to find GUI based reviews, ratings, walkthrough, 30 s—stories based advertisement/real-time feed which is similar to stories on Instagram and Facebook. This will help the users choose the best options for their learning needs within their vicinity.

The proposed solution aims at overcoming the challenges of the existing e-learning platforms like disconnectedness, i.e., lack of one-to-one interaction with the instructors, time and schedule conflicts, technological issues, and rigid approach to learning. It is evident from the following figures that users are open to experimenting newer methods to online-learning which include a mix of online and offline classes, flexible structure of learning, time, and location suitable to learner and instructor. This learning model can offer all these features to its users only through a hyperlocal collaboration. This model describes an approach to infer the location of a social media post at a hyperlocal scale based on its content, conditional to the knowledge that the post originates from a larger area such as a city or even a state [7].

Users were asked if they were open to being part of a system which allowed for both online and offline means of learning depending on their respective interests. Figure 6 clearly shows that 85% of the users were open to being part of an idea that is being proposed in this paper.

On being asked if users would like a model that allowed them to be able to go to classes at their own time and location than being confined to a schedule set by the organizers, 57.5% agreed that a model that let them control their schedules were more optimal (Fig. 7). On further being asked if they would benefit better from courses

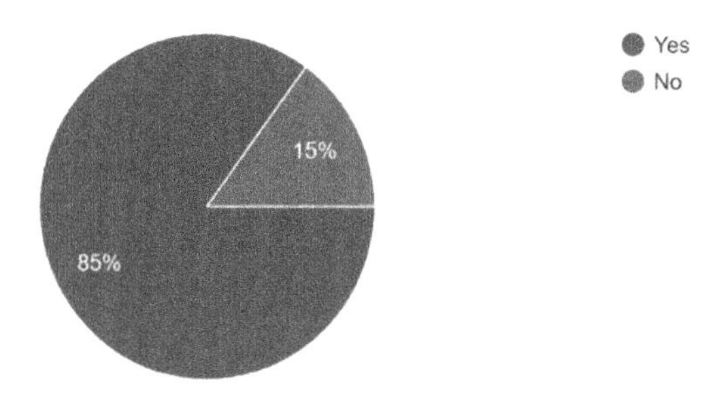
Are you open to the idea of being part of a course that is held both online and offline?
207 responses
Yes
No
15%
85%

Fig. 6 Users opinion on being part of a course that is a mix of both online and offline

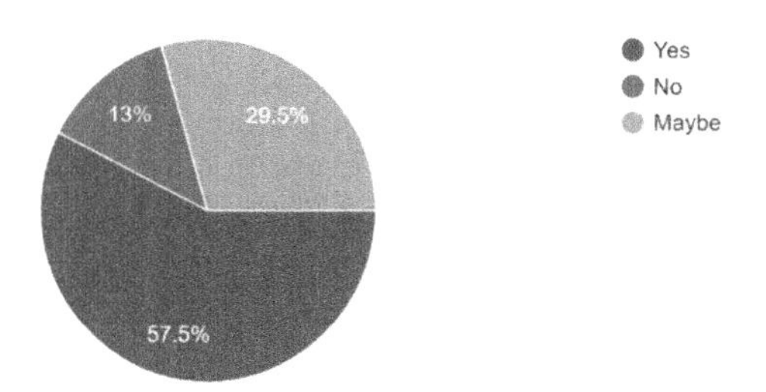
Would you like being able to go to class at times and locations that you choose instead of being tied to a set time and place?
207 responses
Yes
No
Maybe
13%
29.5%
57.5%

Fig. 7 Attending classes at times and locations that is picked by the user vs. being made to follow a fixed time and location schedule

that were more flexible and open to ones learning needs, 81.2% of the users agreed to a model that was tailor made to user needs (Fig. 8).

3.6 Algorithm

Reinforcement learning (RL) is the science of making optimal decisions using experiences. The process of reinforcement learning involves these simple steps:

- Observation of the environment
- Deciding how to act using some strategy
- Acting accordingly
- Receiving a reward or penalty
- Learning from the experiences and refining our strategy
- Iterate until an optimal strategy is found.

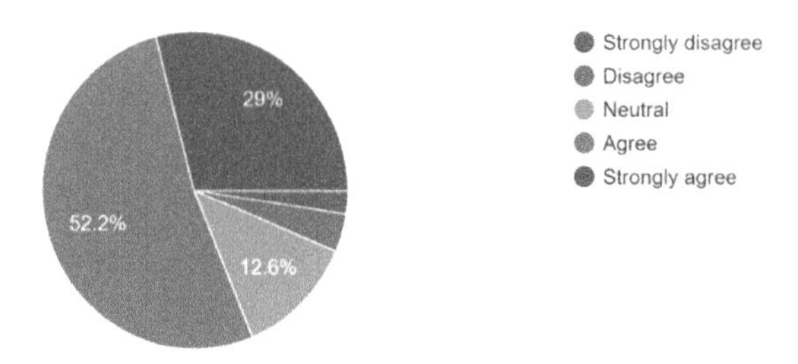

Fig. 8 Users' opinion on benefitting from flexible and open course structures

The objective of RL is to maximize the reward of an agent by taking a series of actions in response to a dynamic environment (Figs. 9 and 10).

Ant colony optimization (ACO) algorithms have been used successfully in the past to tackle combinatorial optimization problems in dynamic environments as its inbuilt mechanisms allow it to adapt to new environments [8]. Our proposed model uses the Q-learning algorithm. Q-learning is a model-free reinforcement learning algorithm. It is a values-based learning algorithm. Value-based algorithms update the value function based on an equation (particularly Bellman equation). However, the other type policy-based estimates the value function with a greedy policy obtained from the last policy improvement.

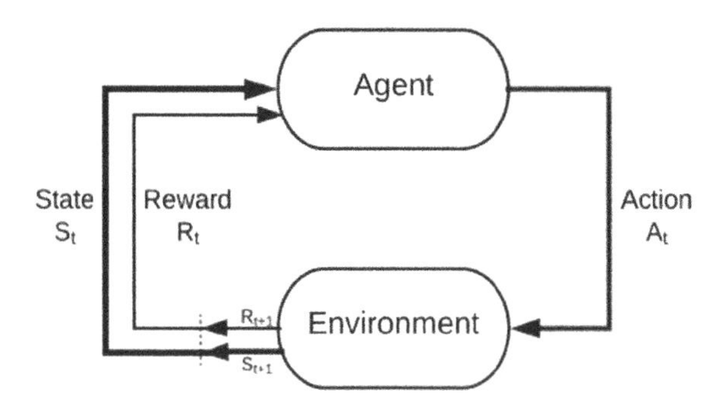

Fig. 9 Four basic components in reinforcement learning: agent, environment, reward, and action

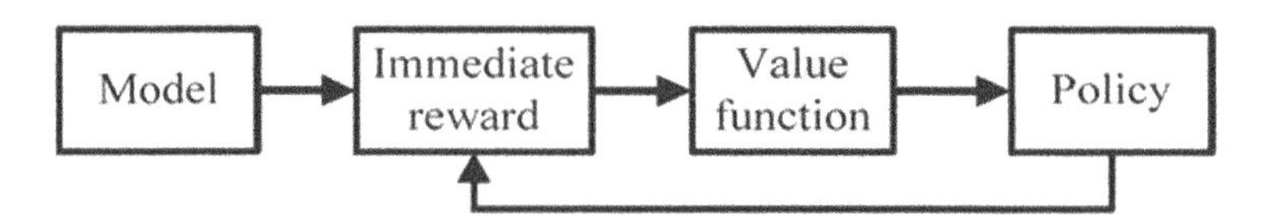

Fig. 10 Reinforcement learning system

Q-learning is also an off-policy learner. It is capable of learning the value of the optimal policy independent of the agent's actions. On the other hand, an on-policy learner learns the value of the policy being carried out by the agent, including the exploration steps, and it will find a policy that is optimal, taking into account the exploration inherent in the policy. The 'Q' in Q-learning stands for quality. Quality here represents how useful a given action is in gaining some future reward.

Q-learning can be defined as:

- Q * (s, a) is the expected value (cumulative discounted reward) of doing a in state s and then following the optimal policy;
- Q-learning uses temporal differences (TD) to estimate the value of Q * (s, a). Temporal difference is an agent learning from an environment through episodes with no prior knowledge of the environment;
- The agent maintains a table of Q [S, A], where S is the set of states and A is the set of actions;
- Q[s, a] represents its current estimate of Q * (s, a).

The focus of this paper is to:

- Propose a mathematical online community-based dynamic customization model
- Explain its practical mechanism
- Solve its dynamic trade-off challenge [9].

In order for Q-learning algorithm to be implemented, it is necessary to have a Q-Table. A Q-Table is the data structure used to calculate the maximum expected future rewards for action at each state. Basically, this table will guide us to the best action at each state. To learn each value of the Q-table, Q-learning algorithm is used (Fig. 11).

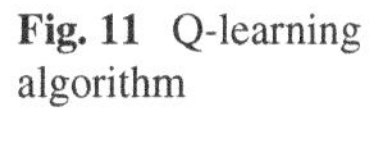
Fig. 11 Q-learning algorithm

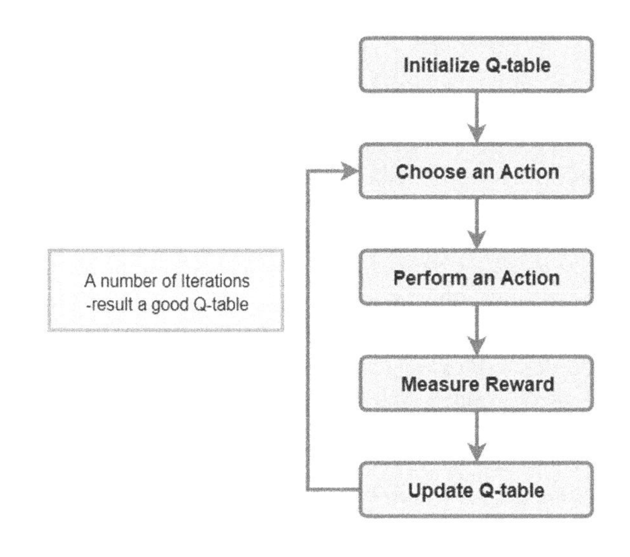

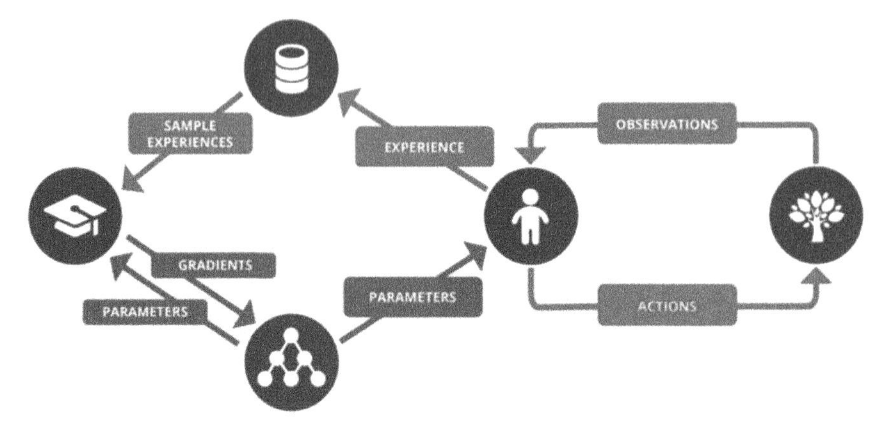

Fig. 12 User experience using the Q-learning algorithm

When Q-learning is applied to the proposed system, it is expected to efficiently enhance the user experience by the means of a constant observation-action change method. This automatically allows the machine to deliver the most appropriate results to enable the users to have a seamless experience. The focus is on information diffusion within users of social media to share data within a targeted audience of either a known friend or cluster of locally connected users. As large heaps of information are globally shared within seconds, the online presence for business, marketing, and customer services can be benefited by our idea and implementation of our proposal [10] (Fig. 12).

4 Conclusion

Users need an instantaneous gratification and ease to move between physical and online experiences. However, the current online learning experiences are disengaged and non-personalized. The proposed model offers a location-based technology to improve the user's learning experience. Owing to a change in the user outlook, to provide a hyperlocal experience, the use of location-based technology needs to be considered. This enables interaction with user's in proximity to enhance their experiences and increase learning efficiency. The advantage of the solution is that it is cost effective since it concentrates only on the region surrounding the user. In addition to providing advantages to the learning community, it also offers benefits to the individuals who have not undergone formal training to teach/instruct but are passionate about sharing their knowledge and expertise with the community. One of the disadvantages of this model could be related to the privacy of users since the results are optimized based on their geographical and demographical parameters. This could be overcome with future enhancements such as an advanced verification system that can be incorporated during the onboarding process. The system could also serve genres that currently do not have any online engagement such as offline photography clubs, trekking, and biking clubs. One other aspect that would enable more footprint toward the proposed model is hyperlocal marketing and monetization. Local community solutions and services could easily use the proposed system for quick outreach programs that could enable more traffic toward their offering. This also enable more freelancers and other on-demand B2C services to restructure their monetization methods.

5 Future Enhancements

The proposed model mainly focuses on providing a hyperlocal service that caters to vocational and independent learning. The same could be scaled up and customized to be applied on mainstream learning at schools and colleges that have a rigid curriculum and structure. It should qualify to provide an on-demand skill-based learning, offered by the local communities.

Another enhancement is to design a model that allows toddlers to be a part of the system. Here, the parents can benefit from finding pre-schools and day-care services offered within the vicinity of either parents' workplace. The motive here is to help connect small-time players who offer quality services in the industry to their potential opportunities.

Lastly, a hyperlocal service that is specially made for kids that require special care and attention in the society. Many a times, these kids are simply denied leading and to experience a normal life for no fault of theirs. To help kids and parents overcome the issues they face in terms of finding activities specially meant for the small group, the proposed model could help identify therapy sessions and other activities that should help in bringing about a difference in how these children go about their routine otherwise.

References

1. West RE (2018) Foundations of learning and instructional design technology—Chapter 12. In: Sociocultural perspectives of learning. Current Trends Press
2. Berge Z, Collins M (1995) Computer-mediated communication and the online classroom in distance learning. In: Collins MP (eds) Computer-mediated communication and the online classroom, vol III, pp 91–104. Hampton Press Inc., Cresskill
3. Harasim LN, Hiltz SR, Teles L, Turoff M (1995) Learning networks: a field guide to teaching and learning online. The MIT Press, Cambridge
4. Gabrielle DM (2001) Distance learning: an examination of perceived effectiveness and student satisfaction in higher education. In: Proceedings of society for information technology & teacher education international conference (SITE), pp 183–188. AACE, Orlando, FL
5. Hiltz SR (1997) Impacts of college-level courses via asynchronous learning networks: some preliminary results. J Asynchronous Learn Netw 1(2):1–19

6. Narayanan V, Rehman R, Devassy A, Rama S, Ahluwalia P, Ramachandran A (2014) Enabling location based services for hyperlocal marketing in connected vehicles. In: International Conference on Connected Vehicles and Expo (ICCVE), pp 12–13. IEEE Press, Vienna. https://doi.org/10.1109/ICCVE.2014.7297526
7. McClanahan B, Gokhale SS (2015) Location inference of social media posts at hyper-local scale. In: 3rd international conference on future internet of things and cloud, pp 465–472. Elsevier Publishing Company, Rome, Italy. https://doi.org/10.1109/FiCloud.2015.71
8. Zhang Z, Cheng W, Song L, Yu Q (2009) An ant-based algorithm for balancing assembly lines in a mass customization environment. In: International workshop on intelligent systems and applications. IEEE, Wuhan, China, pp 1–4. https://doi.org/10.1109/IWISA.2009.5072706
9. Wang Y, Wu J, Lin L, Shafiee S (2020) An online community-based dynamic customisation model: the trade-off between customer satisfaction and enterprise profit. In: International Journal of Production Research. Taylor and Francis Ltd, Milton Park
10. Das U, Basu A, Mohapatra GP, Chowdhury S (2019) Hyperlocal based support and information diffusion in social media. In: International conference on information technology (ICIT), pp 406–411. IEEE CS-CPS, Bhubaneswar, India. https://doi.org/10.1109/ICIT48102.2019.00078

Subsequent Technologies Behind IoT and Its Development Roadmap Toward Integrated Healthcare Prototype Models

Priya Dalal, Gaurav Aggarwal, and Sanjay Tejasvee

Abstract Internet of things (IoT) is a propelled research region that gives further automate, examination, and combination of the physical world into electronic devices and PC frameworks through system foundation. It permits collaboration and participation between a massive assortment of unavoidable objects over remote and wired associations to accomplish explicit objectives. As IoT has some imperative properties like dissemination, receptiveness, interoperability, and dynamicity, their creating presents an incredible test. So, it is an advancement of the Internet and has been increasingly expanded consideration from scientists in both scholarly and modern situations. Progressive mechanical improvements make the advancement of savvy frameworks with a high limit with regard to correspondence and information assortment conceivable giving a few chances to various IoT applications. To permit everybody to encounter the IoT by observing and feeling the possibilities of primary use cases through iterative prototyping. At last, to empower people, networks, and associations to think to envision, and the question "What's Next?" This explicitly concerns the administrations and applications made on the head of these regular use cases, to construct a significant IoT for humans. This paper presents the current status of the quality of IoT structures with attention on the innovations, possibilities, technologies behind IoT, a roadmap of IoT improvement, gateways of IoT, and manageability for model design, particularly in the recent healthcare sector. Moreover, this archive integrates the current assortment of information and distinguishes consistent ideas for new immense significance and direction of future research.

Keywords IoT · Automate · PC framework · Interoperability · Roadmap health care

P. Dalal
Department of IT, MSIT, New Delhi 110058, India
e-mail: priya@msit.in

G. Aggarwal (✉)
Department of Computer Science, JIMS, Bahadurgarh 124507, India
e-mail: gaurav.aggarwal@jagannathuniversityncr.ac.in

S. Tejasvee
MCA Department, Government Engineering College Bikaner, Bikaner 334004, India

A. Choudhary et al. (eds.), *Applications of Artificial Intelligence and Machine Learning*, Lecture Notes in Electrical Engineering 778,
https://doi.org/10.1007/978-981-16-3067-5_6

1 Introduction

Today IoT uses various tools for optimization and automation. The arrangements of optimization and automation show practically interoperability abilities to produce explicit communications. Besides, as IoT single roof like an umbrella wraps up an exclusively whole application in several fields, apparently improvement sequence and modernization transformation incredibly. As an outcome vertical and segregated arrangements rise while just a progressively flat methodology, where application storehouses share a typical specialized establishing and usual design standards, could, in the long run, lead to a full-fledged Internet of things [1, 2]. So, as to keep up and improve individuals' life quality in all times of life, however, especially for more seasoned grown-ups, surrounding helped living stays a multi-disciplinary field that is carefully identified with a biological system of various advances and applications for individual healthcare services checking and unavoidable and pervasive processing [3, 4]. The ascent of the advanced economy, as the majority of the seismic innovation shifts in the course of recent hundreds of years, has on an elementary level changed innovation as well as a business also. The very idea of the "digital economy" keeps on advancing. Where once it was only a segment of the economy that was based on advanced innovations, it has developed getting practically unclear from the "conventional economy" and extending to incorporate almost any change, for example, portable, the Internet of things, cloud computing, and expanded knowledge [5]. There is a core of the digital economy the fundamental need to interface different information regardless of where it lives. This has prompted the ascent of utilization combination, the need to associate various applications and information to convey the best knowledge to the individuals and frameworks who can follow up on it [6].

IoT incorporation implies making the blend of new IoT devices, IoT information, IoT stages, and IoT applications—joined with IT resources (business applications, heritage information, versatile, and SaaS [Software as a service]) function admirably together with regards to executing start to finish IoT business arrangements [7, 8]. So IoT is an umbrella term that incorporates numerous various classes of advances, for example, wireless sensor/actuator systems, Internet-associated wearable, low force inserted frameworks, RFID empowered following, sensors, shrewd cell phones, devices that interface employing Bluetooth-empowered cell phones to the Internet, smart homes, associated vehicles, and some more. The outcome is that no single engineering will suit every one of these territories and the prerequisites every region brings. In any case, measured adaptable engineering that supports including or taking away abilities, just as supporting numerous necessities over a wide assortment of these utilization cases is naturally helpful and essential. It gives a beginning stage to engineers hoping to make IoT arrangements just as a solid reason for a new turn of events [9, 10].

IoT is an environment of associated physical articles that are open through the Internet. As per the Gartner report, by 2020 associated gadgets overall advancements will arrive at 20.6 billion. According to the Cisco report, IoT will produce $14.4

trillion in esteem over all enterprises in the following decade. The entirety of this will prompt better openings for work [11, 12] (Source: Cisco IBS, 2011 Cisco and/or its affiliate and Press Releases Gartner, Inc. (NYSE: IT), Egham, U.K.).

This paper communicates the perspectives on innovations behind the IoT and IoT way of development with reference/model design. The design must cover numerous aspects including the cloud or server-side engineering that permits us to screen, oversee, associate with and process the information from the IoT gadgets the systems administration model to speak with the devices. The specialists and code on the gadgets themselves, just as the prerequisites on what kind of gadget can bolster this reference engineering.

2 Technologies Behind IoT

Mainly, IoT primarily utilizes networking technologies. In-depth, major enabling technologies and protocols used by IoT are RFID, NFC, Bluetooth, Wi-Fi, low-energy radio protocols, LTE-A, etc. [13, 14]. These technologies support the specific networking functionality needed in an IoT system in contrast to a standard uniform network of widespread systems. The following entities provide a diverse technology environment and are examples of technologies used behind IoT [15].

2.1 IoT Development Boards and IoT Hardware

IoT hardware includes a wide scope of devices and tools that forgive the direction of the sensors and accumulators. The hardware behind the IoT has the capacities of the actuation of the framework, provides security, action details, communication, and location of help explicit objectives and activities. Arduino Uno, Beagle Board, XDK 110 Bosch, ARM-MBed, Intel Edison, Intel Galileo, Raspberry Pi, and wireless SoC are the most widely recognized utilized equipment for the IoT model turn of events.

2.2 IDE (Integrated Development Environment)

IDE is the platform for programming gadgets, firmware, and APIs. The visual development tools are intended to be used as an IDE to assist the development cycle of the enhancement of the IoT framework [16]. The devices that have been introduced depend on a conceptual IoT design and guarantee a negligible coding condition for improving the IoT solution [17].

2.3 Protocols

The conventions are used on constant information stream systems to permit mashups of the information streams, include activities, and so forth. The last mashups are deployable in a conveyed situation. The framework particularly centers around the synthesis of information streams from Internet administrations and IoT gadgets with Internet interfaces. RPI, CoAP, RESTful, HTTP, MQTT, XMPP (extensible informing, and nearness convention) are the key conventions utilized for IoT-based designs. These IoT correspondence conventions take into account and meet the particular useful prerequisite of an IoT framework.

2.4 Network Spine

A core or center is a component of PC structure that integrates various device bits, allowing data to be exchanged between different LANs or inter-networks such as IPv4 (Low-Power Wireless Personal Area Network4), IPv6, UDP, and 6LowPAn.

2.5 Internetwork Cloud Stages/server

For mechanical organizations, ThingWorx is one of the key IoT levels, which gives devices easy connectivity. IoT Suite on Microsoft Azure. IoT Platform for Google Cloud, IBM Watson IoT Platform, AWS IoT Platform, and Cisco IoT Cloud Link.

2.6 Software

Programming comprises of two components: programming at IoT gadget and programming the Server RIOT OS, Contiki OS Thing square Minst Firmware Eclipse IoT, and so on (Fig. 1).

3 IoT Gateway Device

An IoT gateway is a response to empower IoT correspondence, generally gadget-to-gadget interchanges or to cloud interchanges of gadgets. The door is normally a programming device lodging equipment gadget that performs fundamental tasks [19]. As an equipment gadget or a virtual programming code, the IoT gateway acts

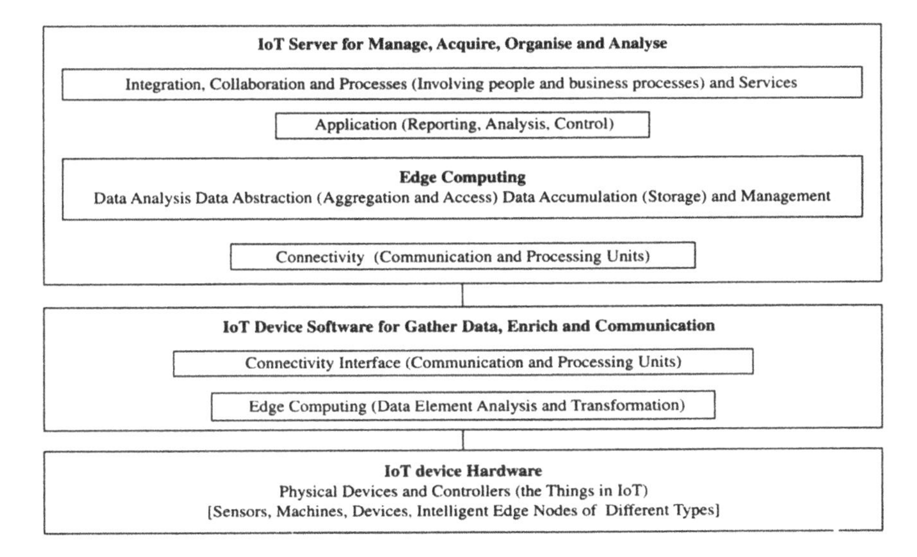

Fig. 1 IoT S/W components device hardware [18]

as a link between both the IoT sensor nodes and the database server on the cloud [20].

Important IoT Gateway Device Errand: Facilitate similarity over IoT Organization. IoT gateway guarantees this by endorsing various communications conventions such as Zigbee, 6lowpan, Bluetooth, Wi-Fi, LoRA, Zwave All gadgets to be monitored or regulated have appropriate sensors (temperature, mugginess, closeness, or specific sensors) mounted on them. These sensors are IP-based; the IoT gateway deals with the cloud server network of these sensors (and thus genuine physical gadgets) [21, 22]. A versatile IoT gateway can perform any of the following functions:

(a) Store, buffer, and wipe data
(b) They facilitate correspondence with heritage or gadgets associated with non-web applications
(c) Building up some information
(d) Preparation, purification, sifting, and forwarding of data
(e) Diagnostic systems
(f) Device design to managers
(g) Information representation and critical information analysis using applications for IoT gateways
(h) Device correspondence device/M2M
(i) Security monitors the highlights of client access and device security
(j) Short-term professional information history highlights
(k) Highlights networking and live information facilitation.

4 IoT Development Roadmap and Healthcare Prototype

From the earliest starting point of the turn of events and usage of IoT, the multi-faceted nature of IoT items and frameworks is brought up in the extent of advanced business change and the circle of shrewd advances. IoT framework is identified with the meaning of an innovation stack, so the administration procedure appears to at the same time control numerous various advancements [23]. To coordinate all parts of an IoT framework (equipment, programming, conventions, stages, and so forth), it is important to introduce the IoT mechanical roadmap into a smooth utilitarian arrangement [24]. The very broadness of the idea and the guide depends on the more broad strides of organization and assessment of tech-stacks to deliberately characterize the innovation activity that aids the way toward establishing IoT items/environments including innovative work. As the IoT item/biological system contains IT/web advancements and different hierarchical procedures, the guide should assist us with defining the period in which the activities of mechanical procedures will be executed. At that point, it is significant in addressing the subject of naming the business objective that will be accomplished, for example, how innovation, applications, forms, and so forth will help the development of an association or framework that executes the innovation guide. Looking from the IoT framework and item viewpoint, an IoT innovation guide is an activity plan which needs to propose fixings and conduct inside the framework at the end of the day, a computerized change process with criticism dependent on information insight going with differing IoT items/resources, tasks, and individuals as clients just as workers inside advanced frameworks (Fig. 2).

Notwithstanding this intelligent division inside the IoT guide, it is basic to include another measurement that decides the quality of the IoT environment timescale. The course of events is significant as exact perceptions have been made in the past period

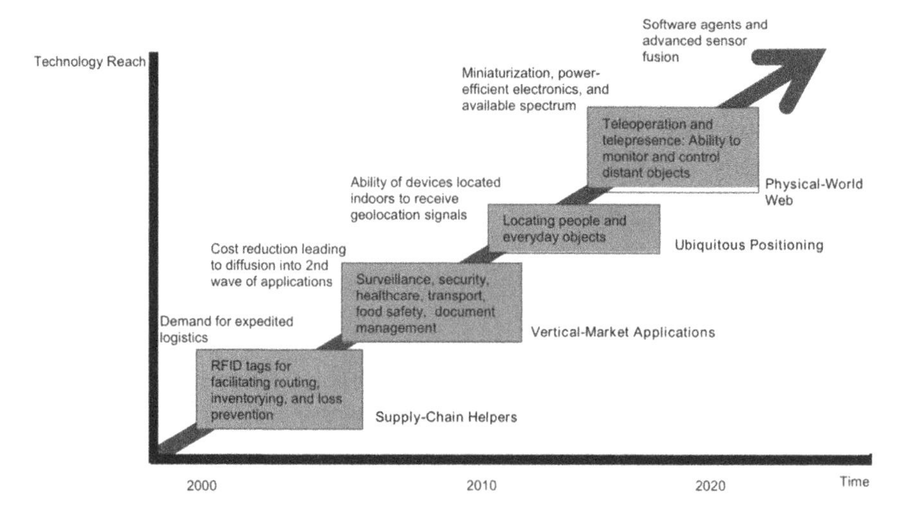

Fig. 2 Technology roadmap: IoT (*Source* SRI consulting Business Intelligence)

just as expectable gauges for the not so distant future. Such expectation can likewise be the reason for a more profound review of the IoT innovation stack around the current IoT biological systems and stages on which they are to a great extent based (Fig. 3).

A portion of the realized merchants advance their corporate dreams for the IoT guides, for example, Windows (Fig. 4).

At that point, AWS IoT way to deal with its guide is concentrated on genuine tech-stack with center around the current proposal of gadget programming, control administrations, and information administrations. At the majority of the solicitation, an IoT framework will be a working arrangement and consider all layers of the IoT innovation stack. Regardless of whether it is a business, mechanical, or associated home resale, the components that supplement IoT tech-stack, regardless of which kind of reference engineering is utilized at the base level are (Fig. 5).

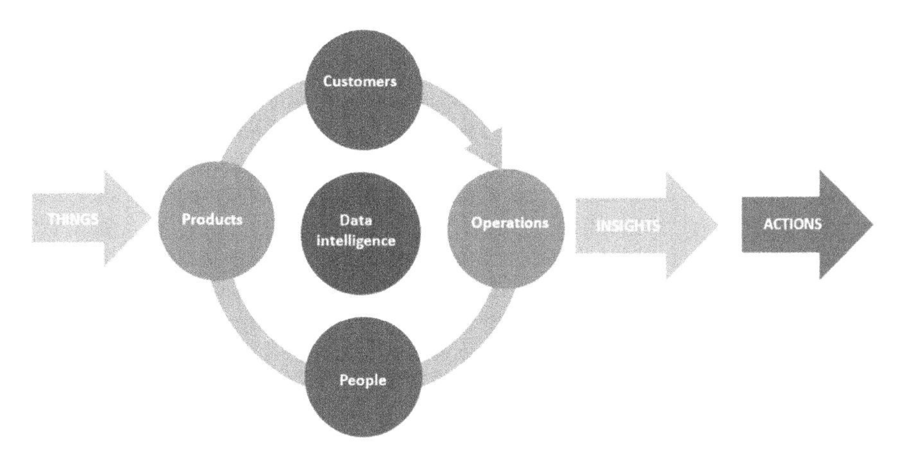

Fig. 3 Inside IoT action plan

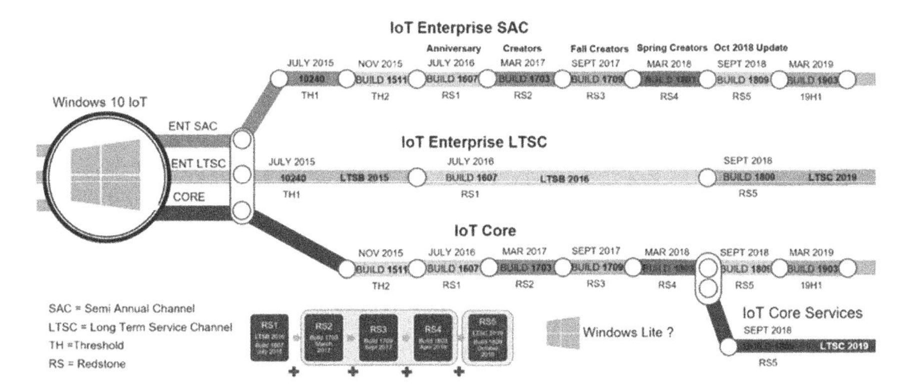

Fig. 4 Corporate visions for IoT roadmap

Fig. 5 Base levels of IoT-based prototype model (*Source* danielelizalde.com/iot-primer)

Spoken to components of the tech-stack has a place with the IoT reference engineering and its fixings, for example, sensors, actuators. Nonetheless, working of an elevated level IoT environment requires, notwithstanding these fundamental components, some more assets to make the IoT guide total and usable. A few distinct partners need to connect with to completely comprehend the standard and usefulness of things to come IoT biological system, for example, administrators, sales, marketing, and engineering must cooperate on the meaning of tech-stack. Moreover, it included the consideration of the expressive component, supposed story planning that brings a novel into a guide that improves comprehension of the subtleties, and finished the degree of IoT tech-stack. The mix of story planning in IoT tech-stack furnishes visual item accumulation with enough data that is required for everyday work. Likewise, by discharging and tending to vital layers, the IoT modeler can upgrade the method of building obstructs into a framework guide with watch out for the master plan yet with center around littler subtleties and therefore improve usefulness. Another progression of expansion IoT tech-stack to improve innovation guide is on including IoT choice system. This is identified with improved access to organize addressing and route through specific layers and to settle on better choices. The components around which the system is based are as per the following:

(a) **User Experience**
In this part, we are not considering specialized subtleties yet just client needs, attempt to comprehend an ideal way which is the client.

(b) **Data**
This speaks to the information decision area, a meaning of by and large information methodology and information move through all layers on an innovation guide. Choices of gadget information types are sort of examination and information volume for cloud activities.

(c) **Business Choices**
In this part, the business choices will be resolved; for example, the monetary capability of the IoT item/framework, as such, how to adapt the arrangement.

(d) **The Technologies**
This choice of territory will give a solid diagram of building nature about equipment needs, gadget programming, geography that can be utilized, correspondence conventions, and sensor portrayals. Additionally, if some cloud arrangement will be included, the choice components for that and propose necessities for a total specialized arrangement.

(e) **Security**
Decisions on how each layer could be undermined, how to react to if any of the gadgets are hacked, and what innovation to execute on making sure about tech-stack for the total guide.

(f) **Standards and Regulations**
Area to explain choices on guidelines and norms which can be influenced to each layer dependent on the kind of item, industry source, and client direction. Likewise some extra components will be announced, for example, similarity as indicated by gadget security, law to go along at each layer of the guide. At every individual layer, an assessment of every choice component is fundamental.

5 Conclusion

The Internet of things (IoT) stage can be used to gather data related to a specific geographic region using far away watching mechanical assemblies and perform examination so early reprimands of a disaster can be found. IoT can have a huge impact concerning far off prosperity checking. Notwithstanding the way that the patient is advancing, his vitals can be continually sent to the pros with the help of the IoT stage. The decision of fitting structure is essential to consider the business needs with the goal that the guide is done, and the last course of action is described. This together braces the accessibility of all accomplices during the time spent executing the IoT condition, smoothes out procedure seeing during progress, and testing similarly as extraordinary completed outcome execution which together brings customer reliability. IoT joins man-made cognizance, accessibility, sensors, dynamic responsibility, and little contraption use. These contraptions control, manage the major tasks, for instance, structure incitation, action subtleties, security, correspondence, and distinguishing proof to support unequivocal targets and exercises. In any case, focal points of IoT length over each zone of lifestyle and business by improving progression of advancement, decrease time, and update data variety methods yet some multifaceted nature as for security, assurance consistency notwithstanding everything exit. IoT has applications overall endeavors, markets, and sharp work especially use full to make keen healthcare applications with astute devices for the continuous situation for prosperity from COVID-19 furthermore. It navigates customer packs from the people who need to diminish imperativeness use in their home to enormous affiliations who need to streamline their exercises. IoT gives strong techniques for checking various strategies, and certified straightforwardness makes a progressively significant detectable quality for improvement openings. It exhibits supportive, yet practically essential in various undertakings as development advances, and we move toward the pushed motorization imagined in the far away future IoT limits in a relative and progressively significant way to current development, assessment, and enormous data. Existing advancement accumulates unequivocal data to make related estimations and models after some time, in any case, that data normally needs significance and precision. Thus, IoT enhances and develops this by studying and observing even more procedures and activities, and with amazement investigating them.

References

1. Universal open platform and reference specification for ambient assisted living. https://www.universaal.info/. Retrieved as on 9 July 2020
2. Junaid M, Abhinav T, Ocneanu AF, Colin J, Chung HL, Andy A et al (2014) Internet of things: remote patient monitoring using web services and cloud computing. In: 2014 IEEE international conference on internet of things (iThings 2014). Green computing and communications (GreenCom 2014), and cyber-physical-social computing (CPSCom 2014); 978-1-4799-5967-9/14. IEEE, pp 256–226. https://doi.org/10.1109/iThings.2014.45
3. Dohr A, Moore-Opsrian R, Drobics M, Hayn D, Schreier G (2017) The IoT for ambient assisted living. In: Proceedings
4. Vanya Z (2015) Identity management approach in the internet of things. Master Thesis, Aalborg University
5. Stankovic JA (2014) Research directions for the IoT. IEEE Internet Things J 1(1):3–9. https://doi.org/10.1109/jiot.2014.2312291
6. Tyagi S, Agarwal A, Maheshwari P (2016) A conceptual framework for IoT-based healthcare system using cloud computing. In: 2016 6th international conference—cloud system and big data engineering (confluence), Noida, pp 503–507. https://doi.org/10.1109/CONFLUENCE.2016.7508172
7. Agile integration architecture using lightweight integration runtimes to implement a container-based and microservices-aligned integration architecture. https://www.ibm.com/downloads/cas/J7E0VLDY. Retrieved as on 11 July 2020
8. Sethi P, Sarangi SR (2017) Review article internet of things: architectures, protocols, and applications. Hindawi J Electric Comput Eng. Article ID 9324035, 25p. https://doi.org/10.1155/2017/9324035
9. Whitmore A, Agarwal A, Da Xu L (2015) The internet of things—a survey of topics and trends. Inf Syst Front 17(2):261–274
10. Mhlaba A, Masinde M (2015) An integrated internet of things based system for tracking and monitoring assets—the case of the Central University of Technology. https://wso2.com/whitepapers/a-reference-architecture-for-the-internet-of-things/#04. Retrieved as on 14 July 2020
11. https://www.gartner.com/en/information-technology/glossary/iotintegration. Retrieved as on 12 July 2020
12. Evans D (2011) Cisco internet business solutions group the internet of things how the next evolution of the internet is changing everything. https://www.cisco.com/c/dam/en_us/about/ac79/docs/innov/IoT_IBSG_0411FINAL.pdf. Retrieved as on 12 July 2020
13. Weyrich M, Ebert C (2016) Reference architectures for the internet of things. IEEE Softw 33(1):112–116
14. Jeong Y-S, Shin S-S (2017) An efficient healthcare service model using IoT device and RFID technique in the hospital environment. J Adv Res Dyn Control Syst 10(Special Issue). ISSN 1943-023X
15. https://www.tutorialspoint.com/internet_of_things/internet_of_things_technology_and_protocols.htm. Retrieved as on 14 July 2020
16. Akhila V et al (2017) An IoT based patient health monitoring system using Arduino Uno. Int J Res Inform Technol 1(1):1–9
17. Kefalas N, Soldatos J, Anagnostopoulos A, Dimitropoulos PA (2015) Visual Paradigm for IoT solutions development. In: Interoperability and open-source solutions for the internet of things. Springer, Berlin, pp 26–45

18. http://www.mhhe.com/rajkamal/iot. McGraw-Hill TMH, IoT Architecture and Design principles by RajKamal. Retrieved as on 14 July 2020
19. Kleinfeld R, Steglich S, Radziwonowicz L, Doukas CG (2014) Things: a mashup platform for wiring the internet of things with the internet of services. In: Proceedings of the 5th international workshop on web of things, Cambridge, MA, USA, 8 October 2014. ACM, New York, pp 16–21
20. Niharika K (2017) IoT architecture and system design for healthcare. In: Systems international conference on smart technology for smart nation. IEEE, 978-1-5386-0569
21. Kumar R, Rajasekaran MP (2016) An IoT based patient monitoring system using raspberry Pi. In: International conference on computing technologies and intelligent data engineering (ICCTIDE), pp 1–4
22. https://openautomationsoftware.com/blog/what-is-an-iot-gateway/. Retrieved as on 15 July 2020
23. https://www.embitel.com/blog/embedded-blog/understanding-how-an-iot-gateway-architecture-works. Retrieved as on 16 July 2020
24. https://www.novatec-gmbh.de/blog/iot-technology-roadmap/. Retrieved as on 17 July 2020

Big Data and Data Mining

Bug Assignment-Utilization of Metadata Features Along with Feature Selection and Classifiers

Asmita Yadav

Abstract In open source software bug repository, lots of bugs are filed or reported in a single days and to handle these with manually is difficult and time consuming process. To build an automatic bug triager is a good way to resolve these bugs efficiently with the use of machine learning based classifiers. This proposed work is built the triager by utilizing the bug metadata features like product name, keywords, bug summary and component name which are extracted after applying the feature selection algorithms that play the important role in triaging process. To measure the prediction accuracy, ML based classifiers like NB, SVM and DT are applied on the dataset. Four open source projects based datasets (Eclipse, Netbeans, Firefox and Freedesktop) are castoff to experiment the En-TRAM triager and achieved the improved outcomes from the state-of-art that goes to approximately 19%. The proposed work has empirically proved that selective inclusion of highly ranked metadata fields improves the bug triaging accuracy.

Keywords Bug classification · Triaging process · Machine learning algorithms · Feature selection · Bugs

1 Introduction

Software bugs are inevitable in open source software projects which occurred in any software programs in the form of mistake or error [1]. Thus, fixing of bugs is of upmost priority when they are reported in the system although daily numbers of bugs are reported in the open source repositories (like- Bugzilla, Mozilla, Eclipse or Jira) by the end user or developers. So, handle these bugs are difficult as well as time consuming process. These software bugs are stored in the software bug repositories that contained all the important information about the software bugs [2, 3]. Typical format of bug report is depicted in Fig. 1 that contains the details about it in terms of various metadata features like: Bug-id, summary, bug product name, bug component name, platform name, bug priority and bug severity, assignee and

A. Yadav (✉)
ABESEC, Ghaziabad, India

A. Choudhary et al. (eds.), *Applications of Artificial Intelligence and Machine Learning*, Lecture Notes in Electrical Engineering 778,
https://doi.org/10.1007/978-981-16-3067-5_7

Fig. 1 Example of BugzillaMozillae Bug Report (#253,291)

bug history [4, 5]. These bug reports provide vital information to the assignee or developer in understanding the bug problem faced by the end user, while using the software. The contents of the reports are also useful in fixing the bugs.

As shown in Fig. 1, bug reports contain several metadata features which are filled by the reporter at the time of bug reporting. Few of the bug features are generated during the bug life cycle that contains finer details about the bug which are hidden in the form of hyperlink. The details of few bug report metadata (namely: Bug-id, summary, product, component, platform, assignee, reporter, severity, priority, status, etc.) from BugzillaMozilla repository as demonstrated below in Table 1. These bug features help the developers in well understanding of bug-report-like bug priority

Table 1 Bug metadata feature description

Feature name	Description
Bug-Id	Represent an exclusive number id to the bug report
Assignee	Developer who finally resolve the bug issue
Component	After classification, second-level categories are components; each belongs to a particular product
Product	Bugs are categorized into products and components. Select a classification to narrow down this list
Creation date	Bug reported timestamp
Keywords	A unique and distinct work to represent the bug summary
Reporter	Any of the end user or developer who filed this bug report
Status	Represent the bug state in the form of- unconfirmed, new, assigned, resolved, closed and reopened etc.
Resolution	The resolution field indicates what happened to the respective bug report
Summary	The bug summary is a short description of the problem
Priority	Prioritize bugs, like: P1,P2,P3,P4,P5, according to its resolution urgency

Table 2 Classifier prediction accuracy with and without feature selection extraction technique

	Recall	Precision	F-score
WOFS algorithm	0.642	0.628	0.599
WFS algorithm	0.759	0.749	0.699

level tells about the urgency of bug, summary contains the brief detail about the bug issue.

Previous works is not utilizing the feature selection (FS) or extraction techniques to build better bug triager. Although, state-of-art used the various bug metadata features to train the classifiers without applying any FS techniques. In this approach, we only worked with those features which are extracted from the FS algorithms namely: Bug report summary that tells the brief description about the bug, keyword that tells a unique word that explain the bug in short, bug product and bug component name and the name of the developer who resolve or report the various types of bugs [6–12]. Some of the important the bug metadata are briefly explained in Table 1.

To validate the importance of the selection of bug metadata features, we have done few experiments on four open source projects in order to rank and select the best metadata feature from the data corpus to train the classifiers. Unlike the most of the work where feature selection was done by using one technique, we empirically selected the best metadata using three feature selection techniques: (i) reliefAttribute, (ii) gainratioAttributeEval and (iii) CFS SubsetEval [7]. These FS algorithms are reduced the noisy bug features from the datasets and ranked the selected features. To measure the importance of FS algorithms, we tested the triager likelihood correctness over without using FS algorithm (WOFS) and with FS algorithms (WFS) is shown Table 2. In primary step, we considered all the reported bug metadata features, where WOFS as name implies it took entire bug metadata features to train the classifiers, on the other hand WFS only used selected high impact bug metadata. These three classifiers are used to measure the prediction accuracy on both conditions, i.e., WFS and WOFS. The means results of all three classifiers produced better accuracy with the use of bug metadata feature selection as compare to WOFS based classifiers results and this is happened due to the removal of unnecessary data from bug reports that only creates the noise in the dataset instead to play a role for accuracy improvement.

We managed this paper in 5 sections, where Sect. 1 contains the brief introduction of the paper, Sect. 2 explained the state-of-art in the form of literature survey with the bug triaging process, En-TRAM working is explained in Sect. 3 and its experimental evaluation and result discussion is prepared in Sect. 4. Conclusion is accomplished in Sect. 5.

2 Literature Survey

Machine learning techniques stand the popular way to triage the bug reports. In this section, we reviewed few of the state-of-art. A bug tossing based method is projected by Jeong et al. [13], and this triager is built with the used of metadata features. This approach is refined in another work by Pamela et al. [14, 15]. This paper worked on multi-feature tossing graph by taking into account bug product name, bug component name and developer activity. Version (source) repository is also utilized for fixing the bug reports in the proposed work of Kagdi et al. [16] and not received as much as good results and shown 50% negative outcome for finding the relevant developers due to the noisy data. Ramin et al. [17] are also worked with source code repositories to extract the noun extraction term which are handled by the developers to find his/her expertise. Jifeng et al. [18] work is focused to reduce the noisy data to achieved good prediction accuracy. They built an approach with the combination of instance and features selection algorithms and successfully reduced 50% of data from the datasets. TRAM [19] paper is based on bug report metadata features which are extracted through VSM and tested the approach on four different open source projects and the triager prediction accuracy suffered owing to less used of bug features. In another paper, bug priority is also taken considered as a major factor to build bug triager [15]. In bug history, few of the important bug features are hidden-like bug timestamp. Calvalcanti et al. [20] and Ramin et al. [21] triager are utilized the bug timestamp metadata features and incorporate it with tf-idf based similarity measure. These method are based on the last activity of the developer on a particular term and tested on four classifiers- SVM, NB, VSM and SUM and achieved upto 45% improved accuracy with the state of work.

All the above approaches are built the bug triager based on various metadata features which are chosen as randomly or with the taken the inspirations from the literature survey. Here, we applied FS algorithms to extract the important bug metadata features from the bug datasets and picked up finest features with topmost ranks.

3 Proposed En-TRAM Triager

Here, we are describing the methodology of En-TRAM Triager. The Framework and flowchart of En-Tram is depicted in Figs. 2 and 3. This En-Tram is divided into two phases: Modeling (Training) and Testing Phase (Fig. 4).

1. In modeling phase, first we have to applying the pre-processing on bug metadata features. Pre-processing is done on unstructured bug dataset to remove irrelevant features. To diminish the noise of the bug reports, the text pre-processing is performed for selection of bug metadata features. Text pre-processing removes all the whitespace, punctuation, numbers and stopwords from the bug reports. Stemming is applied on the preprocessed bug reports. In next step, a term frequency (TF) matrix is built by using most discriminating terms from the bug reports. In Table 2, we shown that a bug report features, in which only most important features are extracted by using Feature selection algorithm, followed by Ranker algorithm. These selective features are ranked according to their weight score. Bug product name, bug summary, bug component and

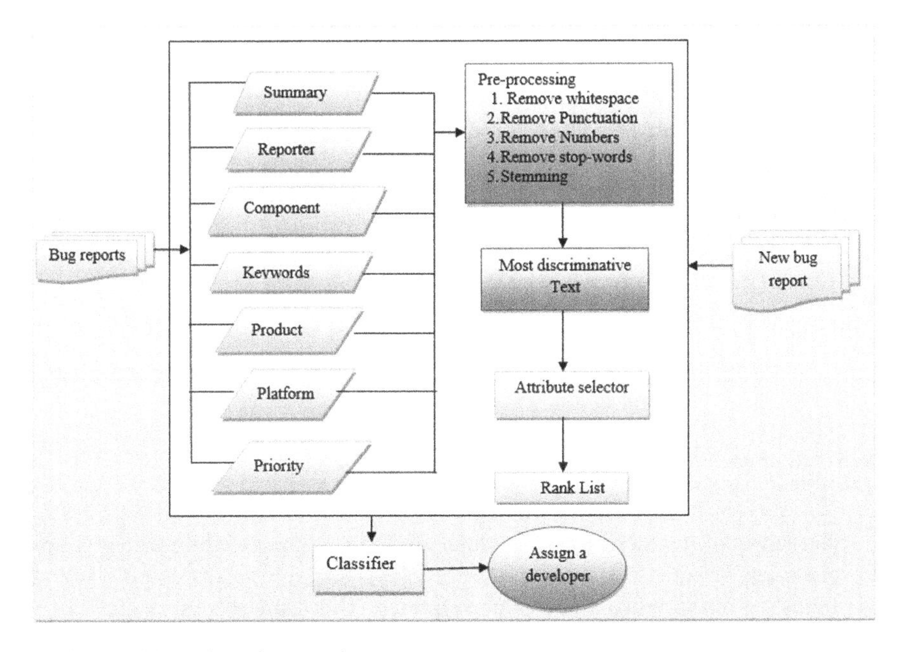

Fig. 2 En-TRAM triager framework

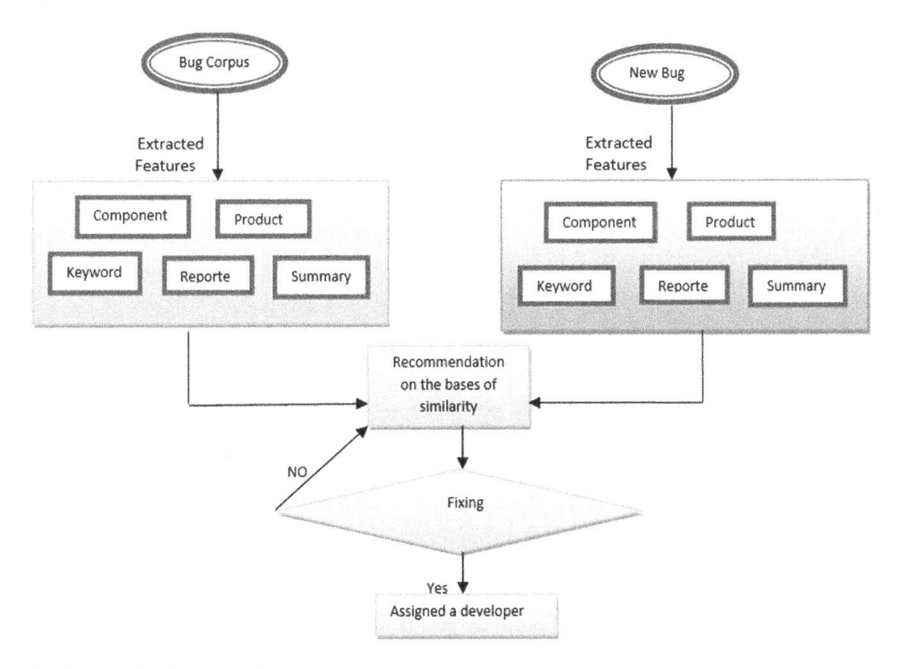

Fig. 3 En-TRAM triager flowchart

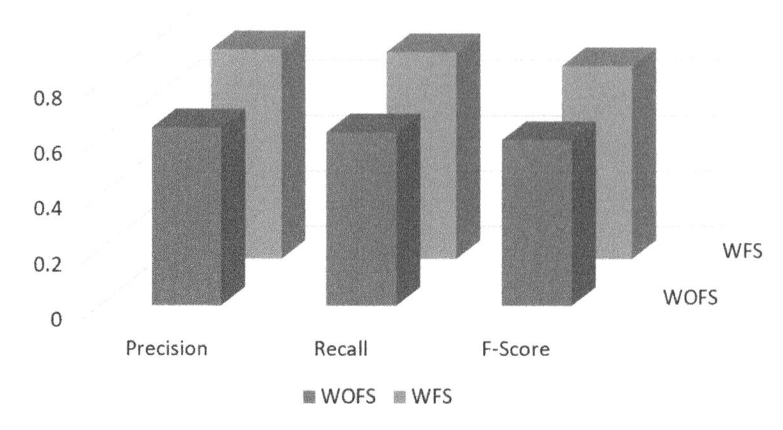

Fig. 4 Triager performance-WFS versus WOFS

bug keyword are those features which are used as input to train the En-Tram classifier.

2. In testing phase, when a new bug is reported. This same process is done on it in terms of applying pre-processing and selection of bug metadata features by using features selection algorithms. Machine learning based classifiers are used to build the predictive model. This predictive model is successfully generate the rank list of the relevant and appropriate developers. This bug also entered in the re-assignment phase if could not be handled by the assigned developer until not get fix.

Unlike TRAM, we are using features extraction/selection algorithms to collect most important bug metadata features from the bug report [22, 23]. En-Tram is a dynamic approach for the selection of bug metadata features. Table 3 depicted the Bug metadata features ranked weigh or score after applying three different feature

Table 3 Weight score of feature after applying FS Techniques

Features	GRE ranked weight	CFS ranked weight	RA ranked weight
Product	–	0.38132	0.758
Components	–	0.048396	0.721
Keywords	–	0.09835	0.479
Summary	–	0.06709	0.452
Reporter	–	0.06652	0.288
Priority	–	0.06334	0.272
Platform	–	0.03612	0.205
Importance	–	0.03569	0.226
Bug-id	–	0.00863	0.013

Table 4 Collection of selected metadata fields after applying FS algorithms

FS algorithms	Selected Bug Metadata features after applying FS algorithms
GainRatioAttributeEval (GRE)	Component, summary, reporter, keywords, priority and product
CFS SubsetEval	Reporter, product and summary
ReliefAttributeEval (RA)	Reporter, keywords, product, summary and component

selection algorithms [24–27]—CFS SubsetEval, Relief attributeEval and GrainRationAttributeEval. These approach are applying nine bug features, and these metadata features are graded in respect its weighted score and finally selected bug features are shown in Table 4 which used in further classification task. This evaluation process is done with the use of Weka 3.7 tool. In which, we simply used the respective FS algorithms over the datasets and collect the all relevant bug metadata features with the weight score.

4 Experimental Setup and Results

4.1 Experiment Setup

A metadata based En-TRAM approach is experimented on four datasets which are collected commencing the bug repositories: Firefox, Eclispe, Netbeans and Freedesktop. In this experiment setup, we only collect those bug reports which are in the status—closed, fixed, verified and resolved. These bugs are contains the details about the whole life cycle of the bug reports and provide the useful information for bug resolution. We completely ignore the duplicate bugs to avoid the redundancy in the dataset. Summary of the datasets is shown in Table 5, and the time span for the collect of bugs are from Jan 1, 2004 to Dec 31, 2012 that contains approximately 32,330 bug reports. We applied pre-processing on the bug reports and remove the present noise from it. The summary of datasets features is also shown in Table 6 that contain the total numbers of bugs, total components and product name, used bug keywords and total developer and reported names.

Table 5 Summary of the four datasets

Project	Initial count of bugs	From	To	Final count of bugs ()
Eclipse	7898	Jan 1, 2004	Dec 31, 2012	7339
Freedesktop	9881	Jan 1, 2004	Dec 31, 2012	8583
Firefox	5101	Jan 1, 2004	Dec 31, 2012	3856
Netbeans	9450	Jan 1, 2004	Dec 31, 2012	8947

Table 6 Summary of dataset features

Project name	# of Bugs	# of Reporter	# of Component	# of Product	# of Keywords	# of Developers
Eclipse	7339	1019	29	06	63	61
Netbeans	8947	1677	256	35	144	55
Firefox	3856	719	40	04	289	41
Freedesktop	8583	3249	237	32	137	73

4.2 Results

The above datasets is used to experiment the En-TRAM approach by divided the dataset into two parts (training and testing datasets) and K-fold method is applied to avoid biased and overlapping results. This is rummage-sale to split the datasets in the form of subsets which is used to train the classifier as well as for bug testing process. Each of the subset is used to train the triager, and rest of the dataset part is applying for testing. We evaluated the previous state-of–art approaches as well as our proposed En-TRAM in this dataset. The performance of these approaches is measured on various measures or parameters as presented in Table 7. As we go through this table, Baseline approach is suffered low prediction accuracy and our En-TRAM is achieved good results in all prediction measures as compared to others that goes to approximately 68% improvement in triager prediction performance.

As we compared En-TRAM to an ABS-Time-tf-idf [9], accuracy on Eclipse and Netbeans dataset is improved upto 62.4% by using SVM classifier and the same improvement in respective datasets has been observed upto 51.3% when used NB classifier (as shown in Table 7 and Figs. 5, 6 and 7). In Fig. 7, we compared our En-Tram approach with the existing approach BPR, and we used the same dataset for the experimental setup in all process. These improved results in triager performance are achieved because of the selection of bug metadata features which are used to train the En-TRAM triager. Table 7 presented the comparison among the En-TRAM versus BPR [15], where Naïve bayes and SVM classifier are improved the prediction accuracy upto 15.2% for both datasets Eclipse and Firefox. The improvement for DT in both dataset, Eclipse and Firefox are upto 49.2%.

We also evaluated the En-TRAM triager performance along with the use of bug metadata that are extracted via using different FS algorithms. In Table 8, we are showing the performance comparison of En-TRAM triager which comes after applying the respective features selection algorithm. En-TRAM approach achieved better prediction accuracy when ReliefAttribute + ranker algorithm based extracted features are used in triaging process and successfully remove the noisy data from the dataset. Both later algorithms have not been able to handle redundant or noisy data and produced inaccurate results as compared to ReliefAttribute.

Our proposed En-TRAM triager achieved the good prediction accuracy in all aspects and this is happened with the use of the right combination of bug metadata

Table 7 En-TRAM versus state-of-art results by using reliefattribute feature extraction technique

Project	Method		Precision	Recall	F-Score
Eclipse	Baseline		0.296	0.289	0.292
	X2		0.408	0.188	0.257
	TRAM		0.493	0.482	0.487
	EnTRAM	**NVA**	**0.758**	**0.749**	**0.699**
		Decision tree	**0.792**	**0.816**	**0.782**
		SVM	**0.996**	**0.996**	**0.996**
Netbeans	Baseline		0.271	0.271	0.250
	X2		0.495	0.154	0.235
	TRAM		0.666	0.638	0.652
	EnTRAM	**NV**	**0.848**	**0.841**	**0.834**
		Decision tree	**0.933**	**0.928**	**0.928**
		SVM	**0.998**	**0.998**	**0.998**
Firefox	Baseline		0.326	0.318	0.322
	X2		0.365	0.322	0.342
	TRAM		0.532	0.528	0.530
	EnTRAM	**NV**	**0.740**	**0.642**	**0.612**
		Decision tree	**0.977**	**0.989**	**0.983**
		SVM	**0.997**	**0.997**	**0.997**
Freedesktop	Baseline		0.31	0.303	0.306
	X2		0.539	0.251	0.343
	TRAM		0.663	0.623	0.642
	EnTRAM	**NV**	**0.849**	**0.825**	**0.814**
		Decision tree	**0.865**	**0.853**	**0.848**
		SVM	**0.991**	**0.991**	**0.991**

features to build it. These features/attributes from metadata have enhanced the accuracy of the classification. For all four datasets, analysis of classifiers has revealed that SVM and decision tree achieved good prediction accuracy in comparison as presented in Fig. 5.

5 Conclusion

Our experiment shown the proposed En-TRAM triager performance in all aspects, and who efficiently achieved the good prediction results as compared to the state-of–art. This proposed approach is mainly dedicated on the importance of bug metadata features in the building of bug triager. To accomplish this task, we are applying three different feature selection algorithms (CFS SubsetEval, Relief attributeEval and

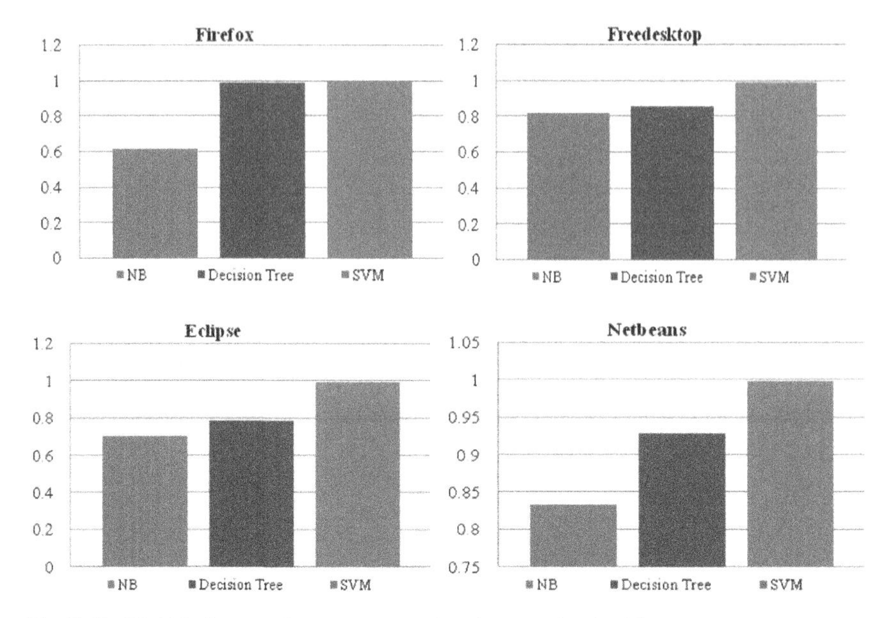

Fig. 5 En-TRAM triager performance comparison between the classifiers

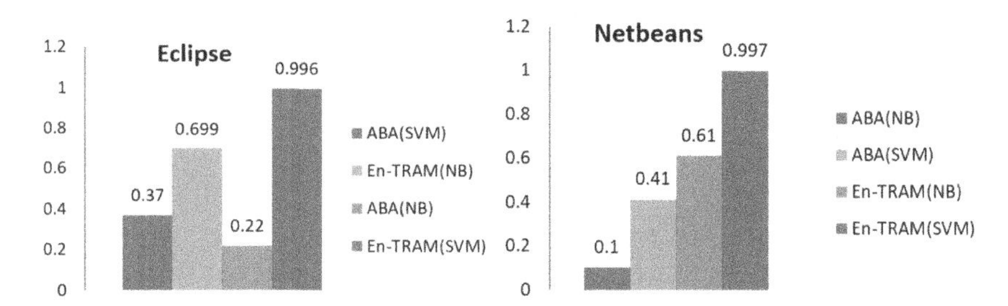

Fig. 6 En-TRAM versus ABA-Time-tf-idf triager performance comparison

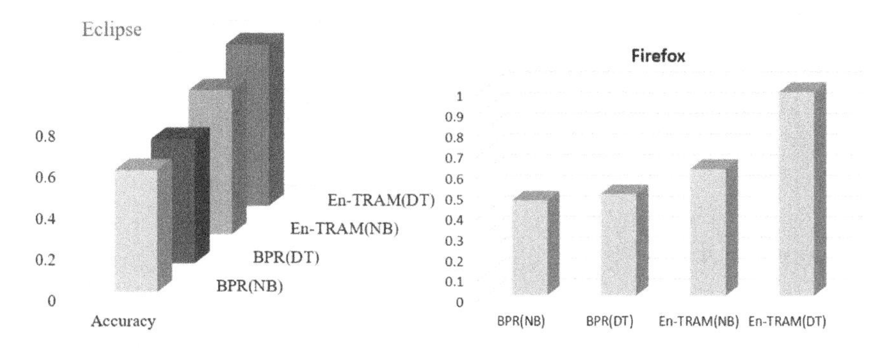

Fig. 7 En-TRAM versus BPR triager performance comparison

Table 8 Triager performance over various FS algorithms

Feature selection approach	Precision	Recall	F-score
ReliefAttributeEval (RA)	**0.404**	**0.568**	**0.470**
CFS SubsetEval (CFS)	0.102	0.244	0.144
GainRatioAttributeEval (GRE)	0.192	0.148	0.167

GrainRationAttributeEval) on the bug reports and extract the most important as well as distinguished bug metadata features from the bug datasets. Three different machine learning based classifiers are used for evaluation process that are NB, SVM and DT and achieved upto 21.2% improved accuracy over TRAM approach and 47.6% improved accuracy over ABA-Time-tf-idf for all four datasets Eclipse, Netbeans, Firefox and Freedesktop.

References

1. Kanwal J, Maqbool O (2012) Bug prioritization to facilitate bug report triage. J Comput Sci Technol 27(2):397–412
2. Alenezi M, Magel K, Banitaan S (2013) Efficient bug triage using text mining. In: Proceeding of 5th international conference and technology
3. Sharma L, Gera A (2013) A survey on recommendation system: research challenge. Int J Eng Trends Technol (IJETT) 4(5)
4. Hassan A (2008) The road ahead of mining software repository. Frontiers of Software Maintenance, Beijing. IEEE
5. Park JW, Lee MW, Kim J, Hwang SW (2011) COSTRIAGE: a cost-aware algorithm for bug reporting system. In: Proceedings of the national conference on artificial intelligence, vol 1
6. Cubranic D, Murphy GC (2004) Automatic bug triage using text categorization. In: Conference: proceedings of the sixteenth international conference on software engineering & knowledge engineering, Banff, Alberta, Canada, June 20–24
7. Yang J, Li YP (2006) Orthogonal relief algorithm for feature selection. Springer, Berlin
8. Vásquez ML, Hossen K, Dang H, Kagdi H, Gethers M, Poshyvanyk D (2012) Triaging incoming change requests: bug or commit history, or code authorship? In: Software maintenance (ICSM), 2012 28th IEEE international conference, pp 451–460, 23–28 Sept 2012
9. Xuan J, Jiang H, Ren Z, Zou W (2012) Developer prioritization in bug repositories. In: Proceedings of the 34th international conference on software engineering, Zurich, Switzerland, pp 25–35
10. Shokripour R, Anvik J, Kasirun ZM, Zamani S (2014) A time-based approach to automatic bug report assignment. J Syst Softw
11. Chen L, Wang X, Liu C (2011) An approach to improving bug assignment with bug tossing graphs and bug similarities. J Softw 6(3)
12. Ahsan SN, Ferzund J, Wotawa F (2009) Automatic software bug triage based on latent semantic indexing and support vector machine. In: 4th international conference on software Engineering Advances, IEEE
13. Anvik J (2007) Assisting bug report triage through recommendation. Ph.D. dissertation, University of British Columbia
14. Bhattacharya P, Neamtiua I, Sheltona C (2010) Automated, highly-accurate bug triage using machine learning. J Syst Softw 85:2275–2292
15. Bhattacharya P, Neamtiua I (2010) Fine-grained incremental learning and multi-features tossing graphs to improve bug triaging. In: Conference: software maintenance (ICSM), vol 85, no 10, pp 1–10

16. Kagdi H, Gethers M, Poshyvanyk D, Hammad M (2012) Assigning change requests to software developers. J Softw: Evolut Process 24(1):3–33
17. Shokripour R, Anvik J, Kasirun ZM, Zamani SA (2014) Time-based approach to automatic bug report assignment. J Syst Softw 102:109–122
18. Xuan J, Jiang H, Ren Z, Zou W (2012) Developer prioritization in bug repositories. In: ICSE 2012 proceedings of the 34th international conference on software engineering, pp 25–35, Zurich, Switzerland
19. Banitaan S, Alenezi M (2013) TRAM: an approach for assigning bug reports using their metadata. Computational intelligent applications in software engineering (CIASE), Beirut, IEEE
20. Cavalcanti YC, Neto PAdMS, Machado IDC, Vale TF, Almeida ES, de Lemos SR (2014) Challenges and opportunities for software change request repositories: a systematic mapping study. J Softw Evaluat Process 26(7):620–653
21. Shokripour R, Anvik J, Kasirun ZM, Zamani S (2013) Why so complicated? Simple term filtering and weighting for location-based bug report assignment recommendation. In: Proceedings of the 10th working conference on mining software repositories, pp 2–11
22. Vinodhini G, Chandrasekaran RM (2013) Effect of feature reduction in sentiment analysis of online reviews. Int J Adv Res Comput Eng Technol (IJARCET) 2(6)
23. Ahmed A, Chen H, Salem A (2008) Opinion analysis in multiple languages: feature selection for opinion classification in web forums. ACM Trans Inform Syst 26(3)
24. Gao L, Li T, Yao L, Wen F (2014) Research and application of data mining feature selection based on relief algorithm. J Softw 9(2)
25. Ting S, Ip W, Tsang AH (2011) Is Naïve Bayes a good classifier for document classification? Int J Softw Eng Its Appl 5(3)
26. Sikonja MR, Kononenko I (2009) Theoretical and empirical analysis of Relief and Relief. Mach Learn J 53:23–69
27. Yadav A, Singh S (2020) DyEnTRAM—dynamically enhanced metadata based approach for bug assignment. Trends Comput Sci Inform Technol 5(1):023–033

Role of Artificial Intelligence in Detection of Hateful Speech for Hinglish Data on Social Media

Ananya Srivastava, Mohammed Hasan, Bhargav Yagnik, Rahee Walambe, and Ketan Kotecha

Abstract Social networking platforms provide a conduit to disseminate our ideas, views, and thoughts and proliferate information. This has led to the amalgamation of English with natively spoken languages. Prevalence of Hindi-English code-mixed data (Hinglish) is on the rise with most of the urban population all over the world. Hate speech detection algorithms deployed by most social networking platforms are unable to filter out offensive and abusive content posted in these code-mixed languages. Thus, the worldwide hate speech detection rate of around 44% drops even more considering the content in Indian colloquial languages and slangs. In this paper, we propose a methodology for efficient detection of unstructured code-mix Hinglish language. Fine-tuning-based approaches for Hindi-English code-mixed language are employed by utilizing contextual-based embeddings such as embeddings for language models (ELMo), FLAIR, and transformer-based bidirectional encoder representations from transformers (BERT). Our proposed approach is compared against the pre-existing methods and results are compared for various datasets. Our model outperforms the other methods and frameworks.

Keywords BERT · ELMO · FLAIR · Hinglish-English code-mixed text

1 Introduction

In the upfront of our social lives, lies the huge platform of social media sites. Social media penetration in India is growing very rapidly with currently over 29 percent of India's population using social media [5]. Due to the rise in usage of social media, contrasting ideology and hateful material on the Internet has escalated. An individual's liberty to free speech is prone to exploitation and hate can be conveyed in

These authors contributed equally to this work.

A. Srivastava · M. Hasan · B. Yagnik
Symbiosis Institute of Technology, Pune, India

R. Walambe (✉) · K. Kotecha
Symbiosis Centre for Applied Artificial Intelligence, Pune, India
e-mail: rahee.walambe@scaai.siu.edu.in

A. Choudhary et al. (eds.), *Applications of Artificial Intelligence and Machine Learning*, Lecture Notes in Electrical Engineering 778,
https://doi.org/10.1007/978-981-16-3067-5_8

terms of specific or particular speech acts or disjunctive sets of speech acts—for example, advocating hatred, insulting or defaming, terrorizing, provoking discrimination or violence, accusation [12]. Hate speech has an adverse effect on the mental health of individuals [11]. Twitter has seen about 900% increase in hate speech during COVID-19 pandemic [15]. YouTube reported about 500 Million hate comments from June 2019 to September 2019 [4]. Hate speech has been recognized as a growing concern in the society and numerous automated systems have been developed to identify and avert it.

With the recent progress in pre-trained models across different language models and support for Hindi and other languages, better results could be achieved in Hinglish hate speech classification.

1.1 Previous Work

One of the earliest works in hate speech recognition is reported in [22] which extracted rule-based features to train their decision tree text classifier. A number of studies have been carried out particularly within the context of social media [6, 8, 19]. In [27], authors surveyed various application of deep learning algorithms to learn contextual word embeddings for classification of tweets as sexist or racist or none using deep learning. Mozafari [16] reported better performance of convolutional neural network (CNN) with BERT on [8] dataset. Hateful text classification in online textual content has also been performed in languages like Arabic [17] and Vietnamese [10].

Various approaches and methods for detection of abusive/hateful speech in Hinglish have been reported in the literature. Hinglish is a portmanteau of the words in Hindi and English combining both in a single sentence.[28] Now with increase in the number of youths using this mixed language, not only in urban or semi-urban areas, but also to the rural and remote areas through the social media, Hinglish is achieving the status of a vernacular language or a dialect. Sinha [21], attempted to translate Hinglish to standard Hindi and English forms. However, due to shallow grammatical analysis, the challenge of polysemous words could not be resolved. Mathur et al. [14] created an annotated dataset of Hindi-English offensive tweets (HEOT) split into three labels: hate speech, non-offensive speech, and abusive speech. Ternary trans-CNN model was pre-trained on tweets in English [8] followed by retraining on Hinglish tweets. An accuracy of 0.83 was reported on using this transfer learning approach. Multi-input multi-channel transfer learning using multiple embeddings like GloVe, Word2vec, and FastText with CNN-LSTM parallel channel architecture was proposed by [13] that outperformed the baselines and naïve transfer learning models. If the text classification approaches designed for English language are applied to such a code-mixed language like Hinglish, it fails to achieve the expected accuracy. Some of these standard models were applied on the Hinglish dataset and the results obtained have been reported in Sect. 5.

In English language, there was an extensive work on hate speech datasets but there were no datasets focusing on Indian Social media texts, so we have focused on developing a hate speech classifier based on Indian audience. The previous methods explained by researchers in Hinglish language did not implement pre-trained transformers embeddings considering the barrier of Hindi pre-trained language models. With the advances in pre-trained language models like multilingual BERT [23] and support for Hindi and other languages, we were able to implement models like BERT, ELMo and FLAIR to exceed the previously obtained results in English and Hinglish. In summary, the contributions of this work are:

- Design of custom web-scraping tools for different social media platforms and development of an annotated hate speech dataset specific to Indian audiences for both English and Hinglish languages.
- Development of separate generic pre-processing pipelines for Hinglish and English languages which are validated on benchmark datasets.
- Demonstration of use of BERT for Hinglish embeddings using pre-trained multilingual weights.
- Demonstration and validation of a novel approach based on the ELMO and implementation of FLAIR framework for Hinglish text classification.

The paper is divided into the following sections. In Sect. 2, the data collection and annotation scheme for both English and Hinglish datasets have been illustrated. In the Sect. 3, classification systems including BERT, ELMO, and FLAIR have been summarized, and the pipelines followed for both English and Hinglish datasets have been discussed in Sect. 4. In Sect. 5, results are presented and compared. In the last section, the conclusion and future scope is discussed.

2 Corpus Creation and Annotation

The datasets provided by Davidson [8] and in Hinglish by Bohra [3] and [13] have been used for validation and demonstration of the proposed approach. In addition to these datasets, primary part of this work was focused on collection and creation of the datasets us for both languages with the intention of encompassing a wider spectrum of issues across various verticals of the society.

The data was collected from Twitter, Instagram, and YouTube comments dated from November 2019 till February 2020 associated with domains such as political, cultural, and gender. The major portion of the corpus was from trending news headlines and hashtags which went viral on social media platforms as they may gather more hate.

Custom web-scraping tools were designed for YouTube and Instagram using selenium [20] while "Twitter_scrapper" was used to scrape data from Twitter [24]. The comments that were collected were saved into CSV file format.

2.1 Annotations

The comments collected were non-homogeneous, i.e., they were a mix of all three languages, namely English, Hindi, and Hinglish. Hence, the segregation of the comments into their respective languages was important as different languages required different pre-processing pipelines. This was carried out by assigning separate labels for each language. The comments were then annotated based on the definition of hate speech stated by the United Nations [26]. The labels of the two classes were "Hate" and "Not Hate". Table 1 shows the examples of hate speech and subsequent annotations from our dataset.

Annotations were performed by the authors who are proficient Hindi as well as English speakers. The inter-annotator agreement [7] was observed to be 0.89. There were around 8000 (32.5%-YouTube, 47.9%-Instagram, 19.5%-Twitter) comments collected which were then divided into 2 datasets, 1683 comments in Hinglish text, while 6160 comments in English. The comments in Hindi were transliterated to Hinglish using a Unidecode [25]. Followed by dataset creation, the development of classifiers has been discussed in the next section.

3 Methodology

Textual data in its raw form cannot be understood by computers and hence embeddings are used in NLP to convert text into multi-dimensional vectors. The recent developments in state-of-the-art language models have proven to be extensive in giving excellent results with small amount of training data. In this section, we demonstrate the of use BERT for English and Hinglish embeddings using pre-trained weights and validate a novel approach based on the ELMO and implementation of FLAIR framework for Hinglish text classification.

Table 1 Examples of hate speech and subsequent annotations

	English	Hinglish
Generic	The most illiterate, ill-mannered, psychopath just barks whatever shit out of his mouth @euler12 😂	Bikao kutta sala. 2 kodi ka insaan. Tujhe sataye hue ke baddua lagenge harami 🤬
Political	We have a stupid government, divide and rule is their motto. #BJP	Tuje toh kutta bhi nahi bol sakta wo bhi wafadar hota hai tu toh desh gad-dar hai sale madarchod
Gender based	Hoes what do you expect. Make money by show in body	Chutiya orat … kisse aajadi chahiye tuje …randapa krne ki aajadi bol?
Religion based	As Muslim will never leave their ideology of killing non-Muslims and now other communities have as well given up on living together peacefully	Madarchod musalman rape hindu girls

3.1 *Bidirectional Encoder Representations from Transformers (BERT)*

BERT is a deep bidirectional transformer-based language representation model [9] which is designed fine-tuning with shallow neural networks to obtain state-of-the-art models. $BERT_{base-uncased}$ ($BERT_{BU}$) and $BERT_{multilingual-uncased}$ ($BERT_{MU}$) were used for this work. $BERT_{BU}$ contains 768 hidden layers and is trained on English Wikipedia and the Book Corpus. $BERT_{MU}$ is similar but trained on lower-cased text in top 102 Languages in Wikipedia. So, this becomes useful when encoding text in Hinglish language. BERT was then fine-tuned using our datasets.

3.2 *Embeddings from Language Models (ELMo)*

ELMo [18] embeddings developed by AllenNLP include both word level contextual semantics and word level characteristics. It is character based, i.e., the model forms a vocabulary of words that are not present in the vocabulary and captures their inner structure. It uses bidirectional LSTMs to create word representations according to the context of the words they are used in. The forward pass contains words before the target token while the backward pass contains words after the target token. This forms the intermediate word vector which acts as input to the next layer of bidirectional language model. The final output (ELMo) is the weighted sum of the intermediate word vectors and the input word vector.

3.3 *Flair*

Flair [2] library consists of models such as GloVe, ELMo, BERT, character embeddings, etc. This interface allows stacking-up of different embeddings which gives a significant increase in results. "Flair embeddings" [1] which are unique to the Flair library uses contextual string embeddings that capture latent syntactic-semantic information and are contextualized based on neighboring text leading to different embeddings for the same words depending on the context. Flair includes support for pre-trained multilingual embeddings including "hi-forward" and "hi-backward" for Hindi.

The standard CNN [29] and Bi-LSTM frameworks are used with the above-specified embeddings. In the next section, we implement distinctive pre-processing approaches along with the above-mentioned methodologies.

4 System Design

4.1 Data Pre-Processing

Separate generic pre-processing pipelines for Hinglish and English languages were constructed and validated on the benchmark datasets.

English
The social media data tends to be casual and informal, leading to a decrease in the ability of a language model to understand the corpus, hence, performing extensive pre-processing on the data became necessary. Figure 1 represents the diagrammatic pipeline for pre-processing in English language. The initial step was expansion of contractions and abbreviations into their standard notation. Common pronouns, conjunctions, articles, and prepositions in English vocabulary generally add no contextual meaning to a sentence and are ignored by search engines thus were removed from the corpus. Removal of URLs, hashtags, mentions, and punctuations was also carried out as a part of pre-processing. The emojis were replaced with the text denoting its meaning.

Lastly, to map the words into their root form, WordNetLemmatizer (NLTK) was applied. Table 2 shows an example of the pre-processing done on the English corpus.

Hinglish
Prior to performing the pre-processing techniques to Hinglish data, the comments were manually cleaned to remove ambiguity, perform spell check, removal of Hinglish specific stopwords and repetitive comments. For this, the English stopwords list was appended and commonly used Hinglish words along with variations in their spellings were added to the list. For example, words like *teko, terko, tujhe,* etc., were added to the Hinglish stopwords list.

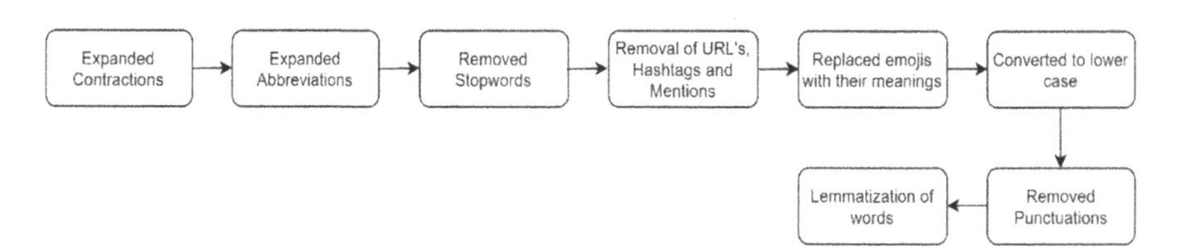

Fig. 1 Pre-processing pipeline for English dataset

Table 2 Example of pre-processing technique for English dataset

Text	Pre-processed text
@amitshah You can't change the minds of such small minded people who are stuck in the past, they just don't understand logics. #India Against CAA 🤬	you cannot change mind small minded people stuck past understand logic face with symbols on mouth

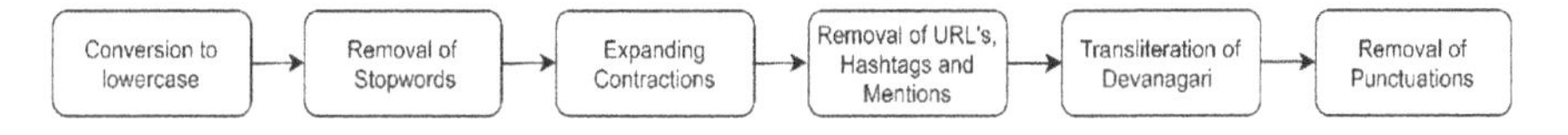

Fig. 2 Pre-processing pipeline for Hinglish dataset

Table 3 Example of pre-processing technique for Hinglish dataset

Text	Pre-processed text
@narendramodi मेरा देश BHAI hate ni pyar phailata ha or jo pyar se nhi manta wo use ache se samjhate hain!! https://twitter.com/4948747235330	meraa desh hate ni pyar phailata pyar nhi manta ache samjhate hain

The text was then converted into lowercase as shown in Fig. 2. The Hinglish data contained a few words in Devanagari script that had to be converted to Roman Script. For this purpose, Unidecode library was used to transliterate the Hindi text to Hinglish. Table 3 shows an example of the pre-processing done on the Hinglish corpus.

4.2 Fine-Tuning Approaches

$BERT_{BU}$+ CNN for English Dataset

The processed text was then converted to tokens using WordPiece Tokenizer which has a vocabulary of 30,523 unique keywords. The outputs were then truncated into 100 tokens along with padding for smaller sequences. The tokens were then used to generate weights of dimensions 1×768 from $BERT_{BU}$. These weights were then fine-tuned using the CNN architecture proposed by [29] that consisted of 6 filters, 2 filters of sizes 2,3, and 4 each followed by a max pooling layer and a softmax layer. The learning rate was set to 1e-3 and the model was trained on NVIDIA DGX Station with 32 GB RAM.

$BERT_{MU}$+ CNN for Hinglish Dataset

The processed text in Hinglish was converted into tokens using the WordPiece Tokenizer, but this time using a different vocabulary file for $BERT_{MU}$ which has a vocabulary of more than 1 M keywords from 102 languages. The outputs were then truncated to 75 tokens. Sequences lesser than it were padded and then the weights were generated by using the $BERT_{MU}$ language model which were fine-tuned to specific task of hate speech detection using above-mentioned CNN architecture with a learning rate of 1e-4 for the HOT dataset [13] (3500 samples) while learning rate of 1e-3 was used for the dataset annotated by the authors, combined with HOT [13] and hate speech dataset [3] (9000 samples).

ELMo and MLP for Hinglish Dataset

For implementing ELMo, ELMo model 2 was imported using TensorFlow Hub. The processed text was taken as raw input vectors for ELMo embeddings and intermediate vectors were computed. The concatenation of outputs of both the layers was carried out to have one vector for each word with a size of 1024. The learning rate was set to 1e-5 with Adam Optimizer. The training process was carried out on Google Colaboratory GPU.

Flair

The pre-processed data was split into train (80%), test (10%), and dev (10%) sets following the convention of flair text classifier with text and labels represented in the form: label__ < class_n > < text >

Further two approaches were used under flair:

Flair-stacked (WordEmbeddings$_{hi}$ + FlairEmbeddings$_{hi+forward}$ + FlairEmbeddings$_{hi+backward}$).

The input split above is embedded over the stacked pre-trained multilingual embeddings of "WordEmbeddings$_{\text{hi}}$", "FlairEmbeddings$_{\text{hi+forward}}$" and "FlairEmbeddings$_{\text{hi+backward}}$" which were specifically chosen for Hinglish-based applications. The weights were then fine-tuned to explicit detection using Bi-LSTM with a learning rate of 1e-5 and Adam optimizer. It was trained on Google Colab GPU Tesla P100.

Flair (BERT + BiLSTM).

Two different embeddings were used in this approach. One was BERT$_{\text{BU}}$ while the other was BERT$_{\text{MU}}$. The input split mentioned above is embedded over two separate pre-trained multilingual embeddings of "BERT$_{\text{BU}}$" and "BERT$_{\text{MU}}$". The weights from both the embeddings were then fine-tuned to explicit detection using an Bi-LSTM with a learning rate of 1e-5 and Adam optimizer and trained on Google Colab GPU.

The above-mentioned approaches were used to train the models on the collected datasets. The results of the same have been discussed in the following section.

5 Results and Discussion

We evaluate the outcomes of different fine-tuning approaches on their datasets and compare them with other baseline datasets. Table 4 summarizes the different results obtained from various fine-tuning approaches for English mentioned in Sect. 4. In

Table 4 Results for English datasets

Dataset	Model	Accuracy (%)	Recall	F1-score
Davidson [8] dataset	BERT$_{\text{BU}}$ + CNN	94	0.94	0.93
English Hate dataset	ELMO	73	0.73	0.71
	BERT$_{\text{BU}}$ + CNN	73	0.73	0.70

English, two datasets were considered for this purpose: Davidson [8] dataset which was trained on BERT_{BU} + CNN approach resulting in an accuracy of 94% while the English hate dataset collected by us was trained on ELMO and BERT_{BU} + CNN both yielding a similar accuracy of 73%. The considerable variation between the datasets is mainly because of the shift in domain as the text in Davidson dataset is mainly generic while the dataset collected by us is event-driven, i.e., it contains comments based on political, gender-specific, religious events which in-turn disseminates different hate-based contextual meaning that may not be directed by any specific abusive words.

The pipelines used in English were applied on the HOT dataset in Hinglish and the results obtained are summarized in Table 5 and show that the English pipelines were ineffective and led to acquiring lower accuracy on Hinglish dataset given the complex nature of Hinglish corpus. Thus, designing a different pipeline for Hinglish dataset was necessary. Table 6 depicts that after applying a different pipeline specific to Hinglish dataset, significantly better results were achieved.

The results obtained using BERT_{MU} were at par with the results of BERT_{BU} in terms of accuracy for Hinglish datasets. But BERT_{MU} can be considered preferable

Table 5 Results of English pipeline applied to Hinglish dataset

Dataset	Model	Accuracy (%)	Recall	F1-score
HOT dataset [13] (Hinglish)	BERT_{BU} + CNN	83	0.83	0.83
	ELMO	80	0.80	0.79

Table 6 Results for Hinglish datasets based on Hinglish pipeline

Dataset	Model	Accuracy (%)	Recall	F1-score
HOT dataset [13] (Hinglish)	FLAIR Embeddings $_{\text{hi}}$ + Flair $_{\text{hi-forward+hi-backward}}$	81	0.84	0.91
	FLAIR BERT_{MU} + Bi-LSTM	84	0.86	0.92
	ELMO + MLP	85	0.85	0.85
	BERT_{MU} + CNN	86	0.86	0.86
	FLAIR BERT_{BU} + Bi-LSTM	88	0.89	0.94
Hinglish Hate dataset + HOT [13]	ELMO + MLP	71	0.71	0.70
	BERT_{MU} + CNN	82	0.82	0.82
	FLAIR BERT_{MU} + Bi-LSTM	82	0.84	0.91
Hinglish hate dataset + HOT [13] + [3] dataset	ELMO + MLP	67	0.67	0.66
	BERT	67	0.67	0.67
	FLAIR BERT_{MU} + Bi-LSTM	73	0.79	0.87

as it has a broader vocabulary including words from many languages, and hence it was able to tokenize words in a more contextual form. For example, "kaam karna he" would be generated as "[CLS] Ka ##am ka ##rna He [SEP]" using BERT_{BU} while using BERT_{MU} it would be "[CLS] kaam karna he [SEP]". Also, this leads to loss of context in case of BERT_{BU} as it contains words from the English vocabulary and tokenizes out-of-vocabulary words into multiple sub-word tokens.

Tables 7 and 8 compare the obtained results with the baseline results observed previously on Davidson dataset [8] in English and HOT dataset [13] in Hinglish.

F1-score of 0.93 was achieved on Davidson dataset [8] using the BERT_{BU} fine-tuned with CNN which exceeds the previous results obtained from [16]. For this, task around 30 K samples from the Davidson dataset [8] were used.

Along with the English dataset, the best results for Hinglish were yielded using BERT_{MU} + CNN and FLAIR BERT_{BU} + Bi-LSTM approach, attaining F1-scores of 0.86 and 0.94, respectively, which exceeds the baseline results obtained by [13]. Figure 3 depicts the confusion matrices for BERT_{BU} on Davidson [8] dataset and ELMO applied on Hinglish dataset [21], respectively.

6 Conclusion

In this work, we have proposed to improve the hate speech detection in English as well as Hinglish languages with specific focus on the social media. Language modeling and text classification in English language are comparatively a well-explored area. However, for a code-mixed language like Hinglish, which is very common and popular in India, the detection of hate speech is highly challenging. We collected the English and Hinglish datasets manually from various online avenues. Fine-tuning-based approaches for Hindi-English code-mixed language are employed by utilizing contextual-based embeddings such as ELMO, FLAIR, and transformer-based BERT.

Table 7 Comparison for accuracies in Davidson dataset [8]

Method	F1-score
Waseem [27]	0.89
Mozafari [16]	0.92 (LSTM) 0.92 (CNN)
BERT_{BU} -CNN	0.93

Table 8 Comparison for accuracies in HOT dataset

Method	F1-score
Multi-channel CNN-LSTM architecture Mathur [13]	0.89
FLAIR BERT_{MU} + Bi-LSTM dataset	0.92
ELMO	0.85
BERT_{MU} + CNN	0.86
FLAIR BERT_{BU} + Bi-LSTM	0.94

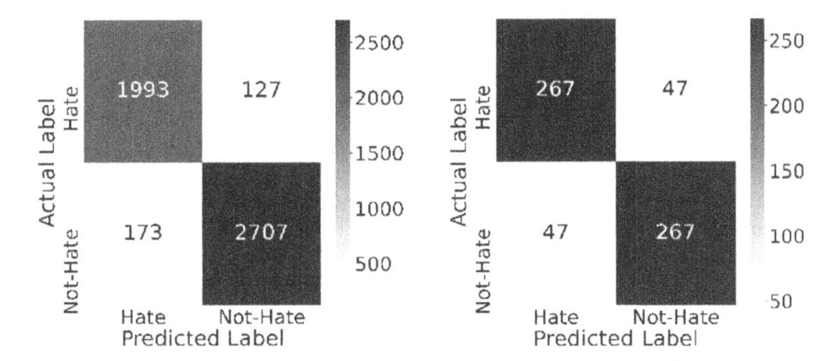

Fig. 3 (Left) Confusion matrix for BERT with Davidson dataset. (Right) Confusion Matrix for ELMO applied on Hinglish dataset

We compared our proposed approach with various existing methods and demonstrated that our methods are promising and outperform the other approaches. Our major contribution lies in designing the annotated hate speech dataset in Hinglish, developing a pre-processing pipeline which is generic in nature and devolvement and validation of deep learning-based approaches for detection of hate speech in Hinglish text.

6.1 Future Scope

Many application-driven pipelines can be designed for Hinglish embeddings as well as classifiers given the advancement in the field of NLP for multilingual domain and the scope of transfer learning for further accuracy. Enhanced approaches include changing the architecture of neural network classifier or using deeper/multi-layer neural networks with larger corpus for Hindi-English code-mix language. The project can also be extended to multiclass classification under degree of hate or as labeled in Davidson [8] dataset that includes classes like toxic, severe-toxic, obscene, threat, insult and identity-hate rather than bi-class, i.e., Hate and Not Hate. The research presented in this paper can be extended to other code-mix local + English languages from any part of the world. We hope that the dataset and the experimental results encourage further research in multilingual domain as well as hate speech recognition.

References

1. Akbik A, Blythe D, Vollgraf R (2018) Contextual string embeddings for sequence labeling. In: Proceedings of the 27th international conference on computational linguistics, pp 1638–1649
2. Akbik A, Bergmann T, Blythe D, Rasul K, Schweter S, Vollgraf R (2019) Flair: an easy-to-use framework for state-of-the-art nlp. In: Proceedings of the 2019 conference of the North

American chapter of the association for computational linguistics (demonstrations), pp 54–59
3. Bohra A, Vijay D, Singh V, Akhtar SS, Shrivastava M (2018) A dataset of Hindi-English code-mixed social media text for hate speech detection. https://doi.org/10.18653/v1/w18-1105
4. Burch S (2019) YouTube deletes 500 million comments in fight against 'hate speech.' TheWrap. https://www.thewrap.com/youtube-deletes-500-million-comments-in-fight-against-hate-speech/
5. Cement J (2020) Social media: active usage penetration in selected countries 2020. Retrieved from https://www.statista.com/statistics/282846/regular-social-networking-usage-penetration-worldwide-by-country/
6. Chen Y, Zhou Y, Zhu S, Xu H (2012) Detecting offensive language in social media to protect adolescent online safety. In: Proceedings—2012 ASE/IEEE international conference on privacy, security, risk and trust and 2012 ASE/IEEE international conference on social computing, SocialCom/PASSAT 2012. https://doi.org/10.1109/SocialCom-PASSAT.2012.55
7. Cohen J (1960) A coefficient of agreement for nominal scales. Educ Psychol Measur 20(1):37–46
8. Davidson T, Warmsley D, Macy M, Weber I (2017) Automated hate speech detection and the problem of offensive language. In: Proceedings of the 11th international conference on web and social media, ICWSM 2017
9. Devlin J, Chang MW, Lee K, Toutanova K (2019). BERT: Pre-training of deep bidirectional transformers for language understanding. In: NAACL HLT 2019—2019 conference of the North American chapter of the association for computational linguistics: human language technologies—proceedings of the conference
10. Do HTT, Huynh HD, Van Nguyen K, Nguyen NLT, Nguyen AGT (2019) Hate speech detection on Vietnamese social media text using the bidirectional-lstm model. arXiv preprint arXiv:1911.03648
11. Gelber K, McNamara L (2016) Evidencing the harms of hate speech. Soc Ident 22(3):324–341
12. Leets L (2002) Experiencing hate speech: perceptions and responses to anti-semitism and antigay speech. J Soc Issues 58(2):341–361
13. Mathur P, Sawhney R, Ayyar M, Shah R (2019) Did you offend me? Classification of offensive tweets in Hinglish language. https://doi.org/10.18653/v1/w18-5118
14. Mathur P, Shah R, Sawhney R, Mahata D (2019) Detecting offensive tweets in Hindi-English code-switched language. https://doi.org/10.18653/v1/w18-3504
15. Mehta I (2020) Twitter sees 900% increase in hate speech towards China because coronavirus. The Next Web. https://thenextweb.com/world/2020/03/27/twitter-sees-900-increase-in-hate-speech-towards-china-because-coronavirus/
16. Mozafari M, Farahbakhsh R, Crespi N (2020) A BERT-based transfer learning approach for hate speech detection in online social media. Stud Comput Intell. https://doi.org/10.1007/978-3-030-36687-2_77
17. Mubarak H, Darwish K, Magdy W (2017) Abusive language detection on Arabic social media. https://doi.org/10.18653/v1/w17-3008
18. Peters ME, Neumann M, Iyyer M, Gardner M, Clark C, Lee K, Zettlemoyer L (2018). Deep contextualized word representations. In: NAACL HLT 2018—2018 Conference of the North American Chapter of the association for computational linguistics: human language technologies—proceedings of the conference. https://doi.org/10.18653/v1/n18-1202
19. Raisi E, Huang B (2017) Cyberbullying detection with weakly supervised machine learning. In: Proceedings of the 2017 IEEE/ACM international conference on advances in social networks analysis and mining, ASONAM 2017. https://doi.org/10.1145/3110025.3110049
20. SeleniumHQ/selenium (2007) GitHub. https://github.com/SeleniumHQ/Selenium
21. Sinha RMK, Thakur A (2005) Machine translation of bi-lingual Hindi-English (Hinglish) Text. In: 10th machine translation summit (MT Summit X).
22. Spertus E (1997) Smokey: automatic recognition of hostile messages. In: Innovative applications of artificial intelligence—conference proceedings
23. Turc I, Chang M-W, Lee K, Toutanova K (2019) Well-read students learn better: the impact of student initialization on knowledge distillation. *ArXiv*.

24. Twitter Scraper (2017) Github. https://github.com/taspinar/twitterscraper
25. Unidecode (2019) PyPI. https://pypi.org/project/Unidecode/
26. United Nations (2020) UN strategy and plan of action on hate speech. https://www.un.org/en/genocideprevention/hate-speech-strategy.shtml
27. Waseem Z, Hovy D (2016) Hateful symbols or hateful people? Predictive features for hate speech detection on Twitter. https://doi.org/10.18653/v1/n16-2013
28. Wikipedia contributors (2020) Hinglish. Wikipedia. https://en.wikipedia.org/wiki/Hinglish
29. Zhang Y, Wallace B (2015) A sensitivity analysis of (and practitioners' guide to) convolutional neural networks for sentence classification. arXiv preprint arXiv:1510.03820.

From Web Scraping to Web Crawling

Harshit Nigam and Prantik Biswas

Abstract The World Wide Web is the largest database comprising information in various forms from text to audio/video and in many other designs. However, most of the data published on the Web is in unstructured and hard-to-handle format, and hence, difficult to extract and use for further text processing applications such as trend detection, sentiment analysis, e-commerce market monitoring, and many others. Technologies like Web scraping and Web crawling cater to the need of extracting a huge amount of information available on the Web in an automated way. This paper starts with a basic explanation of Web scraping and the four methodologies—DOM tree parsing, semantic–syntactic framework, string matching, and computer vision/machine learning-based methodology—developed over time based on which scraping solutions and tools are formulated. The paper also explains the term Web crawling, an extension of Web scraping and introduces Scrapy, a Web crawling framework written in Python. The paper describes the workflow behind a Web crawling process initiated by Scrapy and provides with the basic understanding on each component involved in a Web crawling project, built using Scrapy. Further, the paper dives into the implementation of a Web crawler, namely confSpider that is dedicated to extract information related to upcoming conferences and summits from the Internet and may be used by educational institutions to promote student awareness and participation in multi-disciplinary conferences.

Keywords Web scraping · Web crawling · Data extraction · Scrapy

1 Introduction

The World Wide Web is a rich source of 'big data' for many applications including business intelligence [1, 2], competitive intelligence [3, 4], sentiment analysis [5–7], agriculture and food economics [8], bioinformatics [9, 10], analyzing human

H. Nigam (✉) · P. Biswas
Jaypee Institute of Information Technology, Noida, India

P. Biswas
e-mail: prantik.biswas@jiit.ac.in

A. Choudhary et al. (eds.), *Applications of Artificial Intelligence and Machine Learning*, Lecture Notes in Electrical Engineering 778,
https://doi.org/10.1007/978-981-16-3067-5_9

behavior at social Web level [6, 11, 12] and many more data-driven applications [7, 13]. However, about 70% of this Web data is published in some unstructured and hard-to-handle format like PDFs. This becomes a challenge for professionals involved in data extraction roles who are left with the manual technique of copy-and-paste so as to convert the data into suitable format to perform further analysis. Web scraping and Web crawling cut down this manual job into an automated way of extracting data and converting it into an easy-to-use format like CSV, JSON, or store it into a central local database.

Many research works and surveys [14–18] have been conducted on Web scraping applications and tools available in many different forms with many different features on a case-by-case basis. Briefly, a Web scraper can be designed using libraries and frameworks provided in different programming languages, or can come as a Web site extension or a more powerful desktop application.

This paper highlights the use of Scrapy in the task of Web scraping on multiple web pages that is often termed as Web crawling. Web crawling allows scraping of data on multiple web pages by following all the links on a web page. A past few studies [19–21] deal with effective and scalable Web crawlers. The paper starts with the explanation of four general methodologies working behind Web scraping tools and solutions. The later sections dive into a common understanding of Web crawling and implementation of an application-based Web crawler using Scrapy framework.

2 Methodologies Involved in Web Scraping

In order to understand the working of various Web scraping tools, it is significant to know the technical perspectives involved in Web scraping. The theoretical methods that are implemented in practical scenarios via Web scraping tools have significance in knowing and carrying out further research on them and improving them. The following is a broad classification of common methodologies working behind a Web scraping process at core.

2.1 DOM Tree Parsing

A huge amount of Web data is published on numerous Web pages on multiple sites. However, these unstructured Web pages are commonly made with structured HTML and CSS languages. These Web pages comprise HTML elements or tags and data between those elements that is targeted in Web scraping processes for extraction and use in further analysis and applications [22–24]. This method broadly consists of two steps: first is to gain access to the site by implementing the client side of the HTTP protocol and second is the building of DOM tree of the HTML document behind the concerned Web page and parsing through it while extracting data between the HTML elements using Xpath matching or CSS selector matching. This approach is

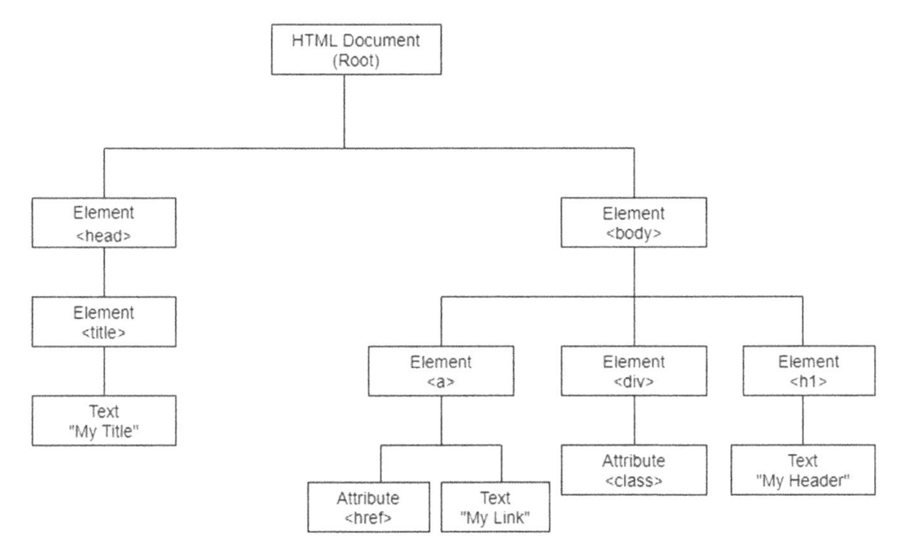

Fig. 1 DOM tree structure of a simple HTML document

basically based on creating a DOM tree structure of the Web page and searching for specific elements in the tree. Figure 1 shows a DOM tree structure of a simple HTML document.

DOM or document object model tree (https://dom.spec.whatwg.org/) is a tree-based hierarchy for representing a Web document. Moreover, HTML parsers such as AngleSharp and HAP in.Net, jsoup and Lagarto in Java, html5lib and lxml in Python, and many more, used in Web scraping tasks are also based on DOM tree.

Pros:

- Many libraries and frameworks support this approach.
- A hybrid model can be built combining this method with the machine learning model to predict appropriate elements based on the features extracted via DOM-based approach [25, 26].

Cons:

- Frequent changes in HTML of a Web page make this approach less reliable.
- Web scraper based on this approach needs to be maintained continuously.
- Time cost of building and parsing the DOM tree is more.

2.2 *String Matching*

The time-consuming DOM tree construction can be eliminated using less time-consuming string matching algorithms. These string matching algorithms address one of the most critical issues in Web data extraction that is searching process. Uzun

[24] experimented with seven pattern matching algorithms on Web pages and index of methods in popular programming languages such as Java, c#, Python, and JavaScript.

Uzun et al. [27] present a lightweight parser called SET parser which, unlike DOM tree parsers, searches in a text of a Web page without performing the creating process of a DOM tree, thus allowing fast extraction as compared to the DOM-based parsers. DOM tree-based approach takes the entire Web document into account for searching which increases the time cost of extraction step, while on the other hand, the string-based methods use an appropriate extraction pattern matching algorithm which provides time efficiency. Using appropriate indexes on the Web page as a starting index for a search algorithm also proves to improve time efficiency. Most search algorithms consist of two main phases: the preprocessing phase and the searching phase. The preprocessing step of an algorithm collects some information like character ordering and statistical data about a pattern and building a model to search that complements the next phase of searching where the algorithm finds the index of a pattern in a text.

To give a better insight, these extraction patterns are HTML tags such as < div > , < section > , < p > , < span >, and others that are appropriately chosen for Web data extraction in a more time-efficient way.

Pros:

- Consumes less time than the DOM-based method.
- Scraping tools like ScrapeBox [28, 29] includes keyword harvesting that uses string matching algorithms to find matching strings based on keywords.

Cons:

HTML structure of a Web page changes frequently that may force users to change the relevant extracting pattern.

2.3 Semantic Framework

Semantics, as the word suggests, deals with the meaning and relationship between entities like words, phrases, signs, or symbols. The World Wide Web is a large collection of data, most of which is not machine-readable, lacks structure, and is difficult to extract in an automated way. With the rising demand of data, the online resources have to be exploited for data-driven tasks. This necessity has marked the evolution of the World Wide Web to the 'semantic Web'. According to the W3C, '*The Semantic Web provides a common framework that allows data to be shared and reused across application, enterprise, and community boundaries.*' The semantic Web is therefore regarded as an integrator of Web data and systems and applications.

This parallel Web or an extension of WWW encompasses machine-readable Internet data that has been semantically encoded to represent metadata in a formal way. Technologies like resource description framework (RDF) and Web ontology language (OWL) enable the encoding of semantics with the data, thus creating a semantic Web markup for various domains. Ontologies can describe the concepts and relationships between entities and categories. OWL (https://www.w3.org/OWL/) is a family of knowledge representing languages formulating ontologies.

RDF (https://www.w3.org/RDF/) is a metadata data model used for conceptual description and modeling of knowledge data in various domains that is further used in knowledge management applications. It is implemented on Web resources using a variety of syntax resources and data serialization formats. It is an abstract model facilitating linked data structure and data interchange on the Web. Fernández-Villamor et al. [30] show the semantic–syntactic framework that uses DOM-based method and RDF-based model to map HTML fragments (extracted via DOM-based method) to the semantic Web resources.

Pros:

- Semantic Web resources facilitate an efficient management of knowledge and information of an organization.

Cons:

- Time-consuming

2.4 *Machine Learning-Based Web Scraping*

Recent studies deal with machine learning techniques applied on training dataset for the extraction of main content on a Web page, setting aside the irrelevant parts of a Web page consisting of advertisements. These methods are used for information extraction and Web search, link analysis, Web usage mining and opinion mining [31]. These techniques are usually used along with the mention of HTML tags of a Web page as features wherein segmentation of the page is improved using various visual cues, as compared to DOM-based method of taking the entire HTML document into consideration. The extraction of main content of the Web page or better page segmentation process is based on visual features such as font size, background color and styles, layout of Web page, text density and text length in different segments of a Web page that serve as features for a learning model. There may be supervised approaches [25, 26, 32–34] and unsupervised or statistical methods [35–40] for content extraction but any of these methods just simulates how a user understands web layout structure based on his visual perception. An implementation of such vision-based Web scraping is seen in Diffbot [41].

Pros:

- Not affected by any change in HTML structure as long as the visual representation of the Web page remains the same.

Cons:

- One needs to have requisite knowledge before applying these methods.
- Most of the machine learning-based Web scraping processes are restricted to certain types of Web pages.

3 Web Crawling

Having said about the methodologies working behind Web scraping processes and tools, this section throws light on the concept of Web crawling commonly used in scraping Web data from multiple Web pages. Web crawling can be seen as an extension of Web scraping that not only scrapes the data from a Web page but also follows the links on the Web page and jumps to multiple pages performing Web data extraction simultaneously. Web scraping is limited to scraping a certain Web page, while Web crawling scrapes the data from all of the Web pages which can be accessed from a certain page, thus huge amounts of data can be scraped using a crawler in a single automated process. There are two ways of following the links that a Web crawler may opt—one way is to let the crawler hop from one link to the other and another way is to add a 'robot.txt' file (Fig. 2) in the root directory of a Web page that tells the crawler, the links allowed to be visited.

An example of a Web crawler that performs data scraping at a large scale can be a 'Google search engine.' A search engine, also called search engine spiders, or search engine robots are human written software programs that can automatically and constantly visit millions of Web sites every day and include what they find, into search engine databases. This process is called crawling or spidering. Working of these search engines can be understood by a simple example. For instance, you have a Web site, and it has to be indexed by search engines so that users can find you through a search engine search. It happens when the search engine visits your Web site, the first file it looks for is 'robots.txt' which should be located in your Web root directory, and looks like, http://www.yoursite.com/robost.txt.

'Robots.txt' file gives you control of the following:

1. Crawlers allowed to visit your site.
2. The part of the Web site allowed to be visited and the parts from which crawlers should stay away.

However, in the absence of this file, crawlers follow all the links freely on the Web site without any restrictions. The restrictions on crawling and techniques to bypass those restrictions are discussed in brief in the later section. The following subsections

```
User-agent: *
Disallow: */s?k=*&rh=n*p_*p_*p_
Disallow: /dp/product-availability/
Disallow: /dp/rate-this-item/
Disallow: /exec/obidos/account-access-login
Disallow: /exec/obidos/change-style
Disallow: /exec/obidos/dt/assoc/handle-buy-box
Disallow: /exec/obidos/flex-sign-in
Disallow: /exec/obidos/handle-buy-box
Disallow: /exec/obidos/refer-a-friend-login
Disallow: /exec/obidos/subst/associates/join
Disallow: /exec/obidos/subst/marketplace/sell-your-collection.html
Disallow: /exec/obidos/subst/marketplace/sell-your-stuff.html
Disallow: /exec/obidos/subst/partners/friends/access.html
```

Fig. 2 Robots.txt for amazon.in. (*Source* https://www.amazon.in/robots.txt)

enable the readers to construct their own Web crawler using Scrapy framework along with the theoretical understanding of the workflow behind all the actions performed by Scrapy.

3.1 Scrapy and Spider

Scrapy is an application framework written in Python specifically for crawling purposes. Scrapy is written with 'Twisted,' a popular event-driven network programming framework for Python. 'Twisted' projects contain asynchronous (non-blocking) code for concurrency that provides for a fast crawling process. It also supports TCP, UDP, SSL/TLS, IP multicast, UNIX domain sockets, and many other protocols. Scrapy requires no dependencies and ensures that it works in all operating systems. Briefly, in Scrapy, robots or spiders or crawlers are defined as classes inherited from BaseSpider class, which defines a set of 'start_urls' and a 'parse' function called at each Web iteration. The following subsections deal with the workflow and components, basically the architecture (Fig. 3) behind the processing done by the Web crawler made using Scrapy.

Data Flow The following figure guides us through the data flow happening behind the hood.

Basic Terminology This section introduces some basic terms used in the above data flow (Fig. 3) and the terms that will be used later in the paper.

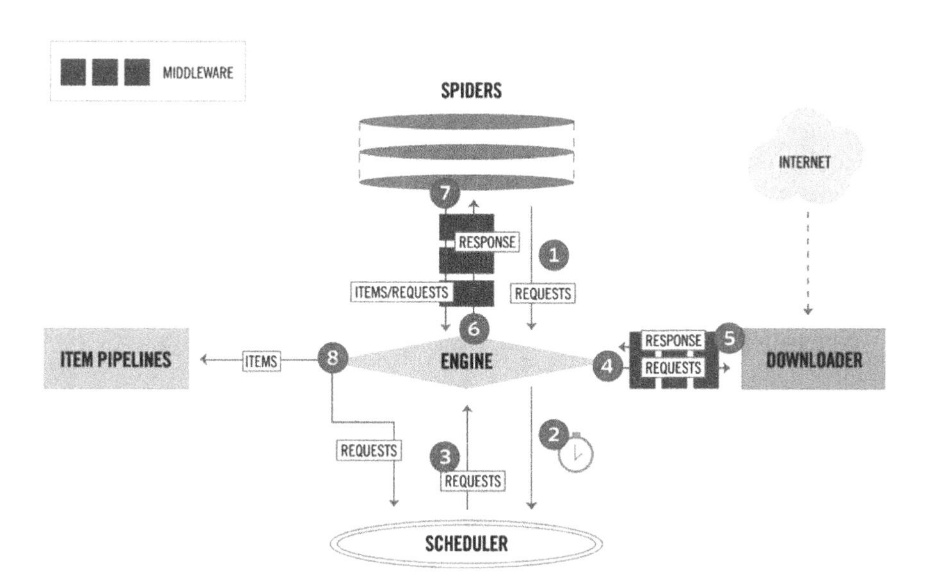

Fig. 3 Data flow in Scrapy. (*Source* https://docs.scrapy.org/)

Engine. It forms the main core of the entire workflow, establishing communication among all the other components and making appropriate action as per the request. The basic functionality includes getting requests to crawl from 'Spider'; scheduling requests in a 'Scheduler' and asking for next requests to crawl; sending requests to the 'Downloader', passing through 'Downloader Middlewares'; receiving response from the 'Downloader' and sending it to the 'Spider' for further action and receiving response from 'Spider' and communicating it with 'Item Pipelines'

Scheduler. Engine schedules request on the 'Scheduler' and continues taking up the follow-up requests of crawling from 'Spider' and schedules them onto 'Scheduler.' The workflow continues until there are no more requests in the 'Scheduler.'

Downloader. It performs the client side of HTTP protocol for accessing and downloading the Web page (whose links are put in 'start_url' variable in 'Spider') upon the request of 'Engine'. Engine takes up the response of 'Downloader' that is the source HTML code of the Web page and forwards it to 'Spider' for scraping.

Spider. These are the custom classes coded by the user to perform scraping of data using CSS selector or Xpath expressions (or by using any library like BeautifulSoup for it) and taking up follow requests. These classes have a 'start_url' global variable that contains the links to the Web pages to be scrapped. The parsing of the 'response' (source code HTML of the concerned Web page downloaded by the 'Downloader') and extraction of data happen in the 'parse' function. These classes are inherited from the BaseClass 'scrapy.Spider.'

Item Pipelines. Items are temporary containers that store the data extracted by the 'Spider' in an organized form. Thereafter, the data in the items are pushed into a database rather than directly storing the extracted data into the database which might cause problems when working on big/multiple projects with many spiders. This is handled by the 'items.py' file explained in the later sections.

Middlewares. In simple words, some changes or modifications can be made to the requests or responses flowing between the 'Engine' and the 'Spider' via 'Spider Middlewares' and between the 'Engine' and the 'Downloader' via 'Downloader Middlewares.'

4 Scraping Conference Data[1]

Every year, numerous conferences and summits are held in various fields. Ignorance of these opportunities can cause a very good opportunity for a student to present his/her work, slip by. Awareness among students on information related to multidisciplinary conferences and summits held on different dates can be significantly increased by collecting the data such as date, location and topics covered and storing them in a single database and communicating them with the students in some relevant

[1]Source code for this project is available on GitHub under the MIT License at https://github.com/NightmareNight-em/Scrapy-for-Web-Crawling.

way. Web crawling can help us in scraping such data in just a few seconds and also allows us to store it in desired format or in a local database. Educational institutions can use this way to communicate with the students such data that is scraped once in a month and can be utilized for many upcoming months.

One such example of a Web crawler, namely 'ConfSpider' is built using Scrapy for this particular task. This would enable the readers to build their own Web crawlers possibly for various different applications. The following subsections explain each.py file that contributes in making a Web crawler using Scrapy. These files are automatically formed by Scrapy when you start a Scrapy project, giving an instruction on Command prompt, 'scrapy startproject project_name'.

Make sure to have Scrapy installed in your system environment [42].

4.1 Settings.py

This default file contains lines of code that are commented, and the user has to uncomment or can add as per the demand the lines of code which a project requires in the crawler. It starts with the BOT_NAME which, in this case, is 'confSpider.' There are some modules and then comes the USER_AGENT along with the instruction of obeying the 'robot.txt' file. The following few lines among others have been made active in this file.

```
BOT_NAME = 'confSpider'.
SPIDER_MODULES = ['confSpider.spiders'].
NEWSPIDER_MODULE = 'confSpider.spiders'.
USER_AGENT = 'confSpider (+http://www.yourdomain.com).
ROBOTSTXT_OBEY = True.
```

Whenever we visit a Web site like google.com, we have to identify ourselves as to who we are (that is done by our system's IP address). The Web site asks the browser to identify itself exactly. If we are scraping, a more responsible way is to give ourselves (the bot or User-Agent) a name so that the Web site who is being requested can identify us. Most Web sites put restrictions on scraping their data, therefore one way to fool them or allow scraping their data freely is using such a User-Agent that cannot be blocked from crawling like a Google bot. We will discuss this later when we will deal with the restrictions put on Web crawling and will know how to bypass it using Scrapy.

Further in 'settings.py,' we have uncommented the item_pipelines so that we could configure pipelines according to our need in 'pipelines.py' file.

```
ITEM_PIPELINES = {
'confSpider.pipelines.ConfspiderPipeline': 300,
}.
```

One can uncomment 'Downloader Middlewares' and 'Spider Middlewares' as per his needs. Two more terms that are worth mentioning are 'Concurrent Requests' and 'Autothrottle,' wherein 'Concurrent Requests' refers to the number of requests made to the Web site server to open up so as to scrape the data, and 'Autothrottle' makes sure that the Web site we are scraping does not get overloaded. To prevent overloading and server getting down, Concurrent Requests are kept at 16 by default, however we can change it in 'settings.py' but a high value is not recommended for it.

4.2 Spider

This is the custom class made by the user that inherits from Scrapy's BaseSpider class 'scrapy.Spider'. The 'parse' method deals with the parsing of the response (i.e., the HTML source code) and extracting relevant data using CSS selectors or/and Xpath expressions. We, in our method, have mostly used CSS selectors to extract desired content corresponding to the CSS syntax (or CSS selector) used in the HTML response source code. CSS selectors can be identified manually by inspecting the HTML code of the concerned Web page or we can use an extension in Google Chrome by the name of 'Selector Gadget' [43] to get the CSS selector of the selected part of the Web page (Fig. 4).

Before the parse method, we had defined 'name', 'start_urls' and 'page' global variables that refer to the bot name, link to the Web page to be scraped [23] and pages to follow up on crawling, respectively. Thereafter comes the parse method where we have scraped 'name,' 'date,' 'location,' 'topicCovered,' 'conflink' (link to the official Web site of the conference), and 'organizer' of the conference. Looping through every division that corresponds to a single conference data, we have extracted data using CSS selector some of which are defined in Table 1.

Now, we put the data extracted, in 'items' that is an object of 'confSpiderItem' class defined in 'items.py' file. The Web site that we are scraping contains ten pages which we followed in our crawling process to extract every page data in multiple

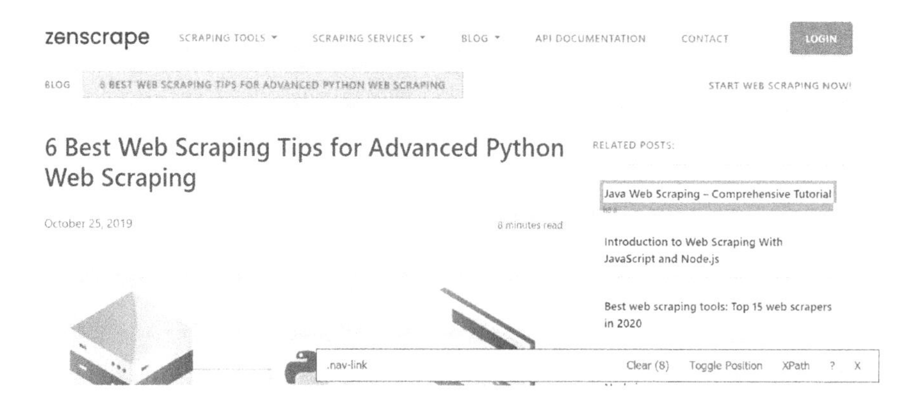

Fig. 4 This extension gives the CSS and Xpath in the taskbar at the right bottom corresponding to the selected (highlighted in yellow color) text (*Source* https://zenscrape.com/)

Table 1 Examples of CSS selectors used for extracting corresponding data

Data to be extracted	CSS Selectors
Name	.css('.conflistspan:nth-child(1)')
Location	.css('span.div_venue')
confLink	.css('a.conflist').xpath('@href')

Note that these selectors are different for different Web pages because the HTML tags behind a Web page changes with the layout of a Web page. This very methodology used in extracting the desired data is the DOM tree-based approach wherein the HTML of a Web page is used to build a DOM tree (Fig. 1) having a hierarchy of tags and content enclosed within tags. This DOM tree is parsed by our spider extracting the content which matches the CSS syntax that we give as an argument for extraction (Table 1).

iterations. The next page variable contains the URL that gets generated when one goes to the next page in the Web site, thus accordingly changes are made in the link to follow up.

4.3 *Items.py*

This file is used to define various fields of data extracted. Herein, we define the fields of 'name,' 'date,' 'location,' 'topicCovered,' 'organizer,' and 'conflink'. These fields are stored in the item container in an organized manner.

4.4 *Pipelines.py*[2]

Web scraped data are handled properly and exported either as.csv/. json/.xml directly using Scrapy or can be stored in a SQL file or MySQL database or mongoDB database. We do not have to mess up with this file if we need to output our scraped content in.csv/. json/.xml format.

The command, *'scrapy crawl confSpider -o yourFile.json,'* starts the process of crawling and stores the scraped data in 'yourFile.json' file. Note that this file must be present in the folder where all project-related files are present. Output can be in.csv and.xml file as well. Scrapy also allows us to change the pipelines.py file according to our needs. For example, storing data in a MySQL database is quite easy with Scrapy with the use of Python's MySQL library[2].

[2]The complete dataset in.csv and.json format as well as pipelining through MySQL database is available at https://github.com/NightmareNight-em/Scrapy-for-Web-Crawling.

4.5 *Middlewares.py*

It allows for cookies and session handling, handling HTTP features, User-Agent spoofing and handling crawl depth restrictions. It particularly handles the work of 'Spider Middlewares' and 'Downloader Middlewares.' We have not messed up with it, however middlewares have significant use in pulling down the restrictions on Web scraping by the Web sites, explained below.

5 Restrictions Imposed on Web Scraping

The below subsections describe ways in which Scrapy helps Python Web scraper to handle restrictions and measures such as user authentication and black listing imposed by the Web sites to secure its data from scraping.

5.1 *Bypass Restrictions Using User-Agents*

When some browser visits a Web site, it is asked for its identity and that identity is termed as 'User-Agent.' The User-Agent may have such an identity that may fool a Web site and could crawl it multiple times without being black listed from server side. Such a User-Agent can be a Google bot or Google User-Agent that has to be allowed to crawl by the Web sites in order to get indexed by its search engine. Thus, defining our User-Agent as Google User-Agent in the 'setting.py' file can bypass the restrictions.

```
USER_AGENT = 'Mozilla/5.0 (compatible; Googlebot/2.1; + http://
www.google.com/bot.html)'.
```

One can find Google bot user agents at,

https://developers.whatismybrowser.com/useragents/explore/software_name/googlebot/

Another way to get access to Web scraping may be using fake user agents in rotation such that the Web site gets fooled into believing that a lot of Web browsers are visiting the Web site and not just one [44].

5.2 *Bypass Restrictions Using Proxies*

Here, proxies refer to the IP addresses that are not of our own PC. Some Web sites like amazon.com may ban the IP of our system from accessing the site because of a lot of scraping done through it. However, using proxies or IP addresses other than ours, in rotation, could bypass this trouble easily.

Note that this is not illegal as of now, and it does not come under identity theft because the proxies used do not belong to other computers. 'Scrapy-proxy-pool' library can help us generate and use proxies [44].

5.3 Bypass Restrictions by Logging in the Web site

Some Web sites put restrictions on data content by hiding the content behind the login page. A user then has to log into the Web site to gain access to the content. Such Web sites require user authentication in order to open up its data to the user.

A user performing Web scraping would need an automated way of logging into such Web sites. This is where Scrapy can help. We just need to change the 'start_urls' variable from our Spider class and give it the link to the login page of the Web site. 'scrapy.http' provides a library 'FormRequest' that helps logging into the Web site and starts scraping freely again [44].

6 Conclusion

Web scraping has been an evident and significant way of collecting Web information for a long time. Professionals use it in many B2B and B2C use cases to integrate data into innovative applications which offer additional values and novelty. Start-ups love it because it is cheap and a powerful way to gather data without the need of partnerships. Big companies use Web scraping for their own benefit but at the same time do not want their data to be scraped by others. Thus, there always has existed a question on the legality of Web scraping. While courts try to judge this technology as legal or illegal, companies tend to develop anti-bots to stop bots from extracting their data. Rising demands of data require the handling of such anti-bot practices.

Our future work would include performing advanced Python Web scraping to handle the measures taken by Web sites to foil Web scraping attempts. Such measures include honeypot trap, building dynamic, and user-friendly Web sites, different layouts in different pages and captchas and redirects among others.

Acknowledgements We would like to thank 'The Dexterity Global Group' for their support and encouragement throughout our endeavors. Additionally, we would like to thank the entire Jaypee Institute of Information Technology fraternity for their support.

References

1. Ferrara E, De Meo P, Fiumara G, Baumgartner R (2014) Web data extraction, applications and techniques: A survey. Knowl-Based Syst 70:301–323. https://doi.org/10.1016/j.knosys.2014.07.007
2. Baumgartner R, Frölich O, Gottlob G, Harz P, Herzog M, Lehmann P, Wien T (2005) Web data extraction for business intelligence: the lixto approach. In: Proceedings 12th conference on Datenbanksysteme in Büro. Technik und Wissenschaft, pp 48–65
3. Anica-Popa I, Cucui G (2009) A framework for enhancing competitive intelligence capabilities using decision support system based on web mining techniques. Int J Comput Commun Control 4:326–334
4. Chen H, Chau M, Zeng D (2002) CI Spider: a tool for competitive intelligence on the Web. Decision Supp Syst 34(1):1–17. https://doi.org/10.1016/S0167-9236(02)00002-7. ISSN 0167–9236
5. Lin L, Liotta A, Hippisley A (2005) A method for automating the extraction of specialized information from the web. In: Hao Y et al (eds) Computational intelligence and security. CIS 2005. Lecture notes in computer science, vol 3801. Springer, Berlin, Heidelberg
6. Suganya E, Vijayarani S (2020) Sentiment analysis for scraping of product reviews from multiple web pages using machine learning algorithms. In: Abraham A, Cherukuri A, Melin P, Gandhi N (eds) Intelligent systems design and applications. ISDA 2018 2018. Advances in intelligent systems and computing, vol 941. Springer, Cham
7. Priyadarshini R, Barik R K, Dubey H (2018) Deepfog: fog computing-based deep neural architecture for prediction of stress types, diabetes and hypertension attacks. Computation. 6:62 https://doi.org/10.3390/computation6040062
8. Hillen J (2019) Web scraping for food price research. British Food J ahead-of-print. https://doi.org/10.1108/BFJ-02-2019-0081
9. Glez-Peña D et al (2013) Web scraping technologies in an API world. Briefings in Bioinformatics Advance Access. https://doi.org/10.1093/bib/bbt026, published April 30, 2013
10. Stein L (2002) Creating a bioinformatics nation. Nature 417(6885):119–120. https://doi.org/10.1038/417119a
11. Catanese SA, De Meo P, Ferrara E, Fiumara G, Provetti A (2011) Crawling facebook for social network analysis purposes. In: Proceedings of the international conference on web intelligence, mining and semantics (WIMS '11). Association for Computing Machinery, New York, NY, USA, Article 52, 1–8. https://doi.org/10.1145/1988688.1988749
12. Traud AL, Kelsic ED, Mucha PJ, Porter MA (2008) Comparing community structure to characteristics in online collegiate social networks. SIAM Rev 53(3):17
13. Barik RK, Misra C, Lenka RK et al (2019) Hybrid mist-cloud systems for large scale geospatial big data analytics and processing: opportunities and challenges. Arab J Geosci 12:32. https://doi.org/10.1007/s12517-018-4104-3
14. Laender AH, Ribeiro-Neto BA, Da Silva AS, Teixeira JS (2002) A brief survey of web data extraction tools. SIGMOD Rec 31(2):84–93. https://doi.org/10.1145/565117.565137
15. Laender AHF, Ribeiro-Neto BA, da Silva AS, Teixeira JS (2002) A brief survey of web data extraction tools. ACM SIGMOD Rec 31(2):84. https://doi.org/10.1145/565117.565137
16. Singrodia V, Mitra A, Paul S (2019) A review on web scraping and its applications. In: 2019 international conference on computer communication and informatics (ICCCI). Coimbatore, Tamil Nadu, India, pp 1–6. https://doi.org/10.1109/ICCCI.2019.8821809
17. Vanden Broucke S, Baesens B (2018) Practical Web scraping for data science, 1st edn. Apress, Berkeley, CA. https://doi.org/10.1007/978-1-4842-3582-9
18. Castrillo-Fernández Q (2015) Web scraping: applications and tools. European Public Sector Information Platform Topic Report No. 2015
19. Heydon A, Najork M (1999) Mercator: a scalable, extensible Web crawler. World Wide Web 2(4):219–229. https://doi.org/10.1023/A:1019213109274

20. Chakrabarti S, Berg M, Dom B (2000) Focused crawling: a new approach to topic-specific Web resource discovery. Comput Netw 31(1623):1640. https://doi.org/10.1016/S1389-1286(99)00052-3
21. Menczer F, Pant G, Srinivasan P (2004) Topical web crawlers: evaluating adaptive algorithms. ACM Trans Internet Techn 4:378–419
22. Kumar A, Paprzycki M, Gunjan VK (eds) (2020) ICDSMLA 2019. In: Lecture notes in electrical engineering. https://doi.org/10.1007/978-981-15-1420-3
23. Zheng X, Gu Y, Li Y (2012) Data extraction from web pages based on structural-semantic entropy. In: proceedings of the 21st international conference on world wide web (WWW '12 Companion). Association for Computing Machinery, New York, NY, USA, 93–102. https://doi.org/10.1145/2187980.2187991
24. Uzun E (2020) A novel web scraping approach using the additional information obtained from web pages. IEEE Access 8:61726–61740. https://doi.org/10.1109/ACCESS.2020.2984503
25. Uzun E, Agun HV, Yerlikaya T (2013) A hybrid approach for extracting information content from Webpages. Inf Process Manage 49(4):928–944
26. Uzun E, Güner ES, Kılıçaslan Y, Yerlikaya T, Agun HV (2014) An effective and efficient Web content extractor for optimizing the crawling process. Softw Pract Exper 44(10):1181–1199
27. Uzun E, Yerlikaya T, Kurt M (2011) A lightweight parser for extracting useful contents from web pages. In: proceedings of 2nd international symposium computer science engineering (ISCSE). Kuşadasi, Turkey, pp 67–73
28. Jason Mun Personal website, https://www.jasonmun.com/using-scrapebox-for-good-not-evil/. Last Accessed 22 May 2020
29. ScrapeBox homepage, http://www.scrapebox.com/. Last Accessed 10 June 2020
30. Jose CAIMG, Fernandez-Villamor I, Blasco-Garcia J (2012) A semantic scraping model for web resources. Applying linked data to web page screen scraping. In: ICAART 2011—proceedings of the 3rd international conference on agents and artificial Intelligence, 2, 451–456
31. Ioan D, Moisil I (2008) Advanced AI techniques for web mining
32. Mashuq M, Michel, Zhou Z Web content extraction through machine learning
33. Nguyen-Hoang B-D, Pham-Hong B-T, Jin J, Le PTV (2018) Genre-oriented web content extraction with deep convolutional neural networks and statistical methods. PACLIC
34. Cai D, Yu S, Wen JR, Ma WY (2003) Extracting content structure for web pages based on visual representation. 406–471. https://doi.org/10.1007/3-540-36901-5_42
35. Gottron T (2008) Content code blurring: a new approach to content extraction. In: Proceedings 19th international conference database expert system applications (DEXA), pp 29–33
36. Weninger T, Hsu WH, Han J (2010) 'CETR: content extraction via tag ratios. In: proceedings 19th international conference of world wide web (WWW), pp 971–980
37. Gupta S, Kaiser G, Neistadt D, Grimm P (2003) DOM-based content extraction of HTML documents. In: Proceedings 12th international conference on worldwideweb, pp 207–214
38. Finn A, Kushmerick N, Smyth B (2001) "Fact or fiction: content classification for digital libraries. In: Proceedings of joint DELOS-NSF workshop, personalization recommender system digital libraries, [Online]. Available: http://citeseerx.ist.psu.edu/viewdoc/citations;jsessionid=8E0FC70BEE7 DFA696487A2F7C6B622FA?
39. Adam G, Bouras C, Poulopoulos V (2009) CUTER: An efficient useful text extraction mechanism. In: International conference on advanced information networking and applications (AINA), pp 703–708
40. Gunasundari R (2012) A study of content extraction from Web pages based on links. Int J Data Mining Knowl Manage Process 2(3):230–236
41. Diffbot homepage, https://en.wikipedia.org/wiki/Diffbot. Last Accessed 10 June 2020
42. Scrapy Installation Guide, https://docs.scrapy.org/en/latest/intro/install.html. Last Accessed 22 June 2020
43. SelectorGadget, Chrome web store, https://chrome.google.com/webstore/detail/selectorgadget/mhjhnkcfbdhnjickkkdbjoemdmbfginb?hl=en. Last Accessed 25 June 2020

44. Python Web Scraping and Crawling using Scrapy, https://www.youtube.com/watch?v=ve_0h4Y8nuI&list=PLhTjy8cBISEqkN-5Ku_kXG4QW33sxQo0t. Last Accessed 22 June 2020
45. Thomas DM, Mathur S (2019) Data analysis by web scraping using Python. In: 2019 3rd international conference on electronics, communication and aerospace technology (ICECA). Coimbatore, India 2019, pp 450–454. https://doi.org/10.1109/ICECA.2019.8822022
46. Feng Y, Hong Y, Tang W, Yao J, Zhu Q (2011) Using HTML tags to improve parallel resources extraction. In: 2011 international conference on Asian language processing. Penang, pp 255–259. https://doi.org/10.1109/IALP.2011.23

Selection of Candidate Views for Big Data View Materialization

Akshay Kumar and T. V. Vijay Kumar

Abstract Big data is a large volume of heterogeneous data, which can be structured, semi-structured, or unstructured, produced at a very rapid rate by several disparate data sources. Big data requires processing using distributed storage and processing frameworks to answer big data queries. Materializing big data views would facilitate faster, real-time processing of big data queries. However, there exist large numbers of possible views and, from among these, computing a subset of views that would optimize the processing time of big data queries is a complex problem. This paper addresses this problem by proposing a framework that reduces the large search space of all possible views by computing comparatively smaller set of candidate views for a given query workload of a big data application. The proposed framework uses the big data view structure graph, which represents the structure of big data views and their dependencies, to compute a set of candidate views and alternate query evaluation plans for big data queries.

Keywords Big data · View materialization · Query processing

1 Introduction

The information technology revolution has transformed the ways of living and working of the vast majority of people around the globe. This has resulted in generation of very large amount of low integrity and low value heterogeneous data, also called big data, which is being produced at a very rapid rate. This data is collected from many disparate sources and then integrated, stored, processed, analyzed, and presented in different visual forms to create useful information for various applications. A big data application can be deployed in different areas, like scientific,

A. Kumar (✉) · T. V. Vijay Kumar
School of Computer and Systems Sciences, Jawaharlal Nehru University, New Delhi 110067, India
e-mail: akshay@ignou.ac.in

A. Kumar
School of Computer and Information Sciences, Indira Gandhi National Open University, New Delhi 110068, India

A. Choudhary et al. (eds.), *Applications of Artificial Intelligence and Machine Learning*, Lecture Notes in Electrical Engineering 778,
https://doi.org/10.1007/978-981-16-3067-5_10

knowledge discovery, commerce, governance, learning, healthcare, disaster management, etc. Two most recent and challenging applications of big data are modeling the spread of the worldwide pandemic, coronavirus disease (*COVID-19*) using big data geographic information system [5] and the development of a framework for studying the environmental sustainability of smart cities [4]. Big data applications should generate correct, precise, consistent, and timely information, which helps in making timely decisions.

Big data is voluminous and heterogeneous and has high data generation rate, low integrity, and low value [10, 14, 18, 21, 35]. A big data application is required to store and process large volumes of structured, semi-structured, and unstructured data reliably, which was not be achieved using the traditional file processing and database management systems. Thus, alternative technologies were required, which supported reliable storage and fault-tolerant distributed processing of very large data. This led to the design of high-performance distributed file systems (*DFS*) like Hadoop distributed file system (*HDFS*), which store data reliably in large big data blocks of size 64 MB and/or 128 MB or more [7, 15, 16]. In addition, fault-tolerant frameworks like MapReduce, Apache Spark, and *NoSQL* databases were developed to process big data, which is stored on *DFS* [7, 8, 21].

A big data application is required to process big data queries efficiently. This can be achieved by materializing the big data views over structured, semi-structured, and unstructured data. Big data applications have a large number of possible views for materialization. However, all these views cannot be materialized due to update processing overheads and storage space constraints. Therefore, a subset of candidate views, which minimizes the query processing time of a big data application for a given query workload, is selected for materialization. One of the prerequisites for big data view materialization is the identification of the candidate big data views from a very large number of possible views. This paper proposes a graph-based approach for identifying such candidate big data views for a given query workload of a big data application. The paper also proposes a theoretical framework and an algorithm to create a big data view structure graph, which represents big data views and their interrelationships [23]. This big data view structure graph can be used to identify candidate views for materialization for a given set of workload queries. The algorithm also creates alternate query evaluation plans, which are useful in computing the optimal query processing cost for selected sets of materialized big data views.

The rest of the paper is organized as follows. Section 2 presents big data view materialization in the context of structured, semi-structured, and unstructured data. Section 3 proposes a theoretical framework for candidate view selection using big data view structure graph. Section 4 presents an algorithm for selecting candidate views and alternate query evaluation plans for a given query workload. An example illustrating the use of the proposed algorithm is discussed in Sect. 5. Section 6 is the conclusion.

2 Big Data View Materialization

View materialization has been studied for various database management systems. In the context of structured data, view materialization was studied for relational database management systems (*RDBMS*), data warehouses, and object-oriented database management systems (*OODBMS*). View materialization results in the reduction of time of query processing, as the time to access materialized views is significantly lower than the time to compute the view query. There can be a large number of possible views in *RDBMS* and data warehouse. However, only some of these views can be materialized, as the materialized views are required to be maintained. The problem of materialization of views was formally defined for *RDBMS* and data warehouse in [30] and is given below.

Given a set R of n relations, represented as $R = \{r_1, r_2, \ldots r_n\}$ and set of m queries on these relations, represented as $Q = \{q_1, q_2, \ldots q_m\}$; the view materialization problem selects a set of p views for materialization (V_m), represented as $V_m = \{v_1, v_2, \ldots v_p\}$, from among large numbers of candidate views, which results in minimizing the cost of query processing at a minimal cost of view maintenance and storage space.

The problem of view materialization is a complex problem, which grow exponentially with increase in the number of dimensions of a data warehouse [17]. In order to solve the view materialization problem, a set of candidate views for materialization is created. These candidate views are represented using a framework of views and view defining query expression dependencies. A number of such frameworks have been designed for data warehouses. Harinarayan et al. [17] proposed a lattice structure, where a node represents a view created using a group by clause, and a link from node V_1 to V_2 exists if V_2 can be directly derived from the node V_1. Thus, a query on a data warehouse that can be answered by view V_2 can also be answered by view V_1, but not vice versa. Gupta [13] represented the views using a directed acyclic AND-OR view graph, which represents the views and the dependencies, in terms of common query sub-expressions. These AND-OR view graphs can be used to create alternate query evaluation plans for query processing in a data warehouse. The view selection problem for *RDBMS* and data warehouse has been solved using several different approaches, viz. greedy approaches [13, 17], empirical approach [3], and evolutionary approaches [20]. Authors [6, 32] presented the theoretical framework on view materialization. However, as big data views and their interrelationships are more complex in nature due to the presence of large heterogeneous data, newer frameworks are needed to represent the big data view structure.

The data in databases is also updated frequently. These updates should also be communicated in real time to the related materialized views. Kenneth et al. [19] proposed to find common sub-expressions among the materialized views to create additional materialized views in order to minimize the view maintenance cost. Authors [12, 31] presented a framework and algorithm to solve the view materialization problem considering the query processing and maintenance costs in the context of a data warehouse. The approach to find common sub-expressions may be extended to find the candidate views for big data view materialization, where several smaller views may be considered for materialization. However, the expected minimum size of a big data view should be at least one block of big data, which will require merging of several related smaller views to a candidate big data view.

Kuno and Rundensteiner [28] presented view materialization for the *OODBMS* and represented object-oriented views as schema graphs of base and virtual classes. A node in the schema graph represents a class, and a link represents a common inherited property of classes. A virtual class is created in the schema graph, as a result of a query operator, thus representing a view. Therefore, views are created without making a copy of the original objects. However, classification of objects into materialized view classes has to be dynamic, as changes in data may change this classification, and a view may be assigned to a new materialized view class, if needed. The view materialization, therefore, in *OODBMS* can minimize maintenance overheads. Big data views may use such a concept when semi-structured data is in the form of objects.

View materialization has also been studied for semi-structured data using the ordered tree structure [1, 2, 33]. The basic idea was to identify the query sub-tree from the ordered tree of the semi-structured data. A greedy algorithm for the selection of views for materialization on XML data was proposed in [33]. For processing updates on the views, an incremental update model was proposed for semi-structured data in [9]. An interesting case of view materialization on semi-structured data was presented in [27, 29], which proposes to materialize web-views for a database-driven web application. A web-view was defined as a component of a web page, which can be reused in several web pages. Kumar and Vijay Kumar [27] also presented an algorithm, which dynamically selects web views for materialization based on web page accesses and relation update requests. The graph-based framework of semi-structured data is needed to be extended to big data view structure representations for the identification of the candidate big data views.

View materialization has not been specifically studied for unstructured data. However, [10] defined the importance of unstructured data in the context of big data and presented methods for efficient handling of unstructured data. An interesting work on query processing was presented in [34], which suggested the processing of queries on unstructured data using three different techniques, viz. linking the unstructured data to a relational model, creating a separate model for the unstructured data, or extracting attributes from the natural language queries and processing the unstructured data using these extracted attributes. These models can be used in the context of big data for the identification of query attributes in unstructured data.

Big data view materialization was proposed in [11] in the context of Hive, a big data warehousing tool. Goswami et al. [11] proposed the big data view materialization, as a multi-objective optimization problem and used the genetic algorithm to solve the problem for a standard dataset and queries. However, the paper did not propose any model for identifying big data candidate views. Kumar and Vijay Kumar [22] defined the big data view materialization problem, which considered the big data characteristics including heterogeneity, volume, rate of data generation, and trustworthiness. Big data view materialization was defined as a bi-objective optimization problem in [23], having the objectives—minimization of query evaluation cost for a query workload and the minimization of the update processing cost of the materialized views. The query evaluation cost was computed using query frequency, change in the size of big data, and alternate query evaluation plans, while update processing cost was computed using the size of the data required to update the materialized views and view integrity. Since the two objectives of the bi-objective optimization problem were conflicting, multi-objective evolutionary algorithms were used to select big data views for materialization. Authors [23–26] used the vector evaluated genetic algorithm (*VEGA*), multi-objective genetic algorithm (*MOGA*), strength Pareto evolutionary algorithm-II (*SPEA-II*), and non-dominated sorting genetic algorithm-II (*NSGA-II*), respectively, to solve the bi-objective optimization problem. However, all these evolutionary algorithms require a set of workload queries, candidate views for materialization, and alternate query evaluation plans as input. In order to identify the candidate views for materialization, [22] suggested the use of query attributes (Q_A), which are used to retrieve specific information from the voluminous big data. [23] suggested to use query attributes and their interrelationships to create directed graphs, which may help in the identification of candidate views. This paper formally defines the big data view structure graph, which were proposed in [23] and extends these graphs to create a list of candidate views and alternate query evaluation plans. This is presented in the next section.

3 Big Data View Structure Graph

The views on big data are generated as a result of a query on big data. A query on big data involves the use of query attributes (Q_A), which are used to produce a directed acyclic view structure graph [23]. This big data view structure graph along with the set of workload queries can be used to identify the candidate big data views, which form the input for the big data view selection algorithm. A sample big data view graph, which uses a set of query attributes $S_{QA} = \{A, B, C\}$, is shown in Fig. 1. This figure also assumes that the query attribute *A* is having a dependence relationship on the query attribute *C*, which is represented as $C \rightarrow A$. The node of this graph represents a big data view consisting of structured data (*SD*), semi-structured data (*SSD*) and unstructured data (*UD*), which has been created for a specific set of query attributes; whereas, the links represent the interrelationships among these views. For example, a view *Node(C, B)* has been created using the query attributes (*C*, *B*) and a link from view *Node*(*C*, *B*) to view *Node*(*C*) indicates that the view *Node*(*C*) can be derived from the view *Node*(*C*, *B*). In addition, a link can also be created due to dependencies of the query attributes. For example, the query attribute dependence $C \rightarrow A$ will result in the creation of links *Node*(*C*, *B*) to *Node*(*B*, *A*) and *Node*(*C*) to *Node(A)* as shown in Fig. 1. The big data view structure graph can be represented by the following graph representation equation, where *VG* represents the big data view structure graph, *VN* represents the view node, and *DE* represents the directed edges:

$$VG = (VN, DE) \tag{1}$$

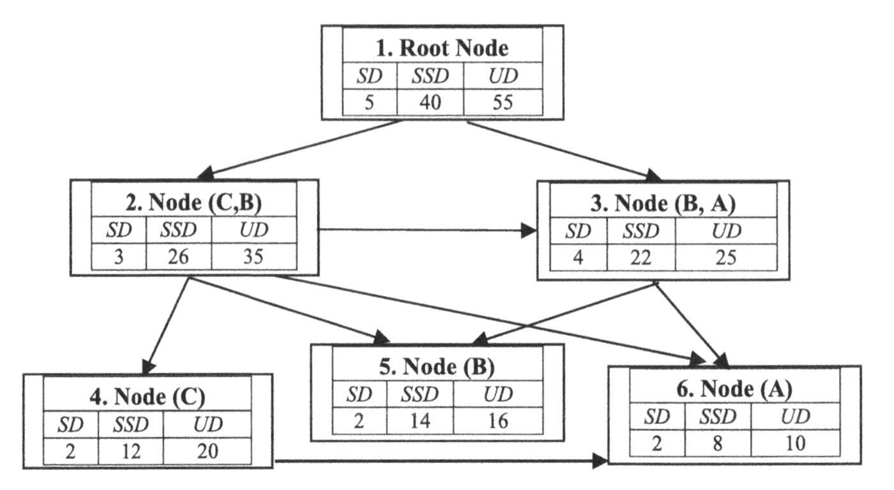

Fig. 1 Big data view structure graph for $Q_A\{A, B, C\}$ with dependence $C \rightarrow A$

VN, which represents a set of candidate views created using Q_A, is defined below:

$$VN = \{VDn | VDn \in \mathcal{P}(Q_A) \wedge (\sim (Q_{Ai} \rightarrow Q_{Aj})(i \neq j)\} \tag{2}$$

VDn is a big data view node, which is created for a member of the power set ($\mathcal{P}(Q_A)$) of the set of query attributes (S_{QA}). A view node is created for every member of the $\mathcal{P}(Q_A)$, except the nodes that have been created using the pair of dependent query attributes (please refer to Eq. (2)). For example, in Fig. 1, the *Node(C, A)* has not been created due to the query attribute dependence $C \rightarrow A$. It may be noted that any multi-level dependence of query attributes may be modeled using the pair-wise binary dependence. In general, the number of nodes in the big data view graph will be of the order of the size of the power set, which is $O(|\mathcal{P}(Q_A)|) = O(2^{|S_{QA}|})$.

The links or the edges in the directed big data view structure graph are defined below:

$$\begin{aligned} DE = \{&(x, y) | \left(x \in Q_A^k, y \in Q_A^{k-1}, \quad \text{where } Q_A^{k-1} \subset Q_A^k\right) \\ &\text{or } \left(Q_{A_i}^k \rightarrow Q_{A_j}^k\right)(i \neq j \text{ and all other } Q_A \text{ are similar}) \\ &\text{for } k = n \text{ down to } 2, \quad \text{where } n = |S_{QA}|\} \end{aligned} \tag{3}$$

In order to define the edges, the term *level of nodes* (k) is defined. A level of the node is the number of query attributes that are used to define the node. Thus, the level of the root node, which includes all the query attributes, will always be n, where $n = |S_{QA}|$, i.e., cardinality of the set of the query attributes. A directed edge is created between two big data view nodes, which are at adjacent levels, provided the query attributes at the lower level are a proper subset of query attributes at the next higher level. This is the first condition of Eq. (3). A directed edge can also be created if the two view nodes are at the same level, having the same query attributes except the one having the dependence relationship. For example, *Node(C,B)* and *Node(B,A)* in Fig. 1 are related as both have a common query attribute *B*, and the other query attribute has a dependence relationship $C \rightarrow A$, and thus, a directed edge will be created from *Node(C, B)* to *Node(B, A)*. The value of k in Eq. (3) varies from n down to 2.

Figure 1 also shows the sample size of different types of data at various levels of the views. The dependence relationship can also be used to identify alternate query evaluation plans. For example, the query attribute dependence $C \rightarrow A$ also implies that a query, which can be answered by a view created on *Node(A)*, can also be answered by the view on *Node(C)*, but *not* vice versa. This big data view graph can be used to generate candidate views. However, it may be noted that a candidate view need not include all types of data. For example, as per the link structure of Fig. 1, a query that can be answered by a view on *Node(A)* can also be answered by *Node(C, B)* or *Node(B, A)* or *Node(C)*, provided they are on the same type of data and are created using the same query operators. In addition, a view node in Fig. 1 can be used to create more than one candidate view on different types of data (see Fig. 2). For example, two different views can be created using *Node(A, B)*, e.g., one on semi-structured data and the other on structured data. It may be noted from Fig. 2 that the root node is used to create three big data views, one each for structured, semi-structured, and unstructured data. All these three views created using the root node are always assumed to be materialized. A node, therefore, can be used to create many different candidate views of different types of data (see Fig. 2). It may be noted that candidate views generated using a single node, sometimes, are combined to create additional candidate views. For example, suppose there are two queries in an application. Suppose the first query can be answered by the semi-structured data

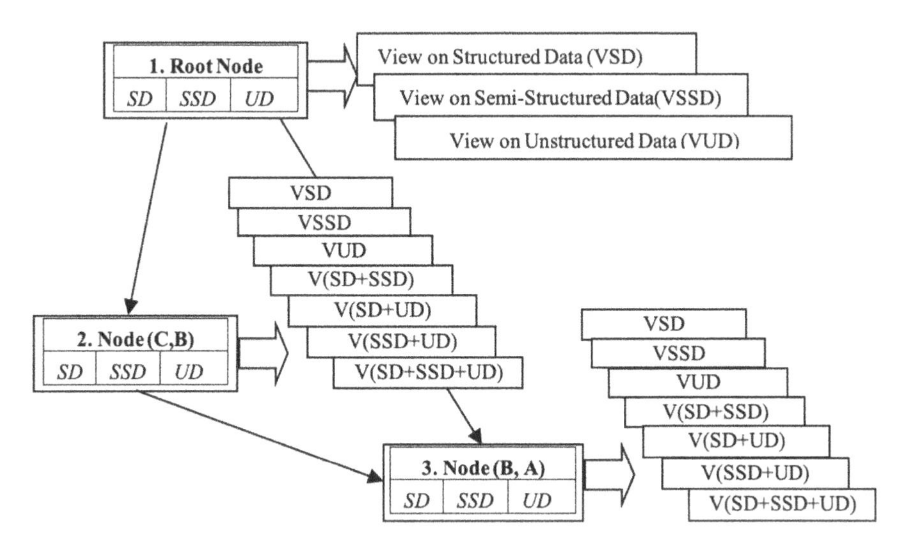

Fig. 2 View node architecture expanded on type of big data

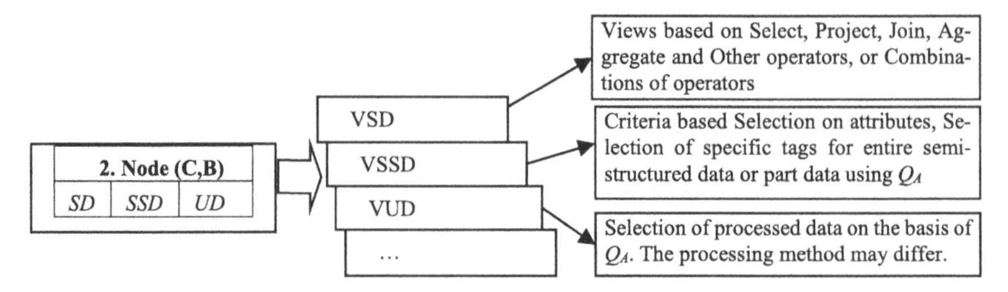

Fig. 3 View creation on the basis of query operators

in node *Node (A, B)*, which will result in a candidate view on the semi-structured data. The second query is on the unstructured data in *Node(A, B)* which will result in a candidate view on unstructured data. In addition, a third candidate view can be created on both semi-structured and unstructured data in *Node(A, B)*. Similarly, if candidate views are on the same node or data, but created using different operators, then a view formed by the union of the two operators can also be considered as a candidate view for materialization. Thus, the big data view structure graph shown in Fig. 1 needs to be further expanded to compute such additional set of candidate views for materialization. Figure 2 shows the expansion of the *root node*, *Node(C, B)* and Node*(B, A)* of Fig. 1 using structured, semi-structured, and unstructured data. Figure 3 shows the expansion of Fig. 2 based on the different operators used for creating individual candidate views.

Thus, the number of views that can be created using the big data view structure graph of Fig. 1 can be very large. The graph of Fig. 1 itself has about $2^{|Q_A|}$ nodes. Thus, the problem of computing the candidate view is of exponential complexity.

An algorithm $BD_{CV\&QP}$ for computing the set of candidate views for materialization and alternate query evaluation plans for the query workload for a big data application is proposed and is discussed in the next section.

4 Algorithm $BD_{CV\&QP}$

The algorithm $BD_{CV\&QP}$ takes the set of queries (*Q*) in a query workload for a big data application, as input and produces a list of candidate views for materialization and alternate query evaluation plans for the query workload in a big data application, as output. The algorithm $BD_{CV\&QP}$ is given below:

Input: The list of Queries (Q) in the query workload for a Big data application
Output: List of candidate views for materialization (CV_m), Alternate query evaluation plans (QEP) for query workload (Q)
Variables: Query Attribute (Q_A); Data accessed by Query (Q_{data});Query operator ($Q_{operator}$) Selection operation S_A (key : value) pair; Display attributes D_A(key : value) pair
Procedure:
Initialization: $Q_A = \emptyset$;$S_A = \emptyset$; $D_A = \emptyset$; $CV_m = \emptyset$
Method:
1. Collection Phase of Q_A
For each query $q \in Q$
For structured data: Collect the Selection operation <key : value> pair ($S_{k,v}$) as <selection attributes (k) : operator and attribute value (v)> ;
$S_A = S_A \cup S_{k,v}$
Identify the projection pair ($D_{k,v}$) as <key of data : attribute list> of the query
$D_A = D_A \cup D_{k,v}$
Identify aggregate function and add <aggregation attributes : function> pair in selection attributes:
$S_A = S_A \cup < aggregation\ attributes : function >$
For semi-structured data
Identify semi-structured data attributes pair ($SS_{k,v}$)<attribute name : selected values>
$S_A = S_A \cup SS_{k,v}$
Identify the tags to be kept in the result pair ($SD_{k,v}$) of <document : tag list>
$D_A = D_A \cup SD_{k,v}$
For Unstructured data
Identify the operation and content object identifiers of results as the pair of unstructured data ($UD_{k,v}$) <content object identifiers : operation, other data>
$D_A = D_A \cup UD_{k,v}$
In case, content object identifier(k) not in S_A
then add <content object identifier : data selection criteria>
$S_A = S_A \cup UD_{k,v}$
2. Creation of Q_A, Q_{data} and $Q_{operator}$ for all pairs in S_A
For each value pair <k, v> in S_A
collect all <k, v> with similar k values in a set
If the size of view > Big data block size, put k in Q_A and value list v in $Q_{operator}$
else take union of values list to create v', where v' is union of the two value lists.
For each pair <k, v> in D_A related to Q_A so created
Create the list of view data relating to Q_A (Q_{data})
3. Creation of Big data View Structure Graph
Create a Power set of $Q_A(\mathcal{P}(Q_A))$
n=Cardinality of set of Q_A
At each level k of view node create a link from n down to 2
if i^{th} node at level k-1 $\subset j^{th}$ node at level k, create a link from *node i to node j*
For every Q_A use Q_{data} to identify related *query attribute dependencies*
if $(Q_{data})_j$ of $(Q_A)_j \subset (Q_{data})_I$ of $(Q_A)_i$
then create a query attribute dependence as $(Q_A)_i \rightarrow (Q_A)_j$

For each *query relationships* of the type ($A \rightarrow B$)
Create a link at each level from ($A \rightarrow B$) if all other Q_A are same for the nodes
Delete the node (A,B) and merge the links of the node containing a hierarchy with the identifying attributes
The output of this step is the Big data View Structure Graph
4. Create the Candidate Views and Query Evaluation Plans
Using the *Big data View Structure Graphs,* $Q_{operator}$ and Q_{data}
For every node identify candidate views, their data type and operator
Create related sub-graphs of View structures
Each node in each sub-graph is listed as candidate views (CV_m)
a candidate view is a triple <node number, data types, query operator>
Combine candidate views to create additional candidate views.
For each Query
To identify the candidate views in each view sub graphs to answer the query
Use the *View Structure Graph,* $Q_{operator}$ and Q_{data}
perform depth first traversal listing all the nodes having same $Q_{operator}$ and Q_{data}
Generate all possible combinations to form alternative query evaluation plans
Output all the Candidate views and Evaluation plans for each query

In the first step, all the queries are explored to create a set of attributes on which an operation has been performed. S_A records the attributes on which the operation has been performed (k), the operator, and the value lists of the operation (v). For each of these $S_{k,v}$ pair, the values of the displayed attributes, e.g., attribute of projection operation in *RDBMS* (<key of data: attribute list >) or tags in semi-structured data (<document: tag list >), are added to $D_{k,v}$. The aggregate functions are also recorded in $S_{k,v}$. These operations are also performed for semi-structured data. For unstructured data, if the selection attribute is not in $S_{k,v}$, then $UD_{k,v}$ is stored in the $S_{k,v}$. In addition, the related values of selected data are included in $D_{k,v}$. After completing the collection phase, at second step, Q_A list is created based on the value of k in the $S_{k,v}$. In addition to collecting the query attributes, the operators that are applied on the query attributes are identified. If the operator produces a view with size less than one big data block, then such view is merged with other views created on the same data using different operators. The output of this step is the set of query attributes Q_A, the values (Q_{data}) of the query attributes (Q_A), and operators (Q_{operator}) that were performed on these query attributes (Q_A).

In the next step, a big data view structure graph is created by creating the power set of the set of query attributes (Q_A). A view node is created for each member of the power set, and the directed edges are created for the adjacent levels of node (Refer to Eq. 3). In order to identify the dependencies among the candidate views at the same level, query dependence relationships are identified using Q_A and Q_{data}. These dependence relationships between query attributes are used to create the links between the candidate views that are at the same level in the view structure graph. Further, the redundant view nodes are removed, and links are accordingly adjusted.

In the final step of $BD_{CV\&QP}$, candidate views for materialization and query evaluation plans for each query are computed. The big data view structure graph along with Q_{data} and Q_{operator} is used to create subgraphs of candidate views. Next, depth first traversal in each subgraph is performed to compute a set of candidate views for materialization and alternate query evaluation plans for the given query workload.

An example illustrating the use of algorithm $BD_{CV\&QP}$ is discussed next.

5 An Example

Consider that a group of banks maintains big data for a banking loan application system of its customers. The loans are issued by various bank branches. The big data of the banks includes structured data, such as the customer accounts and loans; semi-structured data, such as various assets of the customers, the value of assets as per market indexes, mortgage of assets, industries owned by customers, and yearly turnover of industry, etc.; and the unstructured data, such as web based financial statements of companies, complaints, feedback, or any other information about the customers or their assets. Suppose the following queries are raised on this big data application:

Query 1: What is the current status of a customer's assets and their liabilities, as per records of a specific branch?

Query 2: Are there any complaints against the customers, who have taken loan?

Query 3: List of customers who are bank defaulters.

The scope and queries of this application can be vast. However, in this paper, this big data application is used to illustrate the working of the algorithm $BD_{CV\&QP}$. The three queries clearly involve the three main query attributes, viz. the branch, its customers, and the bank. These three Q_As are corresponding to the query attributes A, B, and C in Fig. 1 with A being the branch, B being the customer, and C being the bank. It can also be observed that the query attribute dependence bank → branch is the same as dependence C → A, as shown in Fig. 1. The identification of these attributes by the algorithm $BD_{CV\&QP}$ is given below:

The input to the procedure would be the abovementioned three queries. Using step 1 of the algorithm $BD_{CV\&QP}$, Q_As will be collected as follows.

For Query 1: For the structured data:

<Customer : Bank Balance, Loan balance in a Branch>

For the semi-structured data

<Customer : Assets, liability details in a branch>

For Query 2: For unstructured data

<Customer: Search and selection of complaint against customers, assets>

For Query 3: For structured data

<Branch: Selection on defaulted customer>

<Bank: Common defaulting customer on various branches of banks>

Thus, the set of attributes S_A will include the key value pairs of branch, customer, and bank, as shown above, whereas attributes data D_A will include the customer data that would be displayed as the output of the queries. For bank branches, the data may include the branch ID, customer ID, asset details, etc., and for the bank, the data may include bank-related details in addition to the branch data. It may also be noted that the selection operator would be used for the above-mentioned three queries. Using step 2, query attributes will be created from S_A and D_A. The < key: value > pairs generated in step 1 have a key value (k) customer, which is related to structured data, viz. bank balance, loan balance; semi-structured data, viz. assets and liabilities; and unstructured data, viz. complaint against customers. These < key: value > pairs are used to identify the query attribute customer, which is related to structured data (bank and loan balance), semi-structured data (assets and liability), and unstructured data (complaints). Similarly, the other two query attributes branch and bank will be identified using this step. Using step 3, first the power set of these three query attributes will be created. This power set will result in the generation of the root node with Q_A (A, B, C) and other possible nodes Q_A (A, B), Q_A (A, C), Q_A (B, C), Q_A (A), Q_A (B), and Q_A (C). The null set of the power set will not be selected. The root node is at level 3, which is the cardinality of set Q_A {A, B, C}. Next, the links from the level 3 to level 2 nodes Q_A (A, B), Q_A (A, C), and Q_A (B, C), will be created. In addition, links will be created from the level 2 node Q_A (A, B) to level 1 nodes Q_A (A) and Q_A (B); level 2 node Q_A (A, C) to level 1 nodes Q_A (A) and Q_A (C); and level 2 node Q_A (B, C) to level 1 nodes Q_A (B) and Q_A (C). Since the data of the query attribute—bank (C) is the superset of the query attribute—branch (A), the query dependence C $\rightarrow$ A will be identified. This query dependence will cause deletion of the node Q_A (A, C), and its links will be merged with the node Q_A (B, C), which in turn would result in the creation of a link from Q_A (B, C) to nodes Q_A (A, B) and Q_A (A). In addition, a link from node Q_A (C) to node Q_A (A) will be created, thus creating the view structure graph, as shown in Fig. 1. Using step 4, the query data output and query operator shall be used to generate the set of candidate views. For customers, the possible views are on the structured and the semi-structured data; or are on structured or unstructured data; or are on all the three structured, semi-structured, and unstructured data, as illustrated in Fig. 2. Since a selection operator is used, no further decomposition of views will be needed, as illustrated in Fig. 3. Likewise, the possible views can be identified for the query attributes branch and bank. For the root node, three views, one each for structured (*V1*), semi-structured (*V2*), and unstructured data (*V3*), will be created. Finally, from the view structure graph and queries in the query workload, alternate query evaluation plans will be computed. For query 1, structured and semi-structured data of the query attributes customer (B) and branch (A) is needed to answer the query. The algorithm $BD_{CV\&QP}$ will create the alternate query evaluation plans using the depth first search, as views *V1* and *V2* from the root node; or view *V4*, created using the structured and semi-structured data of node (*B, A*); or views *V5* and *V6* created using the structured and semi-structured data of node *B* and node *A*, respectively. In this way, algorithm $BD_{CV\&QP}$ is used to compute the candidate views for materialization along with alternate query evaluation plans for the query workload.

6 Conclusion

A big data application is required to process big data queries efficiently. Materializing big data views would improve the processing time for big data queries. However, big data applications have a large number of possible views, and selecting, from among these, big data views that would reduce the processing time of big data queries is a complex problem. This paper has addressed this problem and proposed a view structure graph framework that reduces the large search space of all possible views by computing comparatively smaller set of candidate views for a given query workload. The proposed framework uses the big data view structure graph, query attributes, data accessed by the queries, and query operators to compute a set of candidate views and alternate query evaluation plans for big data queries. These candidate views provide a smaller search space from which reasonably good quality big data views can be selected for materialization.

References

1. Abiteboul S, Goldman R, McHugh J, Vassalos V, Zhuge Y (1997) Views for Semi-structured Data. Tech. Report. Stanford InfoLab, Workshop on Management of Semi-structured Data, Tucson, AZ
2. Abiteboul S (1999) On views and xmL, SIGMOD Record 28(4)
3. Agrawal S, Chaudhari S, Narasayya V (2000) Automated selection of materialized views and indexes in SQL databases. In: 26th international conference on very large data bases (VLDB 2000). Cairo, Egypt, pp 486–505
4. Bibri SE (2018) The IoT for smart sustainable cities of the future: an analytical framework for sensor-based big data applications for environmental sustainability. Sustain Urban Areas 38:230–253
5. Chenghu Z, Fenzhen S, Tao P, Zhang Y, Du A, Luo B, Cao Z, Wang J, Yuan W, Zhu Y, Song JC, Xu J, Li F, Ma T, Jiang L, Yan F, Yi J, Hu Y, Liao Y, Xiao H (2020) COVID-19: challenges to GIS with big data. Geograp Sustain 1(1):77–87
6. Chirkova R, Halevy AY, Suciu D (2001) A formal perspective on the view selection problem. In: Proceedings of the 27th VLDB conference. Roma, Italy
7. Dean J, Ghemawat S (2012) MapReduce: a Flexible data processing tool. Commun ACM 53(1)
8. Dezyre (2019) Hadoop ecosystem components and its architecture, from the web site dated 04 Jun 2015. Last accessed on August 08, 2019 from the website: https://www.dezyre.com/article/hadoop-ecosystem-components-and-its-architecture/114
9. El-Sayed M, Rundensteiner EA, Mani M (2006) Incremental maintenance of materialized xquery views. In: Proceedings of 22nd international conference on data engineering (ICDE'06). Atlanta, GA, USA
10. Gandomi A, Haider M (2015) Beyond the hype: Big data concepts, methods, and analytics. Int J Inf Manag 35(2):137–144
11. Goswami R, Bhattacharyya DK, Dutta M (2017) Materialized view selection using evolutionary algorithm for speeding up big data query processing. J Intell Inf Syst 49(3):407–433
12. Gupta A, Mumick IS (1995) Maintenance of materialized views: problems, techniques, and applications. Data Eng Bull 18(2)
13. Gupta H (1996) Selection of views to materialize in a data warehouse. In: Afrati F, Kolaitis P (eds) ICDT 1997 Lecture notes in computer science, vol 1186. Springer, Heidelberg

14. Gupta R, Gupta H, Mohania M (2012) Cloud computing and big data analytics: what is new from database perspective? In: Proceedings of big data analytics-first international conference. Springer
15. Hadoop (2012) http://hadoop.apache.org/. Last Accessed December 2012
16. Hadoop Documentation (2008) on http://hadoop.apache.org/docs/r0.17.0/mapred_tutorial.html. Last Published on May 2008, Last Accessed Oct 2012
17. Harinarayan V, Rajaraman A, Ullman JD (1996) Implementing data cubes efficiently. In: Jennifer W (ed) Proceedings of the 1996 ACM SIGMOD international conference on Management of data (SIGMOD '96). ACM, New York, NY, USA, 205–216
18. Jacobs A (2009) The pathologies of big data, communication of ACM, vol 52(8). ACM
19. Kenneth R, Srivastava D, Sudarshan S (1996) Materialized view maintenance and integrity constraint checking: trading space for time. In SIGMOD international conference on management of data
20. Kumar S, Vijay Kumar TV (2018) A novel quantum inspired evolutionary view selection algorithm. J Sadhana Springer Indian Acad Sci 43(10) 166
21. Kumar A, Vijay Kumar TV (2015) Big data and analytics: issues, challenges, and opportunities. Int J Data Sci 1(2):118–138
22. Kumar A, Vijay Kumar TV (2021) View materialization over big data. Int J Data Anal 2(1):61–85
23. Kumar A, Vijay Kumar TV (2021) A Multi objective approach to big data view materialization. Int J Knowl Syst Sci 12(2):17–37
24. Kumar A, Vijay Kumar TV (2021) Multi objective big data view materialization using MOGA. Int J Appl Metaheuristic Comput 13(3) (Accepted)
25. Kumar A, Vijay Kumar TV (2021) Multi-objective big data view materialization using improved strength pareto evolutionary algorithm. J Inf Technol Res 15(6) (Accepted)
26. Kumar A, Vijay Kumar TV (2021) Multi-objective big-data view materialization using NSGA-II. Inf Resour Manag J 34(2):1–28
27. Kumar A, Vijay Kumar TV (2020) Dynamic web-view materialization. Adv Commun Comput Technol Lecture Notes Electr Eng 668:605–616
28. Kuno HA, Rundensteiner EA (1995) Materialized object-oriented views in multiview. In: ACM research issues data engineering workshop, pp 78–85 (1995)
29. Labrinidis A, Roussopoulos N (2000) WebView materialization. ACM SIGMOD Record 29(2):367–37. In: Proceedings of the ACM SIGMOD conference. Dallas, Texas, USA
30. Mami I, Bellahsene Z (2012) A survey of view selection methods, SIGMOD record 41(1)
31. Mistry H, Roy P, Sudarshan S, Ramamritham K (2001) Materialized view selection and maintenance using multi-query optimization. In: Proceedings of the ACM (SIGMOD) conference on the management of data, 307–318
32. Roussopoulos N (1998) Materialized views and data warehouses, SIGMOD record
33. Tang N, Xu YJ, Tang H (2009) Materialized view selection in xml databases. Database systems for advanced applications, vol 5463
34. Yafooz W, Zainal A, Siti Z, Omar N, Idrus Z (2013) Managing unstructured data in relational databases. In: Proceedings—2013 IEEE conference on systems, process and control. ICSPC 2013, 198–203
35. Zikopoulos P, Eaton C (2011) Understanding big data: analytics for enterprise class hadoop and streaming data 1st edn. McGraw-Hill Osborne Media

A Machine Learning Approach to Sentiment Analysis on Web Based Feedback

Arnav Bhardwaj and Prakash Srivastava

Abstract The advent of this new era of technology has brought forward new and convenient ways to express views and opinions. This is a major factor for the vast influx of data that we experience every day. People have found out new ways to communicate their feelings and emotions to others through written texts sent over the Internet. This is exactly where the field of sentiment analysis comes into existence. This paper focuses on analyzing the reviews of various applications on the Internet and to understand whether they are positive or negative. For achieving this objective, we initially pre-process the data by performing data cleaning and removal of stop words. TF-IDF method is used to convert the cleaned data into a vectorised form. Finally, the machine learning algorithms: Naïve Bayes, Support Vector Machine and Logistic Regression are applied and their comparative analysis is performed on the basis of accuracy, precision and recall parameters. Our proposed approach has achieved an accuracy of 92.1% and has outperformed many other existing approaches.

Keywords Sentiment analysis · Machine learning · Review text · Tf-idf

1 Introduction

The most imperative aspect of any organization/institution is the feedback given by the people/users. With the onset of the technological era, this process of feedback submission has also been revolutionized, and people today provide their feedback through online services such as social networking sites, interactive applications or by electronic mails. Since the written texts take away our ability to understand the feelings and emotions of the users, it becomes extremely important to understand them through their texts. But since we cannot read and review all the feedbacks from the users, we make the machines do this task. Sentiment analysis is the way of

A. Bhardwaj (✉)
Department of Computer Science and Engineering, Amity University, Noida, India

P. Srivastava
Department of Computer Science & Engineering, KIET, Ghaziabad, India

A. Choudhary et al. (eds.), *Applications of Artificial Intelligence and Machine Learning*, Lecture Notes in Electrical Engineering 778,
https://doi.org/10.1007/978-981-16-3067-5_11

understanding the texts which contain emotions, sentiments and other human aspects such as sarcasm, irony etc. These datasets are used for machine learning applications [1].

The organizations/institutes collect the feedback from the users, but it is very difficult to segregate them into positive and negative feedback. For this purpose, they ask the user to provide the rating with their feedback, so as to make the process of training and prediction more accurate. Very strong words must be included in the texts so that the summary of the feedback gets correctly categorized. Even with a lot of data available to train the machines to predict the sentiment, the complexities and construct of the language makes it difficult to predict the sentiment [2]. Sentiments of the texts in social sites, interactive applications or electronic mails are classified into positive or negative on the basis of certain features that further help in classifying the unseen texts [3]. For example, "I loved this application. It was a good choice" has words "loved" and "good choice". These words will be mapped to the positive aspects of the feedback. Similarly, "this was the worst application ever. Made a bad choice" has words "worst" and "bad'. These would be mapped to the negative aspects.

Two of the approaches to sentiment analysis are text mining which builds upon the hard facts such as occurrence of the words, and opinion mining prioritizes the emotions and attitudes in the text [4]. The words that are considered to be positive in one context, can be treated as negative in another context, and can also be treated as neutral/no opinion in some other context [5]. This challenge is faced in sentiment analysis, where sentiments are prioritized over the text, unlike the traditional text mining where preference is given to the words themselves. In sentiment analysis, a phrase is considered different from another slightly different phrase, irrespective of the difference. This mechanism works on the sentences individually to process the sentiments. Also, this mechanism works well with concrete sentences with a definite structure, but fails to recognize sentences that a normal human is capable of but does not have any concrete information [6]. For example, let us consider the sentence, "This application is same as the previous one". Here the user does not provide any concrete information, and the polarity of the review depends upon the reviews of the previous application.

Machine learning can be applied by either supervised learning or unsupervised learning [7]. In our approach, we have performed supervised learning, using Naïve Bayes, Logistic Regression and SVM classifiers. The data is cleaned, transformed into a vector, and then the vectorised data is fed to the classifiers for their training. 90% of the dataset is used for training purpose and 10% for testing, which then generates the results for performance measures. Our achieved accuracy is 92.1%, which is comparable to other similar approaches.

2 Literature Review

Sentiment analysis deals with the analysis of the emotions of the users regarding products, services, organizations, institutions, people or topics. This concept of sentiment analysis was developed in the beginning of 2000s, and since then it is evolving, taking more and more aspects into consideration and becoming more diverse in its approach to unseen reviews.

Sharma et al. [8] have determined the polarity of the reviews of movies at a document level. According to their work, Document Based Sentiment Orientation System has better results when compared to a similar approach of AIRC Sentiment Analyzer for the movie domain, achieving 63% accuracy with their model. He successfully analyzed various feature selection schemes and their impact on sentimental analysis. The classification clearly shows that linear SVM gives more exact results compared to Naive Bayes classifiers. Although many other previous works have also shown that SVM is an improved method for sentiment analysis, it differs from previous work in terms of comparative studies of classification approaches with different patterns of characteristic selection.

Tripathi and Naganna [9] have performed sentiment analysis by applying different algorithms along with various feature selection methods. They were able to analyze different feature selection methods and how they impacted in the field of sentiment analysis. Their results have shown that Support Vector Machine was more successful than Naïve Bayes classifier in this field. Other previous works have also shown the same results, but their approaches were different than mentioned in this Kudo and Matsumoto [10] paper, as their feature selection methods were different.

Khan and Baharudin [11] have proposed a method which categorizes the sentences. The blogs and reviews are categorized into objective and subjective, and the subjective sentences are considered for calculating the semantic score, which in turn is further used to classify the sentence as positive, negative or neutral. Their results have shown that their proposed method was more effective than the machine learning approach, with an accuracy of 87% for feedbacks.

Liu et al. [12] have developed and designed a method to perform movie review and feedback summarization. Feature selection methods have been used to generate a condensed summary of the review. They have used the Latent Semantic Analysis approach for feature selection. Furthermore, the product summary can be reduced by Latent Semantic Analysis approach and the method can be extended to any review system efficiently.

Zedah et al. [13] in their paper have devised a new model named Tensor Flow Network (TFN), which learns both inter and intra modality of the data and categorizes it into unimodal, bimodal or trimodal interactions. Their proposed model produced state-of-the-art performance in comparison to other multi modal approaches.

Ramachandran and Gehringer [14] showed how varied pre-processing steps can help in making the sentiment analysis method more efficient in approach. They have categorized the sentences on the basis of Latent Semantic Analysis and Cosine Similarity to study the results of their text classification done during pre-processing. Their method provides sentiment analysis on the student reviews on different aspects.

Agarwal et al. [15] have considered phrases as an invariant factor in the process of sentiment analysis. The have applied Parts of Speech rule to determine the sentiment rich phrases and dependency relation, followed by Point-wise Mutual Information method to calculate the semantic orientation of the texts in the document. Finally, the semantic orientation are classified into polar reviews to generate outputs.

Zhu et al. [16] have studied the opinion polling method of unlabeled text of reviews of a Chinese restaurant. They identify aspects of the text to perform aspect based bifurcation on the text and finally generate the model to perform sentiment analysis of reviews. Interesting feature of the model is that it does not require labelled data to train the model. This Rambocas and Gama [4] paper follows unsupervised learning approach and achieves 75.5% accuracy. Their model can be extended to other languages and fields other than restaurant reviews also.

Neviarouskaya et al. [17] have generated a method, SentiFul, which scores the sentiment lexicons and expands itself through antonyms, synonyms and compounding with other known lexicons in the grammar. They have categorized four types of lexicons and the role they play: propagate, intensify, weaken and reverse. These provide the basis for expansion of the lexicon set to improve the sentiment analysis process.

Liu et al. [12] have illustrated in their paper feature level sentiment analysis method. They have used fuzzy domain sentiment ontology tree extraction algorithm for sentiment analysis, which automatically constructs a fuzzy tree with inputs as product reviews.

Hemalatha et al. [18] have shown that emoticon used as training set for different machine learning algorithms can achieve great accuracies for sentiment analysis by this mechanism. Even though Twitter comments and reviews are different from other fields and domains, this method proves itself to be equally effective either ways.

Kudo and Matsumoto [10] suggest that the sentiment analysis goes beyond the scope of bag-of-words approach and must be performed in more sophisticated and structured manner. The machine learning algorithms should be able to detect the structures observed in the text. For this, they proposed Bosting algorithm that recognizes sub-structures in the text. They further illustrated the similarity in their approach and Support Vector Machine algorithm.

Jebaseeli and Kirubakaran [19] performed a survey that depicts and underlines the imperative role of sentiment/opinion analysis for the organizations/corpora. They state that one of the major challenges that are faced in the field of sentiment analysis is the process of feature extraction, because feature extraction governs the whole machine learning process and its final outcome.

3 Proposed Methodology

In our approach, we have collected Amazon's application review database and analyzed it. The analysis is based upon labelled data using a combination of unigram, bigram and trigram feature extraction technique. We have applied pre-processing to the texts to make it more concise for feature extraction, followed by Machine

Learning algorithms: Naïve Bayes, Support Vector Machine and Logistic Regression. The complete process along with the methods used is described in detail the next sub-section.

3.1 Data Collection

The paper uses Amazon's application review dataset. The dataset has been collected from http://jmcauley.ucsd.edu/. In this dataset, we have a total of 10,000 reviews, where 5000 are positive reviews and 5000 are negative reviews. The dataset's 'reviews' field has been changed to 'positive' and 'negative' respectively, for labelling purpose.

3.2 Pre-Processing Data

This step breaks down the complete data into individual sentences in a form such that only relevant information is present along with their reviews. The dataset contains a lot of reviews that consist of words and symbols which are useless for the Machine Learning purpose. In order to improve the quality of the data, we pre-process it as follows.

3.2.1 Data Cleaning

It is the process of removing or correcting the words that are of little or no use to the dataset. This removes incomplete and inaccurate parts of the dataset, which is then replaced by complete information, or removed completely from the dataset. Our dataset was checked and cleaned of all the username mentions, hyperlinks, HTML tags, numbers, sitelinks and special characters. This is followed by replacing the abbreviations such as 'isn't' by 'is not' etc.

3.2.2 Stopwords Removal

Stopwords are those words that are very frequently used in a language and are omitted from the data before or after the processing of data in Natural Language Processing. For example, words such as 'I', 'me', 'myself', 'we', 'our' etc. are considered stopwords.

3.3 Feature Extraction

After the pre-processing of the dataset, the dataset gets converted into a concise form, from where we can extract certain properties of the text which are then used to train the model. These properties (adjectives) can be chosen in the form of unigram or n-gram. In the later stages, these features are used for determining the polarity of the unseen texts to determine the sentiments of the texts. We have used a combination of unigram, bigram and trigram features, for feature extraction purposes.

Text Frequency Inverse Document Frequency (TF-IDF) method has been used to determine features in the text and then implement vectorization of the features. It determines how important a certain portion of text is in the whole document. This method determines the frequency of a text in the document and inverse of the frequency of occurrence of the text in a number of documents.

3.4 Training and Classification

Sentiment analysis can be performed by either document, sentence or text analysis, considering each of these as a level. In our study, we have used the document level analysis approach to perform sentiment analysis. This method is companied by three Machine Learning algorithms: Naïve Bayes, Support Vector Machine and Logistic Regression to implement sentiment analysis on the application review dataset. Classification of the dataset is done by TF-IDF algorithm where the features are extracted into a vectorised form, which then is fed into the Machine Learning algorithms.

3.4.1 Naïve Bayes

Naïve Bayes classifier method is one of the simplest algorithm that can work efficiently on a small dataset, generating high accuracies. This classifier algorithm uses Bayes Theorem which assumes all the variables independent of each other. This algorithm is naïve because it assumes that the variables are independent and there is no way to know about a particular variable, given another variable. Bayes theorem is given as follows in Eq. (1):

$$P(A|B) = \frac{P(B|A)P(A)}{P(B)} \tag{1}$$

where A and B are two events independent of each other. The same concept and formula is used while applying Naïve Bayes classifier to the variables. Regardless of its naïve nature, it has proven itself highly efficient as a Machine Learning algorithm.

3.4.2 Support Vector Machine

Support Vector Machine algorithm is used for both classification and regression problems, but is highly used for classification problems for its capability to classify objects efficiently. In this algorithm, we plot the data obtained. The plot is done in an n-dimensional space where n is the number of features of the input. The value of each feature is the coordinate value in the n-dimensional plot. The main objective is to find a hyperplane that separates the two classes of classification problem as distinctly as possible, and the margins of each class should be as far as possible. This classifier is based upon margin distance of a hyperplane, rather than probability as in the case of Naïve Bayes.

3.4.3 Logistic Regression

Logistic Regression is a classification algorithm which is very useful in applications of classification problems. The variables in Logistic Regression can take only discrete values for any given feature of the problem. The algorithm calculates the probability of the variable to lie in a particular category and uses the Sigmoid function, given in Eq. (2), for its calculations.

$$g(x) = \frac{1}{1 + \mathrm{e}^{-x}} \tag{2}$$

3.4.4 Term Frequency—Inverse Document Frequency

Inverse Document Frequency: Term Frequency—Inverse Document Frequency computes the weight of the certain text which is used to determine the importance of that text in the whole corpus. The weight of a certain text increases with the increase in the occurrence in a document, but decreases with the decrease in the occurrence per document in the corpus.

Term Frequency: It is the count of number of times a word occurs in the document and is calculated as given in Eq. (3).

$$\mathrm{TF} = \frac{(\text{Number of times term } 't' \text{ appears in document})}{(\text{Total number of terms in document})} \tag{3}$$

Inverse Document Frequency: It determines how important a word is in the complete given corpora and is calculated as given in Eq. (4)

$$\mathrm{IDF} = \log\_e \frac{(\text{Total number of documents})}{(\text{Number of documents with term } t \text{ in it})} \tag{4}$$

4 Implementation

We, in our study, have used Python and Natural Language Tool Kit to train the models and perform sentiment analysis on unseen texts. The dataset consists of 10,000 texts along with their reviews as 'positive' and 'negative', out of which 9000 is used for training the model and 1000 is used for testing the models. The following diagram shows the flow of process of our study (Fig. 1).

The complete description of the approach of our process is given in the form of a pseudocode as follows:

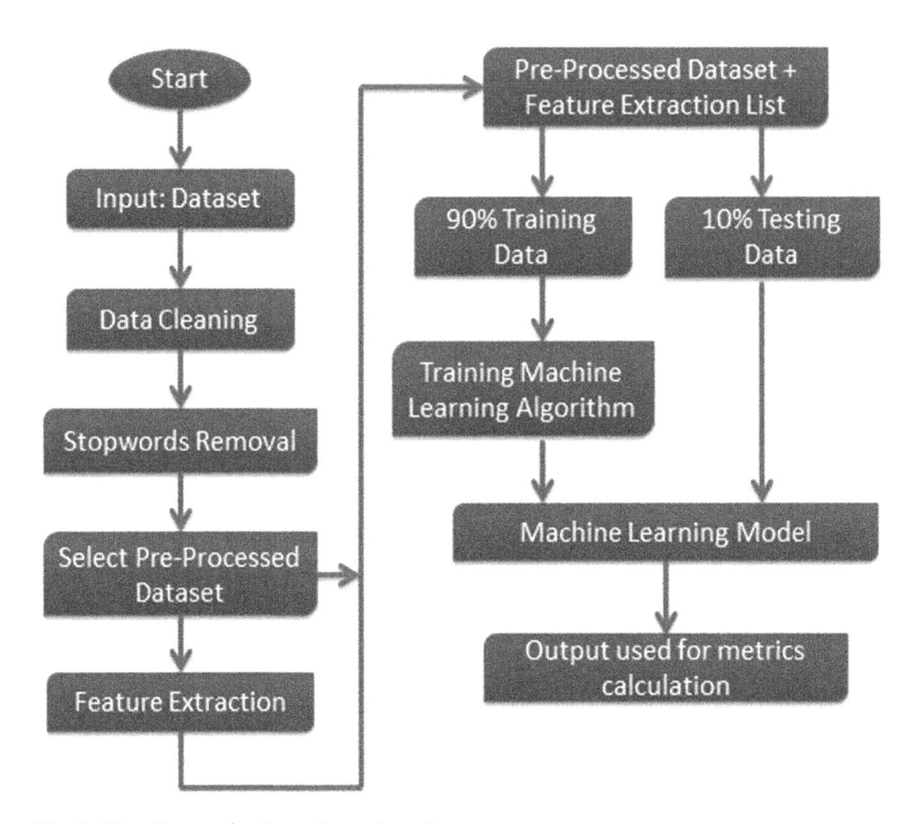

Fig. 1 Flow diagram for the study conducted

```
STEP 1: Import necessary files and required Dataset
STEP 2: Cleaning of Dataset
        Define function Data_Cleaning(Dataset)
                        For (W in Dataset)
                        W = Remove_Mentions (W)
                        W = Remove_Hyperlinks (W)
                        W = Remove_HTML_Tags (W)
                        W = Remove_Special_Characters (W)
                        W = Convert_LowerCase(W)
                        W = Convert_NegativeWords(W)
                Append W to Cleaned_Dataset
                Return Cleaned_Dataset
STEP 3: Removal of Stopwords
        Define function Stopwords_Removal(Cleaned_Dataset)
                        For (W in CleanedDataset)
                        If (W not in Stopwords)
                        Append to Stopwords_Removed
                Return Stopwords_Removed
STEP 4: Convert Reviews to Numbers for Classifiers
        Define function ReviewLabel(Stopwords_Removed)
                        For (W in Dataset)
                        Convert 'positive' of W to 1
                        Convert 'negative' of W to 0
                Return Stopwords_Removed
STEP 5: Vectorise the Dataset
        Define function Vectorise_Dataset (Stopwords_Removed)
                        For(W in Stopwords_Removed)
                        Features = words in Stopwords_Removed
                Return Features
STEP 6: Train the Classifier models with Features
STEP 7: Test the Accuracy, Precision and Recall
STEP 8: Plot the Results
```

The detailed description of our method applied is as follows:

The dataset is imported and preprocessed. The preprocessing includes data cleaning, which further includes removal of hyperlinks, mentions, special keywords, HTML tags and conversion of the text to lower case. Further in preprocessing, we remove stopwords such as 'I', 'me', 'our' etc. from the text. After pre-processing the data, the reviews are converted into numbers; 1 for 'positive' and 0 for 'negative'. This is followed by vectorization of the texts by using TF-IDF method. After the vectorization, the dataset is divided into 90% training set and 10% test set. The training set is fed into our classifiers and the performances are generated on the basis of results of the test set.

5 Result and Discussion

The effectiveness of a Machine Learning model is determined by analyzing its accuracy, precision and recall. These measures are determined by finding True Positive (TP), True Negative (TN), False Positive (FP) and False Negative (FN) through the confusion matrix. In this section, we determine and compare the performance of our Classifiers: Naïve Bayes, Support Vector Machine and Logistic Regression.

Accuracy is calculated as given in Eq. (5):

$$\text{Accuracy} = \frac{\text{TP} + \text{TN}}{\text{TP} + \text{FP} + \text{TN} + \text{FN}} \tag{5}$$

The precision for positive values (positive precision) and negative values (negative precision) is calculated as given in Eqs. (6) and (7):

$$\text{Positive Precision} = \frac{\text{TP}}{\text{TP} + \text{FP}} \tag{6}$$

$$\text{Negative Precision} = \frac{\text{TN}}{\text{TN} + \text{FN}} \tag{7}$$

Similarly, the recall for positive values (positive recall) and negative values (negative recall) is calculated as given in Eqs. (8) and (9):

$$\text{Positive Recall} = \frac{\text{TP}}{\text{TP} + \text{FN}} \tag{8}$$

$$\text{Negative Recall} = \frac{\text{TN}}{\text{TN} + \text{FP}} \tag{9}$$

Tables 1, 2 and 3 show the Precision and Recall of our three classifiers: Naïve Bayes, Support Vector Machine and Logistic Regression (Table 4).

Figure 2 shows the positive and negative precision comparison of the three classifiers. As seen, Naïve Bayes has highest precision. Positive precision is higher than negative precision, implying that greater number of actual positives have been predicted correctly out of the predicted positives as compared to the same for negatives.

Table 1 Performance measures of Naïve Bayes classifier

Performance measures of Naïve Bayes	Percentage (%)
Positive precision	92.87
Negative precision	91.43
Positive recall	90.33
Negative recall	93.70

Table 2 Performance measures of support vector machine classifier

Performance measures of support vector machine	Percentage (%)
Positive precision	93.95
Negative precision	87.36
Positive recall	84.87
Negative recall	95.03

Table 3 Performance measures of logistic regression classifier

Performance measures of logistic regression	Percentage (%)
Positive precision	89.62
Negative precision	91.50
Positive recall	90.75
Negative recall	90.45

Table 4 Accuracy comparison

Accuracy	Percentage (%)
Naïve Bayes	92.10
Support vector machine	90.20
Logistic regression	90.60

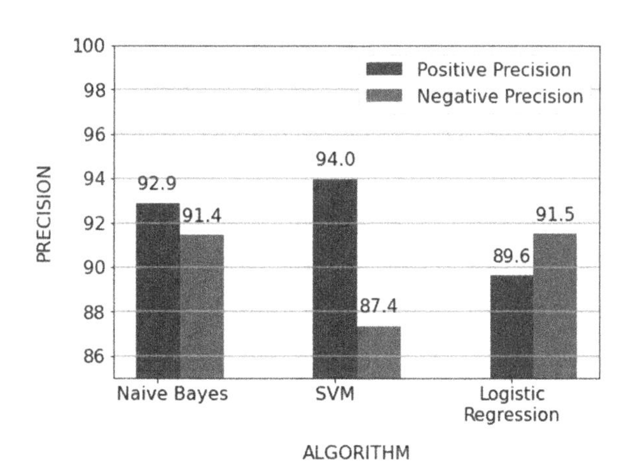

Fig. 2 Measurement of positive and negative precision of the classifiers

Figure 3 shows the comparison of recall of the three classifiers. For the recall, negative recall is greater than positive recall, implying that the ratio of predicted negatives and actual negatives is greater than the same for predicted positives and actual positives.

The result analysis shows 92.1% accuracy for Naïve Bayes, 90.2% for SVM and 90.6% for Logistic Regression. The high accuracy is attributed to the simplicity in the approach and the ease in applicability of the classifiers to the dataset. The results

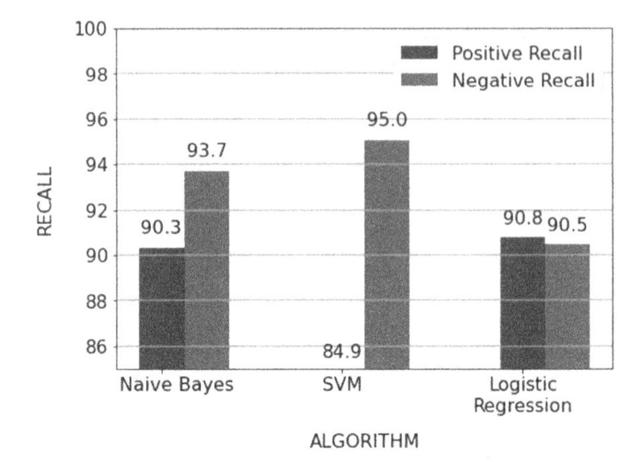

Fig. 3 Measurement of positive and negative recall of the classifiers

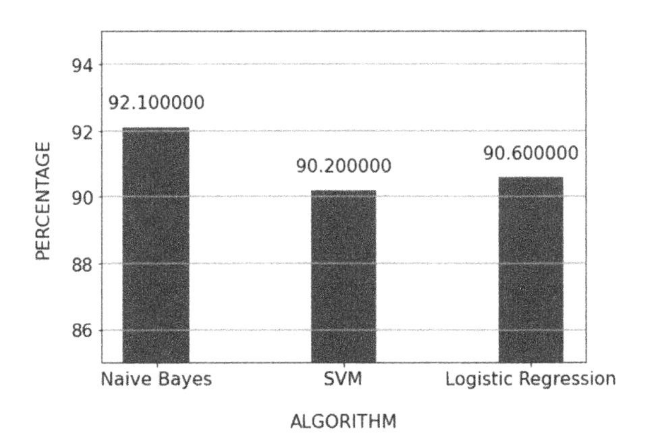

Fig. 4 Accuracy measurement of the classifiers

show that Naïve Bayes has the highest accuracy as well as precision and recall, followed by SVM and Logistic Regression respectively (Fig. 4).

6 Conclusions and Future Scope

In this paper, we have aimed to create a model using TF-IDF vectorizer and Naïve Bayes, Logistic Regression and SVM classifiers that provides a precise analysis of the texts and helps the users to classify reviews as positive and negative. We see that Naïve Bayes has the highest performance of 85% in comparison to SVM and Logistic Regression. The use of collection of unigram, bigram and trigram models improves the results as more phrases can be included for training the model. Thus we can say that Naïve Bayes can be successfully used along with TF-IDF to precisely analyze application reviews.

As a part of future work, we plan to incorporate a neural network model with TF-IDF as classifier to further improve the performance measures.

References

1. Feldman R (2013) Techniques and applications for sentiment analysis. Commun ACM 56(4):82–89
2. Chen P, Sun Z, Bing L, Yang W (2017) Recurrent attention network on memory for aspect sentiment analysis. In: Proceedings of the 2017 conference on empirical methods in natural language processing, pp 452–461
3. Kalaivani P, Shunmuganathan KL (2013) Sentiment classification of movie reviews by supervised machine learning approaches. Indian J Comput Sci Eng 4(4):285–292
4. Rambocas M, Gama J (2013) Marketing research: The role of sentiment analysis (No. 489). Universidade do Porto, Faculdade de Economia do Porto
5. Pang B, Lee L, Vaithyanathan S (2002) Thumbs up? Sentiment classification using machine learning techniques. arXiv preprint cs/0205070
6. Soleymani M, Garcia D, Jou B, Schuller B, Chang SF, Pantic M (2017) A survey of multimodal sentiment analysis. Image Vis Comput 65:3–14
7. Singh Y, Bhatia PK, Sangwan O (2007) A review of studies on machine learning techniques. Int J Comput Sci Secur 1(1):70–84
8. Sharma R, Nigam S, Jain R (2014) Opinion mining of movie reviews at document level. arXiv preprint arXiv:1408.3829.
9. Tripathi G, Naganna S (2015) Feature selection and classification approach for sentiment analysis. Mach Learn Appl Int J 2(2):1–16
10. Kudo T, Matsumoto Y (2004) A boosting algorithm for classification of semi-structured text. In: Proceedings of the 2004 conference on empirical methods in natural language processing, pp 301–308
11. Khan A, Baharudin B (2011) Sentiment classification using sentence-level semantic orientation of opinion terms from blogs. In: 2011 National postgraduate conference. IEEE, pp 1–7
12. Liu CL, Hsaio WH, Lee CH, Lu GC, Jou E (2011) Movie rating and review summarization in mobile environment. IEEE Trans Syst Man Cybern Part C (Appl Rev) 42(3):397–407
13. Zadeh A, Chen M, Poria S, Cambria E, Morency LP (2017) Tensor fusion network for multimodal sentiment analysis. arXiv preprint arXiv:1707.07250.
14. Ramachandran L, Gehringer EF (2011) Automated assessment of review quality using latent semantic analysis. In: 2011 IEEE 11th international conference on advanced learning technologies. IEEE, pp 136–138
15. Agarwal B, Sharma VK, Mittal N (2013) Sentiment classification of review documents using phrase patterns. In: 2013 international conference on advances in computing, communications and informatics (ICACCI). IEEE, pp 1577–1580
16. Zhu J, Wang H, Zhu M, Tsou BK, Ma M (2011) Aspect-based opinion polling from customer reviews. IEEE Trans Affect Comput 2(1):37–49
17. Neviarouskaya A, Prendinger H, Ishizuka M (2011) SentiFul: A lexicon for sentiment analysis. IEEE Trans Affect Comput 2(1):22–36
18. Hemalatha I, Varma GS, Govardhan A (2013) Sentiment analysis tool using machine learning algorithms. Int J Emerging Trends Technol Comput Sci (IJETTCS) 2(2):105–109
19. Jebaseeli AN, Kirubakaran E (2012) A survey on sentiment analysis of (product) reviews. Int J Comput Appl 47(11)
20. Liu L, Nie X, Wang H (2012) Toward a fuzzy domain sentiment ontology tree for sentiment analysis. In: 2012 5th international congress on image and signal processing. IEEE, pp 1620–1624

Forecasting of Stock Price Using LSTM and Prophet Algorithm

Neeraj Kumar, Ritu Chauhan, and Gaurav Dubey

Abstract Advancement in new era of computational techniques has offered a wide opportunity to develop and deploy efficient and faster algorithmic solutions to the extensive research problem in various application domain. Although, they are wide application domain accessible but financial forecasting is among the most desirable area of research due to its broad attainment around the globe. In, context with same stock price prediction is quite essential for any organization in respect to financial gain. Mostly, all financial assets are intense to identify the next move of the share market to attain the maximum profit from it. In past, various machine learning and regression techniques have been applied to detect stock price prediction but they are unable to publish the significant results. In current study of approach, we have implemented the LSTM (Long Short-Term Memory) and Prophet algorithm over the stock market data. The financial time series data has been analyzed for last six years to perform the future forecasting and comparative results shows that LSTM out performs the Prophet algorithm for stock market prediction.

Keywords Long short-term memory · Prophet · Deep learning · Neural network · Recurrent neural network · Financial data

1 Introduction

The prediction modeling of stock forecasting has been a challenging task for the researchers and analysts around the globe. Investors are mostly interested in gaining the maximum profit which is constantly relied more on stock price prediction. In general, mostly all investors are keen to know about future prediction of the stock market. In such, scenario an effective prediction system for stock market can benefit

N. Kumar · R. Chauhan (✉)
Amity University, Uttar Pradesh, Noida, India

N. Kumar
e-mail: nkumar8@amity.edu

G. Dubey
Abes Engineering College, Uttar Pradesh, Ghaziabad, India

A. Choudhary et al. (eds.), *Applications of Artificial Intelligence and Machine Learning*, Lecture Notes in Electrical Engineering 778,
https://doi.org/10.1007/978-981-16-3067-5_12

traders, investors and analysts by providing supportive information like diffuse of value among the certain stocks. We can discuss stock market prediction as a process to determine the future value of company stock or other financial instrument trade on an exchange. However, the successful prediction of stock's future price could yield significant profit. In general, prediction is trivial process in financial systems as there exist lot of complicated financial indicators which also termed as the fluctuation of the stock market which are very violent. However, as a technology is getting advanced, we tend to develop opportunity which can gain insights of stock market with constant informative indicators to make a better prediction.

Usually, stocks tend to be an equity investment which represents the part of ownership in company or corporation. However, Stocks entitle you to be part of a company operation and assets. In general, the organizations are only considerate for prediction of market value, which tends to be a great importance among organizations to generalize or maximize the profits of your stock option purchase and discriminating the low risk patterns. We can say, that the Recurrent Neural Network (RNN) has proven to be the one of the most powerful algorithm for processing sequential data, the Long Short Term Memory (LSTM) is the well-known RNN architectures [1]. The overall working of LSTM consists of the memory cell, a unit of computation that replaces the traditional artificial neurons in the hidden layers of the network. The memory cell networks are capable to effectively associate the memory and input remote in time, hence the suits to grasp the structure of the data dynamically over time with high prediction capacity. LSTM is proven itself one of the best-known algorithms for time series prediction, so checking the quality of results with stock market data is the main motivation of this paper.

On the other end, Prophet Algorithm was discussed firstly by Facebook for prediction in time series analysis. The algorithm tends to predict and discover knowledge for the future price of stock market. It was basically created to automate the time series prediction problem [2, 3]. In the literature study, there are various traditional techniques available for time series forecasting which include statistical techniques and artificial neural networks. Thus, there are various automated issues with current existing techniques such as with artificial neural network, it is very difficult to tune ANN. Prophet is a curve fitting algorithm it is not a machine learning algorithm [2]. Prophet builds the model by finding a best smooth line which can be represented as a sum of these components: Overall growth trend, yearly seasonality, weekly seasonality and holiday effect. There are few benefits of using prophet algorithm over the past traditional approaches. It is not mandatory to have evenly interval time series data, 'NA' or missing data is not a problem for time series forecasting, any type of seasonality can be handled by default, it works well with default settings so minimum parameter tuning is required [2, 4–6]. Prophet forecasting model was designed to work with business time series data of Facebook but as we know stock market data is also a time series data. In current study of approach, we have applied Prophet Algorithm over the stock market data for prediction of future price. Main motivation to use the Prophet algorithm in this paper is just to check the quality of result it returns with very least effort, as it is very simple to implement and very less parameters to tune.

The overall paper is organized in varied sections, related works are discussed in Sect. 2, Sect. 3 discuss the overall stock market prediction, Sect. 4 discuss about the methodology applied, Sect. 5 represents the conclusive results and in last section conclusion is discussed.

2 Related Work

Forecasting in financial research is quiet anticipated area of research from past decades, a lot of research work has been proposed by the researchers for future forecasting. To predict the knowledge, ARIMA model has been applied to forecast Indian stock market. An ARIMA model is discussed as a univariate model for time series forecasting [7]. In context, to same several other techniques are available for forecasting which include Artificial Neural Network i.e. MLP, CNN and LSTM were used to predict the stock market price and found LSTM outperform other two techniques [8]. A complex neuro fuzzy system (CNFS) has been integrated with ARIMA for prediction of stock price movement and depicted positive results [9]. An FGL (Financial loss or gain) model has been proposed for the prediction of future electricity price of GenCo company, an integrated approach using Silhouette criterion and k-means clustering technique were applied to improve the prediction results [10].

A framework has been proposed using "Two-Stream Gated Recurrent Unit" and "Sentiment Analysis" for prediction of short-term trend prediction of stock market [11]. Over- fitting of data is very common problem of Deep Neural Network, a novel technique Dropout is introduced to deal with overfitting problem [12]. A Fully Convolutional Network (FCN) is augmented with LSTM and a model is introduced FCN-LSTM for prediction of time series data and found the results better than some state of art algorithms [13].

A fuzzy method algorithm was applied for prediction of time series data; it is basically inherited the features of Japanese Candlestick theory used for assisting financial prediction [14]. A comparative study of Discreet Wavelet Transform (DWT), ARIMA & RNN for predicting the traffic over computer network [15]. A novel Prophet Algorithm is introduced to work with time series data for future prediction, this algorithm considers various type of seasonality in future prediction [2]. A new method based on convolutional neural network to simplify noisy-filled financial temporal series via sequence reconstruction by leveraging motifs is used for prediction of stock price and found that the results are 4–7% better than the traditional signal processing methods [16]. A new integrated approach which combines the CNN and LSTM and makes a new technique Conv1D-LSTM, it is used to make prediction on the stock price two Indian origin companies TCS and MRF and found the good results [1]. Another research that uses a generalized regression neural network (GRNN), which was applied to automate the time series prediction process [17]. Again, different deep learning techniques, auto encoder and restricted Boltzmann machine are applied over Chinese stock market and compare the results with other machine learning algorithms in the same area [18]. A similar approach related to the Prophet algorithm for prediction of future sales in retail sector was concluded to represent that Prophet Algorithm works well with this kind of data [19]. The series of deep learning techniques, in this paper author compares two deep learning techniques CNN and LSTM over Indian stock market data and found that LSTM outperforms CNN with time series data [20].

3 Stock Market Prediction

The Stock market plays a major role for overall economic standardization of country; it works as an economic indicator. We can say that, if the stock price of companies goes up to higher extend then the linearly, we can correlate and prediction the extensive growth of the country. In India there are two major stock exchanges National Stock Exchange (NSE) and Bombay Stock Exchange (BSE). BSE is an older stock exchange as compare to NSE; BSE started its operation in 1875 whereas NSE starts its operation much later then BSE around mid of 1994. There are around 5000 companies listed in BSE on the other end NSE is new so companies listed are comparatively less to BSE. Around 2000 companies are listed under NSE. There are two separate indexes Sensex and Nifty which used for these two stock exchanges. Sensex is used for BSE and Nifty is used NSE. There are top 30 companies of BSE which represents the Sensex and top 50 companies of NSE represents Nifty. These top companies are generally belonging to the different sectors of business.

Stocks price prediction plays very important role in stock market. An investor can determine the future price of a particular stock in advance, so it can wisely decide while investing in any stock. Generally, brokerage firms are there to do the investment in stock market on behalf for their investors. Now the big question is how we can identify the future price of any stock? As per studies there are various types of factors which may affect the stock price of a company such as: human re- sources, plan and policies, political effect, climate condition, positive or negative news in the market, economic condition of country, condition of global market of that sector and other factors. However, all these factors come under the fundamental factors which may affect the stock price. Hence, while predicting over stock data we have to keep all these factors in our mind, but practically it is not possible to map all these factors inclusively for future prediction.

There is other applicable method which are also known as technical analysis. In technical analysis future price is predicted on the basis of the past price of that stock. There are other techniques available for forecasting of data which may include regression, machine learning, neural network techniques such as: ARIMA, ARCH, GARCH, Random Forest, Neuro-Fuzzy, SVM and many more techniques have been applied to read the pattern from past data and predict the future price [3, 21–23]. In thin current study of approach, we have applied two techniques LSTM and Prophet algorithm over the Indian stock market data for prediction of future price.

4 Methodology

In current study of approach, we have utilized the LSTM and Prophet algorithm over the stock market data of SBI using python script language. The wide associated techniques are discussed in this section to identify the recognizable parameters.

4.1 Artificial Neural Network (ANN)

ANN is a combination of artificial neurons and these artificial neurons integrate or combine themselves together to communicate with each other. We can say that, Artificial neurons are just like the replica of the biological neurons in human mind, it works in the similar context as human brain works. Similarly, like humans it can correlate with the patterns of past events and take the decision, in the comparable way artificial neurons are also trained to do the same. Past data has been provided to train the ANN, once the network has trained completely it can be applied to do a particular task. An ANN with more than one hidden layer is known as deep neural network, number of hidden layers may in- crease as per the nature of the problem. Figure 1 represents the artificial neural network model.

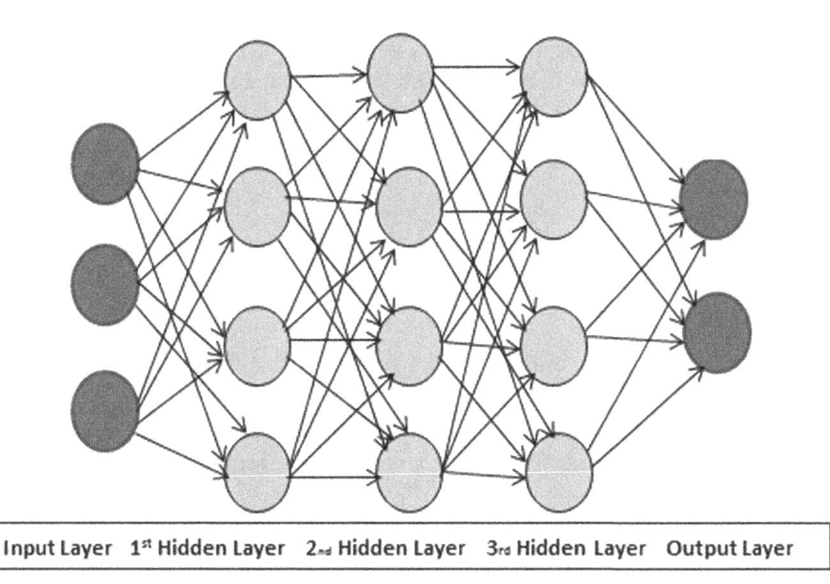

Fig. 1 Artificial neural network

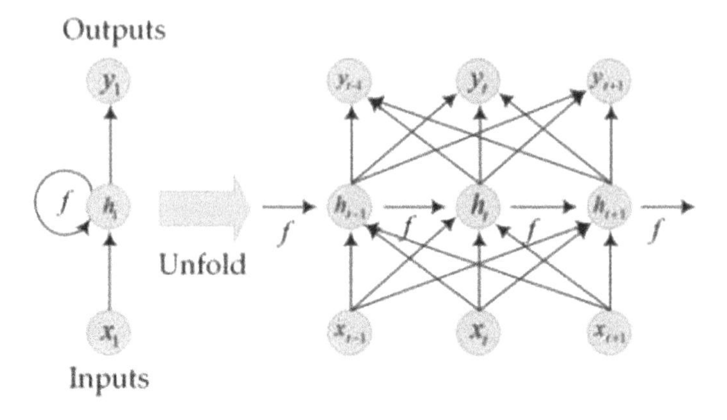

Fig. 2 Structure of recurrent neural network

4.2 Recurrent Neural Network

MLP was the first successful ANN architecture, it works very well with application of big data for learning patterns with high dimensionality but MLP was not the perfect choice for time series data. MLP was not capable to learn from the past sequence of the same data. To overcome this drawback of MLP recurrent neural network has been proposed. RNN's have associative memory with it that can mimic human memory in the network. In Eq. (1) RNN keeps a hidden vector h that updates after the time step t. tanh is hyperbolic function and W is a weight matrix of neurons in RNN whereas I represent a projection matrix. The structure of RNN is discussed in Fig. 2.

$$h_t = \tanh(Wh_{t-1} + Ix_t) \tag{1}$$

$$y_t = \text{softmax}(Wh_{t-1}) \tag{2}$$

$$h_t^l = \sigma\left(Wh_{t-1}^l + Ih_t^{l-1}\right) \tag{3}$$

In Eq. (2). Softmax is used to provide a normal distribution. In Eq. (3) σ is a sigmoid function and W is weight matrix of neurons. Deeper architectures can be created using h.

4.3 LSTM (Long Short Term Memory)

LSTM is a type of RNN (recurrent neural network) and one of the most successful RNN architecture in the field of time series forecasting. As we have already discussed

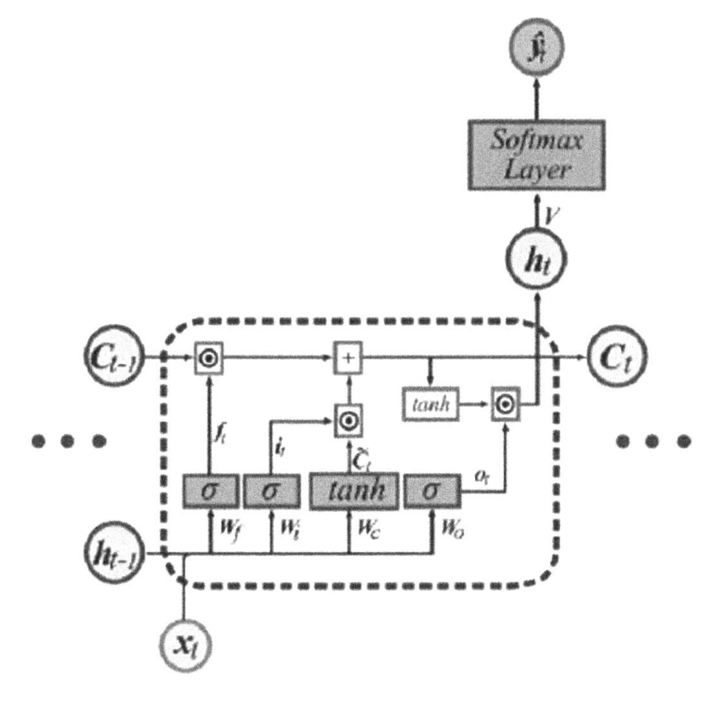

Fig. 3 Structure of LSTM

LSTM uses a memory cell that replaces the artificial neurons in the hidden layers of the network. With these memory cell networks are capable to effectively associate the memory and input remote in time, hence the suits to grasp the structure of the data dynamically over time with high prediction capacity. In Fig. 3 the LSTM architecture is discussed, we have the forget gate, input gate and output gate. LSTM was designed to remove the vanishing and exploding gradient problem. Apart from the hidden state vector each LSTM cell maintain cells state vector and each time stamp the LSTM can choose to reform form it, write to it or reset the cell using gating mechanism. Each unit has three gates of LSTM cell of the same shape each of these are binary gates. Input gate controls whether the memory cell updated or not, forget gate used to reset the memory cell to 0 and the output gate controls whether the information of the current cell state is visible or not. Equations (4), (5) and (6) represents the input, forget & output gate respectively.

$$\text{Input gate}: i^{(t)} = \sigma\left(W^i\left[h^{(t-1)}, x^{(t)}\right] + b^i\right) \tag{4}$$

$$\text{Forget gate}: f^{(t)} = \sigma\left(W^f\left[h^{(t-1)}, x^{(t)}\right] + b^f\right) \tag{5}$$

$$\text{Output gate}: o^{(t)} = \sigma\left(W^o\left[h^{(t-1)}, x^{(t)}\right] + b^o\right) \tag{6}$$

They all have a sigmoid activation, why sigmoid? So they can constitute a smooth curve in the range 0–1 and the model remains differentiable. Apart from these gates we have another vector shows in Eq. (7) (Fig. 3).

$$\mathrm{C}^{(t)} = \sigma\left(W^{c}\left[h^{(t-1)}, x^{(t)}\right] + b^{c}\right) \tag{7}$$

That modifies the cell's state, it has the tanh activation now why tanh here? tan$h$ distributes gradients hence prevents vanishing/exploding. Why LSTM's are constructed in this way and how it mitigates the problem of vanishing or exploding of gra- dients. Each of the gates in LSTM takes the hidden state and the current input x as input, it concatenates the vectors and applied sigmoid C represents a new candidate value can be applied to the cell's state. Now we can apply the gates, like we said earlier input gate controls whether the memory cell updated or not, so its ap- plied to C which is the only vector that can modify the cell state. Forget gate control how much the old state should be forgotten.

$$C^{(t)} = f^{(t)}C^{(t-1)} + i^{(t)}C^{(t)} \tag{8}$$

$$h^{(t)} = \tanh(C^{(t)} \times o^{(t)} \tag{9}$$

Equation (8) applied to the output gate to get the hidden vector. Here, we have three gates per LSTM so we have slow parameters to model. Now we can say LSTM's are basically explicitly designed to avoid long term dependency problem. Remembering information from longer periods of time is practically the default behavior it is not something to struggle to learn. Now we directly move to the stock prediction through LSTM.

4.4 Prophet Algorithm

Prophet is an additive regression model used for time series data which is mainly designed for business forecasting problems [2]. Business forecasting or stock price prediction are interrelated, if business flourish then definitely share price will go up else go down for vice versa, hence both the problems are almost similar. The implementation of prophet algorithm is a very simple and easy to deploy al- gorithm. The prophet algorithm input very a smaller number of parameters that needs to be tuned. Hence, the future forecasting is utilized keeping variable parameters in mind but the most important is the seasonal effects, but most of the methods are not capable enough to map seasonality in future prediction. The ARIMA models are capable to include seasonality factor in the forecasting. According, to this model forecasting can be done on the three points of time series: trend, seasonality and holiday effect [2, 5]. So, this model mainly works within these three points and it can also consider all types of seasonality daily, weekly or yearly. But with stock market data out of these three points holiday effect cannot be considering as stock markets are closed on holidays. This model divides the time series into three major components trend, seasonality and holidays and it can be representing with Eq. (10).

$$y(t) = g(t) + s(t) + h(t) + \in \tag{10}$$

In Eq. (10) $g(t)$ is the trend function, $s(t)$ is seasonality function and $h(t)$ is holiday function, €t is the error term.

Trend Model: There are two variants of trend model 1. Saturating Growth 2.

Piecewise linear model. Saturating growth means how the populations is growing over time so there is no fixed rate of population growth at a particular time that is why it is a nonlinear model.

$$g(t) = \frac{C}{1 + \exp(-k(t - m))} \tag{11}$$

In Eq. (11) C is the carrying capacity, k growth rate, m offset parameter. For forecasting problems that do not show any saturating growth, a linear growth supplies a close and useful model.

$$g(t) = \left(k + a(t)^T \delta\right)t + \left(m + a(t)^T \gamma\right) \tag{12}$$

In Eq. (12) k is the growth rate, δ is rate adjustment and m is offset.

Seasonality: On every time series data there must be some seasonal effects for example: a fireworks company see an upward sale near the festival Christmas & New Year. That upward trend is not for a long period it is because of yearly seasonality. So a time series model must be capable to consider the seasonality effect while doing the forecasting. In Prophet Algorithm Fourier series is used to map the periodic effects. Let P be the time period (i.e. $P = 365$ for yearly data, $P = 7$ for weekly data). Equation (13) shows the Fourier series for calculation of seasonality effect.

$$s(t) = a_0 + \sum_{n=1}^{N} \left(a_n \cos \frac{2\pi nt}{P} + b_n \sin \frac{2\pi nt}{P}\right) \tag{13}$$

Holiday's and Events: holidays and events also affect some businesses time series in positive or negative way. Generally, most of the time series model avoids the effect the holidays on business although it should be part of any time series predictive model. Prophet model provide this function to include all the past and futures events in the model.

5 Results

We have implemented the LSTM and Prophet algorithm over the stock market data of SBI. The last six years' time series data has been collected from 01-01-2013 to 31-12-2018 for training purpose, data has been composed from yahoo finance for training purpose. We have forecasted the price of next sixty days on the basis of training data; we have compared the error percentage after applying both the methods. A representation of SBI stock price for six years in Fig. 4,it shows that there is no clarity of upward or downward trend in the movement of stock price. This is the reason we have taken this data for training purpose, so we can check the pattern reading capability of the models used.

5.1 Implementation of LSTM

In this paper, we have applied univariate LSTM model, further we have chosen Open index of share market data of SBI to train the model. We have also predicted the future price for the Open index only. First step is to preprocess the data, we have chosen Open price column form the columns provided in the dataset['Open','High','Low','Close','Adj Close','Volume'], after that we have

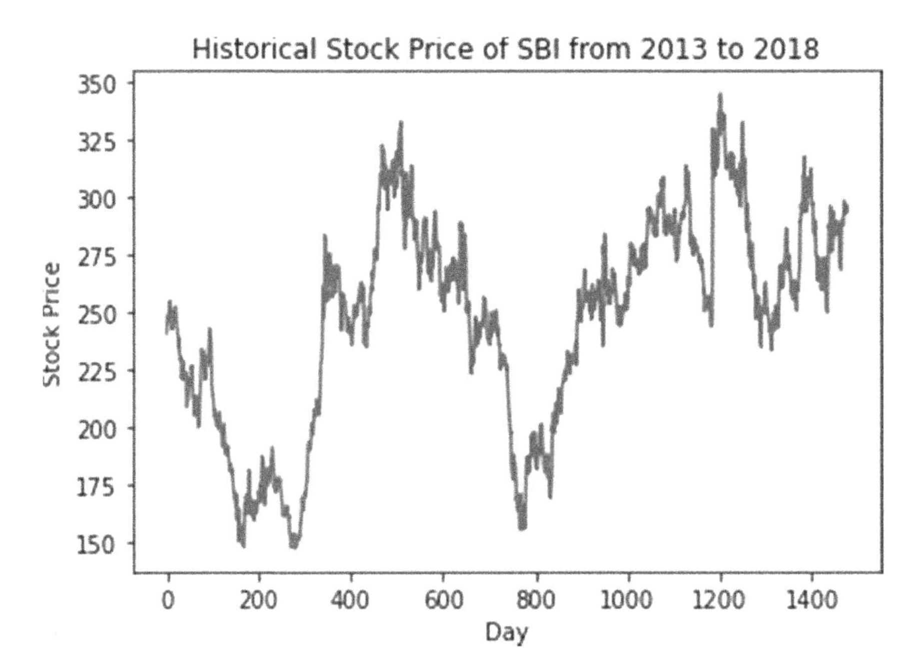

Fig. 4 Graphical representation of SBI stock price

applied MinMax scaler to scale the datafrom range 0 to 1, scaling of data is required for optimal performance.LSTM needs the data to be in specific format generally a 3D array, so we have divide our data into time steps. We are taking 60 time steps as input and 1 feature at each time step. So there will be two separate arrays *X*_Train() & *Y*_Train for time steps and features respectively. Now we have to build the LSTM model, here we are using sequential mode of LSTM with one in-put layer, four hidden layer and one output layer. In each layer we are using 50 neurons to train the model, to prevent the model from overfitting we are using a dropout ratio that 0.2 with each layer, it means the 20% of the neurons will be dropped in each layer during forecasting. In the compilation of LSTM model we have used one of the famous optimizer that is "adam optimizer" and for calculation of loss we used "mean squared error". In the last step to fit the model we used batch size 32 and number of epochs 100.

The Fig. 5 clearly indicates the overall performance of LSTM prediction capability, there is very small difference between the predicted share price and the actual share price. We have calculated performance matrices in Table 1 to show the error rate in the prediction.

Table 1 represents MAE (mean absolute error) is approximately 4 which is very good, MSE (mean squared error) is 25.15, RMSE(root mean squared error) is around 5 and Bias is −0.368.

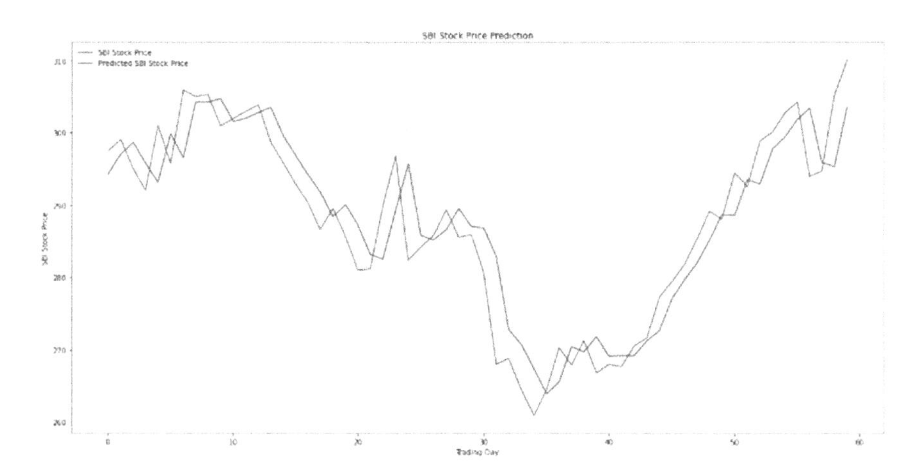

Fig. 5 Actual and predicted stock price of SBI using LSTM

Table 1 Mean absolute error

MAE:	3.965692
MSE:	25.15772
RMSE:	5.015747
Bias:	−0.36869

5.2 Implementation of Prophet Algorithm

Data Preparation: Prophet works good on daily periodicity data, so in this paper we are also considering daily periodicity data of stock market. Prophet is robust algorithm for missing data, so there do not need to handle the missing values, but large amount of missing data may degrade forecasting results. So the data we are using is free from any missing value. Our aim is to predict the future "Open" price of SBI stock using previous data of *N* years. In this paper we are using 5 years previous data of SBI and predicting future prices of next 60 days. To check the effectiveness of this algorithm we have forecasted the different dates in dataset instead of forecasting it on single date. To perform the forecasting, we have trained our model first. As it is a univariate model so in data set, we can have maximum two columns, in our case we have 'Date' and 'Open' price. Prophet always expects two columns in input data frame as 'ds' and '*y*'. Index column must be renamed as 'ds' and other column for which we are doing forecasting should be renamed as '*y*', in our case Date is the index column, so it will be renamed it as 'ds' and Open column as '*y*'. Now our data is ready to use with Prophet algorithm. After we have fit the prophet model over the training data with various parameters of trends and seasonality. After that we have to fit the model on the training data, then the mod- el is ready to do forecasting. Firstly, we have to create a data frame for future dates and then we do forecast for these future dates only on the "Open" price of SBI stock.

In Fig. 6, it clearly represents the prediction for next 60 days, black dots are the actual share prices of last six years and we can see a line between these black dots it shows the prediction. Prophet predicts for entire training period and continues it for

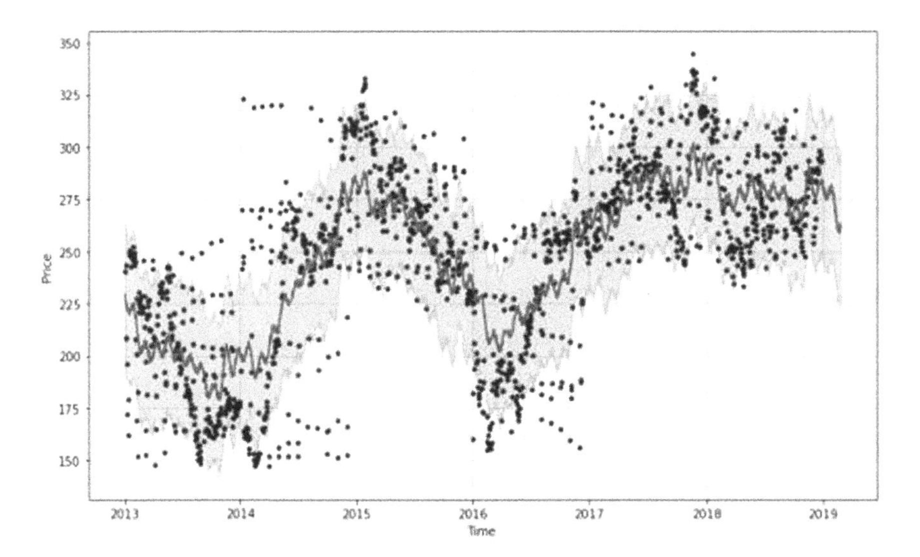

Fig. 6 Prediction of SBI stock price using Prophet algorithm

future forecasting as well. We can see in graph after 2019 there is a prediction line only there are no black dots it shows the prediction for future.

Figure 7 represent various components of trend and seasonality (weekly and yearly). As the data belongs to bank so rarely, we see any seasonal effect over it. Figure 8 shows the comparison between the predicted stock price and the actual stock price. Red line shows the actual share price and green line shows the prediction over it. We can see that predicted price line almost follows the same path as actual price line but somewhere around the mean value of actual price at different time steps. There are two more lines shows the minimum and maximum range of predicted share price. Black line represents the minimum threshold and blue line rep- resent maximum threshold. We can clearly see that the predicted price is within the range of minimum and maximum threshold. So, we can say that prediction quality is very good. We have calculated the performance matrices for prophet algorithm also same

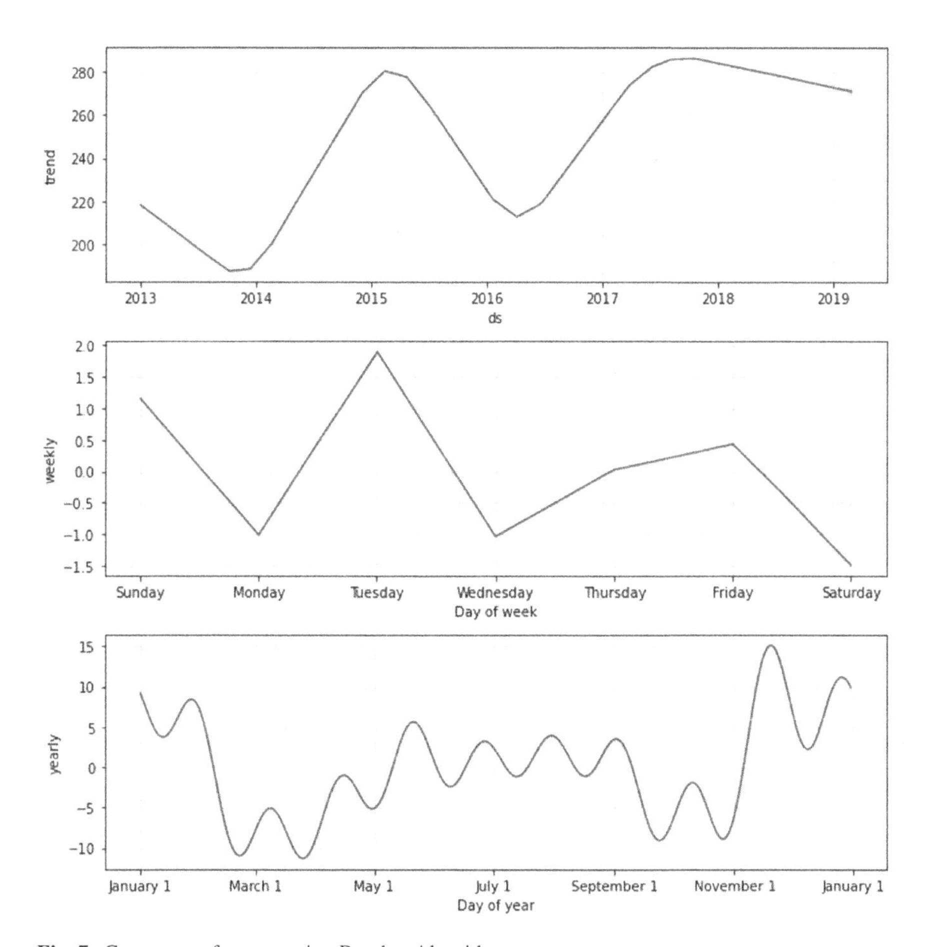

Fig. 7 Component forecast using Prophet Algorithm

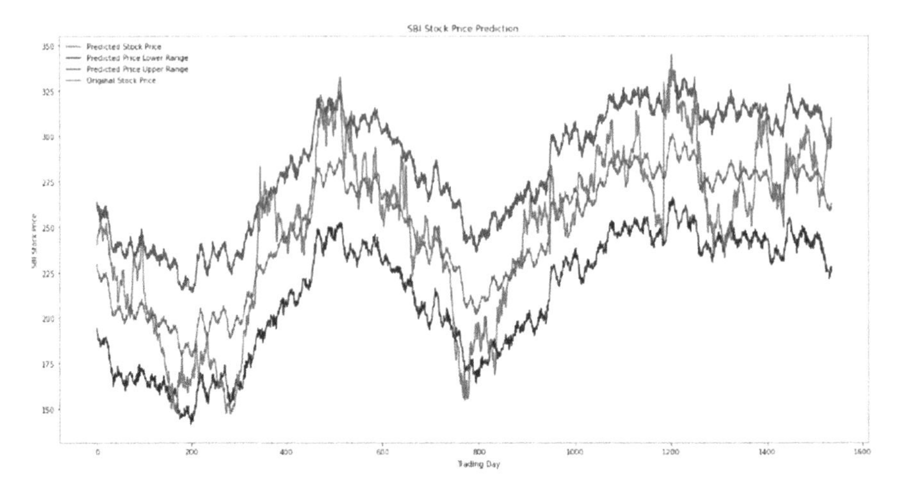

Fig. 8 Comparison graph between actual stock price and predicted stock price

Table 2 Mean absolute error of LSTM

MAE:	17.627316
MSE:	457.361253
RMSE:	21.386006
Bias:	14.060809

as we have calculated for LSTM. In Table 2 we can see the MAE value is 17.62, MSE is 457.36, RMSE is 21.38 and Bias is 14.06.

6 Discussion and Conclusion

We have seen the implementation of LSTM and Prophet Algorithm both and finally we can compare the performance matrices of both the algorithms. If we see the MAE value of LSTM it is approximately 4 and for Prophet, it is 17.62 so we can say LSTM is much better the Prophet is terms of MAE. On the other end if we compare the MSE of both the algorithm LSTM again shows much better results 25.25 is far better then 475.36, definitely same of RMSE also LSTM outperform Prophet in terms of RMSE. If we see the Bias again LSTM beats Prophet. So now we can conclude that in overall performance LSTM is better as compare to Prophet algorithm. But there is one more thing that is to be notice that Prophet algorithm is very easy to implement, very less parameters are there to tune and you get very good results also in very least knowledge of data analytics on the other end LSTM is very accurate but at the same it is not easy to train any neural network there are lots of parameters and very high level network tuning is required to get the better results and it can also face the problems of over fitting and under fitting.

References

1. Jain S, Gupta R, Moghe AA (2018) Stock price prediction on daily stock data using deep neural networks. In: 2018 international conference on advanced computation and telecommunication (ICACAT). IEEE, pp 1–13
2. Taylor SJ, Letham B (2017) Forecasting at scale. PeerJ Preprints
3. Chauhan R, Kaur H, Chang V (2020) An optimized integrated framework of big data analytics managing security and privacy in healthcare data. Wirel Personal Commun
4. https://github.com/facebook/prophet. Accessed Date July 22 2020
5. https://www.kdnuggets.com/2018/11/sales-forecasting-using-prophet.html. Access Date July 25 2020
6. https://blog.exploratory.io/an-introduction-to-time-series-forecasting-with-prophet-package-in-exploratory-129ed0c12112. Access Date Aug 02 2020
7. Idrees SM, Alam MA, Agarwal P (2019) A prediction approach for stock market volatility based on time series data. IEEE Access 7:17287–17298
8. Di Persio L, Honchar O (2016) Artificial neural networks approach to the forecast of stock market price movements. Int J Econ Manage Syst 1
9. Li C, Chiang TW (2012) Complex neurofuzzy ARIMA forecasting—a new approach using complex fuzzy sets. IEEE Trans Fuzzy Syst 21(3):567–584
10. Doostmohammadi A, Amjady N, Zareipour H (2017) Day-ahead financial loss/gain modeling and prediction for a generation company. IEEE Trans Power Syst 32(5):3360–3372
11. Minh DL, Sadeghi-Niaraki A, Huy HD, Min K, Moon H (2018) Deep learning approach for short-term stock trends prediction based on two-stream gated recurrent unit network. IEEE Access, 6:55392–55404
12. Srivastava N, Hinton G, Krizhevsky A, Sutskever I, Salakhutdinov R (2014). Dropout: a simple way to prevent neural networks from overfitting. J Mach Learning Res 15(1):1929–1958
13. Karim F, Majumdar S, Darabi H, Chen S (2017) LSTM Fully convolutional networks for time series classification. IEEE Access 6:1662–1669
14. Lee CHL, Liu A, Chen WS (2006) Pattern discovery of fuzzy time series for financial prediction. IEEE Trans Knowl Data Eng 18(5):613–625
15. Madan R, Mangipudi PS (2018) Predicting computer network traffic: a time series forecasting approach using DWT, ARIMA and RNN. In: 2018 eleventh international conference on contemporary computing (IC3). IEEE, pp 1–5
16. Wen M, Li P, Zhang L, Chen Y (2019) Stock market trend prediction using high-order information of time series. IEEE Access 7:28299–28308
17. Yan W (2012) Toward automatic time-series forecasting using neural networks. IEEE Trans Neural Netw Learning Syst 23(7)
18. Chen L, Qiao Z, Wang M, Wang C, Du R, Stanley HE (2018) Which artificial intelligence algorithm better predicts the Chinese stock market? IEEE Access 6:48625–48633
19. Zunic E, Korjenic K, Hodzic K, Donko D (2020) Application of facebook's prophet algorithm for successful sales forecasting based on real-world data. arXiv preprint arXiv:2005.07575.
20. Kumar N, Chauhan R, Dubey G (2020) Applicability of financial system using deep learning techniques. Ambient communications and computer systems. In: Advances in intelligent systems and computing, vol 1097. Springer, Singapore.
21. Chauhan R, Kaur H, Chang V (2017) Advancement and applicability of classifiers for variant exponential model to optimize the accuracy for deep learning. J Amb Intell Hum Comput Springer. https://doi.org/10.1007/s12652-017-0561-x
22. Chauhan R, Kaur H (2017) A feature based reduction technique on large scale databases. Int J Data Anal Techn Strat 9(3):207
23. Chauhan R, Kaur H, Alam AM (2010) Data clustering method for discovering clusters in spatial cancer databases. Int J Comput Appl Special Issue 10(6):9–14

Towards a Federated Learning Approach for NLP Applications

Omkar Srinivas Prabhu, Praveen Kumar Gupta, P. Shashank, K. Chandrasekaran, and D. Usha

Abstract Traditional machine learning involves the collection of training data to a centralized location. This collected data is prone to misuse and data breach. Federated learning is a promising solution for reducing the possibility of misusing sensitive user data in machine learning systems. In recent years, there has been an increase in the adoption of federated learning in healthcare applications. On the other hand, personal data such as text messages and emails also contain highly sensitive data, typically used in natural language processing (NLP) applications. In this paper, we investigate the adoption of federated learning approach in the domain of NLP requiring sensitive data. For this purpose, we have developed a federated learning infrastructure that performs training on remote devices without the need to share data. We demonstrate the usability of this infrastructure for NLP by focusing on sentiment analysis. The results show that the federated learning approach trained a model with comparable test accuracy to the centralized approach. Therefore, federated learning is a viable alternative for developing NLP models to preserve the privacy of data.

Keywords Federated learning · Natural language processing · Privacy · Machine learning

1 Introduction

Large amounts of data need to be aggregated on a centralized platform for training traditional machine learning models [1]. This induces an inherent requirement of trust in external organizations. We see regular data breaches and misuse of collected data by many organizations [2]. With text and speech being our primary means of communication containing sensitive and personal data, it is essential to find alternative ways of using it for getting required insights and personal user recommendations [3].

O. S. Prabhu · P. K. Gupta · P. Shashank · K. Chandrasekaran (✉) · D. Usha
Department of Computer Science and Engineering, National Institute of Technology Karnataka, Surathkal, Mangalore, Karnataka, India
e-mail: kchnitk@ieee.org

A. Choudhary et al. (eds.), *Applications of Artificial Intelligence and Machine Learning*, Lecture Notes in Electrical Engineering 778,
https://doi.org/10.1007/978-981-16-3067-5_13

In this paper, we discuss federated learning [4], which allows us to train a machine learning model on user devices collaboratively. This method ensures that all forms of sensitive data are only kept on user devices. This uses the existing distributed nature of real-world data while allowing us to comply with privacy regulations, thus overcoming the legal and ethical obstacles to sharing data. Federated learning also eliminates the need to store large amounts of data at a single location, thus minimizing the losses incurred during data breaches. It also encourages different data holders to participate to gain benefits from a better machine learning model without directly sharing their data with any external entity.

Federated learning techniques can be broadly classified as single party or multi-party [5]. In single-party systems, there is a single entity responsible for the governance of the flow system. Data has the same structure on the end devices, and the required model can be trained in a federated manner without much changes for the different clients. In multi-party systems, different parties form an alliance for a collaborative environment for training a common machine learning model. Preprocessing is required at client devices to follow a common defined standard.

The approach of using the inherently distributed nature of the data for federated learning has been used in different application domains. Some of them include banks using their data to train a better fraud detection system, using sensitive private data on mobile devices for training keyboard prediction and emoji prediction models. Medical institutes are building federated learning infrastructures to use their medical records and other sensitive data for collaborative training with other organizations.

Text and speech have been the primary mode of communication for humans. The increasing number of mobile devices has led to an upsurge in the amount of such data being collected. Most of the personal data like text messages, emails, and user habits are in the form of such data. Data protection laws are getting stricter after the regular misuse of data by organizations, leading to difficulties in data collection [3]. Moreover, the amount of such data can be massive, making it difficult and costly to accumulate it on a central location. It can also result in data misuse due to data breaches in the government or the companies. This makes it essential for us to find good ways of gaining insights from such data without compromising the privacy of users. The federated approach can be used to overcome these challenges and help in developing useful natural language processing (NLP) applications.

Although there is evidence of using federated learning in NLP, there is a lack of proper infrastructure for the same. In this paper, we have developed a federated learning infrastructure and trained a deep neural network architecture to predict the sentiment of texts. The federated learning infrastructure can train on remote devices and can aggregate the model updates. The results show that the test accuracy of the model trained using federated learning is comparable to centralized training. However, federated learning carries an additional overhead of aggregating the individual updates, which can be optimized for better training speeds.

This paper is organized as follows: Sect. 2 describes application of federated learning in different domain; Sect. 3 explains the design of federated learning infrastructure including the communication protocol and the phases and provides implementation details of the same; Sect. 4 presents experimental results and analysis; Sect. 5 concludes the paper and discusses some future directions for this work.

2 Related Works

The rise in usage of mobiles as primary user devices has opened up numerous possibilities of exploiting the inherently distributed nature of devices for collaborative training using federated learning. Federated learning has also found numerous applications in health care due to difficulties in getting sufficiently large and high-quality data sets for training. We examine some applications in these domains where federated learning has been applied.

Hard et al. [6] use mobile clients for next word prediction in virtual keyboards. The authors use a long short-term memory (LSTM) recurrent neural network called the coupled input-forget gate (CIFG) and compared it with a classical stochastic gradient done on a centralized server. The model size and inference time prediction latency are low due to limitations in the computational power of the end client devices. The ratio of the number of correct predictions to the total number of tokens is used as a metric for comparison. Top-1 and top-3 recall predictions are found to be on par or better than the server-based approach because of better availability of locally cached data. The authors showed that the federated approach could outperform the traditional methods in a commercial setting by training on a higher-quality dataset, which was not possible due to privacy concerns with the centralized training.

Beaufays et al. [7] use a transfer learning approach for predicting emoji using a pre-trained language model. The authors use a word-level CIFG-LSTM recurrent neural network for prediction and cache the model state to reduce inference-time prediction latency. They discuss different ways of triggering the emoji predictions like using a single language model for predicting the next words as well as emojis, using a separate binary classifier to predict the likelihood of user typing an emoji after a given phrase. The ratio of accurate top-1 emoji predictions to total examples containing an emoji is used as a metric for the model quality. Area under the ROC curve is used to evaluate the quality of the triggering mechanism used. The results showed that federated learning worked even with sparse data and poorly balanced classes and was found to outperform the traditional centralized approach.

Nadiger et al. [8] discuss a federated reinforcement learning (FRL) technique that is focused on improving the personalization time. The authors use in-game metrics and length of a rally in a ping pong game as metrics indicating the level of personalization. The FRL architecture is divided into a grouping policy, a learning policy, and a federation policy. The authors showed a considerable improvement in the time taken for personalization with different number of agents.

Deist et al. [9] have built an IT infrastructure, named euroCAT, for federated learning across five different clinics located in three European countries (Belgium, Germany, and Netherlands). A server within each clinic's infrastructure hosts its local database and its local learning connector. A central server (master) outside the infrastructure of individual sites is connected to individual learning connectors by a global learning connector. They claim that the discriminative performance obtained in euroCAT is similar to previously published SVM [10, 11] models for this use case.

Jochems et al. [12] solve the problem of predicting dyspnea (a common side effect after radiotherapy treatment of lung cancer) by federated learning. They have used a Bayesian network model for distributed learning on horizontally partitioned data located at five different locations. The results obtained show that the average difference of percentages in the conditional probability tables (CPTs) is 1.6% (0.7%), and the average difference in area under the curve (AUC) is 0.002 (0.002) between the distributed model and local (centralized) model for a dataset size of 100,000.

Brisimi et al. [13] tried to predict hospitalizations due to cardiac events using a federated learning approach. They have developed a federated optimization scheme called cluster primal-dual splitting (cPDS) for solving the sparse support vector machine (sSVM) problem. The data is distributed over hospitals connected through a specific graph topology. The results obtained show that cPDS solves the sSVM problem by building a classifier using relatively few features and that cPDS has improved convergence rate compared to the centralized barrier methods such as the SubGD [14], the IncrSub descent [15], and the LAC scheme [16].

Sheller et al. [17] have demonstrated the use of a federated learning approach to solve the problem of brain tumor segmentation. The results show that the federated learning approach achieves 99% of the model performance of the centralized data-sharing model. They have also compared federated learning (FL) with other types of collaborative learning techniques: institutional incremental learning (IIL), where each institution trains the shared model once in succession, and cyclic institutional incremental learning (CIIL), which is IIL performed in rounds with prescribed numbers of epochs. They find that IIL performs poorly compared to FL and CIIL, while CIIL is less stable and harder to validate than FL.

The existing research papers show several applications of federated learning in mobile and healthcare domains. Although there have been some evidences of using federated learning for NLP applications, it lacks a well-defined infrastructure that can show federated learning to be a viable alternative. Therefore, in this paper, we investigate the suitability of using federated learning for training NLP models.

3 Federated Learning Infrastructure

In this section, we present the design of our system tuned for NLP applications. We describe implementation details along with the experimental environment. Our design is inspired by Bonawitz et al. [18], where they introduce a high-level description of the various components and the protocol in a federated learning infrastructure. We have implemented the services in our system using Go [19] and Python [20] programming languages. gRPC [21] is used as a protocol to define the interfaces for communication between the services.

3.1 Design and Implementation

We follow a simple client–server model with client service acting as end devices in federated learning. The phases of federated learning round (fl round), as depicted in Fig. 1, include selection configuration and reporting. The central server is responsible for the selection of clients, which indicates their availability, configuration: sending the required files to selected clients, aggregation: averaging of weights sent by clients, and updating the global model with the averaging process result. Aggregation is done using federated averaging [4]. Algorithms 1 and 2 describe the states in server and clients during an fl round. Currently, we are not considering any selection criteria for the clients.

However, the server can easily incorporate these changes. The clients train the model received with the data present on them and send the updates to the model (updated checkpoints) after training to the server.

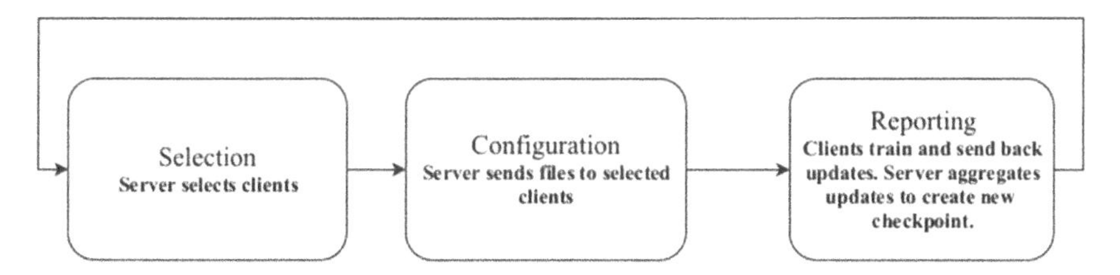

Fig. 1 Sequence of events in a federated learning round

Algorithm 1: Server

1 Wait until *checkInLimit* number of clients have connected. Reject other clients once the threshold is reached.
2 Send model architecture, checkpoint, vocabulary and tokens files to the clients
3 Wait until clients complete training. Receive the checkpoint updates from each client
4 After *checkInLimit* updates are received start the Federated Averaging process.
5 Update the checkpoint

Algorithm 2: Client

1 Send a connection request to the server to participate in the FL round
2 Receive model architecture, checkpoint, vocabulary and tokens files
3 Train the model using available data which is locally stored.
4 Report the new checkpoint created back to the server

To begin the training process, each client needs to receive files such as model file (architecture of the model), checkpoint file (containing the weight/parameter values of the model), vocabulary file (a finite set of words), and token (containing token values for each word in vocabulary). These can be sent only in the first round of federated learning as they are common for all rounds. NLP applications such as sentiment analysis from the text applications explicitly require the vocabulary and token files.

For the implementation, the server has a connection handler function running on a go-routine running in the background, which is responsible for updating count variables so that it is the only routine making updates and hence maintaining consistency. Other routines will communicate with this routine via channels to query or make updates to count variables. For every new client connection, there will be a separate go-routine functioning. After the training process completes, the client calculates the update between checkpoints and sends it along with a weight (number of batches of data) to specify the amount of data it trained on back to the server. Checking whether the count of updates received is greater than some decided threshold, the connection handler starts the federated averaging process, and the global model checkpoint is updated by adding the aggregated values.

For creating a simulation environment, we have used Docker [22]. We containerized the applications for the client and the server and employed three clients and one server docker container instances, as shown in Fig. 2. The global model architecture, checkpoint, vocabulary, and token files are mounted onto the server container, and for the clients, data files are mounted onto them.

3.2 Dataset and Model

The dataset used is the large movie review dataset [23] for demonstrating the proof of concept. This dataset is for binary sentiment classification. We divided the dataset into

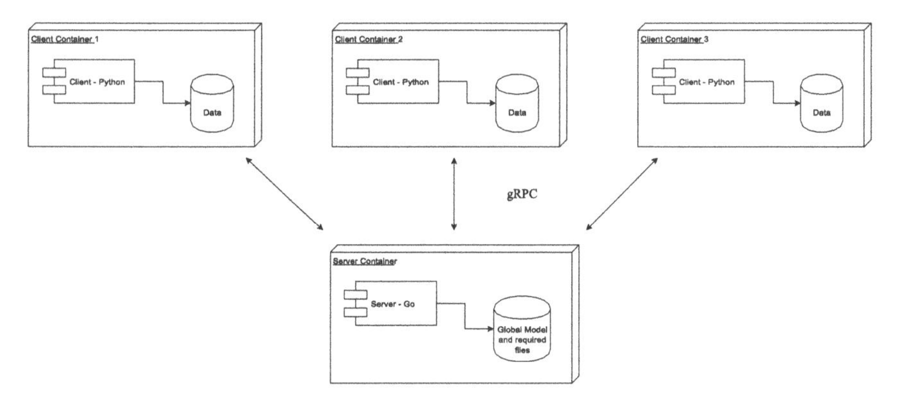

Fig. 2 Container deployment and interaction of client and server

three disjoint subsets of the same size, with each subset containing the same number of positive and negative classes. Each subset contained 4000 positive reviews and 4000 negative reviews. Then these subsets were stored in three different locations to simulate the decentralization of data.

Before training, we preprocess the data, which involves cleaning data to remove punctuations and removing words not present in a preset vocabulary (about 25,000 words). We then tokenize texts (i.e., convert words to unique integer ids) by having a common tokenizer for all devices during both training and testing. We then truncate or pad texts to be of a fixed size to pass it to the input layer of a deep neural network. Our deep learning architecture contains three layers. The first layer is the embedding layer, which learns to map words to fixed-length vectors. The second layer consists of the LSTM (long short-term memory)[24] neural network architecture with dropouts. The third layer is a dense layer with sigmoid activation for binary classification.

4 Results

In this section, we compare the performance of the model trained on our federated learning infrastructure with the model trained using the centralized learning approach.

The performance is measured in terms of train and test accuracy of the model. We provide a brief description of how the experiment was conducted, along with the results obtained.

Federated learning approach: Federated training was performed on three client devices for five rounds. In each round, we used a unique set of 1600 data points with 800 positive and 800 negative classes for training. This methodology was adopted to simulate a real-life scenario where each end device generates or collects varying data continuously at different rounds of federated training. Also, in each round, training was carried out for three epochs. The model was also tested on a separate dataset with 8000 data points with 4000 positive and 4000 negative classes. Test accuracy was calculated just after training as well as after federated averaging in each round. Table 1 shows the accuracy achieved on different end devices in each round. For each client, three kinds of accuracy measures are recorded:

- Train: Accuracy on the training dataset
- Local test: Test accuracy calculated just after training locally on the device
- Fed. Avg. Test: Test accuracy achieved by the model after running federated averaging for the given round.

Centralized learning approach: The model was trained on three epochs with the training dataset formed by collecting and combining the training data present on the three client devices. The train and test accuracy of the centralized approach are as follows.

- Train accuracy: 91.26%
- Test accuracy: 82.46%

From Table 1, we see that the model trained using a federated learning approach improves after each round. After five rounds of federated learning, the test accuracy of the federated learning model reaches a similar value of test accuracy as that of the model trained using a centralized approach. As shown in Fig. 3, the average test accuracy of the model on test data present on the three client devices in the federated learning approach after five rounds is 86.67%, which is higher than the test accuracy of 82.46% of the model trained using the centralized approach.

From the results obtained, we infer that the model trained using the federated learning approach can achieve a similar level of performance as that of the model trained with a traditional centralized approach. Thus, the infrastructure presented in the paper provides a proof of concept for using the federated learning approach to train machine learning models for NLP applications.

5 Conclusion and Future Work

Federated learning infrastructure enables the training of machine learning models while trying to maintain the privacy of data used for training. The quality of a model trained by federated learning (FL) infrastructure should be on par with that trained using a traditional centralized approach. Inspired from previous work on federated learning infrastructures, we have designed a federated learning infrastructure tuned for natural language processing (NLP). We demonstrate the usability of this system for NLP applications by solving a problem on sentiment analysis with the Large Movie Review Dataset. The results show the performance of the model trained on our federated learning system after certain number of rounds is on par with the model trained using the centralized approach. There is a great scope for improvement of the proposed federated learning infrastructure. Improvements in training process, optimized aggregation algorithm, incorporation of secure aggregation protocol for

Table 1 Train and test accuracy (in %) on client devices during federated learning

	Client 1			Client 2			Client 3		
Federated tounds	train	Local test	Fed. Avg. test	Train	Local test	Fed. Avg. test	Train	Local test	Fed. Avg. test
Round 1	85.06	78.74	64.55	71.94	67.81	64.8	76	77.1	64.03
Round 2	95.94	80.26	83.99	94.5	78.81	84.08	95	79.14	82.97
Round 3	95.37	81.76	86.21	97.31	83.96	86.38	95.75	80.62	84.76
Round 4	96	85	86.91	96.94	85.25	86.54	95.94	83.79	85.99
Round 5	95.94	85.32	87	95	83.81	86.86	97.13	83.95	86.17

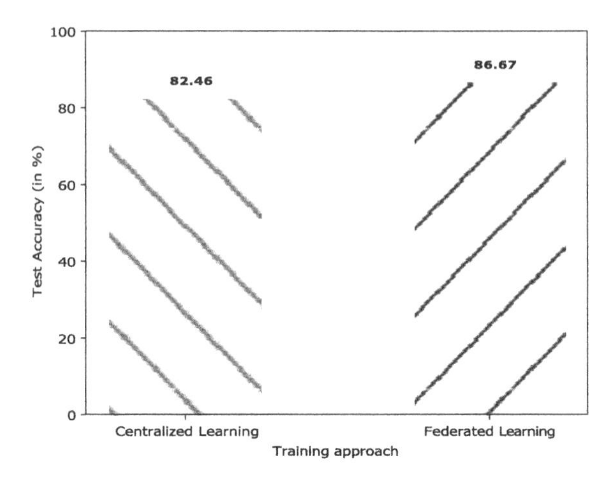

Fig. 3 Comparison of test accuracy between centralized learning and federated learning approaches

securely computing sums of vectors, etc., can be made in the infrastructure. Through this paper, we have shown a proof of concept for a federated learning infrastructure for an NLP application using a dataset for sentiment analysis. We believe that the results presented in this paper will pave the way for future research in using federated learning for NLP applications.

References

1. Roh Y, Heo G, Whang SE (2018) A survey on data collection for machine learning: a big data—AI integration perspective. CoRR, vol. abs/1811.03402, [Online]. Available: http://arxiv.org/abs/1811.03402
2. Karunakaran S, Thomas K, Bursztein E, Comanescu O (2018) Data breaches: user comprehension, expectations, and concerns with handling exposed data. In: Fourteenth symposium on usable privacy and security (SOUPS 2018). USENIX Association, Baltimore, MD, pp 217–234. Available: https://www.usenix.org/conference/soups2018/presentation/karunakaran
3. Mieskes M (2017) A quantitative study of data in the NLP community. In: Proceedings of the first ACL workshop on ethics in natural language processing. Association for Computational Linguistics, Valencia, Spain, pp 23–29. Available: https://www.aclweb.org/anthology/W17-1603
4. McMahan HB, Moore E, Ramage D, Hampson S, Arcas BAY (2016) Communication- efficient learning of deep networks from decentralized data
5. Yang Q, Liu Y, Chen T, Tong Y (2019) Federated machine learning: concept and applications CoRR, vol abs/1902.04885. Available: http://arxiv.org/abs/1902.04885
6. Hard A, Kiddon CM, Ramage D, Beaufays F, Eichner H, Rao K, Mathews R, Augenstein S (2018) Federated learning for mobile keyboard prediction
7. Beaufays F, Rao K, Mathews R, Ramaswamy S (2019) "ederated learning for emoji prediction in a mobile keyboard
8. Nadiger C, Kumar A, Abdelhak S (2019) Federated reinforcement learning for fast personalization. In: 2019 IEEE second international conference on artificial intelligence and knowledge engineering (AIKE), pp123–127
9. .Deist TM, Jochems A, van Soest J, Nalbantov G, Oberije C, Walsh S, Eble M, Bulens P, Coucke P, Dries W, Dekker A, Lambin P (2017) Infrastructure and distributed learning methodology

for privacy-preserving multi-centric rapid learning health care: eurocrat. Clinical Trans Radiat Oncol 4:24–31
10. Dehing-Oberije C, Ruysscher DD, van Baardwijk A, Yu S, Rao B, Lambin P (2009) The importance of patient characteristics for the prediction of radiation-induced lung toxicity. Radiother Oncol 91(3):421–426
11. Nalbantov G, Kietselaer B, Vandecasteele K, Oberije C, Berbee M, Troost E, Dingemans A-M, van Baardwijk A, Smits K, Dekker A, Bussink J, Ruysscher DD, Lievens Y, Lambin P (2013) Cardiac comorbidity is an independent risk factor for radiation-induced lung toxicity in lung cancer patients. Radiother Oncol 109(1):100–106
12. Jochems A, Deist TM, van Soest J, Eble M, Bulens P, Coucke P, Dries W, Lambin P, Dekker A (2016) Distributed learning: Developing a predictive model based on data from multiple hospitals without data leaving the hospital—a real life proof of concept. Radiotherapy Oncol 121(3):459–467
13. Brisimi TS, Chen R, Mela T, Olshevsky A, Paschalidis IC, Shi W (2018) Federated learning of predictive models from federated electronic health records. Int J Med Inf 112:59–67
14. Bertsekas DP (1997) Nonlinear programming. J Operat Res Soc 48(3):334–334
15. Nedic A, Bertsekas DP (2001) Incremental subgradient methods for nondifferentiable optimization. SIAM J Optim 12(1):109–138
16. Olshevsky A (2014) Linear time average consensus on fixed graphs and implications for decentralized optimization and multi-agent control
17. Sheller MJ, Reina GA, Edwards B, Martin J, Bakas S (2018) Multi-institutional deep learning modeling without sharing patient data: a feasibility study on brain tumor segmentation
18. . Bonawitz K, Eichner H,Grieskamp W, Huba D, Ingerman A, Ivanov V, Kiddon CM, Konecˇny´ J, Mazzocchi S, McMahan B, Overveldt TV, Petrou D, Ramage D, Rose- lander J (2019) Towards federated learning at scale: system design 2019, to appear.
19. Go programming language. [Online]. Available:https://golang.org/
20. Python programming language. [Online]. Available: https://www.python.org/about/
21. gRPC: high performance rpc framework. [Online]. Available: https://www.grpc.io/
22. Docker. [Online]. Available: https://www.docker.com/
23. Maas Al, Daly RE, Pham PT, Huang D, Ng AY Potts C (2011) Learning word vectors for sentiment analysis. In: Proceedings of the 49th annual meeting of the association for computational Linguistics: human language technologies. Association for Computational Linguistics, Portland, Oregon, USA, pp 142–150
24. Hochreiter S, Schmidhuber J (1997) Long short-term memory. Neural Comput 9(8):1735–1780

Challenges of Smart Cities Future Research Directions

Analysis of Groundwater Quality Using GIS-Based Water Quality Index in Noida, Gautam Buddh Nagar, Uttar Pradesh (UP), India

Kakoli Banerjee, M. B. Santhosh Kumar, L. N. Tilak, and Sarthak Vashistha

Abstract The current research aims to establish GIS-based water quality index by analysing 51 groundwater samples collected from various sectors of the Gautham Buddh Nagar district of Noida, U.P., India. To determine WQI the groundwater samples were exposed to a detailed physico-chemical experimentation of seven parameters such as pH, Total Hardness, Total Alkalinity, Chlorides, Turbidity, Carbonates and Bicarbonates. The values of the examined samples were compared with the water quality requirements of the Bureau of Indian Standards (BIS). The result of the present study indicates that 84.4% of water samples fell in good quality and that only 15.6% of water samples fell under the poor water category exceeding the acceptable and permissible limits of BIS approved drinking water quality standards (Indian Standard Drinking Water IS:10500–2012). This determined the WQI index for the same and the values ranged from 47.12 to 192.104. Along with that machine learning techniques helped us understand the relation between pH and WQI index, where pH fluctuation is compared to water quality. The study shows that treatment is needed in the region's groundwater before it is used for drinking and other uses.

Keywords GIS · Water quality index (WQI) · Physico-chemical analysis · Noida

K. Banerjee (✉) · S. Vashistha
Department of Computer Science and Engineering, JSS Academy of Technical Education Noida, C-20/1, C Block, Phase 2, Sector 62, Noida, Uttar Pradesh 201309, India
e-mail: kakoli.banerjee@jssaten.ac.in

M. B. Santhosh Kumar (✉) · L. N. Tilak
Department of Civil Engineering, JSS Academy of Technical Education Noida, C-20/1, C Block, Phase 2, Sector 62, Noida, Uttar Pradesh 201309, India
e-mail: santhoshkumar@jssaten.ac.in

L. N. Tilak
e-mail: tilakln@jssaten.ac.in

A. Choudhary et al. (eds.), *Applications of Artificial Intelligence and Machine Learning*, Lecture Notes in Electrical Engineering 778,
https://doi.org/10.1007/978-981-16-3067-5_14

1 Introduction

Groundwater has a major role in water supply system for drinking, irrigation and industrial purposes. Exploitation of groundwater in India collectively surpassed total US and Chinese use, thus holding the world's first position on the list [10]. However, according to the CGWB report, about 245–109 m^3 of groundwater is utilized to meet the demand in India's agricultural sector. About 65% of the world's groundwater is used for drinking purposes; about 20% is used for agriculture purposes and 15% is used for industrial benefits, placing immense stress on this critical resource as demand rises [12]. Human actions such as land use/land cover changes, infiltration from polluted crops, geological formation, change in depletion of rainfall and decrease in precipitation infilteration influence the quantity/quality of groundwater and result in contamination of the groundwater [14]. Regardless of the unsafe and improper disposal of sewage, industrial waste, groundwater pollution is also a major environmental issue in recent times [7]. Groundwater biological, physical, and chemical factors may be affected due to pollution. The water quality index (WQI) assessment of drinking water quality was developed in the 1970s by the Oregon Department of Environmental Quality to evaluate and summarize water quality coditions [1, 3, 8, 16]. Monitoring the quality of water plays a crucial role in protection of environment, humans and marine ecosystem integrity. In this context, the water quality index (WQI) methodology is an important tool that offers policymakers and interested individuals in the study area knowledge about water quality [14]. GIS has been a commonly used method for manipulating multivariate values, processes and spatial mapping outputs, helping to draw a conclusion on the environmental and geological scenario [6]. Due to geogenic and anthropogenic processes, several studies have been published on groundwater quality and pollution in the literature [1, 3–5, 9, 11, 16, 13]. The present research is an attempt on the GIS-based groundwater quality index to determine its adoptability for human use and to communicate for successful management to policy makers and local bodies/people. The aim of this paper is to understand the hydro-geochemical cycle of groundwater using physico-chemical parameters and to determine the groundwater quality of the NOIDA for drinking and other human uses using the WQI and GIS techniques. In fact, the outcome of these investigations provides baseline data on quality of groundwater status in the chosen study area, which benefits in groundwater resources management and the importance of volatility limit the pH should be in to understand its effect on water quality, i.e. WQI. Machine learning techniques such as multivariable linear regression, support vector regression, and decision tree regression were used to predict the pH of the water body based on its geographical location.

2 Study Area

Noida is located in the Uttar Pradesh state's Gautam Buddh Nagar district, near to Delhi's NCR. The geographical location of Noida is between 28° 26′ 39″ N–28° 38″ 10″ N and 77° 29″ 53″ E–77° 17″ 29″ E. Spanning Geographical Area of 203 km^2 and with a population of 642,381 according to the 2011 census. The City's climate is sub-humid, and is marked by hot summer and cold season bracing. In the most part of the year Noida has a hot and humid climate. Annual rainfall averages 642.0 mm. The region typically has flat topography from north-east to south-west (< 1°) with gradual slope. The current study area is 204 m above mean sea level near to Parthala Khanjarpur village situated at the northeastern and with the minimum elevation of 195 m above mean sea level near Garhi Village located at south-west region. Noida city uses water source from Ganga Canal interception at Masoori Dasna, situated in Ghaziabad, and a normal occurrence of groundwater. Noida falls under the Yamuna River catchment area; it is bound by the Yamuna River to the west and south-west, and by the Hindon River to the south-east. In the south, the city is bounded by the two rivers, Yamuna and Hindon, meeting point (Figs. 1 and 2).

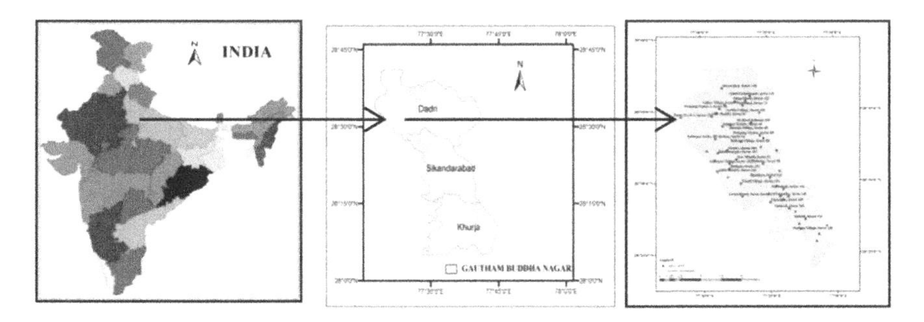

Fig. 1 Study area

3 Materials and Methods

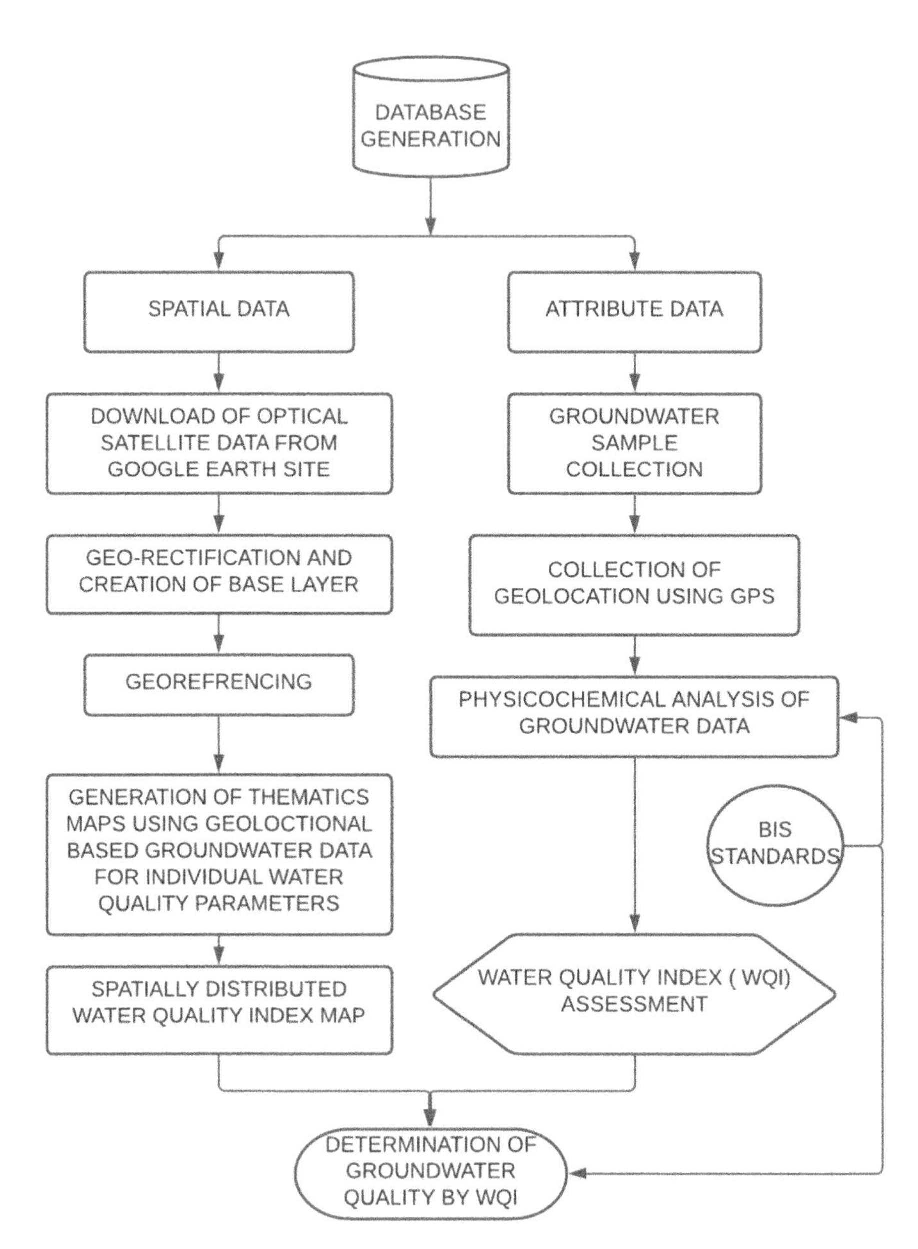

Fig. 2 Flow chart adopted for methodology

3.1 Methodology

The current study focuses on data that are gathered from the field and produced through laboratory research. Physico-chemical analysis of groundwater samples taken from Noida was analysed at environmental laboratory of Civil Engineering Department, JSS Academy of technical education Noida. Results from the physico-chemical analysis of groundwater samples which were acquired from randomly distributed fifty-one different sites, such as public hand pumps, borwells and maintained according to the methods recommended in the American Public Health Association manual (APHA-2320 1999). Using Global Positioning System, geospatial specifics of sample sites were determined. An attempt is made to define the groundwater quality index and compared the results with the values specified in water quality standards such as the Bureau of Indian Standards (BIS), the Indian Standards (IS) [2] and the World Health Organization [15]. All the concentrations of the chemical are measured in mg/l. On an alternate side multivariable linear regression, support vector regression, and decision tree regression are used to predict the pH of locations and using the Bureau of Indian Standards (BIS), the limit of pH we understood and upon applying it as a limiting feature the range of WQI is understood for the locations which are in the pH limits.

3.2 Physico-Chemical Parameters

3.2.1 pH

The pH of the groundwater tested is between 7.1 and 8.9 representing acidic to alkaline groundwater samples, accordingly. After evaluation of the test samples pH value shows that all water samples were in the BIS [2] allowable limits (6.5–8.5) except at eight locations (Fig. 3).

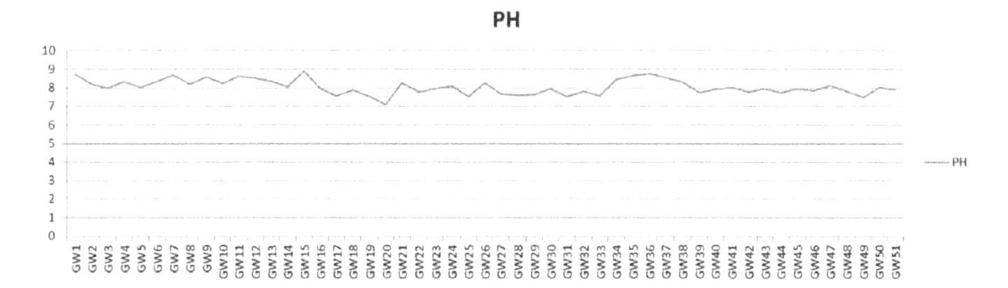

Fig. 3 pH distribution in different groundwater samples

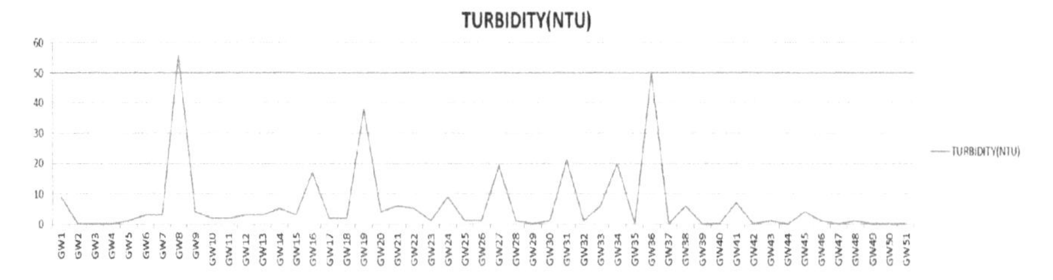

Fig. 4 Turbidity distribution in different groundwater samples

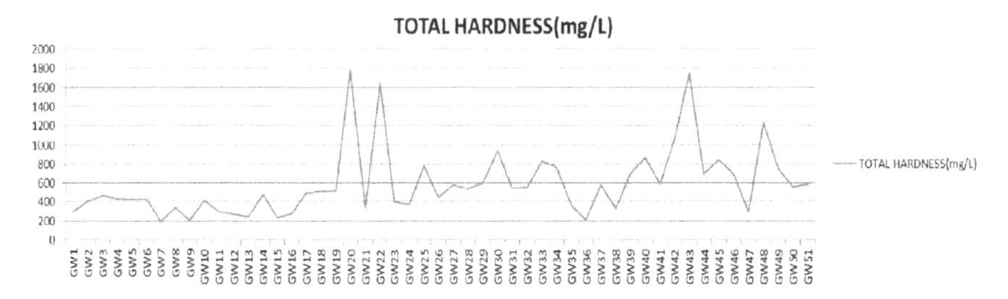

Fig. 5 Total hardness distribution in different groundwater samples

3.2.2 Turbidity

Turbidity is calculation of a liquid's relative visibility. It is an optical characteristic of water that is a measure of the amount of light that is spread in the water by substance when a light is shined through the water sample. Extreme turbidity is aesthetically unattractive in potable water, and may also cause health issue. The GW8 sample shows the highest turbidity with a value of 56 NTU which is preceding the permissible limits for drinking water recommended in [2] (Fig. 4).

3.2.3 Total Hardness

Total Hardness indicates that 68.6% of the groundwater sample lies below the permissible limit (600 mg/l) and 31.4% groundwater are above the permissible limit (600 mg/l) as prescribed in BIS [2]. High TH concentrations above the permissible limit were found at 16 locations in the study region between 600 and 1780 mg/l (Fig. 5).

3.2.4 Chlorides

The chloride level at Baraula Village, Sector 49 is 3288.98 mg/l where the limit is 1000 mg/l and the presence of sewage treatment plant attributing the values is observed using satellite image by visual confirmation. The effect of high Chlorides

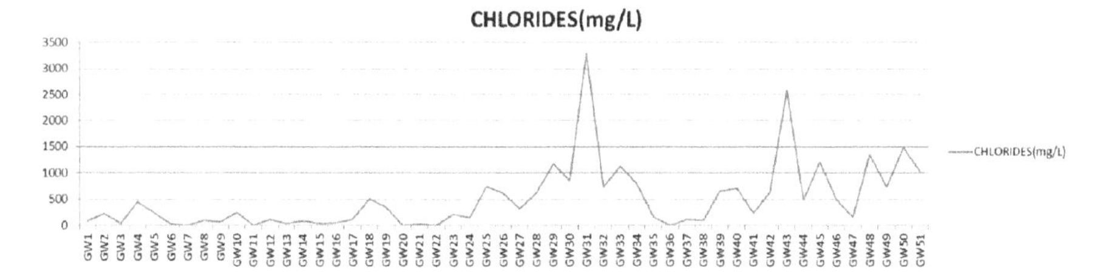

Fig. 6 Chloride distribution in different groundwater samples

levels in groundwater results in indigestion, taste palatability and corrosion of pipes (Fig. 6).

3.2.5 Total Alkalinity

At GW45 the Highest total alkalinity value of 695 mg/l was found and at GW31 the lowest value of 200 mg/l. The alkalinity value was found to be higher than the appropriate level for drinking water as recommended in BIS [2] (Fig. 7).

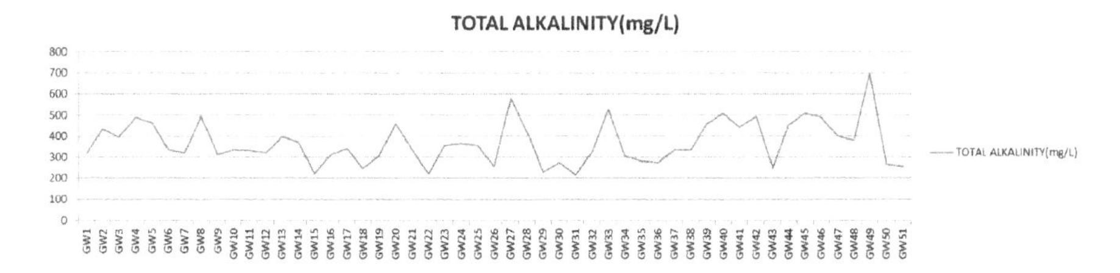

Fig. 7 Total Alkalinity distribution in different groundwater samples

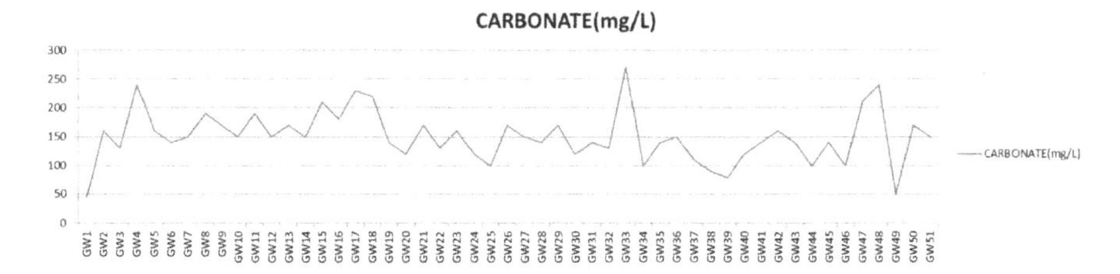

Fig. 8 Carbonate distribution in different groundwater samples

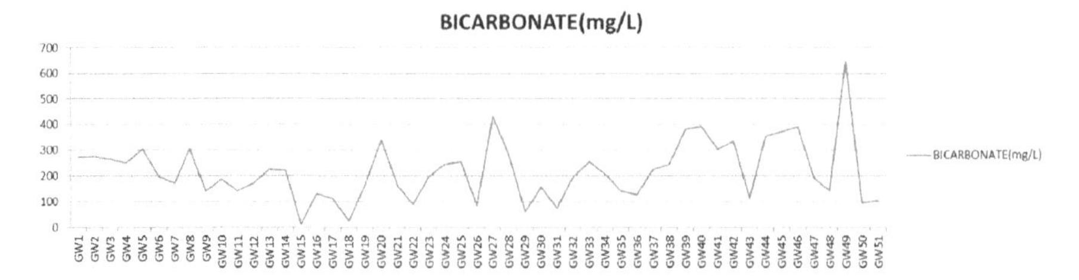

Fig. 9 Bicarbonate distribution in different groundwater samples

3.2.6 Carbonates

Carbonate value of GW33 was found above the allowable limit being as mentioned in BIS [2] for drinking water (Fig. 8) and 270 mg/l was the lowest value of carbonates that was found at GW1 with a value of 45 mg/l.

3.2.7 Bicarbonates

GW49 sample has bicarbonate value higher than the limit prescribed by the BIS [2] for drinking water (Fig. 9) with 645 mg\l and 10 mg\l of bicarbonate was found at GW15.

3.2.8 Water Quality Index Computing

The WQI was measured using the drinking water quality requirements recommended by the World Health Organization (WHO) and the Bureau of Indian Standard (BIS). The weighted arithmetic index method was used in groundwater WQI calculation (Brown et al. 1972). The WQI is computed in 3 steps. In the first step, a weight (*wi*) has been assigned to 07 parameters (pH, TH, Total Alkalinity, Chlorides, Turbidity, Carbonates Bicarbonates) according to its relative significance in the overall quality of drinking water (Table 1) in the first step. 2nd step involves calculating relative weight (*Wi*)

$$W_{\mathrm{i}} = Wii = 1nWi \tag{1}$$

where *Wi* is the relative weight and wi is the weight of each parameter and the total number of parameters is *n*, respectively.

3rd step consists of calculating the quality rating scale (*qi*) for each parameter determined by dividing its standard concentration as specified by the BIS guidelines

$$q_i = (ci/si) \times 100 \tag{2}$$

Table 1 Relative weight of physico-chemical parameters

Parameters	Weighted	Relative weight ($Wi = kS_i$)	BIS [2]
pH	5	0.208333	6.5–8.5
Turbidity (NTU)	3	0.125	1
Hardness (mg/1)	5	0.208333	200
Chlorides (mgl)	4	0.166667	250
Total Alkalinity (mg.l)	4	0.166667	200
Carbonate (mgfl)	2	0.083333	200
Bicarbonate (mg l)	1	0.041667	200
	$Wi = 24$	$Wi = 1.0$	

where qi is the quality rating scale, the concentration of each chemically parameters is ci in each water sample in mg/l, si is the Indian standard water parameters in mg/l as guided by the guidelines of BIS.

Further each physico-chemical parameter, its concentration will be divided by the permissible limits (as mentioned in the BIS) and it will be multiplied by 100.

$$q_i = (ci/si) \times 100 \tag{3}$$

Next the Sli is defined first for each physico-chemical parameter derived by the following equation for the calculation of water quality index.

$$Sl_i = W_i \times q_i \tag{4}$$

$$WQI = \sum_{i=1}^{N} Sli$$

where the sub-index of i th parameters is Sli, the rating of each concentration of i th parameters is qi, and the number of parameters is n. Table 2 displays the measured water quality index in five categories of ranging from > 50 (excellent water) to > 300 (water unsuitable for drinking) range water quality index for drinking purposes.

Table 2 Groundwater classification related to WQI range

WQI	Status
< 50	Esc client water
50–100	Good water
100–200	Poor water
200–300	Very poor water
> 300	Unfit for drinking purpose

3.3 Spatial Analysis

Spatial analysis of different physico-chemical parameters was performed in GIS environment with an open source QGIS programme. The map of the pH, TH, Total Alkalinity, Chlorides, Turbidity, Carbonates and Bicarbonates was prepared for Noida using inverse distance weighted (IDW) interpolation technique.

3.4 Machine Learning Techniques

This machine learning technique is an augmentation to Linear regression. Which, using the linear dataset helps in predicting the required result.

$\{y_i, x_{i1}, \ldots, x_{ip}\}_{i=1}^{n}$ Consisting of n instances.

$$y = \beta_0 + \beta_1 x_1 + \beta_2 x_2 + \cdots + \beta_k x_k + \in \quad (5)$$

where

β_i = slope constants.

$\in$ = Error.

x_i = input variable.

y = output variable.

When the dataset is passed through the algorithm it iterates one by one and finds error and minimizes it with each iteration thus improving the accuracy and the last iteration, yields the best results.

3.4.1 Support Vector Regression

It works on the principle that classes have a separating hyperplane between them which differentiates one class from the other keeping the property of each class in consideration. The procedure in this algorithm invokes the property and minimizes the error by subsequent iterations.

$$\text{minimize } \frac{1}{2}\|w\|^2 \quad (6)$$

$$\text{subject to} |y_i - \langle w, x_i \rangle - b| \le \varepsilon \quad (7)$$

$$y = wx + b \quad (8)$$

where $\underline{x_i}$ = training sample.

$\langle w, x_i \rangle + b$ = prediction of sample.

And ϵ = Threshold free parameter.

The approach for prediction starts by feeding x_i to Eq. (7) and finding the which ensures Eq. (6) is minimized. Then after the result of the hyperplane parameter is used in predicting the y value which is afterwards fed in R^2 method for finding the accuracy.

3.4.2 Decision Tree Regression

It is a tree-like regression algorithm based on decision tree which divides the classes into subsets which are treated as nodes of the tree. The resulting outcomes are considered from the leaf nodes and decision nodes. It priorities the nodes and traversing through the tree gives the prediction result.

Top-down tree construction method is used in this regression. The probabilities of occurring classes are used in this regression.

$$\text{Gini Index}(GI) = 1 - \Sigma Pi2$$

Pi = Probability of occurrence of Pi.

The tree is constructed to ensure that the next class is selected such that the Gini index remains as low as possible. The target is finding the lowest Gini impurity node as the leaf node. The outcome which is predicted is a real number.

4 GIS Statistical Model

The spatial distribution and the spatial modelling in the present study have achieved by inverse distance weighted (IDW) interpolation technique and the groundwater quality index is determined according to BIS [2]. The IDW is an interpolation method that represents difference and continuous type of spatial attribute in the region [17]. The value measured by IDW interpolation is a weighted average of the ground sampling points that are neighbouring it. Weights are determined by reversing the distance from the origin of an observation to the position of the predicted value [17].

5 Results and Discussions

Noida is an industrial area of Gautham Buddh Nagar District UP. Without adequate treatment, the industrial effluents pollute water bodies, rivers, etc. Untreated effluents from sewage treatment plants percolate into the groundwater making it unfit for drinking and other use.

Alongside the machine learning models gave astonishing results which would help in understanding the relation of the pH to the WQI and understanding the mapping relation between these two (Fig. 10; Table 3).

Fig. 10 Spatial distribution of physico-chemical parameters

Table 3 R^2 scores of pH

Model	R^2 test (Training data)	R^2 test (Test data)
Multivariable linear regression	42.90125594276315	−6.07342695815436
Support vector regression	92.09154744482632	46.97531102030899
Decision tree regression	98.28134611023242	94.7828134623048

Then by segmenting the WQI values it is found that the water quality index values which are said as the "Excellent water" resulted in 5 pH values in the limits prescribed by BIS [2] and only 5 WQI which could be classified as "Excellent water".

6 Spatial Distribution of GIS-Based WQI

The WQI calculated for assessing Noida's groundwater quality. Calculated groundwater samples WQI values ranged from 47.10 to 192.10 (avg. 78.14). The highest value was observed at the groundwater samples collected at the Gulavali sector 162 and lowest WQI value was observed in Momnathal, sector 150. The WQI in the samples could be because of natural and anthropogenic activities. The GIS-based WQI analysis shows the groundwater samples, 9.9% excellent water, 74.5% of good water and 15.6% of poor water. These results reveal that the samples collected at the study area are moderately contaminated and inappropriate for direct use in drinking. It is advised to consider any treatment methods before utilizing it (Fig. 11; Table 4).

7 Conclusions

The GIS-based water quality index analysis reveals that the 84.4% of collected groundwater samples falls in excellent and good water category and 15.6% groundwater samples falls in poor category. The results revealed that water quality index varies from 47.10 to 192.10. The water quality index less than 50 is considered as excellent water, 50–100 is considered as good water, 100–200 is considered as poor water, 200–300 is considered as very poor water and WQI greater than 300 will be considered as unfit for drinking. The highest WQI values founded in high in GW19, GW27, GW31, GW34, GW36, GW43 and GW89 groundwater samples which lie in the poor water category. The study recommends some treatment considering for drinking purpose and locals in that area need to treat the water before usage. The GIS-based analysis suggests sewage treatment plants, industries for considering treatment of water before discharging into the water bodies. The study recommends continuous monitoring of groundwater quality, and implementation of methods and techniques for improving water quality. Further, it is advised to avoid water consumption from bore wells and hand pumps should be treated to avoid unnecessary health disorders.

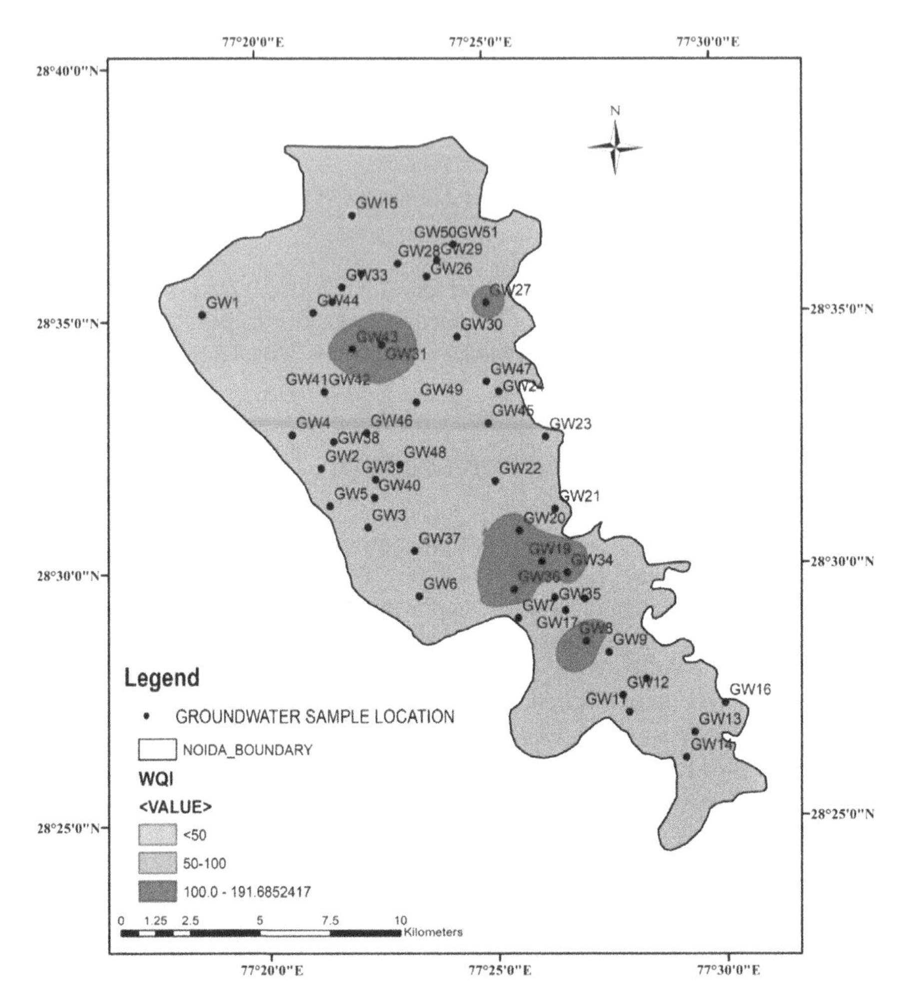

Fig. 11 Spatial distribution of GIS-based WQI

Predicting and mapping the pH value with the WQI gave us insight of the volatility of WQI on the pH, although "good water" in WQI index is also fit for usage but the pH if it is in limit contributes to the "Excellent water".

Table 4 Water quality index for groundwater samples

Location	Sam. No	Lat	Long	WQI values	Description
Pump No. H-1, Sector 15A	GW1	28.58	77.31	66.93783	Good water
Asagarpur Jagir Village, Sector 128	GW2	28.53	77.35	53.82815	Good water
Hindustan Petroleum, Near Jaypee Hospital	GW3	28.51	77.37	50.91777	Good water
Balaji Temple, Sector 126	GW4	28.54	77.34	61.53172	Good water
Ankit Nursery, Sector 131	GW5	28.51	77.35	58.09573	Good water
Green Beauty Farm, Sector 135	GW6	28.48	77.38	55.73013	Good water
Yakootpur, Sector 167	GW7	28.48	77.42	47.48599	Excellent Water
Gulavali, Sector 162	GW8	28.47	77.44	192.1043	Poor water
Jhatta Village, Sector 159	GW9	28.46	77.45	51.10198	Good water
Badauli, Sector 154	GW10	28.46	77.46	56.20924	Good water
Kambuxpur Derin Village, Sector 155	GW11	28.44	77.46	49.13407	Excellent Water
Gujjar Derin, Kambuxpur, Sector 155A	GW12	28.45	77.45	51.59635	Good water
Kondali Bangar, Sector 149	GW13	28.44	77.48	51.69048	Good water
Garhi Samastpur, Sector 150	GW14	28.43	77.47	63.92533	Good water
Momnathal, Sector 150	GW15	28.61	77.36	47.19809	Excellent Water
Shafipur Village, Sector 148	GW16	28.45	77.49	83.99748	Good water
Mohiyapur Village,Sector 163	GW17	28.48	77.43	55.67721	Good water
Nalgadha, Sector 145	GW18	28.48	77.44	60.49112	Good water
Ideal Industrial Training Institute, Sector 143	GW19	28.49	77.43	148.5878	Poor water
Shahdara, Sector 141	GW20	28.50	77.42	105.9785	Poor water
Hindon Flood Plain, Kulesara, Sector 140	GW21	28.51	77.43	60.00302	Good water
Illahabas, Sector 86	GW22	28.52	77.41	97.21823	Good water
Sai Dham Colony, Sector 88	GW23	28.54	77.43	52.57598	Good water
Kakrala Village, Sector 80	GW24	28.55	77.41	71.1939	Good water
Gijhor Village, Sector 53	GW25	28.59	77.36	73.28433	Good water
Sarfabad, Sector 73	GW26	28.59	77.39	58.56977	Good water
Sorkha Village, Sector 118	GW27	28.58	77.41	112.9936	Poor water
Pumping Station 3, Sector 71	GW28	28.59	77.38	65.4666	Good water
Pump House, Sector 122	GW29	28.59	77.39	68.28567	Good water
19, Block H,Sector 116	GW30	28.57	77.40	79.08585	Good water
Baraula Village, Sector 49	GW31	28.57	77.37	153.1029	Poor water
Pumping Station, Sector 35	GW32	28.58	77.35	65.09853	Good water

(continued)

Table 4 (continued)

Location	Sam. No	Lat	Long	WQI values	Description
Pumping Station 3, Sector 34	GW33	28.59	77.36	101.1179	Poor water
Peerbabaji, Sector 144	GW34	28.49	77.43	121.8443	Poor water
Dallupura Village, Sector 164	GW35	28.48	77.43	47.10549	Excellent Water
Dostpur, Mangrauli, Sector 167	GW36	28.49	77.42	164.1895	Poor water
Nangli Village, Sector 134	GW37	28.50	77.38	55.28453	Good water
Bakhtawarpur, Sector 127	GW38	28.53	77.35	60.53784	Good water
Sultanpur Village, sector 128	GW39	28.52	77.37	70.43717	Good water
Shahpur, Sector 131	GW40	28.52	77.37	79.41056	Good water
Sadarpur, Sector 45	GW41	28.55	77.35	77.74496	Good water
Chhalera, Sector 44	GW42	28.55	77.35	84.53748	Good water
Sanatan Temple, Sector 41	GW43	28.56	77.36	135.2111	Poor water
Shiv Mandir, Sector 31	GW44	28.58	77.35	67.55356	Good water
Nagla Charan Dass, Noida Phase-2	GW45	28.54	77.41	97.53002	Good water
Nursery, Sector 104	GW46	28.54	77.37	71.27371	Good water
Pumping Station, Sector 80	GW47	28.55	77.41	47.86249	Excellent Water
Shiv Mandir, Sector 93	GW48	28.53	77.38	101.6343	Poor water
Salarpur Village, Sector 102	GW49	28.55	77.38	80.97442	Good water
Garhi Chaukhandi, sector 121	GW50	28.60	77.39	74.07227	Good water
Pumping Station, Block-G, Sector 63	GW51	28.60	77.39	66.5801	Good water

Acknowledgements Authors are thankful to The JSS MAHAVIDYAPEETA, Management, Principal and Heads of Computer Science & Engineering and Civil Engineering Departments of JSS Academy of Technical Education, Noida for providing constant support and motivation and extend humble gratitude to all faculties and staff of both the departments.

Funding The authors thank Dr. A.P.J AKTU, Lucknow, for providing financial assistance through Collaborative Research and Innovation Program (CRIP).

References

1. Adimalla N, Li P, Venkatayogi S (2018) Hydrogeochemical evaluation of groundwater quality for drinking and irrigation purposes and inte- grated interpretation with water quality index studies. Environ Process 5(2):363–383
2. BIS (2012) Bureau of Indian Standards. New Delhi, 2–3
3. Bashir N, Saeed R, Afzaal M, Ahmad A, Muhammad N, Iqbal J, Khan A, Maqbool Y, Hameed S (2020) Water quality assessment of lower Jhelum canal in Pakistan by using geographic information system (GIS). Groundw Sustain Dev 10:100357

4. Batabyal AK, Chakraborty S (2015) Hydrogeochemistry and water quality index in the assessment of groundwater quality for drinking uses. Res Water Environ Res 87:607–617
5. Brindha K, Elango L (2012) Groundwater quality zonation in a shallow weathered rock aquifer using GIS. Geo-Spatial Inf Sci 15:95–104
6. Brindha K, Rajesh R, Murugan R, Elango L (2011) Fluoride contamination in groundwater in parts of Nalgonda District, Andhra Pradesh, India. Environ Monit Assess 172:481–492
7. Deepesh M, Madan KJ, Bimal CM (2011) GIS-based assessment and characterization of groundwater quality in a hard-rock hilly terrain of Western India. Environ Monit Assess 174:645–663. https://doi.org/10.1007/s10661-010-1485-5
8. Karanth KR (1987) Ground water assessment: development and management. Tata McGraw-Hill Education
9. Magesh NS, Krishnakumar S, Chandrasekar N, Soundranayagam JP (2013) Groundwater quality assessment using WQI and GIS tech-niques, Dindigul district, Tamil Nadu, India. Arab J Geosci 6:4179–4189
10. Munesh K, Rai SC (2020) Hydro geochemical evaluation of groundwater quality for drinking and irrigation purposes using water quality index in Semi-Arid Region of India. J Geol Soc India 95:159–168
11. Saleem M, Hussain and Gauhar Mahmood (2016) Analysis of groundwater quality using water quality index: a case study of greater Noida (Region), Uttar Pradesh (UP), India Civil & Environmental Engineering. Research Art Cogent Eng 3:1237927
12. Salehi S, Chizari M, Sadighi H, Bijani M (2018) Assessment of agricultural groundwater users in Iran: a cultural environmental bias. Hydrogeol Jour 26:285–295
13. Schot PP, Van der Wal J (1992) Human impact on regional groundwater composition through intervention in natural flow patterns and changes in land use. J Hydrol 134:297–313
14. Singh DF (1992) Studies on the water quality index of some major rivers of Pune Maharashtra. Proc Acad Environ Biol 1:61–66
15. Swati B, Mazhar Ali Khan M (2020) Assessment of ground water quality of Central and Southeast Districts of NCT. J Geol Soc India 95:95–103 Assessment of Ground Water Quality of Central and Southeast Districts of NCT of Delhi
16. Vaiphei SP, Kurakalva RM, Sahadevan DK (2020) Water quality index and GIS-based technique for assessment of groundwater quality in Wanaparthy watershed, Telangana India. PMID 32779065 https://doi.org/10.1007/s11356-020-10345-7
17. WHO (2008) World Health Organisation guidelines for drinking-water quality: second addendum. vol 1, Recommendations

An Artificial Neural Network Based Approach of Solar Radiation Estimation Using Location and Meteorological Details

Amar Choudhary, Deependra Pandey, and Saurabh Bhardwaj

Abstract This paper proposes a methodology to estimate monthly average global solar radiation using Artificial Neural Network. Due to abundancy, solar energy is at the cutting edge for a long time among other types of renewable energy. It has several applications in several fields. Solar energy devices completely depend on the amount of solar radiation that is to be received. So, optimization of solar energy is possible only when solar radiation is estimated well in advance. This is challenging since solar radiation is the location and seasonal-dependent. The current study addresses the issue of scarce meteorological stations which in turn limits the radiation measuring devices at the location of interest of the researchers. In this study, an ANN-based solar radiation estimation model is proposed using Levenberg–Marquardt training algorithm here. The study is performed on the six stations of Bihar, India. The Neural Fitting Tool (NF Tool) of MATLAB R2016a is used for simulation purposes. The data set is collected from the FAO, UN. The proposed model shows the R values of 0.9974, 0.97909, 0.90589, 0.9925, slope (m) values of 0.99, 1.0, 0.87 and 0.99, and intercept (c) values 0.0086, 0.021, 0.021 and 0.0066, for 'training', 'validation', 'testing', and 'all', respectively. The mean square error (MSE) is found to be 0.07727, 0.08092, and 0.08076 for 'training', 'validation', and 'testing', respectively.

Keywords Renewable energy · Solar radiation · Artificial neural network · Sustainable energy · Machine learning · Artificial intelligence

A. Choudhary (✉) · D. Pandey
Department of Electronics and Communication Engineering, Amity School of Engineering and Technology, Amity University, Lucknow, Uttar Pradesh 226010, India

D. Pandey
e-mail: dpandey@lko.amity.edu

S. Bhardwaj
Department of Electronics and Instrumentation Engineering, Thapar Institute of Engineering and Technology, Thapar University, Patiala, Punjab 147001, India
e-mail: saurabh.bhardwaj@thapar.edu

A. Choudhary et al. (eds.), *Applications of Artificial Intelligence and Machine Learning*, Lecture Notes in Electrical Engineering 778,
https://doi.org/10.1007/978-981-16-3067-5_15

1 Introduction

The well-known drawbacks of non-renewable energy resources, renewable energy resources are fastly becoming a point of attraction of policy makers. Renewable energy resources are non-pollutant and they are available in plenty. Among all types of renewable energy resources, solar energy attracts the systems developers due to large abundancy [1–3]. In the early twenty-first century, only low-power electronic gadgets were using the solar energy but nowadays researchers are looking towards its industrial utilization to meet huge electricity requirements. Abrupt fluctuations and inconsistency are the major challenges behind the optimization of solar energy [4]. The well-known fact is that solar energy depends upon site/location and seasonal changes which makes it difficult to estimate well in advance. Still, then large investments are being planned for the establishment of solar power plants, solar grid to establish solar energy as an alternative source of conventional energy. These efforts draw the attention of researchers towards the estimation of solar energy. Solar radiation is measured by the devices installed at meteorological stations which are scarce and it may not be the location of interest of the researchers [5]. In the early twenty-first century, solar radiation used to be estimated by mathematical models establishing a mathematical relationship between seasonal variables and solar radiation. But these models were very prone to errors. Since the last 20 years, computer-aided solar radiation estimation is mainly going on. Artificial Intelligence techniques such as ANN, CNN, Deep Neural Network, Fuzzy logic, SVM, etc. became the integral methodology of solar radiation estimation. Each technology has its advantages and disadvantages as well. Several researchers have proposed an ANN-based estimation model for the locations of India with the suitability of their model [6–9].

In this study, the ANN-based solar radiation estimation model is proposed. LM algorithm is used here since it fastly executes minimizing statistical errors. ANN technology is briefed in Sect. 2. Section 3 is for data collection and development of the estimation model. In Sects. 4 and 5 simulation results and conclusions are placed.

2 Artificial Neural Network Based Solar Radiation Estimation

The mathematical Architecture of ANN is shown in Fig. 1. It consists of input, output, weight, bias, transfer function, and activation function. The network output of Fig. 1 is determined by Eq. 1. Its parallel structure enables it for fast learning and computing from the data set [10] and ensures the fastest way to establishing a linear relationship between input and output. ANN has been widely used by a number of researchers in solar radiation estimation [11–19].

$$o_j = \sum_{i=1}^{n} x_i w_i + \phi\theta_i \tag{1}$$

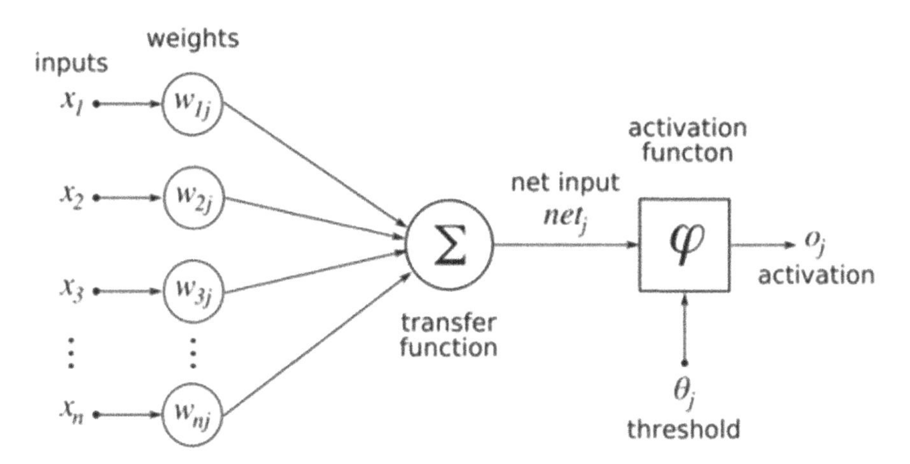

Fig. 1 Mathematical representation of ANN

Here,

O_j is the output of ANN, x_i denotes the input of ANN, w_i denotes the corresponding weights of ANN, θ_i denotes the threshold and $\emptyset$ denotes the activation function.

The neural network is trained before it is used for the estimation of data and testing for accuracy. In the next section development of the estimation model is briefed.

3 Solar Radiation Estimation Modeling

In the present analysis, stations of Bihar (second-most populous state of India) are considered. The stations under consideration are shown by red dots in Fig. 2. The location detail and meteorological data of all six available stations are obtained from the Food and Agriculture Organization (FAO), United Nations. CLIMWAT 2.0 and CROPWAT 8.0 are the two tools by which these data are collected.

Table 1 represents the location details of the stations while Table 2 represents annual averaged meteorological and solar radiation detail of all six available stations.

The 3D climatic plots of all the six stations are shown in Figs. 3, 4, 5, 6, 7 and 8. In those plots, monthly variations of Min Temp, Max Temp, Humidity, Wind speed, Sunshine hours, and Solar Radiation are shown by blue, orange, violet, green, yellow, and pink color respectively. From the plain observation of plots, it is clear that the seasonal/monthly variation of solar radiation is almost following the trend of sunshine hours.

The location parameters (latitude, longitude, and altitude) and meteorological parameters (temperature, humidity, wind speed, sunshine hour, and solar radiation)

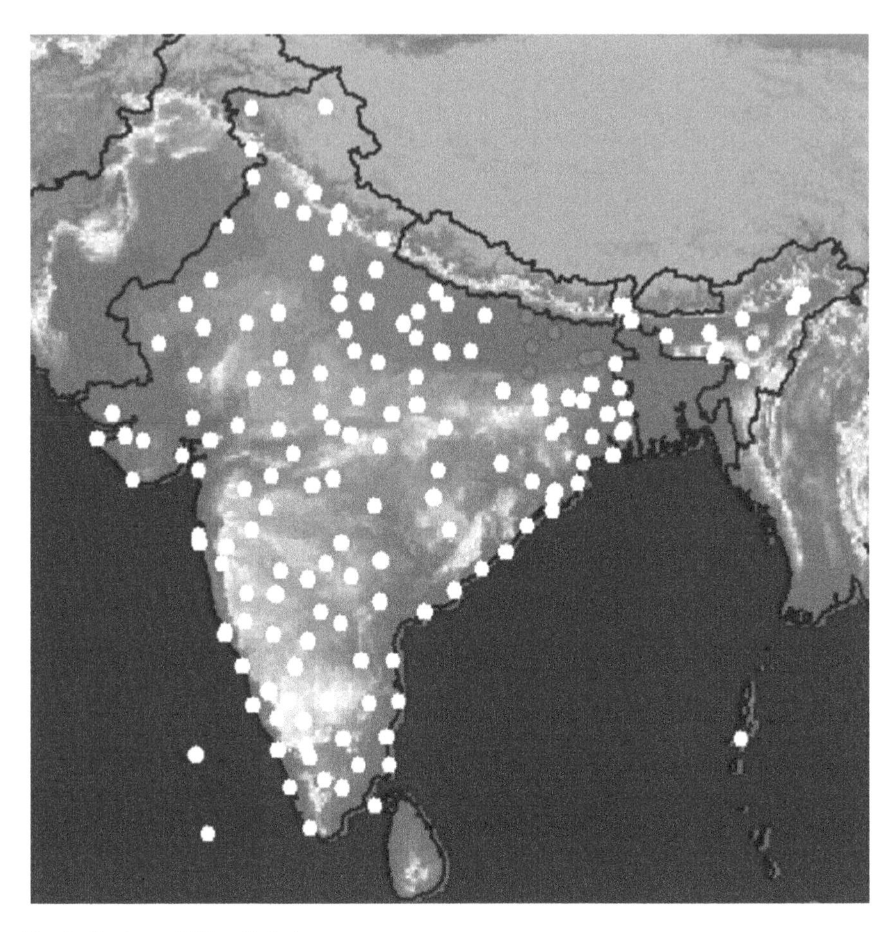

Fig. 2 Stations of Bihar (India)

Table 1 Geographical details of stations under observation

Stations	Latitude (°N)	Longitude (°E)	Altitude (m)
Darbhanga	26.16	85.90	49
Gaya	24.75	84.95	116
Motihari	26.66	84.91	66
Patna	25.60	85.10	60
Purnea	25.26	87.46	38
Sabaur	25.23	87.06	37

Table 2 Annual average meteorological details and solar radiation data of stations

Stations	Maximum temp. (°C)	Minimum temp. (°C)	Humidity (%)	Wind speed (km/day)	Sunshine (h)	Solar radiation (MJ/m^2/day)
Darbhanga	18.2	30.7	68	71	7.4	17.9
Gaya	20.8	32.4	65	204	8.5	19.7
Motihari	18.8	30.6	72	63	7.7	18.2
Patna	20.8	31.2	67	110	7.6	18.2
Purnea	19	30.8	69	78	7.1	17.7
Sabaur	19.2	31.1	68	129	6.6	16.9

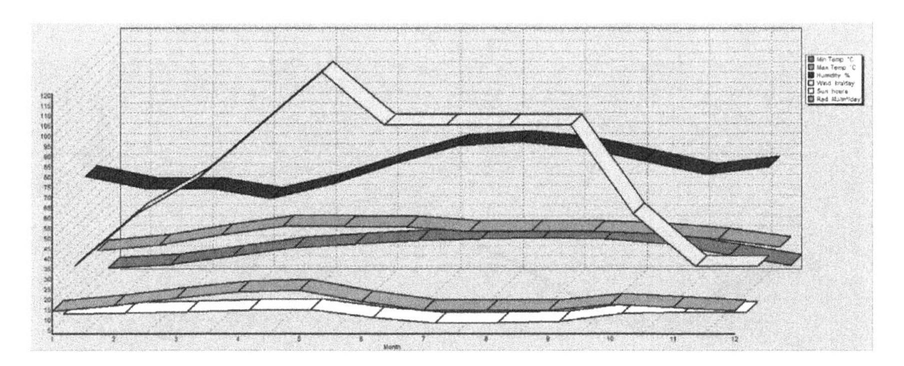

Fig. 3 Climatic plot of Darbhanga

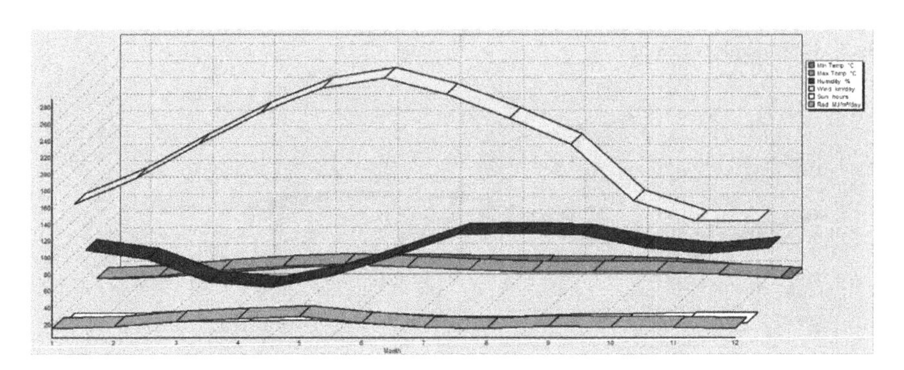

Fig. 4 Climatic plot of Gaya

of Table 2 have different measuring units and scales. These data are normalized before giving input to the model. Here, max–min normalization is executed as per Eq. (2):

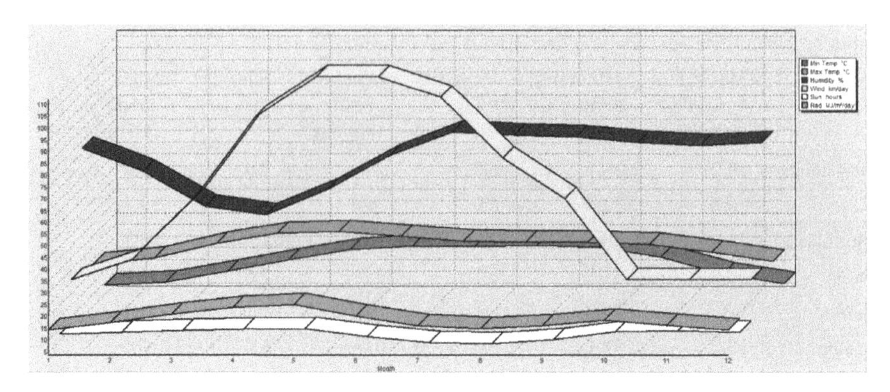

Fig. 5 Climatic plot of Motihari

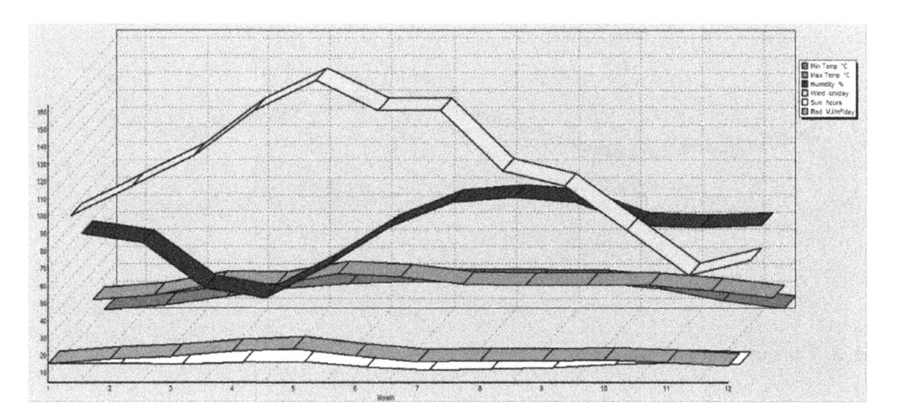

Fig. 6 Climatic plot of Patna

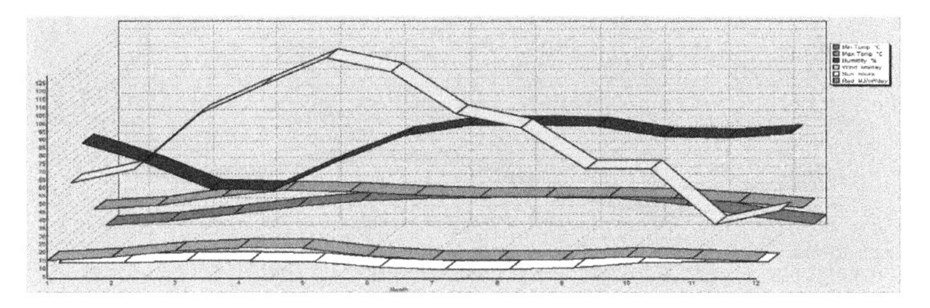

Fig. 7 Climatic plot of Purnea

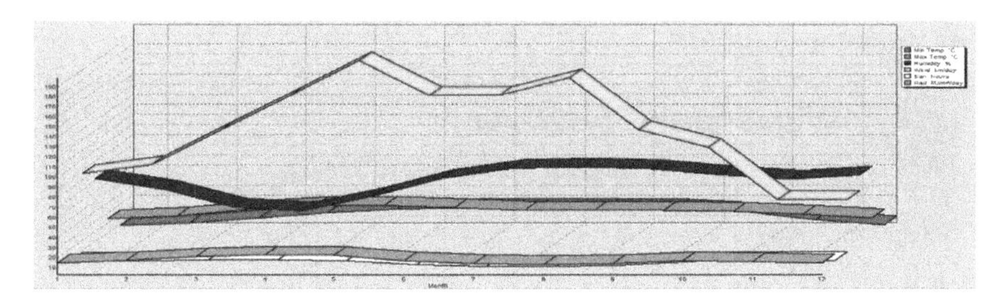

Fig. 8 Climatic plot of Sabaur

Table 3 Customization of neural fitting tool

S. No.	Particulars	Configuration details
1	Network type	Feed forward back propagation
2	Training algorithm	TRAINLM
3	Error function	MSE
4	Number of hidden layers	02
5	Transfer function	TANSIG
6	No. of neurons	10
7	Training parameters	Epochs: 1000, max_fail: 6
8	Data division	Random (dividerand)
9	Performance	Mean squared error (MSE)
10	Calculation	MEX
11	Plot Interval	1 epochs

$$x_{\text{normalised}} = \frac{x - x_{\text{minimum}}}{(x_{\text{maximum}} - x_{\text{minimum}})} \tag{2}$$

Once data is normalized, simulation starts using the Neural Fitting Tool (NF Tool) of MATLAB R2016a. The NF Tool is customized as per Table 3. The network architecture is shown in Fig. 9. Here 9 inputs are applied with 10 hidden layers and

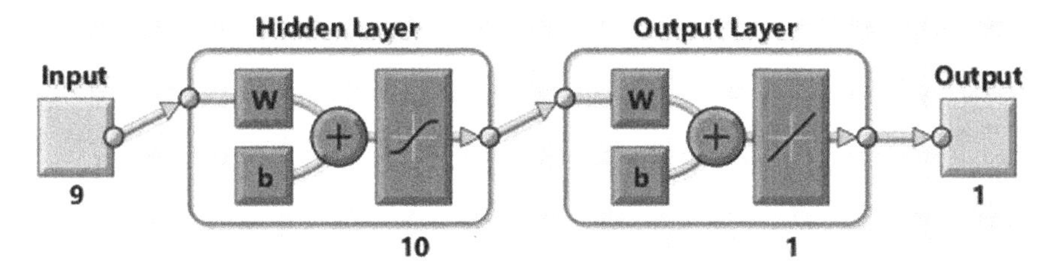

Fig. 9 ANN architecture

Fig. 10 Data division

1 output layer. The training algorithms used in this proposed model is the Levenberg–Marquardt (LM) algorithm. LM is assumed to be the fastest back propagation algorithm.

The normalized data set are divided into three parts as per the default value of NF Tool as shown in Fig. 10. Out of 72 samples, 70% (50 samples) are used for training, 15% (11 samples) for validation, and the rest 15% (11 samples) for testing the neural network.

Once the simulation is performed, the output is obtained. After this, Mean Square Errors (MSE) are found as per Eq. 3. These errors indicate the suitability of the proposed model.

$$\text{MSE} = \left[\left(\frac{1}{n}\right)\sum_{i=1}^{n}\left(SR_{i\,(\text{predicted})} - SR_{i\,(\text{estiamted})}\right)^2\right] \tag{3}$$

Here, n is the number of input, SR is the solar radiation.

4 Results

The training environment using the LM algorithm is shown in Fig. 11. The best network performance is obtained after 9 iterations and 5 validation checks.

The performance of the network is shown in Fig. 12. In this figure, MSE decreases as the epoch increases. The best validation performance of 0.0016254 is obtained at epoch 4.

Figure 13 is for gradient plot in which Gradient of 0.00154, Mu of 0.000006, and Validation Checks of 5 are obtained at epochs 09.

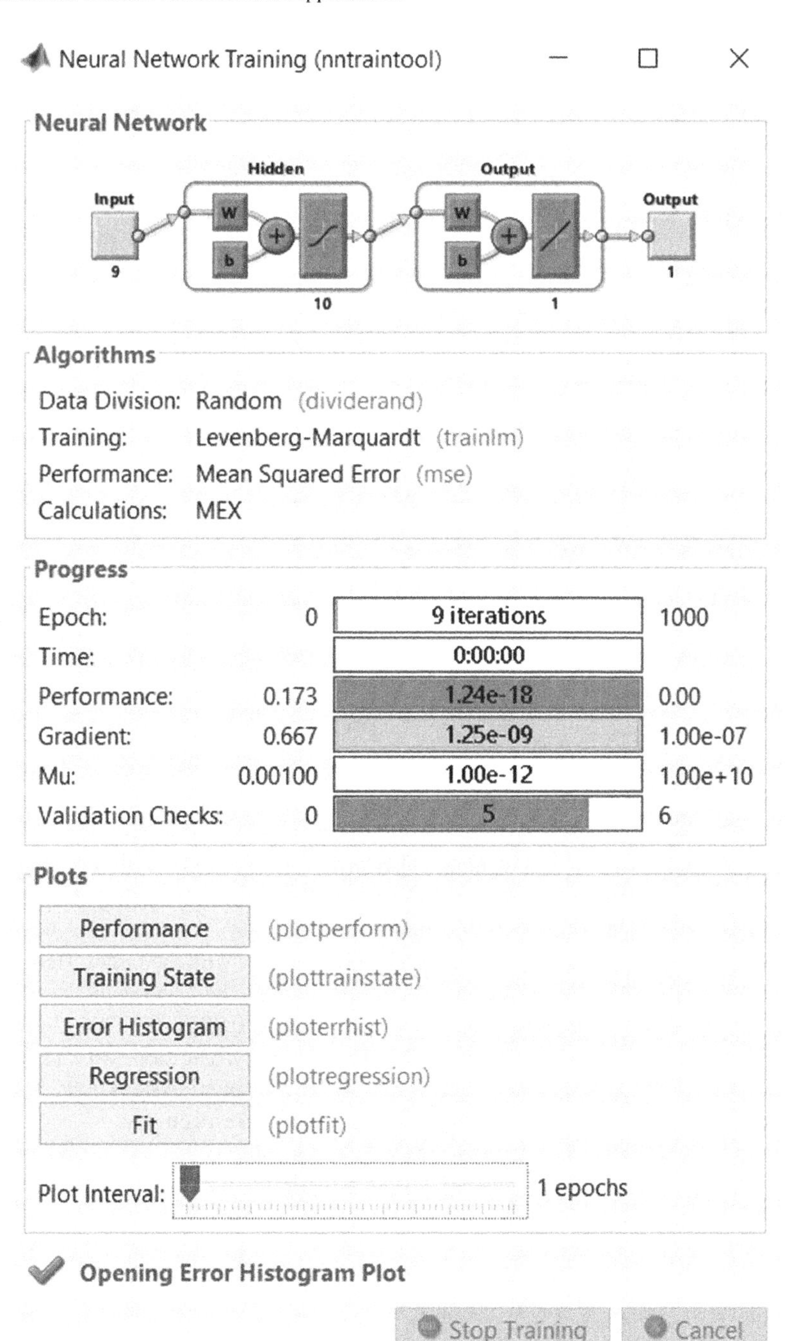

Fig. 11 Training environment

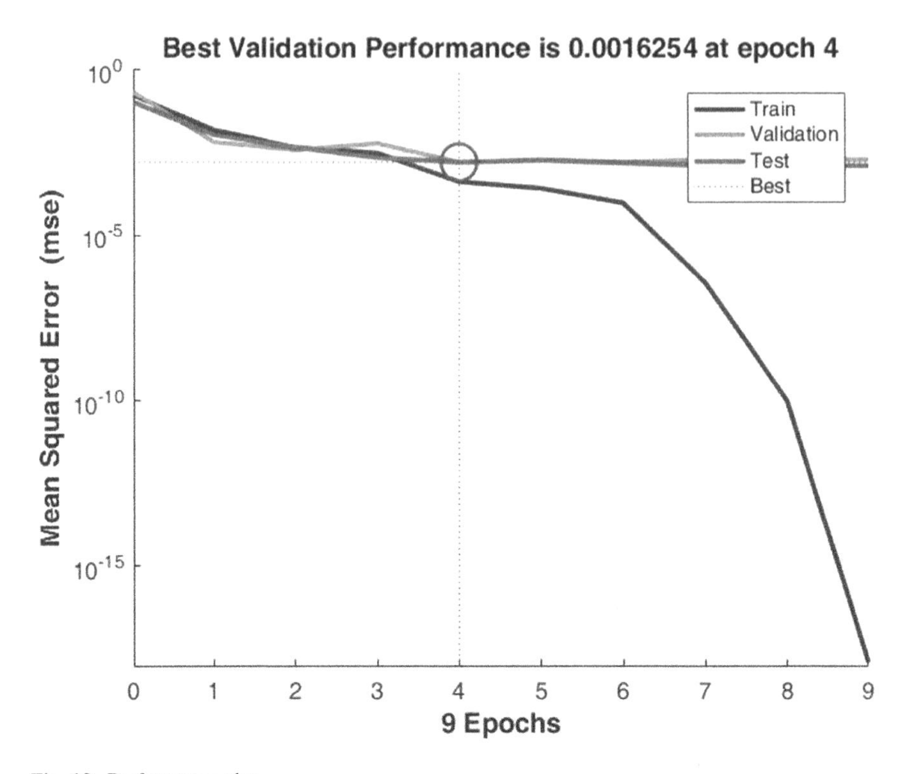

Fig. 12 Performance plot

The plot between error and instances (Error Histogram) is shown in Fig. 14. It is a plot for training data. It has 20 Bins. Here most of the data are below Zero Error line which supports the quality of the simulation.

The regression value determines the variation of output with the target. Here, it is represented in Fig. 15. The best value of R is considered to be 1 and in this case, it is almost obtained. The regression value (R) obtained to be 0.9974, 0.97909, 0.90589, 0.9925 for training, validation, testing, and all respectively. The slope (m) are found to be 0.99, 1.0, 0.87 and 0.99 whereas intercept (c) are found to be 0.0086, 0.021, 0.021 and 0.0066 for training, validation, testing and all, respectively.

Figure 16 is for Mean Square Errors (MSE) during training, validation, and testing and it is found to be 0.07727, 0.08092, 0.08076 respectively. These values are quite satisfactory.

Comparing the obtained results with methodologies such as CVNN, ECWN, and CWN [20], the proposed model shows better accuracy and other values.

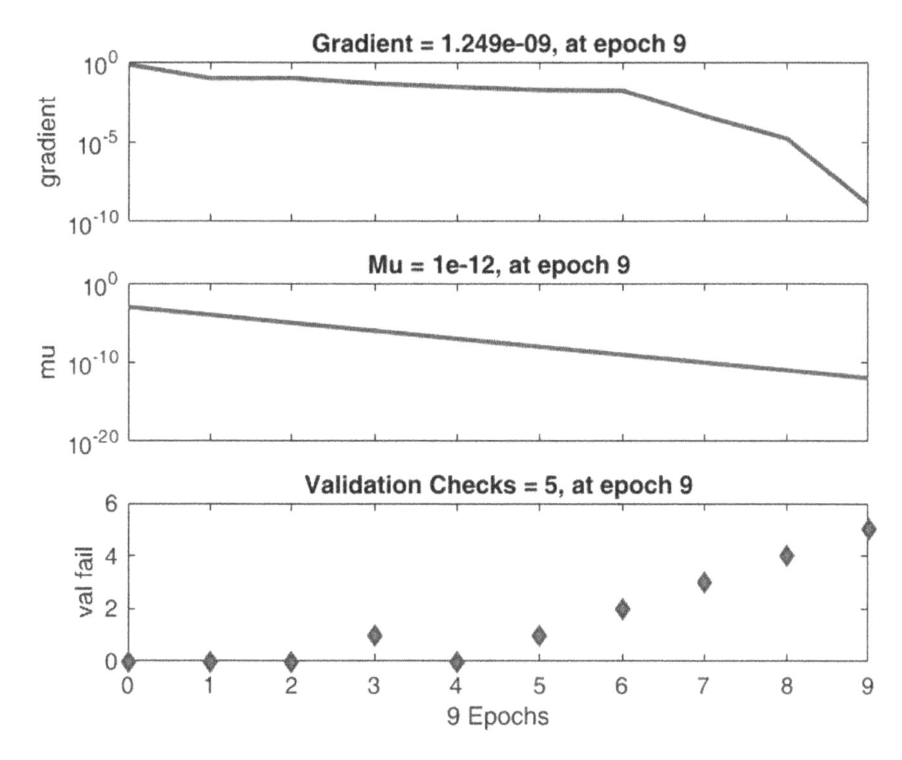

Fig. 13 Gradient plot

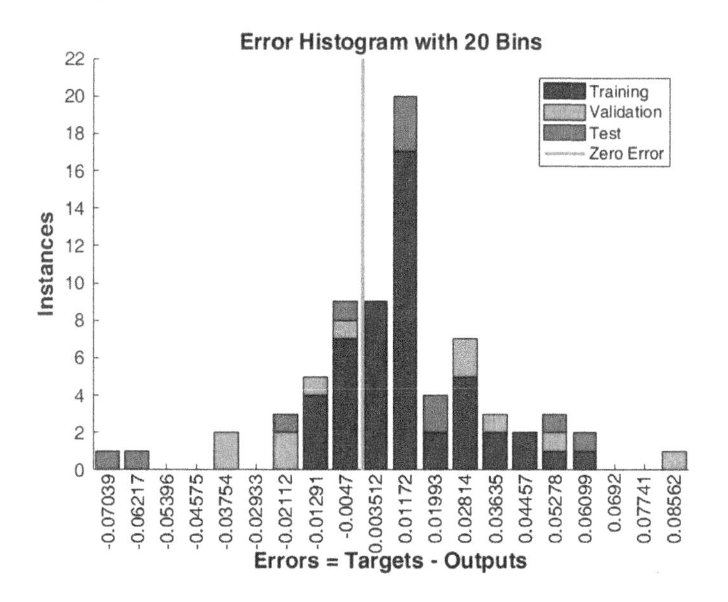

Fig. 14 Error histogram

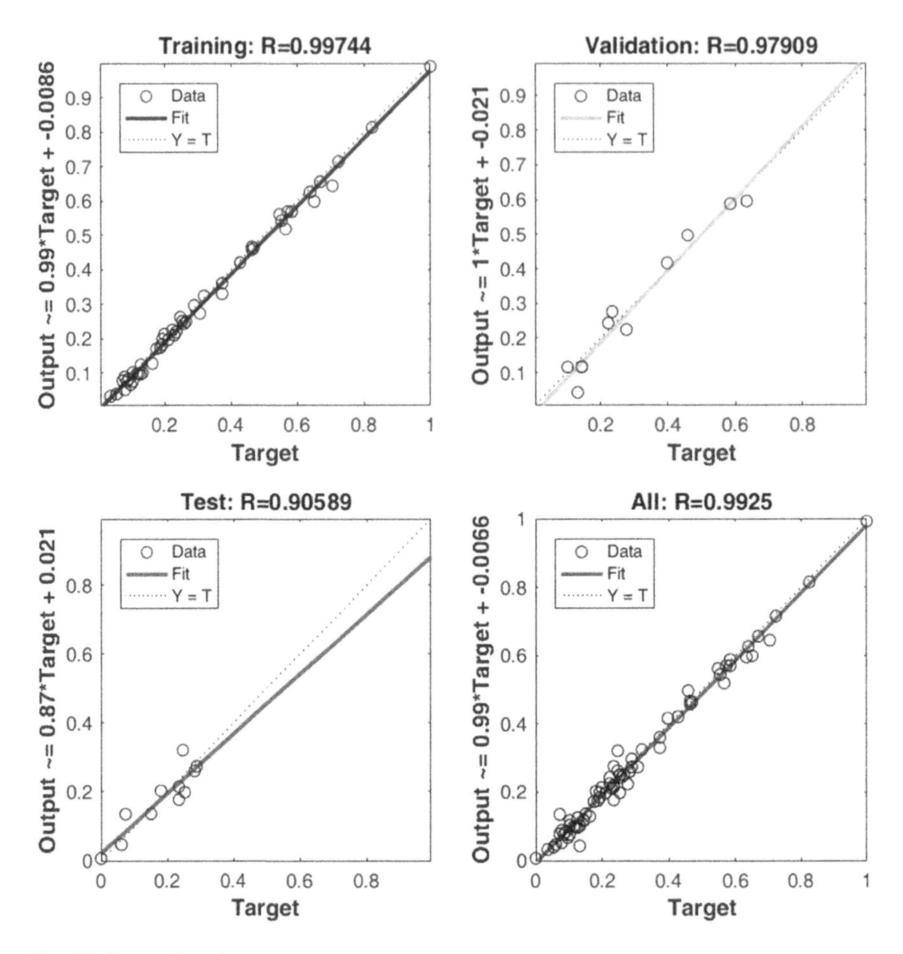

Fig. 15 Regression plot

Fig. 16 Obtained MSE

Results

	Samples	MSE
Training:	50	4.21909e-4
Validation:	11	1.62535e-3
Testing:	11	1.62223e-3

5 Results

This paper briefly gives the idea and need for solar radiation estimation. The proposed models based on ANN for solar radiation estimation are briefed in this paper. The model is proposed with 09 input parameters and it is tested for 06 stations of Bihar, India. The model gives significantly better R values, minimum MSE, better values of slope (m), better intercept values (c), and better performance. The LM algorithm again shows good performance. This model may be used to estimate solar radiation in remote areas where solar radiation measuring devices are scarce. As stations of plain area are used in this simulation and as the estimation of solar radiation is location dependent so, this model may not perform very well in high hilly areas. In the future, improvement may be executed in this regard.

References

1. Deependra P, Amar C (2018) A review of potential, generation and factors of solar energy. J Thermal Eng Appl STM J 5(3). ISSN: 2349–8994 (Online)
2. Kapoor K, Pandey KK, Jain AK, Nandan A (2014) Evolution of solar energy in India: a review. Renew Sustain Rev 40:475–485
3. Jemaa ABEN, Rafa S, Essounbouli N, Hamzaoui A, Hnaien F, Yalaoui F (2013) Estimation of global solar radiation using three simple methods. ScienceDirect, Energy Proc 42:406–415
4. Goh SL, Chen M, Popović DH, Aihara K, Obradovic D, Mandic DP (2006) Complex-valued forecasting of wind profile. Renew Energy 31(11):1733–1750. https://doi.org/10.1016/j.renene.2005.07.006
5. Mubiru J, Banda EJKB (2008) Estimation of monthly average daily global solar irradiation using ANN. ScienceDirect (Elsevier) Sol Energy 82:181–187
6. Amit YK, Chandel SS (2012) Artificial neural network based prediction of solar radiation for Indian stations. Int J Comput Appl 50(9):0975–8887
7. Hasmat M, Sidhharth G (2019) Long-term solar irradiance forecast using artificial neural network: application for performance prediction of Indian cities. Adv Intell Syst Comput 697. https://doi.org/10.1007/978-981-13-18221-1_26
8. Abhishek K, Ravi K (2019) Solar energy prediction using backpropagation in artificial neural networks. In: International conference on advanced computing networking and informatics, advances in intelligent systems and computing, vol 870. https://doi.org/10.1007/978-981-13-2673-8_4
9. Deependra P, Amar C, Suarabh B (2020) Overview of solar radiation estimation techniques with development of solar radiation model using artificial neural network. Adv Sci Technol Eng Syst J 5(4):589–593. https://doi.org/10.25046/aj050469
10. Nitta T (2009) Complex-valued neural networks: utilizing high-dimensional parameters
11. Benghanem M, Mellit A, Alamri SN (2009) ANN-based modeling and estimation of daily global solar radiation data: a case study. Energy Convers Manage 50:1644–1655
12. Amit YK, Chandel SS (2013) Solar radiation prediction using artificial neural network techniques: a review. Renew Sustain Energy Rev. https://doi.org/10.1016/j.rser.2013.08.055
13. Rajesh K, Aggarwal RK, Sharma JD (2012) Solar radiation estimation using artificial neural network: a review. Asian J Contemp Sci 1:12–17
14. Choudhary A, Pandey D, Kumar A (2019) A review of various techniques for solar radiation estimation. In: 3rd international conference on recent developments in control, automation and power engineering (RDCAPE), NOIDA, India, pp 169–174
15. Muammer O, Mehmet B, Besir S (2012) Estimation of global solar radiation using ANN over Turkey. Expert Syst Appl 39(5):5043–5051
16. Sfetsos A, Coonick AH (2000) Univariate and multivariate forecasting of hourly solar radiation with artificial intelligence techniques. Sol Energy 68(2):169–178

17. Behrang MA, Assareh E, Ghanbarzadeh A, Noghrehabadi AR (2010) The potential of different artificial neural network (ANN) techniques in daily global solar radiation modeling based on meteorological data. Sol Energy 84(8):1468–1480
18. Cao J, Xingchun L (2008) Study of hourly and daily solar irradiation forecast using diagonal recurrent wavelet neural networks. Energy Convers Manage 49(6):1396–1406
19. Benghanem M, Mellit A, Alamria SN (2009) ANN-based modeling and estimation of daily global solar radiation data: a case study. Energy Convers Manage 50(7):1644–1655
20. Saad Saoud L, Rahmoune F, Tourtchine V, Baddari K Fully complex valued wavelet neural network for forecasting the global solar radiation

Applications of Machine Learning and Artificial Intelligence in Intelligent Transportation System: A Review

Divya Gangwani and Pranav Gangwani

Abstract Due to the tremendous population growth in the country, the use of vehicles and other transportation means has increased which has led to traffic congestion and road accidents. Hence, there is a demand for intelligent transportation systems in the country that can provide safe and reliable transportation while maintaining environmental conditions such as pollution, CO2 emission, and energy consumption. This paper focuses on providing an overview and applications of how Artificial intelligence (AI) and Machine Learning (ML) can be applied to develop an Intelligent Transportation system that can address the issues of traffic congestion and road safety to prevent accidents. We will then re-view various ML approaches to detect road anomalies for avoiding obstacles, predict real-time traffic flow to achieve smart and efficient transportation, detect and prevent road accidents to ensure safety, using smart city lights to save energy, and smart infrastructure to achieve efficient transportation. Next, we review various AI approaches such as safety and emergency management system to provide safety to the public, autonomous vehicles to provide economical and reliable transportation. We then propose smart parking management and how it can be used to find parking spaces or spots conveniently, incident detection which detects the traffic incidents or accidents in real-time provides a report. Finally, we conclude with predictive models and how the algorithms utilize sensor data to develop an Intelligent Transportation System.

Keywords Artificial intelligence · Machine learning · Smart transportation · Intelligent transportation system

D. Gangwani (✉)
Florida Atlantic University, Boca Raton, FL 33431, USA
e-mail: dgangwani2017@fau.edu

P. Gangwani
Florida International University, Miami, FL 33174, USA
e-mail: pgang002@fiu.edu

A. Choudhary et al. (eds.), *Applications of Artificial Intelligence and Machine Learning*, Lecture Notes in Electrical Engineering 778,
https://doi.org/10.1007/978-981-16-3067-5_16

1 Introduction

Artificial Intelligence (AI), also called machine intelligence, is a wide-ranging branch of computer science that makes machines function like a human brain. It solves various problems that are difficult to address using conventional computation techniques. AI research was first conducted at Dartmouth College at a workshop in 1956. John McCarthy coined the term "Artificial Intelligence" to differentiate the area from cybernetics, which is the science of automatic control systems and communication in both living things and machines. By the middle of the 1960s, the Department of Defense heavily funded AI research in the U.S and established laboratories all around the world [1]. In the 1980s the development of practical Artificial Neural Network (ANN) technology was enabled due to the development of metal-oxide-semiconductor (MOS) very-large-scale integration (VLSI), in the form of complementary MOS (CMOS) transistor technology. AI began to be used in many areas such as medical diagnosis, logistics, data mining, and other areas in the late 1990s and early twenty-first century [2].

Access to a huge volume of data, faster computers, and algorithmic improvements enabled advances in perception and Machine Learning (ML). In 2012, deep learning algorithms started to dominate accuracy benchmarks [3]. Architectures of deep learning including deep belief networks, convolutional neural networks, deep neural networks, and recurrent neural networks have been applied to various application areas. These include speech recognition, audio recognition, computer vision, natural language processing, machine vision, and many more [4].

The most distinguished AI method is ANN, which is used in various applications. ANNs consist of artificial neurons which are a collection of nodes or connected units that loosely resemble the biological brain's neurons. Each connection can transmit a signal to other neurons just like the synapses in a biological brain. The received signal is then processed by the artificial neuron and signals the neurons connected to it. The output of each neuron is the "signal" at a connection, which is a real number and is computed by some non-linear function of the sum of inputs.

Transportation difficulties can become a major challenge especially when the network and users' activities are too difficult to predict and model the patterns in travel. Thus, to overcome the challenges of increasing travel demand, environmental degradation, safety concerns, and CO_2 emissions, AI, and ML are deemed to be a perfect fit for transportation systems. In developing countries, the steady growth of urban and rural traffic due to the increasing population is the main cause of these challenges. In Australia, by 2031, the population is expected to increase to 30 million, and therefore the cost of congestion is expected to reach 53.3 billion [5]. In the twenty-first century, several researchers are attempting to achieve a smarter and reliable transportation system or in other words, Intelligent Transportation Systems (ITS). ITS will have less detrimental effects on the environment, and people will be using AI and ML techniques that are more reliable and cost-effective [6].

This paper is structured as follows: Sect. 2 elaborates on the various Applications of ML in ITS. This section is subdivided into Road Anomaly Detection, Traffic

Flow Detection and Travel Time Prediction, Accident Detection and Prevention, Smart City Lights, and City Infrastructure. Section 3 discusses the Applications of AI in Smart Transportation. This section is subdivided into Safety and Emergency Management System, Autonomous Vehicles, Smart Parking Management, Incident Detection, and Predictive Models. Section 4 discusses the Challenges of AI and ML in ITS. Section 5 discusses Future Trends in Intelligent Transportation Systems. Finally, Sect. 6 concludes this paper review.

2 Applications of ML in ITS

In this section, we cover the major challenges faced in ITS and ML techniques used to solve the problems faced in the transportation system. The applications of ITS are quite broad, but we focus on a few of the important applications and challenges faced in ITS which are crucial for societal development. Table 1 covers all the applications of ITS and lists all the algorithms used in AI and ML.

2.1 Road Anomaly Detection

The condition of the road makes a huge impact on the traffic and can lead to traffic accidents, delays, and cause damage to the vehicles. Hence, road anomaly detection plays an important role in the development of ITS. The main objective of the road anomaly detection system is to detect bumps and potholes on the road and notify

Table 1 ITS applications in AI and ML

S. No.	ITS applications	AI and ML algorithms	References
1	Road anomalies detection	Support vector machine (SVM), Naïve Bayes, K-means clustering	[7–9]
2	Traffic flow detection and travel time prediction	k-Nearest Neighbor (k-NN), support vector regression (SVR), SVM, long short-term memory (LSTM)	[10–14]
3	Accident detection and prevention	Classification and regression tree (CART), k-NN, SVM	[15, 16]
4	Smart city lights		[17–20]
5	City infrastructure	CNN	[21–23]
6	Safety and emergency management system	ANN	[25–27]
7	Autonomous vehicles	CNN, SVM, K-means clustering, k-NN	[28–31]
8	Smart parking management	Genetic algorithm (GA)	[32–35]

the driver [7] so that the vehicle can be controlled and prevented from any damages. Various ML techniques can be used to detect road anomalies.

Road surface monitoring like detecting potholes can be done by using smart mobile devices equipped with GPS and accelerometer, which is used to collect acceleration data [8]. This data can then be used to analyze the road condition for future use. Supervised and unsupervised ML techniques are used to detect road anomalies. Support vector machines (SVM) and Naïve Bayes are the most common ML techniques used to detect and label road conditions such as 'Smooth' or 'Pothole'. Authors in [9] used k-means clustering and SVM to detect road anomalies and labeled the road conditions as 'Smooth', 'Bumps', and 'Potholes'. These ML techniques are capable of detecting road anomalies and make it easy not only for the drivers to avoid the route but also for the government to take necessary action to rectify the road condition.

2.2 Traffic Flow Detection and Travel Time Prediction

It is important to detect traffic flow information in an accurate and timely manner [10]. Over the years, with the increasing population, traffic flow has also increased. This leads to road accidents and delays in arrival time. ML techniques have proved to be capable of solving traffic flow patterns and have contributed to the development of ITS [11]. ML approaches such as k-Nearest Neighbors (k-NN) [12] and Support Vector Regression (SVR) is used to address traffic flow detection problems. Other ML approaches like SVR and SVM are also used to predict the travel time of the road segment.

SVR is an adaptation of the SVM algorithm which is used for regression problems like detecting traffic flow and predicting speed at randomly selected roads [13]. SVR has proven to be successful in solving regression problems and "overcomes the shortcoming of traditional ML algorithms". Hence, SVR has been successfully applied in traffic flow problems. Travel time detection of road segments is also an important contribution when developing an intelligent transportation system. Public transportation such as trains and busses can be utilized effectively if there is an efficient system in place to detect traffic flow and travel time estimation. Several ML techniques such as SVR, SVM, and deep learning methods such as Long Short-Term Memory (LSTM) [14] have been applied in the area of travel time prediction. Deep Learning methods have attracted a lot of attention nowadays as its methods are showing tremendous results in the transportation network.

2.3 Accident Detection and Prevention

Prevention and detection of accidents in a road segment are crucial for the development of a smart transportation system. Creating a smart system to prevent accidents can save human lives. Drivers can avoid common mistakes and concentrate on the road to prevent accidents from happening. An accident prevention system can notify the driver of any critical scenarios and can alert the driver to act on time and avoid accidents. Similarly, prompt detection of accidents on the road can help the driver to avoid taking that particular road and give an update regarding the time delay expected to reach a specific location. ML algorithms have been used extensively in the field of accident detection and prevention. ML techniques can detect road accidents and identify any pattern which can help avoid new accidents from happening.

Nearest neighbor is one of the most common classification techniques that can be used for pattern recognition and classification problems. Classification and Regression Tree (CART) is another classifier that uses entropy or class variance information to detect such ML problems [15]. k-NN and SVM classifiers have been used in combination to detect traffic accidents. A deep learning study has been conducted in the development of accident prevention systems. "An Inception Neural Network was used in [16] to detect accident-prone areas".

2.4 *Smart City Lights*

To develop a smart transportation system, it is beneficial to implement smart streetlights for the city. Major energy expenditure comes from the streetlights being ON the whole night [17]. With the implementation of smart city lights, energy consumption can be reduced dramatically. Smart city lights are implemented based on IoT infrastructure. Smart sensors, GPS, and wireless communication is used to control the mechanism of streetlights. These sensors help in detecting motion which automatically controls the LED lights to be ON or OF depending upon the street utilization at night [18]. The GPS is used as a centralized system to monitor the lights and capture its location and city information so that the maintenance can be easily managed and repair any damages to the lights.

Another approach which the authors implemented in [19] is to use a Raspberry Pi as a microcontroller that controls the lights connected to Infrared sensors. These sensors detect sunrise and sunset and automatically switch the lights ON and OF accordingly.

In [20] the authors proposed a technique in which each city light is used as a Wi-Fi hotspot which helps in transferring information about the number of vehicles passing by the street and severe weather conditions. The light post will be equipped with cameras and sensors to detect changes in the weather condition, damaged vehicles stranded on the street, and detect vehicles and pedestrians passing by the street. All this information will be beneficial in controlling the movement of city lights so that energy conservation can be done efficiently and help in the faster maintenance and development of smart cities.

2.5 City Infrastructure

A smart city transportation system also requires having a smart infrastructure in place. This will benefit modern transportation in many ways. Changes in the infrastructure provide a huge impact on the public transportation system and provide efficient ways of communication among different vehicles. In [21] the authors proposed a method in which vehicles can communicate with each other and can get notified about their speed, location, and other travel information. This vehicle-to-vehicle communication is done by using a GPS device to detect the vehicle's position and speed and other information. This information is uploaded to a server which is then used to alert the driver about the nearby vehicle information. This will help in avoiding accidents and collisions by notifying the drivers about other vehicle speed, movement, and even traffic congestion beforehand [22]. Another approach that requires ML algorithms can be used to control vehicle movement and steering speed [23]. A Convolutional neural network (CNN) is used to control the steering speed and movement of the vehicles for the development of smart infrastructure.

3 Applications of AI in Smart Transportation

This section provides applications of AI in the Intelligent Transportation Systems. The need for ITS is rising linearly with the increase in the population. Hence, the transportation system should be safer, faster, reliable, environment friendly, and cost-effective [24]. Figure 1 shows the overall structure of ITS applications. We will cover a few of the important applications of AI which will be beneficial for the development of ITS.

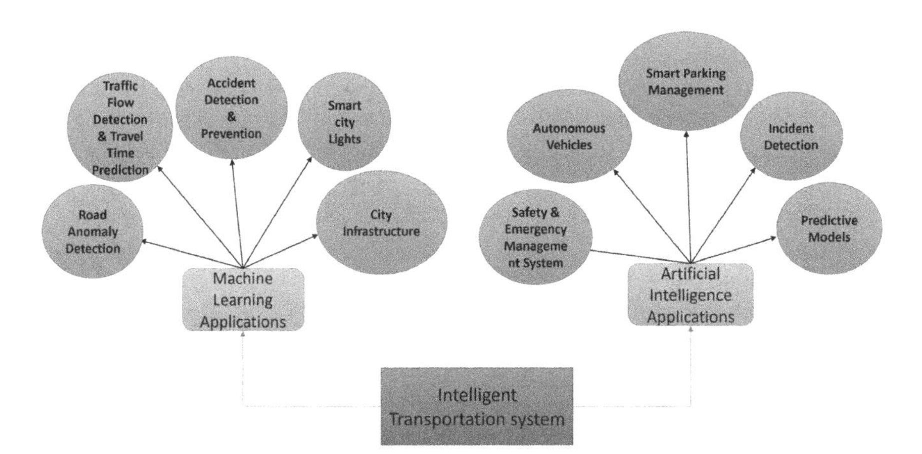

Fig. 1 Block diagram of AI and ML applications

3.1 Safety and Emergency Management System

With the number of accidents on the road, there is always a risk to the people who are traveling as well as the pedestrians on the street. Even with the latest technology in vehicles, safety is a major concern in the development of ITS [25]. Artificial Intelligence plays an important role in the safety and emergency management of smart transportation. AI techniques are used to develop a dependable and comprehensive database of accident data that can provide necessary information for trends and analysis of traffic accidents. ANN is used for road planning and safety management. This information and analysis can then be used to provide safe mobility to people. AI can also provide fast and effective development in the traffic analysis which can be useful in emergencies and provide safety to the people and protect their health [26]. AI is also useful in providing an emergency management system to people which can ensure safe and faster travel from one destination to another [27].

3.2 Autonomous Vehicles

Advancement in AI has given a rise to the introduction of self-driving cars for people who can take advantage of this latest technology. Autonomous vehicles or self-driving cars use a combination of the latest technologies like a sensor, camera, radar, and AI techniques to move around from one location to another [28]. Autonomous vehicles require a lot of trust from the people since there is always a question of safety involved with these vehicles. Therefore, a combination of ML, deep learning, and AI techniques need to prove the safety and reliability of such vehicles. The main algorithm used for self-driving cars is CNN, which is a building block for object detection in a self-driving car database [29]. "This technology works by teaching the vehicle how to drive while maintaining safety and lane discipline" [30]. Several ML algorithms like k-means clustering, SVM, and k-NN are also used in autonomous vehicles. With the use of self-driving cars, major changes have been noticed in the transportation system and their impact on traffic safety and congestion due to their prediction by AI. These cars can tremendously reduce traffic congestion by providing options for ride-sharing [31]. Ridesharing options can be used by many private businesses and provide safe travel to people. This technique can also reduce energy consumption along with providing safety to people.

3.3 Smart Parking Management

AI algorithms have benefited in the development of smart parking systems. A smart parking management system is crucial for the growth of smart cities. Many universities, colleges, and public places face challenges in organizing a systematic parking system so that people don't waste their time to find a parking space. As a result, a well-structured system is needed which can detect available parking spots and notify the driver beforehand. A smart parking management system uses AI technology to detect available parking spaces as well as notify the driver and provide a status update about the availability of a parking space [32]. The most used AI algorithm called Genetic Algorithm (GA) is based on the biological evolution concept. This AI algorithm has now been used to solve many transportation problems of optimization.GA can be used in urban design problems that require optimum utilization of space to create a smart parking management system [33]. Most of the smart cities are equipped with parking lot monitoring sensors which can assist the drivers to find an available parking location [34]. Nowadays with the development in smart cities, multistory parking lots are equipped with digital sensors that can be utilized and managed productively. Drivers can get an online notification about which parking floor is empty and can be directed towards that particular floor. This can help save time and energy and reduce carbon emissions by minimizing the number of drivers searching for parking spaces [35]. With the advancement in these technologies, it is possible to develop an efficient and productive transportation system that will help make traveling safe, smooth, and faster for the public.

3.4 Incident Detection

There have been many attempts to identify, the location, time, and severity of an incident to support professional traffic managers which work to mitigate congestion. All these attempts range from traditional reports to automated algorithms to neural networks. Humans write these traditional reports which can create a delay in detecting traffic incidents. On the other hand, flow characteristics can be measured by algorithms before and after the incident by using the data collected from the sensors along the road. Incident detection algorithms were first implemented using statistical techniques, for instance, California Algorithm. However, in the case of arterial roads, it is difficult to use an algorithm because of the traffic signals and parking on the street. Hence, neural networks have been developed to overcome these challenges. An ANN algorithm was evaluated which performed classification to detect the occurrence of an incident on a freeway [36]. Another research focused on finding a suitable package of the software which detects all objects of vehicles on a real-time basis [37]. The AdaBoost software package was proposed in this research for accurate image detection.

3.5 Predictive Models

ITS is rapidly developing and has emphasized the need to predict traffic information using advanced methods. For the success of ITS subsystems, these methods play an important role. These subsystems include advanced traffic management, commercial vehicle operations, advanced traveler information systems, and advanced public transportation systems. Sensors are attached to the roads and the historical data is extracted from them which is then utilized to develop intelligent predictive systems. Machine learning and AI algorithms utilize this data to make short-term, real-time, and long-term predictions [38]. Short-term flow prediction was the focus of past research which used a simple feedforward neural network [39]. Another research focused on integrating a neural network system to the overall urban traffic control system consisting of one hidden layer.

4 Challenges of AI and ML in ITS

AI and ML have shown promising results in the applications of ITS. When applied to the transportation system, these technologies can improve the quality of lifestyle and provide safety along with the ease of quicker transportation services to people [40]. AI and ML techniques have also been used in the overall development of smart cities. However, these technologies face certain challenges in the development of transportation systems that need to be taken care of by the researchers and developers.

To get better accuracy of the ML algorithm, a very large training data set is needed which provides information such as vehicle speed, the distance between two vehicles, location, miles traveled, etc. [41]. This data will then provide better results in the transportation system because a larger training dataset will provide more accurate results. Therefore, it is important to collect a large volume of transportation data [42].

In the development of ITS, security is a major concern when developing smart techniques using ML. With the advancement in technology and the development of smart infrastructure and smart transportation systems, there is an ever-growing need to protect the security of the system [43]. For example, a security breach at smart parking management or traffic detection can affect the communication between the vehicles and can cause delays in response time. In turn, more advanced techniques like big data analytics are needed to ensure cyber safety and security in the smart transportation system [44].

Autonomous vehicles have proven to be reliable and an efficient way to commute to various locations. AI advancements have changed the way humans think about technology today [45]. The autonomous vehicles such as self-driving cars have been improving and advancing with the recent developments in AI. Researchers are still trying to improve the technology in self-driving cars and make it safer for travel. Even though a great number of models have been proposed and improvisations have been made several times, there is still no proper method to validate the correctness of the model. With the rapid implementation of smart transportation systems, there is an expectation for the faster development of autonomous vehicles [46]. Shortly, as the data keeps growing, it will be possible to collect enough data to minimize the error rate to use self-driving cars more commonly in the real world.

Another challenge is in the security and privacy issue of the people traveling from one location to another [47]. To develop a smart transportation system, there is a need to access a user's smartphone which contains information about their travel. This information is important as it helps to analyze the average travel time of the vehicle and get the information about the busiest location at a particular time of the day. The data can also be used to collect information about the nearest vehicle and help in developing a smart system with sensors that can detect a collision and avoid accidents between the vehicle by giving them an indication of how close the vehicles are to each other [48]. Hence, the data collected by the user's smartphone give very important and useful information that will help in the development of a smart city and transportation system. There is a need to protect the data which is being captured by sensors, actuators, and backend servers [49]. The data must be implemented securely and it should follow all the privacy rules and regulations [50].

A smart transportation system requires a need for a suitable machine learning algorithm that is lightweight and accurate in capturing and detecting traffic anomalies and generating real-time traffic information. This ML algorithm will be capable of detecting real-time traffic information and hence it will be conforming with the privacy rules and regulations as the data will not be transferred to a cloud platform. Implementing this type of ml algorithm that can generate real-time information is challenging and thus require efficient sampling and a scalable model [51].

5 Future Trends in Intelligent Transportation System

In the previous section, we have covered various applications of AI and ML in ITS which will benefit society in the development of a smart transportation system. These applications will provide smart and convenient travel options to people. With the continuous advancements in AI and deep learning algorithms, it is expected that in upcoming years, AI will colonize the entire world. As the data for the transportation system keeps growing, AI can perform and do wonders in the transportation sector. The authors in [52] provided facts of how AI algorithms can improve the future trends in the business and economy of the transportation sector. In the future, AI can provide effective use of roads for faster means of travel. New and improved algorithms will be in use which will provide different and unique route options for each group of travelers. For example, mail delivery vehicles will have a unique route option as compared to common travelers. This will reduce the fuel cost tremendously and also minimize the travel time [53].

It is expected that the percentage of self-driving cars will increase in 2030. By that time AI deep learning algorithms will become smarter and will have much more capability to self-learn and handle safety and emergencies appropriately [54]. With the increase in the use of self-driving cars, the percentage of traffic congestion will likely decrease to a high degree. Over the years, issues related to cybersecurity will likely be much lower. For this reason, the perception of the common public would also change and their trust in AI will grow [55].

Pattern recognition and Natural Language Processing (NLP) algorithms in ML are being widely used in the transportation sector. Within a few years, these algorithms should be able to detect and predict traffic congestion a few days before the actual travel date. For example, with the help of the social network and public comments, these algorithms should be able to detect and predict traffic congestion if there is a huge concert or a football match in the city. It should be able to extend its finding and provide additional details like expected clearance time of the traffic congestion, estimated train delay, and reasons for the delay must also be provided [56]. With more data being available for researchers over the years, a better analysis can be done on the traffic data, which can then provide additional details and advancements in the transportation sector [57].

6 Conclusion

This paper presents a review of AI and ML applications to develop Intelligent Transportation Systems that can tackle problems related to transportation. Our review explains how the applications of AI and ML deal with real-time transportation issues such as road anomalies, road accidents, the energy expenditure of streetlights, improper infrastructure, safety, traffic congestion, and availability of parking space. This paper focuses mainly on applications of the road network for the development of ITS and highlights the ml algorithms which have been widely used for the analysis of traffic accidents and safety concerns. This paper provides a baseline for other authors and researchers to focus on each aspect of the development of smart transportation. Finally, we conclude by analyzing the challenges currently faced by AI and ML in ITS.

References

1. Kyamakya K (2006) Artificial intelligence in transportation telematics. OGAI J. (Oesterreichische Gesellschaft fuer Artif Intell 25:2–4
2. Vittoz EA (1990) Analog VLSI implementation of neural networks. Proc IEEE Int Symp Circ Syst 4:2524–2527. https://doi.org/10.1887/0750303123/b365c79
3. Ciregan D, Meier U, Schmidhuber J (2012) Multi-column deep neural networks for image classification. Proc IEEE Comput Soc Conf Comput Vis Pattern Recognit:3642–3649. https://doi.org/10.1109/CVPR.2012.6248110
4. Gangwani P, Soni J, Upadhyay H, Joshi S (2020) A deep learning approach for modeling of geothermal energy prediction. Int J Comput Sci Inf Secur 18:62–65
5. Infrastructure Australia (2015) Australian Infrastructure Audit, vol 1, no. April. Australian Infrastructure Audit, Australia
6. Abduljabbar R, Dia H, Liyanage S, Bagloee SA (2019) Applications of artificial intelligence in transport: an overview. Sustainability 11. https://doi.org/10.3390/su11010189
7. Zantalis F, Koulouras G, Karabetsos S, Kandris D (2019) A review of machine learning and IoT in smart transportation. Futur Internet 11:1–23. https://doi.org/10.3390/FI11040094

8. Silva N, Shah V, Soares J, Rodrigues H (2018) Road anomalies detection system evaluation. Sensors (Switzerland) 18. https://doi.org/10.3390/s18071984
9. Omar S, Ngadi AH, Jebur H (2013) Machine learning techniques for anomaly detection: an overview. Int J Comput Appl 79:33–41. https://doi.org/10.5120/13715-1478
10. Li C, Xu P (2020) Application on traffic flow prediction of machine learning in intelligent transportation. Neural Comput Appl 6. https://doi.org/10.1007/s00521-020-05002-6
11. Yuan T, Borba W, Rothenberg C, Yuan T, Borba W, Rothenberg C, Obraczka K, Barakat C, Yuan T, Neto R, Rothenberg CE (2019) Harnessing machine learning for next-generation intelligent transportation systems: a survey to cite this version: HAL Id: hal-02284820 harnessing machine learning for next-generation intelligent transportation systems: a survey. https://doi.org/10.13140/RG.2.2.14242.79043
12. Chang H, Lee Y, Yoon B, Baek S (2012) Dynamic near-term traffic flow prediction: system-oriented approach based on past experiences. IET Intell Transp Syst 6:292–305. https://doi.org/10.1049/iet-its.2011.0123
13. Bratsas C, Koupidis K, Salanova JM, Giannakopoulos K, Kaloudis A, Aifadopoulou G (2020) A comparison of machine learning methods for the prediction of traffic speed in Urban places. Sustainability 12:1–15. https://doi.org/10.3390/SU12010142
14. Duan Y, Lv Y, Wang FY (2016) Travel time prediction with LSTM neural network. IEEE Conf Intell Transp Syst Proc ITSC:1053–1058. https://doi.org/10.1109/ITSC.2016.7795686
15. Ozbayoglu M, Kucukayan G, Dogdu E (2016) A real-time autonomous highway accident detection model based on big data processing and computational intelligence. Proc 2016 IEEE Int Conf Big Data:1807–1813. https://doi.org/10.1109/BigData.2016.7840798
16. Ryder B, Wortmann F (2017) Autonomously detecting and classifying traffic accident hotspots. UbiComp/ISWC 2017—Adjun. Proc 2017 ACM Int Jt Conf Pervasive Ubiquitous Comput Proc ACM Int Symp Wearable Comput:365–370. https://doi.org/10.1145/3123024.3123199
17. Assane U (2017) Ziguinchor S. De, Physique, D. De: monitoring the performance of solar street lights in Sahelian environment: case study of Senegal, pp 2–7. https://doi.org/10.1109/DeSE.2017.43
18. Mohandas P, Sheebha J, Dhanaraj A, Gao X (2019) Artificial neural network based smart and energy efficient street lighting system: a case study for residential area in Hosur 48. https://doi.org/10.1016/j.scs.2019.101499
19. Jia G, Han G, Li A, Du J (2018) SSL: smart street lamp based on fog computing for smarter cities. IEEE Trans Ind Inform 14:4995–5004. https://doi.org/10.1109/TII.2018.2857918
20. Tripathy AK, Mishra AK, Das TK (2018) Smart lighting: intelligent and weather adaptive lighting in street lights using IOT. 2017 Int Conf Intell Comput Instrum Control Technol ICICICT 2017:1236–1239. https://doi.org/10.1109/ICICICT1.2017.8342746
21. Chowdhury DN, Agarwal N, Laha AB, Mukherjee A (2018) A vehicle-to-vehicle communication system using Iot approach. Proc 2nd Int Conf Electron Commun Aerosp Technol ICECA 2018:915–919. https://doi.org/10.1109/ICECA.2018.8474909
22. Segal ME (2016) The intelligenter method (I) for making " smarter " city projects and plans 55:127–138. https://doi.org/10.1016/j.cities.2016.02.006
23. Mohammadi M, Al-fuqaha A (2018) Enabling cognitive smart cities using big data and machine learning: approaches and challenges, pp 94–101
24. Agarwal PK, Gurjar J, Agarwal AK, Birla R (2015) Application of artificial intelligence for development of intelligent transport system in smart cities. JTETS 1:20–30
25. Nikitas A, Michalakopoulou K, Njoya ET, Karampatzakis D (2020) Artificial intelligence, transport and the smart city: definitions and dimensions of a new mobility era, pp 1–19
26. Xiao L, Wan X, Lu X (2018) IoT security techniques based on machine learning, pp 41–49. https://doi.org/10.1109/MSP.2018.2825478
27. Hahn DA, Munir A, Member S, Behzadan V Security and privacy issues in intelligent transportation systems: classification and challenges, pp 1–15
28. Smart mobility enhancing autonomous vehicles with commonsense (2017). https://doi.org/10.1109/ICTAI.2017.00155
29. Bailey DE, Erickson I (2019) Selling AI: the case of fully autonomous vehicles, pp 57–61

30. Deep learning for autonomous vehicles, p 2378 (2017). https://doi.org/10.1109/ISMVL.2017.49
31. Vishnukumar HJ (2017) Machine learning and deep neural network—artificial intelligence core for lab and real-world test and validation for ADAS and autonomous vehicles, pp 714–721 (2017)
32. Yamani M, Idris I, Razak Z (2009) Car park system : a review of smart parking system and its technology. https://doi.org/10.3923/itj.2009.101.113
33. Pala Z, Inan N Smart parking applications using RFID technology
34. Shin JH, Kim N, Jun HB, Kim DY (2017) A dynamic information-based parking guidance for megacities considering both public and private parking. J Adv Transp 2017. https://doi.org/10.1155/2017/9452506
35. Alam M, Moroni D, Pieri G, Tampucci M, Gomes M, Fonseca J, Ferreira J, Leone GR (2018) Real-time smart parking systems integration in distributed ITS for smart cities 2018
36. Dia H, Rose G (1997) Development and evaluation of neural network freeway incident detection models using field data. Transp Res Part C Emerg Technol 5:313–331. https://doi.org/10.1016/S0968-090X(97)00016-8
37. Stojmenovic M (2006) Real time machine learning based car detection in images with fast training. Mach Vis Appl 17:163–172. https://doi.org/10.1007/s00138-006-0022-6
38. Zhang J, Wang FY, Wang K, Lin WH, Xu X, Chen C (2011) Data-driven intelligent transportation systems: a survey. IEEE Trans Intell Transp Syst 12:1624–1639. https://doi.org/10.1109/TITS.2011.2158001
39. Ledoux C (1997) An urban traffic flow model integrating neural networks. Transp Res Part C Emerg Technol 5:287–300. https://doi.org/10.1016/S0968-090X(97)00015-6
40. Voda AI, Radu L (2019) How can artificial intelligence respond to smart cities challenges? Elsevier Inc.
41. Applications BS (2019) Machine learning adoption in the challenges, and a way forward. IEEE Access 1. https://doi.org/10.1109/ACCESS.2019.2961372
42. Ullah Z, Al-turjman F, Mostarda L, Gagliardi R (2020) Applications of artificial intelligence and machine learning in smart cities. Comput Commun 154:313–323. https://doi.org/10.1016/j.comcom.2020.02.069
43. Janssen M (2019) Challenges for adopting and implementing IoT in smart cities an integrated MICMAC-ISM approach 29:1589–1616. https://doi.org/10.1108/INTR-06-2018-0252
44. Lin Y, Wang P, Ma M (2017) Intelligent transportation system (ITS): concept, challenge and opportunity. In: Proceedings of 3rd IEEE international conference on big data security cloud, BigDataSecurity 2017, 3rd IEEE international conference on high performance smart computing HPSC 2017, 2nd IEEE international conference on intelligence data security, pp 167–172 (2017). https://doi.org/10.1109/BigDataSecurity.2017.50
45. Nishant R, Kennedy M, Corbett J, Laval U, Prince PP (2020) Artificial intelligence for sustainability: challenges, opportunities, and a research agenda. Int J Inf Manage 53:102104. https://doi.org/10.1016/j.ijinfomgt.2020.102104
46. Jo JH, Sharma PK, Costa J, Sicato S, Park JH (2019) Emerging technologies for sustainable smart city network security: issues. Challenges Countermeasures 15:765–784
47. Mahdavinejad MS, Rezvan M, Barekatain M, Adibi P, Barnaghi P, Sheth AP (2018) Machine learning for internet of things data analysis: a survey. Digit Commun Netw 4:161–175. https://doi.org/10.1016/j.dcan.2017.10.002
48. Conde ML, Twinn I (2019) How artificial intelligence is making transport safer, cleaner, more reliable and efficient in emerging markets. IFC-World Bank Gr. 2023, pp 1–8
49. Chang MC, Wei Y, Song N, Lyu S (2018) Video analytics in smart transportation for the AIC'18 challenge. IEEE Comput Soc Conf Comput Vis Pattern Recognit Work:61–68. https://doi.org/10.1109/CVPRW.2018.00016
50. Chang M-C, Wei J, Zhu Z-A, Chen Y-M, Hu C-S, Jiang M-X, Chiang C-K (2019) AI city challenge 2019-city-scale video analytics for smart transportation. Openaccess.Thecvf.Com, pp 99–108

51. Tizghadam A, Khazaei H, Moghaddam MHY, Hassan Y (2019) Machine learning in transportation. J Adv Transp 2019. https://doi.org/10.1155/2019/4359785
52. Sumalee A, Wai H (2018) Smarter and more connected: future intelligent transportation system. IATSS Res 42:67–71. https://doi.org/10.1016/j.iatssr.2018.05.005
53. Intelligent transportation systems driving into the future with ITS (2006)
54. Camacho F, Cárdenas C, Muñoz D (2018) Emerging technologies and research challenges for intelligent transportation systems: 5G, HetNets, and SDN, pp 327–335. https://doi.org/10.1007/s12008-017-0391-2
55. Kala R, Warwick K (2015) Intelligent transportation system with diverse semi-autonomous vehicles 8:886–899
56. Sil R, Roy A (2019) Artificial intelligence and machine learning based legal application: the state-of-the-art and future research trends. Int Conf Comput Commun Intell Syst:57–62
57. Boukerche A, Tao Y, Sun P (2020) Artificial intelligence-based vehicular traffic flow prediction methods for supporting intelligent transportation systems. Comput Netw 182:107484. https://doi.org/10.1016/j.comnet.2020.107484

Analyzing App-Based Methods for Internet De-Addiction in Young Population

Lakshita Sharma, Prachi Hooda, Raghav Bansal, Shivam Garg, and Swati Aggarwal

Abstract With recent advancements in technology and the excessive use of smartphones, all internet-based applications like WhatsApp, Facebook, Netflix, etc. are one tap away, thereby resulting in increased internet usage on an average, especially among the young population. This has affected the cognitive and affective processes of the users and has caused various problems like loss of focus, fatigue, and burning sensations in the eyes, severe harm to mental health, reduction in response to events happening around, and many more. An unconventional method of recovering from internet addiction could be the use of mobile applications that help users monitor their usage and motivate them to have better self-control. There are a number of such applications, henceforth called apps, available that claim to help recover from internet addiction. However, their efficacy in curbing internet use has not been studied previously. This study is primarily based on assessing the efficiency of these app-based recovery methods from internet addiction. Using statistical analysis and polynomial regression, it was found that these apps do help in lowering internet use. This effect is largely seen in the first week of app use, after which significant reduction is not observed.

Keywords Internet addiction · App-based recovery · Screen-time · Mobile usage

1 Introduction

Behavioral addictions, that is, the compulsive use of non-chemical substances, such as gaming disorder, have been a popular area of research [1–4]. These addictions involve compulsive engagement of stimuli, despite harmful consequences and are also associated with withdrawal symptoms [5, 6]. With the coming in of smartphones and faster internet connections, there has been an upsurge in internet usage, especially among the young population [7]. This pathological use of the internet is a huge health concern [8]. However, pathological gambling and internet gaming disorder are the

L. Sharma · P. Hooda · R. Bansal · S. Garg · S. Aggarwal (✉)
Computer Engineering Department, Netaji Subhas University of Technology, Delhi, India

A. Choudhary et al. (eds.), *Applications of Artificial Intelligence and Machine Learning*, Lecture Notes in Electrical Engineering 778,
https://doi.org/10.1007/978-981-16-3067-5_17

only 2 formally recognized disorders [9, 10] and as such, there is an urgent need for research in this domain.

With tremendous advancement of smartphones, and owing to their availability and portability, self-help healthcare apps and smart monitoring systems are popular and readily available [11, 12]. A number of applications have been made with the aim of monitoring and verifying internet addiction [13]. As opposed to traditional methods of recovery, these apps provide various features and functionalities to tap information regarding internet use and to help users employ self-control in recovering from this addiction. These are more available, flexible, and cheaper as compared to the conventional methods like de-addiction centres. Some of the features include self-monitoring feature, notification feature, manual limit and block feature, automatic limit feature, and the reward feature. With these features, the apps attempt to capture and target cognitive and affective processes like attention, social interaction, emotional laden behavior, response to happenings around, etc. Many such apps are available on Google's Play Store and Apple's App Store, however, their effectiveness in curbing internet use has not been studied, nor has there been any comparative analysis of these apps. This paper aims to understand whether these apps can be used as a viable method of recovery from this enormous problem of internet addiction.

2 Prior Work

A number of diagnostic criteria have been put forward to capture the Internet Addiction Disorder. Of these, Young's Internet Addiction Test (YIAT) [14] is the oldest and one of the most utilized diagnostic instruments for internet addiction. Its features include:

1. It consists of 20 questions that measure mild, moderate, and severe levels of Internet Addiction.
2. Each question has a response on the scale of 0–5.
3. The score is calculated by adding the responses to given 20 questions set.
4. Table 1 lists the result calculated on the basis of the scale provided by Young.

Previous research in the domain of internet de-addiction includes pilot study that was conducted for the internet-addicted college students in China which involved the development of an online expert system named Healthy Online Self-helping Center (HOSC) as an intervention tool to help those who wish to reduce online usage [15]. The study also explored the effectiveness of HOSC for college student's internet addiction behavior. This was one of the first uses of online expert systems to manage

Table 1 Scale for young's internet addiction test

Range	0–30	31–49	50–79	80–100
Result	Normal	Mild	Moderate	Extreme

internet addiction, and was found to be moderately effective in reducing internet usage of the users.

A study carried out in Malaysia that aimed to carry out an investigation on online intervention factors for effective management of Facebook addiction in higher education found six addiction features—relapse, conflict, salience, tolerance, withdrawal, and mood modification—out of which relapse is the most important factor and mood modification is the least important factor [6]. This study also discovered five intervention features (notification, auto-control, reward, manual control, and self-monitoring) out of which notification was the most important intervention feature, whereas self-monitoring was the least important feature [6].

Another comprehensive system called SAMS (Smartphone Addiction Management System) was developed for objective assessment and intervention. Through the system operation verification and the pilot data study, the reliability of the SAMS was verified and also showed examples of its efficacy [13].

These studies demonstrate the use of online expert systems or management systems in recovery from internet addiction. However, no previous research has been done on already existing self-help or self-monitoring mobile applications, to assess their efficacy in reducing internet usage among the addicted users. Hence, the purpose of this study is to assess the efficiency of such commercially available self-control usage tracker apps to facilitate research in the domain of recovery from Internet Addiction.

3 Methodology

The participation of all the participants in the study was kept anonymous. Approval from any board or committee for this research was neither applied nor received as this was a techno-behavioral general study that was non-medicinal, non-intrusive, and non-clinical in nature. Also, all the authors are affiliated to a technology university which has no internal committee related to research on human subjects. Informed consent was obtained from all individual participants included in the study. None of data collected from the participants has been disclosed in the study or elsewhere.

In order to serve the purpose of study, i.e. assessing app-based recovery methods for internet de-addiction, requirements included target population with volunteers addicted to the internet to be surveyed, internet addiction diagnostic instruments, mobile applications to be studied, methods of analysis.

3.1 Mobile Applications

To assess the efficiency of app-based recovery methods, first and foremost requirements were the apps. Since Apple's iOS and Android are the two most popular mobile OS, top three highest-rated and freely available apps were chosen on the basis of their compatibility with both IOS and Android mobiles. These were Screen Time (App A), Space (App B), and Stay Focused (App C).

3.2 The Basic Methodology

The basic methodology followed is shown in Fig. 1.

Step 1: 158 students of age group 17-23 years took the YIAT so that the addicted population could be identified. To stress the gravity of the issue of internet addiction,

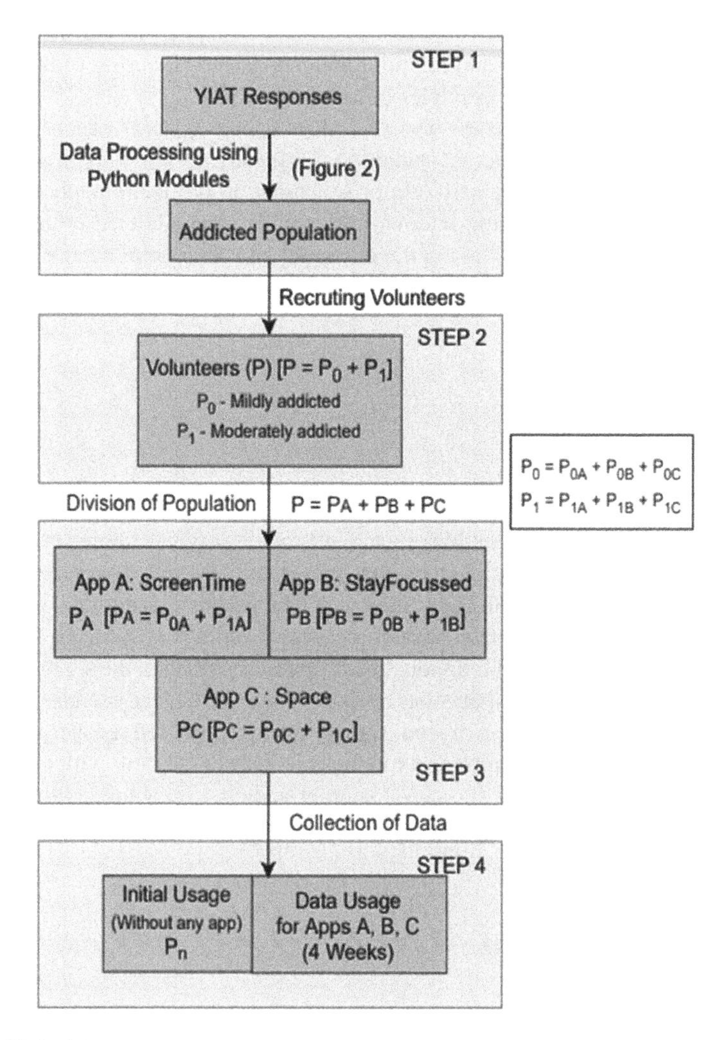

Fig. 1 Methodology

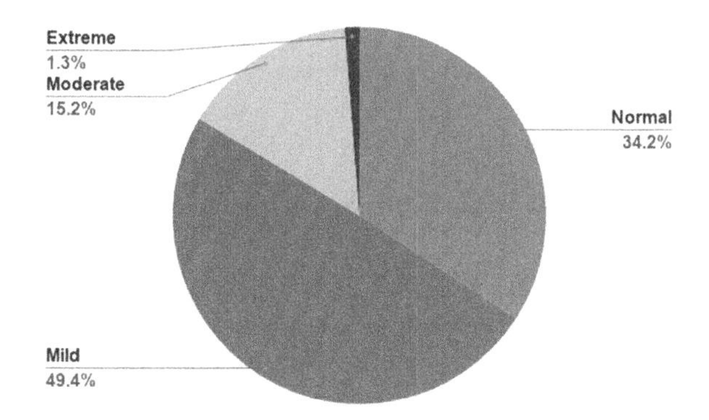

Fig. 2 Population distribution after YIAT test

out of these 158 participants who took the test, 104 were found to be addicted to the internet (mildly, moderately, or extremely) which is about 65.82% of the sample (Fig. 2).

Step 2: The addicted population was then informed about the next stage of the study in which they were supposed to download one of the tracking applications and provide their internet and smartphone usage data over a period of four weeks. After this stage of informed consent, volunteers who agreed for the study were recruited ($P = 30$). Of these, 18 were Mildly addicted ($P_0 = 18$) to the internet and 12 were moderately addicted ($P_1 = 12$). Hence, $P = P_0 + P_1$.

Step 3: In order to study the efficacy of the three identified apps, the volunteers were divided into three groups with approximately equal representation of mildly and moderately addicted individuals. This division was as follows:

App A: $P_A = 10$

Mildly addicted: $P_{0A} = 6$
Moderately addicted: $P_{1A} = 4$

App B: $P_B = 10$

Mildly addicted: $P_{0B} = 6$
Moderately addicted: $P_{1B} = 4$

App C: $P_C = 10$

Mildly addicted: $P_{0C} = 6$
Moderately addicted: $P_{1C} = 4$

No-App: $P_n = 10$

Initial usage(without using monitoring app)

Step 4: The volunteers then downloaded and used the apps and their usage data was captured weekly for 4 subsequent weeks.

4 Results and Analysis

This section discusses the methods used for analysis and the results observed. The analysis can be divided into three steps. The first step consists of getting a consolidated curve for each app so that the performance of each app can be visualized graphically. The second step consists of finding whether these apps are really effective or not. This is done by comparing the internet usage statistics of the population after four weeks of using these monitoring apps and their initial usage data. Finally, the third step consists of comparing these apps against each other and to find out the most efficient app in reducing internet addiction.

4.1 Step 1: Polynomial Regression

To visualize the performance of the apps during four weeks, a consolidated curve was needed which could depict the average trend of usage in the population using each app. For this, polynomial regression was used as it models a non-linear relationship between the independent variable x and the dependent variable y as an nth degree polynomial in x, which can be plotted to get the curve. This curve can be of varying degrees, but the one that most accurately represents the relationship between dependent and independent variables must be used. To find the most suited polynomial degree to get an accurate curve, RMSE (Root Mean Square Error) and R^2 metrics were used to be used in this model. The R^2 metrics denotes the proportion of the variance for a dependent variable that's explained by an independent variable. Whereas RMSE indicates the absolute fit of the model to the data i.e. how close the observed data points are to the model's predicted values. Therefore, a higher R^2 value and a lower RMSE value denotes an increase in accuracy of curve.

The RMSE and R^2 values of the 3 apps for different polynomial degrees are given in the Tables 2, 3 and 4. As observed, the RMSE decreases from degree 1 to degree

Table 2 RMSE and R^2 metrics for App A

Degree	RMSE	R^2
1	11.84	0.023
2	11.75	0.029
3	11.60	0.037
4	11.66	0.041
5	11.66	0.042

Table 3 RMSE and R^2 metrics for App B

Degree	RMSE	R^2
1	10.88	0.053
2	10.80	0.060
3	10.75	0.068
4	10.71	0.075
5	10.70	0.076

Table 4 RMSE and R^2 metrics for App

Degree	RMSE	R^2
1	10.35	0.042
2	10.25	0.059
3	10.18	0.068
4	10.11	0.074
5	10.11	0.075

4 and then it becomes constant on further increase in degree whereas the R^2 value increases from degree 1 to degree 4 and then becomes constant, for all the 3 apps. This shows that polynomial of degree 4 depicts the relationship more accurately than lower degree polynomial. Hence, degree 4 was used to fit the dataset because the error is least at degree 4 and on further increase in degree, it becomes more or less constant.

The consolidated curve, obtained after applying polynomial regression, for each app can be seen in Fig. 3. Here, the green color curve represents App A, blue represents App B and red represents App C. From this figure, it can be inferred that the

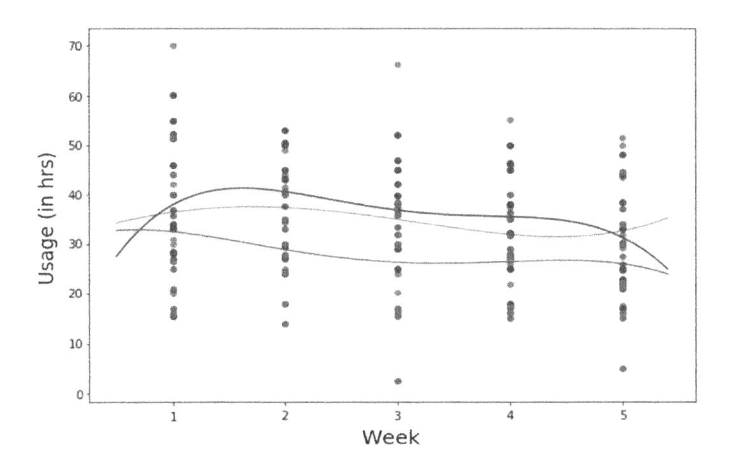

Fig. 3 Comparison of performance of 3 apps using polynomial regression

Table 5 Descriptive statistics (dependent variable)

Apps	Mean	Standard deviation	*N*
App A	32.680	11.7465	10
App B	31.350	10.1271	10
App C	28.175	9.5815	10
Non App	44.060	14.2310	10
Total	34.066	12.6619	40

internet usage of the users who have been using these apps tend to decrease over 4 weeks which shows the effectiveness of the apps in internet de-addiction.

4.2 Step 2: ANOVA

To statistically determine the effectiveness of these apps, the mean difference in the usage data of the population while using these apps was compared with the usage data without using any of these apps. Therefore, ANOVA was used to analyze the differences among these group means.

When ANOVA was used to compare four group means (App A, App B, App C, and No-App) as shown in Table 5, statistically significant results indicate that not all of the group means are equal, p value < 0.05. Therefore, LSD (Least Significant Difference) was further used as a post hoc test to determine exactly which means were significantly different from each other.

According to the Least Significant Difference (LSD), the non-app data is significantly different from other 3 apps data, p values (0.034, 0.019, 0.004) < 0.05 as shown in Table 6. Also, the mean difference (I–J) between non-app and app data is positive for all the 3 apps which means that the usage of non-app users was more than the users with apps installed.

These tests showed that there is significant difference in the internet usage while using these apps and without using any of these apps and also the usage while using these apps is lower than that of without using any app. Hence, it was inferred that these apps are indeed effective in reducing mobile usage.

4.3 Step 3: Multivariate Analysis of Variance (MANOVA)

To compare the efficiency of these apps against each other, the usage data of each app was compared over a period of four weeks. To compare data for multiple weeks i.e. multiple dependent variables, MANOVA was used.

Before applying MANOVA, certain assumptions must be met, like, homogeneity of variance, homogeneity of covariance, and absence of multicollinearity. For testing the assumption that the dependent variables would be correlated with each other in the

Table 6 Least significant difference (LSD)

(*I*) App	(*J*) App	Mean difference (*I*–*J*)	Std. error	Sig. (*p*)	Lower bound (95% C.I.)	Upper bound (95% C.I.)
App A	App B	1.330	5.1714	0.799	– 9.158	11.818
	App C	4.505	5.1714	0.389	– 5.983	14.993
	Non App	– 11.380*	5.1714	0.034	– 21.86	– 0.892
App B	App A	– 1.330	5.1714	0.799	– 11.81	9.158
	App C	3.175	5.1714	0.543	– 7.313	13.663
	Non App	– 12.710*	5.1714	0.019	– 23.19	– 2.222
App C	App A	– 4.505	5.1714	0.389	– 14.99	5.983
	App B	– 3.175	5.1714	0.543	– 13.66	7.313
	Non-App	– 15.885*	5.1714	0.004	– 26.37	– 5.397
Non-App	App A	11.380*	5.1714	0.034	0.892	21.868
	App B	12.710*	5.1714	0.019	2.222	23.198
	App C	15.885*	5.1714	0.004	5.397	26.373

*The mean difference is significant at the 0.05 level

moderate range, Pearson correlations were performed between all of the dependent variables. The results showed absence of multicollinearity where all correlations are positive and below $r = 0.90$, suggesting the appropriateness of MANOVA. To examine the homogeneity of variance, Levene's Test gave non-significant results, $p > 0.05$, which indicates equal variances between groups. In addition, the Box's *M* value of 49.63 had a nonsignificant association with a *P* value of 0.213. Therefore, this indicates the equality of covariance between the groups. Hence, the prerequisite assumptions were met.

Next, MANOVA was performed on the data of four weeks for the three apps to test the hypothesis that one or more mean differences would exist between the groups (App A, App B, App C) of the independent variable with regard to the dependent variables (Table 7). Among the different statistical tests [(a) Pillai's Trace, (b) Wilks' Lambda, (c) Hotelling's Trace, and (d) Roy's Largest Root] which differ in their statistical power, Roy's Largest Root test was used as it had the most power in the results. It provided a statistically significant effect, Roy's Largest Root = 0.593, $F = 2.845$, $p < 0.05$, partial eta squared = 0.372 with observed power = 0.742 which indicates that the null hypothesis is rejected and there is significant difference in the performance of the apps used. Therefore, Tukey HSD was used as a post hoc test to determine where the significant difference lies while reducing Type 1 error.

The post hoc test statistics show that there is a significant difference between App B and App C data in the week 1 as $p < 0.05$ (Table 8). The individual t-tests about the mean difference were conducted in order to investigate the specific mean difference between App B and App C. The results showed a significant difference in the usage in Week 1 for App B ($M = 40.61$, SD = 9.95) and App C ($M = 30.04$, SD = 9.07); $t(18) = 2.47, p = 0.023$. These show that App C fared better than App B as its mean

Table 7 Descriptive statistics (between 3 Apps)

Week	App	Mean	Standard deviation	*N*
Week 1	App A	37.4100	9.30489	10
	App B	40.6100	9.95852	10
	App C	30.0450	9.07948	10
	Total	36.0217	10.17217	30
Week 2	App A	35.0200	13.38895	10
	App B	36.8600	10.68584	10
	App C	28.7750	9.01114	10
	Total	33.5517	11.34269	30
Week 3	App A	31.990	11.6811	10
	App B	35.550	10.7907	10
	App C	27.460	9.0195	10
	Total	31.667	10.7269	30
Week 4	App A	32.6800	11.74647	10
	App B	31.3500	10.12711	10
	App C	28.1750	9.58150	10
	Total	30.7350	10.33614	30

Table 8 Tukey HSD

Week	(*I*) App	(*J*) App	Mean difference (*I–J*)	Std. error	Sig. (*p*)	Lower bound (95% C.I.)	Upper bound (95% C.I.)
Week 1	App A	App B	– 3.2000	4.2284	0.732	– 11.876	5.476
		App C	7.3650	4.2284	0.208	– 1.311	16.041
	App B	App A	3.2000	4.2284	0.732	– 5.476	11.876
		App C	10.565*	4.2284	0.048	1.889	19.241
	App C	App A	– 7.3650	4.2284	0.208	– 16.041	1.311
		App B	– 10.565*	4.2284	0.048	– 19.241	– 1.889

*The mean difference is significant at the 0.05 level

is significantly lower in the first week. The main reason behind better performance of App C is the distinguishing feature of 'blocking' which is missing in other 2 apps. This feature is used to block the mobile applications for which the usage exceeds the set limit by the user on a daily basis. This feature targets the cognitive process of attention and the affective process of social interaction and response to happenings around.

5 Conclusion

The results of the statistical analysis of mobile usage data show that there is a significant difference between the internet usage hours of users who used these recovery apps and that of those who didn't use any of these apps. The number of internet usage hours of the users who have these apps installed on their devices tends to decrease within four weeks. Further, the performance of the three apps was compared and App C (Stay Focused) was found to be the best among them because of its 'blocking' feature.

Hence, it was concluded that the apps that target the cognitive process of attention and restrict the user's internet usage prove to be efficient. The results of the study reaffirm that the app-based recovery methods are a good way of recovering from internet addiction and can be used efficiently to monitor and control a user's internet usage successfully.

References

1. Griffiths M (2009) Internet addiction—time to be taken seriously? Addict Res 8:413–418
2. Villella C, Martinotti G, Di Nicola M et al (2011) Behavioural addictions in adolescents and young adults: results from a prevalence study. J Gambl Stud 27:203–214
3. Leeman RF, Potenza MN (2013) A targeted review of the neurobiology and genetics of behavioural addictions: an emerging area of research. Can J Psychiatry 58(5):260–273
4. Albrecht U, Kirschner NE, Grüsser SM (2007) Diagnostic instruments for behavioural addiction: an overview. Psycho-social Med 4:Doc11
5. Grant JE, Potenza MN, Weinstein A, Gorelick DA (2010) Introduction to behavioral addictions. Am J Drug Alcohol Abuse 36(5):233–241
6. Dogan H, Norman H, Alrobai A, Jiang N, Nordin N, Adnan A (2019) A web-based intervention for social media addiction disorder management in higher education: quantitative survey study. J Med Internet Res 21(10):e14834
7. Lenhart A, Purcell K, Smith A, Zickur K (2010) Social media and young adults. Pew Research Center, Washington, DC
8. Pan C, Zheng L (2008) Pathological internet use and its correlation with college students' personalities and mental health. J Ningbo Univ (Educ Sc Edn)
9. American Psychiatric Association (2013) Diagnostic and statistical manual of mental disorders (DSM-5®). American Psychiatric Pub, Washington, DC
10. Brailovskaia J, Margraf J (2017) Facebook addiction disorder (FAD) among German students—a longitudinal approach. PLoS ONE 12(12):e0189719
11. Schueller SM, Neary M, O'Loughlin K, Adkins EC (2018) Discovery of and interest in health apps among those with mental health needs: survey and focus group study. J Med Internet Res 20(6):e1014
12. Livingston NA, Shingleton R, Heilman ME et al (2019) Self-help smartphone applications for alcohol use, PTSD, anxiety, and depression: addressing the new research-practice gap. J Technol Behav Sci 4:139–151
13. Lee H, Ahn H, Choi S et al (2014) The SAMS: smartphone addiction management system and verification. J Med Syst 38:1

14. Young KS (1998) Caught in the net: how to recognize the signs of internet addiction—and a winning strategy for recovery, 1st edn. Wiley, USA
15. Su W, Fang X, Miller J, Wang Y (2011) Internet-based intervention for the treatment of online addiction for college students in China: a pilot study of the healthy online self-helping center. Cyberpsychology, behavior and social networking, pp 497–503

Revolution of AI-Enabled Health Care Chat-Bot System for Patient Assistance

Rachakonda Hrithik Sagar, Tuiba Ashraf, Aastha Sharma, Krishna Sai Raj Goud, Subrata Sahana, and Anil Kumar Sagar

Abstract Chat-Bot is like our personal virtual assistants which can conduct a conversation through textual methods or auditory methods. One of the important tools of AI is Chabot's which can interact directly with humans and provide them with a considerate solution to their problem. These kinds of AI programs are designed to simulate how a human behaves as a conversational partner, after passing the Turing test. Chabot's are generally accessed via public virtual assistants such as Google Assistant, Microsoft Cortana, Apple Siri, or via various individual organizations' apps and websites. The process of building a Chat-Bot involves two tasks: understanding the user's intent and predicting the correct solution. In the development of Chat-Bot, the first task is to understand the input entered by the user. The prophecy is made depending on the first task. The main feature one can use of Dialog Flow API is follow-up intent. By using follow-up intent, one can create a decision tree that will help in the prediction of a disease by the Chat-Bot. Through this project, one can get insights into the actual understanding of Chat-Bot, explore various NLP techniques, and understand how to harness the power of NLP tools (Bennet Praba et al in Int J Innov Technol Explor Eng 9:3470–3473, (2019) [1]). One can thoroughly enjoy building their Chat-Bot from scratch and learning the advancements in the domain of machine learning and general AI.

Keywords AI · Tensor flow · ML · Dialog flow · Chat-bot · API · LUIS

R. H. Sagar (✉) · T. Ashraf · A. Sharma · K. S. R. Goud · S. Sahana · A. K. Sagar
Department of Computer Science Engineering, School of Engineering and Technology, Sharda University, Greater Noida 201310, India
e-mail: 2018004036.hrithik@ug.sharda.ac.in

T. Ashraf
e-mail: 2018003670.tuiba@ug.sharda.ac.in

A. Sharma
e-mail: 2018005582.aastha@ug.sharda.ac.in

K. S. R. Goud
e-mail: 2018006226.krishna@ug.sharda.ac.in

A. Choudhary et al. (eds.), *Applications of Artificial Intelligence and Machine Learning*, Lecture Notes in Electrical Engineering 778,
https://doi.org/10.1007/978-981-16-3067-5_18

1 Introduction

Chat-Bots are the software program applications that use Artificial intelligence and natural language processing that is used to apprehend what a human wish within the specific field and guides them to get the desired outcome which also makes use of existing consumer conversations to offer better outputs in destiny. Chat-Bots is commonly used a lot in customer interaction, marketing on social network sites, and instant messaging the client. Chat-Bots provides accurate and efficient information based on the user's requirement. Chat-Bots are utilized in conversation machines for diverse functions together with customer service, request routing, or for information collecting [2]. At the same time as a few Chat-Bot packages use substantial world-class techniques, NLP, and sophisticated AI, others truly experiment for trendy keywords and generate responses using commonplace phrases obtained from a related library of databases. Chat-Bots has the technology, which are computer programs that mimic human communication through voice commands or text chat [3]. Smaller Chat-Bots Sometimes chatterbots are an artificial intelligence feature that may be embedded and used in any primary messaging programs which include FB messenger, Viber, WeChat, WhatsApp, Coursera, and so on [4].

Some of the examples of Chat-Bots are Apple has Siri, Google has Google Assistant, and Microsoft has its Cortana for windows [5]. These are the most popular Chat-Bots with many features and high compatibility generally, Chat-Bots intelligence depends on how humans like but are made of pattern-matching technology [6]. They train using predefined patterns are and how good the text from the user is understood moreover some Chat-Bots functions are in an advanced manner using machine learning What is done behind working of Chat-Bots are:

1. User request analysis.
2. Returning the response.

The first task that a Chat-Bot performs is User request; it analyzes the user's request to identify intents and to extract relevant entities after identifying user requests and understanding those responses of a Chat-Bot are predefined and generic text [7].

Textual content retrieved from know-how base that carries unique answers contextualized records primarily based on statistics the consumer has furnished records this is saved in employer machine Movements Results are defined by the way that a Chat-Bot is interacting with backend application with one or more the disambiguating question that allows the Chat-Bot to properly apprehend the purchaser's request. The proposed medical Chat-Bots can interact with the user, giving them a realistic experience with a medical professional. The Chat-Bot will be trained on the dataset which contains categories, patterns and responses from, medical sector [8].

Artificial Intelligence (AI) is a field of study in intelligent agents where the main aim is to make human's work easy with the help of highly intelligent machines. One can achieve it to a few extents over the past few decades. Moreover, it was amusing to know the most precious feature of Dialog Flow which can easily deploy the model to cloud on voice command. For that, it can say, "Deploy NLP app" to Google Assistant on our phone and the code will be deployed on the cloud platform.

1.1 History of Chat-Bots

The first Chat-Bot was introduced even before the launch of PC by MIT Artificial Intelligence Laboratory by Joseph Weizenbaum in 1966 and was named "**Eliza**". Eliza in 2009 a Chinese company created a more advanced Chat-Bot called WeChat which conquered the hearts of many users who demonstrated unwavering loyalty to it. Chat-Bots are a highly thriving social media platform [9].

After Eliza, it became "Alice" which has evolved in 1995 with the aid of Richard Wallace, not like Eliza this Chat-Bot became able to use herbal language processing which allowed greater for sophisticated conversations changed into open for open supply. Developers can use synthetic intelligence mark-up language to create their very own Chat-Bots powered by Alice.

After "Jabberwacky" was created by British programmer Rollo Woodworker. It intends to "simulate herbal human chat in an interesting, exciting and funny way". It's an early try to develop synthetic intelligence via human interaction. The reason for the task changed into creating synthetic intelligence that can pass the Turing check. Jabberwacky has become designed to imitate human interaction and to perform conversations with users. It wasn't designed to perform every other capability [10]. In contrast to greater conventional ai packages, the gaining knowledge of the era is meant as a shape of entertainment instead of being used for computer support structures or company illustration. Modern-day developments to this Chat-Bot do permit managed approach to take a seat atop the overall conversational ai, aiming to carry together the quality of both approaches, and used inside the fields of income and marketing is underway [11]. The closing intention is that software passes from a textual content-primarily based gadget to be voice operated-mastering at once from sound and other sensory inputs. Its creators accept as true that it can be included in items which are around the house such as robots or talking pets, intending each to be useful and interesting, preserving human's agency [12].

Next, it's "Mitsuku" which is made out of AIML generation with the aid of Steve Warwick. It claims to be 18-year-old girl Chat-Bots from Leeds, England. Mitsuku carries all of Alice's AIML files, with many additions from person-generated conversations, and is constantly a work in development [12]. Its intelligence consists of the ability to reason with specific items. For instance, if a person asks "can you devour a residence? Mitsuku looks up the homes for "residence". Find the fee of "crafted from" is about "brick" and reply "no", as a house is not suitable for eating. Mitsuko can play video games and do magic tricks at the user's request. In 2015 Mitsuku conversed, on average, in extra a quarter of a million instances day by day [13].

This healthcare Chat-Bot gadget will assist hospitals to offer healthcare support on number 24 * 7 in Fig. 1. It also gives the basic medications to the customers and if needed it suggests the customers visit the doctors in case of emergency [14].

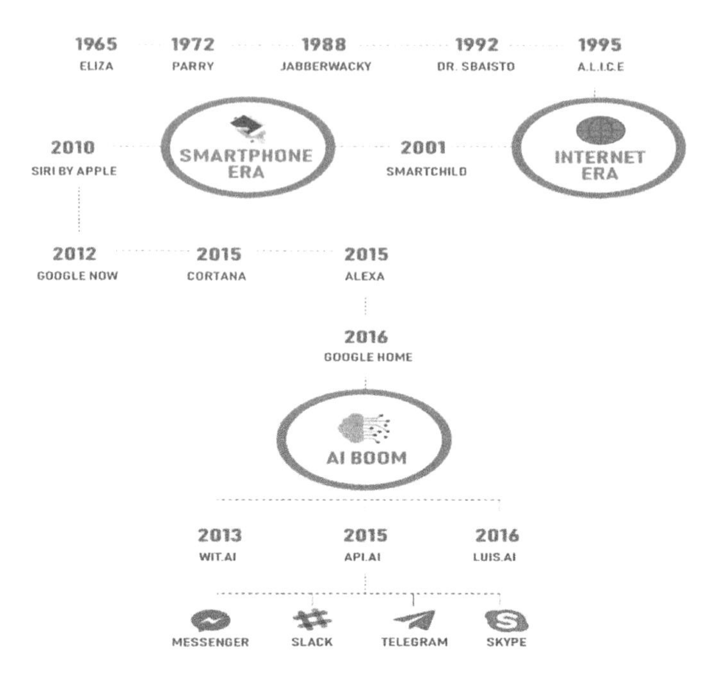

Fig. 1 The history of Chat-Bots from 1965 to 2016

1.2 Ease of Use

Faithful assistant within the pocket: Chat-Bots integrated with applicable cellular apps assist the sick yet busy and the on-the-move patient to time table appointments, problem reminders, manage check outcomes, statistics on nutritional picks, and even provoke a medicinal drug refill.

Inside the function of a virtual nurse, these Chat-Bots are trained using custom-designed questions and their answers frequently handled through physicians.

Effective: Experts consider that via significant facts analysis machines will quickly expect illnesses better than humans.

First-degree primary care and emergency first useful resource: The important scenario when you require ultimate-minute assistance may not be progressed by looking forward to a message reaction from a busy doctor or by the way of planning an emergency visit to a medical institution placed protracted away. An attempted and examined Chat-Bot carrier, specially trained in dealing solely with such emergencies, is frequently a heaven-dispatched solution at such junctures [15].

2 Literature Review

The contemporary paper gives facts about a healthcare Chat-Bot that's useful for clients to achieve what they want precisely. Query answering structures may be recognized as data gaining access to structures which attempt to solve natural language queries by way of giving appropriate answers creating use of attributes to be had in herbal language strategies [16]. The proposed machine of Chat-Bot takes a text from the user as entered and answers for educated scientific questions output via a qualified user is the output. The motive of the Chat-Bot is to provide a frequent way to this problem [10].

The modern-day paper facilitates in recognizing the fact in texts and giving the past content for developing a communique that is utilized in middle-college CSCL eventualities. A clever Chat-Bot for medical customers by using the software as a provider which analyses the message of every utility server [17]. It'll help the user to clear up the difficulty by imparting a human way to interplay the use of LUIS and cognitive offerings that are applied on AWS public cloud-admin feeds the input to the system so that gadget can discover the sentences and take a decision itself as a reaction to a question [18, 19].

The conversations may be executed so that it might add a few expertise to the database as it has no longer been modeled before. Data is remodeled if the entered sentences didn't match with the database [19]. This utility can be developed by using google conversation-flow. The current paper makes use of AI that's expecting the diseases primarily based on the signs and delivers the list of "to be had treatments". It can additionally facilitate us to determine the trouble and to validate the answer [1].

3 Proposed System

Some of the chat bots are compact medical reference books which might be beneficial for doctors, patients, persons, and so forth and for individuals who want to examine something about fitness. The customers might sense that they're included inside the method of their fitness. Patients, who might feel included, which's interacting via Chat-Bots with the healthcare gadget, will stay with the system, and this is crucial for them and the healthcare issuer. The vintage Chat-Bots which are currently available are consumer communications systems and their exceptional attempt is a question and solution web page on a website however Chat-Bots can facilitate to get the not unusual fitness related query and prediction of disorder without human interference. Purchaser pride is the principal problem for growing this system. The welfare of the Chat-Bot is to facilitate the people by using given right steerage regarding the good and wholesome living. For some of the reasons that among the people now do not have fundamental recognition of bodily conditions. Several human beings stay for years debilitating but they no longer pay a whole lot attention to signs and symptoms without a doubt due to the fact they think they don't require a physician. The operating of the cutting-edge device is as follows.

1. Simply logon to Chat-Bot site and begin asking your questions.
2. You may ask some of the questions regarding a few healthcare. And it's related to textual content- text communique. Using Google API for interconversion of textual content-textual content.
3. You may get clinical info on medication names and you can even ask about medicinal drug-related information on the idea of medication names.

Chat-Bots are created using platforms such as wit.ai, Dialog Flow, Microsoft Language Understanding Intelligent Service (LUIS), TensorFlow, etc. Among the various platforms, Dialog Flow was chosen for this project because the platform is very intuitive, the ease of creation of the Chat-Bot was better as compared to other platforms, provides integration with various third-party apps and it provides support for an enormous set of languages [20].

The architecture of creating a Chat-Bot can be divided into design, building, analytics, and maintenance. The Chat-Bot development is the process where developing Chat-Bot as the purpose of interaction between the user and the Chat-Bot. Chat-Bot developers define Chat-Bot personality, the questions that will be asked to the patients (end-users), and the overall interaction. An important part of the Chat-Bot development is also centered around user testing. Dialog Flow provides out of the box techniques for testing and training [17].

3.1 Processing of Chat-Bot

First, it tries to get input from the user then process input and return the value that generated the highest confidence value from the logic adapters then return the response for the input in Fig. 2.

- The general process of building a Chat-Bot is:
    1. Preparing the dependencies.
    2. Import classes.
    3. Create and train the Chat-Bot.
    4. Communicate with python Chat-Bot.
    5. Training the python Chat-Bot with a corpus of data.

Intents: Intents refer to goals of the customer; Entities: Entities used to add values to search; Candidate response generator generates a response for the user to give to respected questions; Context refers to collaborative act; Response refers to Chat-Bot replies [21].

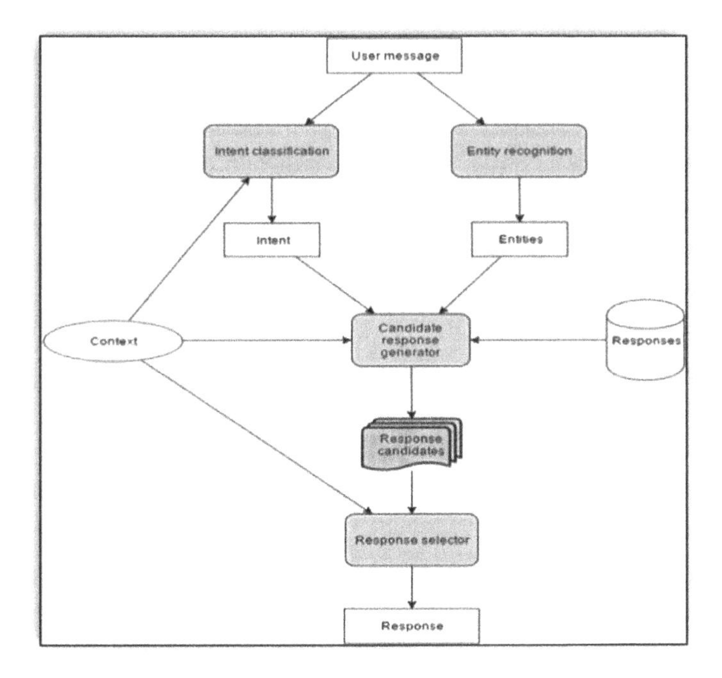

Fig. 2 This is the architecture of Chat-Bot

3.2 Some Common Mistakes

Most common problems faced by current generation Chat-Bots are:

i **Making Chat-Bot Affordable**

Mainly whilst looking to sell to the board of directors, irrespective of enterprise advantages of integration of Chat-Bot into any commercial enterprise model, it relies on who can pay it. It wouldn't forget building a Chat-Bot ourselves using loose software. This not best will result in an inferior product, the hidden fees will quickly pile up, too [22].

ii **Making a secure Chat-Bot**

A Chat-Bot needs to have strong protection from day one. Clients count on high-quality safety protocols along with https and HTTP metadata linked to any net channel that makes use of their non-public facts. Anything which isn't always a relaxed web page and clients will flat-out refuse to apply Chat-Bot.

Need to make sure that safety features are in location, such as stop to-forestall encryption, component authentication, biometric authentication, and authentication timeouts. Behavior finding out of Chat-Bot through manner of jogging penetration exams and API safety tests.

Being updated with the cutting-edge developments in Chat-Bot cybersecurity packages. But, this type of pressure calls for expert information [23].

iii **Getting human beings to love Chat-Bot**

It required a kind of Chat-Bot that actions beyond the 'magic' tag many tech gadgets suffer from and offers clients a few aspects they'll want to interact with—be it text-based Chat-Bot or voice-based Chat-Bot. It does not necessarily have to be 'almost human'—customers might as a substitute have interaction with a Chat-Bot that offers superb, applicable responses than one that's overly pleasant [24]. The 5 key factors to getting human beings to apply a bot:

1. Chat-Bot should be useful.
2. Chat-Bot should be relatable.
3. Chat-Bot should be accurate.
4. Chat-Bot wishes to be truthful.
5. Chat-Bot wishes to be likeable.

iv **Deciding on among text-primarily based Chat-Bots and voice-activated Chat-Bots**

The fact that anyone can choose out among specific forms of Chat-Bots offers us a bargaining chip while seeking to convince different activities' interior agency that a smart Chat-Bot which might pass through possible channels through which has in a position to connect to our clients.

Whilst clients might not be equipped to take a bounce and expand a voice-activated Chat-Bot just but, reluctant events might be more willing to provide entertainment to the concept of a textual kind of content-based Chat-Bot rather, particularly if it has given already robust virtual footsteps.

1. Chat-Bots activated through VOICE
   i. Generally, these are greatly suitable for static use—if our client base desires a "circulated-to" factor or integrated with a special Internet of technology era in conjunction with a domestic hub.
   ii. These are generally expensive to grow and maintain and have great personality quota and move generational of their enchantment.
2. Chat-Bots activated through TEXT:
   i. These are satisfactory for mobile gadgets and records particular responses consisting of online banking or financial statistics.
   ii. The usage of devices that your clients are already sociable with tablets and Phones and people are much less high priced to increase and hold.

v **Ensuring Chat-Bot offers actual cost to customers**

There may be no point handing over attractive, technologically superior Chat-Bots if it does not do something in reality! Chat-Bots should have a USP to make them appealing to customers, and top to that listing of a Chat-Bot makes their lives much less difficult in some kind manner. As a route to be presenting Chat-Bots with instantaneous access to personal statistics, giving them beneficial statistics that enriches them enjoy (or perhaps makes their day better) or assisting them in their paintings.

Thinking cautiously, who might be using the Chat-Bot, and what they need; does it have a USP?

How much better is it than other Chat-Bots?

Determine whether need a Chat-Bot with particular competencies such as tailor-made to our audience [25].

vi **Getting to know our Chat-Bot audience**

Technical challenges of developing Chat-Bot may be without difficulty triumph over by using experienced designers, but it nonetheless needs to make certain that if the Chat-Bot has some person to talk to as soon as it's out there. To try this, it must recognize our target market.

Attempt to research how Chat-Bot is deployed to your unique industry. Communicate to Chat-Bot builders about whether they have any case research that could provide insight into how a specific consumer base has reacted to Chat-Bot deployment in the beyond. Carry out some demographic analytics of the present-day target market through a short questionnaire, online survey, or electronic mail advertising marketing campaign [26].

vii **Constructing the bot**

Technical elements of constructing Chat-Bot depend upon what sort of Chat-Bot that want to grow. Merely textual content-primarily based bot will be massless difficult to increase than a voice-activated model most of the instances as beginners [27].

For any enterprise, there should be a hard and fast approach that any Chat-Bot undergoes through, so it's crucial to extend a 'wise' platform that:

1. Reveals out the intention of character by using the usage of asking the set of questions and then trying to respond to the solutions coherently.
2. Collect an applicable record from the user.
3. Techniques the records and uses its evaluation to respond to the goal of the person.
4. Shops record in the database so that if the same question arises yet again, it may use the information to shape a much accurate reaction than it is normally known as device getting to know.

3.3 Building Bot Using Dialog Flow API

There are various platforms available to create Chat-Bot forex as msg.ai, wit.ai, Dialog flow, Microsoft Language Understanding Intelligent Service (LUIS), etc. Dialogue flow was chosen for this project because the platform is very intuitive, the ease of creation of the Chat-Bot was better when compared to other platforms, provides support for a large set of languages and provides integrations to various third-party apps. But most of the platforms follow the same concepts and terminologies [28].

A Dialog device or conversational agent (CA) is a laptop gadget intended to communicate with a human, with a coherent shape. Chat-Bots are usually being used in scenarios like dialog structures for different kinds of practical functions and use sophisticated NLP structures. The criterion of intelligence relies upon the capacity of a laptop software to impersonate a human in a real-time written verbal interchange with a human decide, sufficiently proper that the device is not able to distinguish reliably among this system and an actual human [22].

Dialog Flow (formerly API.AI) is Google-owned developer of human laptops interplay technology-based totally on herbal language conversations. Dialog Flow allows us to construct herbal and rich conversational studies that offer users new approaches to interact with the product by constructing attractive voice and text-primarily based conversational interfaces powered by way of AI. Dialog Flow contains Google's device studying know-how and merchandise which includes google cloud speech-to-textual content and is backed via google cloud platform which can easily scale millions of customers. Using years of domain information and NLP, Dialog Flow analyses and is familiar with the user's rationale and responds in the maximum useful way [15].

In Fig. 3. The technique a Dialog Flow follows from invocation to fulfilment is similar to a person answering a question, with a few liberties taken of the path.

So, one can start a verbal exchange with an agent, the user wishes to invoke the agent. A user does this by asking to speak with the agent in a manner specified utilizing

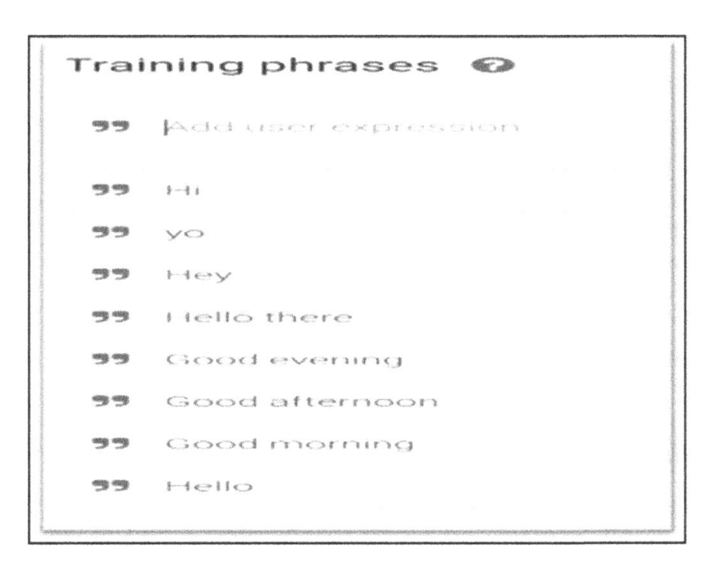

Fig. 3 An explanation of the training phase

Table 1 Table of threshold representation and disease

Name of disease/illness	
Tags	
Threshold value	
Symptoms	
	Score = Value Symptom Check Question Recommended Medication
	Med 1
	Med 2
	Etc.
Recommended Remedy	
	Precaution
	Remedy
	Etc.

the agent's developer but the agent is trained using NLP and machine learning techniques such as word embedding, user data, lexical synonyms, bag-of-words, synonym detection, regular expressions, tokenization, tags, supervised learning, etc. and hence was able to detect what the user said.

3.4 *Tabular Design of Disease Record*

See Tables 1 and 2.

3.5 *Decision Tree of Chat-Bot*

Agents work based on decision trees and at each step of the conversation flow, it can have multiple paths to proceed and determines the best path based on the previous outcome and input as in Fig. 4. By making a decision tree by using the data of symptoms and its illness and then using the follow-up feature of intent. The picture below is the small snip of how we made it:

Figure 5 shows a small part of the decisions taken by our bot. This figure covers a portion of the illness such as common cold, diabetes, gastric problem, COVID-19, and interesting activities.

Table 2 Example table of threshold representation

Migraine
TAGS: Headache, Head pain, pain, eye pain, eye
Threshold
Symptoms
Score = 2 Do you experience pain in your eyes
Recommended Medication
ibuprofen
Aspirin Recommended Remedy
Stay away from light
Close your eyes and relax
Wash your eyes with cold water
Score = 1
Do you feel like vomiting?
Recommended Medication
ibuprofen
Aspirin
Recommended Remedy
Drink warm water
Stay in cold place

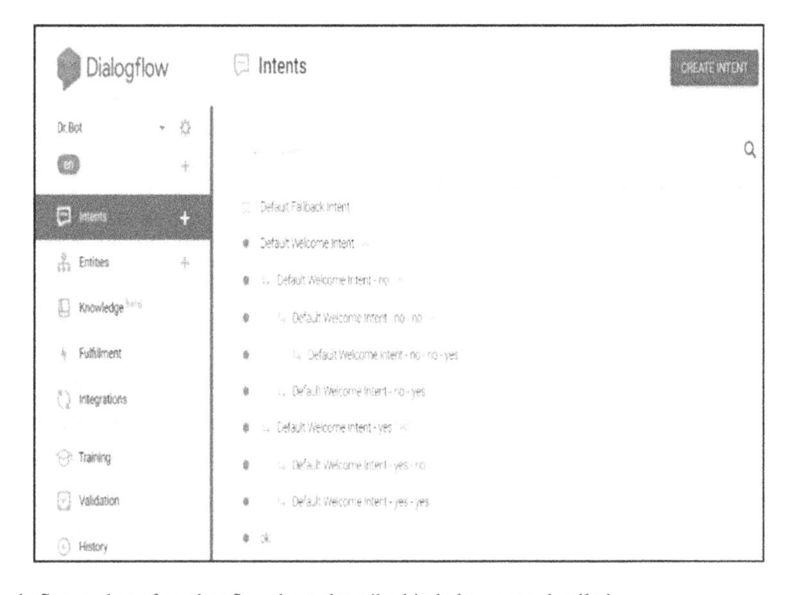

Fig. 4 Screenshot of our bot flowchart, described in below page detailed

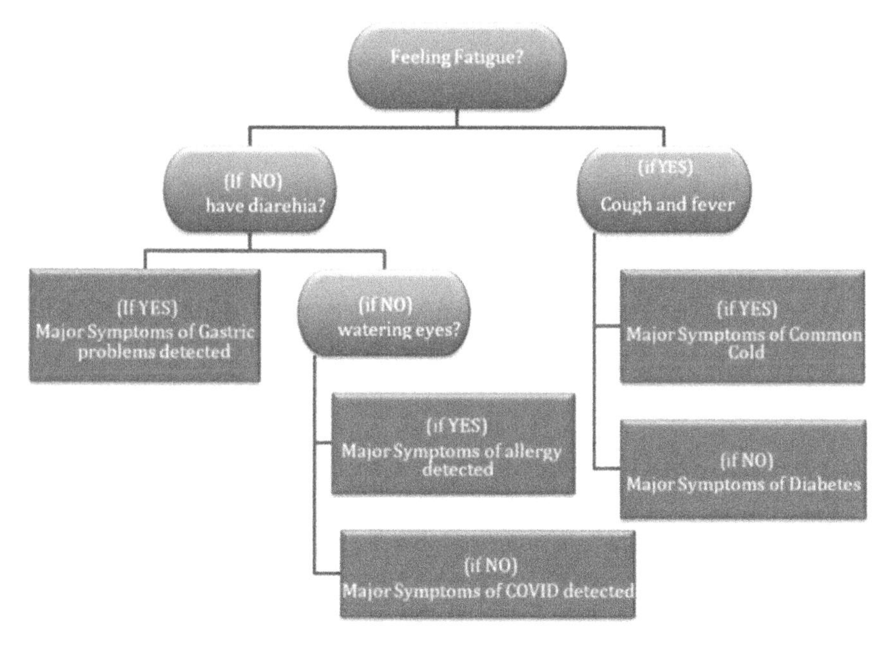

Fig. 5 Decision tree diagram for having fatigue

3.6 Developing Chat-Bot

Construction of Chat-Bots that integrate within a digital gadget mainly in case Chat-Bot is a covered device that makes use of both online portals and IVR telephony is a mission that calls for a big quantity of information.

With the useful resource of ways the exceptional choice to be had is to paint with an expert to create a bespoke tool in particular tailored to your corporation. At the same time as there are structures that allow us to build set up bots, they lack the nuance of tailor-made designs, are not likely to offer our customers with a quality level, and might be tough to combine into our wider channel shift approach.

3.7 Normalizing the Tech Inside Present Digital Channels

Integrating Chat-Bots into present digital channels has to be pretty sincere, mainly if both us and our clients are already acquainted with right away messaging protocol.

The network of a Chat-Bot to most of the other additives of our set-up is the trouble of embedding type of links to most of the online content cloth, ensuring that there aren't any protocol which conflicts between channels and that theirs is an easy course map for users to take to the Chat-Bot and lower again all over again.

To honestly acquire the blessings of our Chat-Bot, it enables us to combine it into an Omni-channel gadget that makes full use of ivy, email, mobile web sites, and SMS integration. This requires an entire channel switch method that experienced customer service era providers will help us to put in force.

1. Get users to get signed up for the Chat-Bot provider by the use of signposting them to the carrier via exclusive channels.
2. Get builders to Trojan horse-take a look at thoroughly earlier than liberating the Chat-Bot so that there are not any coding or protocol conflicts. Channel moving users in the direction of the Chat-Bot via the use of incentives together with precedence responses; get right of entry to big codes that unencumbered one of a kind deals, and so forth.

3.8 Ensuring the Chat-Bot Suits in with Emblem Identity

Prolonged-term emblem identification is the maximum important advertising device. Whether or not we're an industrial agency or a public-sector enterprise, it's how human beings recognize us. Our Chat-Bot wants to supplement our emblem.

Consequently, it needs to look for a Chat-Bot expert who is inclined to work intently to extend a Chat-Bot that represents and respects our unique identity. Whether or not it's adapting scripts to satisfy our particular wishes and needs or providing solutions to client inquiries which are precise to our line of exertions, our Chat-Bot partners will need to paintings issue-via way of-aspect with our institution to make certain the Chat-Bot presentations our brand.

If our logo identification is a laugh and mild-hearted, our Chat-Bot must suit that by being pleasant, chatty, and customized.

To achieve a high-end business-orientated Chat-Bot, which is refined, professional language is quite essential with an entire absence of emojis!

Voice-activated Chat-Bots: The voice that it has picked has to be a representation of our business, which may also show whether or not you pick a male or lady voice.

3.9 Handling Chat-Bot

Even as most of the Chat-Bots are on occasion portrayed as in large part self-assisting, there wishes to be a few tweaking every so often to make certain exceptional performance and for the various folks who move down the direction of building a free Chat-Bot themselves, that is where prices begin to add up.

As each person knows, a negative carrier drives ability client away and ensures that they're not going to ever return. This is a very good-sized hazard with poorly designed, loose Chat-Bots.

As an alternative, it has endorsed partnering with a Chat-Bot expert that.

Is communicative and available 24/7, if there are issues with Chat-Bot protection otherwise you need to make modifications to the script, this need to be finished as soon as viable.

Affords 12 months upkeep services and guarantees the right functioning of the bot.

Try and provide a reporting provider that feeds lower back beneficial person facts for your management team.

3.10 Coping When Matters Go Wrong

Not anything is ever ideal, but with the aid of careful management and thoughtful creation on the front quit, it will minimize capability problems in a while. However, if matters do move incorrect, want a contingency plan. Or, higher but, we need if you want to leap at the smartphone and get a Chat-Bot provider working day and night time on a restore.

Make clarified that its groups are updated with the programming language that it's made with and protocols that bot makes use of.

Have strong safety features in location so that failure isn't always the result of a virus or a hack.

Take survey (feedback) remarks from users on board—in case the bot isn't calculating up then react quicker in place of later to conform the programming and make it more consumer-pleasant.

3.11 Knowing When a Human Need to Take Over

Trouble comes whilst a successful integration of virtual technology like Chat-Bots is the idea that humans could continually as a substitute speaks to a human rather than a system.

That is a commonplace misconception that for example in a current survey achieved with the aid of Hub Spot, fifty-five % of customers stated they have been interested in the usage of a Chat-Bot to engage with an enterprise. Still not convinced, k what about this, mind browser's 2017 survey observed that ninety-five% of users trust the customer service component of business will, in the long run, be more advantageous via Chat-Bots.

But there are usually those cases where a Chat-Bot really cannot really resolve a question, so constantly make certain that there may be always a man or women who could take it over if vital.

1. Most of the time makes sure that there is a choice for persons to switch from bot to a human operator.

2. Always try including a ‘back door’ that customers can access if the bot is not interacting properly for them.
3. Try assisting the user feel valued through imparting them with vanity codes which could permit them to skip automated telephony structures and cross instantly over to a human operator.

3.11.1 Understanding How Successful the Chat-Bot is

Generally, Chat-Bots are most famous with clients, with 50% of the users in a Hub Spot ballot saying they could as a substitute have interaction with a commercial enterprise via the Chat-Bot than every other approach of touch. But most of the customers having the identical revel in?

Analytics shall say how many interactions the Chat-Bot is dealing with, the commonplace interaction duration and whether a person is suffering a better-than-not unusual dropout rate. The maximum Chat-Bot gives integrated analytics modules and documents ordinary usage reports for the one’s individuals of your corporation that need them. Not surely this however directs customer comments are likewise very critical, so:

1. Ballot individuals and ask those at once how they’re locating their Chat-Bot experience.
2. Constantly try to tackle board feedback on how a Chat-Bot can step forward and, genuinely as importantly on what you’re doing right.
3. Evaluate the Chat-Bots channel to different digital touch factors to check if there are special, extra productive routes it must blend into the bot’s protocol.

3.11.2 Word Order Similarity Among Sentences

The order of words is vital too due to the fact change so as of words might bring an incorrect output, as an instance:

“he can’t swim however dogs can” it could trade its order: “dogs can’t swim but the bird can” hence the which means of both sentences is special consequently, it should shape word order vectors for each sentence, specifically $r1$ and $r2$.

For the very first sentence, that is carried out by using doing the subsequent for each word w in the joint phrase set:

If the first sentence carries v, it fill the access in “$r1$” with the corresponding index of v within the first sentence.

If the primary sentence does not comprise v, it finds the word from the first sentence that is much like V. This phrase is denoted as $\sim v$. If the similarity of these is greater than the pre-set threshold, it fills the first sentence’s vector access with the corresponding index of $\sim v$ within the first sentence. If the similarity between them is not greater than the threshold, it fill the vector’s access with zero.

The threshold could be very critical due to the fact are calculating the phrase similarity of various phrases, and consequently the similarity measures may be very low. Since that means "*S*" the phrases aren't comparable, it doesn't need to introduce such noise into our calculation. While it booms the brink, it would probably introduce more noise to our calculations, which isn't always appropriate. It'll repeat the manner for each sentence, so it achieves phrase order vectors for both sentences. The very last price of/for the phrase order similarity measure is evaluated using the following formulation:

$$Sr = 1 - \frac{\|r1 - r2\|}{\|r1 + r2\|} \tag{1}$$

Word order similarity is measured amongst two sentences and is calculated as a normalized differentiation of word order. The measure is very touchy to the gap between words of the word pair. If the distance will increase then the degree decreases.

$$H = \frac{\sum_{i=0}^{n} (\text{score})}{\text{Threshold value}} \tag{2}$$

H	It's the decision parameter which is used to check if the threshold level was hit.
Σ (Score)	It's the total score of all the symptoms that the user claims to experience.
I	Random number starting from first.
Threshold Value	It's the upper limit value until which the Chat-Bot can handle.
Condition for triggering	if $H \geq 1$ the Chat-Bot connects the user to the doctor [9].

4 Results

The above screenshot shows the outputs of our Bot responding to medical healthcare questions of customers/users. The link to the bot is provided below: https://dialogflow.cloud.google.com/#/agent/bot-decision-tree-fxssat/intents The intention of this paper becomes to introduce healthcare Chat-Bot, a machine designed to enhance the health paradigm through the usage of a Chat-Bot to simulate human interplay in scientific contexts based totally on gadget gaining knowledge of and synthetic intelligence strategies, our bot is in a position to triumph over the drawback of classical human-gadget interaction, consequently removing bias and allowing the patient to an unfastened and natural communique (Fig. 6).

The Chat-Bot is efficaciously designed to work as a supporting device in health practitioner-patient conversation, however, it should be a global magazine of fashion in clinical studies and development emphasizing that it must be paintings primarily importantly as complement and not a substitute.

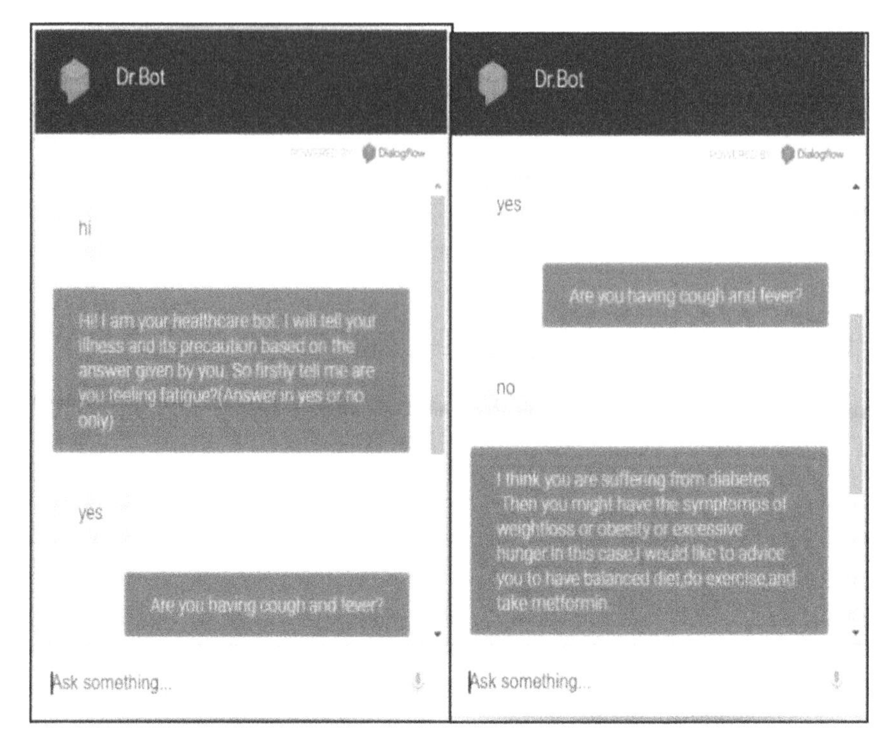

Fig. 6 Screenshots of our bot, when it starts a conversation it'll introduce itself and asks if a person is having any fewer or cold

5 Future Work

Nowadays technology is moving at a very frighteningly fast pace, so it's crucial to avoid the uncertainty of discontinuance by future-proofing Chat-Bot even though incorporated synthetic intelligence represents the next technology of conversation, it may not be the bot that hype revolves, but also 0 the hardware used by the customers. If the included system learns protocols into Chat-Bot then it ought to be able to preserve pace with outside technology advances.

It has to additionally paintings to update our tech and hold us abreast of tendencies but we also can make certain:

This shall rent the very modern-day technology of NLP software program and keep updating it regularly.

It has to plot in advance for brand spanking new structures which include subsequent-generating hardware holding a tight watch on bot's improvement and also if our Chat-Bot is on a valid supported platform, make sure the provider issues everyday updates.

Along the time and more deep knowledge, we can add more features to our Chat-Bot like:

- Adding more training data for more accurate prediction.
- Adding one more function of booking appointments to the doctors according to the patient's location.
- Can provide this Chat-Bot with a more segregated and user-friendly interface.
- For making it a personal healthcare doctor, we can add a function of keeping the record of patient's data so that Chat-Bot doesn't have to ask the same questions every time a patient visits the platform.
- Taking remarks from users on board if the bot isn't measuring up then to react quicker as opposed to later to evolve the programming and make it greater person-pleasant.

It has applied various Machine Learning algorithms in the background for effective text analysis which was provided by Google and also performed Intent and entities-based analysis for implementing smooth Chat-Bot conversation.

As the future work, our team is also working on adding more intents and better specification of entities to cover more symptoms, to make Chat-Bot able to diagnose more diseases. Currently, our bot responds to text thus we would like to add voice input and output.

Face recognition would also be added in future and webcam support so that the user can explain his/her problems to professionals when professional support is added.

In the conclusion of our healthcare Chat-Bot, it was successfully able to understand user/patient's healthcare-related queries and lead the conversation to final diagnosis by an effective text-based diagnostic technique and knew when to handover to the human.

6 Conclusion

Given the huge population, not every person gets to meet a doctor as per WHO doctor to patient ratio in India in the year 2017 was 1:1000. In the USA, "doctor shortage" was reported from year 2010 to 2016 as there are approximately 277 to 295 patients per doctor. It is obvious to us that it is not possible for every human being to have access to a doctor whenever they have a known health problem but not knowing existing health issues is also a major concern. Hence, considering this situation we have planned to create virtual doctor assistance for everyone with no cost and just using Internet.

Chat-Bots are easily accessible and understood by anyone mainly when it is in their own language. Moreover, Chat-Bots interface emulate patient and doctor conversations, thus, we have chosen Chat-Bot as the primary interface.

The assessment proves that the usage of a Chat-Bot is consumer-friendly based on a survey conducted within the community. The usage of a Chat-Bot also does not depend on how far a person is making location an irrelevant concern.

The intelligence of a Chat-Bot can be extended by gathering and growing the database so that the Chat-Bot can cover different types of queries about each sickness/health concern. Hence, considering all additional features not present in the already existing Chat-Bots makes our Chat-Bot more efficient for users and has wider access range.

References

1. Bennet Praba MS, Sen S, Chauhan C, Singh D (2019) Ai healthcare interactive talking agent using Nlp. Int J Innov Technol Explor Eng 9:3470–3473
2. Pal SN, Singh D (2019) Chat-Bots and virtual assistant in Indian banks. Industrija 47:75–101
3. Lawry T (2020) AI in health: a leader's guide to winning in the new age of intelligent health systems. CRC Press
4. Kar R, Haldar R (2016) Applying Chat-Bots to the internet of things: opportunities and architectural elements. arXiv:1611.03799
5. Chaudhari ST, Chandsarkar R, Bodke S, Sule T (2010) Virtual theatre. In: Proceedings of 3rd international conference on emerging trends engineering technology ICETET 2010, pp 312–316. https://doi.org/10.1109/ICETET.2010.127
6. Devi KS (2018) Survey on medical self-diagnosis Chat-Bot for accurate analysis using artificial intelligence. Int J Trend Res Dev 5:2394–9333
7. Conradi P, Heinkel U, Wahl M (1997) Introducing multimedia in teaching of digital system design. In: Proceedings of IEEE international conference on microelectronic systems education MSE, pp 89–90. https://doi.org/10.1109/mse.1997.612560
8. Dharwadkar R, Deshpande NA (2018) A medical Chat-Bot. Int J Comput Trends Technol 60:41–45
9. Battineni G, Chintalapudi N, Amenta F (2020) AI Chat-Bot design during an epidemic like the novel coronavirus. Healthcare 8:154
10. Rarhi K, Bhattacharya A, Mishra A, Mandal K (2018) Automated medical Chat-Bot. SSRN Electron J. https://doi.org/10.2139/ssrn.3090881
11. Sirocki JA (2019) Conversational agent model based on stance and BDI providing situated learning for triage-psychologists in the helpline of 113 suicide prevention
12. Amato F et al (2017) Chat-Bots meet eHealth: automatizing healthcare. In: WAIAH@ AI* IA, pp 40–49
13. Janarthanam S (2017) Hands-on Chat-Bots and conversational UI development: build Chat-Bots and voice user interfaces with Chatfuel, Dialogflow, Microsoft Bot Framework, Twilio, and Alexa Skills. Packt Publishing Ltd.
14. Bobrow DG et al (1977) GUS, a frame-driven dialog system. Artif Intell 8:155–173
15. Fadhil A, Gabrielli S (2017) Addressing challenges in promoting healthy lifestyles: the al-Chat-Bot approach. In: Proceedings of the 11th EAI international conference on pervasive computing technologies for healthcare, pp 261–265
16. Bueno D, Chacón J, Carmona C (2008) Learning to teach sports to handicapped people using games. In: Proceedings of 8th IEEE international conference on advanced learning technology ICALT 2008, pp 590–591. https://doi.org/10.1109/ICALT.2008.181
17. Parker AG (2013) Designing for health activism. Interactions 20:22–25
18. de Clunie GT et al (2011) Developing an android based learning application for mobile devices Desarrollo de una aplicación educativa basada en Android para dispositivos móviles. In: 2011 ieee international conference on technology for education T4E, pp 262–265
19. Wachtel A, Fuchß D, Schulz S, Tichy WF (2019) Approaching natural conversation Chat-Bots by interactive dialogue modelling and Microsoft LUIS. In: 2019 IEEE international conference on conversational data and knowledge engineering (CDKE), pp 39–42

20. Ashcroft M, Kaati L, Meyer M (2016) A step towards detecting online grooming-identifying adults pretending to be Children. In: Proceedings of 2015 European intelligence and security informatics conference EISIC 2015, pp 98–104. https://doi.org/10.1109/EISIC.2015.41
21. Klüwer T (2011) From Chat-Bot s to dialog systems. In: Conversational agents and natural language interaction: techniques and effective practices. IGI Global, pp 1–22
22. Fadhil A (2018) Beyond patient monitoring: conversational agents role in telemedicine and healthcare support for home-living elderly individuals. arXiv:1803.06000
23. Mugoye K, Okoyo H, Mcoyowo S (2019) Smart-bot technology: conversational Agents role in maternal healthcare support. In: 2019 IST-Africa week conference (IST-Africa), pp 1–7
24. Wang YF, Petrina S (2013) Using learning analytics to understand the design of an intelligent language tutor–Chat-Bot lucy. Editor Pref 4:124–131
25. Chandan AJ, Chattopadhyay M, Sahoo S (2019) Implementing Chat-Bot in Educational Institutes
26. Elliott A (2019) The culture of AI: everyday life and the digital revolution. Routledge
27. Stricke B (2020) People v. Robots: a roadmap for enforcing california's new online bot disclosure act. Vanderbilt J Entertain Technol Law 22
28. Zhang B, Wang Q et al (2019) Outfit helper: a dialogue-based system for solving the problem of outfit matching. J Comput Commun 7:50

Air Quality Prediction Using Regression Models

S.K. Julfikar, Shahajahan Ahamed, and Zeenat Rehena

Abstract Due to the urbanization, the cities are under pressure to stay livable all over the world. The quality of air in modern cities has become a remarkable concern nowadays. Air pollution is defined as the presence of harmful substances in the atmosphere that are adverse affected to the health of humans and other living beings. Thus, it is necessary to monitor air quality constantly of a city to provide a smart and healthy environment to citizens. Any air quality monitoring system first collects information about the concentration of air pollutants from the environment then evaluates those raw data. After evaluation, the system provides an assessment or prediction of air pollution of that particular area. In this work, different regression models have been used on the pollutants to find out the suitable model for predicting the air quality. Here, multiple linear regression, support vector regression, and decision tree regression are used for prediction. The simulation results showed that decision tree regression is the best model amongst these three models. It generates a good quality output, and it can be used for predicting the air quality in any air quality monitoring system.

Keywords Air quality prediction · Regression models · AQI · Machine learning

1 Introduction

Not only in India, but also all the cities throughout the world are facing tremendous air pollution. It is one of the major concerns for humans as well as for animals, plants, oceans, and aquatic life worldwide. As air pollution has a direct impact on human health, thus there has been increased public awareness about the same in our country. According to World Health Organization (WHO) [1], air pollution levels are severely very high in many region of the world. The statistics shows that 9 out of 10 people breathe air having high concentration of pollutants. WHO also claims that almost 7

S.K. Julfikar (✉) · S. Ahamed · Z. Rehena
Department of Computer Science and Engineering, Aliah University, Kolkata, West Bengal700160, India

A. Choudhary et al. (eds.), *Applications of Artificial Intelligence and Machine Learning*, Lecture Notes in Electrical Engineering 778,
https://doi.org/10.1007/978-981-16-3067-5_19

million people die every year due to air pollution. A safe and healthy environment is the most important factor for people especially children under age of 5 years are vulnerable to air pollution, hazardous chemicals, havoc change in climate. These environmental risks take lives of 26% [1] of children under 5 year every year.

Further, in most countries, air quality monitoring systems in urban areas are based on centrally located fixed monitoring stations. So, predicting air pollution, the researchers have tried to design and develop models to monitor pollutants in air.

This paper is motivated to predict the air quality using all those labeled data of much known harmful pollutants in our country. This prediction system will help the people to be aware about the air quality and its effect, and also the researcher may get help from this system to find some solutions for this air pollution phenomenon.

This paper studied several regression models to find out the best suitable model of air quality prediction systems. It discusses how the air quality can be predicted by collecting the concentration of pollutants from stations where sensors are available. Multiple linear regression, support vector regression, and decision tree regression models are developed to predict the air quality.

The rest of the paper is organized as follows: Sect. 2 highlights the detailed survey of the literature study on air quality prediction models. Section 3 describes the working procedure of the work presented here. It also demonstrates a brief description of the chosen models. Section 4 shows the simulation results and evaluates the performance of the models. Finally, the paper is concluded in Sect. 5.

2 Related Work

Currently, the whole world is badly facing and going through under different types of toxic waste due to urbanization, population growth, and other changes in environment. Among them, air contamination is one of the major causes of ailing the whole world. The air pollution may occur by the emission of CO_2, CO from vehicles, factories, and industries. Many researchers have done different works in this area.

The authors [2] reviewed various measures which had been used for the prediction of air pollution pollutants. These approaches are deep learning, machine learning, feedforward neural network. In [3], the authors proposed an approach which utilizes the information relevant to the unlabelled data to perform the interpolation and the prediction. They had performed features selection also. In [4], a mechanism has constructed for monitoring the air quality using context-aware computing of urban areas. In [5], the authors worked with time series data using deep learning approaches. They used RNN and LSTM as a framework for estimating the air pollution in South Korea. The authors in [6] compared four unpretentious machine learning algorithms: linear regression, naive Bayes, support vector machine, and random forest to predict air pollutants levels ahead of time. In [7], the authors proposed a system called UH-BigDataSys which monitors urban air quality. They first developed a method by collecting air quality data from multiple sources. Their system finally provides a health guideline of surrounding environments for citizens. In another work, the authors [8] proposed an air quality index prediction model. It is based on historical data, and it predicts the air quality of coming years. The authors proposed a regression model [9] to predict the air quality index (AQI) in Beijing and other pollutant like NOX in an Italian city. Their regression model used machine learning algorithms like support vector regression (SVR) and random forest regression (RFR) for prediction. They compared SVR and RFR models, and they claimed RFR model that shows better result than SVR model.

In [10], the authors have investigated machine learning-based techniques for air quality prediction. In their work, supervised machine learning techniques like logistic regression, random forest, K-nearest neighbors, decision tree, and support vector machines have been used. They compared different pollutants' value using these methods.

In this work, we studied the background work related to prediction of air quality. Three prediction models are thoroughly studied and implemented to find a good model which predicts the air quality in urban region.

3 Overview of the Proposed Work

In this work, different regression models, namely multiple linear regression, support vector regression, and decision tree regression, are studied and developed to understand the best suitable model for predicting air quality in smart city. First, the working steps are discussed and the brief overview of the above-mentioned three models are described later. Figure 1 shows the working procedure of air quality monitoring. These steps are briefly described as follows.

1. **Data Collection**: It is the initial phase of the monitoring process. Required dataset for this work has been collected from Central Control Room for Air Quality Management [11] from the station Anand Vihar, Delhi—DPCC (Delhi Pollution Control Committee).
2. **Sub-index calculation**: Sub-index calculation is the second phase and very important part of this work, because without calculating sub-index, there is no other way to get the air quality index (AQI). The sub-index calculation procedure has discussed later.
3. **AQI calculation**: An air quality index (AQI) is used by government agencies. They communicate to the public regarding the how much pollution in the air. Increase in AQI increases the risk of public health. So it is important to calculate the AQI value. IT can be calculated by comparing the sub-index of all pollutants. The highest sub-index will be the AQI of those pollutants. In this work, AQI is calculated for those parameters which are mentioned in [12].
4. **Creating Dataset**: Once the AQI has been computed, there is no need to keep the sub-indices for predicting the AQI. So, the dataset has been created using the pollutants and the AQI, and all the sub-index has been deleted.

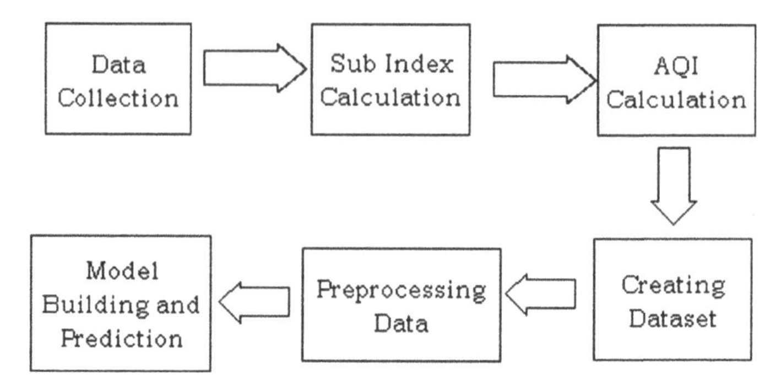

Fig. 1 Working procedure of proposed work

5. **Preprocessing of data**: Preprocessing of data means rectification of data before feeding it into the model. The raw data may have inconsistency, redundant value, missing value, etc. So, preprocessing is used to convert the raw data into a clean dataset by removing redundant values and filling null values. This is done for getting better prediction accuracy.
6. **Model building and prediction:** It is the final step. In this step, model is built using predefined algorithm according to the outcome needed. In this work, three regression models are used for prediction.

3.1 Multiple Linear Regression

Multiple linear regression (MLR) [13] is a statistical technique. It uses two types of variables, namely (i) explanatory variables and (ii) response variable. MLR allows estimating the relation between a response variable and a set of explanatory variables. The explanatory variable is called independent variable and the response variable is called dependent variable. In other words, it can be defined as a multiple linear regression analysis is carried out to predict the values of a dependent variable, say d and a given a set of n explanatory variables $(x_1, x_2, ..., x_n)$. MLR is defined using the following equation.

$$d_i = \gamma_0 + \gamma_1 x_{i1} + \gamma_2 x_{i2} ... + \gamma_n x_{\text{in}} + \epsilon \tag{1}$$

where

i n observations
d_i dependent variables
x_i independent variables
γ_0 d-intercept
γ_n slope coefficient for each independent variable
ϵ error

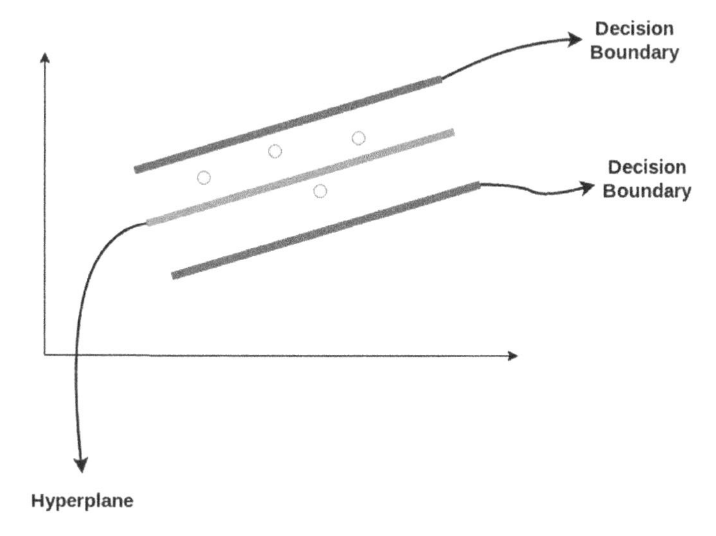

Fig. 2 Support vector regression working functionality

The following assumptions are taken for MLR:

1. A linear relationship exists between the dependent variables and the independent variables.
2. The independent variables are not too highly correlated with each other.
3. Random and independent selection of d_i observations are done from the dataset.

3.2 Support Vector Regression

Support vector machines (SVMs) [14] are very popular and widely used for classification problems in the domain of machine learning. Support vector regression (SVR) is the most common application form of SVMs and uses the same theory as SVM. The problem of regression is defined as a function that approximately maps from an input data in to real data on the basis of a trained data. Figure 2 shows the working functionality of SVR.

In the above figure, consider the two red lines as the decision boundary and the green line as the hyperplane. The objective is to consider the points that are inside the decision boundary line.

3.3 Decision Tree Regression

The decision tree [15] is very simple and the most strong algorithm in machine learning. It belongs to the family of supervised learning algorithms. The decision tree algorithm can be used for solving regression and classification problems as well. It breaks down a dataset into smaller subsets, and as a result, the decision tree is incrementally developed. Thus, the final resultant tree contains decision nodes and leaf nodes. Decision nodes represent a condition, whereas leaf nodes represent the output of the algorithm.

In addition, decision tree used for two categories, namely (i) categorical output problem and (ii) continuous output problem. Decision tree regression is used for the continuous output problem. It selects features of an object and trains a model in the structure of a tree. The trained model helps to predict data in the future to generate a required output. Figure 3 shows the working functionality of decision tree regression.

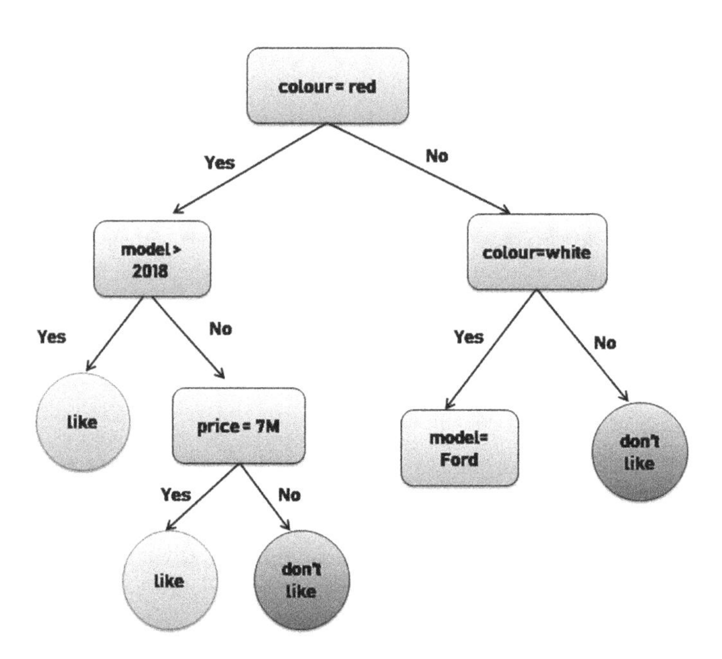

Fig. 3 Working process of DTR

4 Experimental Setup and Result

This section represents experimental setup and simulation process. Simulation and experiments have been done using Python and Jupyter as a tool. The experiment is divided into two primary phases:

1. **Training phase**: The model is trained by using the dataset and is ready for testing.
2. **Testing phase**: In the testing phase, the model is given with the input data for testing. The accuracy is checked in this phase.

The whole dataset has divided into two parts in 80 and 20%. The 80% dataset are for trained the models, and rest of 20% of dataset are used for testing the model and find the accuracy. Below are the following steps which are taken for the experiment.

4.1 Dataset

Initially, the lists of the primary pollutants are formed. The air pollution data of hourly manner are collected from the Central Control Room for Air Quality Management [11] for the following location:

- Station Name: Anand Vihar, Delhi—DPCC.
- Address: Delhi
- Latitude: 28.646835, Longitude: 77.316032

Only eight pollutants have been chosen to calculate the AQI as guided by the India Environment Portal [16]. The dataset of eight pollutants is represented in Fig. 4. These are PM10, NO_2, PM2.5, SO_2, NH_3, CO, O_3, C_6H_6.

After getting the input data, the sub-index of every pollutant has been calculated by the following equation:

$$I_p = [\{(I_{HI} - I_{LO})/(BP_{HI} - BP_{LO})\} * (CO_p - BP_{LO})] + I_{LO} \quad (2)$$

where

I_p	The sub-index of a particular pollutant concentration CO_p
BP_{HI}	Breakpoint concentration higher or equal to given concentration
BP_{LO}	Breakpoint concentration lower or equal to given concentration
I_{HI}	AQI value corresponding to BP_{HI}
I_{LO}	AQI value corresponding to BP_{LO}

Once the sub-indices are formed, AQI can be calculates as AQI = $\text{Max}(I_p)$ as shown in Fig. 5. After doing all this steps, final dataset has been created using seven pollutants as shown in Fig. 6. In this dataset, total number of rows are 8276 of the given dataset. Here, a small portion of it is depicted.

PM2.5	PM10	NO2	NH3	SO2	CO	Ozone	Benzene
361.5	525.5	156.9	62.5	18.13	3.88	1.4	6.08
348	514	142.12	57.52	19.17	2.45	1.7	9.93
298	466.5	121.45	52.55	15	1.8	6.42	9.97
276.5	429.5	113.48	53.95	14.27	1.7	9.98	8.7
295	435.5	108.18	53.2	14.28	1.92	10.5	7.43
346	483.5	88.02	44.25	17.45	1.8	6.9	6.2
359.5	502.5	72.75	40.53	15.48	1.78	7.38	4.02
351.5	501	77.95	43.43	17.1	1.77	3.62	3.58
404.5	581.5	111.57	44.83	20.15	2.4	4.2	3.9
451	636	125.4	49.77	36.6	1.77	9.6	4.18
424	583	103.22	50.82	44.1	1.4	26.98	3.6
369.5	520.5	136.67	69.95	31	1.25	46.4	4.55
329	475.5	135.55	70.38	30.52	1.13	68.95	6.8
309	447.25	118.12	62.78	31.25	0.97	79.7	5.52
239	365.5	90.47	46.75	31.72	0.5	66.17	4.07
127.25	220.75	83.77	45.68	28.95	0.62	77.35	3.6
136	223.5	78.75	34.65	32.9	0.7	56.22	3.43
171.5	295.5	68.92	30.15	35.3	0.82	41.28	1.2
180.75	331.5	83.43	32.9	28.17	1.23	21.07	1.03
221.25	342	123.65	37.43	25.92	1.9	3.58	2.43
316	480.25	97.23	36.8	23.75	1.82	2.5	2.5
300.75	448.75	114.25	47.2	23.27	2.33	1.83	3.22
377.25	519.25	140.85	65.25	21.08	2.43	1.67	6.15
442.25	601.5	140.8	64.85	17.5	2.35	1.67	12.15

Fig. 4 Collected data of eight pollutants

```
In [28]: def calculate_AQI(SI,NI,PMI2_5,PMI10,CI,OI,NHI):

             AQI=0
             if(SI>=NI and SI>=PMI2_5 and SI>=PMI10 and SI>=CI and SI>=OI and SI>=NHI):
                 AQI=SI
             if(NI>=SI and NI>=PMI2_5 and NI>=PMI10 and NI>=CI and NI>=OI and NI>=NHI):
                 AQI=NI
             if(PMI2_5>=SI and PMI2_5>=NI and PMI2_5>=PMI10 and PMI2_5>=CI and PMI2_5>=OI and PMI2_5>=NHI):
                 AQI=PMI2_5
             if(PMI10>=SI and PMI10>=NI and PMI10>=PMI2_5 and PMI10>=CI and PMI10>=OI and PMI10>=NHI):
                 AQI=PMI10
             if(CI>=SI and CI>=NI and CI>=PMI2_5 and CI>=PMI10 and CI>=OI and CI>=NHI):
                 AQI=CI
             if(OI>=SI and OI>=NI and OI>=PMI2_5 and OI>=PMI10 and OI>=CI and OI>=NHI):
                 AQI=OI
             if(NHI>=SI and NHI>=NI and NHI>=PMI2_5 and NHI>=PMI10 and NHI>=CI and NHI>=OI):
                 AQI=NHI

             return (round(AQI))
         data['AQI']=data.apply(lambda x:calculate_AQI(x['SI'],x['NI'],x['PMI2_5'],x['PMI10'],x['CI'],x['OI'],x['NHI']),axis=1)
```

Fig. 5 Calculating AQI

Figure 7 shows correlation factors. It is seen that all the features that are used for the prediction are correlated to each other. Now, this values can be used to train the model.

Data splitting was done as 80% for training and 20% for testing in this work.

4.2 Preprocessing of Data

Data preprocessing has been done (e.g., removing null values, filling null values, etc.), and standard scalar is used as feature selection as shown in Fig. 8.

PM2.5	PM10	NO2	NH3	SO2	CO	Ozone	AQI
361.50	525.50	156.90	62.50	18.13	3.88	1.40	408.0
348.00	514.00	142.12	57.52	19.17	2.45	1.70	403.0
298.00	466.50	121.45	52.55	15.00	1.80	6.42	353.0
276.50	429.50	113.48	53.95	14.27	1.70	9.98	327.0
295.00	435.50	108.18	53.20	14.28	1.92	10.50	345.0
346.00	483.50	88.02	44.25	17.45	1.80	6.90	396.0
359.50	502.50	72.75	40.53	15.48	1.78	7.38	407.0
351.50	501.00	77.95	43.43	17.10	1.77	3.62	402.0
404.50	581.50	111.57	44.83	20.15	2.40	4.20	436.0
451.00	636.00	125.40	49.77	36.60	1.77	9.60	466.0
424.00	583.00	103.22	50.82	44.10	1.40	26.98	448.0
369.50	520.50	136.67	69.95	31.00	1.25	46.40	413.0
329.00	475.50	135.55	70.38	30.52	1.13	68.95	379.0
309.00	447.25	118.12	62.78	31.25	0.97	79.70	359.0
239.00	365.50	90.47	46.75	31.72	0.50	66.17	289.0
127.25	220.75	83.77	45.68	28.95	0.62	77.35	188.0
136.00	223.50	78.75	34.65	32.90	0.70	56.22	193.0

Fig. 6 Final dataset

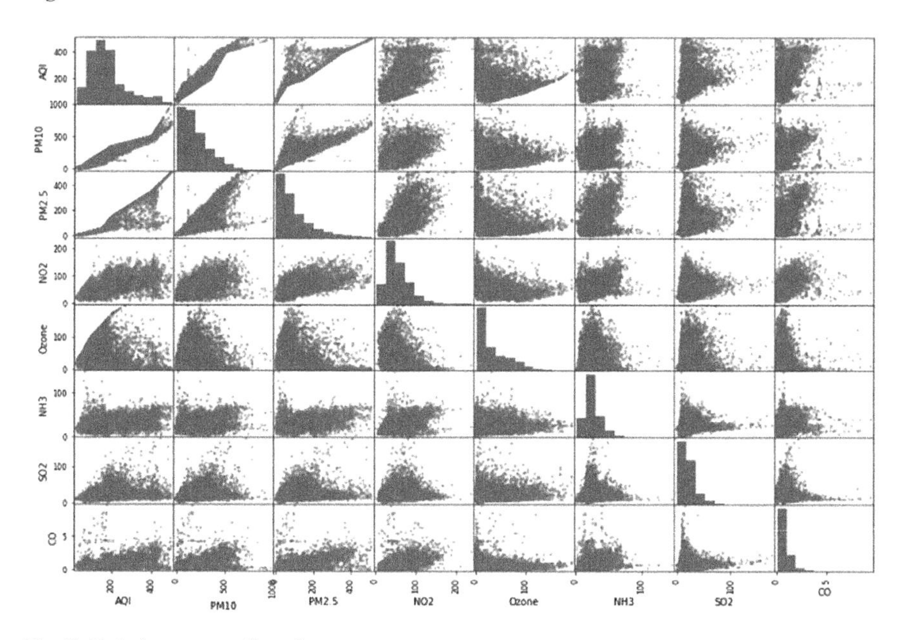

Fig. 7 Relation among all attributes

```
In [44]:  from sklearn.preprocessing import StandardScaler
          sc_X = StandardScaler()
          X_train = sc_X.fit_transform(X_train)
          X_test = sc_X.transform(X_test)
          sc_y = StandardScaler()
          y_train = sc_y.fit_transform(y_train.reshape(len(y_train),1))
          y_test = sc_y.fit_transform(y_test.reshape(len(y_test),1))
```

Fig. 8 Procedure for the feature selection process

Table 1 AQI prediction for three prediction models.

Model	RSME	MSE	Score
Multiple liner regression	22.76	517.87	0.94
Support vector regression	30.61	936.69	0.89
Decision tree regression	9.26	85.74	0.99

The standard score of a sample x is calculated as: $z = (x - u)/s$ where u is the mean of the training samples. s is the standard deviation of the training samples.

RSME and MSE are calculated for each of the model. Table 1 shows the AQI prediction for the three models.

4.3 Results

Comparison of the three models is shown in Fig. 9. It shows the boxplot of AQI of the used models. When compared to other machine learning models applied on the chosen dataset, decision tree regression (DTR) shows the best for this system with the mean accuracy and standard deviation accuracy to be 0.99 (99%). It also can be seen in the figure that the original value and the predicted values of DTR are very close to each other. Figure 10 also depicts the RMSE and MSE values of all the three models. From the figure, it is clear that DTR has the lowest value of RSME and MSE among these. Hence, to predict AQI value from a given sample, the decision tree regression could be used.

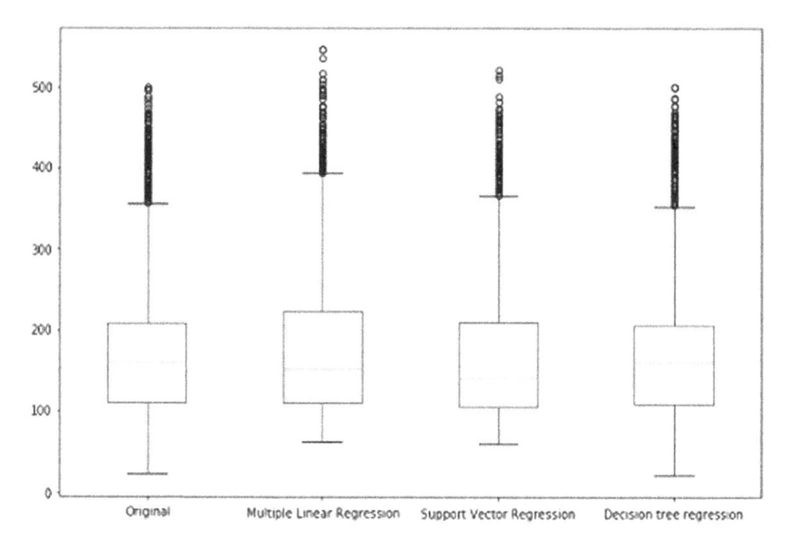

Fig. 9 Comparison of all models using boxplot

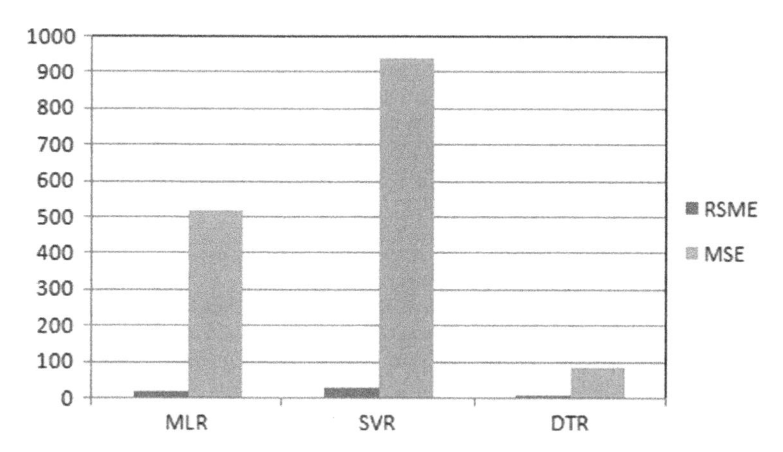

Fig. 10 Comparison of the models using RSME and MSE value

5 Conclusion

Rapid urbanization puts pressure on controlling the air pollutant levels low, and it is becoming one of the most important tasks. Air quality prediction system is important to know the pollution of the surrounding environment. In this work, a thorough study of several prediction models, mechanisms are studied. Here, mainly three prediction models like multiple linear regression, support vector regression, and decision tree regression are studied and implemented. These three models are compared with each other by comparing RSME and MSE value. The results show that decision tree regression can be efficiently used to detect the quality of air and predict the level of AQI in the future. This study will help the researcher as well as the government organizations to use the decision tree regression to develop and form policies for an air quality management system. A good air quality monitoring system is required to develop where this prediction model need to be incorporated as a future direction of this work.

References

1. Air Pollution, WHO (2020) https://www.who.int/health-topics/air-pollution#tab=tab_1. Last accessed 20 March 2020
2. Raj PP, Dewangan S, Harne S (2019) Prediction and optimization of air pollution-a review paper. Int J Res Appl Sci Eng Technol
3. Qi Z, Wang T, Song G, Hu W, Li X, Zhang Z (2018) Deep air learning: interpolation, prediction, and feature analysis of fine-grained air quality. IEEE Trans Knowl Data Eng 30(12):2285–2297
4. Ghose B, Rehena Z (2020) A mechanism for air health monitoring in smart city using context aware computing. Proc Comput Sci
5. Bui TC, Le V, Cha SK (2018) A deep learning approach for forecasting air pollution in South Korea Using LSTM
6. Nemade S, Mankar C (2019) A survey on different machine learning techniques for air quality forecasting for urban air pollution. Int J Res Appl Sci Eng Technol
7. Chen M, Yang J, Hu L, Hossain MS, Muhammad G (2018) Urban healthcare big data system based on crowdsourced and cloud based air quality indicators. IEEE Commun Mag
8. Soundari AG, Jeslin JG, Akshaya AC (2019) Indian air quality prediction and analysis using machine learning. IJAER
9. Liu H, Li Q, Yu D, Gu Y (2019) Air quality index and air pollutant concentration prediction based on machine learning algorithms. Appl Sci 9(19):4069
10. Mahesh Babu K, Rene Beulah J (2019) Air quality prediction based on supervised machine learning methods. Int J Innov Technol Explor Eng
11. CCR (2020) Central control room for air quality management. https://app.cpcbccr.com/ccr/#/caaqm-dashboard-all/caaqm-landing. Last accessed 20 March 2020
12. Central Pollution Control Board (Oct 2014) Air quality index. http://www.indiaenvironmentportal.org.in/files/file/Air. Last accessed 20 March 2020
13. Multiple Linear Regression (MLR) (2020) https://www.investopedia.com/terms/m/mlr.asp. Last accessed 20 June 2020
14. Support Vector Regression (SVR) (2020) https://www.analyticsvidhya.com/blog/2020/03/support-vector-regression-tutorial-for-machine-learning. Last accessed 20 June 2020
15. Decision tree regression in machine learning. https://medium.com/@chughkashish12/decision-tree-regression-in-machine-learning-4f117158cdcc. Last accessed 20 June 2020
16. India Environment Portal (2020). http://www.indiaenvironmentportal.org.in/. Last accessed 20 June 2020

Anomaly Detection in Videos Using Deep Learning Techniques

Akshaya Ravichandran and Suresh Sankaranarayanan

Abstract People's concern for safety in public places has been increasing nowadays and anomaly detection in crowded places has become very important due to this aspect. This paper provides an approach towards identifying suspicious behaviors automatically in a crowded environment. In light of this, we have validated two powerful deep learning based models namely CNN and VGG16. So, for this purpose, we have collected a number of CCTV videos for detecting and differentiating between both normal and anomalous activities. Now based on videos collected, they have been trained using VGG16 and CNN model towards achieving the best accuracy

Keywords CNN · Deep learning · Anomaly detection · VGG16

1 Introduction

Public places, such as airports, train stations, theme parks, and shopping centers are usually crowded. Every year the number of people is increasing due to urbanization. These places should be monitored constantly to avoid such dangerous situations and any unusual activity that is being spotted should be reported immediately. Public health and security are primary considerations for study of crowds. Human operators observe the visual screens continuously to detect any event of interest that becomes challenging to tackle for long durations. This process seems to be more taxing than it sounds and therefore, researchers are developing an automatic system that allows the user to monitor the scene.

The computer vision and signal processing departments have performed extensive research in the area of crowd anomaly identification. Recently, there have been many attempts to exploit deep learning frameworks to avoid some complex approaches for

A. Ravichandran (✉)
Sigma Soft Tech Park, Whitefield Main Road, Bangalore, India

S. Sankaranarayanan
Department of Information Technology, SRM Institute of Science and Technology, SRM Nagar, Kattankulathur Campus, Chennai, India
e-mail: sureshs3@srmist.edu.in

A. Choudhary et al. (eds.), *Applications of Artificial Intelligence and Machine Learning*, Lecture Notes in Electrical Engineering 778,
https://doi.org/10.1007/978-981-16-3067-5_20

the retrieval and delivery of hand-designed software. In [1], computer vision guided anomaly detection studies have first been performed and it proposes various model-based approaches for anomaly detection such as Dirichlet Process Mixture Models (DPMM), Bayesian Mixture Model, etc. In [2], a fully convolutional feed-forward auto encoder to learn both the local features and the classifiers as an end-to-end learning framework has been proposed for this same problem of anomaly detection. In [3], Hochreiter's 1991 analysis of this problem has been discussed, then it has been addressed by introducing a novel, efficient, gradient-based method called Long Short-Term Memory" (LSTM). In this paper [4], a general end-to-end approach to sequence learning that makes minimal assumptions on the sequence structure has been presented. In this paper [5], precipitation nowcasting as a spatiotemporal sequence forecasting problem has been formulated in which both the input and the prediction target are spatiotemporal sequences. This has been done by extending the fully connected LSTM (FC-LSTM). In this paper [6], a comprehensive survey of several representative CNN visualization methods, including Activation Maximization, Network Inversion, Deconvolutional Neural Networks (Deconv Net), and Network Dissection based visualization has been provided. In [7], Adam, an algorithm for first-order gradient-based optimization of stochastic objective functions, based on adaptive estimates of lower-order moments has been discussed. A key insight of the paper [8] is that if anomalies are local optimal decision rules are local even when the nominal behavior exhibits global spatial and temporal statistical dependencies. This insight helps collapse the large ambient data dimension for detecting local anomalies. In [9], a novel unsupervised deep learning framework for anomalous event detection in complex video scenes has been discussed. In [10], a novel framework for anomaly detection in crowded scenes has been discussed. In [11], for each category i.e. anomaly or normal, a basic anomaly detection technique has been discussed, and then how the different existing techniques in that category are variants of the basic technique has been discussed. In [12], an unsupervised approach for anomaly detection in high-dimensional data has been discussed. There are numerous shortcomings in all the above-proposed work, such as the availability of ground truth, the type of anomaly, etc., despite comprehensive research work and advancement in this area. The computer vision group also faces a variety of obstacles in designing successful solutions for detecting anomalies. It covers the unavailability of sensors, severe weather situations, night vision issues, etc. The researchers are unable to detect abnormal scenarios often because of these obstacles. To fix these issues, smartphones are suitable alternatives for other issues, as discussed. Increasing number of mobile phones has helped the network of signal processers to evaluate and appreciate crowd dynamics. The data obtained from these devices may be used to reveal information regarding human behaviors and community dynamics within a crowd.

Inside mobile phones, sensors such as accelerometer, gyroscope, Bluetooth, proximity sensor, webcam, ambient light sensor, etc. aid to take more precise images. These sensors acquire data that helps understand the behavior of the people in crowds. More research can be done relating to the identification of single human action structures. Such systems can catch events such as walking, sitting, biking, swimming, etc. The key purpose of the device developed and suggested by this paper is to identify suspicious events in the crowd by concurrently analyzing the non-visual data collected synchronously from a bunch of individual smartphones. It is the first research to our knowledge that uses non-visual data from handheld accelerometers and gyroscopes to identify crowd-based community anomaly.

One of the main inputs to this paper is the usage of a smartphone equipped with rich sensors to gather data together and track suspicious activity in crowds. This paper presents a comprehensive collection of data coupled with video-based ground reality, collected from accelerometers, gyroscopes, and several fixed cameras that may be used for analysis purposes to examine individual behaviors in crowds. The rest of the paper is organized as follows: Sect. 2 analyzes the state of the art, while Sect. 3 explains the methodology of the new idea. Section 4 reflects on the results of the new program and, ultimately, Sect. 5 summarizes and addresses potential plans for more technology enhancements.

2 Anomaly Detection Recent Surveys

Deep learning algorithms allow the systems to track the behavior based on actual data. It is possible to develop algorithms that allow computers to view behaviors learned from the trained model. Deep learning algorithms are used in a given video to study abnormal instances. The video dataset is trained with multi-data algorithms. This concept is used to identify and track real-time anomalies. Deep learning is important for training and studying anomalous behavior.

A few important findings in this area of study that has been published over the last ten years or so. The authors in [13] used visual trajectories to analyze the identification, tracking, scenario analysis, and behavioral perception. The study addressed surveillance, knowing the behavior, and detecting incidents from the video. The study discussed is the first work covering strategies for anomaly detection. To find phenomena, it included individuals, including detection processes, learning processes, and scenario simulation. From object identification, monitoring, and analysis of actions detailing the success of the last decade of works, this was portrayed with an object-oriented approach. A multi-camera study on surveillance in multi-camera setup is presented in the research. Authors addressing events that are regarded as a type of anomalous event need urgent intervention, arising accidentally, rapidly, and spontaneously.

The work discussed in [14] addresses machine vision-related implementations from the perspective of intelligence and law enforcement. The analysis described in [1] addresses the characteristics of human behavior and mechanisms for conduct awareness. The authors of [15] describe the task of knowing human nature through human actions and experiences. Intelligent camera systems have been studied in [16], covering facets of analytics. Surveillance systems were introduced in [17], with different implementation areas. Datasets used for the identification of phenomena are covered in [18]. In [19], Scientists presented anomalous human behavior recognition analysis with an emphasis on task representation and emulation, object extraction approaches, classification and behavior analysis models, performance assessment techniques, and video surveillance system sample repositories. The paper summarizes major computer-based vision studies undertaken over the last 10 years and also summarizes significant computer-based vision research conducted in the past 10 years. In our work, we have concentrated in particular on the anomaly detection studies. Anomalies are by their very definition subjective. The principles used in detections of phenomena cannot be uniformly generalized across various situations of operation. From the application viewpoint, we examine the capabilities of anomaly detection approaches utilized in road camera surveillance.

3 Proposed Work

We have used CNN and VGG 16- special CNN architecture in this work to identify the videos of an anomaly. The model used are convolution neural normal networks (CNN), and VGG16—specific type of CNN architecture. We train the pattern in this project with videos of a real-time anomaly. We have constructed both CNN and VGG16 models independently and trained them using the same dataset. This has been done in order to compare the performance of the models and make use of the model which performs best.

3.1 CNN Model

A convolution neural network (CNN) is a deep learning algorithm capable of collecting an input image, assigning meaning (learnable weights and biases) to specific aspects/objects in the image, and separating one from another. In CNN the need for pre-processing is far smaller than other classification algorithms. In CNN, we have convolution layer followed by pooling layer where Relu activation is used. Finally, after flattening the layer, it is constructed as fully connected layer where SoftMax activation function applied for output (Fig. 1).

3.2 VGG 16 Model

VGG16, introduced by K, is a blueprint of the convolution neural network. Simonyan, A. Zisserman, of Oxford University, in the article "Very Broad Convolutionary Object Recognition Networks", the algorithm reaches 92.7% of the top-5 measurement accuracy in ImageNet, which is a dataset of over 14 million images belonging to 1000 categories. This was one of the ILSVRC-2014 best-recognized concepts that had been received. Building on Alex Net, it replaced large kernel filters (11 and 5

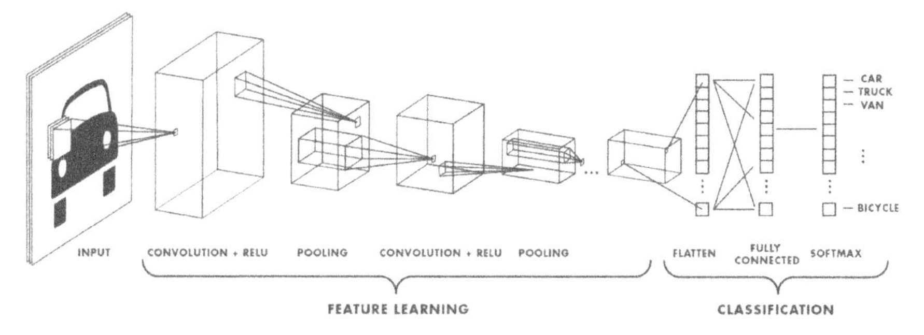

Fig. 1 CNN structure [20]

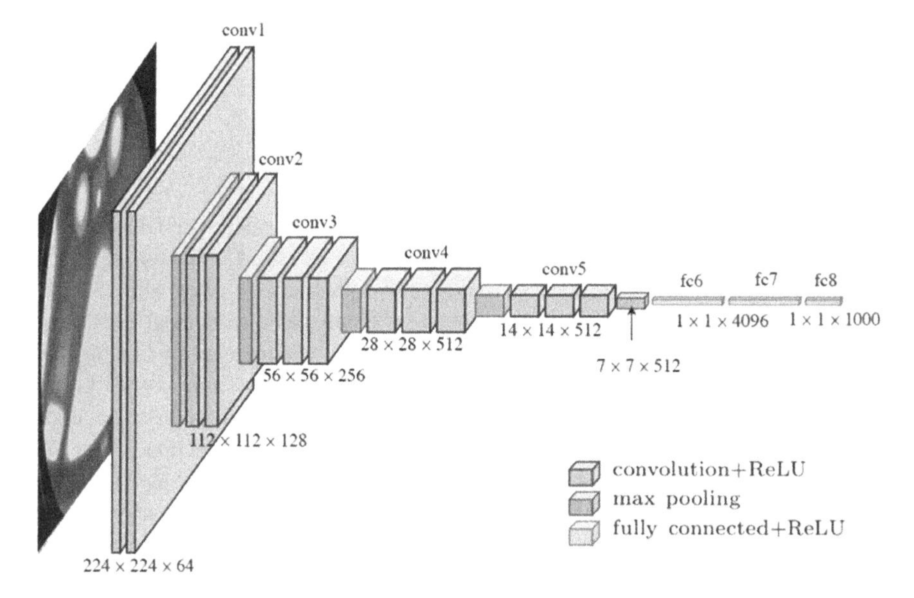

Fig. 2 VGG 16 structure [21]

in the first and second convolution layers respectively) with miniature, one by one, 33 kernel-sized filters. VGG16 was primed for weeks and was using NVIDIA Titan Black GPUs (Figs. 2 and 3).

4 Implementation

This system is developed using Python programming utilizing Anaconda framework. Datasets have been collected retrospectively from the UCF Center for Research in Computer Vision. Videos have been labeled according to their relevant classification

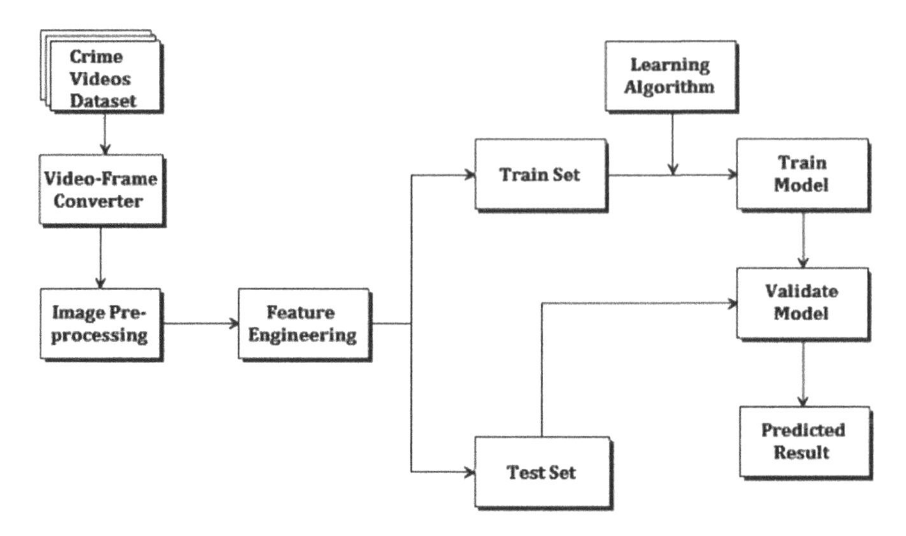

Fig. 3 Anomaly detection system flow

and then fed into the system. Videos have been trained and tested with CNN Model and VGG 16. We have gathered the dataset from the website Center for Research in computer vision [22]. We have classified our data into five categories: ***Arson, Robbery, Fire, Protection, and Routine***. We have 25 videos in each category making up 125 videos on the whole. Surveillance recordings can capture a variety of plausible phenomena. Within this article, we recommend investigating phenomena with the aid of manipulating both anomalous and daily images. To stop annotating the anomalous fragments or excerpts in educational films that can be very time-consuming, we recommend learning anomaly by more than one sample ranking system by exploiting weakly classified educational films. The training marks (anomalous or normal) are at the video stage, instead of the clip-point. In our approach, we find normal and anomalous motion pictures as bags and video segments as instances in a few cases that gain knowledge of (MIL), and periodically evaluate a profound anomaly ranking model that predicts high anomaly rankings for anomalous video segments. Also, inside the ranking fault function, we integrate sparsity and transient smoothness constraints to consider anomaly higher all by planning.

4.1 UCF-Crime Dataset

To test our strategy, we compile a brand new, huge-scale dataset, dubbed UCF-crime. This contained long unregulated surveillance films covering 13 real-world anomalies, including violence, detention, attack, road crashes, burglary, fire, fighting, theft, firing, stealing, shoplifting, and vandalism. Such results are chosen as they have an immense effect on public safety. Here is a summary of each anomalous event.

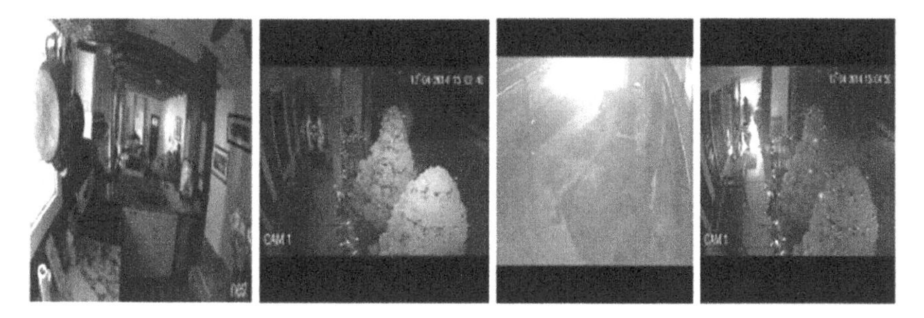

Fig. 4 Video frames for Arson [22]

Fig. 5 Video frames for Burglary [22]

4.1.1 Video Frames for Arson

Figure 4 shows the Arson video where people deliberately setting fire to property.

4.1.2 Video Frames for Burglary

Figure 5 shows the video frames illustrating burglary. This burglary video shows thieves barging into a residence and robbing it. It no longer consists of exerting pressure on humans.

4.1.3 Video Frames for Fighting

Figure 6 shows frames from a fighting video, which shows two or more people attacking each other.

Fig. 6 Video frames for fighting [22]

4.1.4 Video Frames for Explosion

Figure 7 illustrate an Explosion video, showing the destructive event of something blowing apart. This case does not contain images in which a human intentionally causes a fire or starts an explosion.

Fig. 7 Video frames for explosion [22]

Fig. 8 Video frames for normal event [22]

4.1.5 Video Frames for Normal

Figure 8 illustrate a Normal Event. It contains a video where no crime has occurred. These videos include both indoor and outdoor scenes (such as a shopping mall), as well as day and night scenes.

Classification in deep learning refers to a problem of predictive modeling where a class mark is predicted for a given example of input data. Types of classification problems include: Classify whether it is Anomaly or not, provided an example.

4.2 VGG16 & CNN Model Implementation

After collecting the dataset, dataset is arranged based on the classification and has made it ready for the training process. The frames have been extracted from videos using the OpenCV library for the training process. Even though all the frames have been extracted it doesn't mean that the frame size of all pixels are the same. Therefore, all the frames have been resized into (64 * 64 * 3) using OpenCV. After this process, datasets are made ready and model built.

The uniqueness of the VGG16 implementation is that, instead of using a large number of hyper parameters as we do in CNN, more focus has been given in building convolutional layers of 3×3 filter with stride 1. Also, the same padding and maxpool layer of 2×2 filter of stride 2 has been used. Convolutional as well as maxpool layers have been arranged in a consistent manner throughout the architecture. We have used two fully connected layers followed by a SoftMax layer for output.

In VGG16 implementation, firstly all the libraries imported which would be required for implementation. Then, sequential model has been created. After this, Image Data Generator from keras. Preprocessing has been imported. The objective of Image Data Generator is to import data with labels into the model. After this, the data is passed into the neural network. After the creation of the convolutional layer, data has been passed onto the dense layer and ReLu activation function has been used for this layer. Then after the SoftMax layer has been created, the model is finally prepared and then compiled. Followed by this, Adam's optimizer has been used to reach the global minima while training the model.

5 Results and Analysis

Anomaly detection in videos is one of the emerging requirements in this jiffy. Using deep learning techniques, namely CNN and VGG 16 we have built a model to classify anomaly in videos and the results indicate that VGG16 performs slightly better compared to CNN with an accuracy of 90%. CNN achieved an accuracy of 88%. The reason CNN performs better than other previously implemented models is because it detects the important features without any human supervision. The reason VGG16 has performed slightly better is because it uses a more basic architecture with no residual blocks. In terms of Precision-Recall, VGG16 give precision of 0.91 and recall of 0.90 as compared to CNN which shows the maximum videos classified correctly. Also, VGG16 gives a $F1$ score of 0.90 as compared to CNN which indicates the test accuracy. These are shown in Figs. 9, 10, 11 and 12.

```
print(classification_report(y_true=y_true,y_pred=y_predict))

              precision    recall  f1-score   support

           0       0.98      0.80      0.88       148
           1       0.93      0.93      0.93        75
           2       0.91      0.92      0.92       135
           3       0.87      0.89      0.88        85
           4       0.61      0.97      0.75        40

    accuracy                           0.88       483
   macro avg       0.86      0.90      0.87       483
weighted avg       0.90      0.88      0.89       483
```

Fig. 9 CNN results

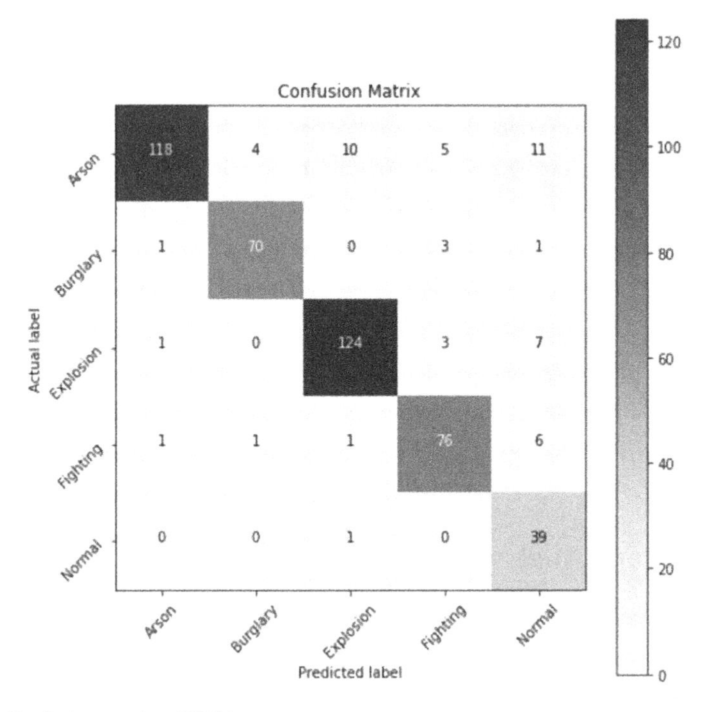

Fig. 10 Confusion matrix of CNN

```
print(classification_report(y_true=actual,y_pred=y_predict))
              precision    recall  f1-score   support

           0       0.95      0.91      0.93       148
           1       0.95      0.93      0.94        75
           2       0.83      0.99      0.90       135
           3       0.99      0.86      0.92        85
           4       0.78      0.62      0.69        40

    accuracy                           0.90       483
   macro avg       0.90      0.86      0.88       483
weighted avg       0.91      0.90      0.90       483
```

Fig. 11 VGG16 results

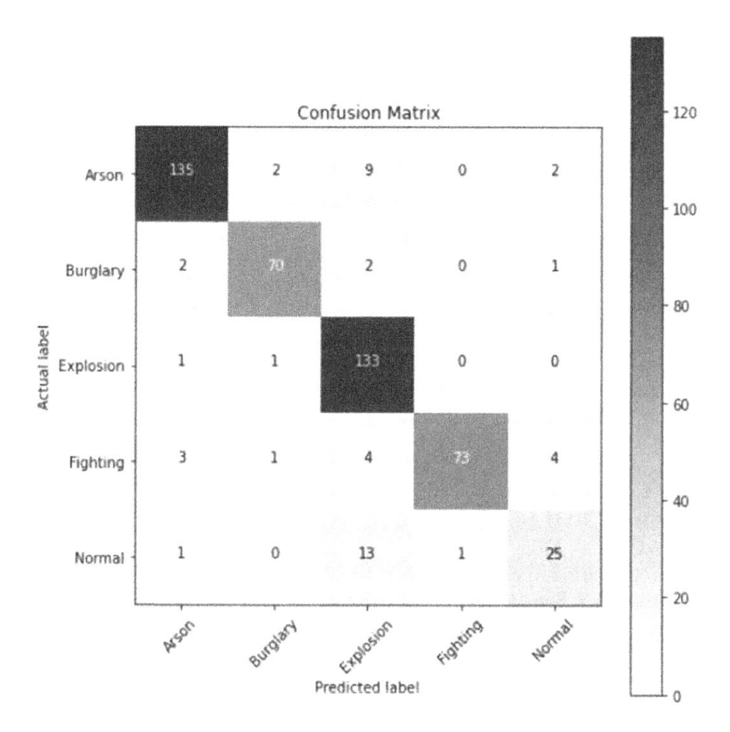

Fig. 12 Confusion matrix of VGG16

6 Conclusion and Future Scope

In this paper, deep learning-based models namely CNN & VGG16 have been implemented for detecting anomalies in surveillance videos. To train and test the models

125 videos have been used. The result shows that VGG 16 performs slightly better than the CNN model. Video classification is much different from Image classification and also it is complex. Also, achieving 90% accuracy in video classification shows that our model works perfectly.

These models have been created and tested using Python on desktops. However, actual usage of video anomaly detection is usually required in CCTV camera systems. Hence, in future, these models can be inserted in hardware chips and added in hardware systems to provide Artificial Intelligence (AI) such that when an anomaly is detected the hardware itself sends an alert notification to the corresponding departments. Also, in future we aim to implement more datasets for better results and comparison.

References

1. Tan H, Zhai Y, Liu Y, Zhang M (2016) Fast anomaly detection in traffic surveillance video based on robust sparse optical flow. In: 2016 IEEE international conference on acoustics speech and signal processing (ICASSP), pp 1976–1980, Mar 2016
2. Hasan M, Choi J, Neumann J, Roy-Chowdhury AK, Davis LS (2016) Learning temporal regularity in video sequences. In: 2016 IEEE conference on computer vision and pattern recognition (CVPR), pp 733–742, June 2016
3. Hochreiter S, Schmidhuber J (1997) Long short-term memory. Neural Comput 9(8):1735–1780
4. Sutskever I, Vinyals O, Le QV (2014) Sequence to sequence learning with neural networks. In: Proceedings of the 27th international conference on neural information processing systems, vol 2, Ser NIPS'14. MIT Press, Cambridge, MA, USA, pp 3104–3112
5. Shi X, Chen Z, Wang H, Yeung DY, Wong WK, Woo WC (2015) Convolutional LSTM network: a machine learning approach for precipitation nowcasting. In: Proceedings of the 28th international conference on neural information processing systems, vol 1, Ser NIPS'15. MIT Press, Cambridge, MA, USA, pp 802–810
6. Zeiler MD, Krishnan D, Taylor GW, Fergus R (2010) Deconvolutional networks. In: 2010 IEEE computer society conference on computer vision and pattern recognition, pp 2528–2535, June 2010
7. Kingma DP, Ba J (2014) Adam: a method for stochastic optimization. CoRR. abs/1412.6980
8. Saligrama V, Chen Z (2012) Video anomaly detection based on local statistical aggregates. In: 2012 IEEE conference on computer vision and pattern recognition, pp 2112–2119, June 2012
9. Xu D, Ricci E, Yan Y, Song J, Sebe N (2015) Learning deep representations of appearance and motion for anomalous event detection, BMVC. BMVA Press, pp 8.1–8.12
10. Mahadevan V, Li W, Bhalodia V, Vasconcelos N (2010) Anomaly detection in crowded scenes. In: 2010 IEEE computer society conference on computer vision and pattern recognition, pp 1975–1981, June 2010
11. Chandola V, Banerjee A, Kumar V (2009) Anomaly detection: a survey. ACM Comput Surv
12. Ramchandran A, Kumar Sangaiah A (2018) Unsupervised anomaly detection for high-dimensional data-exploratory data analysis. In: Computational intelligence for multimedia big data on the cloud with engineering applications
13. Zhou S, Shen W, Zeng D, Fang M, Wei Y, Zhang Z (2016) Spatial-temporal convolutional neural networks for anomaly detection and localization in crowded scenes. Signal Process Image Commun 47(C):358–368
14. Piciarelli C, Micheloni C, Foresti GL (2008) Trajectory-based anomalous event detection. IEEE Trans Circuits Syst Video Technol 18(11):1544–1554

15. Li F, Yang W, Liao Q (2016) An efficient anomaly detection approach in surveillance video based on oriented GMM. In: 2016 IEEE international conference on acoustics speech and signal processing (ICASSP), pp 1981–1985, Mar 2016
16. Tang X, Zhang S, Yao H (2013) Sparse coding based motion attention for abnormal event detection. In: 2013 IEEE international conference on image processing, pp 3602–3606, Sept 2013
17. Kim J, Grauman K (2009) Observe locally infer globally: a space-time MRF for detecting abnormal activities with incremental updates. In: 2009 IEEE conference on computer vision and pattern recognition, pp 2921–2928, June 2009
18. Krizhevsky A, Sutskever I, Hinton GE (2012) Imagenet classification with deep convolutional neural networks. In: Advances in neural information processing systems, vol 25. Curran Associates Inc. pp 1097–1105
19. Wang L, Qiao Y, Tang X (2015) Action recognition with trajectory-pooled deep-convolutional descriptors. In: 2015 IEEE conference on computer vision and pattern recognition (CVPR), pp 4305–4314, June 2015
20. A comprehensive guide to convolution neural networks-EL5 way. Available from https://towardsdatascience.com/a-comprehensive-guide-to-convolutional-neural-networks-the-eli5-way-3bd2b1164a53
21. VGG-16. Available from https://www.geeksforgeeks.org/vgg-16-cnn-model
22. UCF anomaly detection dataset. Available from https://visionlab.uncc.edu/download/summary/60-data/477-ucf-anomaly-detection-dataset

Unsupervised Activity Modelling in a Video

Aman Agrawal

Abstract In today's modern era, human activity recognition is widely used in video surveillance for various purposes like safety and security. This proposed work is also used in health care system, entertainment environment. In the unsupervised activity recognition, it recognize the activity in current frame and compares it to the previous frame. If there is any sudden change in activity then it gives the current frame number. The proposed system is basically divided into three phases: Pre-processing, Feature extraction, and Recognition. Gaussian Mixture Model is used to recognise moving object by using background subtraction. Then I apply some rules to recognize the different activities. I create two types of datasets i.e. two activity and three activity dataset. This approach shows better results in both the datasets.

Keywords Activity recognition · Gaussian mixture model · Centroid · Speed · Precision · Recall · *F*-measure

1 Introduction

Over the last decade, the use of surveillance cameras increases very rapidly. The demanding task is to understand the Human Activities from videos. The main functionalities of video system are automatically detecting the action of humans from video sequence. Video surveillance is important for observing the safety and security of elderly individuals in the home domain. Video surveillance is also used to analyze uncommon movements of patients in the hospital.

Today's video surveillance is leading research areas and also used in various fields like safety and security of common people in public places. It provides important information about various crimes and terrorist attacks and also used to provide good management in transportation and other public services. It detects various activities of humans and categorized into normal, abnormal and suspicious activities. There are two approaches for learning are as follows:

A. Agrawal (✉)
Department of Computer Engineering and Applications, GLA University, Mathura, India
e-mail: aman.agrawal@gla.ac.in

A. Choudhary et al. (eds.), *Applications of Artificial Intelligence and Machine Learning*, Lecture Notes in Electrical Engineering 778,
https://doi.org/10.1007/978-981-16-3067-5_21

In supervised surveillance systems, it requires label data for training. Possibly, all the activities cannot process with one-time training so this approach is not suitable for large-scale deployment. Hence, there is need for unsupervised surveillance. It learns the activities by its own from unlabelled data, i.e. it does not require any labelled data. System is able to learn itself, easy to deploy and make system possible for large-scale monitoring. Therefore, latest research tries to adapt or build the automatic system in unsupervised manner. In activity recognition system, there are various challenges such as occlusion, intraclass variability, illumination changes, noise, etc. For foreground detection there are various techniques for background subtraction are Kalman filter, Gaussian Mixture Model (GMM), HMM, Histogram similarity, Incremental PCA, etc.

The main goal of this approach is to develop a framework for unsupervised activity detection. It gives the frame number when the activity changes in a video sequence. It detects the changes in activity automatically. It does not require any trained data and perform a task in unsupervised manner.

2 Related Works

Lots of research are done in the field of activity recognition. Different researchers give various methods in the field of activity recognition. Some of them are discussed in this section: Ke et al. [1] proposed a survey on human activity recognition in a video. They describes three levels in human activity recognition as follows: core technology, activity recognition system and various applications. In core technology, they discussed about segmentation, feature extraction and classification of the activity. In recognition system, they discussed mainly three types which are single person, multi-person and crowd behaviour. Finally, various applications are discussed like health-care systems, entertainment and surveillance system which also gives the various challenges related to the applications. They gave a fine survey on human activity recognition in a video. Vishwakarma et al. [2] give the review paper on activity recognition in a video. They discussed about simple and complex activities and its various applications. It covers all aspects of general framework of activity recognition and summarized recent research progresses related to general framework and also provides the overview of various datasets of activity recognition. Ning et al. [3] proposed a framework of object tracking by active coutour segmentation and joint registration. It deals with non-rigid shape changes of target. Singh et al. [4] proposed human activity tracking which is capable of detecting people in both indoor and outdoor environment. On the basis of space and time feature, the behaviour of the object in a video sequence is easily detected. The temporal information is useful for motion detection in video sequence that's why Cheng et al. [5] proposed an approach for event detection by temporal sequence.

In this, video is represented by sequence of visual words and apply sequence memorizer to capture dependencies in video sequence in temporal context and this model is further integrated with the classifier for classification of the activities.

This approach is applied to two challenging datasets. For object detection, various approaches are proposed: Deepak et al. [6] gives the implementation of bounding box for object detection. It separates the moving object from background and detect all the objects from video sequence. They design a bounding box for human detection in the presence of crowd. Bounding box is used to track the object in each frame and help for recognising the various activities in a video sequence. Hossen et al. [7] proposed activity recognition in thermal infrared video.

Semisupervised approach is the combination of both supervised and unsupervised. Initially, it needs manpower to train and then it increments after every iteration. So, this approach is not suitable and requires large memory. Zhang et al. [8] give semisupervised approach using HMM for detection of unusual events. In unsupervised, different researchers gave different approaches. Nater et al. [9] gave temporal relation in video for unsupervised activity analysis. In this approach, it encoded the temporal relation by discriminative slow features between consecutive frames and automatically segment the activities in hierarchical manner. It is used to analyse the unseen data activities which are modelled in a generative manner. This approach is purely feature-independent and data-driven. Yeo et al. [10] proposed unsupervised coactivity detection using Markov Chain. In this, it constructs a complete multipartite graph where vertices in subsequence of video and edges are connective between those vertices and weight is directly proportional to the similarity of two end vertices. This algorithm identifies the subset of subsequences as a co-activity by estimating absorption time. The main advantage of this algorithm is that it can handle two videos naturally. Wang et al. [11] proposed unsupervised activity perception in crowded and complicated scenes. Author uses hierarchical Bayesian model to connect three elements that are visual feature, atomic activities and interactions. Author proposes three Bayesian models which is Latent Dirichlet Allocation (LDA) mixture model, Hierarchical Dirichlet Processes (HDP) and Dual Hierarchical Processes (Dual-HDP) model.

Based on Hidden Markov Model Trabelsi et al. [12] gave an unsupervised approach for automatic activity recognition. They proposed the method which is based upon joint segmentation of multidimensional time series and it uses expectation maximization algorithm where no activity is labelled. For abnormality activity recognition, lots of research is done. Hu et al. [13] give unsupervised activity detection in crowded scenes using semiparametric scan statistics. Here, Training datasets are not needed before detection. The abnormality is measured by likelihood ratio test statistics. Si et al. [14] give unsupervised learning of events using and-or grammar. It uses predefined set of unary and binary relation for each frame and their co-occurrences are cluster into atomic actions.

Najar et al. [15] this paper proposed finite multivariate generalized Gaussian mixture model to recognize human activities using unsupervised learning. They develop algorithm by using fixed point covariance matrix combined with Expectation-Maximization algorithm. This recognition applies both images and videos.

3 Proposed Methodology

Unsupervised activity detection system developed to make surveillance system intelligent and smart. In recent years, when the criminal activities are increases. We need a good surveillance system. This approach is helpful to detect activity as well as any mis-happening in a video. In this work, activity is detected and classified but it does not give the name of the activity. When the person suddenly changes its activity then it gives the alert and frame number when activity is change. This proposed approach is work well for real-time videos. This approach used an unsupervised learning method and it does not require any training database unlikely as supervised requires training database.

Previously, many researchers work in the field of unsupervised activity detection. Initially, it does not require any training database. When new activity arrived, it added to the database and incremented. When next time activity is arrived it match to the database. This approach required more memory to save the activities. This method is not suitable for large videos and real-time videos.

In this approach, Gaussian mixture model is used to detect the human in a video. After foreground detection, different features are extracted like centroid, height-width ratio, distance, and speed. When the person change its activity it gives a frame number of the video.

The framework of proposed work is shown in Fig. 1. There are three phases in proposed framework: Pre-processing, Feature Extraction, and recognition phase. In pre-processing phase, first we extract frames from video, then detect the moving object and then noise removal is done. In feature extraction, calculate different features like centroid, speed and dimensions. Finally, in recognition phase implement rules based on that it classifies different activities and gave frame number when activity changes as an output.

3.1 Pre-processing

The Pre-processing phase consists of frame extraction from input video, foreground detection and noise removal. For foreground detection, Gaussian Mixture Model (GMM) is used. Gaussian Mixture Model (GMM) [16] is a popular method that has been employed to tackle the problem of background subtraction. In this, each pixel in a frame is modelled into Gaussian distribution. Each pixel is divided by its intensity value and compute probability for each pixel. According to the probability, each pixel is decided whether it belongs to foreground or background pixel. Some snapshot of moving object detection is shown in Fig. 2. By using the following Eq. (1):

$$P(Z_t) = \sum_{i=1}^{K} w_{i,t}.\eta\big(Z_t, \mu_{i,t}, \Sigma_{i,t}\big) \tag{1}$$

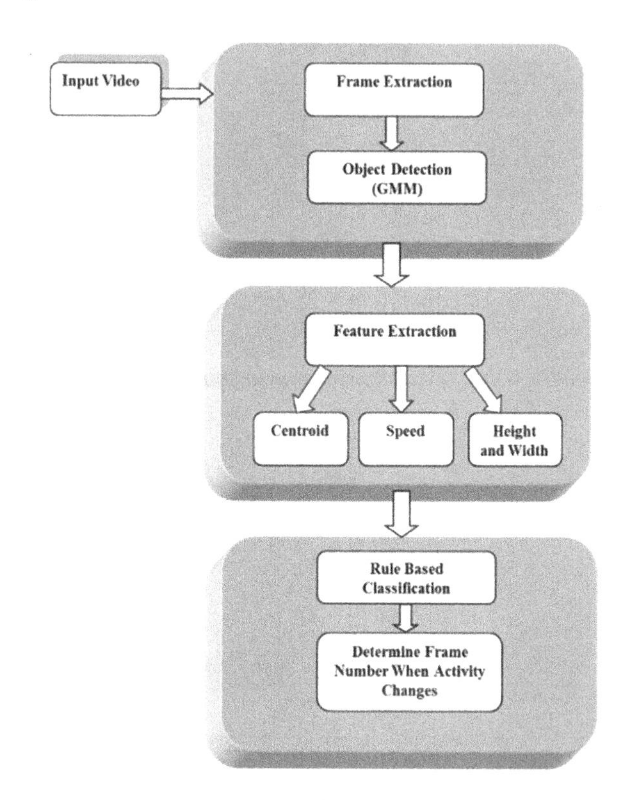

Fig. 1 Framework of proposed system

where

Z_t Current pixel in frame t,
K Number of distributions in the mixture.
$w_{i,t}$ The weight of the k_{th} distribution in frame t.
$\mu_{i,t}$ The mean of the k_{th} distribution in frame t.
$\sum_{i,t}$ The standard deviation of the k_{th} distribution in frame t.

And $\eta(Z_t, \mu_{i,t}, \Sigma_{i,t})$ is probability density function which is defined as Eq. (2):

$$\eta = \frac{1}{(2\pi)^{\frac{n}{2}}|\Sigma|^{\frac{1}{2}}} \exp^{\frac{-1}{2}(Z_t-\mu)\Sigma^{-1}(Z^t-\mu)} \quad (2)$$

3.2 *Feature Extraction*

Feature extraction is an important step in recognition of different activities. After object detection, different features are extracted. For classification of different

Fig. 2 Snapshots of moving object detection

activities, we consider different parameters like Centroid, Distance, Speed and Measurements. It is defined as follows:

Centroid: Centroid is a centre of mass of an object. Centroid is computed for both x and y directions. It is formulated as:

$$\text{Centroid in} x\text{-direction} = \frac{\text{width of blob}}{2} \tag{3}$$

$$\text{Centroid in} y\text{-direction} = \frac{\text{height of blob}}{2} \tag{4}$$

Distance: It is defined as distance covered by an object in two consecutive frames. Distance is calculated by Euclidean formula.

$$\text{Distance} = \sqrt{(x_2 - x_1)^2 + (y_2 - y_1)^2} \tag{5}$$

Speed: It is defined as distance travelled by object in consecutive frames divided by time taken by that object. To distinguish between walking and running speed is calculated. It is formulated as:

$$\text{Speed} = \frac{\text{distance covered by object}}{\text{total time taken by object}} \tag{6}$$

Measurements: It the measurement of the ratio of height of blob to width of blob. It is calculated by using a formula:

$$\text{Height Width ratio} = \frac{\text{height of blob}}{\text{width of blob}} \tag{7}$$

3.3 Recognition

In the proposed framework, I define some rules to classify different activities. This method is unsupervised so it does not require any labelled training database. In this method, define rules to differentiating between different activities and every frame is compared with previous frame on the basis of different features. It checks whether object changes its activity or same as previous frame. When object changes, its activity gives the frame number of the video.

In Fig. 3, there are multiple frames which shows different activities in a single video. In the starting frames, person is walking. After some time, person changes its activity from walk to sit in next frames. Again, person changes its activity from sit to walk in further frames. So, in this particular video, person changes its activity two times.

Fig. 3 Different activities in a single video

4 Experimental Results

The entire examinations and approval with different measurements performed in the proposed approach. The test comes about performed on the dataset used to incite the best outcomes. Different measurements applied to demonstrate that our outcomes are predominant and analyzed.

4.1 Dataset

There is no availability of standard dataset. We make our own dataset. Our dataset consists of 165 videos of different activities. Our dataset consists of 4 activities: walk, run, sit and stop. Our dataset categorised into two categories are as follows: 1. two activity dataset 2. Three activity dataset.

In two activity dataset, there are five combinations which are as follows: walk-stop, run-stop, walk-run, walk-sit, and run-sit.

In three activity dataset, there are six combinations which are as follows: run-sit-run, run-sit-walk, run-stop-run, walk-stop-walk, walk-sit-walk, and run-stop-walk. Some snapshots of our dataset are shown in Fig. 4.

Fig. 4 Snapshot of our dataset

4.2 Result Analysis

The performance of our proposed framework is evaluated for two activities dataset and three activities dataset. I have calculated precision and recall scores for quantitative analysis for both categories.

The precision-recall can be calculated by the equation given in Eqs. (8) and (9)

$$\text{precision} = \frac{\text{tpr}}{\text{tpr} + \text{fpr}} \tag{8}$$

$$\text{recall} = \frac{\text{tpr}}{\text{tpr} + \text{fnr}} \tag{9}$$

where, tpr is the true positive rate, fpr is the false positive rate and fnr is the false negative rate.

The Accuracy is for the proposed method and to validate the performance of our proposed method. I have assumed the general computation in this measure to determine flexibility.

F-measure is a harmonic mean of precision and recall, which is defined as

$$F\text{-measure} = \frac{2.\text{precision}.\text{recall}}{\text{precision} + \text{recall}} \tag{10}$$

Table 1 shows the detection rate and accuracy of two activity dataset. Proposed framework correctly classified the videos and gives the frame number when activity changes and also classified as true positive and false negative. I take the 75 videos of two activity which consists 5 combination of activities. Each combination has 15 videos. Table 2 shows the detection rate and accuracy of three activity datasets. I take 90 videos of three activity dataset and make 6 combinations. Each has 15 videos and both tables show the average of various measurements.

In Fig. 5, shows the PR curve of both two and three activity dataset. In this, I performed precision-recall curve (PR Curve) to validate our results. If the approach

Table 1 Detection rate of two activities dataset

Classes of two activity videos	Total number of videos	True positive	False positive	False negative	Precision	Recall	F-measure
Walk–Stop	15	12	2	1	0.85	0.92	0.88
Run–Stop	15	11	2	2	0.84	0.84	0.84
Walk–Run	15	12	1	2	0.92	0.85	0.88
Walk–Sit	15	12	1	2	0.92	0.85	0.88
Run–Sit	15	11	3	1	0.78	0.91	0.84
Total	75	58	9	8	0.86	0.88	0.86

Table 2 Detection rate of three activities dataset

Classes of three activity videos	Total number of videos	True positive	False positive	False negative	Precision	Recall	*F*-measure
Run–Sit–Run	15	10	3	2	0.77	0.83	0.80
Run–Sit–Walk	15	12	2	1	0.86	0.92	0.89
Run–Stop–Run	15	11	2	2	0.84	0.84	0.84
Run–Stop–Walk	15	11	2	2	0.84	0.84	0.84
Walk–Sit–Walk	15	13	1	1	0.93	0.93	0.93
Walk–Stop–Walk	15	13	1	1	0.93	0.93	0.93
Total	90	70	11	9	0.86	0.89	0.87

Fig. 5 Graph show PR curve of **a** two activity dataset, **b** three activity dataset

correctly given the frame number when activities are changed and compared with the ground truth then it adds to the True Positive Rate. If it gives more activity changes when compared with the ground truth then it adds to the True Negative Rate and if it gives no activity change then it adds to False Negative Rate.

In Fig. 6, graph shows the average *F*-measure of two activity and three activity videos. The accuracy computed for three activity videos is slightly larger than the two activity videos as shown in the graph. The *F*-measure has been computed for both

Fig. 6 Graph shows the accuracy of two activity and three activity dataset

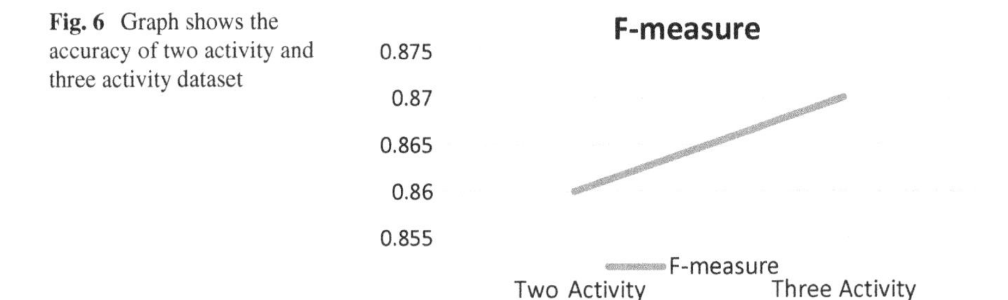

two and three activity datasets. The analysis in this graph determines the computation for both datasets. Our approach has found better results compare with the ground truth on both of the datasets.

5 Conclusion and Future Work

In this proposed approach the unsupervised activity recognition in a video surveillance and create our own dataset with different scenes, person, illumination changes and natural scenes. Various researchers have done valuable researches in unsupervised activity recognition and proposed different kinds of approaches and techniques which accomplish the goals of activity recognition successfully The proposed system is basically divided into three phases: (1) Pre-processing (2) Feature extraction and (3) Recognition. In pre-processing, we extract frames from video then we apply GMM for moving object detection. In feature extraction, we extract different features like centroid, speed and height-width. In recognition phase, with the help of features extracted it detect different activities and give frame number when activity changes. We have verified the proposed approach on our own created dataset. We make two datasets that are two activity and three activity dataset. We calculate Precision, Recall and F-measure to validate our results in both datasets. The aim to extend the proposed work could be as follows: To detect the activity of multiperson in a video and we can also detect more activities in a video.

References

1. Ke S-R, Thuc HLU, Lee Y-J, Hwang J-N, Yoo J-H, Choi K-H (2013) A review on video-based human activity recognition. Computers 2(2):88–131
2. Vishwakarma S, Agrawal A (2013) A survey on activity recognition and behavior understanding in video surveillance. Vis Comput 29(10):983–1009
3. Ning J, Zhang L, Zhang D, Yu W (2013) Joint registration and active contour segmentation for object tracking. IEEE Trans Circuits Syst Video Technol 23(9):1589–1597
4. Singh D, Yadav AK, Kumar V (2014) Human activity tracking using star skeleton and activity recognition using HMMS and neural network. Int J Sci Res Publ 4(5):9
5. Cheng Y, Fan Q, Pankanti S, Choudhary A (2014) Temporal sequence modeling for video event detection. In: Proceedings of the IEEE conference on computer vision and pattern recognition, pp 2227–2234
6. Deepak P, Krishnakumar S, Suresh S (2014) Human recognition for surveillance systems using bounding box. In: International conference on contemporary computing and informatics (IC3I). IEEE, pp 851–856
7. Hossen J, Jacobs EL, Chowdhury FK (2015) Activity recognition in thermal infrared video. In: SoutheastCon 2015. IEEE, pp 1–2
8. Zhang D, Gatica-Perez D, Bengio S, McCowan I (2005) Semi-supervised adapted hmm for unusual event detection. In: IEEE computer society conference on computer vision and pattern recognition. CVPR 2005, vol 1. IEEE, pp 611–618
9. Nater F, Grabner H, Van Gool LJ (2011) Temporal relations in videos for unsupervised activity analysis. BMVC 2:8

10. Yeo D, Han B, Han JH (2016) Unsupervised co-activity detection from multiple videos using absorbing markov chain. In: AAAI, pp 3662–3668
11. Wang X, Ma X, Grimson WEL (2009) Unsupervised activity perception in crowded and complicated scenes using hierarchical Bayesian models. IEEE Trans Pattern Anal Mach Intell 31(3):539–555
12. Trabelsi D, Mohammed S, Chamroukhi F, Oukhellou L, Amirat Y (2013) An unsupervised approach for automatic activity recognition based on hidden markov model regression. IEEE Trans Autom Sci Eng 10(3):829–835
13. Hu Y, Zhang Y, Davis L (2013) Unsupervised abnormal crowd activity detection using semiparametric scan statistic. In: Proceedings of the IEEE conference on computer vision and pattern recognition workshops, pp 767–774
14. Si Z, Pei M, Yao B, Zhu S-C (2011) Unsupervised learning of event and-or grammar and semantics from video. In: 2011 IEEE international conference on computer vision (ICCV). IEEE, pp 41–48
15. Najar F, Bourouis S, Bouguila N, Belghith S (2019) Unsupervised learning of finite full covariance multivariate generalized gaussian mixture models for human activity recognition. Multimedia Tools Appl 78(13):18669–18691
16. Nurhadiyatna A, Jatmiko W, Hardjono B, Wibisono A, Sina I, Mursanto P (2013) Background subtraction using gaussian mixture model enhanced by hole filling algorithm (GMMHF). In: 2013 IEEE international conference on systems, man, and cybernetics (SMC). IEEE, pp 4006–4011

Performance Comparison of Various Feature Extraction Methods for Object Recognition on Caltech-101 Image Dataset

Monika, Munish Kumar, and Manish Kumar

Abstract Object recognition system helps to find the label of the object in an image. The identification of the object depends on the features extracted from the image. Features play a very important role in the object recognition system. The more relevant features an object has, the better the recognition system will be. Object recognition system mainly works in two major phases—feature extraction and image classification. Features may be the color, shape, texture, or some other information of the object. There are various types of feature extraction methods used in object recognition. These methods are classified as handcrafted feature extraction methods and deep learning feature extraction methods. This article contains a comprehensive study of various popular feature extraction methods used in object recognition system. Various handcrafted methods used in the paper are scale invariant feature transformation (SIFT), speeded-up robust feature (SURF), oriented FAST and rotated BRIEF (ORB), Shi-Tomasi corner detector, and Haralick texture descriptor. The deep learning feature extraction methods used in the paper are ResNet50, Xception, and VGG19. In this article, a comparative study of various popular feature extraction methods is also presented for object recognition using five multi-class classification methods—Gaussian Naïve Bayes, k-NN, decision tree, random forest, and XGBoosting classifier. The analysis of the performance is conducted in terms of recognition accuracy, precision, $F1$-score, area under curve, false positive rate, root mean square error, and CPU elapsed time. The experimental results are evaluated on a standard benchmark image dataset Caltech-101 which comprises 8677 images grouped in 101 classes.

Keywords Object recognition · Feature extraction · Image classification · Deep learning

Monika
Department of Computer Science, Punjabi University, Patiala, India

M. Kumar (✉)
Department of Computational Sciences, Maharaja Ranjit Singh Punjab Technical University, Bathinda, Punjab, India

M. Kumar
Department of Computer Science, Baba Farid College, Bathinda, Punjab, India

A. Choudhary et al. (eds.), *Applications of Artificial Intelligence and Machine Learning*, Lecture Notes in Electrical Engineering 778,
https://doi.org/10.1007/978-981-16-3067-5_22

1 Introduction

Object recognition system is a key research area in the field of artificial intelligent and computer vision. This system works on identifying objects in the real world from an image in the world. The system receives an image as an input and provides the correct label of the object in image as an output. Humans can recognize objects effortlessly and instantaneously but it is very difficult for a machine to do the same task. There is a need to train the machine by storing the relevant information of the images with their associated label. The trained data enables the system to assign a label for the input image/test data. Object recognition system mainly works in two phases—feature extraction and image classification.

Feature extraction is one of the most important phases in object recognition system. It describes the relevant features of an image. It is a kind of feature reduction algorithm. An input image in the recognition system contains a lot of information which is too large to process and needs a huge memory to store. Even this huge information may cause the problem of overfitting during classification. So there is a need to extract the most relevant features from the image that will help to identify the label of the object in the image. Feature extraction algorithms involve the procedure of transforming the large dimensional data into the reduced dimensional data by accepting important data and discarding unnecessary data. An image contains various types of features like color, texture, shape, or some other information. The efficiency of recognition system depends on the selection of features and depends on the choice of particular feature extraction method. This phase forms the feature vector of extracted features that are then used by the classifier to recognize the image. There are various types of feature extraction methods used in object recognition. In the beginning of computer vision, a lot of work has been done using various handcrafted (conventional) feature extraction methods. Nowadays, feature extraction through deep convolutional neural network has made a tremendous response for object recognition. Specifically, features are extracted using handcrafted and/or deep learning feature extraction methods. In the paper, the authors have considered some well-known feature extraction methods such as scale invariant feature transform (SIFT), speeded-up robust feature, oriented FAST and rotated BRIEF (ORB), Shi-Tomasi corner detector, Haralick texture descriptor, ResNet50, Xception, and VGG19. The authors have evaluated the performance of each method in terms of recognition accuracy, precision, $F1$-score, area under curve (AUC), false positive rate (FPR), root mean square error (RMSE), and CPU elapsed time. The experiment is conducted on one of the most challenging image dataset Caltech-101. This dataset contains 101 object classes and 1 background. Each class contains numerous images in the range of 40–800. This is an unbalanced and multi-class dataset. The authors have experimented their work on 101 object classes. So for classification, the authors have considered multi-class classification methods and evaluated the results using multi-class performance evaluation measures.

The rest of the paper is organized in various sections; Sect. 2 presents the literature review related to the feature extraction methods. Section 3 comprises the detail of various feature extraction methods used in the work. In Sect. 4, the authors have given the description on various classification algorithms used in the work. Section 5 reports the comparative analysis of various feature extraction methods over various classification methods. Finally, the paper is concluded in Sect. 6.

2 Related Work

Affonso et al. [1] proposed a combined approach of feature extraction to find the quality of wood boards based on their images. They combined deep learning and Haralick texture descriptor for feature extraction. Then, they applied various classification algorithms to predict the quality of the board. They represented the comparison of experimental results achieved using various classification algorithms. Agarwal et al. [2] performed a comparative study of SIFT and SURF algorithms for object recognition. They evaluated the performance of both algorithms on different objects under different background conditions. At the end, they evaluated that SIFT can work for noisy and illuminated images but SURF is not able to extract features for noisy images. SURF executes faster than SIFT. Routray et al. [3] presented the performance analysis of three popular feature extraction techniques—SIFT, SURF, and HOG over noisy images. The experiment was evaluated on three types of noisy images such as Gaussian, salt and pepper, and speckle. Three levels of variance were considered for Gaussian, salt and pepper, and speckle noisy images. The performance of these methods was presented by considering the number of correct matches between original and noisy images. The paper concluded that SIFT extracts more features and matching points than SURF and it is slow. Rather SURF is faster than SIFT and extracts more relevant information from images. However, HOG focuses on textural information of the image. He et al. [4] experimented with a modified SIFT feature extraction method for content-based image retrieval (CBIR). They compared the performance of the proposed method with some other feature extraction methods such as SURF and HoG. The experiment proved the performance of the proposed system for image retrieval. Xiao et al. [5] presented a comparison among three pre-trained convolutional neural network (CNN) models—Inception V3, ResNet50, Xception, a convolutional neural network with three layers and traditional hand-crafted feature extraction methods. They proposed a combined approach using three pre-trained CNN models on breast ultrasound images. Gupta et al. [6] performed an analysis on two feature extraction techniques, i.e., SIFT and ORB for object recognition. In the paper, they used feature selection and dimensionality reduction algorithms using k-means clustering and locality preserving projection, respectively. They made a comparative analysis on three-dimension sizes of feature vectors—8, 16, and 32 dimensions and observed highest results for a feature vector of size 8. Finally, they proposed a combined approach using both feature extraction methods for object recognition system. Tropea and Fedele [7] presented a comparative analysis of various machine learning classifiers on a pre-trained convolutional neural network Google Inception v3 model. The classifiers used in the experiment were multi-class logistic regression (MLR), Gaussian Naïve Bayes, k-nearest neighbor (k-NN), random forest, and support vector machine (SVM). The experiment was conducted on a benchmark dataset Caltech-256 where the first 100 images of each category were chosen. The authors also reported the performance results on varied dataset size for training and test dataset such as 50:50, 60:40, 70:30, 80:20, 90:10.

3 Feature Extraction Methods

Features are defined as the properties of an object that helps to identify it. Feature extraction is the procedure to select a subset of important features from the whole dataset of an image. Object recognition system uses many types of features such as color, edges, corners, texture, or some other information. These features are categorized in local, global, or deep features. Local features represent the small portion of the image such as corner, edges, and shape. Some of the local feature extraction methods are scale invariant feature transform (SIFT), speeded-up robust feature (SURF), oriented FAST and rotated BRIEF (ORB), Shi-Tomasi corner detector, Haralick texture descriptor, etc. Global features are usually some characteristics of the image such as area, perimeter, and moments. It describes the image as a whole. Some of the global features are Hough transform, Zernike moments, Hu moments, discrete cosine transform (DCT), etc. Deep features are the features derived from a set of neural network layers. There are various popular pre-trained deep models used for feature extraction such as VGG19, ResNet50, and Xception. The detail of local and deep feature extraction methods is described in this section.

3.1 Local Features

Local feature extraction methods refer to the low level information of an image. In this subsection, the authors have described various local feature extraction methods used in the work.

3.1.1 Scale Invariant Feature Transform (SIFT)

Scale invariant feature transform (SIFT) provides a robust mechanism for feature extraction in object recognition. This method is more robust to translation, scale, rotation, and illumination. SIFT is proposed by Lowe [8]. It also works for low resolution images efficiently. SIFT performs in four computational stages—scale space extreme detection, keypoint localization, orientation computation, and keypoint descriptors.

- *Scale-space Extreme Detection*—In the first step, Difference-of-Gaussian (DoG) is computed to identify the keypoints. The computed keypoints are invariant to scale and orientation.
- *Keypoint Localization*—This step performs the selection of important keypoints. This is done with the use of threshold. The intensity of the keypoints computed in the first step is compared with the threshold. If the intensity is less than the selected threshold, then that keypoint is discarded. This removes the low contrast keypoints and unnecessary edges.
- *Orientation Computation*—In this step, various rotations are assigned on each keypoint. An orientation histogram of 36 bins covering 360° is created. It also computes the direction of gradient which makes the image invariant to rotation.
- *Keypoint Descriptors*—Here, a 16×16 neighborhood around each keypoint is computer, which are further divided into 16 sub-blocks of 4×4 size. For each sub-block, an orientation of 8 bins is applied. Finally, it computes a feature vector of 128 dimensions.

3.1.2 Speeded-Up Robust Feature (SURF)

Speeded-up robust feature (SURF) is a local feature detector and descriptor proposed by Bay et al. [9]. SURF, which is motivated by SIFT, is much faster than SIFT. It works in two stages—feature detection and feature description that is described as:

- *Feature Detection*—Here, features are extracted using convolution of Laplacian of Gaussian with box filter on integral images. SURF follows the space scale by keeping the same size of an image and increasing the size of the filter. Further, a non-maximum suppression in a $3 \times 3 \times 3$ neighborhood is applied to localize the keypoints in an image.
- *Feature Description*—Here, features are described by constructing a square region aligned to the selected orientation. A rotation of angle 60° is done for the scanning region and wavelet responses are recomputed. This results in a feature vector of 64 dimensions.

3.1.3 Oriented FAST and Rotated BRIEF (ORB)

Oriented FAST and rotated BREIF (ORB) is also a feature detector and descriptor used in object recognition developed by Rublee et al. [10]. It is built on two algorithms—FAST keypoint detector and BRIEF keypoint descriptor. Some modifications are further introduced in the method to enhance the performance. It works in the following steps.

- First it uses features from accelerated segments test (FAST) algorithm to find the keypoints.
- Then, it applies Harris corner measure to find the top N points among them.
- For scale invariance, it uses a multiscale image pyramid. For rotation invariance, moments are computed with x and y. From these moments, centroid for each

patch is computed and orientation of the patch is computed to make it rotation invariance.

- Then, binary robust independent elementary feature (BRIEF) descriptor is used to create the feature vector. BRIEF is not robust to rotation so ORB uses a steered version of BRIEF. Finally, a feature vector of length 64 is created for object recognition.

3.1.4 Shi-Tomasi Corner Detector

Shi-Tomasi corner detector is used to detect the corners of the object which helps in the identification of the object. It is proposed by Shi and Tomasi [11]. This method is based on Harris corner detector but it has given more accurate results than Harris algorithm. A modification has been done in the selection criteria (R) of Harris corner detector which has made it better than the original. The value of R is compared with a certain predefined value. If it exceeds, it can be marked as a corner.

3.1.5 Haralick Texture

Haralick texture feature is a local feature extraction method proposed by Haralick et al. [12]. It extracts the features of an object based on the texture. For this, it uses the gray level co-occurrence matrices (GLCM) to count the co-occurrence of gray levels in the image. GLCM contains the information about how image intensities in pixels with a certain position in relation to each other occur together. This method computes fourteen features from the elements of GLCM. These features are angular second moment (ASM), contrast, correlation, variance, inverse difference moment, sum average, sum variance, sum entropy, entropy, difference variance, difference entropy, information measures of correlation, and maximal correlation coefficient.

3.2 Deep Features

In the current era, deep learning is widely used for object recognition and in other fields of research. Deep learning models are designed on the architecture of the human brain. There are various pre-trained convolution neural networks (CNN) used for feature extraction. Further, these features are classified using machine learning or deep learning classifiers. Specifically, CNN performs both feature extraction and classification automatically (see Fig. 1). By removing the last layer(s) from the model, it can be used for feature extraction. The models are trained on a predefined huge dataset that may be implemented on other datasets. In the study, the authors have used three popular pre-trained CNN models—ResNet50, Xception, and VGG19. These models are already trained on ImageNet dataset which contains 1000 categories and

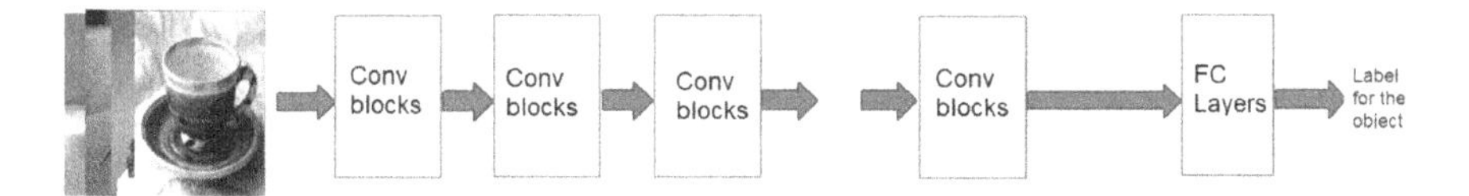

Fig. 1 Basic diagram of convolutional neural network (CNN) model

millions of images. In the experiment, the authors have reduced the size of the feature vector by 8 dimensions. This section describes the architecture of the models as.

3.2.1 ResNet50

ResNet50 proposed by He et al. [13] is designed in four stages. It considers an image of size $224 \times 224 \times 3$. Initially, it performs convolutional and max-pooling using 7×7 and 3×3 kernel sizes, respectively. Afterward, Stage 1 starts consisting of 3 residual blocks containing 3 layers. Then, Stage 2 starts by performing an operation on the first residual block with stride 2. It reduces the size of input but doubles the channel width. Continuing it, the next 3 residual blocks perform operation. Stage 3 has 1 residual block with stride 2 and 5 normal residual blocks. Stage 4 has 3 residual layers out of which first performs stride 2. Each residual layer contains 3 layers of convolution of 1×1, 3×3, and 1×1 in sequence. Finally, there is an average pooling layer followed by a fully connected layer.

3.2.2 Xception

Xception was developed by Chollet [14]. The architecture of the model is based on depthwise separable convolution layers. It consists of three sections—entry flow, middle flow, and exit flow. Entry flow section initially starts with 2 convolution layers, and afterward, contains 3 Residual blocks. At the start of each residual block, there is a convolution layer with stride 2 to reduce the size of input by half. Then, each of three Residual blocks contains 2 separable convolution layers followed by a max-pooling layer. The middle flow section contains one residual block having 3 separable convolution layers. Exit flow section again has 2 residual blocks which starts with a convolution layer with stride 2. First residual block consisting of 2 convolution layers ends with max-pooling layer. Second residual block comprises of 2 convolution layers and ends with global averaging pooling. Finally, a fully connected layer is added to the end of the model.

VGG19

VGG is named after Visual Geometry Group at Oxford's and is proposed by Simonyan et al. [15]. VGG19 contains 5 residual blocks followed by a max-pooling layer. First two residual blocks consist of 2 convolution layers of 3×3 kernel size.

Next three residual blocks consist of four convolution layers of 3×3 kernel size. Next VGG19 has three fully connected layers used for classification.

4 Classification Methods

Classification is used to classify the images according to the similar features of an image. It helps to predict a label for a given image in object recognition. This is a supervised approach which is accomplished by training a part of the dataset and then applying the test data (image) for the recognition of the object. There are various machine learning and deep learning classifiers used in object recognition system. In the article, the authors have made a study on various well-known machine learning classifiers such as Gaussian Naïve Bayes, k-NN, decision tree, random forest, and XGBoosting classifier. The authors have presented a comparative analysis of these classifiers on various feature extraction methods mentioned in Sect. 3. The results of comparison are reported in Sect. 5. In this section, a detail of various classifiers used in the experiment is described as follows.

4.1 Gaussian Naïve Bayes

Gaussian Naïve Bayes is a supervised machine learning algorithm used for classification of images. It is a variant of Naïve Bayes classifier. Gaussian Naïve Bayes classifier assumes that the value of one feature does not depend on values of other features. It supports the continuous data and distributes the data according to the normal (or Gaussian) distribution. It uses the probabilistic approach to predict the class label for the object.

4.2 K-Nearest Neighbor (k-NN)

K-NN is a very simple and easy to interpret method used for supervised machine learning classification. The performance of the algorithm depends on the value of k. This algorithm assigns the correct label to the object through majority voting by considering the distance between its k-nearest neighbors closest to the training data. Various distance measures are used to find the distance for k-NN algorithm such as Euclidean, Manhattan, Minkowski, and Hamming distance. Most classifiers use Euclidean distance. For categorical variables, Hamming distance must be considered. After evaluating the distance between k neighboring points, a majority voting scheme is applied to find the class for a test data.

4.3 Decision Tree

Decision tree is a supervised machine learning algorithm where the features of the images are represented through a decision tree [16]. The tree is created by splitting the training data (or features) of the images according to a certain criterion. Root of the tree represents all the features of the images. Each node of the tree represents a specific attribute, and a branch connects one node to the other by splitting the features into further nodes. The leaf of the tree specifies the label (or class) of the object. Class for the test data is obtained by comparing the features of the test to each node of the tree in a specific path. Various measures are used to determine the best way to split the data such as Gini index, entropy, and misclassification error. The tree is grown by considering impurity, Gini index, and information gain. These parameters are used to create an efficient structure of the tree by removing the impurities in the tree.

4.4 Random Forest

Random forest is a supervised and ensemble machine learning method used for image classification. It is introduced by Kleinberg [17]. It consists of several decision trees that are created during training of the dataset. Random forest overcomes the limitations of decision tree such as overfitting problem and dependency on one criterion. In random forest algorithm, a forest is created by considering several decision tree (CART) in parallel using some form of randomization. It also adds the concept of bootstrapping which ensures that each individual decision tree in the forest is unique. Randomness is applied during both steps of classification—training and testing of data. During training, randomness enables to grow the decision tree using a different subset. In testing phase, it helps to select the correct nodes that matches the attributes of the test data. It follows the top-down approach to build the forest. For splitting of training dataset, Gini index is used that minimizes the impurities in the forest. The class label is predicted using the majority voting scheme of all the trees. As compared to the decision tree, it works slow due to the large number of trees but gives more efficient results.

4.5 XGBoosting Classifier

XGBoosting classifier proposed by Chen et al. [18] is an ensemble machine learning classifier that is designed using three components—bagging, stacking, and boosting. Bagging or Bootstrap Aggregation helps in classification during training phase and predictive phase. In the training phase, it separates the training dataset into different random subsamples, and in the predictive phase, it uses the majority voting for all the subsamples. Stacking helps to improve the accuracy and reduce overfitting. Boosting helps to convert the weaker trees into stronger trees. This reduces the bias from the tree and improves the accuracy of the recognition system. XGBoosting classifier works in two steps. First, it regularizes learning objective to convert the weaker trees into stronger trees. Next, it uses gradient tree boosting to train the data in an additive manner. XGBoosting classifier has many advantages over other classifiers as it is highly scalable, fast, and more accurate. Hyperparameter for this algorithm improves the recognition results of the system.

5 Experimental Results and Discussion

This section comprises about the dataset used in the study and results of the experiment.

5.1 Dataset and Data-Partitioning Methodologies

Caltech-101 is one of the most challenging datasets used for supervised learning. It contains 102 categories out of which 101 categories represent different objects such as airplanes, faces, ant, chair, pizza, wrench, and 1 background scene. Each class contains a collection of images in the range of 40–800. The images in the dataset are noisy, rotated at different angles, scaled in different sizes. In the experiment, the authors have used 101 class categories which comprises a total of 8677 images. A standard data-partitioning strategy, i.e., 70:30 is used to train the data where 70% images of each category are used as training dataset and remaining 30% are used for testing purpose.

5.2 Performance Evaluation

In this subsection, the authors have made a comparative analysis of various hand-crafted feature extraction techniques, i.e., SIFT, SURF, ORB, Haralick texture descriptor, Shi-Tomasi corner detector, and various pre-trained convolutional neural network models, i.e., ResNet50, Xception, and VGG19 model. Because of multi-class dataset used in the experiment, various multi-class classification algorithms, i.e., Gaussian Naïve Bayes, k-NN, decision tree, random forest, and XGBoosting classifier are applied on the extracted features. The performance of these feature extraction methods is computed using various classifiers in terms of recognition accuracy, precision, false positive rate (FPR), $F1$-score, area under curve (AUC), root mean square error (RMSE), and CPU elapsed time. The algorithms are implemented on Intel Core i7 processor having Windows 10 operating system and 8 GB RAM. Programming language Python 3.6, Keras and OpenCV image processing library are used for coding. Table 1 depicts the recognition accuracy of each feature extraction method using various classifiers. It shows that pre-trained model VGG19 has achieved maximum accuracy (85.64%) for object recognition using XGBoosting classifier. Table 2 represents the comparative view among various feature extraction techniques using precision rate. Table 3 presents $F1$-score of various feature extraction techniques. Comparison of other parameters like AUC, FPR, RMSE, and CPU elapsed time are presented in Tables 4, 5, 6 and 7, respectively. For all parameters, pre-trained model VGG19 has shown best results as comparison to other feature extraction methods. The experiment exhibits that feature extraction using pre-trained model VGG19 using XGBoosting classifier is performing best among all the classifiers used except for FPR, RMSE, and CPU elapsed time. The experiment shows that in the same case, XGBoosting algorithm takes more time (2.63 min) as compared to others but other parameters like precision (84.76%), $F1$-score (84.78%), and AUC (92.73%)

Table 1 Feature extraction technique-wise recognition accuracy for object recognition

Feature extraction algorithm/classifier	Classifier-wise recognition accuracy (%)				
	Gaussian Naïve Bayes	k-NN	Decision tree	Random forest	XGBoosting
SIFT	54.20	53.89	55.17	57.43	64.43
SURF	48.32	49.72	47.84	51.85	59.58
ORB	56.68	57.79	57.96	61.04	72.01
Haralick texture	61.00	17.94	69.26	55.93	80.73
Shi-Tomasi corner Detector	50.17	51.84	52.53	58.62	64.84
ResNet50	50.19	49.97	51.01	54.69	60.25
Xception	54.66	52.46	54.73	56.92	65.27
VGG19	63.45	75.23	74.71	82.06	85.64

Table 2 Feature extraction technique-wise precision for object recognition

Feature extraction algorithm/classifier	Classifier-wise precision (%)				
	Gaussian Naïve Bayes	k-NN	Decision tree	Random forest	XGBoosting
SIFT	52.23	54.94	54.56	56.41	63.52
SURF	45.32	50.92	46.00	50.22	57.66
ORB	54.70	60.63	57.97	60.54	70.68
Haralick texture	60.20	20.93	70.39	57.58	81.38
Shi-Tomasi corner Detector	47.74	55.60	51.94	57.99	63.63
ResNet50	47.24	51.35	50.02	52.99	59.09
Xception	53.12	52.69	53.83	55.35	64.12
VGG19	61.85	77.31	73.36	83.47	84.76

Table 3 Feature extraction technique-wise $F1$-score for object recognition

Feature extraction algorithm/classifier	Classifier-wise $F1$-score (%)				
	Gaussian Naïve Bayes	k-NN	Decision tree	Random forest	XGBoosting
SIFT	52.32	52.43	54.35	56.41	63.60
SURF	45.77	47.98	46.26	49.90	57.59
ORB	54.85	57.36	57.03	59.90	70.72
Haralick texture	59.00	17.40	68.82	55.61	80.31
Shi-Tomasi corner Detector	48.05	51.37	51.64	57.42	63.35
ResNet50	47.78	48.87	49.36	52.51	59.16
Xception	52.86	50.73	53.80	55.26	64.19
VGG19	61.58	74.91	73.49	81.51	84.78

Table 4 Feature extraction technique-wise area under curve (AUC) for object recognition

Feature extraction algorithm/classifier	Classifier-wise AUC (%)				
	Gaussian Naïve Bayes	k-NN	Decision tree	Random forest	XGBoosting
SIFT	52.48	76.65	52.59	52.66	81.96
SURF	50.16	67.28	50.19	50.24	79.53
ORB	53.68	78.63	53.57	54.02	85.82
Haralick texture	80.36	57.08	84.52	77.83	90.30
Shi-Tomasi corner Detector	52.72	75.62	52.64	52.77	82.20
ResNet50	50.14	74.70	50.34	50.36	67.27
Xception	58.55	75.73	58.34	59.04	82.42
VGG19	50.22	87.49	50.33	50.43	92.73

Table 5 Feature extraction technique-wise false positive rate (FPR) for object recognition

Feature extraction algorithm/classifier	Classifier-wise FPR (%)				
	Gaussian Naïve Bayes	k-NN	Decision tree	Random forest	XGBoosting
SIFT	0.57	0.60	0.56	0.53	0.50
SURF	0.58	0.58	0.61	0.54	0.53
ORB	0.47	0.54	0.46	0.41	0.38
Haralick texture	0.27	0.60	0.21	0.28	0.12
Shi-Tomasi corner Detector	0.53	0.60	0.53	0.45	0.45
ResNet50	0.55	0.57	0.55	0.50	0.49
Xception	0.48	0.51	0.49	0.45	0.42
VGG19	0.50	0.26	0.29	0.17	0.17

Table 6 Feature extraction technique-wise root mean square error (RMSE) for object recognition

Feature extraction algorithm/classifier	Classifier-wise RMSE (%)				
	Gaussian Naïve Bayes	k-NN	Decision tree	Random forest	XGBoosting
SIFT	30.02	32.97	30.88	29.58	28.60
SURF	33.31	33.90	33.47	32.46	31.20
ORB	29.71	32.79	29.24	2763	26.14
Haralick texture	18.67	31.56	17.09	20.36	11.76
Shi-Tomasi corner detector	30.24	34.85	30.93	29.39	28.72
ResNet50	31.53	33.31	31.07	29.98	30.42
Xception	26.85	29.48	27.55	26.86	24.56
VGG19	30.37	20.03	23.00	17.80	16.82

Table 7 Feature extraction technique-wise CPU elapse time for object recognition

Feature extraction algorithm/classifier	Classifier-wise CPU elapsed time (min)				
	Gaussian Naïve Bayes	k-NN	Decision tree	Random forest	XGBoosting
SIFT	0.00	0.01	0.00	0.17	2.95
SURF	0.00	0.01	0.01	0.31	3.26
ORB	0.00	0.01	0.01	0.26	2.67
Haralick texture	0.00	0.02	0.01	0.25	2.49
Shi-Tomasi corner detector	0.00	0.01	0.01	0.25	2.89
ResNet50	0.00	0.01	0.01	0.25	3.04
Xception	0.00	0.01	0.01	0.22	2.90
VGG19	0.00	0.01	0.00	0.15	2.63

are maximum as compared to others. But the results obtained for FPR and RMSE using VGG19 model are not satisfactory as compared of Haralick texture descriptor method. Haralick texture method achieves FPR as 0.12% and RMSE as 11.76%. The experimental study on various feature extraction methods using various classifiers might help other researchers for the selection of object recognition technique.

6 Conclusion

In this article, the authors have presented a comparison among various feature extraction techniques and multi-class classification methods for object recognition. The authors have taken some well-known handcrafted methods, i.e., SIFT, SURF, ORB, texture, and Shi-Tomasi corner detector method. An analysis of various deep feature extraction methods, i.e., ResNet50, Xception, and VGG19 is also made along the handcrafted methods. The study is done on a benchmark dataset Caltech-101. The comparison shows that pre-trained CNN model VGG19 are performing best among other feature extraction methods and XGBoosting classifier performed best among various classifiers like Gaussian Naïve Bayes, k-NN, decision tree, and random forest. If false positive rate and root mean square error are considered, then Haralick texture descriptor has proved best results. The authors have shown the performance of each algorithm providing a comparison among them in order to show the best feature extraction method and the best classifier on the basis of accuracy and time. In future, more feature extraction methods will be explored for object recognition and a study using deep learning classifiers will be done to represent a comparison between machine learning and deep learning.

References

1. Affonso C, Rossi ALD, Vieira FHA, de Leon Ferreira ACP (2017) Deep learning for biological image classification. Expert Syst Appl 85:114–122
2. Agarwal A, Samaiya D, Gupta KK (2017) A comparative study of SIFT and SURF algorithms under different object and background conditions. In: 2017 international conference on information technology (ICIT), Bhubaneswar, pp 42–45. https://doi.org/10.1109/ICIT.2017.48
3. Routray S, Ray AK, Mishra C (2017) Analysis of various image feature extraction methods against noisy image: SIFT, SURF and HOG. In: Second international conference on electrical, computer and communication technologies (ICECCT), pp 1–5
4. He T, Wei Y, Liu Z, Qing G, Zhang D (2018) Content based image retrieval method based on SIFT feature. In: 2018 international conference on intelligent transportation, big data and smart city (ICITBS), Xiamen, pp 649–652. https://doi.org/10.1109/ICITBS.2018.00169
5. Xiao T, Liu L, Li K, Qin W, Yu S, Li Z (2018) Comparison of transferred deep neural networks in ultrasonic breast masses discrimination. Biomed Res Int.https://doi.org/10.1155/2018/4605191
6. Gupta S, Kumar M, Garg A (2019) Improved object recognition results using SIFT and ORB feature detector. Multimedia Tools Appl 78(23):34157–34171
7. Tropea M, Fedele G (2019) Classifiers comparison for convolutional neural networks (CNNs) in image classification. In: 2019 IEEE/ACM 23rd international symposium on distributed simulation and real time applications (DS-RT), pp 1–4
8. Lowe DG (2004) Distinctive image features from scale-invariant keypoints. Int J Comput Vision 60(2):91–110
9. Bay H, Ess A, Tuytelaars T, Van Gool L (2008) Speeded-up robust features (SURF). Comput Vis Image Understanding 110(3):346–359
10. Rublee E, Rabaut V, Konolige K, Bradski G (2011) ORB: an efficient alternative to SIFT or SURF. In: Proceedings of the IEEE international conference on computer vision (ICCV), pp 1–8
11. Shi J, Tomasi C (2004) Good features to track. In: Proceedings of the IEEE computer society conference on computer vision and pattern recognition (CVPR), pp 593–600
12. Haralick RM, Shanmugam K, Dinstein IH (1973) Textural features for image classification. IEEE Trans Syst Man Cybern 6:610–621
13. He K, Zhang X, Ren S, Sun J (2016) Deep residual learning for image recognition. In: 2016 IEEE conference on computer vision and pattern recognition (CVPR). Las Vegas, NV, pp 770–778. https://doi.org/10.1109/cvpr.2016.90

14. Chollet F (2017) Xception: deep learning with depthwise separable convolutions. In: Proceedings of the IEEE conference on computer vision and pattern recognition, pp 251–1258
15. Simonyan K, Zisserman A (2014) Very deep convolutional networks for large-scale image recognition. arXiv:1409.1556
16. Quinlan JR (1986) Induction of decision trees. Mach Learning 1(1):81–106
17. Kleinberg EM (1996) An overtraining-resistant stochastic modeling method for pattern recognition. Ann Stat 24(6):2319–2349
18. Chen T, Guestrin C (2016) XGBoost: a scalable tree boosting system. In: Proceedings of the 22nd ACM SIGKDD international conference on knowledge discovery and data mining, pp 785–794

Leukemia Prediction Using SVNN with a Nature-Inspired Optimization Technique

Biplab Kanti Das, Prasanta Das, Swarnava Das, and Himadri Sekhar Dutta

Abstract Blood smear examination is a basic test that helps us to diagnose various diseases. Presently, this is done manually, though automated and semiautomated blood cell counters are also in vogue. Automated counters are very costly, require specialized and proper maintenance to work, and trained manpower. Thus, in small laboratories and in periphery blood smears are mostly being done manually, followed by microscopic evaluation by trained medicos. Though this method is easily available and cost-effective, but there are always chances of variation in the result due to differences in the methods of preparation of slides and experience of the pathologist. To overcome the manual methods of blood smear examination, various studies are being undertaken which are targeted not only to identify different blood cells but also to specifically identify blast cells which are the cornerstone of diagnosis of acute leukemia by automated methods. This study proposes a leukemia detection method using salp swarm optimized support vector neural network (SSA-SVNN) classifier to identify leukemia in initial stages. Adaptive thresholding on LUV transformed image was used to perform segmentation of the preprocessed smear. From the segments, the features (shape, area, texture, and empirical mode decomposition) are extracted. Blast cells are detected using the proposed method based on the extracted features. The accuracy, specificity, sensitivity, and MSE of the proposed method are found to be 0.96, 1, 1, and 0.1707, respectively, implies that compared to other methods—KNN, ELM, Naive Bayes, SVM, there is improvement in leukemia detection.

Keywords Support vector neural network · Leukemia · SSA · Blast cells

B. K. Das (✉) · P. Das
Calcutta Institute of Technology, Uluberia, India

S. Das
Indian Institute of Technology, Kharagpur, Kharagpur, India

H. S. Dutta
Kalyani Government Engineering College, Kalyani, India

A. Choudhary et al. (eds.), *Applications of Artificial Intelligence and Machine Learning*, Lecture Notes in Electrical Engineering 778,
https://doi.org/10.1007/978-981-16-3067-5_23

1 Introduction

Blood investigations are one of the most used investigations that we undergo every now and then. In routine blood smear examination, we examine the cellular components of blood. They are RBC, WBC, and platelets [1]. Depending on the suspected disease, target cell is selected. The most commonly looked one is WBC as these cells change both in number and morphology in various diseases. There are five subtypes of WBC—neutrophil, lymphocyte, monocyte, eosinophil, and basophil. Depending on the underlying diseases their number, proportional count, size, shape changes, and these changes guide us to accurate diagnosis of the disease. The origin of these cells is in the bone marrow, and they reach peripheral blood after the maturation process completes. Naturally, any abnormality in the bone marrow or of the maturation process will also be reflected by the presence of abnormal progenitor cells in the peripheral smear. One such disease is leukemia, which involves abnormal reproduction of WBCs from affected bone marrow with unsuppressed cell growth [2]. Two types of acute leukemia are acute lymphocytic leukemia (ALL) and acute myeloid leukemia (AML). Leukemia diagnosis procedure requires an analysis, involving the detection of abnormal WBCs. Two analyses are performed, i.e., classification and counting of blood cells. ALL affects progenitor cells of WBC lineage, so the presence of blast cells (abnormal progenitor) in the smear is an indicator. The percentage of blasts in the smear and morphology both are important for diagnosis. Naturally, a good blood smear and experienced pathologist both are prerequisites for this analysis. Starting from manual preparation of smear up to the examination of smear by a trained person; at every step, there is a possibility of human error which may be due to ignorance, inadequate knowledge, fatigue, and inter-observer variation; even there are chances of using defective staining material methods. Therefore, an automatic segmentation method is used for effective WBC identification using microscopic smear images. In the case of morphological analysis, the blood sample is not necessary, as it is achieved using a single image [3]. Edges, texture, geometric, statistical features, and histogram of gradients (HOG) [4] are the features by which the classification of images is done. In computer vision, to detect leukemia using the smear image, involves five stages, i.e., image acquisition, preprocessing, segmentation, feature extraction, and detection of leukemia cells [3].

The goal of the paper is to propose a classifier, named salp swarm algorithm (SSA) [5]-based support vector neural network (SVNN) classifier, to spot the blast cells.

1.1 The Novelty of This Paper

The proposed SSA-based support vector neural network is used to detect the presence of leukemia based on blast cell WBCs. The proposed SSA-based technique is applied to optimize the weights of the classifier. This paper is structured as Sect. 1 gives an introduction to leukemia identification. The literature review of older methods is shown in Sect. 2. The proposed SSA-based SVNN classifier for leukemia detection is dealt with in Sect. 3. Section 4 discusses the results obtained, and the paper is concluded in Sect. 5.

2 Motivation

This section describes previous methods that are associated with the identification of leukemia and their limitations. The proposed work has been inspired by the following methods.

2.1 Literature Survey

A method developed by Putzu et al. [6] to automatically detect and identify leucocytes showed high accuracy, but it provides very low specificity and the sensibility for different lighting conditions and resolution. To get the total number of lymphocytes, Le et al. [7] developed a system by applying the watershed algorithm, which shows high accuracy with low execution time and little computational complexity. The shortcoming of the process is that counting of the cell cannot be applied in this method when the clinical testing procedure is done. Chatap and Shibu [8] designed a model using SVM and nearest neighbor to detect leukemia. The classification was reliable and also efficient with less processing time. This model gives high accuracy, less error, and cost. But its limitation is its inability to find lymphoblast subtypes. An automatic detection system with the counting of lymphoblasts was developed by Shankar et al. [9]. This model provides higher speed, improved accuracy, and added scope for the early identification of leukemia. But the limitation of it is poor accuracy. A light-induced optical biosensor with CNTs was developed by Gulati et al. [10] to detect leukemia early. The best part of this system is its simplicity and cost-effectiveness. The main disadvantage is that it produces poor performance. An approach was developed by Soni et al. [11] to synthesize the polyaniline-nanotubes by using manganese oxide. This method shows improved redox activity and conductivity. However, it was not up to the mark in point of view of diagnostics. For selective and ultrasensitive identification of Leukemia, Khoshfetrat and Mehrgardi [12] developed an electrochemical biosensor. It was robust with high sensitivity and simplicity. But it was not suitable to identify other cancer-affected cells. Analytical methods were developed by Yazdian-Robati et al. [13] such as electrochemical and optical aptasensors to identify leukemia. But it was not considered as a valuable tool by clinicians. A coupling algorithm based on glowworm swarm optimization and bacterial foraging algorithm was developed by Wang et al. [14] to solve multi-objective optimization problems. But the computational time is high.

3 Proposed Method of Leukemia Detection

The paper's main objective is to propose an SSA-based SVNN classifier to detect the existence of leukemia. The block diagram is displayed in Fig. 1 to the detection of leukemia. Initially, microscopic images of blood smear were used as input. These images are passed through a preprocessing phase to enhance certain aspects of the input image. Using adaptive thresholding on LUV transformed image, the preprocessed image is then segmented. Feature extraction is done on these segmented images. In the feature extraction step, different features—area, shape, empirical mode decomposition, and texture—are extracted. Salp swarm algorithm (SSA)-SVNN classifier to detect the blast cells is applied on generated features of WBCs.

3.1 Preprocessing of the Input Image

To enhance the microscopic smear image, the images pass the preprocessing phase and are followed by other successive phases like segmentation and classification. The required proper area is identified from the input image, and the resulting preprocessed image is denoted by K_j.

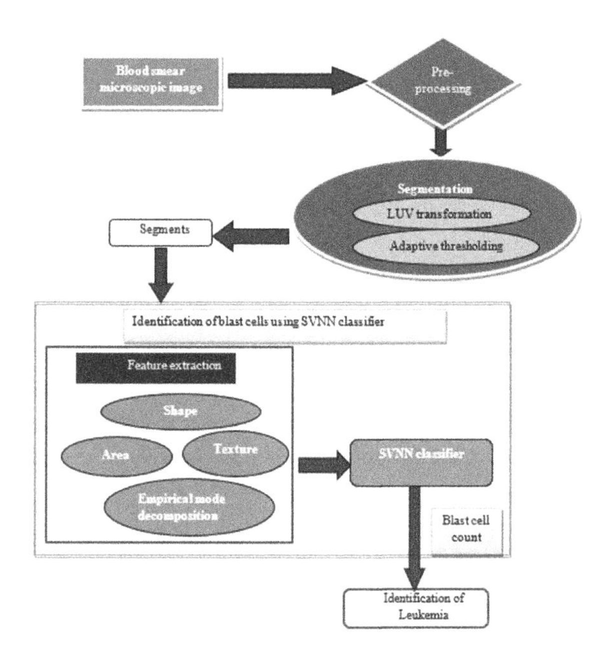

Fig. 1 Block diagram for leukemia detection

3.2 Blast Cell Segmentation

Segmentation generates segments from the preprocessed images. Through this process, meaningful patterns are produced for the simplification of analysis.

LUV transformation of the preprocessed image. The preprocessed image undergoes transformation to generate the LUV image. The jth RGB image (A_j^{RGB}) is changed into the LUV space (K_j^{LUV}).

Adaptive thresholding of LUV transformed image. From the binary LUV region (K_j^{LUV}) just the 'U' space is taken to generate thresholding, after generating image segments from input (RGB) image segment A_j. If the value of the pixel is less than the average of $(u \times u)$ the neighboring window of the integral image, the pixel is black and white otherwise. Equation (1) shown below is for segments generated by using adaptive thresholding.

$$T^j = \left\{T_1^j, T_2^j, \ldots, T_i^j, \ldots T_n^j\right\} \tag{1}$$

where T_i^j indicates the ith segment of the jth image, and n is the total number of generated segments from the jth image.

3.3 Identification of WBCs

To identify WBCs among the RBCs is a complex process. The geometry of the cells is extracted and used for classification, to minimize the difficulty in the detection process. The steps for identification are:

Feature extraction process. To extract features, a suitable feature set with fewer dimensions is defined from a sparse distribution of needless data. The features which are extracted consist of area, texture, shape, and the EMD of the cells.

Extraction of grid-based shape features. Grid-based features are used to represent the shape of the segment. Scanning is performed from top-left to bottom-right to cover the shape fully or partially is assumed '1', and the area which is outside the shape boundary is treated '0'. Thus, a binary boundary is formed to represent the shape. Grid-based features form a set of size $[1 \times 64]$, each representing a point on the boundary, and is denoted by $q_{i,j}^1$.

Texture-based features extraction using LGP. Gradient values are calculated by subtracting intensities of the center pixel and its neighboring pixels. With the implementation of LGP [15], the gradient values of eight-neighbor pixels with respect to the center pixel are generated. Hence, the variation of neighboring pixel w.r.t. the center pixel is the texture. Equation (2) represents the LGP operator

$$l_{i,j}^{2} = \mathrm{LGP}_{\mu,\tau}^{r^j,s^j} = \sum_{\tau=0}^{\mu-1} \alpha(\gamma_\tau - \gamma^*) * 2^y \tag{2}$$

$$\alpha \begin{cases} 0; & u < 0 \\ 1; & \text{Otherwise} \end{cases} \tag{3}$$

γ^* represents the average gradient values and γ_τ represents the center pixel's gradient value, at (r^j, s^j).

$$\gamma_\tau = |\delta - \delta^*| \tag{4}$$

The threshold is applied to the absolute intensity difference using adaptive LGP local threshold ϕ^* to generate the code as in Eq. (2). Depending on gradient differences, LGP produces invariant patterns and the LGP feature dimensions are [1 × 64]. If there occur local color variations, then also LGP is not affected.

Area. All non-zero pixels are counted from the image region of the original image, to get segment area. $a_{i,j}^{3}$ represents the area features with size as [1 × 1].

Employment of empirical mode decomposition. Intrinsic mode functions (IMFs) [16] comprising a set of various frequency components are obtained by using an adaptive method named EMD from the decomposition of non-stationary and nonlinear signals. The main advantage is that there is no requirement of predefined wavelets or filters. To extract bi-dimensional IMFs, the bi-dimensional sifting process is applied. Based on the characteristic multiscale, making use of sifting, the separation was performed on the local mode of the image. The sifting process generates final image residue with the size of [1 × 256].

Feature vector. Equation (5) represents the segment feature vector

$$V = \{b_{i,j}^{1}, b_{i,j}^{2}, b_{i,j}^{3}, b_{i,j}^{4}\} \tag{5}$$

where $b_{i,j}^{1}$, $b_{i,j}^{2}$, $b_{i,j}^{3}$, and $b_{i,j}^{4}$ represent area, texture, shape, and EMD of the segment. Equation (6) shows the features associated with ith segment.

$$V = \{m_1, m_2, \ldots, m_g, \ldots, m_p\} \tag{6}$$

where p represents the total feature count and the gth feature of the segment is represented by m_g. The size of the generated feature vector is [1 × 329].

Classification using the proposed method. The aim of the proposed model is to perform the detection of blast cells, given the [1 × 329] feature set. The extracted features consist of various data that helps to find the blast cells. To identify the WBCs, these features are analyzed by applying the proposed SSA-based SVNN classifier. To model the swarming behaviors and population of salps, two groups are formed—leader and followers, forming the mathematical model of the salp chain. The leader salp will always at the front to lead the group and the rest are followers, who follow its preceding one.

The search space of n dimensions is defined for the position of salp, n being the number of parameters of the problem. A 2D matrix termed x is used to store all the salp positions. It is also presumed that there is a target of the swarm (food source F) in the search space. Similarly, a salp swarm-based algorithm has been used to train the proposed SVNN classifier for the classification to detect WBCs.

Architecture of SVNN classifier. Figure 2 shows the SVNN [17, 18] architecture. It shows three layers, namely input, hidden, and output layer. The feature vector V acts as input to the first-layer (input layer). The weights of the input layer are multiplied with the feature vector. The result is passed as the input to the hidden layer. The weight and the bias values are processed to obtain the output of the hidden layer. After that the hidden layer output is passed as input to the final output layer, where WBCs are identified. The output equation of the proposed classifier is shown in Eq. (7) as

$$S^{\mathrm{SVNN}} = x_3 \times \log\mathrm{sig}\left[\left(\sum_{\substack{j=1 \\ j \in i}}^{4} b_{i,j} * x_2\right) + x_1\right] + x_4 \tag{7}$$

where x_1 denotes bias of hidden layer, x_2 is hidden layer's weight, x_3 is the output layer's weight, and x_4 is the output layer's bias.

Algorithmic steps of the proposed SSA-based method. The algorithm of the proposed SSA-based method is as follows:

Step 1: Salp (leader) positions are set in a random manner, initialized as $H_d(t)$.

Step 2: Assign upper and lower bounds of the search space for all d dimensions, as per current salp-positions, as shown in Eq. (8)

$$ube_d = max(H_d(t)); Ib_d = min(H_d(t)) \tag{8}$$

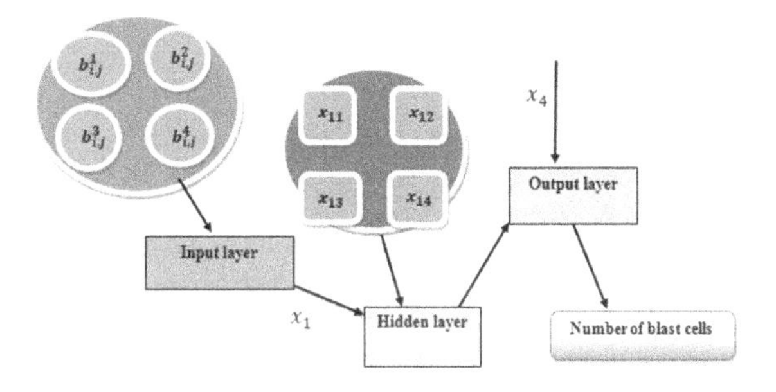

Fig. 2 Architecture of SVNN

where, ube_d and lb_d denote upper and lower bounds of the dth dimension.

Step 3: Regularization factor, loss, and cardinality of the training dataset are used to calculate the fitness value for all positions of the salp are shown in Eq. (9).

$$\text{KF}_d^t = \rho_{\max} + \rho_{\min} + \frac{\text{ER} \times \text{be}}{C} \tag{9}$$

where $\rho_{\min}$ is the minimum, and $\rho_{\max}$ is the maximum values of ρ, respectively.
Where,

$$\rho = \text{Eig}\left(N_{\text{w}} \cdot N_{\text{w}}^{\text{T}}\right) \tag{10}$$

Fitness of the dth salp at tth interval is represented by KF_d^t. Regularization factor denoted by ER, be represents the SVNN-loss, N_{w} refers to the SVNN weight matrix and the cardinality of the training set is denoted by C.

Step 4: Initialize the global best, and the personal best.

Step 5: Next, SSA efficiently initializes the random solutions and helps to converge to the optimized value. Equation (11) represents the standard form of SSA.

$$H_d(t) = \begin{cases} \text{MF}_d + k_1((\text{ube}_d - \text{lb}_d)k_2 + \text{lb}_d), & k_3 \geq 0 \\ \text{MF}_d - k_1((\text{ube}_d - \text{lb}_d)k_2 + \text{lb}_d), & k_3 < 0 \end{cases} \tag{11}$$

where for the dth salp, $H_d(t)$ signifies the location at tth interval (H_d being its personal best), location of food source represented by MF_d (Global best so far), upper and lower bound of the dth dimension denoted by ube_d, lb_d, respectively, k_1, k_2, k_3 denote randomly chosen numbers. The leader changes its position w.r.t. the source of food only that has been shown in Eq. (11). Coefficient k_1, balances exploitation and exploration and is thus the most significant parameter in SSA. It has been defined as:

$$k_1 = 2\mathrm{e}^{-\left(\frac{4l}{L}\right)^2} \tag{12}$$

Step 6: Update, personal best, and the global best position.

Step 7: End of the algorithm.

3.4 Detection of Leukemia

Based on the trained SVNN classifier's output, the input image is classified, and the classification process helps to decide whether the input image is of affected or un-affected blood cells. If the image is normal, then the output is marked as '0', otherwise the level of infection (IL) is calculated by Eq. (13). The level of infection is calculated based on the number of blast cells and pixel count.

$$\mathrm{IL} = \frac{1}{2}\left(\frac{N_{\mathrm{BC}}}{T_{\mathrm{BC}}} + \frac{N_{\mathrm{BP}}}{T_{\mathrm{BP}}}\right) \times 100 \tag{13}$$

where the calculated blast cell count is represented by N_{BC}, calculated blast pixel count is referred by N_{BP}, T_{BC} refers to the actual count of blast cells, and T_{BP} is the actual blast pixels.

4 Results and Discussion

In this section, the results of the classifier have been compared to the conventional classifiers in terms of mean square error (MSE), the ROC curve, sensitivity, specificity, and accuracy. The research was carried out by training our proposed model on the open ALL-IDB1 database (free dataset of microscopic blood sample images in JPG format with resolution 2592 × 1944, 24-bit color depth) [19], and intermediate outputs from random samples have been shown. Two sample images (Fig. 3a, b) were considered for this intermediate analysis. The color transformation is applied to obtain the clarity of the images and is displayed in Fig. 4a, b. Then, preprocessing is applied to the images to enhance the image. Then, WBCs are detected as shown in Fig. 5a, b. Finally, Fig. 6a, b shows the WBCs that were identified from the original images.

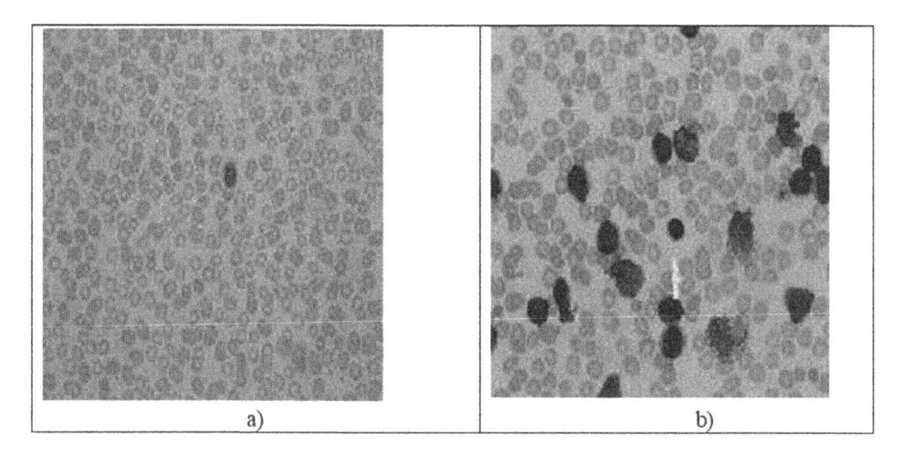

Fig. 3 Sample input images for **a** original image 1 and **b** original image 2

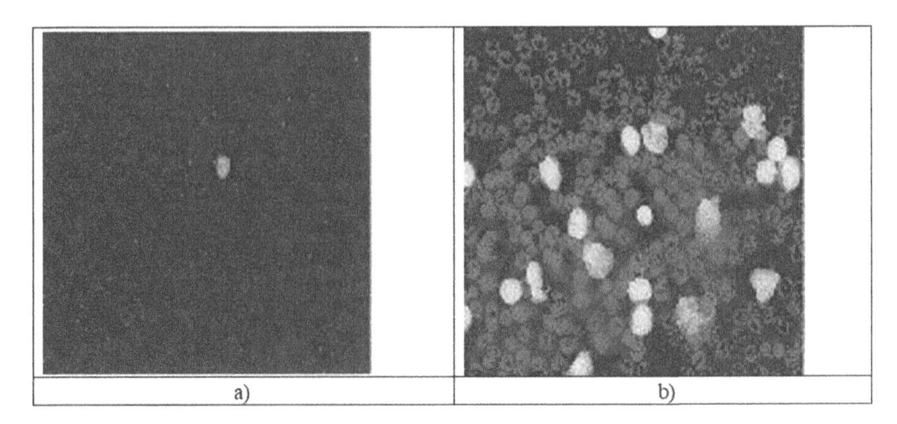

Fig. 4 Color space transformation for original images **a** image 1 and **b** image 2

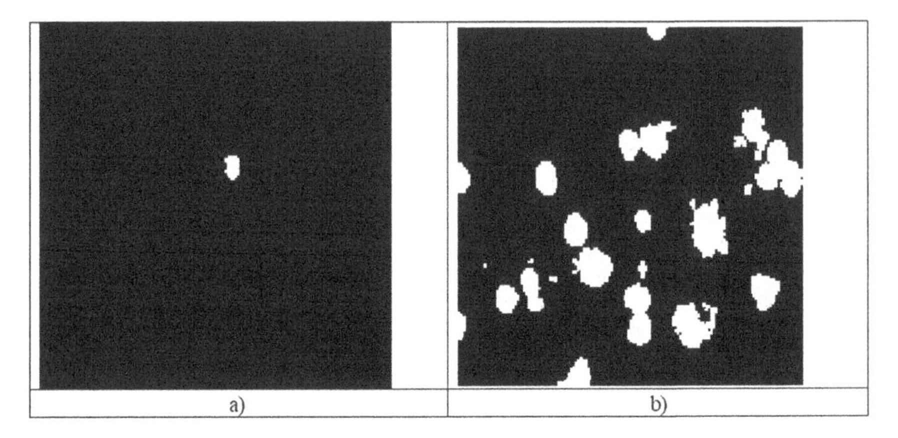

Fig. 5 Cell identification for **a** image 1 and **b** image 2

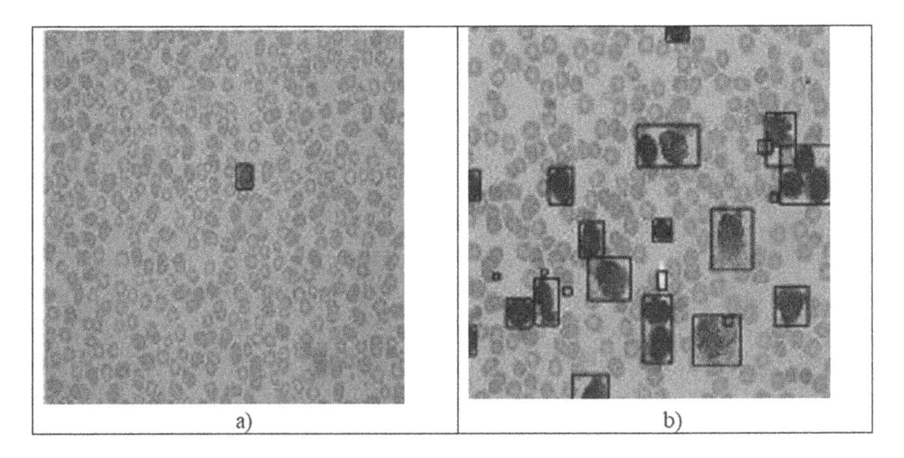

Fig. 6 WBCs identification for **a** for image 1 and **b** for image 2

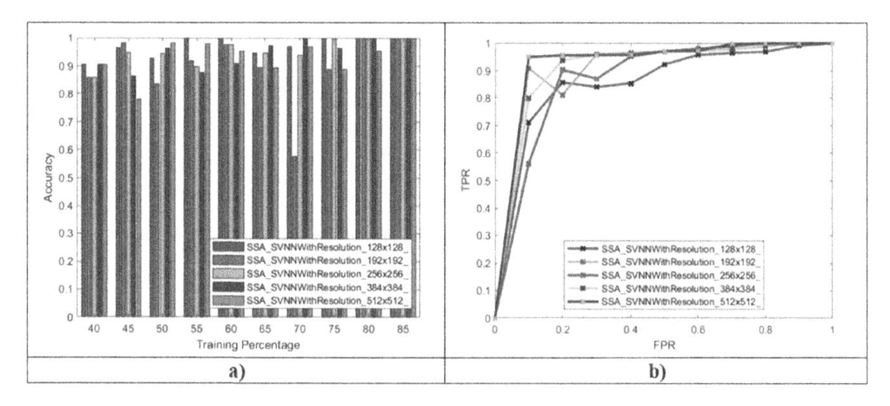

Fig. 7 Performance w.r.t. resolution **a** accuracy and **b** ROC curve

4.1 Performance Analysis

Performance is analyzed with changing resolution. At the training percentage of 50, accuracies of the SSA-based-SVNN classifier with respect to respective image resolutions [128 × 128], [192 × 192], [256 × 256], [384 × 384], and [512 × 512] (Fig. 7a) are 0.9272, 0.8363, 0.9454, 0.96363, and 0.98181. The proposed classifier is observed to have an improved accuracy at higher resolutions generally, and it also shows an increase in accuracy as the training percentage increases. The ROC curve is plotted (Fig. 7b) by observing the values of TPR at the corresponding values of FPR. At 20% FPR, corresponding TPR with respect to respective image resolutions [128 × 128], [192 × 192], [256 × 256], [384 × 384], and [512 × 512] were found to be 85.7, 81.81, 90.33, 93.75, and 95.56%.

4.2 Comparative Analysis

To prove the efficiency of the proposed classifier, it is compared to previous methods, K-nearest neighbors classification, extreme learning machine, Naive Bayes, and support vector machine.

The accuracy of each method at various training percentages is shown in Fig. 8a. At the training percentage of 40, the accuracy of the different methods—KNN, ELM, Naive Bayes, SVM, and the SSA-based-SVNN classifier, respectively, are 0.75, 0.6666, 0.796875, 0.5, and 0.8906. At training percentage of 60, accuracy of these respective classifiers become 0.8636, 0.6944, 0.8636, 0.7924, and 0.9598. The proposed SSA-based-SVNN classifier has improved accuracy, making it better than previous methods; also showing increase in accuracy as the training percentage increases. The MSE of each method at various training percentages is shown in Fig. 8b. As seen in the figure, at the training percentage of 40, MSE values of the

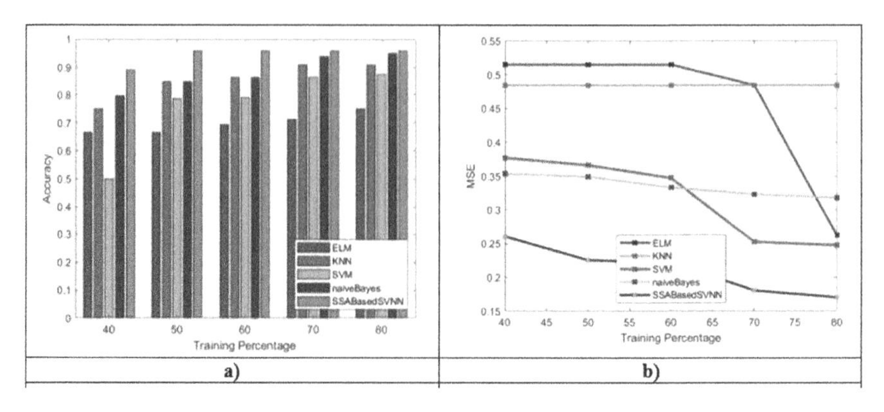

Fig. 8 Comparative analysis based on training percentage, **a** accuracy, **b** MSE

different methods—KNN, ELM, Naive Bayes, SVM, and the SSA-based SVNN classifier, respectively, are 0.4848, 0.5151, 0.3535, 0.3771, and 0.2606. Similar to the improvement as in the case of accuracies, at a training percentage of 60, MSE of these respective classifiers become 0.2626, 0.4848, 0.2474, 0.3181, and 0.1707. Thus, the proposed SSA-based-SVNN classifier has a decreased MSE, making it better than previous methods; also, generally showing a decrease in MSE value as the training percentage increases. Besides having low mean square error and high accuracy, a model should have both high sensitivity and high specificity to ensure the absence of bias or skew, i.e., the classifier performs well for both positive and negative samples. The sensitivities of each method at various training percentages are shown in Fig. 9a. It is observed at the training percentage of 40, sensitivities of the different methods—KNN, ELM, Naive Bayes, SVM, and the SSA-based-SVNN classifier, respectively, are 0.7878, 0.6493, 0.8055, 0.5, and 0.8361. Similar to the improvement as in the case of accuracies, at a training percentage of 60, sensitivities of these respective classifiers become 0.95, 0.75, 0.909, 0.7894, and 0.9673. So, the

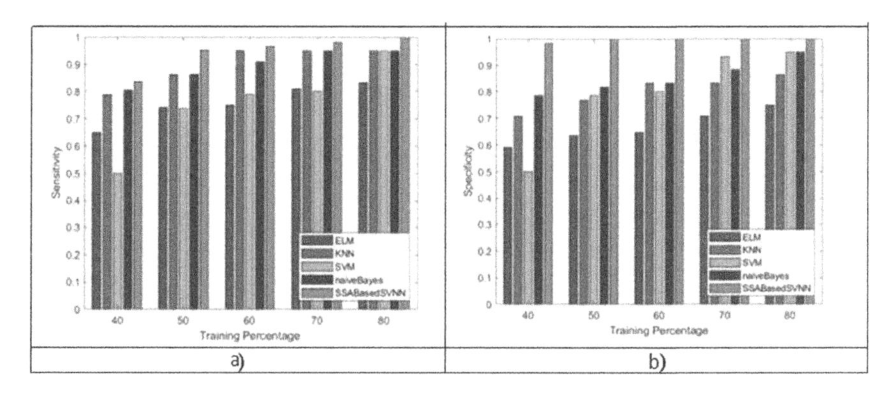

Fig. 9 Comparative analysis based on training percentage, **a** sensitivity, **b** specificity

proposed classifier has an improved sensitivity value, making it better than previous methods; also showing an increase in sensitivity as the training percentage increases. Finally, the specificities of each method at various training percentages are shown in Fig. 9b. At the training percentage of 50, specificities of the different methods—KNN, ELM, Naive Bayes, SVM, and the SSA-based-SVNN classifier, respectively, are 0.7692, 0.6363, 0.8181, 0.7857, and 1. Also, at a training percentage of 70, specificities of these respective classifiers become 0.8333, 0.7096, 0.8823, 0.9333, and 1. The proposed SSA-based-SVNN classifier has a higher specificity value, making it better than previous methods. Therefore, it is clear from the results of this analysis, this novel method involving the proposed SSA-based-SVNN classifier delivers high accuracy, sensitivity, and specificity as well as low MSE, and is thus a better performer in blast cell-based leukemia detection than other existing techniques.

5 Conclusion

The classification is found to be sufficiently capable in estimating the presence of blast cells, to determine whether or not a person is suffering from leukemia. This method involving the proposed SSA-based-SVNN classifier was proven to be effective, when trained on the ALL-IDB1 database, relative to existing methods with respect to evaluation metrics—accuracy, sensitivity, specificity, and MSE. The accuracy, specificity, sensitivity, and MSE of the proposed method are found to be 0.96, 1, 1, and 0.1707, respectively, implies that compared to other methods, and there is an improvement in leukemia detection. Future work should include further variation and take steps to improve accuracy.

References

1. Das BK, Jha KK, Dutta HS (2014) A new approach for segmentation and identification of disease affected blood cells. In: Proceeding of the international conference on intelligent computing applications (ICICA2014), IEEE (CPS), pp 208–212
2. Raje C, Rangole J (2014) Detection of leukemia in microscopic images using image processing. In: International conference on communication and signal processing. IEEE, India, pp 255–259
3. Bagasjvara RG, Candradewi I, Hartati S, Harjoko A (2016) Automated detection and classification techniques of acute leukemia using image processing: a review. In: Proceedings on international conference on science and technology-computer (ICST). IEEE, Indonesia, pp 35–43
4. Harb HM, Desuky AS, Mohammed A, Jennane R (2017) Classification of white blood cells based on morphological features. Int J Comput Appl 165(3):23–28
5. Mirjalili S, Gandomi AH, Mirjalili SZ, Saremi S, Faris H, Mirjalili SM (2017) Salp swarm algorithm: a bio-inspired optimizer for engineering design problems. Adv Eng Softw 114:163–191
6. Putzu L, Caocci G, Ruberto CD (2014) Leucocyte classification for leukemia detection using image processing techniques. Artif Intell Med 62(3):179–191

7. Le DT, Bui AA, Yu Z, Bui FM (2015) An automated framework for counting lymphocytes from microscopic images. In: Proceedings of the international conference and workshop on computing and communication (IEMCON). IEEE, Vancouver, pp 1–6
8. Chatap N, Shibu S (2014) Analysis of blood samples for counting leukemia cells using support vector machine and nearest neighbour. IOSR J Comput Eng (IOSR-JCE) 16(5):79–87
9. Shankar V, Deshpande MM, Chaitra N, Aditi S (2016) Automatic detection of acute lymphoblasitc leukemia using image processing. In: Proceedings of the international conference on advances in computer applications (ICACA). IEEE, Coimbatore, pp 186–189
10. Gulati P, Kaur P, Rajam MV, Srivastava T, Ali MA, Mishra P, Islam SS (2018) Leukemia biomarker detection by using photoconductive response of CNT electrode. Analysis of sensing mechanism based on charge transfer induced Fermi level fluctuation. Sens Actuators B Chem 270:45–55
11. Soni A, Pandey CM, Solankia S, Kotnalab RK, Sumanab G (2018) Electrochemical genosensor based on template assisted synthesized polyaniline nanotubes for chronic myelogenous leukemia detection. Talanta 187:379–389
12. Khoshfetrat SM, Mehrgardi MA (2017) Amplified detection of leukemia cancer cells using an aptamer-conjugated gold-coated magnetic nanoparticles on a nitrogen-doped graphene modified electrode. Bioelectrochemistry 114:24–32
13. Yazdian-Robati R, Arab A, Ramezani M et al (2017) Application of aptamers in treatment and diagnosis of leukemia. Int J Pharm 529(1–2):44–54
14. Wang Y, Cui Z, Li W (2019) A novel coupling algorithm based on glowworm swarm optimization and bacterial foraging algorithm for solving multi-objective optimization problems. Algorithms 12(3):61
15. Zhou W, Yub L, Qiu W et al (2017) Local gradient patterns (LGP): an effective local-statistical-feature extraction scheme for no-reference image quality assessment. Inf Sci 397–398:1–14
16. Huang NE, Shen Z, Long SR et al (1998) The empirical mode decomposition and the Hilbert spectrum for nonlinear and non-stationary time series analysis. Proc R Soc Lond A 454:903–995
17. Ludwig O, Nunes U, Araujo R (2014) Eigen value decay: a new method for neural network regularization. Neurocomputing 124:33–42
18. Parveen, Singh A (2015) Detection of brain tumor in MRI images, using combination of fuzzy C-means and SVM. In: International conference on signal processing and integrated networks (SPIN). IEEE, Noida, pp 98–102
19. ALL IDB database. https://homes.di.unimi.it/scotti/all/. Accessed 23 Nov 2017

Biplab Kanti Das received an M. Tech (IT) degree from Jadavpur University, West Bengal, India, and pursuing Ph.D. in Computer Science and Engineering from Maulana Abul Kalam Azad University of Technology, West Bengal, India. He is presently working as Assistant Professor at the MCA Department of Calcutta Institute of Technology, Uluberia, West Bengal, India. His area of interest in research includes image processing and courseware engineering. He published research papers in several international journals and conferences and reviewed papers for reputed journal. He has also published 4 books.

Prasanta Das received his BCA degree from Prabhat Kumar College, Contai under the Vidyasagar University, India. He has completed an MCA degree from Calcutta Institute of Technology under the Maulana Abul Kalam Azad University of Technology, Kolkata, India. His area of interest in the field of research includes medical image processing and machine learning.

Swarnava Das received his B.Tech degree from Vellore Institute of Technology, Vellore, India. Presently, pursuing M.Tech (CS) from IIT-Kharagpur, West Bengal, India. His research interests include natural language processing, machine learning, and deep learning.

Himadri Sekhar Dutta awarded M.Tech. degree from the University of Calcutta, Kolkata, India, in Optics and Opto-Electronics and Ph.D. in Technology from Institute of Radio Physics and Electronics, Kolkata, India, respectively. He is working as Assistant Professor at the Department of ECE of Kalyani Government Engineering College, Kalyani. His research areas include opto-electronic devices, embedded systems, and medical image processing. He has published more than 70 research papers in various international journals and conferences. He was Chairperson of IEEE Young Professional, Kolkata Section for the years 2016 and 2017.

Selection of Mobile Node Using Game and Graph Theory for Video Streaming Application

Bikram P. Bhuyan and Sajal Saha

Abstract Solving the bandwidth scarcity problem is the need of the hour. In this paper, we propose an architecture that offloads the computational resources from cloud server to edge node to reduce the power consumption and data traffic consumption of the receiving mobile node. The edge node may be local edge server of the ISP or a mobile node. A cooperative game-theoretic framework is being proposed to identify the mobile nodes receiving same video streaming content. Identified mobile nodes create different clusters based on their similarity and then selects a cluster head that acts as the edge node. The edge node receives the video streaming locally from the ISP and distributes it among the other nodes in the cluster. Analytically, the proposed architecture reduces the data traffic consumption significantly.

Keywords Edge computing · Cooperative game theory · Lattice theory · Stable clustering · Graph theory

1 Introduction

Data usage and traffic growth have increased exponentially in the last decade due technological advancement of mobile communication, access technology, and availability of affordable smart phone. Monthly data traffic per smart phone user has grown in many folds in all regions in India. India's data traffic usage has increased to 11 GB/user/month in 2018 as compared to 6.4 GB/user/month in 2015. The Internet traffic gets choked due to the same type of packet that ISP is going to send to all its

B. P. Bhuyan (✉)
Department of Informatics, School of Computer Science, University of Petroleum and Energy Studies, Dehradun, India

S. Saha
Department of Computer Science and Engineering, Kaziranga University, Jorhat, India
e-mail: sajalsaha@kazirangauniversity.in

A. Choudhary et al. (eds.), *Applications of Artificial Intelligence and Machine Learning*, Lecture Notes in Electrical Engineering 778,
https://doi.org/10.1007/978-981-16-3067-5_24

users. Instead of sending the same video streaming packets to all the users, multiple number of time that consumes the bandwidth in many folds, the ISP can stream the videos to a cluster head of all the users which will be selected through a cooperative game-theoretic modeling. The cluster head will distribute the streaming video locally to all other users. In this way, ISPs will get out of the bandwidth scarcity problem; at the same time, users will get the data from the local cluster head with minimum delay to achieve the highest quality of service.

In this paper, we propose a solution that reduces data traffic consumption during video/multimedia streaming by storing the same streaming content in a local server/edge server if the server is available. Otherwise streaming content will be stored in localized mobile device cache and stream the video content from the mobile device. Mobile device is selected through a game-theoretic model.

Let us consider the situation where 'n' number of users that is $m_1, m_2, \ldots, m_n$ request the same video from the Internet service provider (ISP) as shown in Fig. 1. Based on the streaming content type, three clusters 'm, n, o' are created. The streaming content is streamed locally through an edge server to different clusters. If the edge server is not available, each cluster selects the cluster head based on configuration of the mobile. The proposed architecture selects pool of nodes based on their priority. The highest priority node is selected as the cluster head. If the cluster head mobile node leaves the place/cluster, then second priority mobile node is selected as the cluster head. The pool of nodes is being reorganized after each mobile node's entry and exit in the cluster. The mobile nodes, whose video streaming contents are not similar in nature, are kept outside the cluster.

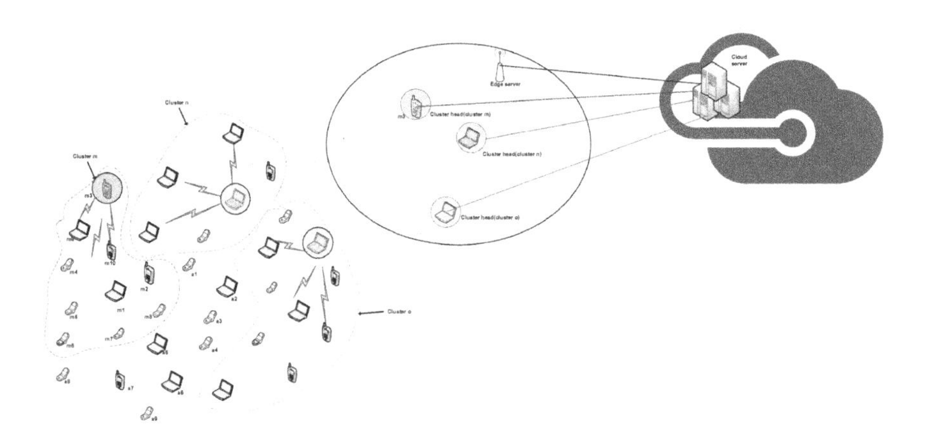

Fig. 1 Proposed offloading architecture

2 Literature Survey

Edge or fog computing is an optimizing method of cloud computing application where some portion of application (data and services) are taken from central cloud server to logical extreme (edge) of the Internet to serve the end user. Edge computing in one hand, significantly decreases the volume of data movement for end user to cloud server; on the other hand, it reduces the traffic flow, transmission costs, shrinking latency, and improving QoS.

Moura et al. [1] survey different applications of game theory in edge computing to reduce the service latency of cloud computing services. In another paper [2], authors considered three tier architecture to offload mobile data and computational resources to cloud servers. First tier consists of user node, middle tier consists of resource constraint local cloudlet (Access Point, etc.), and third tier consists of distant cloud servers which have practically infinite resources. In [3], the author proposes portioning technique for data streaming locally to the mobile nodes. Furthermore, in the paper [4], the authors segregated the optimization problem in mobile edge computing in two sub-problems, namely resource allocation (RA) with fixed task and task offloading (TO) problems. As the problem is computationally hard, convex optimization tool is used for the RA problem; while for the TO problem, a heuristic approach is being proposed. A binary offloading policy is being adopted in [5] to ensure the total task execution in the edge device or at the server end. For the offloading problem (online), the authors made use of reinforcement learning. The problem of multi-access fog computing was further taken in [6], and logic-based benders decomposition technique was cited as a potential solution to the computationally hard problem. In [7], the authors tackled the offloading problem by subdividing it into two sub-problems, namely user association and sub-channel assignment. While a many to one matching game is being proposed for the former, an one to many game is being played for the later with proper simulations. In [8], the authors handled the replication caused because of the offloading procedure by proposing a optimal communication upload download latency with a proper proof of convergence. A proper survey was conducted by [9] to understand the complexity and proposed solutions in the computation offloading problem in edge computing.

Recent articles like [8, 10] focuses on the evolutionary games for specific purposes like linkage detection for edge computing networks.

In this paper, the authors have formulated non-cooperative game-theoretic model to analyze the characteristics of user node usage and determine the data offloading point. The user node (Mobile, Tablet, laptop, etc.) can offload the data to middle tier or to the third tier directly based on stable coalition of pay off matrix (refer Fig. 1).

2.1 Overlapping Coalition Game

In game theory [11], agents (players/videos) form coalitions to perform cooperative jobs (video streaming from respective cloud servers) which produces a payoff vector along with a coalition structure as an outcome. The backup model that we are following is described in [12] which avoids the formation of grand coalition.

Cooperative games are simply categorized into *non-transferable utility* (NTU) and *transferable utility* (TU) games. We are interested in playing an NTU game which avoids any exchange between the players through any means. An NTU game can be formally stated as [13, 14] (Table 1):

Definition 1 [14] A non-transferable utility game can be defined a pair (V, β), where V is a finite set of players (videos) and β is the coalition function denoting the value of the coalition (a mapping defined for each coalition $X \subset V$).
The set $v(X) \subset \mathbb{R}^X$ represents feasible payoff vectors for each X.

Table 1 Notation and symbols used in this paper

Symbol	Description
V	Set of video streaming jobs
A	Set of properties (video streaming characteristics)
(A, Δ)	Meet semi-lattice
λ	Mapping from video to property (λ: $V \rightarrow A$)
X^{∇}	Common properties for a set of videos $X \subseteq V$
d^{∇}	Common videos for a set of properties $d \in (A, \Delta)$
$\perp$	Partial ordered relation between video pattern cluster
$\xi_{v_i v_j}$	Similarity index between videos v_i and v_j
β	Value of the coalition
CS	Coalition structure (list of partial coalitions)
Υ	Videos participating in coalition formation ($\Upsilon \subseteq X \subseteq V$)
c_j^k	Marginal contribution (payoff) of video 'j' in the 'kth' coalition (cluster)
c^k	Payoff vector of each video in $\{V\}$ in the 'kth' coalition (cluster)
support(c^k)	Support index of the 'kth' coalition
$v(c^i)$	Characteristic function of the 'kth' coalition
SC	Stable coalition
U	Set of user (edge) nodes streaming the similar video $\in$ SC
P	Set of processing speed for each node $\in U$
R	Set of memory for each node $\in U$
S	Set of storage space for each node $\in U$
PR	Set of priority for each node $\in U$
$>$	Priority ordering between each node $\in U$

Definition 2 [12, 14] A coalition structure CS on a set of players (videos) Υ where $\Upsilon \subseteq X \subseteq V$; is a finite list of partial coalitions $\text{CS}_{\Upsilon} = (c^1, c^2, \ldots, c^s)$ which satisfies

(i) $c^i \in [0, 1]^n$
(ii) $\text{support}(c^i) \subseteq \Upsilon$ for all $i = 1, \ldots, s$; and
(iii) $\sum_{i=1}^{s} c_j^i \leq 1 \quad \forall j \in \Upsilon.$

where 's' is the size of the coalition structure CS_{Υ} thus, $|\text{CS}_{\Upsilon}| = s$.

3 Basic Terminologies

Each user streams a video of their individual choice. Let $\{V_1, V_2, \ldots, V_n\}$ be the list of videos streaming at time t_i and $\{A_1, A_2, \ldots, A_n\}$ are the properties of the videos. We now define some basic terminologies required for the game-theoretic formulation.

Definition 3 A video pattern can be defined as a triple $(V, (A, \Delta), \lambda)$ where V is the set of videos, (A, Δ) is a complete meet-semi-lattice of the properties of the videos, and λ: $V \rightarrow A$ is a mapping from video to property.

Table 2 shows a video pattern where $V = \{V_1, \ldots, V_{12}\}$. The value associated with each property is shown in Fig. 2.

Table 2 Video pattern

Video	Genre	Connection	Type
1	TV shows	TCP	MP4
2	Web series	UDP	FLV
3	Music video	TCP	AVI
4	Web series	UDP	FLV
5	TV shows	TCP	MP4
6	Music video	TCP	AVI
7	TV shows	TCP	MP4
8	Web series	UDP	FLV
9	Web series	UDP	FLV
10	Live event	UDP	3GP
11	Live event	UDP	3GP
12	Web series	TCP	MOV

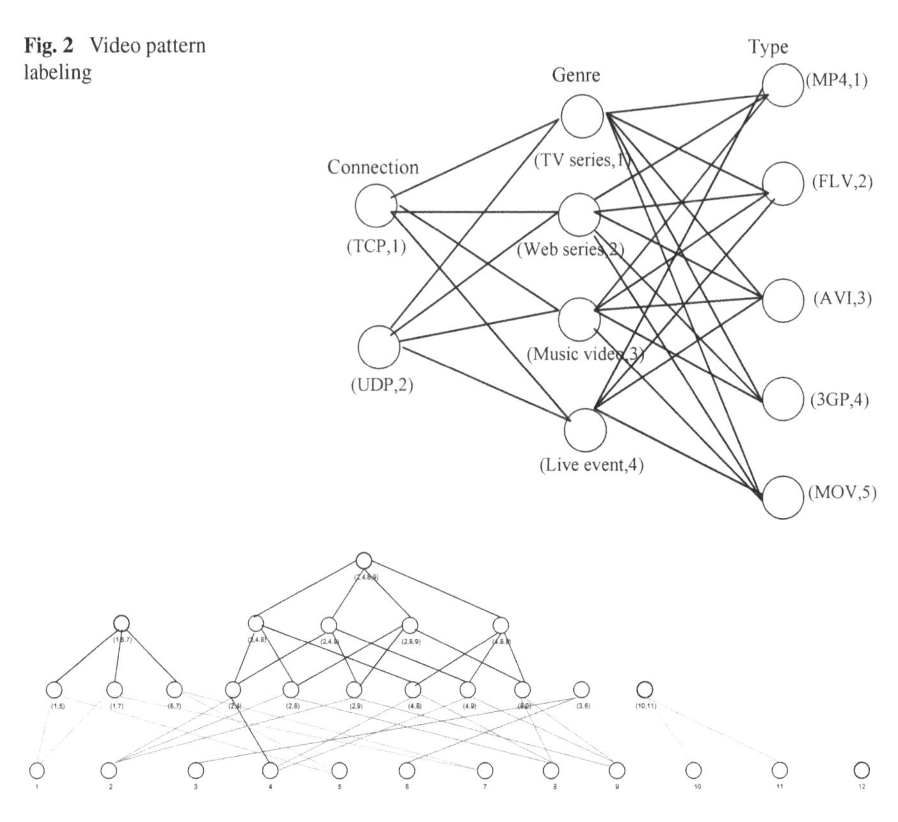

Fig. 2 Video pattern labeling

Fig. 3 Pattern lattice

Definition 4 The common properties for a set of videos can be represented by

$$X^{\nabla} = \Delta_{g \in X} \lambda(g) \quad \text{for } X \sqsubseteq V \tag{1}$$

Definition 5 The common videos for a set of properties can be represented by

$$d^{\nabla} = \{g \in X | d \sqsubseteq \lambda(g)\} \quad \text{for } d \in (A, \Delta) \tag{2}$$

Definition 6 A video pattern cluster can be represented as a pair (X, d) where $X \sqsubseteq V$ and $d \in (A, \Delta)$ such that $X^{\nabla} = d$ and $d^{\nabla} = X$.

Definition 7 A partial ordered relation $\perp$ between the video pattern clusters is defined as $(X_k, d_k) \perp (X_{k+1}, d_{k+1}) \Leftrightarrow X_k \subseteq X_{k+1} \Leftrightarrow d_{k+1} \subseteq d_k$. The pattern clusters together form a complete meet lattice called as pattern lattice.

The pattern lattice for the video pattern in Table 2 is shown in Fig. 3 which is formed by playing a cooperative game as shown in the subsequent section.

4 Cooperative Game-Theoretic Modeling

We first define a similarity matrix. The terms coalition and video pattern cluster are used interchangeably. The definition of stable coalition is maintained on the similarity index. To avoid global coalition, an avoidance parameter is defined. We then find a coalition structure of the videos where we get maximum payoff value for each video which forms a meet lattice.

We define cosine similarity index as

Definition 8 Cosine similarity index ξ between two videos v_i and v_j having properties $\{i_1, i_2, \ldots, i_m\}$ and $\{j_1, j_2, \ldots, j_m\}$, respectively, is defined as

$$\xi_{v_i v_j} = \frac{i_1 j_1 + i_2 j_2 + \cdots + i_m j_m}{\sqrt{i_1^2 + i_2^2 + \cdots + i_m^2}\sqrt{j_1^2 + j_2^2 + \cdots + j_m^2}} \quad (3)$$

The range of ξ is [0, 1].

In the meet lattice, the level of the lattice determines the coalition structure (CS) which is a collection of the set of coalitions (video pattern cluster) present at that level. In simple terms (refer Definition 2), when $\Upsilon = 2$ and $\Upsilon \subseteq X$, $t = \binom{X}{2}$ clusters are formed in a CS. We rewrite CS at level 2 as $CS_2 = (c^1, c^2, \ldots, c^t)$. c^k (the kth cluster) contains the vector $c^k = (c_1^k, c_2^k, \ldots, c_X^k)$ where c_j^k represents the marginal contribution (payoff) of video 'j' in the 'kth' coalition at level 2.

For example, for the coalition between two videos v_i and v_j (level 2), the individual payoff of each video in $\{V\}$ is represented by the vector,

$$c^k = \left(0, 0, \ldots, \frac{\sum_{\forall v_j \in |\Upsilon|} \xi_{v_i v_j}}{\sum_{\forall v_l \in |X|} \xi_{v_i v_l}}, 0, \ldots, \frac{\sum_{\forall v_i \in |\Upsilon|} \xi_{v_j v_i}}{\sum_{\forall v_l \in |X|} \xi_{v_j v_l}}, \ldots, 0\right)$$

Definition 9 Support index of a video pattern cluster in a coalition structure is defined as

$$\text{support}(c^k) = \{v \in V | c_v^k \neq 0\} \quad (4)$$

Definition 10 The characteristic function $v(c^i)$ of a video pattern cluster is defined as

$$v(c^i) = \sum_{\forall v_i \in |\Upsilon|} c_i^k \quad (5)$$

We now formulate the criteria for stable video pattern coalition.

Definition 11 A set of video patterns (S) are said to be stable iff

$$\frac{\sum_{\forall v_l \in |S'|} \xi_{v_i v_l}}{\sum_{\forall v_l \in |X|} \xi_{v_i v_l}} \leq \frac{\sum_{\forall v_j \in |S|} \xi_{v_i v_j}}{\sum_{\forall v_j \in |X|} \xi_{v_i v_j}} \quad (6)$$

where $S \bigcap S' = \phi$.

The above definition states that the videos which are most similar will form a coalition which is stable and no other video belongs to the coalition.

After a set of stable coalition is formed in a lattice at a level, we are now ready to form stable coalitions about other levels.

Theorem 1 *The stable coalition* $\{SC_i\}$ *at level* $(r-1)$ *will form a stable coalition* $S = \{SC_i \bigcup v_j\}$ *at level* (r) *iff for each* $i \in \{SC_i \bigcup v_j\}$*; the following criteria holds,*

$$\frac{\sum_{\forall v_l \in |S'|} \xi_{v_i v_l}}{\sum_{\forall v_l \in |X|} \xi_{v_i v_l}} \leq \frac{\sum_{\forall v_h \in |S|} \xi_{v_i v_h}}{\sum_{\forall v_h \in |X|} \xi_{v_i v_h}} \tag{7}$$

where $v_j \in \{V \setminus SC_i\}$*;* $l \in \{V \setminus \{SC_i \bigcup v_j\}\}$*;* $S = \{SC_i \bigcup v_j\}$*;* $S' = \{SC_i \bigcup v_l\}$*;* $S \bigcap S' = \phi$*;* $v_j \neq v_l \neq v_h$ *and* $\xi_{v_i v_j}$ *is the similarity index between two videos* v_i *and* v_j.

To avoid the formation of global coalition in a coalition structure (CS), we now set up an avoidance parameter.

Definition 12 The avoidance parameter stops the coalition formation game if the following criteria holds true

$$\text{support}(\text{SC}_i) > \text{support}(\text{SC}_j) \tag{8}$$

where SC_i is any stable coalition and $\text{SC}_j \in \text{CS}$.

Algorithm 1 Stable Video Pattern Clustering

Cluster videos with similar properties represented as a triple $(V, (A, \Delta), \lambda)$
Input: Video Pattern $(V, (A, \Delta), \lambda)$
Output: Pattern lattice of coalition structure (CS)

```
1: procedure FORMULATE–LATTICE
2:    Compute the similarity matrix using Definition 8.
3:    for each cluster ∈ ϒ in level 2 do
4:       Form Stable Clusters (SC) (Apply Definition 11).
5:       Apply Avoidance Parameter (Definition 12).
6:    end for
7:    for each SC from level 2 do
8:       Apply Theorem 1 to form larger stable clusters.
9:       Avoid Global Coalition (Definition 12).
10:      Find v(S) (Definition 10).
11:      Form Pattern Lattice (Definition 7)
12:   end for
13: end procedure
```

Algorithm 1 shows the stable clustering of similar videos. The lattice formed is shown in Fig. 3. The bold circles represents a stable coalition. Thus, the set of stable

Table 3 Edge node configuration

Node	Processing speed (GHz)	RAM (GB)
1	1.8	2
2	2	3
3	2	2
4	1.8	3
5	3.6	4
6	1.8	2
7	1.8	2
8	14.4	6
9	1.8	2
10	2.8	3

coalition formed from the video pattern in Table 2 are $\{(1, 5, 7); (2, 4, 8, 9); (10, 11); (12)\}$. A stable cluster in fact denotes a cluster of users streaming the similar type of video.

Now we have to select the user from the list of users involved in a stable cluster bearing different configurations to forward a stable coalition. The user selection and replacement strategy is defined in the next section.

5 Edge Node Replacement Strategy

Let $U = \{U_1, U_2, \ldots, U_n\}$ be the set of user (edge) nodes streaming similar videos forming a stable coalition $c \in \mathrm{CS}$.

Definition 13 Configuration setup of user node can be represented as a tuple (P, R); where $P = \{P_1, P_2, \ldots, P_n\}$ is the set of processing speed and $R = \{R_1, R_2, \ldots, R_n\}$ is the set of device memory (refer Table 3).

A priority set $\mathrm{PR} = \{\mathrm{PR}_1, \mathrm{PR}_2, \ldots, \mathrm{PR}_n\}$ is defined for each of the user nodes.

Definition 14 Priority (PR_i) of an edge node (U_i) is computed as,

$$\mathrm{PR}_i = P_i * R_i \tag{9}$$

Definition 15 A priority ordering $>$ is defined for any edge nodes U_i and U_j as $\mathrm{PR}_i > \mathrm{PR}_j \Leftrightarrow P_i \geq P_j \Leftrightarrow R_i \geq R_j$.

A max heap is formed based on the priority of the user. Algorithm 2 formally defines the above process.

Algorithm 2 selects the node as the head of the cluster (stable coalition) with the highest priority. If the cluster head node stops streaming, it gets deleted from the

Algorithm 2 Edge Node Replacement

Create a max heap of the edge (user) nodes on a stable cluster $c \in CS$ based on priority.
Input: Configuration setup of each edge node represented as a tuple (P, R).
Output: Cluster head

1: **procedure** FORMULATE–MAX HEAP
2: **for** each edge node $U_i \in c$ **do**
3: Compute PR_i (using Definition 14)
4: Insert U_i with value PR_i in the max heap
5: Perform heapify().
6: **end for**
7: head = findmax()
8: Set head as the Cluster head.
9: **if** (U_i stops streaming)
10: Perform deletemax()
11: Perform heapify()
12: head = findmax()
13: Set head as the new Cluster head.
14: **end procedure**

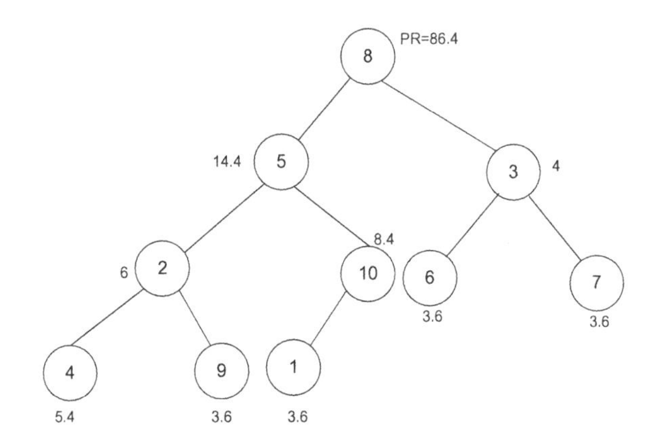

Fig. 4 Max heap

heap using deletemax() operation, and the next priority node present in the max heap becomes the cluster head following the heapify() operation of max heap. The cluster head is used to directly stream videos which will be working as a proxy node for the other edge nodes for streaming.

Table 3 shows the user configuration for a set $U = \{U_1, U_2, \ldots, U_{10}\}$ of users belonging to a stable cluster formed by algorithm 1. A heap, as shown in Fig. 4, is formed by algorithm 2 with user node priority which selects user U_8 as the cluster head. When U_8 stops streaming, U_5 is the potential next cluster head.

6 Discussion and Time Complexity Analysis

Algorithm 1 has a polynomial time complexity in the order of $O(v^L)$, where v is the number of users streaming videos simultaneously at time 't' and L is the size of the pattern lattice. Algorithm 2 also has a polynomial time complexity of $O(n \log n)$, where 'n' is the number of users in a stable coalition.

We discuss algorithm 1 with the help of the video pattern shown in Table 2. The nominal data is first represented with some real data as shown in Fig. 2. For example, (MP4, 1) denotes that MP4 is represented by the real value '1.' The data is now given as an input to algorithm 1, and Fig. 3 shows the output. Each circle in the lowest level of the lattice denotes a video sequence. The bold circles represent stable coalition of videos streamed by their respective users (edge node) based on similarity. Our example creates $\{(1, 5, 7); (2, 4, 8, 9); (10, 11); (12)\}$ as a coalition structure, where each element, for example (2, 4, 8, 9) is a stable coalition.

Now a stable coalition is given as an input to algorithm 2. We suppose that (1, 2, …, 10) is a stable coalition of edge node, and their configuration is shown in Table 3. Algorithm 2 creates the heap as shown in Fig. 4. Here U_8 works as the cluster head at time 't' and when at time t_i, U_8 stops streaming, U_5 is the potential next cluster head.

7 Conclusion

In this paper, we proposed an architecture that offloads the computational resources from cloud server to edge node to reduce the power consumption and data traffic consumption of the receiving mobile node. A cooperative game-theoretic framework is first proposed to cluster the mobile (edge) nodes receiving same video streaming content. Identified mobile nodes create different clusters based on their similarity and then selects a cluster head that acts as the edge node. The edge node receives the video streaming locally from the ISP and distributes it among the other node in the cluster.

The proposed novel algorithm is to be implemented in future and compared with the other architectures in literature to ensure its optimality as discussed.

References

1. Moura J, Hutchison D (2017) Survey of game theory and future trends for applications in emerging wireless data communication networks. CoRR
2. Cardellini V, Persone VDN, Di Valerio V, Facchinei F, Grassi V, Presti FL, Piccialli V (2016) A game-theoretic approach to computation offloading in mobile cloud computing. Math Program 157(2):421–449

3. Yang L, Cao J, Yuan Y, Li T, Han A, Chan A (2013) A framework for partitioning and execution of data stream applications in mobile cloud computing. ACM SIGMETRICS Perform Eval Rev 40(4):23–32
4. Tran TX, Pompili D (2018) Joint task offloading and resource allocation for multi-server mobile-edge computing networks. IEEE Trans Veh Technol 68(1):856–868
5. Huang L, Bi S, Zhang YJ (2019) Deep reinforcement learning for online computation offloading in wireless powered mobile-edge computing networks. IEEE Trans Mob Comput
6. Alameddine HA, Sharafeddine S, Sebbah S, Ayoubi S, Assi C (2019) Dynamic task offloading and scheduling for low-latency IoT services in multi-access edge computing. IEEE J Sel Areas Commun 37(3):668–682
7. Pham QV, LeAnh T, Tran NH, Hong CS (2018) Decentralized computation offloading and resource allocation in heterogeneous networks with mobile edge computing. arXiv preprint arXiv:1803.00683
8. Li Q, Hou J, Meng S, Long H (2020) GLIDE: a game theory and data-driven mimicking linkage intrusion detection for edge computing networks. Complexity
9. Jiang C, Cheng X, Gao H, Zhou X, Wan J (2019) Toward computation offloading in edge computing: a survey. IEEE Access 7:131543–131558
10. Iyer GN (2020) Evolutionary games for cloud, fog and edge computing—a comprehensive study. In: Computational intelligence in data mining. Springer, Singapore, pp 299–309
11. Roger BM (1991) Game theory: analysis of conflict. The President and Fellows of Harvard College, USA
12. Chalkiadakis G, Elkind E, Markakis E, Polukarov M, Jennings NR (2010) Cooperative games with overlapping coalitions. J Artif Intell Res 39:179–216
13. Hart S (2004) A comparison of non-transferable utility values. In: Essays in cooperative games. Springer, Boston, MA, pp 35–46
14. Bhuyan BP (2017) Relative similarity and stability in FCA pattern structures using game theory. In: 2017 2nd international conference on communication systems, computing and IT applications (CSCITA). IEEE, pp 207–212

Attentive Convolution Network-Based Video Summarization

Deeksha Gupta and Akashdeep Sharma

Abstract The availability of smart phones with embedded video capturing mechanisms along with gigantic storage facilities has led to generation of a plethora of videos. This deluge of videos grasped the attention of the computer vision research community to deal with the problem of efficient browsing, indexing, and retrieving the intended video. Video summarization has come up as a solution to aforementioned issues where a short summary video is generated containing important information from the original video. This paper proposes a supervised attentive convolution network for summarization (ACN-SUM) framework for binary labeling of video frames. ACN-SUM is based on encoder–decoder architecture where the encoder is an attention-aware convolution network module, while the decoder comprises the deconvolution network module. In ACN-SUM, the self-attention module captures the long-range temporal dependencies among frames and concatenation of convolution network and attention module feature map result in more informative encoded frame descriptors. These encoded features are passed to the deconvolution module to generate frames labeling for keyframe selection. Experimental results demonstrate the efficiency of the proposed model against state-of-the-art methods. The performance of the proposed network has been evaluated on two benchmark datasets.

Keywords Video summarization · Convolutional neural network · Self-attention · Deep learning · Computer vision

1 Introduction

The accessibility to low-cost video capturing devices, huge storage capacity and high-speed Internet has resulted in exponential growth in videos on our personal devices as well as on social Web sites like YouTube. According To Cisco Annual

D. Gupta · A. Sharma (✉)
University Institute of Engineering and Technology, Panjab University, Chandigarh, India
e-mail: akashdeep@pu.ac.in

D. Gupta
Mehr Chand Mahajan DAV College for Women, Panjab University, Chandigarh, India

A. Choudhary et al. (eds.), *Applications of Artificial Intelligence and Machine Learning*, Lecture Notes in Electrical Engineering 778,
https://doi.org/10.1007/978-981-16-3067-5_25

Report, by year 2022, 82% of the total internet traffic will be consumed by online videos [1]. Such a large volume of videos with limited meta-data make it difficult to index or browse the video content on the web. Video summarization comes in rescue to cope up with the challenges imposed by large-scale video data [2].

Video summarization generates condensed versions of video which is much smaller in size but contains the important and interesting content of the video [2]. Due to fast and efficient processing of short videos, video summarization can also be included as a preprocessing step in various video analyzing tasks like anomaly detection [3] and action recognition [4].

From the summary form perspective, video summarization techniques have been categorized into two classes: static video summarization [5–15, 23, 24] and dynamic video summarization [16–22]. Static video summarizations produce summaries in the form of keyframes [6, 12, 23], mosaic [24], storyboard [7], where temporal information is not included. Dynamic video summarization selects important video shots as a summary upholding temporal aspect also. Static summaries help to provide efficient indexing and retrieval and dynamic summaries support a user decision of watching a video by providing gist of video content. The technique generates video summaries in the form of key-shots. The proposed network assigns a binary label to every frame, specifying its selection, or rejection in video summary.

Earlier work in the field of video summarization was based on heuristic approaches exploiting low-level features to model the visual representativeness [5, 6, 12, 19, 25, 26] and diversity [15, 27–29] for summary generation. The inadequacy of low-level feature in depicting the high-level semantic information of video content leads to poor performance. In quest of user oriented summaries, the video summarization research area has gone through a paradigm shift from heuristic to deep models in the past half-decade. In recent studies, RNN-based approaches [18, 19, 25, 30] have shown better performance in modeling temporal dependencies among video content but the inherent limitation of RNN of processing video frames sequentially in left to right direction hampers parallel processing of frames [31]. This leads to poor utilization of GPU and slows down the process.

The aforementioned limitation of RNN encouraged development of ACN-SUM where main building blocks comprise of fully convolutional layers rather than RNN. ACN-SUM is based upon encoder–decoder framework, where the encoder encompasses a fully convolutional module for temporal information and visual feature extraction, followed by the attention-aware module that assigns importance score to frames for quality video summarization. The work is inspired with [31], where a fully convolutional sequential network was used to generate video summary. In ACN-SUM, the convolution network abstract information and attention scores are combined to get higher level information. Finally, the deconvolution module generates binary labels of frames. Extensive experiments on two benchmark datasets are performed to demonstrate the performance of the method.

The structure of the rest of the paper is as follows: In Sect. 2, review of the related literature is provided. Section 3 presents the detailed description of every component of our model. Section 4 includes experiment results obtained. Section 5 represents the concluding remarks.

2 Related Work

Video summarization research area is devoted to obtaining a short video containing all important information and events while eliminating redundant content from the original video. Depending upon the different applications, different forms of summary can be generated that include keyframe [32], storyboard [7], hyperlapses [33], mosaics [24], video skims [20, 22], etc. Our work is related to keyframe-based video summary generation. Research in video summarization has seen steady growth in the literature. Earlier approaches were based on heuristic criteria [5, 9, 10, 13, 34, 35] with handcrafted features like color [5, 6, 8, 10], motion [14, 36, 37], and HOG [20, 26]. Some techniques focus on diversity [15, 17, 27] of content while others targeted interestingness or importance [14, 16, 22, 38] of frames. High-level features like hand position [39] and object interaction [16] are computed for interpreting the importance of frames. Some research work exploited external information like title [40], web images [41] or EEG recordings [42] for generating semantically rich summaries.

From a learning perspective, recent deep learning model-based work can be categorized into supervised [18, 31, 43–46] and unsupervised [25, 30, 47, 48] techniques. Unsupervised approaches learn the model parameters while optimizing user defined loss functions for diversity, representativeness, and summary length. Zhou et al. [48] proposed reinforcement-based learning optimized with diversity and representative award functions. Mahasseni et al. [25] presented an unsupervised adversarial model where the generator comprises variational autoencoder (VAE) as encoder and discriminator contains LSTM. The unsupervised approaches cope up with the problem of unavailability of ground truth annotated video dataset but their performance is compromised as compared to that of supervised methods.

Supervised methods employ original video and its corresponding summary pair for training of summary generating models. Under supervised method [15] was the first supervised attempt to generate video summary, optimized over representative loss between annotated ground truth and generated summary. Zhang et al. [18] presented LSTM-based supervised encoder–decoder framework for summary generation. Sharghi et al. and Zhang et al. [17, 49] proposed LSTM-based model followed by determinantal point process (DPP) module for incorporating diversity in generated summary.

The use of LSTM in summary generation task shows better performance owing to their nature of memorizing and modeling sequential data. But the LSTM-based model also suffers from two limitations. First, owing to sequential modeling, the processing of video frames gets very slow as one frame can be processed only after previous frame processing. Second, due to limited memory in LSTM, they can restore small-range temporal structure, making them unsuitable for long video summarization. Because of mentioned shortcomings of RNN, we used convolution network as basic building block for summary generation. Our work is inspired by the fully convolutional sequence model used by Rochan et al. [31]. The main issue with FCSN [31] is that, it ignores underlying temporal dependency and treat every frame with importance. But in our proposed model, we made an attempt to improvise the performance by employing the attention mechanism, which assigns a higher score to a frame containing more important information or event. Soft attention-aware mechanism proposed by Bahdanau et al. [50] emphasizes the important region contained inside an image. Attention-based models have shown tremendous performance in machine translation [50] and image caption generation [51]. Attention coefficients are computed as the importance value of the image. This paper incorporates a self-attention module to enhance the frame level feature information to improvise the generated summary quality. As supervised models outperform unsupervised models, this motivates the use of supervised mechanisms for training of underlying modules.

3 Methodology

A video is a collection of frames, and images and frames lying in close proximity contain redundant information. Motivated by this fact, instead of processing all the frames of a video, pre-sampling of frames is performed. The aim of pre-sampling is to reduce the computational cost of a model without incurring any information loss.

Let us assume a video '*V*' contains '*T*' frames. After preprocessing video, '*V*' is represented as a collection of pre-samples '*n*' frames such that $V = \{F_1, F_2, F_3, \ldots, F_n\}$, where F_i represents pretrained CNN-based feature descriptor of ith frame. These feature descriptors are processed by convolution network module (CM) which is followed by self-attention module (SAM) as shown in Fig. 1.

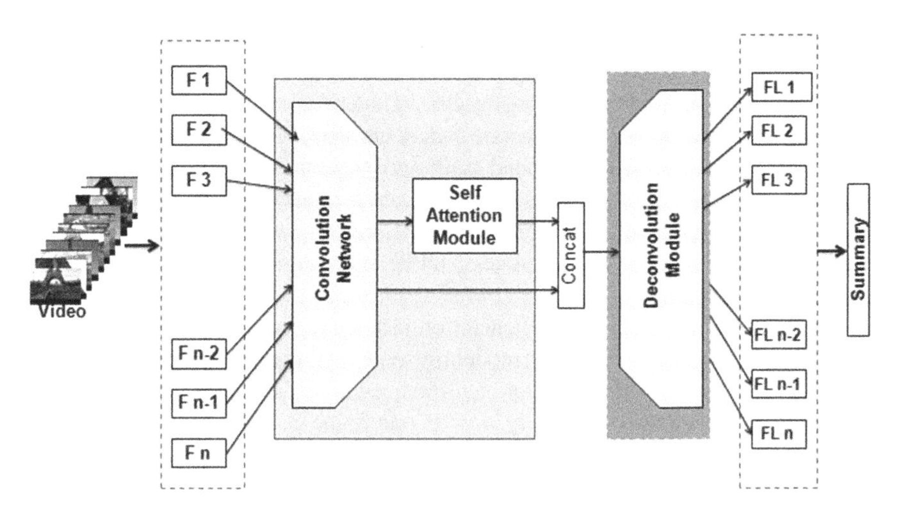

Fig. 1 Attentive convolution network (ACN-SUM)

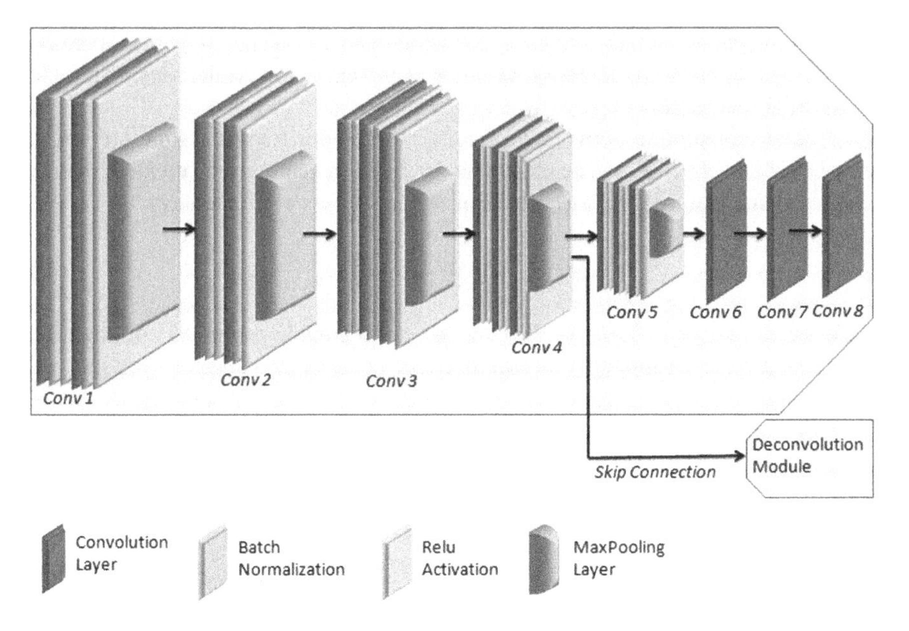

Fig. 2 Convolution network module

3.1 Convolution Network Module

Convolution network has tremendous capability to extract high-level features from input images. Inheriting this property, the convolution module of our approach is stimulated by fully convolution sequence network used for semantic segmentations [52]. Convolution layer structure used is inspired by [52], where the below 5 layers contain multiple cascaded convolution layers, batch normalization, and ReLu activation function. Every layer is followed by a temporal max-pooling layer as shown in Fig. 2. Intermediate pooling layers reduce dimension along the temporal axis. One of the major reasons behind employing convolution networks is their efficiency in inference and backpropagation due to sharing of convolution maps layer by layer. Along with the output of the eighth convolution layer, the output of the Pool 4 layer is also routed to the deconvolution block. This merging results in richer temporal information by combining lower layers coarse features with upper layer fine features.

3.2 Self-attention Module

The consideration of temporal dependencies among video frames plays an important role while generating summaries. RNN variant LSTM-based models capture these dependencies but are limited to local neighborhoods only [53]. Zhao et al. [45] proposed use of hierarchical LSTM to model long-range temporal relationship, which improves the range but fails to confine the whole video sequence. The self-attention component computes the attention score of each frame unit while considering the whole video sequence.

Let after convolution network processing, the output frames feature are represented as (f_1,f_2,f_3 ...$f_{n'}$). The alignment matrix representing self-attention weight between frame feature f_i and whole frame sequence f_j is calculated as:

$$m_i^j = W \tanh\left(W_1 f_j + W_2 f_i\right) \tag{1}$$

where $i \in n'$ and $j = [0, n')$. W, W_1, and W_2 are learnable parameters of attention module. The attention mask coefficient for frame f_i is computed by applying softmax function on m_i^j.

$$v_i^j = \frac{\exp\left(m_i^j\right)}{\sum_{k=1}^{n'} \exp\left(m_i^k\right)} \tag{2}$$

Finally, attention vector is computed as weighted sum of convolution network output descriptor.

$$a_i = \sum_{j=0}^{n'} v_i^j f_i \tag{3}$$

Concatenation of the attention vector and convolution network processed frame feature provides a semantically rich descriptor for each frame.

3.3 Deconvolution Module

Deconvolution module consists of two deconvolution layers that map reduced temporal dimension to their original size. First deconvolution layer accepts concatenated convolution module output and self-attention module output feature map. Output of the first deconvolution layer is merged with conv4 layer output (followed by 1×1 convolution to map to desired output channel) and routed to the second deconvolution layer. This merging implement skip connection to enhance summarization model performance for restoring temporal information. The second deconvolution layer followed by sigmoid generates final prediction for frames.

4 Experiment

During experiments, datasets are processed to generate oracle summaries in the form of frame level labels for training purpose. The videos are processed to extract frame descriptors representing video frames. ACN-SUM model framework is designed and trained using binary cross-entropy loss. The model performance is evaluated using overlap F-measure metric under canonical setup where training and testing videos belong to the same dataset.

In this section, first, the datasets used for evaluation of the model are specified. Then, the implementation details of the experiment are presented followed by evaluation metrics used for performance evaluation. In the last, baseline models and experimental results are discussed.

4.1 Dataset

Two benchmark datasets, SumMe [22] and TVSum [40], are used for performance evaluation of the proposed ACN-SUM model. SumMe dataset comprises 25 videos of diverse nature belonging to three different categories of first person videos—egocentric videos, static camera videos, and moving camera videos. The videos are of short duration ranging from 1 to 6 min, resulting in a total duration of more than one hour. In the dataset, for every video 15–18 human summaries are provided resulting in 390 summaries in total. The summaries are provided as a shot level importance score. TVSum dataset comprises 50 videos belonging to 10 different categories (5 videos per category). 20 user annotations are provided for each video resulting in 2000 annotations in total. Annotations are in the form of frame level importance score.

4.2 Ground Truth Preparation

For training of models, oracle summaries are obtained from available multiple user annotated ground truth scores. An oracle summary contains frame level importance score computed using multiple user annotations with an aim to maximize the F-measure. For comparative evaluation with baseline methods, we need to represent ground truth summaries in the form of key-shot summaries. SumMe dataset already contains key-shot level score, but TVSum dataset contains frame level score in user annotations. So, following [18], frame level scores of TVSum are transformed to shot level score by following 4 step procedure: First, the video is segmented using kernel temporal segmentation (KTS) mechanism specified by [54]. Second, from user annotation, average segment score is computed and assigned to each segment constituent frame. Third, rank all the frames based on their score and forth, select keyframes as per specified summary length, using knapsack algorithm [40]. Thus, keyframe level summaries are obtained from score-based annotations.

4.3 Implementation Detail

Frame Descriptor. For fair comparison with baseline models, pretrained Inception V1 (GoogleNet) descriptors are used to represent each frame. For this, frame representation is obtained from the penultimate layer of GoogleNet network resulting in a 1024 D feature vector. For any video with 'n' sampled frames, $(1 \times n \times 1024)$ dimensional input is passed to the convolution network.

Model Training. For training of model, keyframe-based annotations are used. Under supervised setting, to maximize accuracy, binary cross-entropy loss is used to minimize representation error between selected keyframe and ground truth summary. The network is trained using ADAM optimizer with a learning rate of 0.001 and momentum of 0.9 with batch size equal to 5. For training and testing purposes, the dataset videos are randomly split into 80:20 ratios (80% for training and 20% for testing).

$$\mathcal{L}_{\mathrm{BCE}} = -\frac{1}{n}\sum_{t=1}^{n} y_t \log(\hat{y}) + (1 - y_t)\log(1 - \hat{y}) \tag{4}$$

where y_t represents ground truth label and $\hat{y}$ denotes predicted label for frame f_t.

4.4 Evaluation Metric

Following [18, 22, 25, 45], key-shot-based evaluation metrics are used for performance comparison. Let AS specifies the automatic generated summary and GS specifies the user specified ground truth summary. Precision and recall metrics are used to compute temporal overlap score of AS and GS, as follows:

$$\text{Recall} = \frac{\text{Overlapped duration of AS and GS}}{\text{Duration of AS}} \tag{5}$$

and

$$\text{Precision} = \frac{\text{Overlapped duration of AS and GS}}{\text{Duration of GS}} \tag{6}$$

F-measure is computed using precision (P) and recall (R) as their harmonic mean: So,

$$F\text{-measure} = \frac{2(\text{Recall} * \text{Precision})}{\text{Recall} + \text{Precision}} \quad (7)$$

As specified in Sect. 4.2, model generated keyframe summary is transformed into shot-based summary for evaluation purpose.

4.5 *Baseline Models*

Various supervised methods are considered as a baseline for comparative evaluation of the proposed model. These baseline models are described in this section.

Zhang et al. [18] proposed supervised model vsLSTM that employs bidirectional LSTM in encoder–decoder architecture followed by a multilayer perceptron (MLP) network. Another variant of the same architecture [18]—dppLSTM is presented in which determinantal point process probabilistic model is integrated with LSTM to focus on diverse key frame selection. Ji et al. [46] introduced supervised sequence to sequence attention-aware encoder–decoder-based model where decoder uses attention information to determine the important segments. Two variants A-AVS (additive attention video summarization) and M-AVS (multiplicative attention video summarization) are presented in the study. Yujia et al. [19] proposed dilated temporal relational generative adversarial network (DTR-GAN) where training of generator and discriminator is done in three-player loss manner. A hierarchical LSTM-based approach is proposed in [45], where a two-layer LSTM concept is implemented, first layer LSTM encodes video shots and second layer predicts the shot scores. Finally, [20] proposes an adversarial process that learns a mapping function from raw videos to human-like summaries, based on professional summary videos available online.

5 Experimental Results

Table 1 presents the performance of the proposed ACN-SUM model (in bold) with respect to other state-of-the-art supervised approaches, on SumMe and TVSum dataset. It shows that our model outperforms prior methods for the SumMe dataset. The inflation in result justified the effectiveness of our approach to select important frames from video. Among previous studies, performance of the SUM-FCN [31] is comparable to the proposed approach for SumMe dataset but for TVSum dataset, our method shows better performance with a large margin. One possible explanation may be the capability of our model to capture important region information in a better way for edited videos. The first runners up results are underlined in Table 1.

Table 1 Comparison results on SumMe and TVSum datasets in terms of F-measure in % age

Dataset	Model	F-measure
SumMe (80% training + 20% testing)	vsLSTM [18]	37.6
	DPPLSTM [18]	38.6
	DTR-GAN [19]	44.6
	A-AVS [46]	43.9
	M-AVS [46]	44.4
	SUM-FCN [31]	47.5
	H-RNN [45]	44.3
	ACN-SUM (proposed)	**47.9**
TVSum (80% training + 20% testing)	DTR-GAN [19]	59.1
	A-AVS [46]	59.4
	M-AVS [46]	61
	SUM-FCN [31]	56.8
	H-RNN [45]	62.1
	ACN-SUM (proposed)	**62.8**

5.1 *Ablation Study*

The ablation study is carried out to analyze the impact of the attention module in the model. Under ablation study, ACN-SUM model without attention module is evaluated for both of the datasets. The ACN-SUM without attention module contains temporal convolution module and deconvolution module. The resulting model lacks in modeling temporal relationship between the video frames. Table 2 specifies behavior of the proposed model with and without self-attention module is computed.

The results, in bold, indicate the significance of the attention module in the proposed model. Attention module models long-range temporal relationships over whole frame sequences and thus generate diverse summaries. Inclusion of the attention layer improves results for both of the datasets.

Table 2 Ablation study of ACN-SUM with and without attention module

Dataset	Model	F-measure
SumMe (80% training + 20% testing)	ACN-SUM_w/o_attention	43.6
	ACN-SUM	**47.9**
TVSum (80% training + 20% testing)	ACN-SUM_w/o_attention	56.9
	ACN-SUM	**62.8**

6 Future Scopes and Conclusion

This paper specifies a novel supervised attention-based convolutional network framework for video summarization. The convolution networks reduce the temporal dimension, to emphasize the subset of frames. Attention module computes the attention score to weight the frame's importance. Deconvolution layers result in binary labeling of frames. Experimental results show that the proposed model outperforms the recent state-of-the-art models. The experiments are carried out on two diverse nature datasets.

The embedding of attention in temporal convolution network has shown promising results. It shows a promising direction to include different natures attention and at different stages of the convolution network. Also, the global diverse attention might be helpful in boosting the efficiency of summarization model.

References

1. Cisco (2020) Cisco annual internet report (2018–2023). Cisco, pp 1–41
2. Truong BT, Venkatesh S (2007) Video abstraction: a systematic review and classification. ACM Trans Multimed Comput Commun Appl 3(1):3:1–3:37
3. Feng Y, Yuan Y, Lu X (2017) Learning deep event models for crowd anomaly detection. Neurocomputing 219:548–556
4. Tejero-de-Pablos A, Nakashima Y, Sato T, Yokoya N, Linna M, Rahtu E (2018) Summarization of user-generated sports video by using deep action recognition features. IEEE Trans Multimed 20(8):2000–2011. https://doi.org/10.1109/TMM.2018.2794265
5. Mundur P, Rao Y, Yesha Y (2006) Keyframe-based video summarization using Delaunay clustering. Int J Digit Libr 6:219–232
6. Avila S, Brandaolopes A, Luz A, Araujo A (2011) VSUMM: a mechanism designed to produce static video summaries and a novel evaluation method. Pattern Recogn Lett 32(1):56–68
7. Furini M, Geraci F, Montangero M, Pellegrini M (2010) STIMO: STIll and MOving video storyboard for the web scenario. Multimed Tools Appl 46:47. https://doi.org/10.1007/s11042-009-0307-7
8. Mahmoud KM, Ismail MA, Ghanem NM (2013) VSCAN: an enhanced video summarization using density-based spatial clustering. In: Petrosino A (eds) Image analysis and processing—ICIAP 2013. Lecture notes in computer science, vol 8156. Springer, Berlin, Heidelberg
9. Wu J, Zhong S, Jiang J, Yang Y (2017) A novel clustering method for static video summarization. Multimed Tools Appl 76:9625–9641. https://doi.org/10.1007/s11042-016-3569-x
10. Kumar K, Shrimankar DD, Singh N (2018) Eratosthenes sieve based key-frame extraction technique for event summarization in videos. Multimed Tools Appl 77:7383–7404
11. Shroff N, Turaga SP, Chellappa R (2010) Video précis: highlighting diverse aspects of videos. IEEE Trans Multimed 12(8):853–868
12. Mahmoud KM, Ghanem NM, Ismail MA (2013) VGRAPH: an effective approach for generating static video summaries. In: 2013 IEEE international conference on computer vision workshops, Sydney, NSW, pp 811–818. https://doi.org/10.1109/ICCVW.2013.111
13. Iparraguirre J, Delrieux C (2013) Speeded-up video summarization based on local features. In: 2013 IEEE international symposium on multimedia, Anaheim, CA, pp 370–373. https://doi.org/10.1109/ISM.2013.70

14. Srinivas M, Pai MM, Pai RM (2016) An improved algorithm for video summarization a rank based approach. Procedia Comput Sci 89:812–819
15. Gong B, Grauman K (2014) Diverse sequential subset selection for supervised video summarization. Adv Neural Inf Process Syst 3:2069–2077
16. Lu Z, Grauman K (2013) Story-driven summarization for egocentric video. In: 2013 IEEE conference on computer vision and pattern recognition, Portland, OR, pp 2714–2721. https://doi.org/10.1109/CVPR.2013.350
17. Sharghi A, Borji A, Li C, Yang T, Gong B (2018) Improving sequential determinantal point processes for supervised video summarization. In: Lecture notes computer science (including Lecture notes artificial intelligence. Lecture notes bioinformatics) LNCS, vol 11207, pp 533–550
18. Zhang K, Chao WL, Sha F, Grauman K (2016) Video summarization with long short-term memory. In: ECCV. Springer, pp 766–782
19. Yujia Z, Kampffmeyer M, Zhao X, Tan M (2019) DTR-GAN: dilated temporal relational adversarial network for video summarization. In: ACM TURC '19: proceedings of the ACM turing celebration conference—China, pp 1–6. https://doi.org/10.1145/3321408.3322622
20. Lei Z, Zhang C, Zhang Q, Qiu G (2019) FrameRank: a text processing approach to video summarization. In: IEEE international conference on multimedia and expo (ICME), Shanghai, pp 368–373. https://doi.org/10.1109/ICME.2019.00071
21. Wei H, Ni B, Yan Y, Yu H, Yang X, Yao C (2018) Video summarization via semantic attended networks. In: Proceeding 22nd AAAI conference, artificial intelligence, pp 216–223
22. Gygli M, Grabner H, Riemenschneider H, Gool NV (2014) Creating summaries from user videos. In: Proceeding European conference on computer vision. Springer, pp 505–520
23. Asadi E, Charkari NM (2012) Video summarization using fuzzy c-means clustering. In: 20th Iranian conference on electrical engineering (ICEE2012), Tehran, pp 690–694. https://doi.org/10.1109/IranianCEE.2012.6292442
24. Viguier R, Lin CC (2015) Automatic video content summarization using geospatial mosaics of aerial imagery. In: 2015 IEEE international symposium on multimedia (ISM), Miami, FL, pp 249–253. https://doi.org/10.1109/ISM.2015.124
25. Mahasseni B, Lam M, Todorovic S (2017) Unsupervised video summarization with adversarial LSTM networks. In: Proceedings of the IEEE conference on computer vision and pattern recognition, CVPR 2017, vol 2017, Jan 2017, pp 2982–2991
26. Garcia A, Boix X, Lim J, Tan A (2012) Active video summarization: customized summaries via on-line interaction with the user. In: Proceedings of the thirty-first AAAI conference on artificial intelligence, pp 4046–4052
27. Anirudh R, Masroor A, Turaga P (2016) Diversity promoting online sampling for streaming video summarization. In: 2016 IEEE international conference on image processing (ICIP), Phoenix, AZ, pp 3329–3333. https://doi.org/10.1109/ICIP.2016.7532976
28. Sharghi A, Borji A, Li C, Yang T, Gong B (2018) Improving sequential determinantal point processes for supervised video summarization. In: Proceedings of the European conference on computer vision (ECCV), pp 517–533
29. Yandong L, Wang L, Yang T, Gong B (2018) How local is the local diversity? Reinforcing sequential determinantal point processes with dynamic ground sets for supervised video summarization. In: European conference on computer vision
30. He X, Hua Y, Song T, Zhang Z, Xue Z, Ma R, Robertson N, Guan H (2019) Unsupervised video summarization with attentive conditional generative adversarial networks. In: Proceedings of the 27th ACM international conference on multimedia (MM '19). Association for Computing Machinery, New York, NY, pp 2296–2304. https://doi.org/10.1145/3343031.3351056
31. Rochan M, Ye L, Wang Y (2018) Video summarization using fully convolutional sequence networks. In: European conference on computer vision (ECCV-2018)
32. Rani S, Kumar M (2020) Social media video summarization using multi-visual features and Kohnen's self organizing map. Inf Process Manag 57(3). https://doi.org/10.1016/j.ipm.2019.102190

33. Kopf J, Cohen MF, Szeliski R (2014) First-person hyper-lapse videos. ACM Trans Graph 33(4):78:1–78:10. [Online]. http://doi.acm.org/10.1145/2601097.2601195
34. Wang J, Wang Y, Zhang Z (2011) Visual saliency based aerial video summarization by online scene classification. In: 2011 sixth international conference on image and graphics, Hefei, pp 777–782. https://doi.org/10.1109/ICIG.2011.43
35. Ejaz N, Tariq TB, Baik SW (2012) Adaptive key frame extraction for video summarization using an aggregation mechanism. J Vis Commun Image Represent 23(7):1031–1040. https://doi.org/10.1016/j.jvcir.2012.06.013
36. Elhamifar E, Kaluza MC (2017) Online summarization via submodular and convex optimization. In: 2017 IEEE conference on computer vision and pattern recognition (CVPR), Honolulu, HI, pp 1818–1826. https://doi.org/10.1109/CVPR.2017.197
37. Ejaz N, Baik S, Majeed H, Mehmood I, Chang H (2018) Multi-scale contrast and relative motion-based key frame extraction. EURASIP J Image Video Process 40. https://doi.org/10.1186/s13640-018-0280-z
38. Ejaz N, Mehmood I, Baik S (2013) Efficient visual attention based framework for extracting key frames. Signal Process Image Commun 28:34–44
39. Lee YJ, Ghosh J, Grauman K (2012) Discovering important people and objects for egocentric video summarization. In: IEEE conference on computer vision and pattern recognition (CVPR)
40. Song Y, Vallmitjana J, Stent A, Jaimes A (2015) Tvsum: summarizing web videos using titles. In: IEEE conference on computer vision and pattern recognition
41. Cai S, Zuo W, Davis LS, Zhang L (2018) Weakly-supervised video summarization using variational encoder-decoder and web prior. In: European conference on computer vision (ECCV-2018)
42. Mehmood I, Sajjad M, Rho S, Baik SW (2016) Divide-and-conquer based summarization framework for extracting affective video content. Neurocomputing 174:393–403
43. Zhang K, Grauman K, Sha F (2018) Retrospective encoders for video summarization. In: European conference on computer vision
44. Fajtl J, Sokeh HS, Argyriou V, Monekosso D, Remagnino P (2019) Summarizing videos with attention. In: Carneiro G, You S (eds) Computer vision—ACCV 2018 workshops. ACCV 2018. Lecture notes in computer science, vol 11367. Springer, Cham. https://doi.org/10.1007/978-3-030-21074-8_4
45. Zhao B, Li X, Lu X (2017) Hierarchical recurrent neural network for video summarization. In: ACM multimedia
46. Ji Z, Xiong K, Pang Y, Li X (2020) Video summarization with attention-based encoder–decoder networks. IEEE Trans Circuits Syst Video Technol 30(6):1709–1717. https://doi.org/10.1109/TCSVT.2019.2904996
47. Zhang Y, Liang X, Dingwen Z, Tan M, Xing EP (2018) Unsupervised object-level video summarization with online motion auto-encoder. Pattern Recogn Lett
48. Zhou K, Qiao Y, Xiang T (2017) Deep reinforcement learning for unsupervised video summarization with diversity-representativeness reward. arXiv preprint arXiv: 1801.00054
49. Zhang K, Chao WLF, Grauman K (2016) Summary transfer: exemplar-based subset selection for video summarization. In: Proceeding IEEE conference on computer vision and pattern recognition (CVPR), Dec 2016, vol 2016, pp 1059–1067
50. Bahdanau D, Cho K, Bengio Y (2015) Neural machine translation by jointly learning to align and translate. In: Proceeding international conference on learning representations. http://arxiv.org/abs/1409.0473
51. Xu K, Ba JL, Kiros R, Cho K, Courville A, Salakhutdinov R, Zemel R, Bengio Y (2015) Show, attend and tell: neural image caption generation with visual attention. In: Proceeding of international conference on learning representations. http://arxiv.org/abs/1502.03044
52. Zhang Y, Qiu Z, Yao T, Liu D, Mei T (2018) Fully convolutional adaptation networks for semantic segmentation. In: 2018 IEEE/CVF conference on computer vision and pattern recognition (CVPR-2018), pp 6810–6818
53. Wang X, Girshick R, Gupta A, He K (2018) Non-local neural networks. In: Proceedings of IEEE conference on computer vision and pattern recognition (CVPR), pp 7794–7803

54. Potapov D, Douze M, Harchaoui Z, Schmid C (2014) Category-specific video summarization. In: ECCV—European conference on computer vision, Zurich, Sept 2014, pp 540–555. https://doi.org/10.1007/978-3-319-10599-4_35

Static Video Summarization: A Comparative Study of Clustering-Based Techniques

Deeksha Gupta, Akashdeep Sharma, Pavit Kaur, and Ritika Gupta

Abstract The spectacular increase of video data, due to the availability of low cost and large storage enabled video capturing devices, has led to problems of indexing and browsing videos. In the past two decades, video summarization has evolved as a solution to cope up with challenges imposed by big video data. Video summarization deals with identification of relevant and important frames or shots for efficient storage, indexing, and browsing of videos. Among various approaches, clustering-based methods have gained popularity in the field of video summarization owing to their unsupervised nature that makes the process independent of the need for the expensive and tedious task of obtaining annotations for videos. This study is an attempt to comprehensively compare various clustering-based unsupervised machine learning techniques along with evaluation of performance of selective local and global features in video summarization. Quantitative evaluations are performed to indicate the effectiveness of global features—color and texture as well as local features—SIFT along with six clustering methods of different nature. The proposed models are empirically evaluated on the Open Video (OV) dataset, a standard video summarization dataset for static video summarization.

Keywords Static video summarization · Keyframe extraction · Clustering algorithms · Machine learning · Computer vision

1 Introduction

The advancement in processing power, storage mechanism, and multimedia technology led to the generation of large amounts of digital content corpus. According to statistics of YouTube [1], every minute, videos of a total 500 h duration are being uploaded by the users. With the exponential growth of video data, it has become

D. Gupta
Mehr Chand Mahajan DAV College for Women, Panjab University, Chandigarh, India

D. Gupta · A. Sharma (✉) · P. Kaur · R. Gupta
University Institute of Engineering and Technology, Panjab University, Chandigarh, India
e-mail: akashdeep@pu.ac.in

A. Choudhary et al. (eds.), *Applications of Artificial Intelligence and Machine Learning*, Lecture Notes in Electrical Engineering 778,
https://doi.org/10.1007/978-981-16-3067-5_26

difficult to implement real-time applications like searching, browsing, and indexing of videos. To cope up with the challenges posted by deluge of video content, a mechanism is required for efficient storing, searching, indexing, and retrieval of videos.

Video summarization is one of the key technologies for meeting the aforementioned requirements. It is the process of representing video with meaningful frames or video skims, removing redundancy to maximum possible extent [2]. The application of video summarization mechanism reduces storage, navigation time, and transmission bandwidth requirements considerably. Depending upon the kind of summary generated, video summarization is categorized into two modes: static video summarization and dynamic video summarization [3]. Static video summarization deals with extraction of a set of important and diverse keyframes from the video while ignoring audio content. Dynamic video summarization results in generation of video skims where a set of important consecutive frames or excerpts are selected along with related audio information [4]. The static video summaries assist a framework to browse, index, access, and retrieve video in an efficient manner [2]. It also facilitates real-time-based applications by providing fast storage and competent analysis of video content. The extracted keyframes may be represented in the form of storyboard [5], mosaic [6], panorama [7], or thumbnail [8]. The condensed version of video allows a peek into video content without watching the entire video, thus saving time for analyzing the video information. Keyframe extraction has been one of the key research areas in content-based video retrieval (CBVR) for the past two decades. Many supervised and unsupervised approaches have been proposed by the researchers for static video summarization. This research paper focuses on unsupervised clustering-based approaches in automatic static summary generation. Although numbers of clustering methods-based approaches [5, 7–14] are available, still there is no consensus on the most suitable method for generation of a video summary. So, quest for the appropriate feature and clustering method for static video summary generation is still an active area of research. This paper is an attempt toward this direction to propose a comparative analysis of behavior of different clustering methods while keeping under the same constraints.

This paper presents an experimental study of various clustering methods on the basis of distinctive global as well as local features like color, texture, and SIFT. The effectiveness of various extractive models is compared on the basis of four quantitative metrics: comparisons of user summaries-accuracy (CUS-A)/recall, comparisons of user summaries error (CUS-E), precision, F-measure.

Major contributions of the study are as follows:

1. An experimental study to assess the effectiveness of various clustering techniques, belonging to different classes, is performed through extensive experiments for keyframe extraction-based video summarization.
2. Scope varied features effectiveness is assessed by considering global scope features like color, texture, and local scope features like SIFT.

3. Comparison of user summary-accuracy (CUS-A), comparison of user summary-error (CUS-E), recall, precision, and *F*-measure are also used for performance evaluation of different models under study.

The underlying structure of paper is as follows: Sect. 2 covers the related work done in the field of video summarization based on clustering techniques. Section 3 presents the methodology of the video summarization process pipeline under study, in which various features and clustering algorithms under experiment are discussed along with various evaluation measures used for generated summary analysis. Section 4 includes the experiment detail and results obtained. Section 5 represents the comparative analysis of results to extract the effectiveness of clustering method with conclusions being the last section.

2 Related Work

In the field of video summarization, various techniques can be broadly classified into two categories: supervised methods and unsupervised methods. Supervised methods show better performance for domain specific applications but fail to give reliable results on general videos especially when little labeled data is available and obtaining labeled data for suitable implementation of supervised algorithms is an expensive practice. Due to aforementioned reasons, traditional unsupervised methods are still preferred. Among unsupervised learning methods, clustering algorithms are one of the popular approaches. In context of video summarization, the choice of clustering method and feature selection is the most critical decision, impacting video summary quality.

For static summarization of videos, Mundur et al. [5] used Delaunay triangulation (DT) clustering where Delaunay diagrams are constructed with frames as data points represented in color feature space. Clusters are obtained from Delaunay diagrams by eliminating the inter-cluster edges. The processing time using DT clustering is 10 times the duration of video, owing to high computational overhead.

Another attempt to extract keyframe from edited videos using K-means clustering algorithm is employed in [11]. Frames are represented with 16 bin Hue histogram to reduce computational complexity. Author has also proposed objective summary evaluation metrics, Comparison of user summaries-accuracy (CUS-A) and comparison of user summary-error (CUS-E). Shroff et al. [15] proposed modified version of K-means clustering where inter-cluster center variance is used to compute diversity and intra-cluster distance is used to promote representativeness of selected keyframes. The diversity term makes the results sensitive to those events which occur infrequently in video.

Fuzzy C-mean clustering algorithm is used in [16], where frames are represented with color component histogram. Frame with cluster-wise maximum membership value is selected as keyframe. VSCAN algorithm [12] generates static video summary using DBSCAN clustering algorithm with Bhattacharya distance as similarity metric.

Combined color and texture features are used for frames description. In VGRAPH approach [17], uniformly sampled frames are clustered using KNN graph clustering with discrete Haar wavelet transforms-based texture feature. For number of keyframe estimation, HSV color space histograms of consecutive frames are compared using Bhattacharyya distance.

Furini et al. [18] proposed STIll and MOving Video Storyboard (STIMO)—a scalable and customized approach based on modified furthest point first (FPF) clustering algorithm with triangular inequality. Wu et al. [19] presented a density-based clustering algorithm—high density peak search (HDPS) for diverse keyframes extraction. In this method, the summarization process is divided into two steps. First, the singular value decomposition method is applied to extract candidate keyframes. Candidate frames are represented using a vector representing SIFT feature distribution in the codebook of bag of word (BOW). Then, VRHDPS clustering algorithm is applied to cluster candidate frames.

Chamasemani et al. [13] proposed a new algorithm for generating static summary of general videos by extracting color, texture, SURF, and energy features. DENCLUE, an adaptive nonparametric kernel density estimation (KDE)-based, clustering algorithm is used to generate the clusters, and the middle core frame of every cluster is selected as a keyframe.

For real-time application, Kumar et al. [20] proposed multi-stage clustering. The video frames are segmented into prime and nonprime indexed frames followed by K-means clustering to generate keyframes. Clustering is performed on feature vectors consisting of special frequency, spectral residual and gray color frame level information. For estimating the optimal number of clusters, Davies–Bouldin Index (DBI) method is adopted.

Zhao et al. [14] used affinity propagation clustering method along with cluster validity index technique. Histogram difference-based high-level Fuzzy Petri net model (HLFPN) is used for early shot detection then, SURF-based fast random sample consensus (FRANSAC) algorithm is used to filter out detected false shots. From each shot, candidate frames are selected based on interestingness score computed using low-level and high-level features. Affinity propagation clustering method with similarity consistency clustering validity index is used to select representative frames from candidate frames.

Ou et al. [21] adopted GMM clustering for intra-view summarization under a multi-view video environment. GMM clustering is also employed in [22] for extraction of key-event candidates for sports video summary generation.

For evaluation, different metrics are proposed and used in previous work. Objective evaluation methods, comparison of user studies (CUS) [11, 13, 14, 18, 23], and F-measure [10, 17, 24] provide quantitative comparisons between the automatic summary and ground truth summary. In our study, CUS and F-measure both metrics are used to analyze the behavior of clustering methods.

3 Methodology and Taxonomy

The spatial similarity of frames along the temporal order results in redundancy in video content. The clustering-based video summarization techniques exploit the frames similarity in spatial domain to reduce the redundancy in the generated summary. If a video containing n number of frames, $V = (f_1, f_2, f_3 \ldots f_n)$ is represented by a matrix of size '$d \times n$' where each frame is represented by a feature descriptor of 'd' dimensional vector. Then the goal of video summarization is to obtain video summary S, such that:

$$S = F_k(V) = \{f_1, f_2, f_3 \ldots f_m\}$$

where $F_k(.)$ represents keyframe extraction procedure, $S \subset V$ is set of keyframes extracted from V and f_i denoted selected keyframe where $i \in n$ and $m << n$.

Figure 1 depicts the video summarization process pipeline used in the study. Our summarization process comprises of three major steps: video preprocessing, feature extraction, and clustering.

3.1 Video Preprocessing

In context of video summarization, video preprocessing is required to reduce the computational complexity of the rest of the pipeline by condensation of frames. Preprocessing step includes selection of candidate keyframes either by using a trivial sampling method or by using some complex shot or segment detection-based approach [25–27] . In our study, down sampling is done to get a reduced set of candidate frames followed by removal of monochrome frames. This step is kept the same for all the models under study.

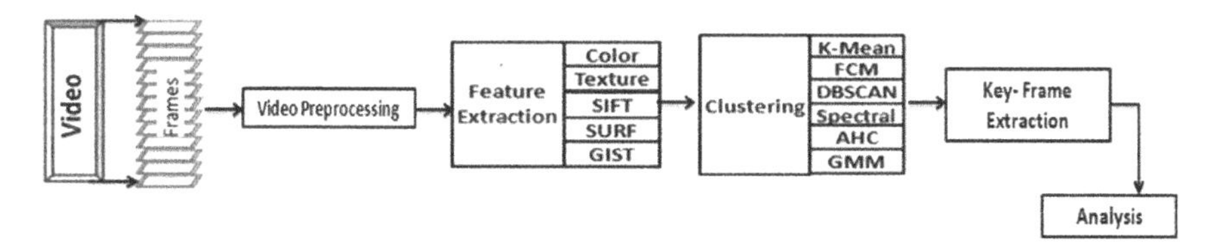

Fig. 1 Video summarization pipeline

3.2 Feature Extraction

The decision on feature selection for representation of frames plays an important role in terms of generated summary quality as well as computational complexity of the algorithm. In our approach, we have used global features (color and texture) as well as local features (SIFT) for clustering purposes.

Color Descriptor. Out of many color models (RGB, CMYK, YUV, HSV etc.), user-oriented color models should be preferred for video summarization tasks. On account of this, hue–saturation–value (HSV) color space is preferred for summarization as it closely aligns with the human vision perception system [5, 11, 12]. Due to hue component dominance in color spectrum over saturation and value components, a normalized 16-dimensional vector representing 16-bin color histogram of hue component only is used as color descriptor for clustering purpose.

Texture. Texture feature is another elementary feature that supports the human image perception system [12, 27]. This study is an effort to observe the role of texture feature independently in video summarization with different clustering models. Statistical approach-based 13-D Haralick texture feature is used to represent frames due to its characteristics of being fast to compute, intuitive in nature, and better in performance [28].

SIFT. In our study, the bag of words approach with SIFT [29] features is used for clustering purposes. We have extracted SIFT local descriptors from every sampled frames of videos and clustered them to create a vocabulary (codebook) of keypoints of size 1024D. The generated codebook represents a frame as a frequency histogram of visual words comprising the codebook. The codebook size is fixed to 1024D to optimize the tradeoff between performance and computational cost [30]. For summary generation purpose, each frame is represented as a normalized frequency histogram using BOVW approach for clustering purpose.

3.3 Clustering

Clustering is one of the machine learning concepts in computer vision techniques, which focuses on intra-cluster homogeneity and inter-cluster heterogeneity among the data objects. Clustering algorithms can be broadly classified into various diverse categories including partition-based, density-based, hierarchical, model-based, soft computing-based, and graph-based clustering [31, 32]. This paper makes an attempt to study behavior and performance constraints of various clustering methods by selecting one method from each family, as indicated in Table 1. The method selection is done on the basis of popularity, applicability, and capability to handle high-dimensional data.

Table 1 Clustering types and selected algorithms

Category	Selected method
Partition-based clustering	K-means
Soft computing-based clustering	FCM (Fuzzy C-mean)
Density-based clustering	DBSCAN (Density-based spatial clustering of applications with noise)
Hierarchical clustering	AHC (Agglomerative hierarchical clustering)
Model-based clustering	GMM (Gaussian mixture models)
Graph-based clustering	SC (Spectral clustering)

Partition-Based K-means Clustering. K-means clustering is based on minimization of cost function specified as the Euclidean distance or squared error distance between cluster members and cluster center. Equation (1) represents the cost function exploited for K-means clustering.

$$\text{Cost Function}\,(J) = \sum_{j=1}^{k}\sum_{i=1}^{n}\left\|x_i^{(j)} - c_j\right\|^2 \tag{1}$$

Here $x_i^{(j)}$ represents the frame feature vector, c_j represents the cluster center, and $\|.\|$ represents the Euclidean distance.

This algorithm requires a priori number of clusters (k) as parameter, and the computational complexity of the algorithm is highly dependent on 'k'. In our study, consecutive video frames difference influenced method is adopted to get approximate number of clusters as specified in Eq. (2), given below:

$$\text{No. of clusters}\,(k) = \alpha\mu - \beta \tag{2}$$

where μ represents the mean of consecutive frame pair distances within a video and α, β are constant. In experiment, $\alpha = 1.6$ and $\beta = 0.03$ are determined empirically, for generating reasonable number of clusters.

Soft Computing-Based Fuzzy C-Mean (FCM) Clustering. FCM clustering [33] allows frames to belong to different clusters based on the membership bound. Due to the aforementioned reason, it is also classified as an overlapping clustering method. Equation (3) represents the cost function for FCM clustering.

$$\text{Cost Function}\,(J) = \sum_{j=1}^{k}\sum_{i=1}^{n}u_{ij}^{m}\left\|x_i^{(j)} - c_j\right\|^2 \quad \text{where } 1.0 < m < \infty \tag{3}$$

Here u_{ij}^{m} represents the membership degree of frame x_i in cluster j with m (fuzziness of membership grade), and it is inversely related to the distance between frame and the cluster center, as given in Eq. (4):

$$u_{ij}^{m} = \frac{1}{\sum_{n=1}^{k} \left(\frac{d_{ij}}{d_{ik}}\right)^{\frac{2}{m-1}}} \tag{4}$$

Here d_{ab} represents the distance between ath frame and bth cluster centers. Like K-means clustering, FCM also requires number of clusters to be specified priori. In experiment, FCM algorithm follows greedy search strategy for membership matrix formulation and centroids selection, as higher the degree of membership of a frame closer it will be to the cluster center, until the algorithm converges with sensitivity threshold $\epsilon = 10^{-3}$. For an approximating number of clusters, the same method is adopted as used in K-means.

Density-Based Clustering: Density-Based Spatial Clustering of Applications with Noise (DBSCAN). DBSCAN is based on the concept that a cluster center has higher neighborhood density than its neighbors and it is sufficiently distant from the frames having equivalent higher neighborhood densities [34]. It requires two parameters to be specified for clustering of data points: neighborhood distance (Eps) and neighborhood density (Minpts). Eps defines the space around the data point to be considered as its neighborhood and Minpts defines the minimum number of neighbor data points within Eps radius. In our experiment, Eps= 0.45 times the maximum distance between video frames pair and Minpts $= 3$ are selected empirically and shared by the models under test. The Euclidean distance metric is used as a frame distance measure. After clustering, from each cluster, the middle frame in sequential order is selected as a keyframe. The main advantage of using this algorithm for frame clustering is that there is no need to make a decision on the number of clusters a priori.

Hierarchical Clustering: Agglomerative Hierarchical Clustering (AHC). Agglomerative hierarchical clustering is one of the popular hierarchical clustering techniques which follow a bottom-up approach for grouping data points [35]. Clusters are formed by employing an affinity matrix, containing inter-frame distance. In this implementation, the affinity matrix for sampled frames is computed using Euclidean distance. The limitation of this algorithm is the same as that of the K-means and FCM algorithm, that is, the need to specify number of clusters to be generated. For this, the same approach as specified under K-mean clustering approach, is followed. The merging of frames to form clusters is done by using Ward's minimum variance method where frames are merged to minimize within cluster variance.

Model Based Clustering—Gaussian Mixture Model (GMM). GMM clustering is based on assumption that data is Gaussian distributed rather than circular distributed [36]. For each cluster formation, the Gaussian distribution parameters (mean and variance) are approximated using an optimization algorithm named as expectation–maximization (EM).

$$P(C_k|f_i) = \frac{P(f_i|C_k) * P(C_k)}{P(f_i)} \tag{5}$$

where

$$P(C_k) = \frac{1}{\sqrt{2\pi\sigma}} e^{\left(\frac{-(f_i - \mu_k)^2}{2\sigma_k^2}\right)} \tag{6}$$

Hence,

$$P(f_i) = \sum_{k=1}^{n} p(f_i|C_k) * p(C_k) \tag{7}$$

This method also requires a number of clusters to be specified a priori for which a common approach is followed in this study. Then clustering starts with random initialization of the Gaussian parameters for each cluster. The probability of each frame belonging to any cluster C_k is computed using Eq. (7).

Graph-Based Clustering: Spectral Clustering (SC). Spectral clustering [37], unlike K-means, generates random shape clusters by exploiting adjacency relation ($A \in R^{n \times n}$) between frame feature space representations. In the present study, affinity matrix is computed using a radial basis function (RBF) Gaussian kernel with Euclidean distance measure followed by Laplacian matrix ($L = I - D^{-1/2}AD^{-1/2}$) construction. Then eigenvalues and eigenvectors from matrix L are computed to represent frames from high-dimensional space to low-dimensional embedding followed by clustering of data points as per pre-specified number of clusters. This technique works without any prior assumptions about number, density, and shape of clusters.

3.4 Evaluation Method

Evaluation metric plays a pivotal role in measuring the performance of a model. In our study, we have used the CUS-A, CUS-E, precision, recall, and F-measure evaluation metric. The purpose of employing numerous evaluation methods is to verify the model performances from various aspects.

Comparison of User Summaries (CUS). Comparison of user summaries (CUS) is an quantitative measure to evaluate automatic summary (AS) and user summary (US). It includes two metrics—CUS-A (CUS-Accuracy) and CUS-E (CUS-Error). CUS-A and CUS-E are defined as follows:

$$\text{CUSA} = \frac{\text{No. of matched keyframe from AS}}{\text{Total no. of keyframes in US}} \tag{8}$$

$$\text{CUSE} = \frac{\text{No. of non-matched keyframe from AS}}{\text{Total no. of keyframes in US}} \tag{9}$$

where $0 \leq \text{CUS-A} \leq 1$, and $0 \leq \text{CUS-E} \leq \gamma$. Where γ = Total number of keyframes in AS/Total number of keyframes in US. An algorithm is stated to be best if CUS-A = 1 and CUS-E = 0, supporting the complementary nature of two metrics.

***F*-measure**. F-measure is another objective video summary evaluation method which is based on two primitive values, recall and precision. F-measure is harmonic mean of recall and precision, where

$$\text{Recall} = \frac{\text{No. of matched keyframe between AS and US}}{\text{Total no. of keyframes in US}} \tag{10}$$

and

$$\text{Precision} = \frac{\text{No. of matched keyframe between AS and US}}{\text{Total no. of keyframes in AS}} \tag{11}$$

So,

$$F\text{-measure} = \frac{2(\text{Recall} * \text{Precision})}{\text{Recall} + \text{Precision}} \tag{12}$$

High value of F-measure corresponds to high performance of model and vice-versa.

4 Experiment and Results

The various models are evaluated on the basis of comparative experimental results. In this section, the experimental environment is discussed followed by performance comparison among all the models.

4.1 Experiment Environment

Dataset. The comparative analysis is performed on video datasets named Open Video [11] specifically designed and a benchmark dataset for static video summarization. OV dataset contains 50 videos selected from Open Video Project [38], covering several genres like documentary, educational, ephemeral, historical, lecture videos. All videos are in MPEG-1 format with 30 fps and 352 × 240 pixel frame size. All videos are 1–4 min long duration resulting in 75 min duration videos in total. In the dataset, every video is annotated with 5 keyframe summaries from 5 different users to deal with the subjective nature of the problem.

4.2 Video Summarization Models (VSM)

In experiment, sampling of frames at the rate of 1 fps is done to reduce complexity of all models under evaluation. All possible combinations of 3 different features (color, texture, and SIFT) and 6 diverse clustering algorithms (AHC, FCM, DBSCAN, GMM, K-means and SC) resulted in 18 video summarization models (VSM). The comparative analysis of models counting, best model and worst model, best performing clustering method and features, most stable clustering model and feature are discussed in the next section.

4.3 Evaluation Results

In this study, all 18 models are evaluated on the basis of generated summary quality computed using four evaluation metrics: CUS-A/recall, CUS-E, precision, *F*-measure. Automatic summary is compared with each of five user summaries provided in the dataset, and mean pair-wise metrics score is computed. In the end, the average of each evaluation metric over all 50 videos is computed and specified in Tables 2, 3, and 4.

Table 2 Color feature-based models evaluation for OV dataset

Model	Model (feature + clustering)	CUS-A/recall	CUS-E	Precision	*F*-measure
VSM1	Color + AHC	0.84	0.90	0.59	0.69
VSM2	Color + FCM	0.83	0.91	0.59	0.69
VSM3	Color + DBSCAN	0.71	0.36	0.71	**0.71**
VSM4	Color + GMM	0.80	0.94	0.56	0.66
VSM5	Color + K-means	0.85	0.89	0.60	0.70
VSM6	Color + SC	0.83	0.87	0.59	0.69

Table 3 Texture feature-based models evaluation for OV dataset

Model	Model (feature + clustering)	CUS-A/recall	CUS-E	Precision	*F*-measure
VSM7	Texture + AHC	0.92	1.70	0.43	0.59
VSM8	Texture + FCM	0.88	1.73	0.41	0.56
VSM9	Texture + DBSCAN	0.74	0.51	0.66	**0.69**
VSM10	Texture + GMM	0.84	1.78	0.39	0.53
VSM11	Texture + K-means	0.80	1.01	0.52	0.63
VSM12	Texture + SC	0.91	1.56	0.46	0.61

Table 4 SIFT feature-based models evaluation for OV dataset

Model	Model (feature + clustering)	CUS-A/recall	CUS-E	Precision	*F*-measure
VSM13	SIFT + AHC	0.82	0.92	0.58	0.68
VSM14	SIFT + FCM	0.32	0.13	0.76	0.45
VSM15	SIFT + DBSCAN	0.78	0.41	0.71	**0.74**
VSM16	SIFT + GMM	0.71	1.03	0.50	0.59
VSM17	SIFT + K-means	0.76	1.00	0.53	0.63
VSM18	SIFT + SC	0.53	0.61	0.59	0.56

Tables 2, 3, and 4 cover 3 classes of models categorized on the basis of 3 features taken into account. The bold value represents the maximum score for the specific evaluation measure in the corresponding feature-wise categorized models.

As shown in Tables 2, 3, and 4, CUS-A or CUS-E are not sufficient for comparative analysis of video summarization models. Provided both CUS-A and CUS-E, still it is difficult to judge performance of two different models in the absence of a single score value. As, while comparing performance of VSM1 (Color + AHC) and VSM7 (Texture + AHC) the computed CUS-A, CUS-E pair scores are (0.84, 0.90) and (0.92, 1.70), respectively. It is difficult to select the best model out of two models, as although CUS-A for VSM7 is higher but higher value of CUS-E will have a negative impact too. Whereas, *F*-measure metric assigns a single score value 0.69 and 0.59 to models VSM1 and, respectively, making VSM1 a clear winner.

The comparison charts, along with their analysis for the obtained experimental results, are discussed in the next section.

5 Comparative Analysis

The analysis of 18 models is done on various facets including feature clustering duo performance with respect to four different evaluation metrics, behavior analysis of local and global features and clustering algorithms consistency across the various feature space. The performance of all 18 models is represented in Figs. 2, 3, and 4 for, CUS-E, CUS-A, and *F*-measure, respectively.

5.1 Best Model and Worst Model

Model performance is judged on the basis of CUS-A, CUS-E, and *F*-measure scores exhibited by it. The experiment result show that best performance for summarizing 50 videos of OV dataset is given by VSM15 (SIFT + DBSCAN) followed by VSM3 (Color + DBSCAN) model with CUS-A, CUS-E, and *F*-measure (0.78, 0.41, 0.74) and (0.71, 0.36, 0.71), respectively. It verifies the fact that density-based

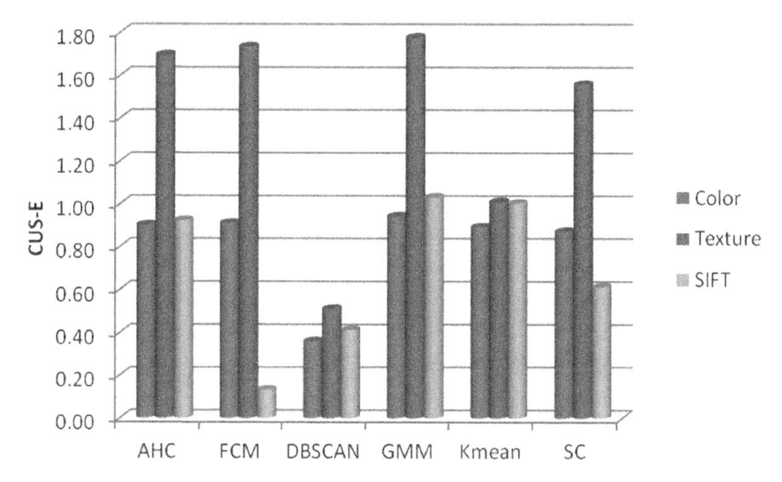

Fig. 2 CUS-E for all models (VSM1–VSM18)

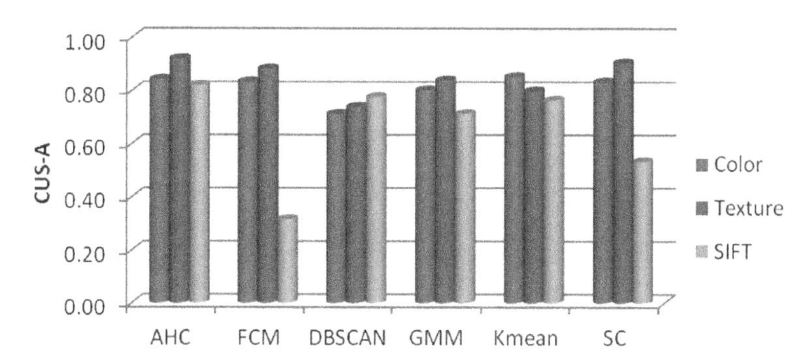

Fig. 3 CUS-A for all models (VSM1–VSM18)

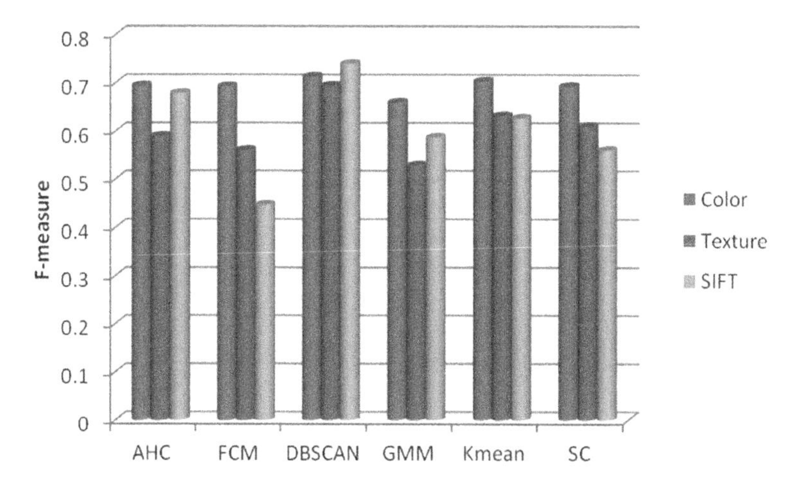

Fig. 4 *F*-measure for all models (VSM1–VSM18)

clustering model is capable of handling high-dimensional data along with irregular shape clusters.

On the other side, from Tables 2, 3, and 4, VSM14 (SIFT + FCM) shows the poorest behavior. The main reason behind this poor performance is due to sensitivity of fuzzy C-mean clustering to outliers and noise present in the image. Also, from Table 3, the evaluation metric values for VSM7, VSM8, and VSM10 represent the incapability of texture features to diversify the data frames in the clustering algorithm and causing large number of keyframes in the summary, which in terms results in high value of CUS-A and CUS-E both as shown in Figs. 2 and 3. The selection of large number of false positive results in lower precision score and lower F-measure scores for texture-based models as shown in Fig. 4.

5.2 Local Versus Global Features

Color features give best results with all clustering algorithms except DBSCAN as depicted in Fig. 4. The performance of color features witnesses their properties of robustness to illumination and high correlation with human vision perception. Texture features selects many keyframes in summary resulting in high value of CUS-E, hence dropping the performance of clustering model. It concludes that during the video summarization task, using a texture descriptor alone is not a reliable decision. While using local features SIFT the performance is highly clustering method dependent. SIFT features with GMM and FCM show average performance while giving the best results with the DBSCAN algorithm.

5.3 Consistency Study of Clustering Method

Here consistency deals with the ability of a clustering algorithm to maintain the performance for summary generation irrespective of feature selection. As interpreted from Fig. 5, DBSCAN clustering is showing the most consistent behavior across various features followed by K-means, while FCM demonstrates the most inconsistent behavior during the experiment. AHC, GMM, and spectral clustering techniques show comparable behavior in terms of consistency.

5.4 Consistency Study of Local and Global Features

Figure 6 summarizes the behavior of features with respect to various clustering algorithms. Global features color show most consistent behavior, while SIFT features show high variation in performance corresponding to clustering techniques. So, if SIFT features are used as frame descriptors, then selection of clustering methods

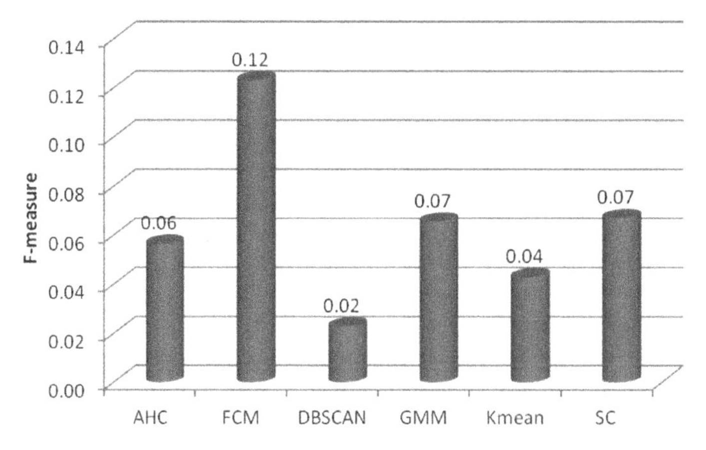

Fig. 5 Standard deviation in F-measure score of clustering methods across features

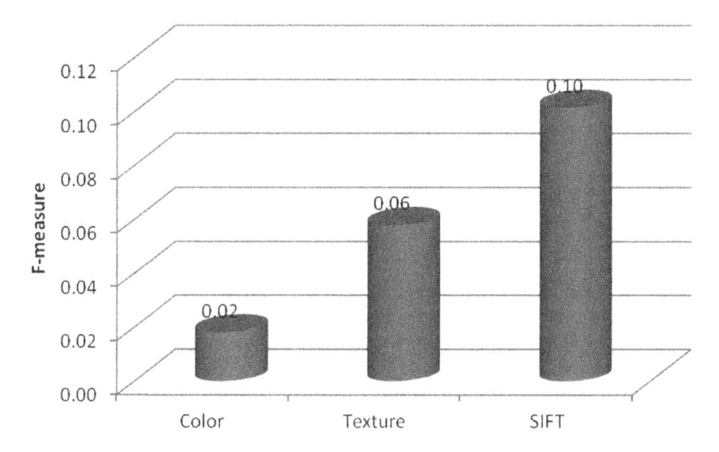

Fig. 6 Standard deviation in F-measure score of features across clustering methods

plays a vital role in performance of the model. Global feature texture also shows unpredictable behavior justifying its sensitivity to the method employed.

6 Conclusion and Future Work

The static summaries obtained can play an important role in various applications ranging from system level browsing and indexing to user level decision making on watching a video or not. In this paper, eighteen clustering-based models with manually designed criteria have been implemented in order to study their behavior under different environmental conditions in context of automatic static summaries generation. Valuable insights are obtained on the performances of different clustering algorithms with diverse nature (local and global) features. DBSCAN shows best results in terms of performance and stability across features. K-means is quite reliable under different environments. Agglomerative clustering and spectral clustering has shown comparable performance. Fuzzy C-mean and Gaussian mixture model require suitable preprocessing of video frames and post-processing of generated summaries to generate comparable results.

Among various features, hue component-based color features manifest the most consistent result across all clustering algorithms. The local features SIFT produces high-quality summary with careful clustering algorithm selection. The texture features indicate poor performance by selecting redundant keyframes. So the summaries generated using textual features need post-processing for redundancy removal.

All the performance measurements are done by exploiting a number of objective metrics. The performance of clustering-based methods is highly dependent on the fine tuning of the parameters for different genre videos. This limitation generates the need for more advanced techniques in order to support a large collection of diverse nature videos. Future study can also focus on bringing deep architectures and machine learning techniques together for better quality summary generation.

References

1. Track YouTube analytics, future predictions, & live subscriber counts—social blade. [Online]. Available: https://socialblade.com/youtube/. Accessed 10 July 2020
2. Truong BT, Venkatesh S (2007) Video abstraction: a systematic review and classification. ACM Trans Multimed Comput Commun Appl 3(1):3:1–3:37
3. Hanjalic A (1999) An integrated scheme for automated video abstraction based on unsupervised cluster-validity analysis. IEEE Trans Circuits Syst 9(8):1280–1289
4. Elharrouss O, Almaadeed N, Al-Maadeed S, Bouridane A, Beghdadi A (2020) A combined multiple action recognition and summarization for surveillance video sequences. Appl Intell 50
5. Mundur P, Rao Y, Yesha Y (2006) Keyframe-based video summarization using Delaunay clustering. Int J Digit Libr 6(2):219–232
6. Viguier R, Lin CC (2015) Automatic video content summarization using geospatial mosaics of aerial imagery. In: 2015 IEEE international symposium on multimedia (ISM), Miami, FL, pp 249–253. https://doi.org/10.1109/ISM.2015.124
7. Trinh H, Li J, Miyazawa S, Moreno J, Pankanti S (2012) Efficient UAV video event summarization. In: Proceedings of the 21st international conference on pattern recognition (ICPR2012), Tsukuba, pp 2226–2229
8. Choi J, Kim C (2016) A framework for automatic static and dynamic video thumbnail extraction. Multimed Tools Appl 75(23):15975–15991
9. Kalita S, Karmakar A, Hazarika SM (2018) Efficient extraction of spatial relations for extended objects vis-à-vis human activity recognition in video. Appl Intell 48(1):204–219
10. Zhou Y, Cheng Z, Jing L, Hasegawa T (2015) Towards unobtrusive detection and realistic attribute analysis of daily activity sequences using a finger-worn device. Appl Intell 43(2):386–396
11. Avila S, Brandaolopes A, Luz A, Araujo A (2011) VSUMM: a mechanism designed to produce static video summaries and a novel evaluation method. Pattern Recogn Lett 32(1):56–68

12. Mahmoud KM, Ismail MA, Ghanem NM (2013) VSCAN: an enhanced video summarization using density-based spatial clustering. In: Petrosino A (eds) Image analysis and processing—ICIAP 2013. Lecture notes in computer science, vol 8156. Springer, Berlin, Heidelberg
13. Chamasemani FF, Affendey LS, Mustapha N, Khalid K (2018) Video abstraction using density-based clustering algorithm. Vis Comput 34:1299–1314
14. Zhao Y, Guo Y, Sun R, Liu Z, Guo D (2019) Unsupervised video summarization via clustering validity index. Multimed Tools Appl 78(7)
15. Shroff N, Turaga SP, Chellappa R (2010) Video précis : highlighting diverse aspects of videos. IEEE Trans Multimed 12(8):853–868
16. Asadi E, Charkari NM (2012) Video summarization using fuzzy c-means clustering. In: 20th Iranian conference on electrical engineering (ICEE2012), Tehran, pp 690–694. https://doi.org/10.1109/IranianCEE.2012.6292442
17. Mahmoud KM, Ghanem NM, Ismail MA (2013) VGRAPH: an effective approach for generating static video summaries. In: 2013 IEEE international conference on computer vision workshops, Sydney, NSW, pp 811–818. https://doi.org/10.1109/ICCVW.2013.111
18. Furini M, Geraci F, Montangero M, Pellegrini M (2010) STIMO: STIll and MOving video storyboard for the web scenario. Multimed Tools Appl 46:47. https://doi.org/10.1007/s11042-009-0307-7
19. Wu J, Zhong S, Jiang J, Yang Y (2017) A novel clustering method for static video summarization. Multimed Tools Appl 76:9625–9641. https://doi.org/10.1007/s11042-016-3569-x
20. Kumar K, Shrimankar DD, Singh N (2018) Eratosthenes sieve based key-frame extraction technique for event summarization in videos. Multimed Tools Appl 77:7383–7404
21. Ou S, Lee C, Somayazulu VS, Chen Y, Chien S (2015) On-line multi-view video summarization for wireless video sensor network. IEEE J Sel Top Signal Process 9(1):165–179. https://doi.org/10.1109/JSTSP.2014.2331916
22. Javed A, Irtaza A, Khaliq Y, Malik H, Mahmood MT (2019) Replay and key-events detection for sports video summarization using confined elliptical local ternary patterns and extreme learning machine. Appl Intell 49:2899–2917. https://doi.org/10.1007/s10489-019-01410-x
23. Ejaz N, Bin T, Wook S (2012) Adaptive key frame extraction for video summarization using an aggregation mechanism. J Vis Commun Image Represent 23(7):1031–1040
24. Wei H, Ni B, Yan Y, Yu H, Yang X (2018) Video summarization via semantic attended networks. In: Proceedings of the thirty-second (AAAI) conference on artificial intelligence, New Orleans, 2–7 Feb 2018, pp 216–223
25. Ejaz N, Baik S, Majeed H, Chang H, Mehmood I (2018) Multi-scale contrast and relative motion-based key frame extraction. J Image Video Process 2018:40. https://doi.org/10.1186/s13640-018-0280-z
26. Dash A, Albu AB (2017) A domain independent approach to video summarization. In: International conference on advanced concepts for intelligent vision systems, Nov 2017. https://doi.org/10.1007/978-3-319-70353-4_37
27. Shanmugam K, Dinstein I (1973) Textural features. IEEE Trans Syst Man Cybern Syst 3(6):610–621
28. Humeau-Heurtier A (2019) Texture feature extraction methods: a survey. IEEE Access 7:8975–9000. https://doi.org/10.1109/ACCESS.2018.2890743
29. Low DG (2004) Distinctive image features from scale-invariant keypoints. Int J Comput Vis 91–110
30. Aldavert D, Rusiñol M, Toledo R, Llados J (2015) A study of bag-of-visual-words representations for handwritten keyword spotting. Int J Doc Anal Recogn 18:223–234
31. Camastra F, Vinciarelli A (2008) Clustering methods. In: Machine learning for audio, image and video analysis, pp 117–148. ISBN 978-1-4471-6734-1
32. Berkhin P (2006) A survey of clustering data mining techniques. In: Kogan J, Nicholas C, Teboulle M (eds) Grouping multidimensional data. Springer, Berlin, Heidelberg. https://doi.org/10.1007/3-540-28349-8_2

33. Tilson LV, Excell PS, Green RJ (1988) A generalisation of the fuzzy C-means clustering algorithm. In: International geoscience and remote sensing symposium, 'remote sensing: moving toward the 21st century', Edinburgh, pp 1783–1784. https://doi.org/10.1109/IGARSS.1988.569600
34. Daszykowski M, Walczak B (2009) Density-based clustering methods. In: Comprehensive chemometrics, vol 2, pp 635–654
35. Davidson I, Ravi SS (2005) Agglomerative hierarchical clustering with constraints: theoretical and empirical results. In: Lecture notes computer science, vol 3721. Springer, Heidelberg, pp 59–70
36. Reynolds D (2009) Gaussian mixture models. In: Encyclopedia of biometrics, no 2, pp 659–663
37. Arias-Castro E, Chen G, Lerman G (2011) Spectral clustering based on local linear approximations. Electron J Statist 5:1537–1587. arXiv:1001.1323. https://doi.org/10.1214/11-ejs651
38. The open video project. http://www.open-video.org. Accessed 1.8.2020

A Review: Hemorrhage Detection Methodologies on the Retinal Fundus Image

Niladri Sekhar Datta, Koushik Majumder, Amritayan Chatterjee, Himadri Sekhar Dutta, and Sumana Chatterjee

Abstract Diabetic retinopathy (DR) is a microvascular symptom where retina is affected by fluid leaks of the fragile blood vessels. Clinically, retinal Hemorrhages are one of the earliest indications of diabetic retinopathy disease. In this contrast, the Hemorrhage count is used to indicate the severity of this disease. The early detection of retinal Hemorrhages obviously prevents the incurable blindness of the DR patients. But, retinal Hemorrhage detection is still a challenging task. Highly reliable, accurate, platform independent retinal Hemorrhage detection method is still an open field. In this research article, we have reviewed the principal methodologies which are used to diagnose the retinal Hemorrhages under the diabetic retinopathy screening operations. This review article helps the researchers to develop a high quality retinal Hemorrhage screening method in future.

Keywords Medical image processing · Diabetic retinopathy · Hemorrhages · Red lesions · Ocular diseases

1 Introduction

Today, diabetic retinopathy (DR) is a foremost reason of adult blindness in the world. According to WHO, 140 million people have suffered from DR and it will increase to near about 400 million by 2030 [1]. The ophthalmologists diagnose the

N. S. Datta (✉)
Department of Information Technology, Future Institute of Engineering and Management, Kolkata, West Bengal, India

K. Majumder
Department of Computer Science and Engineering, Maulana Abul Kalam Azad University of Technology, Kolkata, West Bengal, India

A. Chatterjee · H. S. Dutta
Department of Electronics and Communication Engineering, Kalyani Government Engineering College, Kalyani, West Bengal, India

S. Chatterjee
Department of Ophthalmology, Sri Aurobindo Seva Kendra, Kolkata, West Bengal, India

A. Choudhary et al. (eds.), *Applications of Artificial Intelligence and Machine Learning*, Lecture Notes in Electrical Engineering 778,
https://doi.org/10.1007/978-981-16-3067-5_27

DR based on the presence of microaneurysms (MAs), exudates, Hemorrhages, blood vessel areas, and existing texture of the retinal images [2]. Clinically, DR is categorized in two major types namely non-proliferative diabetic retinopathy (NPDR) and proliferative diabetic retinopathy (PDR). The NPDR is the early stage of DR and in more advance stage, called as PDR. The intermediate stages for NPDR are mild, moderate, and severe [3]. In the mild stage of NPDR, round shape reddish color bulges, i.e., microaneurysms are appeared. This microaneurysms later forms retinal Hemorrhages. According to the medical point of view, retinal Hemorrhages and MAs are the primary symptom for DR disease. In addition with, the grading of DR, more specifically on NPDR is directly related with the count of existing MAs and Hemorrhages of the retinal images. In severe NPDR, the Hemorrhages are observed in all four quadrants. Finally, in PDR, newly developed fragile vessels are bled and produce large size Hemorrhages [4]. As a result, incurable blindness appears to the DR patients. Figure 1 represents the Hemorrhages on DR affected eye. Table 1 represents the grading procedure of NPDR disease as per the count of Hemorrhages and MAs in retinal images [5]. Thus, Hemorrhages detection is one of the important tasks for early detection of this ocular disease. The dimensions of retinal Hemorrhages are irregular and color is similar to background. Hence, the detection of retinal Hemorrhages is one of the difficult tasks under the DR screening. Automated detection of retinal Hemorrhages has drawn severe interest to the researchers as it directly related with DR grading and help the ophthalmologists to investigate and diagnose the disease more effectively. The objective of this research article is to review the existing methods for the identification of retinal Hemorrhages and compare the existing results. This research article is organized as follows: Sect. 2

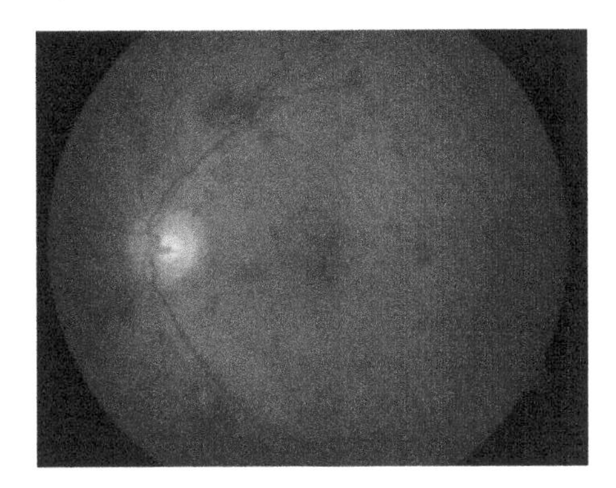

Fig. 1 Hemorrhages on retinal images

Table 1 Hemorrhages count for NPDR grades (M: microaneurysms, H: Hemorrhages)

NPDR grades	
Grade 0/No DR	$M = 0$ and $H = 0$
Grade 1/Mild DR	$1 \leq M \leq 5$ and $H = 0$
Grade 2/Moderate DR	$5 < M < 15$ and $0 \leq H \leq 5$
Grade 3/Severe DR	$M \geq 15$ or $H > 15$

presents the review on Hemorrhages identification methods; performance evaluation is done on Sect. 3. Finally, conclude the paper.

2 Review: Hemorrhage Identification Methods

Retinal Hemorrhages occurs while blood leakage from fragile vessels. Screening of retinal Hemorrhages has become one of the current research activities as there is a still requirement of highly reliable, sensitive, and accurate Hemorrhage identification methods. It is much more complex task compare to MAs or exudates screening as there is no standard database available till date to classify the shape of the retinal Hemorrhages. Thus, lack of research activity observed in this specific area. Few researchers grouped MAs and retinal Hemorrhages as red lesions and diagnose altogether [6–11]. Besides this, many authors have proposed computer-aided diagnoses (CAD) systems for the screening of Hemorrhages on DR affected eye. These research activities only focused on the Hemorrhages identification. The CAD-based Hemorrhages identification scheme consists of two major stages; first: the red lesion candidate's extraction and second: classification. Initially, suitable contrast enhancement scheme and noise removal method are used as a preprocessing step of retinal images. Thereafter, the red surface of the preprocessed image is extracted and segmented as a candidate of the red lesion. To reduce the false detection of retinal Hemorrhages, different blood vessel extracting algorithms are used on the red lesion. Then, the feature analysis is carried out for the selection and extraction of Hemorrhage lesions. Finally, classification algorithm classify the candidate into the abnormal group (i.e., Hemorrhage) or normal group (i.e., non-Hemorrhage). Figures 2 and 3 represent the DR symptoms and the general architecture of Hemorrhage screening procedure, respectively. The review based on main methodologies is stated as.

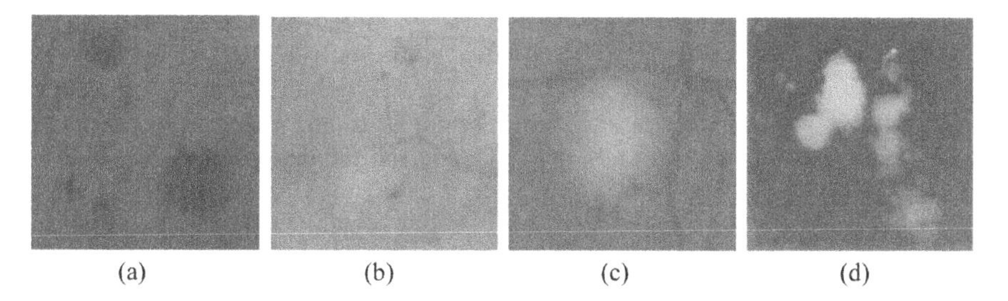

Fig. 2 Clinical symptom of NPDR and PDR **a** Hemorrhages, **b** microaneurysms, **c** hard exudates, **d** soft-exudates

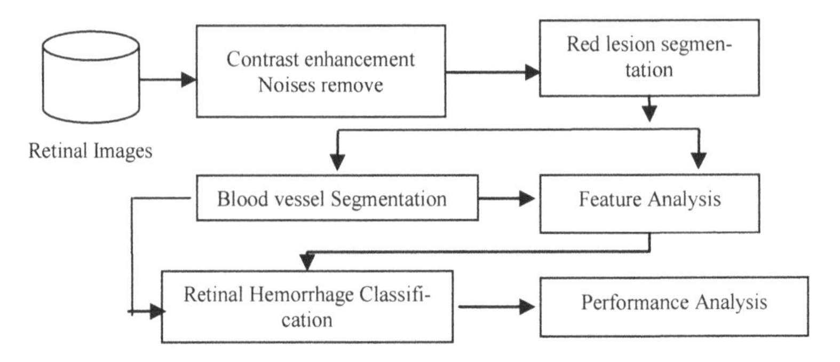

Fig. 3 General architecture of Hemorrhage detection method

2.1 *Mathematical Morphology*

The most commonly used method for Hemorrhage detection is morphology. Pixel by pixel-based erosion, dilation, closing, and opening are basic fundamental concept of it. It is low cost, fast in execution, platform independent method which is easily implemented by the researchers. Jadhav and Patil [12] used this scheme for the extraction and segmentation of Hemorrhages on RGB color plane. Here, two consecutive steps are performed. Initially, contrast enhancement of retinal image is done. Thereafter, build a morphological algorithm to extract blood vessels from red lesions. Sopharak et al. [13] have applied almost same methodology for the detection of Hemorrhages but here green plane of RGB retinal image is used. Shivaram et al. [14] have applied arithmetic-based same topology to suppress the blood vessels. The predictive value for the Hemorrhage detection is reported as 98.34%. Karnowski et al. [15] have reported morphological reconstruction method for the segmentation of red lesion. Here, authors create 'Lesion population' feature vector to classify normal and abnormal classes. Kande et al. [16] have applied morphological Top-Hat transformation method to detect red lesion candidate. This method reported 100% sensitivity and 91% specificity. Matei and Matei [17] have used multi-scale morphological operation to spot the red lesion in image. Here, blood vessels are eliminated by scale-based lesion elimination technique. Using this scheme, 30 retinal images are tested and sensitivity reported as 84.10%. Few authors have reported threshold-based morphological setup to segment the Hemorrhages [18]. Acharya et al. [19] have applied 'ball-shaped' structuring element on the morphological operation. Here, blood vessels are extracted from red lesions and finally Hemorrhages are obtained. Singh and Tripathi [20] have proposed a mixed setup for Hemorrhage identification. Here, mathematical morphology with fuzzy clustering scheme both are applied together for vessel suppression. Finally, segmented image represent the Hemorrhages. A hybrid model is also created by Niemeijer et al. [21]. Here, morphology is used to extract red candidate. Whereas, the KNN classifier is used for classify red lesion. This scheme has reported the sensitivity as 100% and specificity as 87%.

2.2 *Recursive Region Growing*

Region growing is a segmentation method in image processing where same neighbor pixels are grouped in same class. Few researchers have applied this method. These are included here. Sinthanayothin et al. [22] have used 'Moat-Operator' and dynamic thresholding to enhance the contrast of the red lesion. Thereafter, recursive region growing method is applied to extract the vessels from retinal Hemorrhages. Here, blood vessels are tracked by matched filter. Bae et al. [23] have proposed a hybrid scheme where Hemorrhages are segmented by adaptive seed growing method. Here, Hemorrhage lesions are extracted by template matching cross correlation. The sensitivity is found as 85%. Marino et al. [8] have presented a correlation filter-based recursion region growing algorithm to identify the Hemorrhages. Initially, correlation filter is used to detect the red lesion. Thereafter, matched filter has applied to remove false positive pixels.

2.3 *Artificial Neural Network*

Authors have been applied artificial neural network to diagnose the retinal Hemorrhages. Garcia et al. [7] have applied multilayer perception-based neural network to detect the Hemorrhages. This scheme is verified by 50 retinal images which include 29 features to describe the color and shape of the red lesion. This algorithm report average sensitivity as 86.1% with predictive value 71.4%. To detect Hemorrhages backpropagation technique of artificial neural network is applied by Gardner et al. [24]. Here, the entire image set is divided into 30×30 and 20×20 pixels applied as training and testing data. The features from each dataset are described as 'Hemorrhages,' 'blood-vessels,' 'microaneurysms,' etc. The success rate of this proposed scheme is reported as 73.80%. Neural network-based scheme has also employed by Usher et al. [25]. After the preprocessing scheme 'Moat-Operator' is applied here to detect the Hemorrhages. The sensitivity and specificity for the proposed method is reported as 95.10% and 46.30%, respectively. Logistic regression-based classification has reported for the detection to retinal Hemorrhages [6]. Here, sensitivity and specificity are reported as 86.01% and 51.99%, respectively. Recently, Grinsven et al. [26] have presented neural network architecture to detect Hemorrhages. Ground truth evaluation performed in two different databases. They have reflected screening performance in area under curve (AUC) which are 0.97 and 0.98, respectively. Khojasteh et al. [27] have proposed a convolution neural network method for the Hemorrhage detection. Public databases DIARETDB1 and e-Ophtha have been applied for testing the proposed method. Here, the input images are analyzed by probability maps produced by softmax layer's score value. The screening reaches 97.3% and 86.6% accuracy for DIARETDB1 and e-Ophtha datasets, respectively. Very recently, Eftekhari et al. [28] have presented convolution neural network-based DR screening method where Hemorrhages are identified. Here, the neural network scheme uses the Keras library. The testing is done on e-Ophtha-MA dataset and sensitivity is reported as 80%. Shah et al. [1] have proposed an artificial neural network-based method for screening of Hemorrhages. Here, macula centric database MESSIDOR is used to validate the proposed method. The sensitivity and specificity are reported as 99.7% and 98.5%, respectively. Wang et al. [29] have shown a neural network-based method to diagnose and grading of retinal Hemorrhages. Here, a total of 48,996 retinal images are used for testing purpose. The overall sensitivity is reflected as 97.44%.

2.4 Classification

Different classification algorithms have been applied to detect retinal Hemorrhages. Hatanaka et al. [30] have applied Mahalanobis distance classifiers for the identification of retinal Hemorrhages. The sensitivity and specificity both are recorded as 80%. Pradhan et al. [31] have proposed a seed generation algorithm to extract the red lesions. Here, a hybrid classifier which builds up by k-nearest neighbor and Gaussian mixture model has been employed to classify red lesions from retinal images. The sensitivity and specificity are recorded as 87% and 95.53%, respectively. Tang et al. [32] have used splat-based watershed algorithm for screening of Hemorrhages. Here, the KNN classifier is applied on DRIVE and MESSIDOR database for testing the results. The sensitivity and specificity are recorded as 92.6% and 89.1%, respectively. In review, it is observed that authors have extensively used the support vector machine (SVM) classifiers on this field. Zhang and Chutatape [33] have applied SVM classifiers to count the evidence value of the Hemorrhage detection. Here, top-down and bottom-up approaches are applied to detect dark and bright lesions, respectively. The sensitivity and specificity found as 90.6% and 89.1%, respectively. Very recently, Sreeja et al. [34] have proposed a hardware setup using supervised classifiers to detect the DR Hemorrhages. Here, two classifiers are trained by SPLATE features and verified the output by ophthalmologists. The best output is selected for hardware implementation. The sensitivity has observed as 80%. Kurale and Vaidya [35] have also presented SPLATE-based SVM classifier for Hemorrhage identification. The sensitivity and specificity of this scheme is reported as 89% and 88%, respectively. Kumar and Nitta [36] have proposed SVM and GLCM classifiers for retinal Hemorrhages screening. Here, blood vessels are extracted by Kirsch's operator. The sensitivity and specificity are reported as 99.64% and 95.84%, respectively. Godlin and Kumar [37] have applied ANSIF classifiers to segment the retinal Hemorrhages. Here, MRG segmentation scheme is employed to obtain the maximum accuracy level during Hemorrhages detection. The sensitivity is recorded as 92.56% and specificity as 89%.

2.5 *Inverse Segmentation*

Kose et al. [38] have proposed an inverse segmentation scheme to detect the retinal Hemorrhages. Here, low intensity region is segmented from dark region and extract the Hemorrhages from background. This scheme subdivides the retinal surface into low and high contrasted area and use the homogeneity of the healthy surface to detect the Hemorrhages. In testing phase, the sensitivity and specificity is recorded as 93% and 98%, respectively. In review process, it is observed that authors rarely used this scheme.

3 Performance Analysis

In preceding section, we have reviewed principal methodologies to detect retinal Hemorrhages. The existing principal methodologies are followed by three approaches: lesion, image, and pixel based. In lesion-based methodology, the numbers of Hemorrhages are compared in ground truth evaluation. In image-based analysis, the images are verified as normal or abnormal (i.e., Hemorrhages are observed). Whereas, in pixel-based analysis, ground truth evaluation is done pixel by pixel. The widely used scheme is pixel-based scheme. The most of the authors are selected sensitivity and specificity to measure the screening performances. The sensitivity shows the actual positives which are correctly identified and specificity reflects the actual negative which are correctly identified. In Fig. 4, we observe that pixel and image-based analysis reflects high values of sensitivity and specificity altogether. Here, red and blue colored bars indicate sensitivity and specificity, respectively. Whereas, in lesion-based analysis, low specificity is observed in few cases. Thus, to detect the exact position of retinal Hemorrhages, the pixel by pixel analysis will be better choice. Where as to classify only the normal and abnormal (i.e., Hemorrhages present) images, the image-based analysis fulfill the requirement.

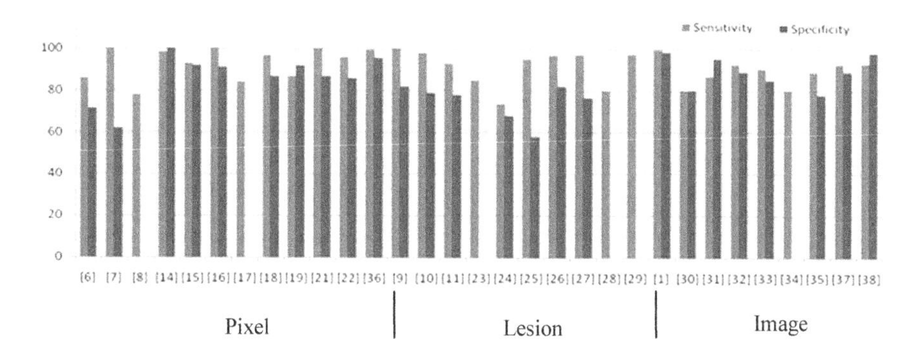

Fig. 4 Performance evaluation for Hemorrhages detection schemes

4 Conclusion

The screening of retinal Hemorrhages is a current research interest. It is one of the significant steps of diabetic retinopathy screening operation. In this contrast, the computer-aided retinal Hemorrhages detection faces several challenges. During screening, segmentation and extraction are tough enough compare to the screening of other existing mellitus of retinal images as there is no standard geometric shape found of Hemorrhages. In this consequence, small Hemorrhages are sometime wrongly detected as Microaneurysms. Thus, retinal Hemorrhage detection is a complicated task. Highly reliable, accurate and platform independent retinal Hemorrhage detection method is still an open field. Thus, there is a future scope to improve the existing retinal Hemorrhage algorithms. This article reviews the existing principal methodologies on this field. Based on this review work, the researchers can develop better algorithms.

Acknowledgements The authors would like to thank The Ophthalmology Department of Sri Aurobindo Seva Kendra, Kolkata, India for their clinical support.

Funding This research activity is financially supported by the R&D Project, sponsored by the Department of Science and Technology and Biotechnology, Government of West Bengal, India (Memo No: 148(Sanc.)/ST/P/S&T/6G-13/2018).

References

1. Shah P, Mishra D, Shanmugam M, Doshi B, Jayaraj H, Ramanjulu R (2020) Validation of deep convolutional neural network-based algorithm for detection of diabetic retinopathy—artificial intelligence versus clinician for screening. Indian J Ophthalmol 68(2):398–405
2. Datta N, Dutta H, Majumder K, Chatterjee S, Wasim N (2019) An improved method for automated identification of hard exudates in diabetic retinopathy disease. IETE J Res. https://doi.org/10.1080/03772063.2019.1618206
3. Wiseng K, Hiransakolwong N, Pothirut E (2013) Automatic detection of retinal exudates using a support vector machine. Appl Med Inform 32(1):32–42
4. Kade M (2013) A survey of automatic techniques for retinal diseases identification in diabetic retinopathy. Int J Adv Res Technol 2:199–216
5. Datta N, Dutta H, Majumder K (2016) Brightness preserving fuzzy contrast enhancement scheme for the detection and classification of diabetic retinopathy disease. J Med Imaging 3(1):014502. https://doi.org/10.1117/1.JMI.3.1.014502
6. Baint A, Andras H (2016) An ensemble-based system for microaneurysm detection and diabetic retinopathy grading. IEEE Trans Biomed Eng 59(6):1720–1726
7. Garcia M, Lopez MI, Alvarez D, Hornero R (2010) Assessment of four neural network based classifiers to automatically detect red lesions in retinal images. Med Eng Phys 32(10):1085–1093
8. Marino C, Ares E, Penedo M, Ortega M, Barreira N, Gomezulla F (2008) Automated three stage red lesions detection in digital color fundus images. WSEAS Trans Comput 7(4):207–215
9. Datta N, Dutta H, Majumder K (2016) An effective contrast enhancement method for identification of microaneurysms at early stage. IETE J Res. https://doi.org/10.1080/03772063.2015.1136573

10. Esmaeili M, Rabbani H, Dehnavi A, Dehghani A (2010) A new curvelet transform based method for extraction of red lesions in digital color retinal images. In: Proceedings of the IEEE international conference 2010, ICIP. IEEE, Hong Kong, pp 4093–4096
11. Kande G, Tirumala S, Subbaiah P, Tagore M (2009) Detection of red lesions in digital fundus images. In: Proceedings of the IEEE international conference 2009, ISBI, Boston, pp 558–561
12. Jadhav A, Patil P (2015) Classification of diabetes retina images using blood vessel area. Int J Cybern Inform 4(2):251–257
13. Sopharak A, Uyyanonvara B, Barman S (2011) Automatic microaneurysm detection from non-dilated diabetic retinopathy retinal images using mathematical morphology methods. Int J Comput Sci 38(3):1–7
14. Shivaram J, Patil R, Aravind H (2009) Automated detection and quantification of haemorrhages in diabetic retinopathy images using image arithmetic and mathematical morphology methods. Int J Recent Trends Eng 2:174–176
15. Karnowski T, Govindasamy V, Tobin K, Chaum E, Abramoff M (2008) Retina lesion and microaneurysm segmentation using morphological reconstruction methods with ground-truth data. In: Proceedings of the IEEE international conference 2008, EMBS, Canada, pp 5433–5436
16. Kande G, Tirumala S, Subbaiah P (2010) Automatic detection of microaneurysms and hemorrhages in digital fundus images. J Digit Imaging 23(4):430–437
17. Matei D, Matei R (2008) Detection of diabetic symptoms in retina images using analog algorithms. Int J Med Health Sci 2(9):323–326
18. Langroudi M, Sadjedi H (2010) A new method for automatic detection and diagnosis of retinopathy diseases in colour fundus images based on morphology. In: Proceedings of the IEEE international conference on bioinformatics and biomedical technology 2010, China, pp 134–138
19. Acharya U, Lim C, Ng E, Chee C, Tamura T (2009) Computer-based detection of diabetes retinopathy stages using digital fundus images. J Eng Med 223(5):545–553
20. Singh N, Tripathi R (2010) Automated early detection of diabetic retinopathy using image analysis techniques. Int J Comput Appl 8(85):18–23
21. Niemeijer M, Ginneken V, Staal J, Suttorp-Schulten A, Abrmoff M (2005) Automatic detection of red lesions in digital color fundus photograph. IEEE Trans Med Imaging 24(5):584–592
22. Sinthanayothin C, Boyce J, Williamson T, Cook H, Mensah E, Lal S (2002) Automated detection of diabetic retinopathy on digital fundus images. Diabet Med 19(2):105–112
23. Bae J, Kim K, Kang H, Jeong C, Park K, Hwang J (2011) A study on hemorrhage detection using hybrid method in fundus images. J Digit Imaging 24(3):394–404
24. Gardner G, Keating D, Williamson T (1996) Automatic detection of diabetic retinopathy using an artificial neural network: a screening tool. Br J Ophthalmol 80(11):940–944
25. Usher D, Dumskyj M, Himaga M, Williamson T, Nussey S, Boyce J (2004) Automated detection of diabetic retinopathy in digital retinal images: a tool for diabetic retinopathy screening. Diabet Med 21(1):84–90
26. Grinsven M, Venhuizen F, Ginneken B, Hoyng C, Theelen T, Sanchez C (2016) Automatic detection of hemorrhages on color fundus images using deep learning. Investig Ophthalmol Vis Sci 57(12):5966–5972
27. Khojasteh P, Aliahmad B, Kumar D (2018) Fundus images analysis using deep features for detection of exudates, hemorrhages and microaneurysms. BMC Ophthalmol 18(1):1–13
28. Eftekhari N, Pourreza H, Masoudi M, Shirazi K, Saeedi E (2019) Microaneurysm detection in fundus images using a two step convolutional neural network. BioMed Eng OnLine 18(67):1–10
29. Wang B, Xiao L, Liu Y, Wang J, Liu B, Li T, Ma X, Zhao Y (2018) Application of a deep convolutional neural network in the diagnosis of neonatal ocular fundus hemorrhage. Biosci Rep 38(6):1–24
30. Hatanaka Y, Nakagawa T, Hayashi Y, Hara T, Fujita H (2008) Improvement of automated detection method of hemorrhages in fundus images. In: Proceedings of the IEEE international conference on EMBS2008, Canada, pp 5429–5432
31. Pradhan S, Balasubramanian S, Chandrasekaran V (2008) An integrated approach using automatic seed generation and hybrid classification for the detection of red lesions in digital fundus

images. In: Proceedings of the IEEE international conference on computer and information technology workshops, Sydney, pp 462–467
32. Tang L, Niemeijer M, Abramoff M (2011) Splat feature classification: detection of the presence of large retinal hemorrhages. In: Proceedings of the IEEE international conference on biomedical imaging: from nano to macro, Chicago, pp 681–684
33. Zhang X, Chutatape O (2005) Top-down and bottom-up strategies in lesion detection of background diabetic retinopathy. In: Proceedings of the IEEE international conference on computer vision and pattern recognition CVPR2005, San Diego, pp 1–7
34. Sreeja K, Kumar S, Pradeep A (2020) Automated detection of retinal hemorrhage based on supervised classifiers and implementation in hardware. In: Proceedings of smart innovation, systems and technologies SIST 2020, Singapore, vol 182, pp 57–67
35. Kurale N, Vaidya M (2017) Retinal hemorrhage detection using splat segmentation of retinal fundus images. In: Proceedings of the IEEE international conference on conference on computing, communication, control and automation ICCUBEA 2017, Pune, pp 1–6
36. Kumar S, Nitta G (2019) Early detection of diabetic retinopathy in fundus images using GLCM and SVM. Int J Recent Technol Eng 7(5S4):17–20
37. Godlin L, Kumar P (2018) Detection of retinal hemorrhage from fundus images using ANFIS classifier and MRG segmentation. Biomed Res 29(7):1–9
38. Kose C, Sevik U, Ikiba C, Erdol H (2012) Simple methods for segmentation and measurement of diabetic retinopathy lesions in retinal fundus images. Comput Methods Programs Biomed 107(2):274–294

A Study on Retinal Image Preprocessing Methods for the Automated Diabetic Retinopathy Screening Operation

Amritayan Chatterjee, Niladri Sekhar Datta, Himadri Sekhar Dutta, Koushik Majumder, and Sumana Chatterjee

Abstract Recent days, diabetic retinopathy (DR) is a principal cause of incurable blindness to the diabetic patients. Manual screening of DR is time-consuming and resource demanding activity. Thus, today, setup of the reliable automated screening is an open issue for the researchers. In that concern, the automated analysis of retinal images, preprocessing stage plays a vital role. The overall success of screening operation is dependent on it completely. Preprocessing the retinal image prior to screening is a common task as noisy image degrades the screening performance. This paper reviews the different DR screening methods and points out the vastly used preprocessing scheme on that field. Finally, it indicates the most effective preprocessing scheme for DR screening as per the data analysis.

Keywords Medical image processing · Diabetic retinopathy · Contrast enhancement · Adaptive histogram equalization · Contrast limited adaptive histogram equalization

1 Introduction

Today, diabetic retinopathy, i.e., DR, is an outcome of diabetic mellitus and foremost reason of blindness across the globe. As per WHO, 150 million people currently have

---

A. Chatterjee (✉) · H. S. Dutta
Department of Electronics and Communication Engineering, Kalyani Government Engineering College, Kalyani, West Bengal, India
e-mail: amritayanchatterjee@gmail.com

N. S. Datta
Department of Information Technology, Future Institute of Engineering and Management, Kolkata, West Bengal, India

K. Majumder
Department of Computer Science and Engineering, Maulana Abul Kalam Azad University of Technology, Kolkata, West Bengal, India

S. Chatterjee
Department of Ophthalmology, Sri Aurobindo Seva Kendra, Kolkata, West Bengal, India

A. Choudhary et al. (eds.), *Applications of Artificial Intelligence and Machine Learning*, Lecture Notes in Electrical Engineering 778,
https://doi.org/10.1007/978-981-16-3067-5_28

suffered from diabetics and it will be 350 million at 2030 [1]. Thus, early detection by automated screening is required to prevent the incurable blindness. In computer-aided diagnose (CAD) screening, the preprocessing is the vital step and the overall success depends on it. Usually, poor contrast and noise as well as uneven illumination degrade retinal image quality. Mostly due to error, noise and angle of capturing device, eye movement, etc., cause the fundus image improper for disease diagnosis. Therefore, appropriate preprocessing method selection for the retinal fundus image is a mandatory step. This preprocessing step improves the quality of input retinal image as well as facilitates to diagnose the disease automatically or manually. The main part of the preprocessing steps is contrast enhancement, shade correction, resizing, etc. After preprocessing steps, the improved image is used to detect the candidate that later used for the detection of the disease. Microaneurysms (MA), hemorrhages, hard and soft exudates, etc., are the clinical sign of DR. MAs appear as the red dot spots in retina and are the early sign of DR. Hard exudates are yellowish-white color and visualize in non-proliferative diabetic retinopathy (NPDR) disease. Soft exudates, i.e., cotton wool spots, are developed by the leakage of blood vessels. In NPDR, for blood clots, retinal hemorrhages are produced [2]. Clinically, identification of MAs, exudates, and hemorrhages forms retinal image treated as DR screening operation. In this research paper, a comprehensive study has been made for different types of preprocessing scheme which are applied in the DR screening operation. So to detect MAs, exudates and hemorrhages from retinal image properly and accurately the fundus image needs to be preprocessed to improve the contrast between the foreground and background. If the original fundus image without preprocessing is used to predict the disease by detecting the candidates, mostly the false candidate detection increases and true lesion candidate decreases which decreases the accuracy and reliability of the automated system. The proper selection of the preprocessing method is necessary for the detection of the disease accurately so that it can detect and predict the stage of the DR properly. Here, point out the most effective preprocessing scheme as per the results appeared. This paper is arranged as in Sect. 2; review report is presented. In Sect. 3, the preprocessing methods like histogram equalization, CLAHE, and bottom hat transformation have discussed. The frequently used filtering methods have also been presented. A brief description of input dataset is given on which research work is carried out. In Sect. 4, experimental results are provided and, finally, conclude the paper.

2 Current State of the Art

Review on the preprocessing steps for automated diabetic retinopathy screening has been carried out. It reflects that low contrast and noisy images obviously degrade the screening performance [3, 4]. As per clinical point of view, near about 20% retinal images are unused for poor image quality. Hence, prior to the DR screening, selection of an effective preprocessing scheme is mandatory [5, 6]. Review works have found that histogram equalization (HE) on green channel and contrast limited adaptive histogram equalization (CLAHE) are the most renowned preprocessing methods for retinal fundus image. Addition with this, different types of image processing filters like median filter, Wiener filter, and bottom hat transformation also used by the researchers. Mengko et al. [7] have used HE to preprocess the retinal image for exudates identification. Wiseng et al. [8] have also applied the above-said contrast enhancement scheme in angiography for retinal images. Agurta et al. [9] opt histogram mean normalization to detect hard exudates in macular region of retinal image. Franklin and Rajan [10] have used HE methods for exudates detection. Here, need to mention that, in case of retinal image contrast enhancement, CLAHE is very common scheme that has opted by the researchers. In this research work, consider the mean filtering scheme to smooth and to improve the quality of retinal image. Need to mention here that, mean filter is incapable to select monotonically decreasing frequency response. But, it is not a suitable filter to smooth the retinal image. Lazar and Hajdu [11] have applied Gaussian mask for smoothing and suppression of retinal fundus images. It produces better quality than mean filter. But cut down the basic information of retinal image is also noticed. Baint and Andras [12] have applied non-uniform illumination for the shade correction at preprocessing stage. Fathi and Nilchi [13] have selected wavelet transform method for the segmentation of blood vessels in retinal surface. Here, few cases are found where inverted green channel uses to reduce the non-uniform illumination. Shanmugam and Banu [14] have proposed 5 by 5 mean filter using Gaussian kernel for the shade correction of input image. Martin et al. [15] have also used the shade correction technique selecting grayscale image. Recently, brightness preservation is a significant criterion for medical image analysis. Popularly known as dynamic histogram equalization (DHE) is a good contrast enhancement method for retinal fundus images but it is incapable to store the mean brightness [16]. Datta et al. [3] have proposed a brightness preserving contrast enhancement scheme for preprocessing of retinal fundus image. Here primarily, green layer extracted from RGB color image. Thereafter, calculation of fuzzy histogram is done. In the next stage, partition of the image histogram and intensity equalization is performed. Finally, normalization is done on output and input images to equalize the input and output image brightness. This retinal image preprocessing method reflects the better results for DR screening on low contrast noisy retinal images. Here, need to mention that authors have employed the quality assessment of retinal image prior DR identification. Survey on this field, the significance of the retinal image preprocessing has realized. Based on this review work, an experiment is conducted to point out the optimal preprocessing method for retinal image.

3 Preprocessing Schemes

A good number of retinal image dataset have required to conduct the research work in this field. Here the research motto is to point out the best suitable retinal preprocessing scheme which is applied on the DR screening operation. Initially, in our research work, three retinal image databases with different image quality are created. All the input images are collected from Ophthalmology Department of Sri Aurobindo Seva Kendra, Kolkata, Multi-specialty Hospital. These images are captured by Canon CR5 3CCD non-mydriatic digital camera with 35-degree field of view (fov). Here, Dataset1 contains forty images with resolution 768 × 584. Whereas, Dataset2 and Dataset3 are containing sixty-one and eighty-seven retinal images maintaining the resolution 700 × 605 and 1440 × 960, respectively. Three different quality retinal image (resolution varied) datasets will indicate the effectiveness of the different preprocessing steps. The following subsections will describe that the very frequently used methodology for retinal image preprocessing.

3.1 Histogram Equalization

The most renowned image processing scheme applied for the preprocessing of retinal fundus image is the histogram equalization (HE) [17]. The HE method is stated as

$$H = \text{floor}(N - 1) \sum \sum_{j=0}^{k} pr(rj) \tag{1}$$

Here, mathematical floor function is used to round of the integer value to its nearby value. Possible intensity level for image is defined by N. Probability of occurrences for the intensity level is

$$P_r(\chi_k) = \frac{Nk}{ef} \tag{2}$$

where e and f are the pixel in the image.

3.2 CLAHE

Image processing point of view, HE over amplifies the image contrast. Thus, noise amplification takes place. CLAHE, i.e., contrast limited adaptive histogram equalization, resolves this problem. Basically, CLAHE is the variant of AHE and it restricted the noise amplification. CLAHE limits the noise amplification by histogram clipping via cumulative distribution function. CLAHE very frequently used by the researchers for preprocessing the retinal image prior to DR screening [17, 18].

3.3 Bottom Hat Transformation

Bottom hat transformation is a morphological image processing operation [2, 17]. Initially, morphological closing is done on input image. Then the output image subtract from the input image. The equation is expressed as

$$B_{\text{hat}}(C) = (C \cdot b) - C \tag{3}$$

Here $C \cdot b$, is the closing operation and C is the input image. The morphological closing operation on C is defined as the dilation of C using the structuring element b followed by the morphological erosion operation. The equation is defined as

$$(C \cdot b) = (C_{\text{dilation}} b)_{\text{erosion}} b \tag{4}$$

The flat disk-type structuring element is selected for contrast enhancement.

3.4 Wiener Filter

Wiener filter has applied several times to remove the noise [17]. This scheme uses the liner estimation technique to erase the noise. It minimizes the mean square error (MSE) of the retinal input image. Wiener filter is able to preserve image edges and existing anatomical structure. The Wiener filter frequency domain is stated as

$$\delta(f_1 f_2) = \frac{h(f_1 f_2), \chi(f_1 f_2)}{\left[\text{square}(h(f_1 f_2)) \cdot \chi(f_1 f_2)\right] + S(f_1 f_2)} \tag{5}$$

where $\chi(f_1 f_2)$, $S(f_1 f_2)$ are the power spectral density of input image and noise, respectively. $h(f_1 f_2)$ is denoted as filter transfer function.

3.5 Median Filter

Median filter is vastly used for removing the noise from retinal image. This filtering scheme is able to preserve edge during preprocessing operation. Generally, it selects the median value for replacing the pixel value [18]. If all adjacent pixels are presented by

$$Q[x, y] = \text{median}\{c[p, q], (p, q) \in A\} \tag{6}$$

where A is the neighborhood centered in $[x, y]$ of the image. Median filter is the effective filtering scheme comparing to the convolution technique.

3.6 Brightness Preserving Preprocessing Scheme

Review on this specific area, it is observed that retinal image preprocessing should have brightness preserving capability. But, existing DHE, i.e., dynamic histogram equalization scheme [16], is incapable to consider the inexact gray level values of input image. Currently, Datta et al. [3] have proposed a preprocessing method which is able to preserved retinal image brightness very effectively. The preprocessing steps are presented as

Step 1: Choose the green plane from RGB colored retinal image.

Step 2: Calculation of fuzzy histogram is performed

$$h = h + \sum_i \sum_j \xi f(x, y)v \tag{7}$$

$\xi f(x, y)v$ is fuzzy membership function defined in [3].

Step 3: Image histograms are partitioned.

Here, to calculate the local maxima points on image histogram, first- and second-order derivations are applied. The local maxima is expressed as

$$\theta_{\max} = \theta \forall h(\theta + 1) \times h(\theta - 1) < 0 \tag{8}$$

where $\overline{\overline{h}}(\theta) < 0$.

Here $\overline{\overline{h}}(\theta)$ is the second-order derivative.

Thereafter, the local maxima points are used to create the partition. If n numbers of maxima found at $\{m_1, m_2 \ldots m_n\}$ in the range $[F_{\min}, F_{\max}]$, then $(n + 1)$ sub-histogram found as

$$\{[F_{\min}, m - 1], [m_1, m - 2] \ldots [m_n, F_{\max}]\} \tag{9}$$

Step 4: The sub-histograms are equalized by dynamic histogram equalization (DHE) method. To create dynamic range, the mathematical expression is

$$R_k = \frac{(L - 1)(h_k - L_k) \log p_k}{\sum_{i=1}^{n-1} (h_i - L_i) \times \log p_i} \tag{10}$$

Here, h_k and l_k are the highest and lowest intensity values for Kth sub-histogram. P_k denotes the pixels in the sub-partitions. The dynamic range is expressed as

$$\text{Start}_k = \sum_{i=1}^{K-1} R_i + 1 \tag{11}$$

Table 1 SSIM comparison for different preprocessing scheme

Image source	HE	CLAHE	Optimal preprocessing method
HOSPITAL DATA SET 1 (No: 40, Res: 768 × 584)	0.69	0.76	0.82
HOSPITAL DATA SET 2 (No: 61, Res: 700 × 605)	0.68	0.74	0.83
HOSPITAL DATA SET 3 (No: 87, Res: 1440 × 960)	0.58	0.78	0.81

Table 2 AMBE analyses for brightness preserving test

Image source	HE	CLAHE	Optimal preprocessing method
HOSPITAL DATA SET 1 (No: 40, Res: 768 × 584)	13.09	8.07	0.012
HOSPITAL DATA SET 2 (No: 61, Res: 700 × 605)	24.06	18.91	0.062
HOSPITAL DATA SET 3 (No: 87, Res: 1440 × 960)	21.31	10.67	0.021

$$\mathrm{Stop}_k = \sum_{i=1}^{k} R_i \tag{12}$$

Step 5: Intensity levels of the sub-histogram are equalized. This is mathematically expressed as

$$\theta = \mathrm{Start}_k + R_k \times \sum_{i=\mathrm{start}_k}^{\theta} \frac{h}{p_k} \tag{13}$$

Here, θ indicates the new intensity range and h indicates the histogram values at kth position.

Step 6: Apply normalization method to equalize the brightness difference between output and input images. The detailed explanation of this scheme is provided at [3]. Performance analysis for the optimally selected preprocessing method is reported in Tables 1 and 2.

4 Experimental Results

The identification of diabetic retinopathy through automatic screening is clearly dependent on the retinal input images quality. The preprocessed retinal image quality is verified by structure similarity index measurement (SSIM). Here, the absolute mean brightness error (AMBE) is also applied to examine the brightness difference of input and output images. If SSIM is 1, the output retinal image is original and nearer to 1 indicates the better quality of retinal images. AMBE compares the brightness preservation on input and output images. Logically, lower AMBE and higher SSIM reflect the better contrast enhancement and able to preserve the mean brightness effectively. Review reflects that histogram equalization and contrast limited adaptive histogram equalization are very frequently used contrast enhancement scheme for the retinal image prior to the diabetic retinopathy detection. In this research paper, an optimal contrast enhancement method is selected and compared the performance with HE and CLAHE methods. Tables 1 and 2 represent the comparative results of SSIM and AMBE, respectively. In SSIM assessment, for each dataset optimal preprocessing shows the highest value which is more than 81% close to 100%, means more accurate compared to the HE and CLAHE which are in the range of 70–75%. On the other hand, for the analysis of brightness preservation scheme, the optimal preprocessing method reflects the lowest value on the same dataset. The HE and CLAHE methods give value between 10 and 30 where the optimal preprocessing method gives value between 0.01 and 0.05 which is far more better compared to the previous two methods as the absolute mean brightness error reduced to a great extent. Testing results here also reflect that for low-resolution images (dataset 1 and dataset 2) HE and CLAHE method not work properly. Figure 1a is the retinal input image selection; Fig. 1b is the outcome of optimally selected contrast enhancement method for retinal images. The fundus image needs to be preprocessed to enhance the contrast between the background and the lesions to detect the true lesions properly. The incorporation of fuzzy logic for contrast enhancement gives the advantage compared to existing HE or CLAHE by preserving the brightness of the image accurately by reducing the AMBE error and increasing the SSIM index. The optimal preprocessing method

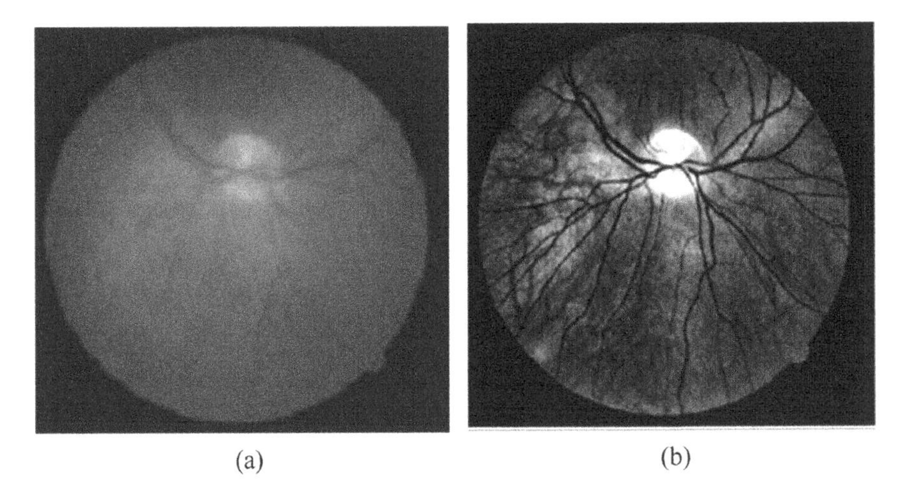

Fig. 1 **a** A sample of retinal input image, **b** optimal choice preprocessing used on sample input for contrast enhancement

gives much better result compared to existing HE and CLAHE in SSIM and AMBE index of measurement that can improve the accuracy of the automated system to detect the DR as using the optimal preprocessing method reduces error and increase the chance to detect the true candidate lesion properly.

5 Conclusion

Preprocessing of retinal image plays a vital role for automatic detection of diabetic retinopathy diseases. In this concern, a lot of preprocessing methods have been used along with different filtering scheme for removing noises. This research paper finds the optimum choice of preprocessing scheme for diabetic retinopathy detection. Datta et al. [3] proposed the brightness preserving preprocessing method of retinal images prior to automatic diabetic retinopathy screening. On survey, HE and CLAHE are the frequently used scheme of automated diabetic retinopathy screening system. But using fuzzy logic in contrast enhancement is a new way in preprocessing steps. After preprocessing steps, the processed image can be used to detect MAs, hemorrhages, exudates, vessels, etc., for accurate detection of the disease. If the preprocessing steps can be improved, the true lesion detection becomes easier in later stages. Hence, optimally selected method compared with HE and CLAHE scheme performs better and produces better image quality with better contrast between background and foreground. Thus, we can conclude that optimally selected preprocessing scheme is the better choice for retinal images as it gives increased value of SSIM index above 81% compared to 70–75% in HE and CLAHE and decreased AMBE value less than 0.01 where HE and CLAHE give very high value 10–30%. In the future, authors will try to detect and segment MAs, hemorrhages, exudates, vessels, etc., using the optimally selected preprocessing scheme discussed in this paper. After extracting the candidate lesions, DR detection and disease stage prediction, i.e., PDR or NPDR, can be made using some machine learning classification method. Deep learning can also be used to detect the DR as it is a new area of research in the field of medical image processing and disease diagnosis.

Acknowledgements The authors would like to thank The Ophthalmology Department of Sri Aurobindo Seva Kendra, Kolkata, India, for their clinical support.

Funding This research activity is financially supported by the R&D Project, sponsored by the Department of Science and Technology and Biotechnology, Government of West Bengal, India (Memo No: 148(Sanc.)/ST/P/S&T/6G-13/2018).

References

1. Shah P, Mishra D, Shanmugam M, Doshi B, Jayaraj H, Ramanjulu R (2020) Validation of deep convolutional neural network-based algorithm for detection of diabetic retinopathy—artificial intelligence versus clinician for screening. Indian J Ophthalmol 68(2):398–405
2. Datta N, Dutta H, Majumder K, Chatterjee S, Wasim N (2019) An improved method for automated identification of hard exudates in diabetic retinopathy disease. IETE J Res. https://doi.org/10.1080/03772063.2019.1618206
3. Datta N, Dutta H, Majumder K (2016) Brightness preserving fuzzy contrast enhancement scheme for the detection and classification of diabetic retinopathy disease. J Med Imaging 3(1):014502. https://doi.org/10.1117/1.JMI.3.1.014502
4. Kade M (2013) A survey of automatic techniques for retinal diseases identification in diabetic retinopathy. Int J Adv Res Technol 2:199–216
5. Datta N, Dutta H, Majumder K (2016) An effective contrast enhancement method for identification of microaneurysms at early stage. IETE J Res. https://doi.org/10.1080/03772063.2015.1136573
6. Sopharak A, Uyyanonvara B, Barman S (2011) Automatic microaneurysm detection from non-dilated diabetic retinopathy retinal images using mathematical morphology methods. Int J Comput Sci 38(3):1–7
7. Mengko T, Handayani A, Valindria V (2009) Image processing in retina angiography: extracting angiographical features without the requirement of contrast agents. In: Proceedings of the machine vision applications 2009. IAPR, Japan, pp 451–457
8. Wiseng K, Hiransakolwong N, Pothirut E (2013) Automatic detection of retinal exudates using a support vector machine. Appl Med Inform 32(1):32–42
9. Agurta C, Murray V, Yu H, Wigdahl J, Pattichis M, Nemeth S, Barriga E, Soliz P (2014) A multiscale optimization approach to detect exudates in the macula. IEEE J Biomed Health Inform 18:1328–1336
10. Franklin S, Rajan S (2014) Diagnosis of diabetic retinopathy by employing image processing technique to detect exudates in retinal images. IET Image Process 8(10):601–609
11. Lazar I, Hajdu A (2012) Retinal microaneurysm detection through local rotating cross section profile analysis. IEEE Trans Med Imaging 32(2):400–407
12. Baint A, Andras H (2016) An ensemble-based system for microaneurysm detection and diabetic retinopathy grading. IEEE Trans Biomed Eng 59(6):1720–1726
13. Fathi A, Nilchi A (2013) Automatic wavelet based retinal blood vessels segmentation and vessel diameter estimation. Biomed Signal Process Control 8(1):71–80
14. Shanmugam V, Banu R (2013) Retinal blood vessel segmentation using an extreme learning machine approach. In: Proceedings of the point of care healthcare technologies 2013, Bangalore, pp 318–321
15. Martin D, Aquino A, Arias M, Bravo J (2011) A new supervised method for blood vessel segmentation in retinal images by using gray-level and moment invariants-based features. IEEE Trans Med Imaging 30(1):146–158
16. Ibrahim H, Kong N (2007) Brightness preserving dynamic histogram equalization for image contrast enhancement. IEEE Trans Consum Electron 53(4):1752–1758
17. Sridhar S (2012) Digital image processing, 2nd edn. Oxford University Press, New Delhi
18. Datta N, Sarkar R, Dutta H, De M (2012) Software based automated early detection of diabetic retinopathy on non-dilated retinal image through mathematical morphological process. Int J Comput Appl 60(18):20–24

FFHIApp: An Application for Flash Flood Hotspots Identification Using Real-Time Images

Rohit Iyer, Parnavi Sen, and Ashish Kumar Layek

Abstract Extreme climate changes have become the new norm in today's world. As a result, flash flood disasters continue to increase. It has become imperative to devise a quick disaster response system for minimizing the magnitude of damage and reducing the difficulties of human life. In this paper, we propose an application using android technology that provides real-time updates and prompt flow of authentic information of flood-ravaged areas to the rescue personnel or common people. The application accepts images belonging to flood-affected regions from rescue personnel, volunteers, etc. It authenticates and then filters those images using deep learning techniques. The severity of floods is estimated and plotted on a map using the associated location information of the images. The data is analyzed by using clustering techniques and visualized on the map. Subsequently, the affected areas of the flash flood are identified. Their peripheries are mapped so that these hotspots can be targeted for immediate relief operations.

Keywords Flood-image detection · Flood hotspots identification · Transfer learning · VGG19 · Clustering · Convolutional neural network

1 Introduction

Recent disasters such as cyclone Aila 2009, Amphan 2020, Nisarga 2020, and many more have shown how human society is helpless when nature's fury has its say. Other than cyclones, monsoonal floods cause mayhem as well. Commuters get stuck in waterlogged streets, open wires pose a grave danger to nescient citizens, and there is an immense loss of livelihood and shelters. The NDRF (*National Disaster*

R. Iyer (✉) · P. Sen · A. K. Layek
Indian Institute of Engineering Science and Technology, Shibpur 711103, India
e-mail: parnavi.sen.ps@gmail.com

A. K. Layek
e-mail: ashish@cs.iiests.ac.in

A. Choudhary et al. (eds.), *Applications of Artificial Intelligence and Machine Learning*, Lecture Notes in Electrical Engineering 778,
https://doi.org/10.1007/978-981-16-3067-5_29

Response Force) personnel and civic volunteers are those warriors who tackle such grave situations and provide a helping hand to the victims. In this age of technology, if we can integrate disaster response systems with cutting edge technology, we would be better equipped to handle these crises. With the support of digital platforms, the information can be quickly communicated to disaster management teams and the general public about the severity of inundation.

There exists several works relating to the early detection of floods. Jyh-Horng et al. [8] utilized video streams to determine the water levels of the main rivers and trigger an alarm. Mean-shift (MS) and region growing (RegGro) algorithms were used to identify flooded areas. Napiah et al. [13] used smartphones and IoT technologies for monitoring and delivering real-time flood information. Sensors were employed to detect water level. A CNN-based model for detecting flood-images from social media images is proposed in the work [10]. It also used possible color filters of floodwater along with CNN-based model to identify flood-images. Another work [3] implemented mask R-CNN as base architecture, for instance, segmentation to estimate the level of flooding. Zhang et al. [18] have documented a comparative study between various conventional image segmentation techniques that can be used for real-time flood monitoring. The techniques adopted were region growing, canny edge detection, and graph cut. The work [5] also used these image segmentation techniques on video streams. Menon and Kala [12] developed a video surveillance system and a mobile application to inform people about the disaster. Work [7] uses the VGG16 network on satellite images to detect inundation.

These works related to flood detection have mainly focussed on various image segmentation techniques and detecting the water level. However, smartphones being ubiquitous nowadays, there is a need for communicating real-time reliable data and providing detailed analysis to the public using these devices. Moreover, most of the works are dependent on social media images. But in the case of social media images, the geolocation from the metadata of those images is stripped off. This makes it difficult for accurate mapping of flood-affected areas. There might also be too many farce images that might be of a different location or a different date leading to inconsistencies and delay in the hazard mitigation process.

In the proposed work, we took the initiative to present a novel *flash flood hotspots identification* (for brevity, we call it as FFHI) application keeping flash flood situation in mind, which can be useful mainly in the crowded cities. It uses android technology to provide a platform for such critical information to be shared in the public domain. The civic volunteers and NDRF personnel would play a major role in this transfer of real-time information. They suppose to upload images of the inundated localities in different pockets of the city to the FFHI application. Sazara et al. [15] proposed a similar technique of floodwater detection on roadways from image data. They used pretrained deep neural networks for flood area segmentation using three different feature extractors. Probably, this work [15] is closest to our proposed work. However, that is a comparative study between various classification techniques and does not provide an end-to-end application that disseminates useful information to the public domain.

In our approach, based on the weather forecast, the server-side application configures an FFHI event in a city where it is very likely that a flash flood situation may occur on a day. By doing so, it aims to identify the number of hotspots of flood over there. The users of the proposed application, which happens to reside in a city where an FFHI event is configured, are notified to contribute images. At first, images posted through the android app by a smartphone user are validated with the date and location of the target city for which the FFHI event is configured. The metadata of the images is used to obtain the date and location information. Then, validated images are checked if they belong to the flood-image category using a deep learning approach by the server-side application. Finally, those remaining candidate images are clustered based on their location, and boundaries are calculated to obtain the hotspots.

Understandably, a large number of volunteers might be reluctant to take part in this crowd-sourcing initiative. But in case of a heavy flash flood situation, even if it gathers a small community of dedicated individuals, the data would be enough to get a clear picture of the hotspots of the disaster. Let us assume the proposed application manages to rope in a hundred dedicated volunteers willing to contribute, and each person uploads ten to twelve images. Out of those images uploaded in total, if the proposed FFHI application finally filters out about seven/eight hundred authentic flood-related snapshots, it would be able to work fine in terms of identifying important flash flood hotspots. Note that based on the severity of the flood condition of a locality, it is expected that more volunteers will respond to the situation and upload more images from those affected areas.

The later part of this paper is organized as follows. In the next Sect. 2, the proposed methodology is discussed in detail. In Sect. 3, the simulation process and experimental results are discussed. Concluding notes are presented in Sect. 4.

2 Proposed Work

The proposed method of flash flood hotspots identification has two major components. The first one is the android application for the smartphone device. In Sect. 2.1, we describe the functionalities involved in the android app. In brief, it does the preprocessing of images which are posted by the users. It also enables the option of displaying identified flash flood hotspots within a city/place. The server-side application is another major component that deals with flood-image detection and performing clustering algorithms on the stream of images uploaded by different users of the android app. This entire technique is discussed in detail in Sect. 2.2.

2.1 Android Application for Smartphone

The android application provides a simple interface to the end user. As a prerequisite, the location information of the smartphone is accessed by the application. In the

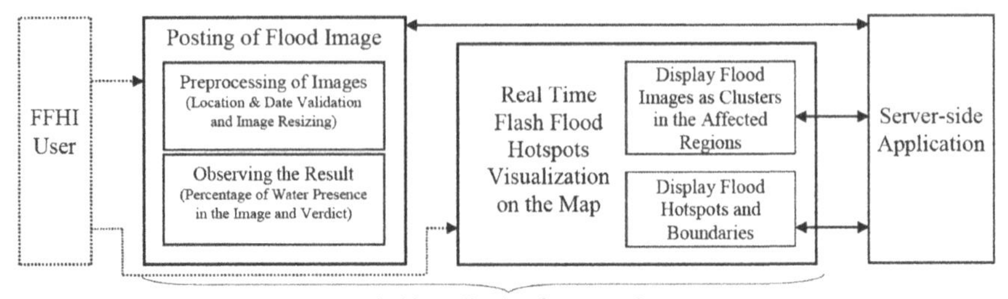

Fig. 1 Android application overview

registration process, the android app exchanges the location information with the server. As a result, it gets to know from the server if there is any configured active FFHI event in that city. If such an event is configured for that city, only then the user gets the option for posting images to the server.

The android application has mainly two important activities that are pictorially shown in Fig. 1. The basic operation one user can perform is the image posting. However, as mentioned, image posting is possible only if there is an active FFHI event for that city. Once the android app receives an image to be posted, it performs some preprocessing steps. There are three steps performed in a sequence, namely (a) date and location verification using image metadata, (b) image resizing to lower resolution, and (c) near-duplicate image detection. At first, it checks whether the date of the image is matching with the current date, otherwise rejected. Then, it checks if the location information associated with the image belongs to the same city from where the user is trying to post. Here, we use the "Geocoding API" provided by Google to carry out this activity. If the image does not belong to the specified city, it is rejected. Further, it examines whether the image is at the periphery of a water body. In general, most of the flood-image detection techniques [10] cannot distinguish accurately between floodwater and a natural water body. However, only those images can be sent to the server that does not contain natural water bodies. If the coordinates of the image are at the coastline or on a riverbank or at the fringes of any lake or pond, it is rejected. In the second step, the image is resized to a lower dimension keeping aspect ratio unaltered. This is done to avoid uploading very high-resolution images to the server. In the resizing process, the height (or width—whichever is more) is fixed at 512 pixels. However, we have taken care that the metadata of the images remains unchanged, and the geolocation of the image is preserved. There is also a possibility that the same user may post duplicate (or near-duplicate) images again and again. Hence, near-duplicate image detection is necessary, and we use a similar type of near-duplicate clustering technique proposed in [19]. After the preliminary processing, the image is sent to the server. Once the image is verified by the flood-image-detector at the server, it returns a response to the android app in an asynchronous manner. The response contains the final verdict on the image, i.e., if it is a flood-image or not. If it is concluded as flood-image, then the percentage of water present in the image is also included in the response.

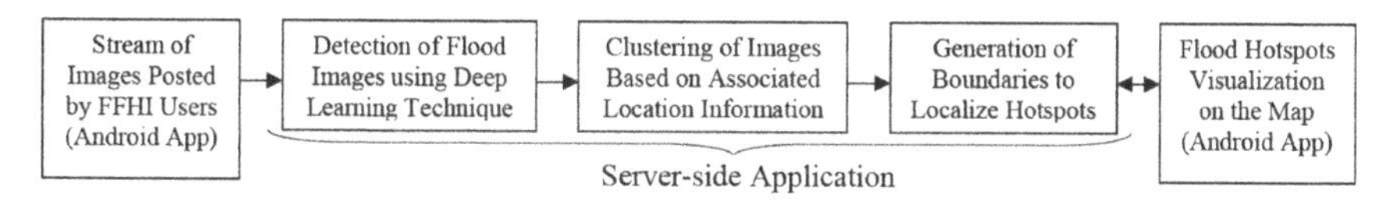

Fig. 2 Server-side application overview

Apart from the image posting, the proposed android app also provides the option of viewing the various clusters of images and hotspots that have suffered havoc by the floods. The user could view these by entering the city name of his choice. The hotspots are identified by mapping the periphery of the areas which need relief aid and should be avoided by people. This aspect of the proposed system is discussed in detail in Sect. 2.2.

2.2 *Server-Side Application*

The server-side application plays a major role to ensure end-to-end functionalities of the flash flood hotspots identification system. At the time of a flood, it receives a stream of images from different users on a real-time basis. The activities of the server-side application are pictorially presented in Fig. 2. Mainly, it performs three major tasks on the images it receives. At first, it checks if the image falls into the category of flood-image using a deep learning technique. We call an image as a flood-image if it has some amount of water presence. If the images are found to be of flood-image category, in the next step, they are clustered based on their associated location information. Finally, it calculates the boundaries and identifies the flash flood hotspots.

2.2.1 Flood-Image Detection

It is observed [10] that several images posted by news agencies or in social media do not convey any pictorial information about the flood event. In flood-related disasters, several images are circulated over the Internet. However, only near about half the number of images are relevant [10] to the event. No doubt the same kind of scenarios might occur in the context of the proposed FFHI application. Without understanding the true motive behind this application, a user may post many images that are not related to flood. Thus, these images, once having been uploaded to the server, undergo a verification process. The images are passed through a deep learning model for classification. Here, we have proposed a flood-image-detector model using the transfer learning approach using a pretrained VGG19 [17] network. The proposed flood-image-detector model is discussed separately in Sect. 2.3.

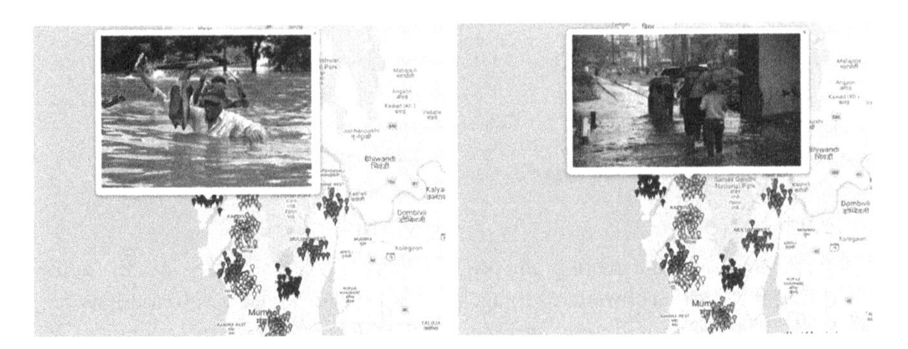

Fig. 3 Examples of image displayed on clicking a marker and corresponding info-box

Once an image is classified as flood-image, then in the next step, those are used to identify flash flood hotspots. Note that an image can be classified as a flood-image based on the amount of floodwater present within the image. The basic assumption here is that the severity of the flood is proportional to the amount of floodwater present in the image. The images which have floodwater content less than a specified threshold are ignored, while the rest can be viewed on the map. If a user wants to view images from different flood-affected regions on the map, the server provides the required information to the android app. Each marker created on the map for an image has an on-click event in the android app, which opens the associated image in an *info-box*. Figure 3 shows examples of a couple of info-boxes, where the respective posted image is displayed on clicking the marker.

2.2.2 Clustering of Images

We carry out the clustering operation to divide the flood-affected areas into various zones. This would facilitate the local authorities to split into respective teams to carry out rapid relief measures. This clustering operation is performed by the server periodically, such that whenever some finite number of new authentic flood-images are posted, the clustering operation is repeated along with all the previous images. In the proposed system, we carry out the process once ten (10) new authentic images are posted. The number of clusters is selected dynamically as we use the DBSCAN algorithm [4] for the clustering. This algorithm does not depend upon the shape of the clusters. For a new cluster to form, the minimum number of samples has to be thirty (30). In our FFHI application where we have worked with geo-coordinates, the parameters used for the DBSCAN algorithm are as follows: (i) Distance measure = "Haversine," (ii) Min-samples = 30, and (iii) Epsilon = 2 (km)/R, where $R = 6373$ km is the radius of the earth. Here, the '*ball tree*' algorithm is used while performing DBSCAN clustering. It reduces the number of candidate points which are required for neighbor search by using the triangle inequality. However, the said val-

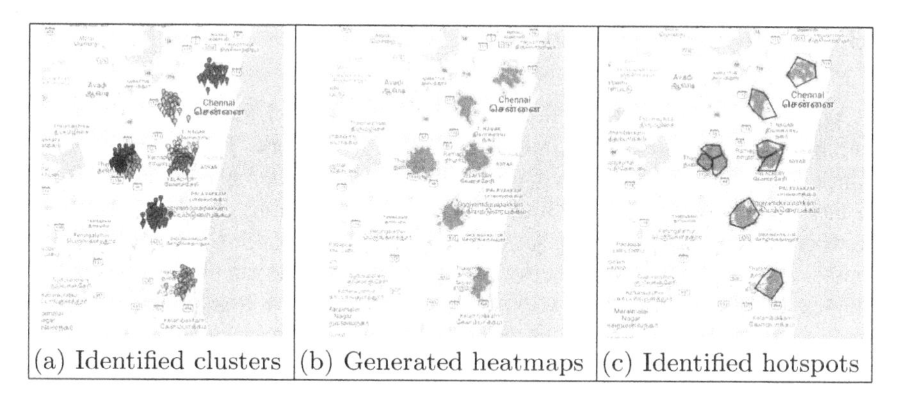

(a) Identified clusters (b) Generated heatmaps (c) Identified hotspots

Fig. 4 Identified clusters, heatmaps, and hotspots for a flood event (plotted on map of Chennai city)

ues for those parameters can be tuned keeping the number of target hotspots in mind. Once the DBSCAN algorithm performs the clustering, it returns the total number of identified clusters. Using this number, further K-means [11] clustering is performed, which gives the cluster centroids. Figure 4a provides a visual representation of the clusters plotted on a map. Note that the cluster centroids are useful to find the center of the affected regions. This enables people to roughly get an idea of the names of the affected colonies or localities within the city. Using *reverse geocoding* method, the set of cluster centroid coordinates is sent to the "Geocoding API" to get the name of the localities.

2.2.3 Generating Heatmap and Boundaries to Discover "Hotspots"

The heatmap of affected regions provides a measure of severity of the catastrophe. By default, areas of higher intensity will be colored red, and areas of lower intensity will appear green. The heatmaps corresponding to the identified clusters (Fig. 4a) are plotted on a map and presented in Fig. 4b. By generating boundaries to the existing clusters of affected regions, it can define certain regions as hotspots. These hotspots can be immediately cordoned off to allow only rescue personnel and health workers to access the area to provide relief. Delineation of flood risk hotspots is a very challenging task for the authorities which have been automated in this work. The detailed boundary can be used as a reference by civic authorities, power and fire department, municipal corporation, and other local authorities. Figure 4c shows the identified hotspots for the respective heatmaps shown in Fig. 4b. Boundaries are created with the *Graham Scan algorithm* [6] which is used to find the convex hull of a set of points. It creates the smallest polygon encompassing all the coordinates.

2.3 Flood-Image-Detector: CNN Model

We propose a CNN-based [9] flood-image-detector model following the transfer learning approach using a pretrained VGG19 [17] network. This CNN model is trained using two different kinds of image samples: *floodwater* and *non-water*. Here, we have used a similar approach to generate several training samples as stated in the work [10]. We use a training sample dimension of $48 \times 48 \times 3$. A total of 70 K samples are generated in this process, out of those half of the samples are *floodwater* samples. While training the network, 50 K samples are used. Remaining 20 K samples are used for validating and testing the flood-image-detector model in an equal ratio.

2.3.1 CNN Model Architecture

Many existing works suggest the use of conventional image segmentation algorithms [5, 18], while some suggest training all the layers of a CNN model [3, 10]. However, we use a transfer learning [14, 16] approach to create our model. With transfer learning, instead of starting the learning process from the very beginning, we start from patterns that have been learned when solving a different problem. This method is faster and computationally inexpensive. Related works [7, 15] suggest the use of pretrained models, especially the VGG network for flood detection. We use the VGG19 [17] pretrained network in our server-side application which performed well in the classification task. However, we did not perform a comprehensive analysis of all the pretrained models, and there might be models that perform the classification task more efficiently. We found out that freezing some layers while unfreezing others performed best. Our model has the first four blocks of convolution layers frozen while the last convolution block (having four convolution layers) unfrozen. Two fully connected layers are added after the stack of convolution layers, the first having 64 channels (chosen empirically). The second fully connected layer has two channels which is followed by the softmax [2] function used for classification into *floodwater* and *non-water*. In the training phase, first, the pretrained weights of the network are loaded. We train the network with training samples of dimension $48 \times 48 \times 3$ over 50 epochs. First, with a high learning rate to quickly converge to a near maximal position and subsequently with a lower learning rate to fine-tune the network. In each epoch, the network is trained in batches of 500 with 100 samples each. After each epoch, the network is cross-validated with 10 K samples. Overall training accuracy achieved in this model is 97%. Table 1 describes the layers in the network. Frozen indicates that the layers are not trainable while unfrozen is trainable.

Table 1 Pretrained VGG19 network with two fully connected layers

Serial no.	Layer type	Filter dimension	Number of filters	Output dimension	Layer status
1	Input	–	–	48 ∗ 48 ∗ 3	Frozen
2	First 12 convolution layers	–	–	3 ∗ 3 ∗ 512	Frozen
3	block5_conv1	3 ∗ 3 ∗ 3	512	3 ∗ 3 ∗ 512	Unfrozen
4	block5_conv2	3 ∗ 3 ∗ 3	512	3 ∗ 3 ∗ 512	Unfrozen
5	block5_conv3	3 ∗ 3 ∗ 3	512	3 ∗ 3 ∗ 512	Unfrozen
6	block5_conv4	3 ∗ 3 ∗ 3	512	3 ∗ 3 ∗ 512	Unfrozen
7	block5_pool (MaxPooling2D)	2 ∗ 2	–	1 ∗ 1 ∗ 512	Unfrozen
8	Fully connected (64)	–	–	1 ∗ 1 ∗ 64	Unfrozen
9	Fully connected (2)	–	–	1 ∗ 1 ∗ 2	Unfrozen
10	Softmax (output)	–	–	1 ∗ 1 ∗ 2	Unfrozen

2.3.2 Flood Region Detection in Image

In this section, the process of identifying flood-images is described. Here, it is checked if an input image has any floodwater presence. It is done mostly in two steps. At first, the response matrix is generated for an input image using a sliding window technique. Finally, it calculates the percentage of floodwater presence in the image. Based on that, we classify the image into a flood-image or any other.

Response Matrix Generation: We use an overlapping sliding window approach to generate the response matrix for the input image. The size of the window is $48 \times 48 \times 3$, and stride used here is eight pixels (both vertical and horizontal). As a preprocessing step, the image is zero-padded with eight pixels on all sides of the images. Steps for response matrix generation process are as follows:
Step-1: The response matrix is a 2D matrix that has the same dimensions as the input image (with zero-padded).
Step-2: Each image patch under sliding window of dimension $(48 \times 48 \times 3)$ is fed to the flood-image-detector model. The output (y_pred) is a prediction of value 1 (*floodwater*) or 0 (*non-water*). Here, we create an output matrix of dimension 48×48, and output value (y_pred) is set to each of its elements. The equation used to calculate each element of the output matrix is as follows:

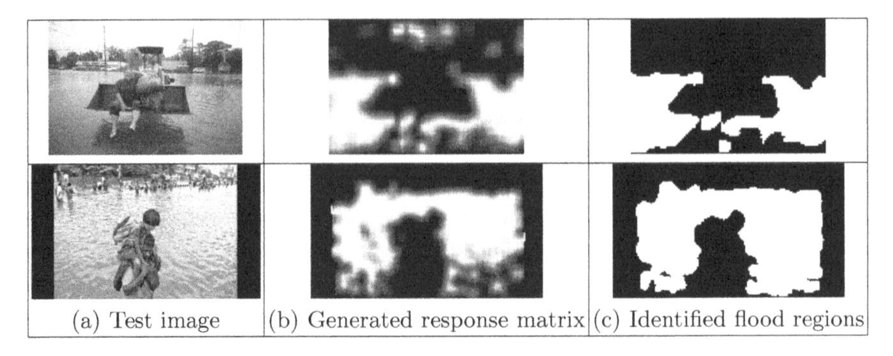

Fig. 5 Pictorial representation of flood-image detection process

$$y\_pred = \begin{cases} 0, & \text{if image patch} = \textit{non-water} \\ 1, & \text{if image patch} = \textit{flood water} \end{cases} \tag{1}$$

Step-3: Each element of the output matrices under the sliding window is cumulatively added to the respective elements of the response matrix.
Step-4: After predicting all the image patches in the sliding window process, the values of the elements of response matrix are normalized using the highest value of matrix elements. The values in the response matrix after this normalization process lie within the range [0, 1]. The padded elements are then discarded to obtain the final response matrix.

The flood-image detection process for two different input images is pictorially illustrated in Fig. 5. Figure 5b shows the generated response matrix for the corresponding input images in Fig. 5a.

Flood Region Identification and Classification: Finally, the response matrix is binarized to obtain the floodwater regions in an input image. Here, we use Otsu's algorithm for image binarization which divides the image histogram into two classes. It uses a threshold in such a way that the in-class variability is very small. Thus, the threshold value is generated dynamically in the binarization process. Figure 5c shows the binarized image for the corresponding input images in Fig. 5a. The white portion of the image is considered as floodwater regions.

After the image binarization, the percentage of the image covered by floodwater is calculated as $\frac{\text{no. of floodwater pixels}}{\text{total no. of pixels}} * 100$. If it is more than some threshold percentage, then we conclude that the input image is of the flood-image category. In this work, empirically, we have considered the value of the threshold percentage as 15%. Once an image is categorized as a flood-image, it is included for further processing, such as clustering and generating boundaries to discover hotspots as discussed in Sect. 2.2.

2.4 System Implementation Strategy

We implemented the android application using android studio. The user is provided with the options of uploading flood-images and viewing the various visualizations by entering the city name which he wishes to view. The *Express Server* is used for developing the server-side application. It runs on the HTTP port to handle the requests from the android app. The JSON format is used for transportation of information between the android app and server-side application. In the proposed system, the flood-image-detector is implemented as a *Python module* which is developed using the *TensorFlow* [1] (version 1.9.0) Python library. The express server acts as a mediator between the android app and the Python module. It receives a stream of images from the android app, those are classified using the flood-image-detector. Once a stream of images reaches the server endpoint, a shared queue of incoming image streams is maintained. An optimal number of processes are spawned depending upon the load in the server. The spawned processes concurrently use the Python module (flood-image-detector) to classify respective images. Once the express server receives the response from the Python module, it returns the response to the android app in JSON format, thus ensuring two way communication.

Apart from the flood-image-detector, the Python module is also responsible for implementing required functionalities related to flood hotspots visualization. This visualization scripts do the jobs of performing different clustering techniques and identifying hotspots boundaries. It also propagates required information to the express server periodically, so that server can respond to the requests from android app related to visualizations (clusters, hotspots, etc.) on map.

3 Simulation and Experiments

We have not found any flood-image dataset where images are having metadata with time and location information intact. Therefore, we simulated a couple of scenarios to test our application. The city of Mumbai and Chennai were selected for this purpose. We prepared two datasets by scraping images of major flood havocs in recent years in those cities. Many images were scraped from the Internet using the *Selenium web driver* and *BeautifulSoup* Python library.

We scraped approximately five hundred flood-images for each of these two cities. However, none of these images have location information present. Therefore, to simulate a real calamity scenario, we tweaked the metadata of the images and assigned them coordinates. We identified certain regions in those cities which are low-lying areas or are prone to floods. For example, in Mumbai, we selected the areas of Andheri, Kurla, Sion, Juhu, Santacruz, etc. Coordinates of these locations were taken as cluster centroids. All the scraped images were distributed randomly to each of the clusters, such that every cluster had a minimum of thirty images and a maximum of seventy images. The centroid of the clusters is defined as cluster_centroid =

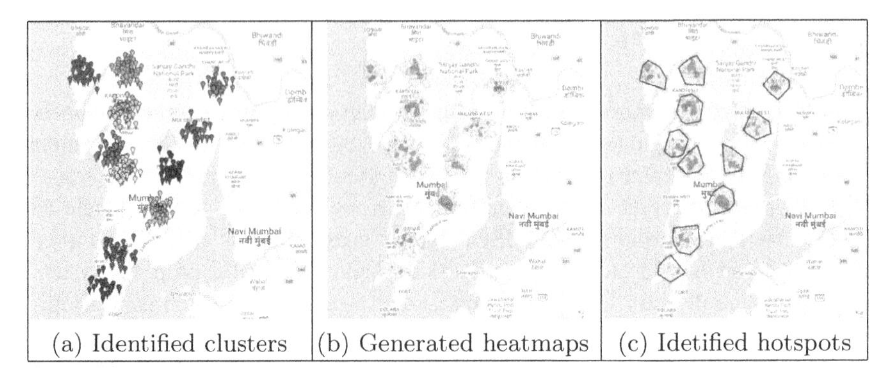

(a) Identified clusters (b) Generated heatmaps (c) Idetified hotspots

Fig. 6 Identified clusters, heatmaps, and hotspots for a flood event (plotted on map of Mumbai city)

[x_coordinate y_coordinate], where x_coordinate is the latitude of the centroid of the cluster, and y_coordinate is its longitude. An image belonging to a particular cluster was assigned coordinates randomly within a square of side 4 km with cluster_centroid as its centroid. The random values for the coordinates were taken from a normal distribution such that each pair of latitude and longitude lied inside the defined cluster.

Once the coordinates were assigned to the images, we ran this simulation process for these two cities (Chennai and Mumbai) separately. The observed clusters, corresponding heatmaps, and hotspots for the Chennai city are already shown in Fig. 4. It can be seen there that a total number of nine hotspots are identified. Similarly, for the city of Mumbai, simulation results are shown in Fig. 6. In this case, the number of hotspots identified is eleven.

4 Conclusion

In this era of social media, unreliable data leads to a lot of panic and rumor. The proposed flash flood hotspots identification application developed in this work effectively filters out the farce images and provides authentic data to the public domain. We have proposed a methodology to evaluate a flood-image using the transfer learning and image segmentation approach. The overall application delivers lucid visualization and detailed analysis of the affected clusters with the help of unsupervised machine learning algorithms. Citizens across the city can be cautioned to restrain themselves from traveling to such disaster affected regions. The localities identified as hotspots by this application would quickly alert the rescue personnel, civic volunteers, and municipal corporation to carry out relief operations in those zones. The ease of use, responsive nature, and prompt flow of real-time information directly from the men on ground makes our FFHI application unique and highly beneficial. The data collected over a period can be analyzed to locate certain flood-prone areas that require robust infrastructure, road maintenance, and stormwater drainage facilities. We can also identify the strategic location of emergency shelter homes. One of our future research intentions is to develop an algorithm for citizens to navigate to a destination by avoiding travel through the flood-affected hotspots, hence ensuring a safe journey. The FFHI application is an endeavor to generate awareness among the public to share information and show eagerness to contribute toward the benefit of society.

References

1. Abadi M et al (2015) TensorFlow: large-scale machine learning on heterogeneous systems. http://tensorflow.org
2. Bishop CM (2006) Pattern recognition and machine learning. Springer
3. Chaudhary P, D'Aronco S, de Vitry MM, Leitão JP, Wegner JD (2019) Flood-water level estimation from social media images
4. Ester M, Kriegel HP, Sander J, Xu X (1996) A density-based algorithm for discovering clusters in large spatial databases with noise. In: Proceedings of the second international conference on knowledge discovery and data mining, KDD96. AAAI Press, pp 226–231
5. Filonenko A, Wahyono, Hernndez DC, Seo D, Jo K (2015) Real-time flood detection for video surveillance. In: IECON 2015—41st annual conference of the IEEE industrial electronics society, pp 004082–004085
6. Graham RL (1972) An efficient algorithm for determining the convex hull of a finite planar set. Inf Process Lett 1:132–133
7. Jain P, Schoen-Phelan B, Ross R (2020) Automatic flood detection in Sentinel-2 images using deep convolutional neural networks. In: Proceedings of the 35th annual ACM symposium on applied computing, SAC 20. Association for Computing Machinery, New York, NY, pp 617–623. https://doi.org/10.1145/3341105.3374023
8. Jyh-Horng W, Chien-Hao T, Lun-Chi C, Lo SW, Lin FP (2015) Automated image identification method for flood disaster monitoring in riverine environments: a case study in Taiwan. https://doi.org/10.2991/iea-15.2015.65
9. Krizhevsky A, Sutskever I, Hinton GE (2017) ImageNet classification with deep convolutional neural networks. Commun ACM 60(6):84–90. https://doi.org/10.1145/3065386
10. Layek AK, Poddar S, Mandal S (2019) Detection of flood images posted on online social media for disaster response. In: 2019 second international conference on advanced computational and communication paradigms (ICACCP)
11. MacQueen J (1967) Some methods for classification and analysis of multivariate observations. In: Proceedings of the fifth Berkeley symposium on mathematical statistics and probability, volume 1: statistics. University of California Press, Berkeley, pp 281–297. https://projecteuclid.org/euclid.bsmsp/1200512992
12. Menon KP, Kala L (2017) Video surveillance system for realtime flood detection and mobile app for flood alert. In: 2017 international conference on computing methodologies and communication (ICCMC), pp 515–519
13. Napiah MN, Idris MYI, Ahmedy I, Ngadi MA (2017) Flood alerts system with android application. In: 2017 6th ICT international student project conference (ICT-ISPC), pp 1–4
14. Sakai T, Tamura K, Kitakami H, Takezawa T (2017) Photo image classification using pre-trained deep network for density-based spatiotemporal analysis system. In: 2017 IEEE 10th international workshop on computational intelligence and applications (IWCIA), pp 207–212
15. Sazara C, Cetin M, Iftekharuddin KM (2019) Detecting floodwater on roadways from image data with handcrafted features and deep transfer learning. In: 2019 IEEE intelligent transportation systems conference (ITSC), pp 804–809
16. Shaha M, Pawar M (2018) Transfer learning for image classification. In: 2018 second international conference on electronics, communication and aerospace technology (ICECA), pp 656–660

17. Simonyan K, Zisserman A (2014) Very deep convolutional networks for large-scale image recognition. https://arxiv.org/abs/1409.1556
18. Zhang Q, Jindapetch N, Duangsoithong R, Buranapanichkit D (2018) Investigation of image processing based real-time flood monitoring. In: 2018 IEEE 5th international conference on smart instrumentation, measurement and application (ICSIMA), pp 1–4
19. Zhou Z, Wu QJ, Huang F, Sun X (2017) Fast and accurate near-duplicate image elimination for visual sensor networks. Int J Distrib Sens Netw 13(2)

Infrastructure and Resource Development and Management Using Artificial Intelligence and Machine learning

An Optimized Controller for Zeta Converter-Based Solar Hydraulic Pump

K. Sudarsana Reddy, B. Sai Teja Reddy, K. Deepa, and K. Sireesha

Abstract Due to advancements in renewable technology, the agricultural sector can independently harvest its energy for running its respective equipments. One such equipment being hydraulic pump that waters the field can be run by solar array. This whole mechanism can be controlled through many present-day control techniques like soft computing techniques. The main aim of the present work is to obtain the best controller technique among fuzzy and genetic algorithm for the fast response of set value according to the climatic conditions and nature of field area. A mathematical model of zeta converter has been provided for studying the performance of the control techniques. This mathematical modelling of the respective converter has been done through state space averaging technique (SSA). This work even contributed a comparative study of zeta and SEPIC converter for its respective performance. These converters are chosen so that the output will be maintained at constant voltage for the range of input voltage. The work has been simulated in MATLAB/Simulink software for the respective study.

Keywords Solar array · Single-ended primary inductor converter (SEPIC) · Zeta converter · Genetic algorithm (GA)-tuned PI controller · Fuzzy logic controller · Hydraulic pump · State space averaging technique (SSA)

1 Introduction

Nowadays, the demand for electricity has increased. So, production has to be increased. Due to reduction in fossil fuels, moving towards renewables is the best alternative [1]. Electricity is generated mainly from solar energy among all renewable

K. Sudarsana Reddy (✉) · B. Sai Teja Reddy · K. Deepa · K. Sireesha
Department of Electrical and Electronics Engineering, Amrita School of Engineering, Amrita Vishwa Vidyapeetham, Bengaluru, India

K. Deepa
e-mail: k_deepa@blr.amrita.edu

K. Sireesha
e-mail: k_sireesha@blr.amrita.edu

A. Choudhary et al. (eds.), *Applications of Artificial Intelligence and Machine Learning*, Lecture Notes in Electrical Engineering 778,
https://doi.org/10.1007/978-981-16-3067-5_30

energies. It is low-maintenance, eco-friendly and has a less operational cost. With the help of photovoltaic cells, they can convert solar energy directly into electrical energy. As a result, the workforce can be reduced compared to conventional energy production technology. Photovoltaic arrays are a combination of several PV cells. Voltage is increased by connecting cells in series, and the current is increased by connecting in parallel. In the present work, hydraulic pump is being supplied by PV array [2–4].

The most efficient way to increase or decrease the DC voltage is by using a DC–DC converter [5]. There are different converters like a buck, boost, buck–boost [6], cuk, SEPIC [7], etc. Depending on the applications, the converters buck or boost the voltage. But some applications require a constant voltage if there is a variable input voltage, and this can be achieved by using the buck–boost converter, but when it is in use this can result in a strong ripple in the output, and this can be reduced by using a cuk converter which inverts output. For overcoming these disadvantages, SEPIC and Zeta converters are into the study which provides output voltage of same polarity as the input voltage. Due to the property of having continuous current in the input capacitor, the input capacitance for lower ripple voltage is reduced. These act as a buck converter when the duty cycle is less than 0.5 and act as a boost converter when the duty cycle is greater than 0.5. These have an efficiency in the ranges of 91–96%. A controller is needed to maintain the constant output from the converter. In the present work, controller is used to control the duty ratio of the zeta converter. Any PI controller [8] takes error as the input and controls the output. The output of the PI controller is the sum of the integration coefficients and the proportion. Increase in proportion constant reduces the steady-state error. If the proportion gain is high, the system may go into unstable state. The control action may go slow if the gain is too less. Proportion gain plays a major role in changing the output value. Magnitude of error and duration of error are proportional to integral term. It is the sum of error over time and gives accumulated error, and this is multiplied to integral gain. Integral gains fasten the process towards the set point. Since it responds to accumulated error, the present values may over shoot. Proper selection of the proportional gain and integral gain makes the system perfect. But tuning of these values manually is a tedious process and may not be perfect, and thereby the process of genetic algorithm [9] is followed to properly tune the gains of the system. The respective response of the system has been tested with the fuzzy logic controller [10–12]. This zeta converter has to be mathematically modelled for studying its controller parameters for different control techniques. This has been done through SSA technique [13, 14].

The aim of the present work is to mathematically model the zeta converter and study its performance in open loop and closed loop with respective to GA-tuned PI controller and fuzzy logic controller, respectively. The best-performed controller is chosen to run solar run hydraulic pump. For Indian climatic conditions, it is required to have a controller which can control the motor speed accordingly has to be chosen by this study. An extension to the present works of a comparative study between SEPIC and zeta has been provided. The organization of the paper has been done in six sections. Section 2 is the system and specifications along with its required modelling. Section 3 details about the control algorithms. Section 4 discusses the

Fig. 1 Block diagram

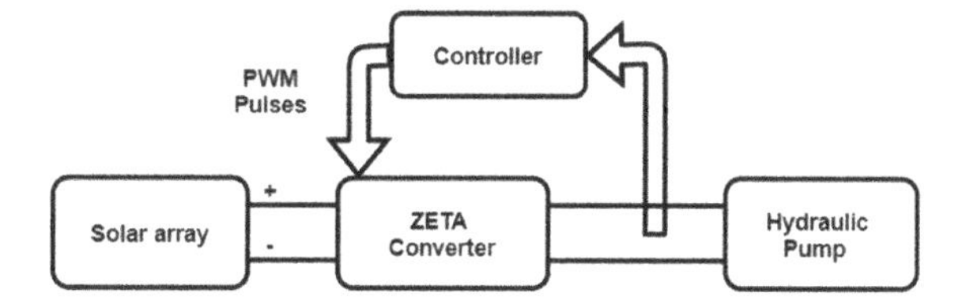

simulation and its results. Section 5 is dedicated for analysing the lookup table for the set values along with comparison of zeta and SEPIC converters. Section 6 concludes the work.

2 System and Specifications

2.1 *Block Diagram of Proposed System*

The proposed block diagram of entire system in which the 800 W hydraulic pump is being supplied by using solar subsystem through zeta converter is shown in Fig. 1. Zeta converter is used to maintain constant voltage at the load with the minimal variation in input. The desired voltage is obtained from the zeta converter by using the controller. In this present work, two controller techniques have been studied for understanding better performance of the system.

2.2 *PV Array*

The 800 W power required to supply load is generated by PV array maintaining 1000 W/m^2 irradiance and 25 °C temperature. Eoplly New Energy Technology EP156M-60-250W module is used which has an open-circuit voltage of 37.26 V, and the short-circuit current of 8.91 A and has the maximum power of 251.346 W. Figure 2 details the MPP curve and V–I characteristics of the PV array at different irradiance. The specifications of the PV array are tabulated in Table 1 (column 1–2).

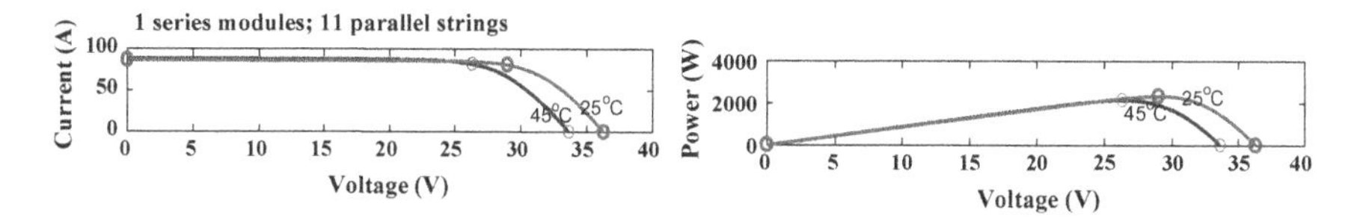

Fig. 2 V–I characteristics and MPP curve of solar array simulated

Table 1 Specifications and parameters of solar array along with zeta converter

Parameters of solar array	PV array values	Parameters of zeta converter	Zeta converter values
Short-circuit current I_{sc} (A)	8.91	Output capacitor (C_2) in F	4.7577×10^{-4}
Open-circuit voltage V_{oc} (V)	37.26	Flyback capacitor (C_1) in F	9.3602×10^{-4}
Irradiance (W m^{-2})	1000	Inductor ($L_1 = L_2$) in H	283.14×10^{-9}
Maximum power (W)	251.346	Load resistance (Ω)	0.1791
Series module per string	1	Maximum duty cycle	0.44
Parallel strings	11	Minimum duty cycle	0.57
Current at MPP I_{mp} (A)	8.15	ΔV_{out} (V)	50
Voltage at MPP V_{mp} (V)	30.84	ΔI_l (A)	26.691

2.3 ZETA Converter

By switching MOSFET, constant output voltage is maintained by transferring energy between inductors and capacitors. The circuit diagram of the zeta converter is shown in Fig. 3a. It can be operated in both buck and boost modes by adjusting duty ratio.

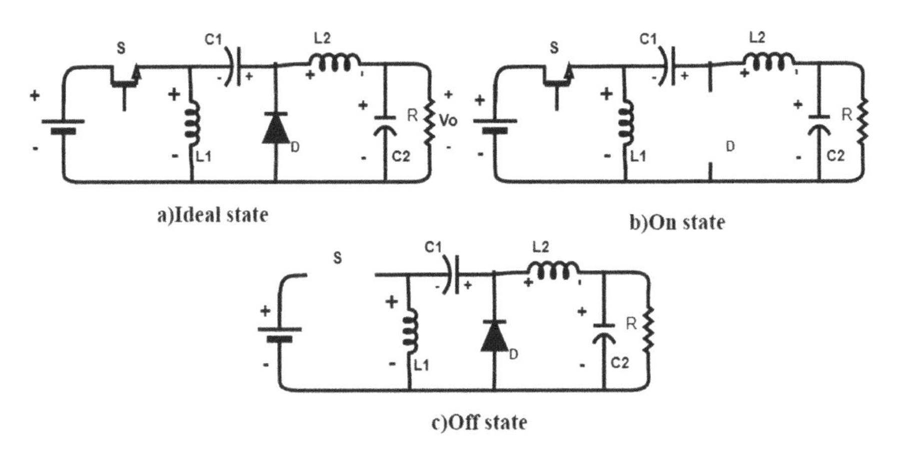

Fig. 3 Different modes of operation of zeta converter

Working and design

When switch S is ON. As the switch is ON, the voltage across the diode D is negative, and it will be in reverse bias. The voltages across the inductors L_1 and L_2 are equal to input voltage Vi. *Both* the inductors are being energized by input voltage, their currents I_{L1} and I_{L2} are increased linearly, and there by switch current also increases. During this time, the output capacitor gets charged and the flyback capacitor discharges. Equivalent circuit of the zeta converter when switch is ON is shown in Fig. 3b.

When switch S is OFF. As the switch is OFF, the diode D is forward biased. The voltage across the inductor is $-Vo$, L_1 supplies energy to $C_{1,}$ and L_2 supplies load. The current linearly decreases in inductors L_1 and L_2. The voltage across the switch will be $Vin + Vo$. Equivalent circuit of the zeta converter when switch is OFF is shown in Fig. 3c. Zeta converter can be operated in continuous conduction mode with given:

$$D = \frac{V_0}{V_0 + V_i}, \quad D_{max} = \frac{V_0}{V_0 + V_{i(min)}}, \quad D_{min} = \frac{V_0}{V_0 + V_{i(max)}} \tag{1}$$

where V_0 is the output voltage, V_i is the input voltage, and D is the duty ratio. D_{max} is the maximum duty ratio, and D_{min} is the minimum duty ratio.

$$\begin{aligned} &\Delta I_{L(p-p)} = 0.3 \times \frac{D}{1-D} \times I_0, \quad I_{L(p-p)} = \frac{V_{i\,max} \times D_{max}}{2 \times L1 \times f_{sw(min)}}, \\ &L_1 = L_2 = \frac{V_i \times D_{v\,min}}{2 \times \Delta I_{L(p-p)} \times f_{sw(min)}} \end{aligned} \tag{2}$$

I_0 is the output current. $\Delta I_{L(p-p)}$ is the output current ripple.

$$C_1 = \frac{I_o \times D_{max}}{\Delta V_{ripple} \times f_{sw(min)}}, \quad C_2 = \frac{I_{L(p-p)}}{8 \times 0.025 \times f_{sw(min)}} \tag{3}$$

where ΔV_{ripple} is taken to be 1% of output voltage. C_2 is the output capacitance, C_1 is the flyback capacitance, and the minimum and maximum switching frequencies are taken as 340 kHz and 460 kHz, respectively. Equations (1)–(3) represent the design of the zeta converter. Parameter values of the zeta converter are specified in Table 1.

Small signal modelling for SSA technique. MOSFET operates in two states. It is ON for dT time and OFF for $(1 - d) * T$ where d is the duty cycle. The steady-state parameters of the circuit when switch is ON and OFF are given by G_1 and G_2, where G can be A, B, V and Z, where $\dot{p}(t)$ represents differentiation of p with respect to t. Weighted average equations are given by (4).

$$\begin{cases} \dot{x}(t) = A_w x(t) + B_w u(t) \\ y(t) = V_w x(t) + Z_w u(t) \end{cases} \tag{4}$$

where $A_w = A_1 d + A_2(1-d)$ and similarly remaining can be found. The averaged equation is nonlinear continuous time equations. Linearized equations can be found by small signal perturbation by considering. $x = X + \hat{x}$ where $\hat{x}$ represents small signal, and X represents DC value. Steady-state solutions can be found by (5), respectively.

$$\begin{aligned} &X = -A^{-1}BU, \quad Y = \left(-VA^{-1}B + Z\right)U, \\ &\hat{x}(s) = \left[(sI-A)^{-1}B \quad (sI-A)^{-1}B_d\right]\begin{bmatrix}\hat{u}(s)\\ \hat{d}(s)\end{bmatrix}, \\ &\hat{y}(s) = \left[V(sI-A)^{-1}B + Z \quad V(sI-A)^{-1}B_d + Z_d\right]\begin{bmatrix}\hat{u}(s)\\ \hat{d}(s)\end{bmatrix} \end{aligned} \tag{5}$$

where $G = G_1 D + G_2(1-D)$, where G can be A, B, V and Z, $B_d = (A_1 - A_2)X + (B_1 - B_2)U$, $Z_d = (V_1 - V_2)X + (Z_1 - Z_2)U$.

The state space equations for ON and OFF states can be written as stated in (6)

$$\begin{aligned} &\dot{i_{L1}} = \frac{V_i}{L_1}\Delta, \; \dot{i_{L2}} = \frac{V_i}{L_2}\Delta, \; \dot{v_{C1}} = \frac{i_{L1}}{C_1}(1-\Delta) - \frac{i_{L2}}{C_1}\Delta, \\ &\dot{v_{C2}} = \frac{i_{L2}}{C_2} - \frac{v_{C2}}{RC_2} - \frac{i_0}{C_2}, \; V_0 = V_{C2}. \end{aligned} \tag{6}$$

The equations in the (6) are written as a function of Δ. $\Delta = 1$ represents the switch is ON, the equations represents ON state equations, and when $\Delta = 0$ the switch is OFF, respective equations represent OFF state equations. The averaged matrices are given by (7)–(9).

$$\begin{aligned} &A = D + A_2(1-D) = \begin{bmatrix} 0 & 0 & \frac{-1}{L_1}(1-D) & 0 \\ 0 & 0 & \frac{D}{L_2} & \frac{-1}{L_2} \\ \frac{1}{C_1}(1-D) & \frac{-D}{C_1} & 0 & 0 \\ 0 & \frac{1}{C_2} & 0 & \frac{-1}{C_2 R} \end{bmatrix} \\ &Z = Z_1 D + Z_2(1-D) = \left[0\,0\,0\right] \end{aligned} \tag{7}$$

$$B = B_1 D + B_2(1-D) = \begin{bmatrix} \frac{D}{L_1} & 0 \\ \frac{D}{L_2} & 0 \\ 0 & 0 \\ 0 & \frac{-1}{C_2} \end{bmatrix}, \; V = V_1 D + V_2(1-D) = \left[0\,0\,0\,1\right], \tag{8}$$

$$\begin{aligned} &B_d = (A_1 - A_2)X + (B_1 - B_2)U = \begin{bmatrix} \frac{V_i}{(1-D)L_1} \\ \frac{-V_i}{(1-D)L_2} \\ \frac{-DV_i - RI_0(1-D)}{(1-D)^2 RC_1} \\ 0 \end{bmatrix} \\ &Z_d = (C_1 - C_2)X + (Z_1 - Z_2)U = [0] \end{aligned} \tag{9}$$

The state space equations can be formulated from (10) and (11).

$$\begin{bmatrix} \dot{i}_{L1}(t) \\ \dot{i}_{L2}(t) \\ \dot{v}_{C1}(t) \\ \dot{v}_{C2}(t) \end{bmatrix} = \begin{bmatrix} 0 & 0 & \frac{-1}{L_1}(1-D) & 0 \\ 0 & 0 & \frac{D}{L_2} & \frac{-1}{L_2} \\ \frac{1}{C_1}(1-D) & \frac{-D}{C_1} & 0 & 0 \\ 0 & \frac{1}{C_2} & 0 & \frac{-1}{C_2 R} \end{bmatrix} \begin{bmatrix} i_{L1}(t) \\ i_{L2}(t) \\ v_{C1}(t) \\ v_{C2}(t) \end{bmatrix} + \begin{bmatrix} \frac{D}{L_1} & 0 & \frac{V_i}{(1-D)L_1} \\ \frac{D}{L_2} & 0 & \frac{-V_i}{(1-D)L_2} \\ 0 & 0 & \frac{-DV_i - RI_0(1-D)}{(1-D)^2 RC_1} \\ 0 & \frac{-1}{C_2} & 0 \end{bmatrix} \begin{bmatrix} V_i(t) \\ I_0(t) \\ d(t) \end{bmatrix}, \tag{10}$$

$$V_0(t) = \begin{bmatrix} 0 & 0 & 0 & 1 \end{bmatrix} \begin{bmatrix} i_{L1}(t) \\ i_{L2}(t) \\ v_{C1}(t) \\ v_{C2}(t) \end{bmatrix} + \begin{bmatrix} 0 & 0 & 0 \end{bmatrix} \begin{bmatrix} V_i(t) \\ I_0(t) \\ d(t) \end{bmatrix} = v_{C2}(t) \tag{11}$$

Duty ratio-to-output voltage transfer function can be found from (10) and (11), and the transfer function is given by (12), respectively.

$$\begin{aligned} T_{dv}(s) &= \frac{V_0(s)}{d(s)} = C(sI - A)^{-1}Bd + Ed \\ &= \frac{\left(r_1 s^2 + r_2 s + r_3\right)(r_4 s + 1)}{(1-D)^2\left(r_5 s^4 + r_6 s^3 + r_7 s^2 + r_8 s + r_9\right)} \end{aligned} \tag{12}$$

$r_1 = V_i(1-D)RL_1C_1$, $r_2 = -V_iL_1D^2 - L_1D(1-D)RI_0$, $r_3 = V_i(1-D)^2R$, $r_4 = 0$, $r_5 = RL_1C_1L_2C_2$, $r_6 = L_1C_1L_2$, $r_7 = RL_2C_2(1-D)^2 + RL_1C_1 + RL_1C_2D^2$, $r_8 = L_2(1-D)^2 + RL_1C_1 + L_1D^2$.

2.4 Motor

DC motor-coupled hydraulic pump is used in agriculture to water the plants. DC shunt motor has been used in, and the results are simulated. The DC motor used is supplied by solar array through zeta converter. The physical parameters for which the design is made have been tabulated in Table 2.

Table 2 Physical parameter values of hydraulic pump

Parameter	value	Parameter	value
Voltage (V)	12–24	Hydraulic power unit (dimensions)	27.5 L * 11.8 W * 12 H
Motor power (kW)	0.8–3	Rotation	Clock wise
Displacement of hydraulic pump GPM	0.825	Relief valve pressure (MPa)	22
Protection degree	IP54	Rated speed (R/min)	2850
Size name	8quart double	Weight (kg)	3

3 Control Technique

3.1 Open Loop

The mathematical model of the zeta converter has been discussed in Sect. 2. The respective transfer function of zeta converter after SSA technique has been provided by (12). These specifications provided in Table 1 (column 2–4) have been substituted in (12), and hence, the open-loop step response of this transfer function has been evaluated with the help of MATLAB software. This respective response can be depicted in Fig. 4a. This underdamped response has settling time of 4 ms with undesirable control parameters like peak overshoot of 60% and steady-state error of 6%. So, a controller is required which can mitigate these undesirable quantities and perform well under any environment. This open-loop plant transfer function which is given by (13) has been connected in unity feedback, and two controller schemes are implemented to control the parameters. The respective controller schemes utilized in this work are genetic algorithm-tuned PI controller and fuzzy logic controller. The comparative study between these schemes with respective to zeta converter has been done, and the better performing has been considered for the further analysis.

$$\frac{\mathrm{Vo}(s)}{d(s)} = \frac{1.837\mathrm{e}-10s^2 - 1.536\mathrm{e}-06s + 0.298}{1.182\mathrm{e}-21s^4 + 7.504\mathrm{e}-17s^3 + 5.977\mathrm{e}-11s^2 + 1.443\mathrm{e}-07s + 0.03312} \tag{13}$$

3.2 Genetic Algorithm-Tuned PI Controller

The plant transfer function has been cascaded with PI controller transfer function whose gains are being continuously tuned by GA through its respective objective function until the termination condition reaches. This corresponding cascaded transfer function has been connected in unity feedback, and the closed-loop response

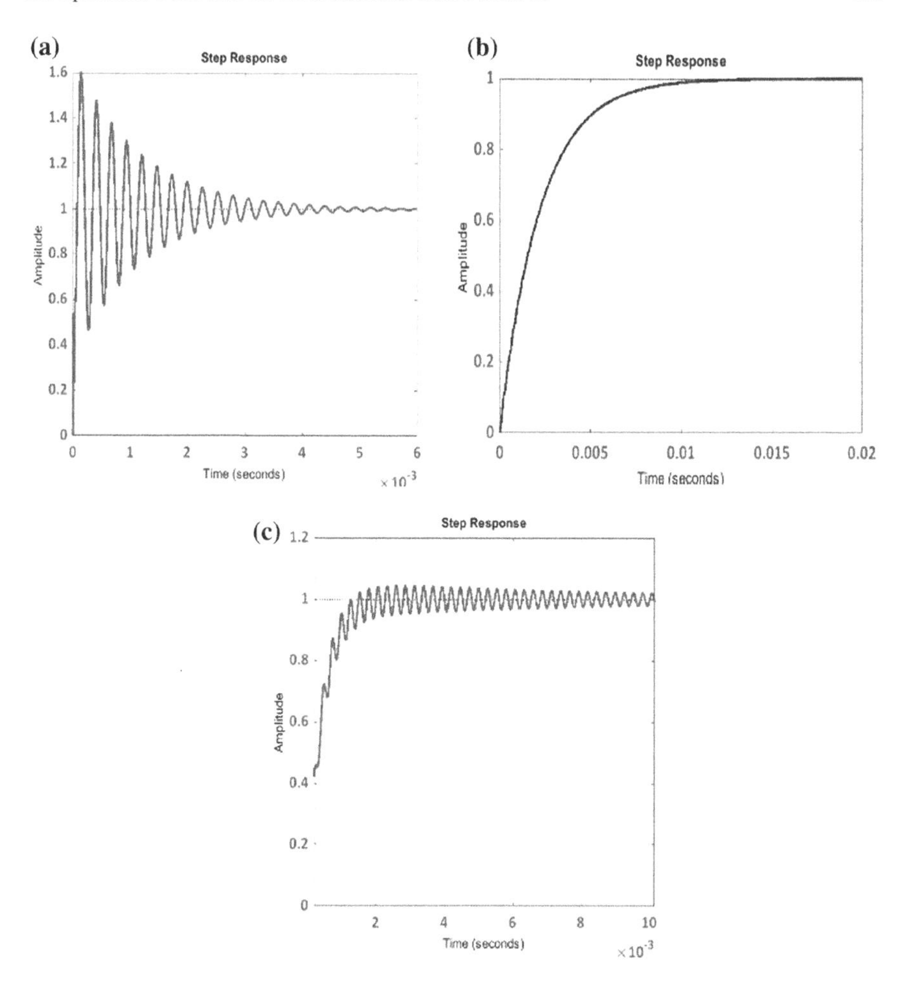

Fig. 4 Step response of **a** open loop, **b** with GA-tuned PI controller, **c** with FLC

of the plant is obtained. The genetic algorithm is one of the search algorithms which optimizes the objective function in the given range with the help of crossover and mutation functions. This respective algorithm will have initial population which gets optimized by the help of Darwin's theory 'Fittest of the Survival'. This algorithm has been implemented in MATLAB to tune gain values of PI through optimization toolbox.

The objective function has been given by integral time absolute error (ITAE) which is given by (14). This function has to be optimized in the range [5, 100] until the termination condition reached. The population size is specified as 150. The initial population is taken as [0, 10] from the given range for proceeding. The scaling function is selected as rank for fitness scaling for evaluating respective fitness value for the population. From all the fitness function values, the population need

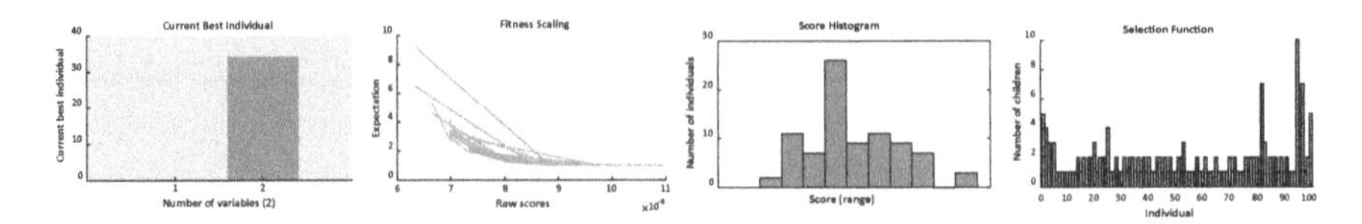

Fig. 5 Parameters of genetic algorithm at 86th iteration

to be selected for further breeding. This has been implemented through stochastic uniform function. The crossover and mutation functions used for further population generation have been taken as constraint-dependent function. The crossover fraction or crossover probability has been taken as 0.6645. The penalty function is the one which penalizes the solutions which are infeasible by reducing their fitness values. The penalty factor is specified as 100 and initial penalty as 10. These both constitute the constraints for this genetic algorithm, and accordingly crossover and mutation functions are selected. The stopping criteria for this algorithm have been specified by using no. of generations, function tolerance, constraint tolerance and time limit which are specified as 200, 10^{-6}, 10^{-3} and infinity, respectively.

$$B = \text{sum}\big(t'. * \text{abs(e)} * \text{d}t\big) \tag{14}$$

where B is objective function, t is time, and abs(e) is absolute error. The genetic algorithm has reached its criteria in 86th iteration. The parameters of the algorithm for the respective iteration are shown in Fig. 5. From Fig. 5, it can be observed that the first variable (K_p) for that iteration (last) has value of 7.9453e−04 and the optimized value for second variable (K_i) is 36, respectively. Figure 5 even shows the plots for fitness scaling, selection function and score histogram, respectively. The fitness scaling plot is expectation for various raw scores. The score histogram is plot for number of individuals for range of scores. The number of children for various individuals is selection function plot.

This GA optimized gain values have been substituted in the controller transfer function, and the respective closed-loop step response for the plant is found. The respective response has been shown in Fig. 4b. The closed-loop transfer function with this controller has been given by (15). The settling time has been noted as 8.7 ms with 0% peak overshoot and steady-state error. Closed-loop transfer function with PI-tuned GA controller is given in (15).

$$\frac{1.459\text{e}-13s^3+7.964\text{e}-09s^2+0.00016s+14.9}{1.182\text{e}-21s^5+7.504\text{e}-17s^4+5.991\text{e}-11s^3+1.523\text{e}-07s^2+0.03328s+14.9} \tag{15}$$

3.3 *Fuzzy Logic Controller*

The plant transfer function has been simulated with controller as fuzzy logic controller (FLC) in unity feedback for finding the control parameters for the plant with this respective controller to compare with GA-tuned PI controller parameters.

The respective crisp input variables of the FLC which are error (E), change in error (CE) are converted to its respective linguistic variables with the help of the membership functions. The respective membership functions and rules defined in the universe of discourse are listed in Table 3 for both input and output fuzzy variables, respectively, where NL and PL are defined using trapezoidal membership functions and remaining using triangular membership functions. The overall graphical representation of the defined FLC is shown in Fig. 6a. The membership functions defined for the input and output are represented in Fig. 7, respectively. Defuzzification has been processed using ‘centroid’. Figure 6b shows the surface of the defined fuzzy rules for the FLC. This whole process has been processed over this plant transfer function, and respective closed-loop response with this controller has been given in Fig. 4c. Settling time is 10.2 ms with peak overshoot of 4.55% and steady-state error of 2.24%, respectively.

Table 4 represents the control parameters like settling time, peak overshoot, steady-state error for the step responses of open loop, closed loop with GA-tuned PI controller and closed loop with FLC, respectively. From Table 4, it can be inferred that if the design considerations are to minimize the peak overshoot and steady-state error, then step response of the plant with GA-tuned PI controller would be the better choice as the respective values are 0%, respectively. The settling time and rise time for the open-loop configuration are less. But the desirable parameters like steady-state error and peak overshoot are 6% and 60%, respectively. The undesirable oscillations are induced in this response. The rise time of closed loop with FLC is better compared with closed loop with GA-tuned PI controller. So, for the remaining work GA-tuned PI controller is chosen for the simulation.

Table 3 Rule base for defined fuzzy logic controller

Change in error	Error						
	PL	PMed	Psm	Zero	NSm	Nmed	NL
PL	PL	PL	PL	PL	PMed	Psm	Zero
PMed	PL	PL	PL	PMed	Psm	Zero	NSm
Psm	PL	PL	PMed	Psm	Zero	NSm	Nmed
ZERO	PL	PMed	Psm	Zero	Nsm	Nmed	NL
NSm	PMed	Psm	Zero	NSm	Nmed	NL	NL
Nmed	Psm	Zero	NSm	Nmed	NL	NL	NL
NL	Zero	NSm	Nmed	NL	NL	NL	NL

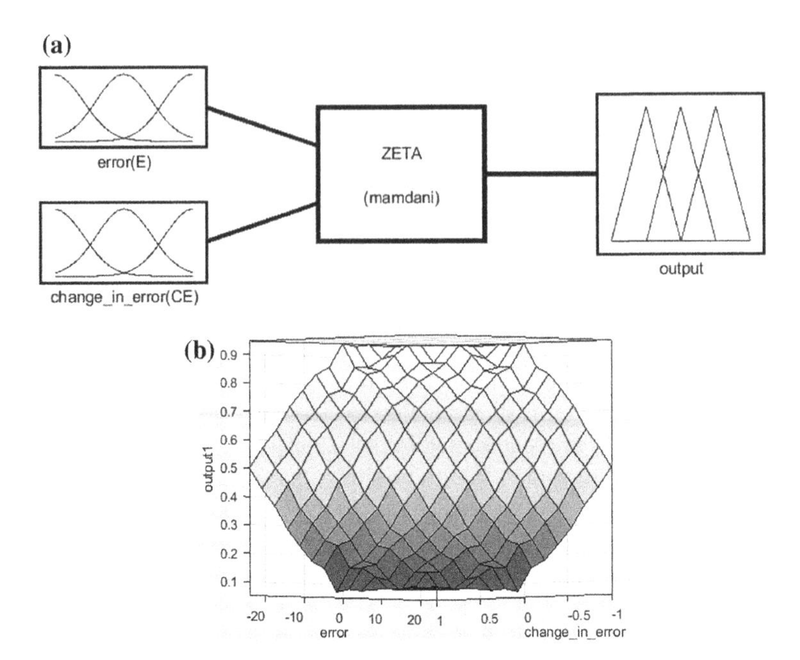

Fig. 6 **a** Overall graphical representation of defined FLC, **b** surface of defined FLC

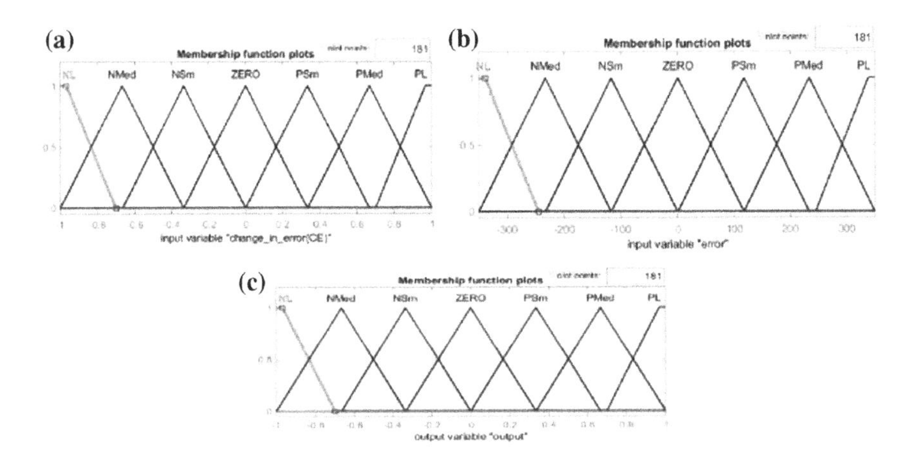

Fig. 7 Membership functions of input variable **a** error, **b** change in error, **c** output

Table 4 Comparative table for control parameters for different configurations

Control parameters ↓	Configuration →		
	Open loop	Closed loop with GA-tuned PI controller	Closed loop with FLC
Rise time (ms)	0.0658	4.9	0.86886
Settling time (ms)	4	8.7	10.2
Peak overshoot (%)	59.93	0	4.5583
Peak value (amplitude)	1.5993	1	1.0456
Peak time (ms)	0.1448	23.5	0.0026
Steady-state error (%)	6	0	4.55

4 Simulation and Results

4.1 Simulation Figure

The proposed system in Fig. 1a has been simulated in MATLAB/Simulink to implement the application stated with best controller scheme as shown in Fig. 8. The required power to run the DC motor-coupled hydraulic pump is supplied by solar system through zeta converter. This zeta converter is used to maintain the constant voltage at the load with the help of controller. The two controller schemes have been discussed in Sect. 3 and were concluded that GA-tuned PI controller will suit for the respective application where steady-state error and settling time need to be as low as possible. So, this overall system has been controlled with the help of GA-tuned PI controller.The voltage and maximum power from solar module is in the range of 9-15 V and 1.5 kW respectively which can run the motor of 0.8kW at constant 12V with help of zeta converter. The controller's set value is 12 V which is continuously

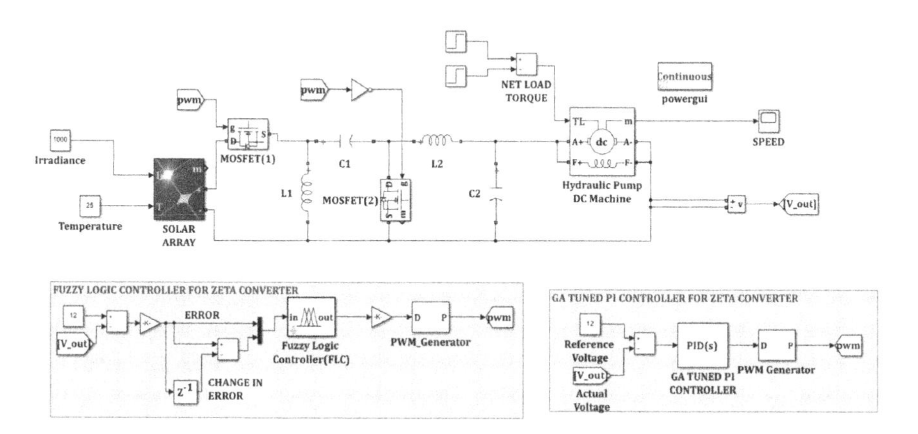

Fig. 8 Simulation figure of the proposed system with both controllers

compared with load voltage, and an error generated is fed as input to the GA-tuned PI controller which produces a duty ratio for controlling the MOSFET(1) of zeta converter. Diode in between C1 and L2 is replaced by a MOSFET(2) aiming at greater efficiency. The load has been varied between 100 and 75% so as to check the variation of the mechanical power. The DC motor utilized here is a shunt motor. The solar array has been maintained at 25 °C temperature and 1000 W m^{-2} irradiance.

Figure 8 shows the implementation of FLC. The error and change in error are given as the inputs to FLC which provides a duty ratio as the output. This duty ratio will be converted to PWM pulses with the help of D-P block. These pulses are fed to MOSFET correspondingly.

4.2 Simulation Results

The simulation Fig. 8 (with diode in place of MOSFET(2)) has been simulated for the two input voltages which are 9 and 15 V. The respective results have been discussed before and after this dynamic change. Figure 9b depicts the output voltage of the zeta converter which is the supply of the motor for dynamic change of the solar voltage. It can be inferred that the input (solar) voltage has dynamically changed from 9 to 15 V at 0.0245 s, and output has been maintained at constant 12 V (desired). Figure 9c depicts the load current maintained at 67 A even after dynamic change. The DC power to the hydraulic pump from zeta converter has been depicted in Fig. 9d. It can be seen that the required 800 W to the motor has been supplied from solar array through zeta converter. Figure 10a, b shows the current and voltage waveforms across the MOSFET switch. It can be inferred from Fig. 10b that the peak amplitude of current waveform across MOSFET has been decreased from 200 to 176 A during the dynamic change. Similarly, the MOSFET voltage has also changed its peak amplitude from 28

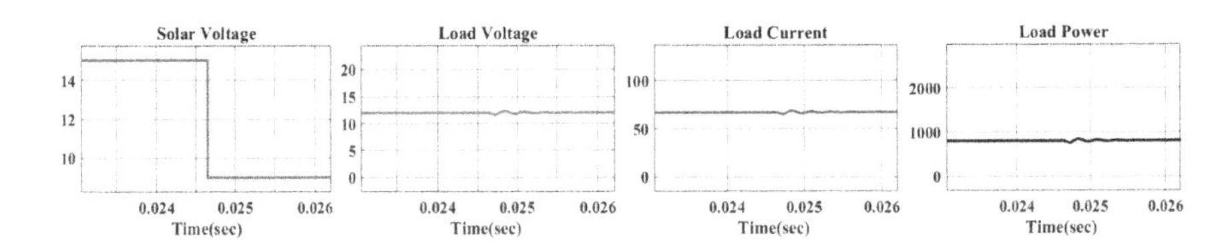

Fig. 9 **a** Input and **b** output voltage in volts, **c** load current in A, **d** load power in W

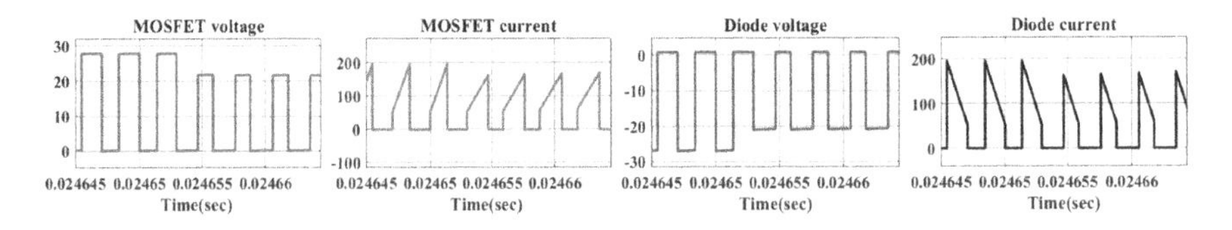

Fig. 10 MOSFET and diode **a**, **c** voltage in (V) and **b**, **d** current (A)

to 22 V during the change of input voltage from 9 to 15 V, respectively. Figure 10c, d shows the diode current and diode voltage, respectively. The diode current will be available when MOSFET current becomes zero and diode current discharges from its peak to zero during its conduction. It can be seen that peak amplitude of both diode current (from 200 to 176 A) and diode voltage (from −27.5 to −21.5 V) has been reduced during transition of input voltage.

From Fig. 11a, b, it can be inferred that the inductor current across L1 has before dynamic change maximum peak of 100 A and minimum value of 0 A. During the 'ON' state of switch, the inductor gets charged to 100 A, and it starts discharging to 0 A during 'OFF' state of the switch. The voltage across L1 can be seen having maximum amplitude of 15 V and with minimum peak being −12 V when input voltage is maintained at 9 V. After the dynamic change, the inductor current changed its maximum peak to 80 A and its voltage maximum peak changed to 10 V, respectively. Figure 11c, d shows the waveforms of current and voltage across the inductor L2. It can be seen that these waveforms follow the waveforms of inductor L1, respectively.

Figure 12a, b shows the capacitor C1 voltage and current waveforms, respectively. It can be seen that average voltage of −11.5 and −11.7 V has been maintained before and after dynamic change with 25 mV ripple. The current waveform has 100 A of maximum peak and −120 A peak as minimum, respectively. Figure 12c, d shows the output capacitor (C2) waveforms. It can be inferred that the maximum peak of current waveform is changing 23–12 A during transition of input voltage from 9 to 15 V and maintaining its minimum peak value at −23 A. Irrespective of changes in the input voltage from 9 to 15 V, the voltage across capacitor C2 (load voltage) has been maintained at 12 V, respectively.

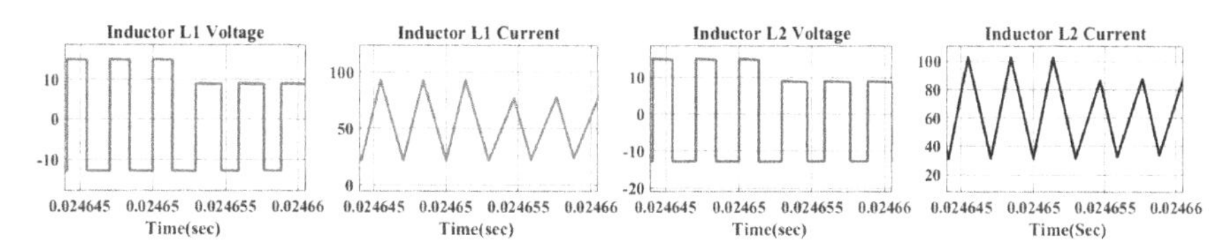

Fig. 11 Inductor L1 and L2 **a**, **c** voltage (V) and **b**, **d** current (A)

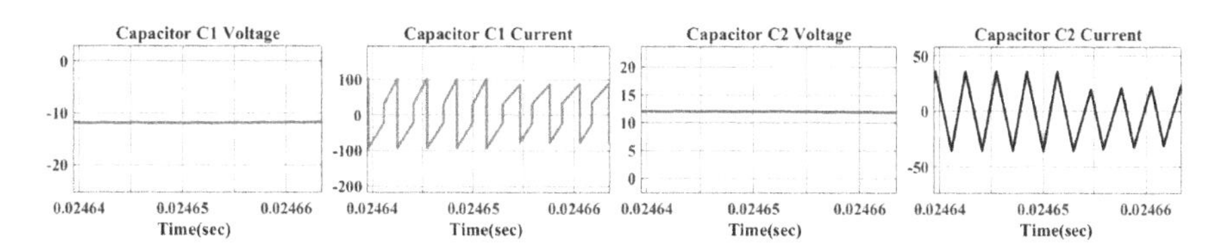

Fig. 12 Capacitor C1 and C2 **a**, **c** voltage (V) and **b**, **d** current (A)

5 Analysis

This section presents a detail analysis on the advantages of zeta converter over SEPIC converter and the reference (Set) value of the controller with respect to the climate and wetness condition of the crop land.

5.1 Set Values for Different Conditions

The climatic conditions can reflect the power input to the converter module as the input supply is from solar array which has greater proportionality with climate in generating power. If the crop field is already wet, it does not require to run the hydraulic pump at its maximum power and vice versa. It can be observed that when climate is sunny and crop is dry, the motor needs to run at its 100% reference value, and when climate being rainy and crop field being dry, then motor needs to run at 0% of its reference value (i.e. motor supply should be turned OFF). When crop field being somewhat wet and climate being sunny, then motor can run at 25% of its value. Similarly, when field being somewhat dry and climate being cloudy, then motor can be run at its 50% of its reference value to wet the land.

5.2 SEPIC Versus Zeta

The zeta and SEPIC converters are of the similar DC–DC converter topology. The similarity lies in the number of passive and active components in both converters. The two converter topologies have been simulated with same specifications and its comparison in the efficiency, conduction and switching losses of switch and diode (*D*) along with voltage and current stress of the diode and switch have been tabulated in Table 5. The respective equations for calculating the same have been provided from (15) and (16). The modified zeta converter is shown in Fig. 8, here the diode of the

Table 5 Comparative table of SEPIC, zeta and modified zeta converter

	SEPIC converter	Zeta converter	Modified zeta converter
P_{in} in W	809.01	774.81	824.76
P_{out} in W	758.8496	708.144	799.8354
η in %	93.79	91.3905	96.97
P_{cond} in W	*S*: 6.928, *D*: 4.895	*S*: 3.969, *D*: 7.303	*S*: 4.71831, *D*: 7.965
$P_{switch\ (overall)}$ in W	38.338	59.394	12.24079
Voltage stress in V	*S*: 26.1, *D*: −24.9	*S*: 21.5, *D*: −20.05	*S*(1): 21.2, *S*(2): −20.9
Current stress in A	*S*: 17,460, *D*: 201	*S*: 202.5, *D*: 202.5	*S*(1): 211.4, *S*(2): 211.6

zeta converter has been replaced by another MOSFET (S) to study the performance. From Table 5, it can be inferred that the efficiency of modified zeta converter is high, and switching losses are very less when compared to SEPIC and zeta converter. The voltage stress across the switches and diodes among all the converters has been similar, whereas the current stress across MOSFET in SEPIC is 17 kA which is very high when compared with zeta and modified zeta converter MOSFET switches. These comparisons prove that the modified zeta converter has better performance over zeta and SEPIC configurations.

$$P_{in} = V_{in} * I_{in},\ P_{out} = V_{out} * I_{out},\ \eta = \frac{P_{out}}{P_{in}}, \tag{16}$$

$$P_{cond} = I_{d|s}^2 * R_{on},\ P_{switch(overall)} = P_{in} - P_{out} - P_{cond(overall)} \tag{17}$$

6 Conclusions

The solar array of maximum 1 kW has been selected to meet the demands of load along with the losses. The corresponding open-circuit and short-circuit test values have been considered to plot V–I characteristics and MPP curve, respectively. The zeta converter has been designed according to the requirements of the proposed system. To study the optimized controller over genetic algorithm-tuned PI controller and fuzzy logic controller, the zeta converter has been mathematically model by using state space averaging technique considering the small signal perturbation for linearizing. The respective open-loop transfer function has been used for finding the controller parameters of open-loop response and closed-loop responses (GA-tuned PI and FLC). It has been concluded the GA-tuned PI controller has 4 ms settling time and 0% steady-state error and overshoot, respectively, which are the best controller parameters for this application. This controller scheme has been considered for controlling zeta-based solar hydraulic pump. The respective proposed model has been simulated using MATLAB/Simulink, and the respective simulation results have been discussed. The analysis for changing the set values in controller according to the climatic and field area conditions has been discussed. The SEPIC converter, zeta converter and modified zeta converter were compared in terms of efficiency and losses, and modified zeta converter where the conventional diode has been replaced with MOSFET has performed well with efficiency of 96.97%.

References

1. Dotzauer E (2010) Greenhouse gas emissions from power generation and consumption in a nordic perspective. Energy Policy 38(2):701–704

2. Porselvi T, Deepa K, Aishwarya S, Bharathraj E, Sairam S (2019) Solar PV fed super-lift LUO converter and modified sepic converter for LED drivers. Test Eng Manag 81:2627–2635
3. Mahalakshmi R et al (2015) Implementation of grid connected PV array using quadratic DC-DC converter & 1ϕ multi level inverter. Indian J Sci Technol 8:1–5
4. Sudarsana Reddy K, Mahalakshmi R, Deepa K (2020) Bio-diesel fed solar excited synchronous generator. J Green Eng 10:1–27
5. Sreelakshmi S, Mohan Krishna S, Deepa K (2019) Bidirectional converter using fuzzy for battery charging of electric vehicle. In: IEEE transportation electrification conference (ITEC-India), Bengaluru, pp 1–6
6. Kim C-K, Rhew H-W, Kim YH (1998) Stability performance of new static excitation system with boost-buck converter. In: Proceedings of the 24th annual conference of the IEEE industrial electronics society IECON '98, Germany, vol 1, pp 402–409
7. Ibrahim NM, Gangadhar JV, Partho M, Deepa K (2016) High frequency SEPIC-WEIBERG converter for space applications. In: International conference on inventive computation technologies (ICICT), Coimbatore, pp 1–6
8. Wang GJ et al (2001) Neural-network-based self-tuning PI controller for precise motion control of PMAC motors. IEEE Trans Ind Electron 48(2):408–415
9. Reddy S, Neppalli Y, Sireesha K (2018) Load optimization and forecasting for microgrids. In: ICICCS, pp 1106–1112
10. Bhukya et al (2018) Fuzzy logic based control scheme for doubly fed induction generator based wind turbine. Int J Emerg Electr Power Syst 34
11. Muruganandam M et al (2009) Performance analysis of FLC based DC-DC converter fed DC series motor. In: Chinese control and decision conference 2009, pp 1635–1640
12. Liu C, Li B, Yang X (2011) Fuzzy logic controller design based on genetic algorithm for DC motor. In: International conference on electronics, communications and control, Ningbo, pp 2662–2665
13. Sarkawi H (2013) Dynamic model of zeta converter with full-state feedback controller implementation. Int J Res Eng Technol 02(08):34–43
14. Vuthchhay E et al (2008) Dynamic modeling of a zeta converter with state-space averaging technique. In: 5th international conference on electrical engineering/electronics, computer, telecommunications and information technology, pp 969–972

Automated Detection and Classification of COVID-19 Based on CT Images Using Deep Learning Model

A. S. Vidyun, B. Srinivasa Rao, and J. Harikiran

Abstract Medical image classification is one of the important areas of application of deep learning. In CT scan images, the structures overlapping against each other are eliminated, thus providing us quality information which helps in classification of images accurately. Diagnosing COVID-19 is the need of the hour and its manual testing consumes a lot of time. Deep learning approach toward COVID CT image classification can reduce this time and provide us with faster results compared to conventional methods. This paper proposes a fine-tuning model, containing a dropout, dense layers, and pretrained model which is validated on publicly built COVID-19 CT scan images, containing 544 COVID and NON-COVID images. The obtained result is compared with different other models like VGG16 and approaches like transfer learning. The experimental result provides us with fine-tuning of the VGG-19 model, which performed better than other models with an overall accuracy of 90.35 ± 0.91, COVID-19 classification accuracy or recall of 92.55 ± 1.25, overall f_1-score of 88.75 ± 1.5, and an overall precision of 88.75 ± 1.5.

Keywords Image processing · Deep learning · Medical image classification

1 Introduction

The COVID-19 pandemic is considered as a World Health Threat of the century [1]. The virus responsible for this disease is SARS-COV2. Current statistics show that the virus has affected more than 5.1 million people [2]. Slower results from testing is the major problem faced by hospitals and other testing centers. It is essential to

A. S. Vidyun (✉) · B. Srinivasa Rao · J. Harikiran
School of Computer Science and Engineering, Vellore Institute of Technology, VIT-AP University, Amaravati, India
e-mail: vidyun.as@gmail.com

B. Srinivasa Rao
e-mail: srinivas.battula@vitap.ac.in

J. Harikiran
e-mail: harikiran.j@vitap.ac.in

A. Choudhary et al. (eds.), *Applications of Artificial Intelligence and Machine Learning*, Lecture Notes in Electrical Engineering 778,
https://doi.org/10.1007/978-981-16-3067-5_31

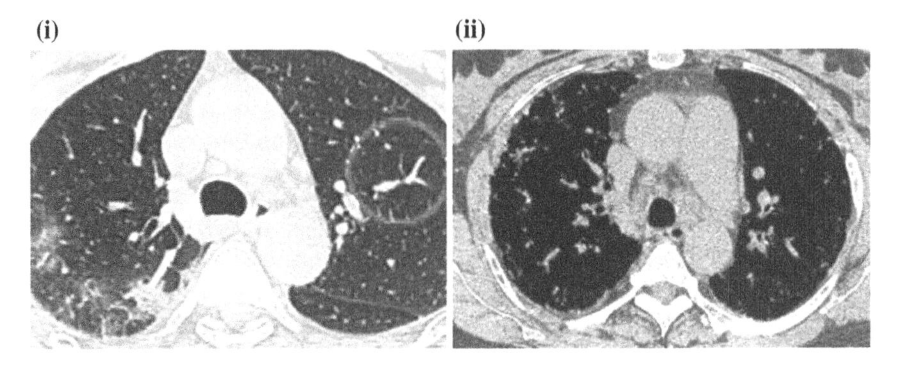

Fig. 1 **i** COVID-19 CT scan image, **ii** NON-COVID-19 CT scan image

diagnose and cure the disease as early as possible. Generally, it takes 24 hour for the results to arrive, and it requires a different kit to detect the virus [3]. This problem can be resolved using artificial intelligence and deep learning techniques. Using deep learning models, COVID-19 is diagnosed using images and they provide accurate and fast results compared to traditional methods.

The manufacture of present COVID-19 diagnosing kit consumes time and money. The number of diagnosing Kits is less, which slows down the diagnosing process [4]. On the other hand, the CT scan machines are present all over and are already well developed. The CT images are accurate with less noise, and it takes a 360° image of the whole organ [5]. Many hospitals in China used CT scans to diagnose the virus [6]. CT scan image dataset is created by Zhao et al. [7] which contains CT scan images of COVID and NON-COVID patients of 216 patients. These CT images are used in training and validating the model. The sample images of the dataset are as shown in Fig. 1.

During validation, accuracy of the model is not the only metric to be taken into consideration. The weightage of error should also be considered. COVID-19 patients misclassified as NON-COVID-19 has more weightage than NON-COVID-19 patients misclassified as COVID-19. Recall of a patient must also perform well to get the best model. This must also be considered during selection of the best model and approach.

Another factor which hinders the research is the number of COVID-19 CT scan images that are available. Most of the COVID-19 CT scan images are made unavailable due to privacy concerns. The CT image which are presently available of low-quality causes less accurate classification of viral and non-viral patients' images. Training the model with 544 images will not provide an efficient model. But with available resources, the best model and approach is found and proposed through this paper.

Kermany et al. [8] proposed a new approach in deep learning for training the dataset with less number of data, called the transfer learning. These are models which are already pretrained with 1.2 million images of 1000 different categories of ImageNet dataset [9], and the weights and biases of the model are optimized.

So, this approach is used to compare our model and approach. There are many pretrained models which are available but only the models which performed well were considered for comparison and models with less accuracy were neglected.

In this paper, we propose the fine-tuning approach [10, 11] where some layers of pretrained model is freezed and other layers are allowed to get trained. Different pretrained models are considered for fine-tuning, and the best model is considered among them. It is then validated using the COVID-19 CT scan image dataset containing 349 COVID-19 images and 195 NON-COVID-19 images created by Jinyu et al. Sample images are shown in the figure. Accuracy and recall of COVID classification are the mainly considered parameters, and the parameters for selecting the best models are precision and f_1 which are considered next after accuracy and recall [12].

2 Materials and Methods

The dataset and methods used for comparing the results with the proposed model are discussed below. Later results of the models are discussed, and also the proposed models are validated on bigger dataset for strengthening the proposed approach to work on future even with big dataset. The colorectal cancer histology image dataset [13] is used for validating the model on bigger dataset.

2.1 Covid CT Image Dataset

The dataset used in this study was built by Zao et al. [7] from the preprints of the virus—medRxiv and bioRxiv—and consisting of 544 CT scan images of COVID-19 and NON-COVID-19 patients. From these images, 349 images belong to COVID-19 positive, and the remaining 195 images are labeled as negative. The dataset is manually split into 295 for training, 102 for validation, and 145 for testing. The split of the data is shown in Table 1.

Table 1 Data split table of COVID CT samples

	COVID-19	NON-COVID-19	Total
Train	191	104	295
Validation	60	42	102
Test	98	49	147
Total	349	195	544

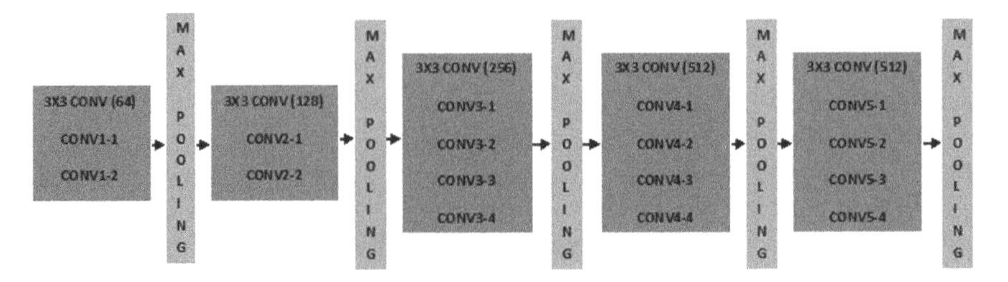

Fig. 2 Architecture of VGG19

2.2 *Transfer Learning*

In our work, we performed a transfer learning approach [14] on the COVID-19 CT dataset [15]. In this technique, a model is already pretrained with the ImageNet dataset by 1.2 million images and 1000 separate object categories. It is then reused, and the last layer is removed and replaced with a layer of neurons equal to the number of classes present in our working dataset, so that the model classifies images only among the provided classes. There are different pretrained models with different architecture and perform differently on different dataset. Here, we used many pretrained models, whose results were compared and VGG16, VGG19 models [16, 17] outperformed other transfer learning models; therefore, these models are used as comparison to our proposed method. The architecture of VGG19 is as shown in Fig. 2.

2.3 *The Proposed Fine-Tuning Method*

With the available number of samples, it is necessary to achieve maximum accuracy. Since the number of samples are less, these pretrained models help us achieve better accuracy since their weights are optimized. In fine-tuning, the pretrained model from transfer learning is taken, which is already trained with ImageNet dataset containing 1.2 million images and 1000 separate categories. Here, we used VGG19 architecture which consists of 16 fully connected and convolutional layers where the initial 12 layers of the pretrained model is freezed and the weights of those layers remain unchanged. The last six layers are allowed to get trained, and their weights are fine-tuned to perform to the provided dataset. This method allows us to train the model efficiently without overfitting as we use a small dataset. The training time for the data is also less compared to training a new model. Since the weights are optimized, better feature extraction from the image is possible. Finally, fully connected layers along with dropout layers [18, 19] are added to prevent the overfitting of models, and neurons in the layers are activated and deactivated while training at the probability of 50%, so the weights in the fully connected layers are equally distributed and prevent models from getting overfit. The last layer is activated with softmax with 2 neurons which returns the probability of the image being COVID positive or negative. The

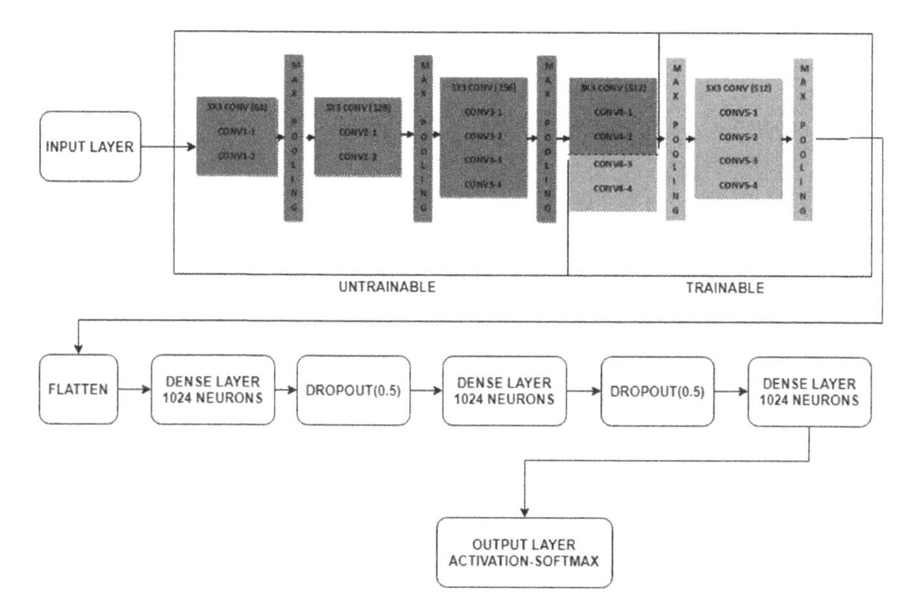

Fig. 3 VGG19 fine-tuning (proposed approach)

model is tested against the test set, and the achieved results are compared with other models. The standard deviation for the accuracy of the model is also calculated to check the consistency. Since the proposed model is only validated against small dataset, it is necessary to validate them on bigger dataset to make sure the model will also work on bigger datasets. The colorectal cancer histology image dataset is used for validating the model on bigger dataset. The dataset consists of 5000 images categorized in eight categories. The proposed fine-tuning model architecture is shown in Fig. 3.

3 Experimental Results

The proposed model and other methods are validated on COVID-19 CT dataset and are trained for five times, accuracy over each time is calculated, and their standard deviation is found to find the consistency of the model. Recall of COVID-19 is considered with high priority as weightage of error for misclassification of COVID-19 as NON-COVID is high. Later the model is validated on bigger dataset to make sure model will work on bigger dataset on future.

Table 2 Performance of different model on COVID-19 dataset

Method	Accuracy	COVID-19-recall	Precision	F_1-score
VGG16-transfer learning	89.28 ± 1.54	90.5 ± 3.14	87.25 ± 2.21	88.25 ± 1.5
VGG19-transfer learning	88.68 ± 2.54	88.9 ± 5.21	86.6 ± 3.04	87.2 ± 2.5
VGG16-fine-tuning	90.33 ± 1.71	92.52 ± 1.94	88.75 ± 1.89	88.75 ± 1.89
VGG19-fine-tuning	90.35 ± 0.91	92.55 ± 1.25	88.75 ± 1.5	88.75 ± 1.5

3.1 Results

On validating the model on a COVID-19 test set which is manually split, the results achieved by different models confirm that the fine-tuning of VGG19 model outperformed other models. The metrics that are considered here are accuracy, recall of COVID-19 classification,f_1 score, and precision in the same order. Recall of COVID-19 classification is considered important because the weightage of error of misclassifying COVID-19 positive patients as negative will be more serious than misclassification of NON-COVID-19 patient classified as COVID-19. Every approach is tested for five times, and standard deviation of the model is also found to help us understand the consistency of model. VGG19 fine-tuning model achieved an accuracy of 90.25 ± 0.91, recall of COVID classification as 92.55 ± 1.25, f_1 score as 88.75 ± 1.5, and precision score of 88.75 ± 1.5. Other models which are considered for comparison are VGG16 fine-tuning, VGG16 transfer learning, and VGG19 transfer learning models. The results obtained on different approaches are given in Table 2. The confusion matrix of the proposed approach is shown in Fig. 4. On validating the same model on colorectal cancer histology image dataset, VGG19 fine-tuning model outperformed other model with 91.2% accuracy, precision score 0.913, f_1 score 0.909, and recall score 0.91. Table 3 gives the result achieved by model on colorectal dataset.

3.2 Discussion

VGG16, VGG19 tansfer learning model performed relatively lesser compared to fine-tuned VGG19, VGG16 models which help us reach the conclusion that in a small dataset, fine-tuning models are more efficient than other models. It is also necessary to consider the weightage of error depending on seriousness of the disease in medical image classification problems. Among VGG16 and VGG19 models, VGG19 performed better when compared to VGG16. The model with lesser standard deviation of accuracy is said to be consistent as it suggests that on every training the model achieves accuracy closer to mean and will not have large deviation from mean, so the model produces consistent output on any run. Therefore, from the results, we can conclude that fine-tuning of VGG19 model contains better accuracy and less

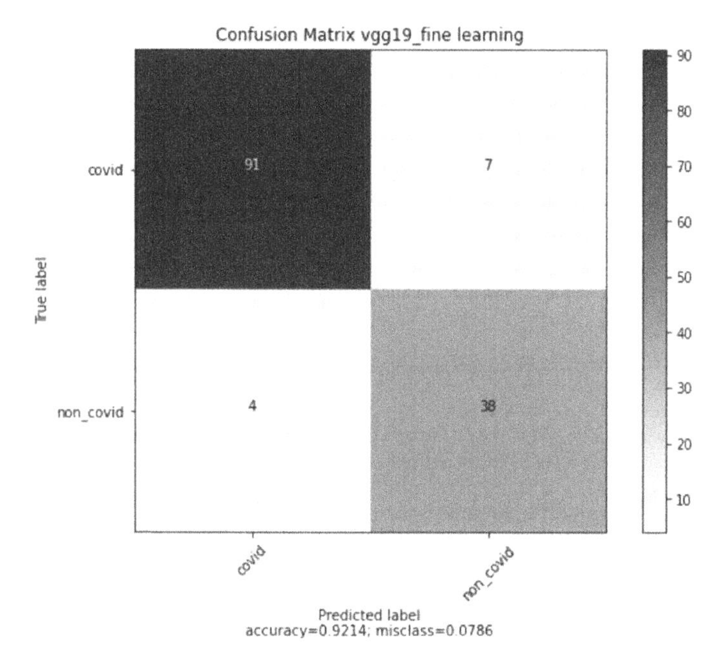

Fig. 4 Confusion matrix of proposed approach on COVID-19 dataset

Table 3 Performance of different model on colorectal dataset

	Accuracy	Precision	F_1 Score	Recall
SVM	65.4	0.657	0.65	0.651
CNN-traditional	85.9	0.857	0.857	0.859
VGG19-transfer learning	88.5	0.884	0.882	0.883
VGG19-fine-tuning	91.2	0.913	0.909	0.91

weightage of error against COVID CT scan image dataset and also on colorectal cancer dataset.

4 Conclusion

This paper proposes a better approach among neural networks toward small dataset containing COVID CT scan images and different ways of evaluating the best model based on seriousness of this disease in medical image diagnosis. Four different approaches were used to compare the performance in terms of accuracy, recall of COVID-19, precision, f_1 score, and confusion matrix. From the result, fine-tuning model has outperformed other approaches, and among them, VGG19 model is considered from other fine-tuned model and achieved an accuracy of 90.25 ± 0.91, recall of COVID classification as 92.55 ± 1.25, f_1 score of 88.75 ± 1.5, and a precision score of 88.75 ± 1.5.

In future, when availability of COVID-19 CT images are increased, the accuracy of the model can be further improved, and different combinations of architecture can be tried. In long run, complete framework for analysis of COVID-19 can be designed.

References

1. https://www.medicalnewstoday.com/articles/covid-19-is-now-a-pandemic-what-next. Pandemic news
2. https://www.cnbc.com/2020/05/21/coronavirus-live-updates.html. Stats
3. https://www.cnet.com/health/coronavirus-test-how-long-will-it-take-to-get-my-covid-19-results-back/
4. https://timesofindia.indiatimes.com/business/india-business/diagnostic-firms-face-shortages-of-kits-experts-urge-for-mass-screening-tests-transparency-in-approvals/articleshow/74923571.cms
5. https://www.envrad.com/difference-between-x-ray-ct-scan-and-mri/. [CT advantage]
6. https://www.wired.com/story/chinese-hospitals-deploy-ai-help-diagnose-covid-19/. CT image
7. Zhao J, He X, Yang X, Zhang Y, Zhang S, Xie P (2020) COVID-CT-dataset: a CT image dataset about COVID-19. arXiv:2003.13865
8. Kermany DS, Goldbaum M, Cai W, Valentim CCS et al (2018) Identifying medical diagnoses and treatable diseases by image-based deep learning. Cell 172:1122–1131
9. Deng J, Dong W, Socher R, Li L-J, Li K, Fei-Fei L (2009) ImageNet: a large-scale hierarchical image database. In: IEEE Conference on computer vision and pattern recognition, pp 2–9
10. https://cv-tricks.com/keras/fine-tuning-tensorflow/. Fine tuning
11. Fine-tuning convolutional neural network on own data using keras tensorflow. https://cv-tricks.com/keras/fine-tuning-tensorflow/
12. https://towardsdatascience.com/accuracy-precision-recall-or-f1-331fb37c5cb9
13. Kather JN et al (2016) Multi-class texture analysis in colorectal cancer histology. Sci Rep 6:27988
14. Alinsaif S, Lang J (2020) Histological image classification using deep features and transfer learning. In: 17th Conference on computer and robot vision (CRV), Ottawa, ON, Canada, pp 101–108. https://doi.org/10.1109/CRV50864.2020.00022
15. Wang C et al (2019) Pulmonary image classification based on inception-v3 transfer learning model. IEEE Access 7:146533–146541. https://doi.org/10.1109/ACCESS.2019.2946000
16. Hijab A, Rushdi MA, Gomaa MM, Eldeib A (2019) Breast cancer classification in ultrasound images using transfer learning. In: Fifth international conference on advances in biomedical engineering (ICABME), Tripoli, Lebanon, pp 1–4. https://doi.org/10.1109/ICABME47164.2019.8940291
17. Kaur T, Gandhi TK (2019) Automated brain image classification based on VGG-16 and transfer learning. In: 2019 International conference on information technology (ICIT), Bhubaneswar, India, pp 94–98. https://doi.org/10.1109/ICIT48102.2019.00023
18. Srivastava N, Hinton G, Krizhevsky A, Sutskever I, Salakhutdinov R (2014) Dropout: a simple way to prevent neural networks from overfitting. J Mach Learn Res 15:1929–1958
19. Dropout neural network layer in keras explained. https://towardsdatascience.com/machine-learning-part-20-dropout-keras-layers-explained-8c9f6dc4c9ab

Comparative Study of Computational Techniques for Smartphone Based Human Activity Recognition

Kiran Chawla, Chandra Prakash, and Aakash Chawla

Abstract Human activity recognition (HAR) has been popular because of its diverse applications in the field of health care, geriatrics care, the security of women and children, and many more. With the advancement in technology, the traditional sensors are replaced by smartphones. The mobile inbuild accelerometer detects the orientation or acceleration, and the gyroscope detects the angular rotational velocity. In this study, computational techniques-based comparative analysis has been carried out on publicly available dataset on human activity recognition using smartphone dataset. Traditional and contemporary computational techniques (support vector machine, decision tree, random forest, multi-layer perceptron, CNN, LSTM, and CNN-LSTM) for HAR are explored in this study to compare each model's accuracy to classify a particular human activity. Support vector machine outperforms in most of the activity recognition tasks.

Keywords Machine learning · Human activity recognition using smartphones · Deep learning

1 Introduction

The smartphone-based human activity recognition is a field of study in which the activity or the movement performed by a person is identified using sensors. These sensors can be motion, accelerometer, gyroscope, proximity, etc. The study of recognizing human activity has become focused as its applications are being explored vastly in eldercare, health care, women security, etc. Along with the earlier methods, one of the most attainable, practical, and convenient sensors that can be used for

---

K. Chawla (✉)
Indira Gandhi Delhi Technical University for Women, Delhi 110006, India

C. Prakash
National Institute of Technology Delhi, Delhi 110040, India

A. Chawla
SupplyCopia Software Services Pvt. Ltd., Noida, Uttar Pradesh 201307, India

A. Choudhary et al. (eds.), *Applications of Artificial Intelligence and Machine Learning*, Lecture Notes in Electrical Engineering 778,
https://doi.org/10.1007/978-981-16-3067-5_32

recording the movements of an individual is mobile sensors. According to a Statista Research Department survey, the global smartphone sales to end-users have tremendously increased worldwide from 0.122 billion units in 2007 to 1.56 billion in the year 2018 [1]. In today's world, mobile phones are not confined to only telephony; instead, they offer various services like mobile banking, e-commerce, entertainment, and sensors. In many smartphones, the sensors are in-built to classify the activities. With the increase in the smartphone user population, using smartphones as a sensor for health monitoring, diagnosis, and activity recognition is the most pragmatic and suitable solution. Any smartphone with embedded accelerometer and gyroscope sensors can be used for collecting the data.

The activity conducted by a human the next moment depends on the action being performed at the current instance. For example, if a person is walking upstairs, it is less probable that he/she will be lying down the next moment or vice versa. Similarly, if a person is laying, it is more probable that he/she would be standing or sitting in the immediate moments compared to walking, walking upstairs, or walking downstairs. So, the actions coming up by a human being at an instance are contingent on the activities performed shortly before. These particular observations can be adopted to discover any deviation from the typical behaviour.

Human activity recognition has been a popular research subject among researchers for a long time. A lot of work has been done on classifying the activities using various datasets. The dataset used in this study is UCI human activity recognition using smartphones. Various researchers have worked on this dataset with conventional and contemporary machine learning models to classify the activities. The most cited study by Anguita et al. [2] used multi-class SVM as the classifier to predict the labels for the activities, which stated 96% overall accuracy. The study by Gaur and Gupta [3] that compared logistic regression with and without cross-validation, decision tree, and random forest concluded that logistic regression with cross-validation is most accurate among the four with an accuracy of 91.9%, whereas Fan et al. [4] and Feng et al. [5] proposed decision tree classifiers and ensemble random forest, respectively.

Since conventional models require defining the features manually, contemporary research is inclined to use deep learning for activity recognition. Sharma et al. [6] used state-of-the-art techniques for HAR; they used neural networks (artificial neural networks) on the UCI dataset to reach an accuracy of 83.96% earlier in 2008. Cho and Yoon [7] and Gholamrezaii and Almodarresi [8] used one-dimensional convolutional neural networks and two-dimensional convolutional neural networks, respectively. LSTMs have become quite popular over time and have emerged as great models for activity recognition, as stated in the study by Chen et al. [9]. Along with that, CNN-LSTM architecture proposed in the study by Kun Xia et al. proposed the results on UCI HAR dataset as 95.78% [10].

The limitation of using a single model for activity recognition is that a model might perform well for classifying a particular activity but might not be as useful to classify the other. This study also explores using multiple ML models that work best for respective human activities. Section 2 discusses the methodology used for the classification of human activities. Section 3 presents the result followed by the discussion and future scope in Sect. 4.

2 Methodology

In this section, the methodology used in this study is presented and is illustrated in Fig. 1. After the collection of accelerometer and gyroscope readings, the data is pre-processed, for noise removal, segmentation, and extraction of the features from the accelerometer and gyroscope readings. The next step is feature selection, which is specific to traditional machine learning algorithms and not for deep learning algorithms. In the feature selection phase, the relevant features from the pool of attributes extracted in the previous step are selected. A feature metric is constructed and is passed on further to the models. Finally, different techniques for activity classification are performed.

2.1 Dataset

The dataset used for the study is human activity recognition using Smartphones Dataset Version 1.0 [11]. The experiment was conducted on 30 volunteers aged between 19 and 49 years. In this particular dataset, the data was collected using Samsung Galaxy S II tied with the volunteers waist. The volunteers' six activities were walking, walking upstairs, walking downstairs, sitting, standing, and laying. The data collected by the accelerometer and gyroscope is 3-axial linear acceleration (x, y, z) and 3-axial angular velocity (x, y, z) at a constant rate 50 Hz, respectively. The data of the volunteers was arbitrarily split in 70:30 ratio into training and testing set. Since there is a lot of noise in the raw data, the data pre-processing was done by applying noise filters. The pre-processed data was then sampled in fixed-width sliding windows of 2.56 s with a 50% overlap, that is, 128 readings/window. 0.3 Hz cutoff frequency Butterworth low-pass filter was used to separate the components of gravity from body motion components of the accelerometer readings since gravitation is assumed to have low-frequency components. The resulting vector of features collected for each separate window was obtained from the time and frequency domain

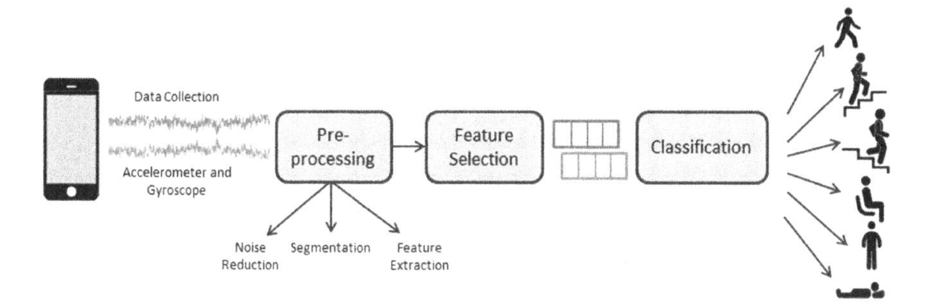

Fig. 1 Methodology used in this study

[12, 13]. For activity recognition, the sensors collect the accelerometers data in three dimensions: along (x, y, z) axes. The gyroscope collects the angular velocity in the three dimensions as well.

2.2 Feature Analysis and Selection

The data is prepared and visualized to get the outline of various attributes of the data. This subsection contains the plots for various relations among the features to understand the data better. The activity distribution lies in the range of 13–19%, and all the activities have almost equal distribution in the dataset.

The kernel density estimation curve shows the distribution of the probability density of a continuous variable. For some features of the data collected by the smartphone sensors, kernel density estimation curves are as shown in Fig. 2a, b. As an example, for the mean of total body acceleration in the X direction, density is maximum in the range 0.0–0.5, which shows that the value of the total body acceleration in the X direction is most probable to lie near to 0.2.

The feature distribution of the median absolute deviation of total body acceleration in the Y direction for each activity is as shown in Fig. 3a. It is observed that the scatter plots for the categorical activities are immediate to each other for static activities (i.e.

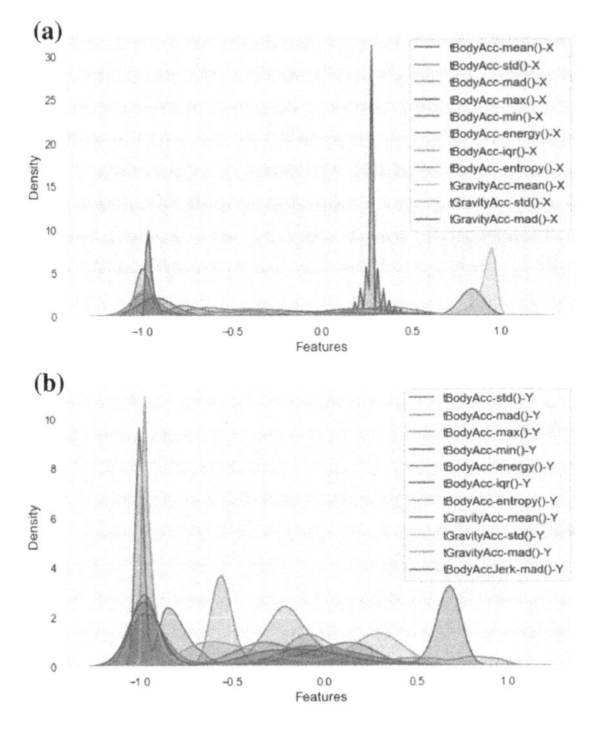

Fig. 2 Kernel density estimation curve of selected features in X, Y, Z direction

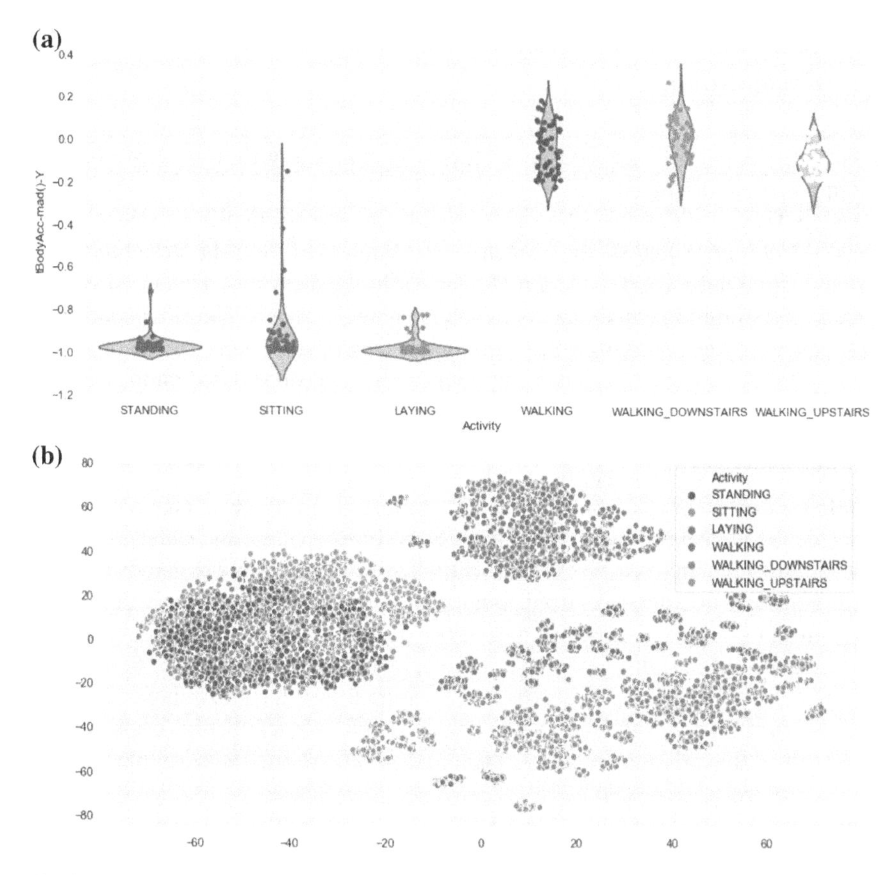

Fig. 3 Feature distribution of the median absolute deviation of total body acceleration for each activity (**a**) in the Y direction (**b**) in the X, Y, Z direction

standing, sitting, and laying) and dynamic activities (i.e. walking, walking upstairs, and walking downstairs). Similar patterns can be observed in Fig. 3b. If we consider a straight line passing through the origin with unit slope, it can be observed that static activities lie on one side and the dynamic on the other.

Since the data collected from smartphones have a significantly large number of features, it is needed to decide which of them should be selected to improve the models' accuracy. The classification models like decision tree, random forest, SVM require manual feature selection. In contrast, the models like LSTM, CNN select the features on their own by multiple iterations and assign weights to the features based on their importance; this is one advantage of deep networks over any conventional architecture. In this study, the data is pre-processed, and three axial features are selected. The features used in this study are mean, standard deviation, median absolute deviation, signal magnitude area, entropy, correlation coefficients of two signals, skewness, etc. [11].

2.3 Computational Techniques for Human Activity Recognition

In this section, the approaches used to classify human activities from sensor data of smartphones are discussed.

Decision Tree is one of the oldest supervised learning algorithms. The rules for the decision are represented by the branches, the features of the data are represented by the internal nodes, and the leaf nodes represent the decisions. The decisions are made at every minute level of the tree. A decision tree is a sound algorithm that gives good accuracy when working with high-dimensional data [4]. For HAR dataset, the decision tree would give the possibility to make the decisions based on the specific conditions using a heuristic approach; that is, the attribute that produces the purest node is chosen. The node selection criteria are majorly dependent on Information Gain, Gini Index, and Gain Ratio.

Random Forest takes the average of the outcome for multiple decision trees merged to decide the final class, which is comparatively more accurate and stable [14]. The prime reason random forest was introduced, when the decision tree is already there, is that the decision trees that are "deep" might undergo over-fitting. In contrast, random forest prevents this from happening by constructing smaller sub-trees and considering the majority votes, resulting in reduced variance. In random forest classifier, hyper-parameters are the number of decision trees in the forest, the maximum number of features taken into consideration when a node is split, the number of levels in each decision tree, the minimum number of leaf nodes, and the criteria to sample the data points (Entropy or Gini) [5].

Support Vector Machine aims to find a plane that divides the data according to the maximum distance between the data points for all the classes. SVM classifier uses the linear decision hyper-planes to classify data points in various classes or labels. SVM supports cost-effective fitting of the high-dimensional data to give precise results. With the increase in the number of training samples, the complexity of the training increases as well [2].

Multi-layer Perceptron uses the feed-forward artificial neural networks for the classification; it passes the extracted features to the input layer [15]. The building blocks or the functioning unit of the artificial neural networks are the neurons. We feed the input to each layer in the network; each layer's outputs are passed to the succeeding layer [16]. The parameters that are passed from one layer to the other have some weight associated with them. Initially, random weights are assigned to the parameters, but as the training is done, model adjusts the weights and learns with multiple iterations and backpropagation.

2-D Convolutional Neural Networks were developed to train the models based on the image data by extracting features from the image in the form of an array of pixels; this process is known as feature learning. In the case of sensor data, the model learns to extract the internal features and make observations from the accelerometer and gyroscope data. In 2-D kernels, there are two convolutional layers, succeeded by the pooling layer and a softmax layer. Since there are a lesser number of layers in 2-D convolutional neural networks, the performance of the model rises, whereas the cost of computation falls [8, 17].

Long Short-Term Memory, popularly known as the LSTM model, is created to eliminate the long-term dependency problem, i.e. when the gap is more between the relevant information, it becomes difficult for recurrent neural networks to learn and predict the information. LSTMs also behave like RNNs, but the structure of the repeating module is different [18]. The horizontal line passing through the top of the cell is called the cell state. This horizontal line passes through all the cells in LSTMs interacting linearly. The LSTMs have the potential to keep or eliminate any information. The gates that are present in LSTMs provide the ability to select or reject a piece of information. There is a sigmoid layer present which returns a value between 0 and 1, which depicts the percentage of selection, that is, the returned value 0 means nothing should get through, whereas the returned value 1 means everything should get through. In human activity recognition, the LSTMs perform quite well because they can remember the information [9].

CNN LSTM CNN LSTM is a combination of convolutional neural networks and LSTM. A prominent feature of convolutional neural networks is that it can extract the features of interest from the data [10, 19]. CNN dives into the fundamental layers of the matrix that are packed together and extracts the pertinent features. Because of this property of CNN, they are added to the front of the architecture. The features extracted by the convolutional neural networks are then passed to the LSTM model, which is proficient in predicting the sequence because of its ability to remember the relevant information. After the LSTM, a dense layer is added to return the output. For human activity recognition, the procedure for classification is demonstrated in Fig. 4.

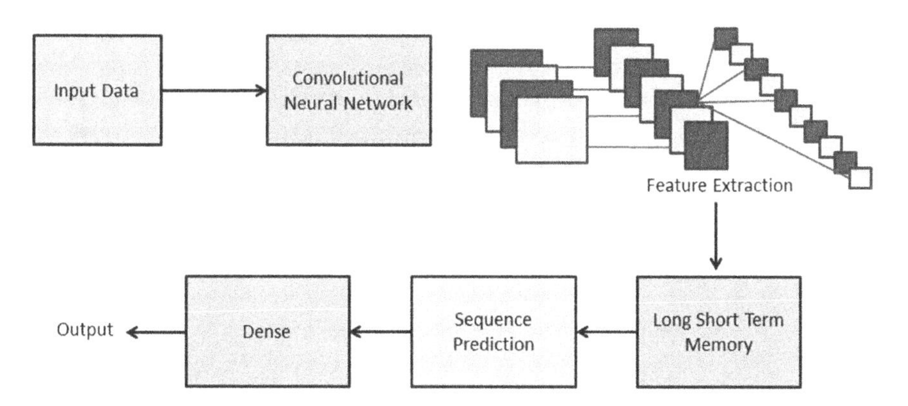

Fig. 4 The process of activity recognition using CNN LSTM

3 Result and Discussion

This section presents the comparative results of the models applied to human activity recognition. Using various algorithms, we can classify human activities based on the extracted features from total acceleration, linear acceleration, and gyroscope readings. Multiple parameters are used in the study to compare the quality of the results.

3.1 Confusion Matrix

The confusion matrix of the model illustrates its performance when tested. The diagonal elements depict the true positive values, i.e. the values that are predicted correctly by the trained model. In this study, the x, y labels in the confusion matrix for all the models represent the subjects' activities. The diagonal values of the confusion matrix for all the algorithms are shown in Table 1. From Table 1, it can be inferred that

Table 1 Confusion matrix of true positive rate for the test data

Activity		Walking	Walking upstairs	Walking downstairs	Sitting	Standing	Laying
Decision tree		0.9455	0.8046	0.6214	0.8146	0.7988	1.0
Random forest		0.9697	0.9278	0.8380	0.8859	0.9774	1.0
SVM	RBF	0.9838	0.9660	0.9404	0.9083	0.9849	1.0
	Linear	0.9919	0.9575	0.9761	0.8940	0.9511	1.0
	Poly	0.9838	0.9554	0.9642	0.9083	0.9812	1.0
2-D CNN		0.9677	0.8853	0.9809	0.8248	0.8966	0.9515
MLP		0.8407	0.8492	0.8500	0.7474	0.8176	0.9199
LSTM		0.9919	0.9575	0.8571	0.8696	0.7819	0.9497
CNN LSTM		0.9516	0.9426	0.9857	0.8594	0.7462	0.9813
Maximum accuracy per activity		0.9919	0.9660	0.9857	0.9083	0.9849	1.0
Model with maximum accuracy per activity		SVM-Linear	SVM-RBF	CNN LSTM	SVM-RBF	SVM-RBF	Decision tree, random forest, SVM

The rows represent the diagonal values of the confusion matrix for each activity with respect to the technique used. The columns represent the activity

Fig. 5 Decision tree for HAR

for different activities, the performance of models varies. One model might perform very well for correctly classifying one human activity, whereas the same might not be very suitable for another activity. The techniques used for human activity recognition show that for accurately classifying sitting, standing, walking upstairs from others SVM when the kernel is radial basis function performs well; walking downstairs is classified by the CNN LSTM model with the highest accuracy, whereas walking is correctly classified by SVM when the kernel selected is linear. Similarly, laying is accurately predicted by decision tree, random forest, and SVM. So, for different activities, different models are performing efficiently.

The decision tree for human activity recognition [20] using CART is as shown in Fig. 5. Minimum of total gravitational acceleration in X-direction is selected as the root node; the condition to make level 1 of the child node is the value of this feature either less than/equal to 0.096 or greater than 0.096. For both the conditions, further decisions are made at each level to classify the activities. It can be observed that the nodes to the right of the decision tree are classifying the static activities, whereas the nodes to the left are classifying the dynamic activities. At every level, the nodes with the Gini Index value greater than 0.0 are explored further. The six classes are predicted based on the number of samples and values. Figure 6a shows the confusion matrix of decision tree for the actual activity labels versus the predicted activity labels. For human activity recognition, in this study, various combinations of hyper-parameters are considered. For example, the criteria can be Gini Index and entropy, and the maximum depth can be 10, 20, 30, 40, 50, ..., etc. Figure 6b shows the confusion matrix of the actual activity labels versus the predicted activity labels using the random forest classifier.

The kernel plays a vital role in SVM classifier [21]. The kernel is a set of mathematical equations which takes the data as input and converts it into a specific format by applying mathematical transformations. Which kernel would be best for SVM on a dataset cannot be decided prior; instead, we need to apply all the kernels and check for the results. For human activity recognition, the kernels used are linear, polyno-

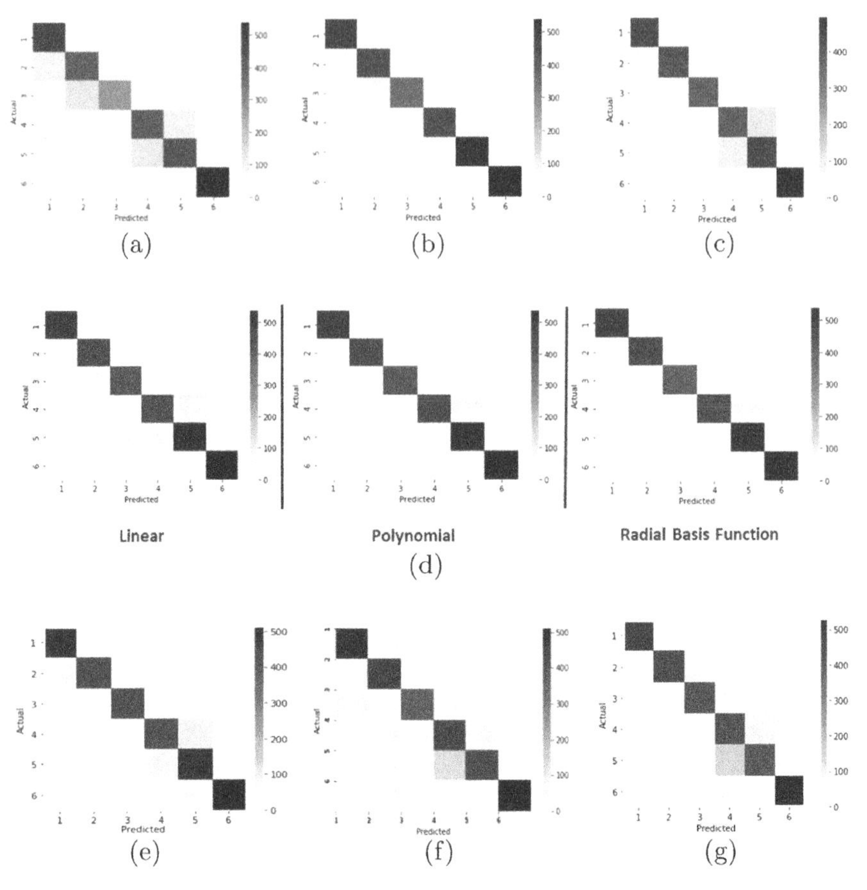

Fig. 6 Confusion matrix—**a** decision tree **b** random forest **c** multi-layer perceptron **d** support vector machine **e** CNN **f** LSTM **g** CNN LSTM

mial, and radial basis function (RBF). The confusion matrix for various kernels for SVM is shown in Fig. 6d.

In this study, the activation function used is ReLu, and the confusion matrix is as shown in Fig. 6c [22]. The confusion matrix for CNN is as shown in Fig. 6e. The confusion matrix for LSTMs is as shown in Fig. 6f. The confusion matrix of CNN-LSTM is as shown in Fig. 6g.

3.2 Evaluation Parameters

Accuracy It is the traditional way to judge the results of a machine learning model. Accuracy is not a very precise term to judge a model since let say false negatives might create trouble in some serious decisions to be made. The calculation is done as follows:

$$\frac{\mathrm{TP}+\mathrm{TN}}{\mathrm{TP}+\mathrm{FP}+\mathrm{TN}+\mathrm{FN}} \tag{1}$$

Precision Precision is described as the proportion of the positive instances over the total positives that the model predicted. Precision gives an idea about the correctness of the model when it claims to be right. The formula to calculate precision is as shown:

$$\frac{\mathrm{TP}}{\mathrm{TP}+\mathrm{FP}} \tag{2}$$

Recall It is the true positive rate, that is, the percentage of true positives to the total actual number of true positives in the dataset. The formula to calculate recall is as follows:

$$\frac{\mathrm{TP}}{\mathrm{TP}+\mathrm{FN}} \tag{3}$$

Specificity It is the percentage of true negatives to the total negatives. It is the parameter that evaluates how different the classes are from each other. The formula to calculate specificity is as shown:

$$\frac{\mathrm{TN}}{\mathrm{TN}+\mathrm{FP}} \tag{4}$$

F_1-score F_1 score is associated with the balance between precision and recall and is calculated as the harmonic mean of precision and recall. The more is the F_1 score, the better is the model. The formula of the F_1 score is as shown:

$$\frac{2}{\frac{1}{\text{precision}}+\frac{1}{\text{recall}}}=\frac{2*\text{precision}*\text{recall}}{\text{precision}+\text{recall}} \tag{5}$$

These evaluation parameters are used to validate the results and choose, in particular, which algorithm performs well. The derived values for the parameters mentioned above for all the techniques used in this study are presented in Table 2. It can be concluded from the results that the support vector machine gives an accuracy of 96.64% when the selected kernel is polynomial, the value of error control hyperparameter, C, is 0.001, and gamma, the deciding factor for the curvature of the decision boundary, equals 1. The lower value of C indicates a lesser amount of error, and a higher value of gamma indicates more curvature.

Table 2 Evaluation parameters for HAR using smartphones

Classifier		Hyper parameters	Accuracy (in %)	Precision	Recall	f_1-score	Specificity
Decision tree		Maximum number of leaf nodes: 9 Minimum split in samples: 3	83.84	0.8467	0.8384	0.8364	0.8383
Random forest		Criterion: GiniMaximum Depth: 30 Maximum features: log2 Number of trees in the forest: 500	93.68	0.9391	0.9368	0.9364	0.9368
SVM	RBF	C: 100 Gamma: 0.01	96.53	0.9664	0.9653	0.9653	0.9653
	Linear	C: 1	96.4	0.9649	0.964	0.9639	0.964
	Poly	C: 0.001	96.64	0.9671	0.9664	0.9663	0.9664
CNN		–	92.26	0.9241	0.9226	0.9226	0.9226
MLP		–	89.54	0.8973	0.8954	0.8948	0.8954
LSTM		–	92.12	0.9166	0.9158	0.9154	0.9261
CNN LSTM		–	90.8	0.9116	0.9111	0.9081	0.9046

4 Discussion and Future Scope

Human activity recognition, particularly with smartphones, plays a vital role in building suitable emergency measures. In this study, using UCI HAR dataset, multiple approaches for recognizing human activities are proposed. Different validation measures prove that the proposed method returns the considerably accurate result to identify human activities. This study also discusses the comparison of various ML models that work best to recognize different activities.

To identify the human activities, the ensemble model of all the applied algorithms with assigned weightage to each model will contribute to the recognition of each particular activity much more efficiently. There is a scope of building up the systems that use these models to strengthen women's security, health care, and eldercare by recognizing the emergency or urgent signals to implement suitable prevention measures.

References

1. O'Dea S (2020) Global smartphone sales to end users 2007–2021
2. Anguita D, Ghio A, Oneto L, Parra X, Reyes-Ortiz JL (2012) Human activity recognition on smartphones using a multiclass hardware-friendly support vector machine. In: International workshop on ambient assisted living. Springer, Berlin, pp 216–223
3. Gaur S, Gupta GP (2020) Framework for monitoring and recognition of the activities for elderly people from accelerometer sensor data using apache spark. In: ICDSMLA 2019. Springer, Berlin, pp 734–744
4. Fan L, Wang Z, Wang H (2013) Human activity recognition model based on decision tree. In: 2013 International conference on advanced cloud and big data. IEEE, pp 64–68
5. Feng Z, Mo L, Li M (2015) A random forest-based ensemble method for activity recognition. In: 2015 37th Annual international conference of the IEEE engineering in medicine and biology society (EMBC). IEEE, pp 5074–5077
6. Sharma A, Lee Y-D, Chung W-Y (2008) High accuracy human activity monitoring using neural network. In: 2008 Third International conference on convergence and hybrid information technology, vol 1. IEEE, pp 430–435
7. Cho H, Yoon SM (2018) Divide and conquer-based 1D CNN human activity recognition using test data sharpening. Sensors 18(4):1055
8. Gholamrezaii M, Mohammad S, Almodarresi T (2019) Human activity recognition using 2D convolutional neural networks. In: 2019 27th Iranian conference on electrical engineering (ICEE). IEEE, , pp 1682–1686
9. Chen Y, Zhong K, Zhang J, Sun Q, Zhao X (2016) LSTM networks for mobile human activity recognition. In: 2016 International conference on artificial intelligence: technologies and applications. Atlantis Press
10. Kun X, Jianguang H, Hanyu W (2020) LSTM-CNN architecture for human activity recognition. IEEE Access 8:56855–56866
11. Anguita D, Ghio A, Oneto L, Parra X, Reyes-Ortiz JL (2013) A public domain dataset for human activity recognition using smartphones. In ESANN, vol 3, p 3
12. Bayat A, Pomplun M, Tran DA (2014) A study on human activity recognition using accelerometer data from smartphones. Procedia Comput Sci 34:450–457
13. El Moudden I, Jouhari H, El Bernoussi S, Ouzir M (2018) Learned model for human activity recognition based on dimensionality reduction. In: Smart application and data analysis for smart cities (SADASC'18)
14. Xu L, Yang W, Cao Y, Li Q (2017) Human activity recognition based on random forests. In: 2017 13th International conference on natural computation, fuzzy systems and knowledge discovery (ICNC-FSKD). IEEE, pp 548–553
15. Talukdar J, Mehta B (2017) Human action recognition system using good features and multilayer perceptron network. In: 2017 International conference on communication and signal processing (ICCSP). IEEE, pp 0317–0323
16. Jain AK, Mao J, Moidin Mohiuddin K (1996) Artificial neural networks: a tutorial. Computer 29(3):31–44
17. Andrey I (2018) Real-time human activity recognition from accelerometer data using convolutional neural networks. Appl Soft Comput 62:915–922
18. Hochreiter S, Schmidhuber J (1997) Long short-term memory. Neural Comput 9(8)
19. Liu YH (2018) Feature extraction and image recognition with convolutional neural networks. J Phys Conf Ser 1087:062032
20. Zubair M, Song K, Yoon C (2016) Human activity recognition using wearable accelerometer sensors. In: 2016 IEEE International conference on consumer electronics-Asia (ICCE-Asia). IEEE, pp 1–5
21. Han S, Qubo C, Meng H (2012) Parameter selection in SVM with RBF kernel function. In: World automation congress 2012. IEEE, pp 1–4
22. Ronao CA, Cho S-B (2016) Human activity recognition with smartphone sensors using deep learning neural networks. Expert syst Appl 59:235–244

Machine Learning Techniques for Improved Breast Cancer Detection and Prognosis—A Comparative Analysis

Noushaba Feroz, Mohd Abdul Ahad, and Faraz Doja

Abstract Breast cancer prevails as the most widespread and second deadliest cancer in women. The typical symptoms of breast cancer are minimal and curable when the tumor is small, hence screening is crucial for timely detection. Since delayed diagnosis contributes to considerable number of deaths, a number of cross-disciplinary techniques have been introduced in medical sciences to aid healthcare experts in the swift detection of breast cancer. A vast amount of data is collected in the process of breast cancer detection and therapy through consultation reports, histopathological images, blood test reports, mammography results etc. This data can generate highly powerful prediction models, if properly used that can serve as a support system to assist doctors in the early breast cancer diagnosis and prognosis. This paper discusses the need of machine learning for breast cancer detection, presents a systematic review of recent and notable works for precise detection of breast cancer, followed by a comparative analysis of the machine learning models covered in these studies. Then, we have performed breast cancer detection and prognosis on three benchmark datasets of Wisconsin, using seven popular machine learning techniques, and noted the findings. In our experimental setup, *K*-Nearest Neighbor and Random Forest perform with the highest accuracy of 97.14% over Wisconsin Breast Cancer Dataset (original). Moreover, random forest displays best performance over all the three datasets. Finally, the paper summarizes the challenges faced together with conclusions drawn and prospective scope of machine learning in breast cancer detection and prognosis.

Keywords Breast cancer · Machine learning · Benign · Malignant · Wisconsin breast cancer dataset

1 Introduction

Cancer is caused by the uncontrolled cell growth in an organ and extending to other tissues around it. Breast cancer develops as cells in the breast tissue split and expand

N. Feroz (✉) · M. A. Ahad · F. Doja
Department of Computer Science and Engineering, Jamia Hamdard, New Delhi, India

A. Choudhary et al. (eds.), *Applications of Artificial Intelligence and Machine Learning*, Lecture Notes in Electrical Engineering 778,
https://doi.org/10.1007/978-981-16-3067-5_33

without usual cell division and cell death controls. The available statistical data indicates that breast cancer is one of the most common forms of cancer globally, but it is also one of the most curable ones, if detected early. The World Research Fund for Cancer (WCRF) has reported that breast cancer is the most widespread cancer in women and the second most prevalent cancer in general [1]. It is one of the leading causes of death in women [2]. Around 2.1 million women are diagnosed with breast cancer annually, with 627,000 deaths reported around the globe in 2018 alone, contributing to nearly 15% of reported cancer related deaths [3]. The "Indian Council of Medical Research (ICMR)" has projected that by 2020, around 1.9 lakh women are likely to be affected with breast cancer in India alone [4]. Since the causes and treatment of breast cancer are still being studied and no preventive measure is available yet [5], the key element for breast cancer regulation is early detection to enhance recovery and survival. Hence, timely detection of breast cancer is a critical issue in the medical sector [6]. "Fine-Needle Aspiration Cytology" (FNAC), "Magnetic Resonance Imaging" (MRI) and mammography are extensively used for breast cancer diagnosis, but these procedures are prone to display false diagnostic results [7]. This may be due to fatigue or inexperience on the part of doctor or technical lapses in the procedures. Mammograms fail to detect about 15% of breast cancer cases [8]. Another common screening method, FNAC, is able to detect about 90% of the cases correctly, hence better diagnosis systems are urgently required [9].

Though the role of an oncologist is considered adequate for local breast cancer diagnosis and treatment, this chronic disease requires systematic therapies for successful treatment necessitating the early diagnosis of the disease. Oncologists require tools to select the medications needed to treat the patients successfully in order to avoid cancer recurrence (prognosis) by minimizing the adverse effects and costs of such therapies. The advances in machine learning, decision support systems and expert systems have led these technologies into diverse domains, including health care. Machine learning classifiers conduct the thorough examination of enormous medical data in shorter time. The retrieval of useful information from historical medical data is one of the critical tasks for accurate diagnosis. These algorithms facilitate learning from past experience, trends and examples [10], and are therefore, being readily used in medical diagnosis. Machine learning involves data collection, model selection, and model training, followed by testing. A machine learning model classifies the data by placing each observation under the right category. In the case of breast cancer prediction, the model classifies whether the tumor is "benign" (non-cancerous) or "malignant" (cancerous) (see Fig. 1).

This work outlines numerous machine learning classifiers to assist doctors in timely diagnosis of cancerous cells in breast cancer patients. This paper is divided into the following sections: Sect. 2 presents a comprehensive review of recent works in machine learning based breast cancer detection and Sect. 3 portrays a comparative analysis of these works. Section 4 presents the experimental setup adopted in the work, Sect. 5 depicts the results obtained, and Sect. 6 discusses the challenges and future scope of machine learning in breast cancer detection.

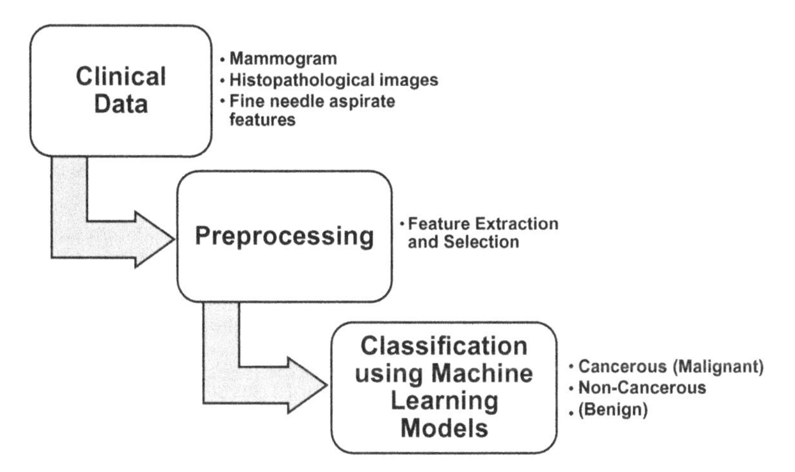

Fig. 1 Series of steps for breast cancer identification in patients

2 Literature Review

The detection of breast cancer by the application of machine learning models/classifiers has been addressed in several previous research studies focused on classifying benign tumors from malignant cancers. This section presents a precise review of 20 recent and notable works in machine learning-based breast cancer prediction. Image data, such as mammography and histopathological images, is mainly used in these studies. The Wisconsin datasets (original and diagnostic) are extensively used in the referred works. The models are evaluated by applying percentage split and K-fold cross validation [11]. The parameter chosen for comparison of performance is prediction accuracy.

A comparative study of NB, C4.5, KNN and SVM for breast cancer prediction, is presented in [12]. The authors have selected the "Wisconsin (original) Breast Cancer Dataset" (WBCD) [13] for analysis. The study has been conducted in WEKA machine learning environment [14]. A fusion of classifiers has been implemented to improve accuracy and it is observed that the combination of "NB + C4.5 + SVM" performs with the maximum accuracy of 97.31%.

In [15], the authors have used genetic algorithms for feature extraction on the original and diagnostic datasets of Wisconsin. The study has been conducted in WEKA environment and the classifier systems analyzed are MLP, SVM, random forest (RF), Bayesian Network, C4.5, RBFN, logistic regression and rotation forest. The performance evaluation shows that GA combined with rotation forest performs exceedingly well with an accuracy of 99.48% for the WBC (diagnostic) dataset.

The authors in [16] have analyzed the performances of 12 models for breast cancer classification. WBCD has been selected as an experimental dataset, and MLP, Decision Table, NB, Lazy K-star, J48, Lazy IBK, Random Tree, LR, AdaBoost-M1, Multiclass Classifier, Random Forest and J-Rip are analyzed. It is observed that

"Tree" and "Lazy" classifiers perform with the highest accuracy of 99.14% while Naïve Bayes has the lowest accuracy of 73.21%.

In [17], the authors have compared SVM and KNN models for early breast cancer detection using the Wisconsin (diagnostic) dataset [18]. The experimental results demonstrate that KNN classifier with "PCA value = 13" exhibits the highest accuracy of 97.489%.

In [19], SVM and KNN have been studied for classifying breast cancer. WBCD (diagnostic) has been taken as an experimental dataset. It is found that the highest accuracy of 98.57% is displayed by SVM.

A novel model "Nested Ensemble (NE)" has been introduced in [20], which is a fusion of multiple ensemble learners and base algorithms (LMT, SGD, BN, REPTree, J48, NB). The authors have used WEKA 3.9.1 tool to develop the proposed model and have chosen WBCD (diagnostic) for analyzing its performance. It is observed that "SV-BayesNet-3-MetaClassifier" and "SV-NaïveBayes-3-MetaClassifier" with $K = 10$, yield the highest accuracy of 98.07%.

The authors in [21] have used genetic algorithm (GA) to identify appropriate biomarkers to detect breast cancer at an early phase. The authors have used the "Breast Cancer Coimbra Dataset" (BCCD) from UCI repository [22] for performing experimentation, containing 116 cases, 9 attributes and no missing values. In BCCD, 64 cases (55.17%) are malignant and 52 (44.83%) are benign. Five ML classifiers have been compared, viz. decision tree, random forest, AdaBoost, gradient boosting and LR. experimental results show that GA combined with gradient boosting classifier exhibits the highest prediction accuracy of 0.7908 (79.08%) and logistic regression performs with the lowest accuracy of 0.6994 (69.94%). From these findings, it is observed that ensemble learners perform better than traditional models.

In [23], the authors have applied J48, LR, NB, random forest, AdaBoost, SVM and Bagging to WBCD and analyzed the findings. The experimental findings show that Random Forest outshines the rest with an accuracy of 95.85% and AdaBoost exhibits the lowest accuracy of 92.60%, when applied to complete dataset. For reduced dataset, both Adaboost and LR exhibit the highest accuracy of 97.92%.

In [24], the authors have proposed "Feature Ensemble Learning" based on "Sparse Autoencoders" and "Softmax Regression" for breast cancer detection. As an experimental dataset, the Wisconsin (diagnostic) dataset is used and "Feature Ensemble based Stacked Autoencoder + Softmax Regression-based Model (FE-SSAE-SM)" and "Stacked Autoencoder and Softmax Regression-based Model (SSAE-SM model)" are analyzed. It is observed that FE-SSAE-SM model (with size of AE1 = 8 and AE2 = 6), performs with the highest accuracy of 98.60%.

In [25], SVM, C4.5, NB and KNN have been evaluated for breast cancer detection on WBCD (original), using the WEKA tool. The results obtained demonstrate that SVM gives the maximum classification accuracy of 97.13%.

The authors have introduced the "Breast Cancer Detection based on Gabor wavelet transform (BCDGWT)" in [26] as an effective method of classifying mammogram lumps. As an experimental dataset, the authors have used the "Digital Database for Screening Mammography" (DDSM) dataset, comprising of 2604 cases [27], and have used "Gabor Wavelet Transform" to obtain features. The paper analyzes five machine learning models: ANN, CHAID, SVM, C5.0 and quest tree on this dataset. It is observed that SVM offers the highest accuracy of 0.968 (96.8%) and ANN offers the lowest accuracy of 0.939 (93.9%).

The authors in [28] have applied four classifiers: NB, J48, KNN and SVM on WBCD and "Breast Cancer Digital Repository" (BCDR). BCDR contains 362 cases, out of which 187 (51.66%) cases are benign and 175 (48.35%) cases are malignant [29]. The authors have applied WrapperSubsetEval [30] method on the datasets for dimensionality reduction. The results show that accuracy is increased in all the four classifiers after dimensionality reduction, with the highest accuracy of 97.9123% shown by SVM on WBCD and 97.9021% by KNN on BCDR.

In [31], the authors have compared the effectiveness of BN, TAN and NB for breast cancer prediction. They have selected seven nominal datasets from the UCI repository [32] for experimentation. It is observed that BN shows the highest accuracy of 74.47%, while TAN shows the lowest accuracy of 69.58%.

The authors in [33] have evaluated the performance of Linear Regression, SVM and KNN classifiers on WBCD (original), for early prediction of breast cancer using minimal features. KNN is found to perform with the highest accuracy of 99.28%.

In [34], the authors have combined mammogram processing and machine learning classifiers for breast cancer prediction, using LR, AdaBoost, gradient boost, decision tree, random forest, KNN and SVM. As an experimental dataset, the "Mammogram Image Analysis Society" (MIAS) dataset [35] has been selected which contains 322 instances. The highest accuracy of 90% is reported in SVM and lowest in Gradient Boost = 52%.

In [36], SVM, NB and Random Forest have been assessed for the classification of malignant and benign tumors in WBCD. The studies have been carried out in R. The experimental results report that the maximum accuracy of 99.42% is exhibited by random forest.

In [37], the performance of three classifiers has been analyzed on the Wisconsin Breast Cancer Dataset (original), viz. Sequential Minimal Optimization (SMO), Best First (BF) Tree and IBK (KNN classifier). The Weka software has been used for simulation, with classification accuracy as the evaluation metric. The experimental findings indicate that Sequential Minimal Optimization (SMO) performs with the highest accuracy of 96.2%.

A comparison of six classifiers has been presented in [38]: SVM, MLP, Gated Recurrent Unit SVM (GRU-SVM), Softmax Regression, Linear Regression and Nearest Neighbor (NN) search. The WBCD (diagnostic) has been used for evaluating the classifiers and the experimental findings reveal that all classifiers performed with an accuracy of over 90%, with MLP demonstrating the highest classification accuracy of 99.04%.

In [39], the authors have provided a holistic analysis of the application of automatic breast cancer detection systems, intending to offer a roadmap for building an advanced breast cancer diagnosis system. A comparative analysis of five machine learning algorithms has been carried out (SVM, MLP, recurrent NN, combined NN and probabilistic NN) on WBCD using MATLAB framework and the classification accuracies noted. The findings show that SVM exhibits the highest diagnostic accuracy of 99.54%.

The authors in [40] have combined three machine learning classifiers (KNN, probabilistic NN and SVM) with feature selection algorithms to distinguish between benign and malignant breast cancer. The authors have used FNAB dataset I (containing 692 samples of FNAB, with 235 malignant instances and 457 benign instances) and gene microarrays dataset II (containing 295 microarrays, with 115 belonging to "good-prognosis" class and 180 belonging to "poor-prognosis" class) for their experimentation. The results indicate that SVM performs with the highest accuracy of 98.80% over dataset I and 96.33% over dataset II.

3 Comparative Analysis of Existing Literature

For better comprehension of the performance of machine learning models covered in the literature review, a chronological tabular representation of the works, datasets used, best performing models and their respective accuracies, is given (see Table 1). Accuracy has been chosen as the measuring parameter for comparing the classifiers covered in the literature review.

4 Experimental Setup

The key aim of this work is to mimic the diagnostic efficiency of human specialists, and extend to exceed human diagnostic performance for breast cancer detection. In addition to diagnostic prediction, this work includes prognostic prediction to assess the clinical effectiveness and likelihood of recovery. The Wisconsin Breast Cancer Dataset (WBCD) has surfaced as a benchmark dataset for the analysis and evaluation of different machine learning techniques. Here, we have applied machine learning classification techniques to three benchmark datasets of Wisconsin. Our analysis is guided toward reaching the best classifier that outperforms the others over all the datasets. The datasets chosen are Wisconsin Breast Cancer Dataset (Original), Wisconsin Breast Cancer Dataset (Diagnostic) and Wisconsin Breast Cancer Dataset (Prognostic) [41]. The features in these datasets are obtained from digitized image of a "fine-needle aspirate" (FNA) of breast tissues of individuals. The first two datasets (original and diagnostic) assign a discrete class (malignant or benign) to new input. The third dataset (prognostic) assigns one of the two discrete classes (recurring or non-recurring) to new input.

Table 1 Comparative analysis of existing literature

S. No.	Author(s) and year	Dataset	Highest accuracy model	Accuracy (%)
1.	Übeyli (2007) [39]	Wisconsin breast cancer dataset	SVM	99.54
2	Osareh and Shadgar (2010) [40]	FNAB dataset I and gene microarrays dataset II	SVM-RBF (for dataset I)	98.80
3	Asri et al. (2016) [25]	WBCD (original)	SVM	97.13
4	Ang et al. (2016) [31]	7 nominal datasets	GBN	74.47
5	Aličković and Subasi (2017) [15]	WBCD (original and diagnostic)	Rotation forest	99.48
6	Islam et al. (2017) [19]	WBCD (diagnostic)	SVM	98.57
7	Elgedawy (2017) [36]	Wisconsin breast cancer dataset	Random forest	99.42
8	Chaurasia and Pal (2017) [37]	WBCD (original)	SMO	96.20
9	Abdar et al. (2020) [20]	WBCD (diagnostic)	SV-BayesNet-3-MetaClassifier ($K = 10$), SV-NaïveBayes-3-MetaClassifier ($K = 10$)	98.07
10	Kumari and Singh (2018) [33]	WBCD (original)	KNN	99.28
11	Agarap (2018) [38]	WBCD (diagnostic)	MLP	99.04
12	Asri et al. (2019) [12]	WBCD (original)	NB + C4.5 + SVM	97.31
13	Mishra et al. (2019) [21]	Breast cancer Coimbra dataset (BCCD)	GBC	79.08
14	Goyal et al. (2019) [23]	Wisconsin breast cancer dataset	AdaBoost, LR	97.92
15	Kadam et al. (2019) [24]	WBCD (diagnostic)	FE-SSAE-SM	98.60

(continued)

Table 1 (continued)

S. No.	Author(s) and year	Dataset	Highest accuracy model	Accuracy (%)
16	Ghasemzadeh et al. (2019) [26]	DDSM dataset	SVM	96.8
17	Shaikh and Ali (2019) [28]	WBCD and breast cancer digital repository (BCDR)	KNN	97.90
18	Kumar et al. (2020) [16]	Wisconsin breast cancer dataset	Lazy (K-star, IBK), Random (tree, forest)	99.14
19	Sadhukhan et al. (2020) [17]	WBCD (diagnostic)	KNN	97.49
20	Kashif et al. (2020) [34]	MIAS cancer dataset	SVM	90.00

4.1 Machine Learning Classifiers Adopted

For this work, we have selected seven popular machine learning classifiers for breast cancer classification. The classifiers are Logistic Regression (LR), *K*-Nearest Neighbor (KNN), Support Vector Machine (Linear), Support Vector Machine (Radial Basis Function), Gaussian Naïve Bayes (GNB), Decision Tree and Random Forest. The classifiers have been selected considering their significant prevalence in the existing literature and notable results in this domain. All the algorithms are supervised machine learning algorithms. A brief description of each classifier is given below:

a. Logistic Regression (LR): LR is based on probability and assigns new data to a discrete group of classes. It may be defined as a "statistical model" that applies a logistic function for modeling a binary target variable.
b. *K*-Nearest Neighbor (KNN): KNN assigns a new data point to one of the available discrete set of classes based on its neighboring "K" data points that are already assigned to groups.
c. SVM (Linear): SVM (Linear) produces a linear function (hyper-plane) for classification problems. This "hyper-plane" partitions the classes explicitly using certain data points known as "support vectors" that participate in obtaining the outcome.
d. SVM (RBF): SVM (RBF) is used when the hyper-plane is not a straight line. RBF is a function with value based on its distance from the origin or a point. In SVM (RBF), the classes are not partitioned with a straight line, but a curve.

e. Gaussian Naïve Bayes (GNB): GNB, contrary to classical Naïve that supports categorical features and models that correspond to a multinomial distribution, supports continuous features and models that correspond to a Gaussian distribution.
f. Decision Tree: Decision tree is a leading classification technique in which a tree is built by partitioning observations in branches recursively in order to enhance the prediction accuracy. It is composed of root, internal nodes and leaf nodes.
g. Random Forest: It is an ensemble form of classifier comprised of several techniques to generate an optimal technique. It is a blend of a number of decision trees that combine to form a forest. To select distinct trees, it uses samples at random with replacement.

4.2 Performance Metrics Adopted

Accuracy has been selected as the criterion used for assessing the performance of the machine learning classifiers covered in this work. Accuracy is a strong indicator of the overall success of the model performance. It is the degree of the correct prediction corresponding to the incorrect predictions. The classification accuracy is obtained from confusion matrix as follows:

$$\text{Accuracy} = \frac{\text{TruePositive} + \text{TrueNegative}}{\text{TruePositive} + \text{FalsePositive} + \text{TrueNegative} + \text{FalseNegative}} \times 100\%$$

4.3 Implementation

We have selected Jupyter Notebook of Anaconda Framework to carry out our experimentation. The implementation has been done using Python language. Given below is the step-wise implementation applied in this work (see Fig. 2).

a. Data Preparation: The first step is to import the essential libraries and datasets into Jupyter. Three Wisconsin Breast Cancer Datasets (original, diagnostic, prognostic) from UCI machine learning repository are selected and imported into Jupyter.
b. Data Exploration: Then, the datasets are examined by obtaining the first few rows of each dataset. Moreover, the datasets can be visualized for better comprehension.
c. Labeling of Categorical Data: The categorical data are converted into numbers so that the models can comprehend them better. Then, the data are split into training and testing subsets. The model learns over the training subset and the performance of the model is tested over the testing subset. 75% data is selected for training the model and 25% is selected for testing the model.

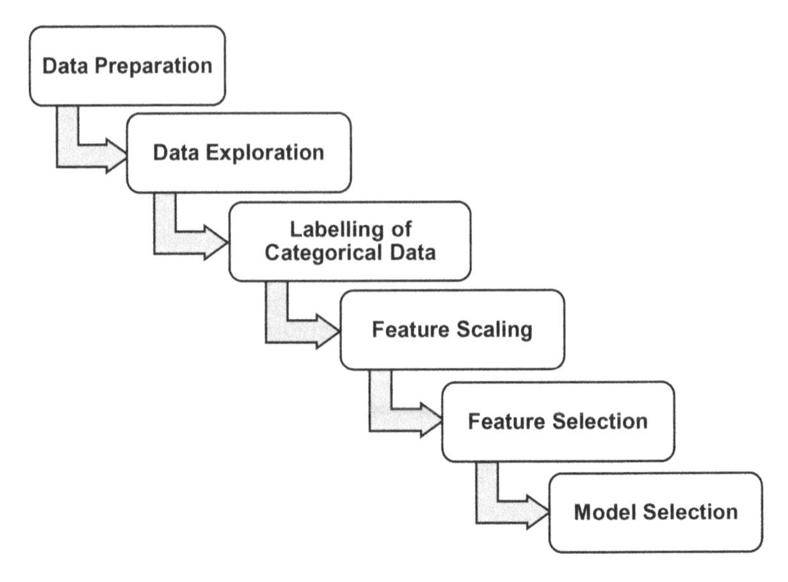

Fig. 2 Implementation

d. Feature Scaling: Feature scaling is performed to get the features within a specific magnitude range. This is done for data normalization.
e. Feature Selection: Feature selection is done to improve the accuracy and predictive power of the classifiers as well as reduce the computation cost. A subset of features is selected that contribute most to the output in order to boost the performance.
f. Model Selection: The models are run over the three datasets and their performance recorded. Then, the model with optimal performance is selected for further implementations.

5 Results

For WBCD (original), the training and testing accuracies of each classifier are given in Table 2.

For WBCD (diagnostic), the training and testing accuracies of each classifier are given in Table 3.

For WBCD (prognostic), the training and testing accuracies of each classifier are given in Table 4.

The highest accuracy reported over all the datasets is 97.14%, achieved by KNN and random forest over WBCD (original). Decision tree achieved over 99% training accuracy over all the three datasets and random forest surfaces as the best performing

Table 2 Training and testing accuracies for WBDC (original)

S. No.	Classifier	Training accuracy (%)	Testing accuracy (%)
1.	Logistic regression	97.32	96.56
2.	KNN	97.51	97.14
3.	SVM (linear)	97.32	96.00
4.	SVM (RBF)	97.32	96.00
5.	GNB	95.41	95.42
6.	Decision tree	99.62	96.00
7.	Random forest	99.42	97.14

Table 3 Training and testing accuracies for WBDC (diagnostic)

S. No.	Classifier	Training accuracy (%)	Testing accuracy (%)
1.	Logistic regression	99.06	94.40
2.	KNN	97.65	95.80
3.	SVM (linear)	98.82	96.50
4.	SVM (RBF)	98.35	96.50
5.	GNB	95.07	92.30
6.	Decision tree	99.62	95.10
7.	Random forest	99.53	96.50

Table 4 Training and testing accuracies for WBDC (prognostic)

S. No.	Classifier	Training accuracy (%)	Testing accuracy (%)
1.	Logistic regression	86.48	82.00
2.	KNN	81.75	82.00
3.	SVM (linear)	87.16	80.00
4.	SVM (RBF)	83.78	76.00
5.	GNB	70.94	78.00
6.	Decision tree	99.62	80.00
7.	Random forest	98.64	82.00

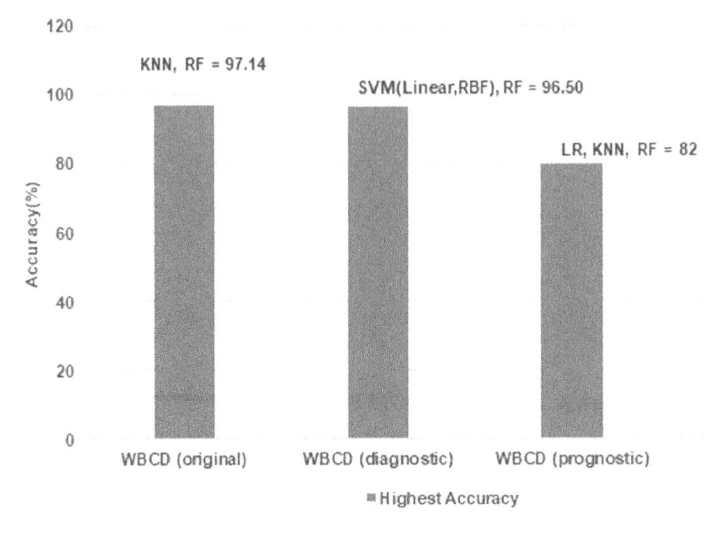

Fig. 3 Highest accuracy reported

classifier over all the three datasets. Moreover, we observe that with feature selection, the results for WBCD (original) and WBCD (prognostic) improve, while no significant improvement is witnessed in the performance of WBCD (diagnostic).

Given below is the highest accuracy reported over each dataset (see Fig. 3).

6 Conclusion, Challenges and Future Scope

A great deal of inter-disciplinary research has been conducted in the domains of medical sciences and machine learning for the detection of malignant and benign breast cancer. This paper analyzes the performance of various classifiers for the diagnosis of breast cancer over different datasets. The findings have demonstrated that the classification efficiency differs with respect to the approach and dataset chosen. A single classifier achieves different accuracies over different datasets. Moreover, the classifiers perform well for diagnostic prediction, but not nearly good enough for prognostic prediction.

This work has been carried out over the three datasets of Wisconsin and the results are restricted to these datasets. In order to obtain a generalization, other diverse benchmark datasets need to be worked upon to ensure that the models produce fairly identical results on other datasets too. The models have been compared based on a single metric (accuracy) and other metrics could be included for better comprehension of their predictive powers.

The development of reliable and efficient classifiers for medical applications remains a major challenge in machine learning domain. Future work could be driven by reduction of time complexity, systematic investigation, and parallel execution of cascaded classifiers, along with larger and diverse datasets to evaluate the models. The recent advances in Deep Learning could be utilized to create a robust automatic breast cancer prediction model for providing enhanced medical services by reducing cost, time and death rate. Since the models and approaches proposed so far in this area operate in offline mode, future work could focus on replicating these systems as online diagnostic systems. Moreover, further research could focus on the development and analysis of classifiers that can uncover models with increased survival rate in datasets.

References

1. World cancer research fund breast cancer statistics. https://www.wcrf.org/dietandcancer/cancer-trends/breast-cancer-statistics. Last accessed 15 Feb 2020
2. Negi R, Mathew R (2018, December) Machine learning algorithms for diagnosis of breast cancer. In: International conference on computer networks, big data and IoT. Springer, Cham, pp 928–932
3. World Health Organization (2020) Cancer: breast cancer. https://www.who.int/cancer/prevention/diagnosis-screening/breast-cancer/en/. Last accessed 15 Feb 2020
4. Cancer cases in India likely to soar 25% by 2020: ICMR. https://timesofindia.indiatimes.com/india/Cancer-cases-in-India-likely-to-soar-25-by-2020-ICMR/articleshow/52334632.cms. Last accessed 20 Jan 2020
5. Suri JS, Chang RF, Giraldi GA, Rodrigues PS (2006, October) Non-extensive entropy for cad systems of breast cancer images. In: 2006 19th Brazilian symposium on computer graphics and image processing. IEEE, pp 121–128
6. Bhardwaj A, Tiwari A (2015) Breast cancer diagnosis using genetically optimized neural network model. Expert Syst Appl 42(10):4611–4620
7. Elmore JG, Wells CK, Lee CH, Howard DH, Feinstein AR (1994) Variability in radiologists' interpretations of mammograms. N Engl J Med 331(22):1493–1499
8. Alarabeyyat A, Alhanahnah M (2016, August) Breast cancer detection using k-nearest neighbor machine learning algorithm. In: 2016 9th International conference on developments in esystems engineering (DeSE). IEEE, pp 35–39
9. Karabatak M (2015) A new classifier for breast cancer detection based on Naïve Bayesian. Measurement 72:32–36
10. Witten IH, Frank E (2002) Data mining: practical machine learning tools and techniques with Java implementations. ACM SIGMOD Rec 31(1):76–77
11. Gutierrez-Osuna R, Nagle HT (1999) A method for evaluating data-preprocessing techniques for odour classification with an array of gas sensors. IEEE Trans Syst Man Cybern Part B (Cybern) 29(5):626–632
12. Asri H, Mousannif H, Al Moatassim H (2019, July) A hybrid data mining classifier for breast cancer prediction. In: International conference on advanced intelligent systems for sustainable development. Springer, Cham, pp 9–16
13. UCI machine learning repository: breast cancer Wisconsin (original) data set. https://archive.ics.uci.edu/ml/datasets/Breast+Cancer+Wisconsin+%28Original%29. Last accessed 10 Jan 2020
14. Hall M, Frank E, Holmes G, Pfahringer B, Reutemann P, Witten IH (2009) The WEKA data mining software: an update. ACM SIGKDD Explor Newsl 11(1):10–18
15. Aličković E, Subasi A (2017) Breast cancer diagnosis using GA feature selection and rotation forest. Neural Comput Appl 28(4):753–763

16. Kumar V, Mishra BK, Mazzara M, Thanh DN, Verma A (2020) Prediction of malignant and benign breast cancer: a data mining approach in healthcare applications. In: Advances in data science and management. Springer, Singapore, pp 435–442
17. Sadhukhan S, Upadhyay N, Chakraborty P (2020) Breast cancer diagnosis using image processing and machine learning. In: Emerging technology in modeling and graphics. Springer, Singapore, pp 113–127
18. UCI machine learning repository: breast cancer Wisconsin (diagnostic) data set. http://archive.ics.uci.edu/ml/datasets/Breast?Cancer?Wisconsin?(Diagnostic). Last accessed 20 Jan 2020
19. Islam MM, Iqbal H, Haque MR, Hasan MK (2017, December) Prediction of breast cancer using support vector machine and *K*-nearest neighbors. In: 2017 IEEE region 10 humanitarian technology conference (R10-HTC). IEEE, pp 226–229
20. Abdar M, Zomorodi-Moghadam M, Zhou X, Gururajan R, Tao X, Barua PD, Gururajan R (2020) A new nested ensemble technique for automated diagnosis of breast cancer. Pattern Recogn Lett 132:123–131
21. Mishra AK, Roy P, Bandyopadhyay S (2019, September) Genetic algorithm based selection of appropriate biomarkers for improved breast cancer prediction. In: Proceedings of SAI intelligent systems conference. Springer, Cham, pp 724–732
22. Breast cancer Coimbra data set: UCI machine learning repository. https://archive.ics.uci.edu/ml/datasets/Breast+Cancer+Coimbra. Last accessed 15 Feb 2020
23. Goyal K, Sodhi P, Aggarwal P, Kumar M (2019) Comparative analysis of machine learning algorithms for breast cancer prognosis. In: Proceedings of 2nd international conference on communication, computing and networking. Springer, Singapore, pp 727–734
24. Kadam VJ, Jadhav SM, Vijayakumar K (2019) Breast cancer diagnosis using feature ensemble learning based on stacked sparse autoencoders and softmax regression. J Med Syst 43(8):263
25. Asri H, Mousannif H, Al Moatassime H, Noel T (2016) Using machine learning algorithms for breast cancer risk prediction and diagnosis. Procedia Comput Sci 83:1064–1069
26. Ghasemzadeh A, Azad SS, Esmaeili E (2019) Breast cancer detection based on Gabor-wavelet transform and machine learning methods. Int J Mach Learn Cybern 10(7):1603–1612
27. Ebrahimpour MK, Mirvaziri H, Sattari-Naeini V (2018) Improving breast cancer classification by dimensional reduction on mammograms. Comput Methods Biomech Biomed Eng Imaging Vis 6(6):618–628
28. Shaikh TA, Ali R (2019) Applying machine learning algorithms for early diagnosis and prediction of breast cancer risk. In: Proceedings of 2nd international conference on communication, computing and networking. Springer, Singapore, pp 589–598
29. Lopez MG, Posada N, Moura DC, Pollán RR, Valiente JMF, Ortega CS et al (2012, July) BCDR: a breast cancer digital repository. In: 15th International conference on experimental mechanics, vol 1215
30. Kohavi R, John GH (1997) Wrappers for feature subset selection. Artif Intell 97(1–2):273–324
31. Ang SL, Ong HC, Low HC (2016) Classification using the general Bayesian network. Pertanika J Sci Technol 24(1):205–211
32. Lichman M (2020) UCI machine learning repository. University of California, School of Information and Computer Science, Irvine, CA. http://archive.ics.uci.edu/ml. Last accessed 20 Jan 2020
33. Kumari M, Singh V (2018) Breast cancer prediction system. Procedia Comput Sci 132:371–376
34. Kashif M, Malik KR, Jabbar S, Chaudhry J (2020) Application of machine learning and image processing for detection of breast cancer. In: Innovation in health informatics. Academic Press, pp 145–162
35. Suckling J, Parker J, Dance D, Astley S, Hutt I, Boggis C et al (2015) Mammographic image analysis society (MIAS) database. https://www.repository.cam.ac.uk/handle/1810/250394. Last accessed 01 Feb 2020
36. Elgedawy M (2017) Prediction of breast cancer using random forest, support vector machines and Naive Bayes. Int J Eng Comput Sci 6(1):19884–19889
37. Chaurasia V, Pal S (2017) A novel approach for breast cancer detection using data mining techniques. Int J Innov Res Comput Commun Eng (An ISO 3297: 2007 Certified Organization), vol 2

38. Agarap AFM (2018, February) On breast cancer detection: an application of machine learning algorithms on the Wisconsin diagnostic dataset. In: Proceedings of the 2nd international conference on machine learning and soft computing, pp 5–9
39. Übeyli ED (2007) Implementing automated diagnostic systems for breast cancer detection. Expert Syst Appl 33(4):1054–1062
40. Osareh A, Shadgar B (2010, April) Machine learning techniques to diagnose breast cancer. In 2010 5th international symposium on health informatics and bioinformatics. IEEE, pp 114–120
41. UCI machine learning repository: breast cancer Wisconsin (prognostic) data set. http://archive.ics.uci.edu/ml/datasets/Breast?Cancer?Wisconsin?(Prognostic). Last accessed 20 May 2020

Multiclass Classification of Histology Images of Breast Cancer Using Improved Deep Learning Approach

Jyoti Kundale and Sudhir Dhage

Abstract Breast Cancer has become a serious threat to women life in the overall word. It is necessary to diagnosis for Breast cancer early to avoid mortality rate. In the era of medical imagining with the advent of Artificial Intelligence systems, it has become easy to detect breast cancer at an early stage. Histopathology image modality is one of the best ways for breast cancer detection as it is easy to store in digital format for a long time. Breast cancer is classified into two main categories one is benign and malignant. These two classes are further divided into subclasses. In this paper, we have proposed an improved deep learning model for breast cancer multiclass classification. This proposed model uses a two-level approach one is blocked based and image-based. The blocked based approach is used to reduce overhead with lower-cost processing. Here ensemble learning approach is used for feature extraction which is a combined approach of the various pre-trained model without negotiating accuracy of the system. Final classification is done based on image based improved deep learning approach into eight classes.

Keywords Histopathology images · Deep learning · Ensemble model

1 Introduction

Even though outstanding modern developments in diagnosis and treatment, now-a-day's cancer is becoming a massive public health problem that leads to mortality around the world [1]. Due to demographic aging and the adoption of bad habits before restricted to industrialized countries are now contributing to an increase in the occurrence of this type of disease. Cancer is dangerous threat to human life. The survey done by World Cancer Research Fund (WCRF) says that there is a 15–20%

J. Kundale (✉)
Ramrao Adik Institute of Technology, D Y Patil University, Navi Mumbai, India

Rajiv Gandhi Institute of Technology, University of Mumbai, Mumbai, India

S. Dhage
Sardar Patel Institute of Technology, University of Mumbai, Mumbai, India
e-mail: sudhir_dhage@spit.ac.in

A. Choudhary et al. (eds.), *Applications of Artificial Intelligence and Machine Learning*, Lecture Notes in Electrical Engineering 778,
https://doi.org/10.1007/978-981-16-3067-5_34

rise in cancer patients every year and there will be chances of 28 million new cases of cancer until the next decade [2]. Cancer is a disease in which division of abnormal cell takes place uncontrollably and destroys other healthy body tissue; finally forms a tumor.

Almost 9.6 million lives are threatened by cancer; unfortunately, India has the largest share in it which is 8.17%. The most worrying thing is that India will have 1.18 million new cancer cases this year and more than 50% of these will be diagnosed in women, said by the World Health Organization (WHO) [2]. Women age range between 21 and 59 years and aged more than 59 years are caused by death by breast cancer [3]. In the early era, cervical cancer was most commonly found in women, but now it is surpassed by breast cancer. There are 14.5 million people who have died because of cancer and in 2030 the count will be above 28 million [4]. In the USA, after lung and bronchus cancer, major death is found by the breast cancer for the woman which is 14% (a total of 41,000 in 2017) study done by the American Cancer Society (ACS) [5].

By overall estimation, at least 1,797,900 women in India may have breast cancer in the year 2020. According to a survey taken at Tata memorial hospital, breast cancer has been reported to occur in 1 woman out of 2000 during the period 1975–78. But today in India, it occurs 1 in 10, so it is required to take preventive steps against this dangerous disease-like breast cancer [6].

Reasons are a delay in bearing a child, absence of breastfeeding, medical usage of hormones, lack of understanding of early symptoms of breast cancer, lack of advanced screening methods, and other factors like absence of diagnostic centers and knowledgeable oncologists. So there is a need to have a domain that will mainly focus on early stage discovery, analysis, and treatment of breast cancer in women.

The main aim to build systems that will detect cancer at an early stage and other is to diagnose and give proper treatment to breast cancer [7]. Currently, no such vaccination is available to prevent breast cancer. Hence proper preventive measures are not present. However, complete curing of breast cancer is possible if it is detected in its earlier stages. The rate of survival will improve by 95% if early detection of breast cancer is done for a patient. Hence, the mortality rate will reduce automatically [8].

There are several ways to detect cancer of breasts such as clinical examination, mammograms, magnetic resonance imaging (MRI), ultrasound, and biopsy. But the main challenge is to detect cancer by the use of imaging. In the biopsy method, a doctor uses a fine needle to take out a tissue from a suspicious area, which will result in the detection of cancer surely. Various types of biopsy techniques, procedures are there such as vacuum-assisted breast biopsy (VABB), fine needle aspiration (FNA), core needle biopsy (CNB) and surgical, (open) biopsy (SOB) stand out. Then, colored dyes are put into tissues, which are observed under a microscope to get useful information about the tissue compositions. Histopathology images play a vital role in proper diagnoses of every type of cancer [9, 10].

Breast cancer for the female population has a large impact on public health; so there is a need to provide tools that will automatically classify breast cancer using digitized images from histological slides, which will support the Pathologists, so that it can be diagnosed early [11].

Organization of the paper as follows: Sect. 2 gives an idea about existing methods for classification. Section 3 helps to understand drawbacks/limitations in existing systems. Section 4 shows the Proposed Novel Framework for multiclass classification followed by the conclusion is given in Sect. 5. Subsequent paragraphs, however, are indented.

2 Review of Existing System

After a lung cancer second most commonly found cancer is breast cancer [1]. In the year 2020, around 8.2 million women may cause death due to cancer said by International Agency for Research on Cancer (IARC), which is part of the World Health Organization (WHO) [2] and a some cases is likely to increase more than 27 million by 2030 [3]. By using histology and radiology images breast cancer can be diagnosed. With the help of radiology images analysis, it is easy to identify the abnormal areas, but difficult to verify the cancerous area [8, 9]. To check whether cancer is present or not, the biopsy method is used in which tissues are studied under an ultra-microscope for confirmation of cancer tissue [10, 11]. The study of tissue related cell structure, distribution and its shape done by Histopathologists, based on which level malignancy is decided [9, 11].

Md Zahangir Alom et al. used Inception Recurrent Residual Convolution Neural Network (IRRCNN) model which combination of Inception Network (Inception-v4), the Residual Network (ResNet) the Recurrent Convolution Neural Network (RCNN). They have used BreakHis and Breast Cancer (BC) classification challenge 2015 [5]. Nadia Brancati et al. used deep learning methods for automatic classification that lead to the detection of invasive ductal carcinoma in breast histological images and sub-types classification of lymphoma. In both the cases they have used Residual Convolution Neural Network which is part of a Convolution Auto encoder Network [12]. P. Sherubha et al. uses Multi-Classed feature selection algorithm for feature selection and adaptive Neuro-Fuzzy k-NN Classification algorithm is used to improve the performance of classification [13]. Nidhi Ranjan et al. make use of multilayer CNN with a hierarchical classifier to deal with inter-class similarity and intra-class variability issues. The accuracy is 79% and 95% for hierarchical classifier 1 and hierarchical classifier 2, respectively [14]. Majid Nawaz et al. make uses of DenseNet CNN model achieved high-processing performances with around 96% of accuracy [15]. In [16] using CNN deep features are extracted and these extracted features are used as input to a traditional classifier. Sara Reis et al. uses multi-scale basic image features (BIFs) and local binary patterns (LBPs), in combination with random decision trees classifier for classification of breast cancer stroma regions-of-interest (ROI) with an accuracy of 84% [17]. Spanhol et al. [18] uses different feature extractor and different classifier for classification of histopathological images of breast cancer which is not handled by a small set of databases. Bayramoglu et al. [19] worked on break his dataset by using a single task convolution neural network for malignancy and multi-task convolution neural network both malignancy and image magnification level simultaneously. In [20] authors have done classification into four classes, i.e., normal tissue, benign lesion, in-situ carcinoma and invasive carcinoma using CNN and support vector machine (SVM). A. Nahid et al. use CNN; it takes raw images as input. In the next step, it extracts global features and classifies the histopathological images [21]. In [22] they suggested a tiny SE-ResNet module that improves the mixture of residual module and Squeeze-and-Excitation block and achieves comparable efficiency with fewer parameters, also suggested a novel learning rate scheduler that can deliver outstanding output. The recommended framework includes several measures including image enhancement, image segmentation, extraction of feature, and classification of images. In the segmentation phase, the suggested framework uses a novel mixture of watershed algorithms and *K*-means clustering which uses ruled based and decision tree classifier to classify images into benign and malignant classes. In [23] author have proposed CNN a deep learning approach with help of fully connected layers to enhance the performance of classification at various image magnifications (40×, 100×, 20× and 400×).

3 Limitations of Existing Systems

The aim is to design a system such that it will efficiently and accurately classify histopathological images into non-cancerous or cancerous tissues in less time, so following relevant issues should be properly addressed

1. Need for Accurate feature extraction using ensemble modeling:
   Histopathological images classification involves one highly challenging step of feature extraction. It is quite difficult to decide upon which features to use for classification. Also, the choice of features used has a great impact on the entire classification process. Current research into histopathological images classification that makes use of features may not be the best way of classification, especially when deep learning methods have been commonly employed recently. So, it will be a great choice if we make use of ensemble modeling for appropriate feature extraction. By considering strength of every different model, it is possible to extract essential features, it will be helpful during multiclass classification.
2. Need to handle massively scaled data without overlapping classes and negotiating accuracy at the same time:
   As the number of patients is more prone to breast cancer, the size of the dataset also increases. There is a chance of overlapping classes that affect the accuracy of the system which leads to misclassification. An efficient system is needed that will able handle scalable data without compromising accuracy during multiclass classification.
3. Need to develop a model based on semi-supervised and unsupervised learning-based algorithms:

In the existing dataset, the most commonly supervised learning approach is used where images are labeled. But in real-time, it is not guaranteed that all the images are correctly labeled. It is a very tedious and time-consuming task to label images by an only specialized pathologist. So there is a need to develop a model which is can work well in case of non-labeled images based on the clustering approach.

4. Lack of comprehensive classification based on the fusion of dataset by various image modalities:
   Till now, the dataset is available for single image modalities for classification. It is needed to have a dataset of different image modalities in the single dataset and the system should able to perform multiclass classification for such types of datasets efficiently. It is necessary so that a patient can get treatment according to breast cancer according to its type and stage.

4 Dual Stage Multiclass Breast Cancer Classification Using Deep Learning Framework

There is a need to develop a model that will comprehensively classify breast cancer histological images automatically, which will helpful for a pathologist. The system will categorize histopathological images into different types of malignant or benign classes. The goal is to achieve high accuracy and reduce false negatives, i.e., ensure that we do not miss cancer detection. Most of the current histological breast cancer image classification work has been done based on whole-slide imaging (WSI). There are various issues present in existing technology mainly, difficulty in handling a large size database, resource-intensive processing, and high cost etc. The aim is to classify the breast cancer histopathological images into different classes using an economical viable process without compromising accuracy. Figure 1 shows proposed framework for dual stage multiclass breast cancer classification using deep learning approach.

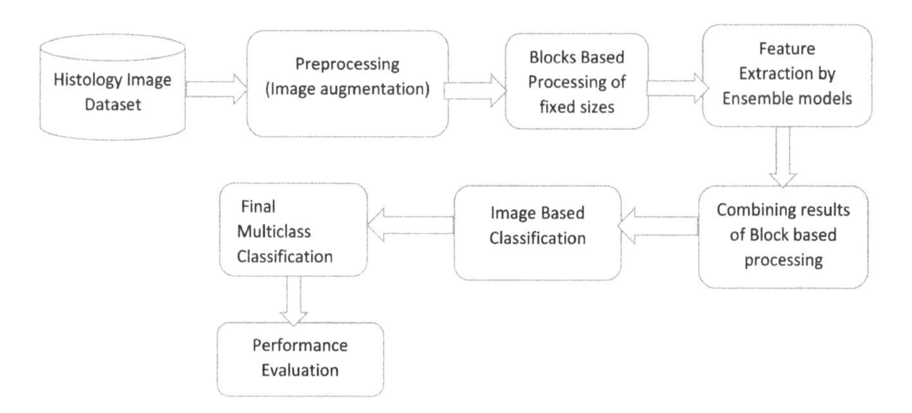

Fig. 1 Dual-stage multiclass breast cancer classification using deep learning framework

Steps for the proposed framework as follows:

1. Histological images are used as input for classification.
2. Preprocessing is done on entire the dataset using various techniques such as data augmentation, staining, normalization.
3. Instead of taking the whole magnification image, we consider image block, which will be reducing overhead.
4. Constant-sized blocks are retrieved by sliding a block-sized L by L and Sr stride over an image.
5. This will help to down sampled total image size. More discriminative features are extracted from image block by using various ensemble models like pre-trained model or other deep learning model used as a feature extractor such as VGGnet, ResNet, and Inception.
6. Thus, we retrieve total number of $[1 + (Ic - L)/Sr] * [1 + (Ir - L)/Sr]$ blocks, Ic and Ir being image columns and rows.
7. It is necessary to select the optimum size of stride Sr as here overlapping of blocks can be considered. Overlapping blocks may contain crucial information regarding features.
8. The result of blocked based processing are combined using majority voting or maximum probability approach and given as input to image-based improved deep learning.
9. Image-based improved deep learning model is trained to perform final classification based on local features given by each block and global feature shared among the different blocks.
10. The final multiclass classification that is into eight subclasses is done by image-wise deep learning through various training and validation.
11. Various performance measure parameters such as Accuracy, Recall, Precision, F_1-score are calculated.

There are various dataset is used by the researcher like WDBC, BACH etc. BreakHis dataset [24] is commonly used by most of the researcher. It consists of 7909 images, having two main classes such as benign and malignant. These two main classes are further divided into eight subclasses. Data augmentation technique is most commonly used as preprocessing technique, where images are flipped horizontally, vertically 90 and 180 degrees. This preprocessing techniques help to avoid problem of over-fitting. Combination of different pre-trained model which can give different relevant feature used as feature extractor on every equal-sized block of images. Result of classification of block-based approach which depends on maximum voting, maximum probability of particular class. These results can be given as further input to image-wise improved deep learning classification model. In this approach image-wise classification is done using combination of CNN and pre-trained model. Finally, images are classified into eight classes, performance of the system can be evaluated using confusion matrix.

5 Experimental Analysis

The experiments are carried out on BreakHis Dataset, which Consists of 7909 images of malignant and benign classes. Each of this classes are subdivided into four more subclasses of each. Histology image is taken from database. To avoid over fitting issue, we have applied the data augmentation techniques to increase the size of the dataset. In preprocessing step, images are rotated by 90, 180°. From the whole image, maximum six non overlapping patches are generated of size $224 \times 224 \times 3$ for each magnification factor ($40\times$, $100\times$, $200\times$, $400\times$). This methods helps to void system overhead due to large size of the image. On each patch pre-trained models are applied like VGG16, VGG19, GoogleNet, ResNet which is used as feature extractor. After extracting feature from every patch by maximum voting results are combined for image based classification. Finally, images are classified into eight classes by support vector machine (SVM).

Table 1 shows comparative analysis of existing method with proposed method. In table, italic values shows highest accuracy for respective to pre-trained model. In $40\times$ magnification factor Resnet50 gives accuracy of 88.23%. ResNet101 gives 89.67% accuracy for $100\times$ magnification factor. 91.00%, accuracy is achieved for $200\times$ magnification factor. In case of $400\times$ magnification factor, accuracy is achieved is 85%. Drawback of existing method is that authors have not considered pre-trained model for feature extraction. Features are extracted using as handcrafted method or CNN from scratch. Therefore, overall performance of model is less in terms of accuracy. Figure 2 shows performance analysis of various Deep Learning (CNN, De-novo model) as well as Pre-trained based model (VGG16, VGG19, GoogleNet, RestNet50, ResNet 101) for magnification factor 40X (zooming factor) along with SVM in terms of Accuracy, Precision, Recall, F1-score. Figure 3 shows performance analysis of various Deep Learning (CNN, De-novo model) as well as Pre-trained based model (VGG16, VGG19, GoogleNet, RestNet50, ResNet 101) for magnification factor 100X (zooming factor) along with SVM in terms of Accuracy, Precision, Recall, F1-score. Figure 4 shows performance analysis of various Deep Learning (CNN, De-novo model) as well as Pre-trained based model (VGG16, VGG19, GoogleNet, RestNet50, ResNet 101) for magnification factor 200X (zooming factor) along with SVM in terms of Accuracy, Precision, Recall, F1-score. Figure 5 shows performance analysis of various Deep Learning (CNN, De-novo model) as well as Pre-trained based model (VGG16, VGG19, GoogleNet, RestNet50, ResNet 101) for magnification factor 400X (zooming factor) along with SVM in terms of Accuracy, Precision, Recall, F1-score.

Table 1 Comparative analysis with different magnification factor for proposed model

Magnification factor	Pre-trained network + classification	Performance measure parameter (percentage)			
		Accuracy	Precision	Recall	F_1 score
40×	CNN + augmented data [25]	83.79	84.27	83.79	83.74
	CNN + SVM [25]	82.89	–	–	–
	De-novo model [26]	89.60	–	–	–
	Combination of CNN from scratch [26]	85.60	–	–	–
	VGG16 + SVM	87.12	85.12	90.00	87.49
	VGG19 + SVM	86.79	89.10	91.00	90.04
	GoogleNet + SVM	81.12	84.23	88.97	86.54
	ResNet50 + SVM	*88.23*	90.12	91.31	90.71
	ResNet101 + SVM	85.45	87.00	92.67	89.75
100×	CNN + augmented data [25]	84.48	84.29	84.48	84.31
	CNN + SVM [25]	80.94	–	–	–
	De-novo model [26]	85.00	–	–	–
	Combination of CNN from scratch [26]	83.50	–	–	–
	VGG16 + SVM	88.12	89.74	92.23	90.97
	VGG19 + SVM	89.15	89.57	92.67	91.09
	GoogleNet + SVM	84.32	85.92	91.47	88.61
	ResNet50 + SVM	86.00	91.53	90.18	90.85
	ResNet101 + SVM	*89.67*	92.12	93.00	92.56
200×	CNN + augmented data [25]	80.83	81.85	80.83	80.48
	CNN + SVM [25]	79.44	–	–	–
	De-novo model [26]	84.00	–	–	–
	Combination of CNN from scratch [26]	83.10	–	–	–
	VGG16 + SVM	87.45	88.42	91.27	89.82

(continued)

Table 1 (continued)

Magnification factor	Pre-trained network + classification	Performance measure parameter (percentage)			
		Accuracy	Precision	Recall	F_1 score
	VGG19 + SVM	86.32	89.15	89.83	89.49
	GoogleNet + SVM	83.12	83.29	90.10	86.56
	ResNet50 + SVM	*91.00*	92.00	91.27	91.63
	ResNet101 + SVM	90.70	93.73	92.45	93.09
400×	CNN + augmented data [25]	81.03	80.84	81.03	80.63
	CNN + SVM [25]	77.94	–	–	–
	De-novo model [26]	80.80	–	–	–
	Combination of CNN from scratch [26]	80.90	–	–	–
	VGG16 + SVM	*85.00*	87.17	90.71	88.90
	VGG19 + SVM	81.61	83.65	89.00	86.24
	GoogleNet + SVM	79.44	82.62	89.79	86.06
	ResNet50 + SVM	83.73	85.79	90.12	87.90
	ResNet101 + SVM	84.13	86.45	88.23	87.33

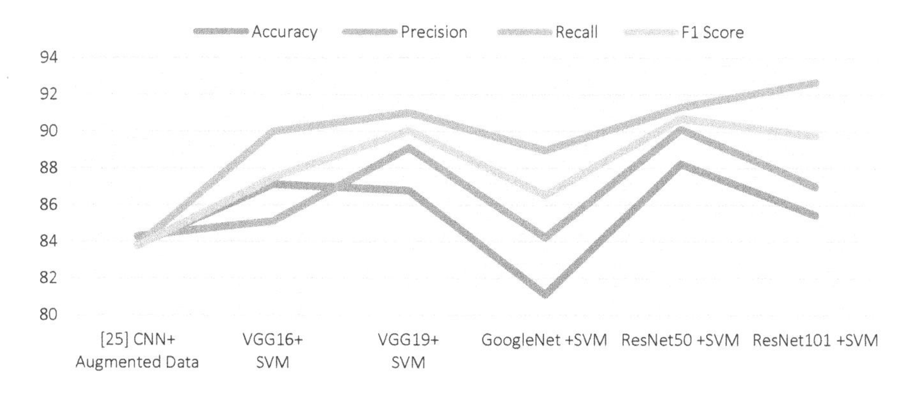

Fig. 2 Magnification factor: 40×

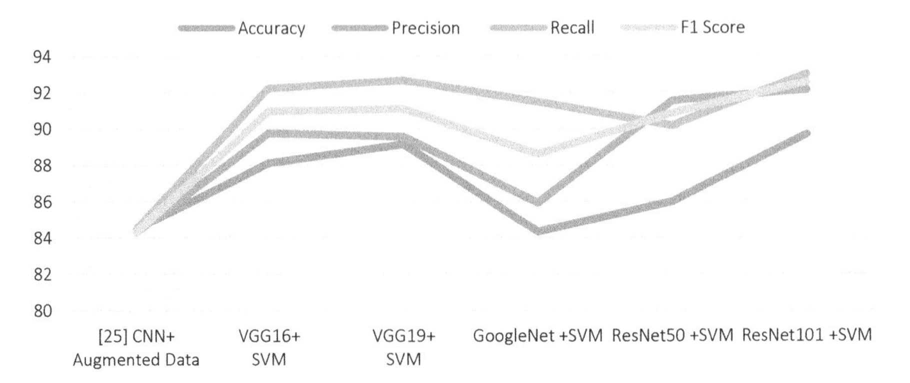

Fig. 3 Magnification factor: 100×

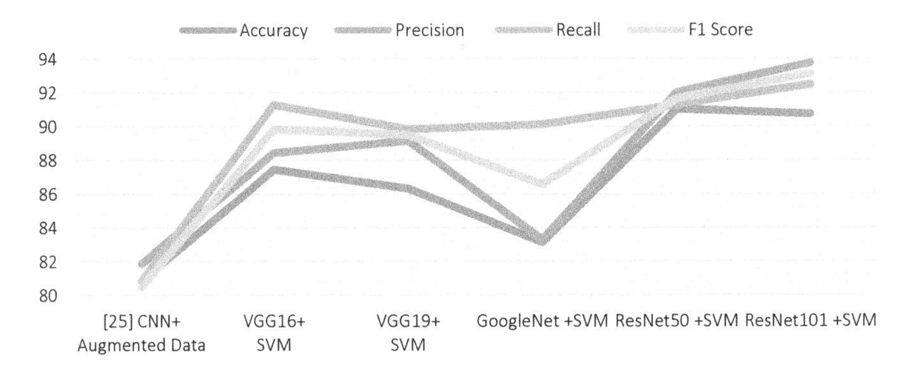

Fig. 4 Magnification factor: 200×

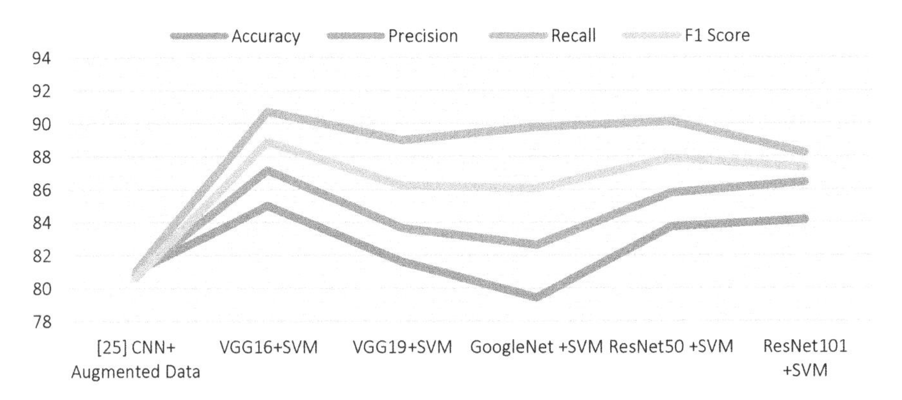

Fig. 5 Magnification factor: 400×

6 Conclusion

This paper represents novel framework for multiclass classification for breast cancer using an ensemble of the model for classification. In ensemble model, one can use different various pre-trained network such as VGG16, VGG19, Inception, ResNet for feature extraction, as features carries relevant information. After extracting feature from various ensemble model, blocked wise classification is done to reduce system overhead. Output of blocked based processing is given to the image-wise classification where improved deep learning model is used for classification to improve the overall accuracy of the system. In case of 40×, 100× and 200× ResNet + SVM gives 88.23%, 89.67%, 91% accuracy. The pre-trained VGG16 + SVM gives 85% accuracy for 400× magnification factor.

7 Future Scope

In future, other preprocessing technique can be applied like normalization, staining. The overall accuracy of system can be improved by fine tuning each layer and applying hybrid approach, while performing multiclass classification. Focus can be given on to number of patch generation with respect to magnification factor. Time complexity is one of the measure aspect should be considered in future, while considering different magnification factors.

References

1. Chun MC (2018) Breast cancer: symptoms, risk factors, and treatment. Medical News Today. https://www.medicalnewstoday.com/articles/37136.php. 10 Mar 2018
2. World Health Organization. http://www.who.int/en/. 10 Mar 2018
3. Boyle P, Levin B (2008) World cancer report. http://www.iarc.fr/en/publications/pdfsonline/wcr/2008/wcr2008.pdf
4. Arau T, Aresta G, Castro E, Rouco J, Aguiar P, Eloy C, Polonia A, Campilho A (2017) Classification of breast cancer histology images using Convolutional Neural Networks. Research Article, 1 June 2017
5. Alom MZ, Yakopcic C, Nasrin MS, Taha TM, Asari VK (2019) Breast cancer classification from histopathological images with inception recurrent residual convolutional neural network. J Digit Imaging
6. Rangarajan B, Shet T, Wadasadawala T, Nair NS, Madhu Sairam R, Hingmire SS, Bajpai J (2016) Breast cancer: an overview of published Indian data. South Asian J Cancer 5(3):86–92. https://doi.org/10.4103/2278-330X.187561
7. Xie J, Liu R, Luttrell IV J, Zhang C (2019) Deep learning based analysis of histopathological images of breast cancer. Front Genet, 19 Feb 2019 [Online]. Available: https://doi.org/10.3389/fgene.2019.00080
8. Joy J, Penhoet E, Petitti D (2005) Saving women's lives. National Academies Press, Washington, DC, USA

9. He L, Long LR, Antani S, Thoma GR (2012) Histology image analysis for carcinoma detection and grading. Comput Methods Programs Biomed 107(3):538–556
10. Nbcf. Biopsy: The National Breast Cancer Foundation. http://www.nationalbreastcancer.org/breast-cancer-biopsy. 10 Mar 2018
11. He L, Long LR, Antani S, Thoma G (2010) Computer-assisted diagnosis in histopathology. Seq Genome Anal Methods Appl 271–287
12. Brancati N, De Pietro G, Frucci M, Riccio D (2018) A deep learning approach for breast invasive ductal carcinoma detection and lymphoma multi-classification in histological images. IEEE Transl Content Min
13. Sherubha P, Banu Chitra M, Narmadha B (2018) Multi-class feature selection algorithm (MCFSA) for breast cancer detection. Int J Pure Appl Math 118(11):301–306
14. Ranjan N, Machingal PV, Jammalmadka SSD, Thenkanidiyoo V (2018) Hierarchical approach for breast cancer histopathology images classification. In: Conference on medical imaging with deep learning, Amsterdam, The Netherlands (MIDL 2018)
15. Nawaz M, Sewissy AA, Soliman THA (2018) Multi-class breast cancer classification using deep learning convolutional neural network. (IJACSA) Int J Adv Comput Sci Appl 9(6)
16. Araújo T et al (2017) Classification of breast cancer histology images using convolutional neural networks. PLoS ONE 12(6):e0177544
17. Reis S, Gazinska P, Hipwell JH, Mertzanidou T, Naidoo K, Williams N, Pinder S, Hawkes DJ (2016) Automated classification of breast cancer stroma maturity from histological images. IEEE Trans Biomed Eng
18. Spanhol FA, Oliveira LS, Petitjean C, Heutte L (2016) Breast cancer histopathological image classification using convolutional neural networks. In: Proceedings of the international joint conference on neural networks (IJCNN), pp 2560–256, July 2016
19. Bayramoglu N, Kannala J, Heikkilä J (2016) Deep learning for magnification independent breast cancer histopathology image classification. In: Proceedings of the 23rd international conference on pattern recognition (ICPR), pp 2440–2445, Dec 2016
20. Bardou D, Zhang K, Ahmad SM (2018) Classification of breast cancer based on histology images using convolutional neural networks. IEEE Transl Content Min
21. Nahid A-A, Kong Y (2018) Histopathological breast-image classification using local and frequency domains by the convolutional neural network. Information 9(1):19
22. Jiang Y, Chen L, Zhang H, Xiao X (2019) Breast cancer histopathological image classification using convolutional neural networks with small SE-ResNet module, 29 Mar 2019
23. Al Rahhal MM (2018) Breast cancer classification in histopathological images using convolutional neural network. Int J Adv Comput Sci Appl 9(3)
24. Spanhol FA, Oliveira LS, Petitjean C, Heutte L (2015) A dataset for breast cancer histopathological image classification. IEEE Trans Biomed Eng
25. Bardou D, Zhang K, Ahmad SM (2018) Classification of breast cancer "based on histology images using convolutional neural networks." IEEE Access 6:24680–24693
26. Spanhol FA, Oliveira LS, Petitjean C, Heutte L (2016) Breast cancer histopathological image classification using convolutional neural networks. In: 2016 International joint conference on neural networks (IJCNN). IEEE, Vancouver, British Columbia, Canada, pp 2560–2567. ISBN 1509006206. https://doi.org/10.1109/IJCNN.2016.7727519

Enhancing the Network Performance of Wireless Sensor Networks on Meta-heuristic Approach: Grey Wolf Optimization

Biswa Mohan Sahoo, Tarachand Amgoth, and Hari Mohan Pandey

Abstract The sensing technology has brought all advancements in the human lives. Wireless sensor network (WSN) has proven to be a promising solution to acquire the information from the remote areas. However, the energy constraints of the sensor nodes have obstructed the widely spread application zone of WSN. There has been a great magnitude of efforts reported for acquiring the energy efficiency in WSN, these efforts varying from conventional approaches to the meta-heuristic method for enhancing the network performance. In this paper, we have presented a comparative evaluation of state of art meta-heuristic approaches that helps in acquiring energy efficiency in the network. We have proposed grey wolf optimization (GWO-P) algorithm with the empirical analysis of the existing methods PSO, GA and WAO that will help the readers to select the appropriate approach for their applications. It is similarly exposed that in different other execution measurements GWO-P beats the contender calculations for length of stability, network lifetime, expectancy and so on.

Keywords WSN · Meta-heuristic · Empirical analysis · Grey wolf optimization · Network performance

---

B. M. Sahoo (✉)
Department of Computer Science and Engineering, Amity University Noida, Noida, Uttar Pradesh, India

B. M. Sahoo · T. Amgoth
Indian Institute of Technology (Indian School of Mines), Dhanbad, India
e-mail: tarachand@iitism.ac.in

H. M. Pandey
Department of Computer Science, Edge Hill University, Ormskirk, Lancashire, UK
e-mail: Pandeyh@edgehill.ac.uk

A. Choudhary et al. (eds.), *Applications of Artificial Intelligence and Machine Learning*, Lecture Notes in Electrical Engineering 778,
https://doi.org/10.1007/978-981-16-3067-5_35

1 Introduction

In computing and communication technology, background intelligence including electric battery driven nano sensors, wireless communication technology is developing a desire for advancement of wireless sensor network. Additionally, nowadays the huge use of wireless sensor system will be the primary target area of investigation wireless sensor system becoming more and more attractive every day for the potential approach of its in ecological monitoring, natural area monitoring, battle field surveillance, natural attack detection in any obscure and ordinary environments [1]. The event is found by a humongous number of small, affordable and minimal powered devices anytime it can feel virtually any earthy movement (pressure, high heat, sound, areas getting some magnetic qualities, vibration, etc.) [2]. Each one of those devices is known as a sensor node. A WSN is composed of thousands or hundreds of inexpensive sensor nodes which may also use a fixed place or randomly deployed for checking purpose [3]. The relaying of information finishes at a unique node identified as base stations (also known as sinks). A base station links the sensor system to the next public network as web to disseminate the sensed information for more processing [4].

Apart from sensing as well as transmitting information, you will find a number of restrictions including power management, distance management, real time difficulties, topological issue, design issue, energy usage, etc. as well as among them the main restriction is the power usage in terminology of the longevity of WSN due to the irreplaceable and limited battery backup of the sensor nodes [5]. Moreover, the nodes close to the base station is the very first one in order to run out of power due to its extra relaying of information of the nodes that are miles away from the BS [6]. To locate an answer of these problems' different studies as radio communication hardware, moderate access management have been studied.

Meta-heuristic optimization algorithms are starting to be increasingly more well-known in engineering apps since they: (i) depend on relatively easy ideas and therefore are not hard to implement; (ii) don't need gradient info; (iii) is able to avoid community optima; (iv) may be employed in a broad range of issues covering various disciplines. Nature-inspired meta-heuristic algorithms solve optimization difficulties by mimicking physical or biological phenomena [7]. They may be grouped in 3 major categories (see Fig. 1): Evolution based, physics based and swarm-based methods. Evolution-based techniques are influenced by the laws of organic evolution. The search process begins with a randomly generated population that is evolved over the following generations. The strength point of these techniques is the fact that the most effective people are constantly coupled together to form the coming generation of individuals. To be able in order to optimize network's lifetime, this grey wolf optimization (GWO) [8] Meta-heuristic algorithm for choosing the effective clustering to data transmitting information period from the sensor node to sink and to enhance energy efficiency to maximize network's lifetime. A few scientific studies are recommended to the literature deal with all the optimization issues such as;

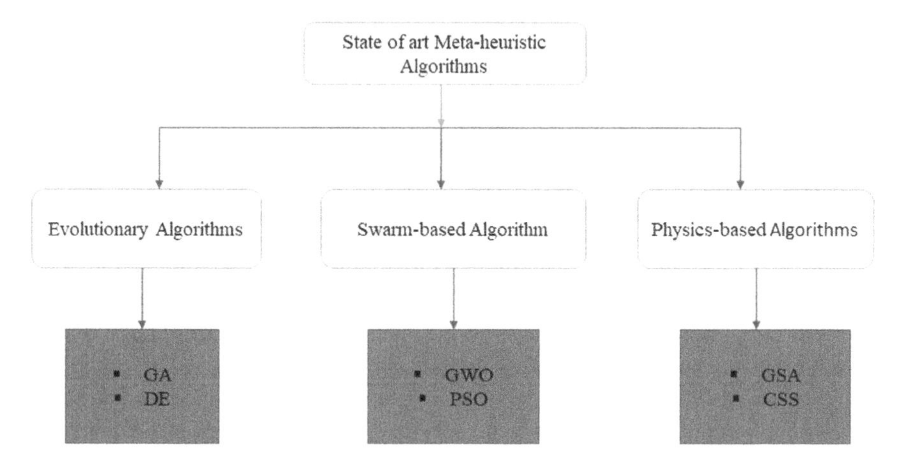

Fig. 1 Categorization of meta-heuristic algorithms

variant of particle swarm optimization (ICRPSO) [9], variant of genetic algorithm (GA-LEACH) [10] and the variant of whale optimization algorithm (WOTC) [11].

Clustering is among the most crucial strategies to enhance the system lifetime in WSNs. Clustering is the procedure of partitioning the whole region into a selection of subregions, known as clusters. Based on the analysis, the use and also the optimization of electricity usage by the sensor in the bunch-based network version can be accomplished far more efficiently. For information transmission clustering is among the well-known and simpler mechanism. In simpler language, a pair of nodes produce a team or even cluster. Each cluster communicates with one another and also works towards the targeted objective. Nodes from each cluster could be included as well as taken out of the bunch whenever and of all the nodes one has long been used as a leader, named group head (CH) [10]. All of the collected information is delivered by the sensor nodes to the head of theirs of the group and after that the CHs aggregate the information and also delivered to the sink or maybe base station (BS) indirectly or directly like through some other CHs. Due to several following effective capabilities clustering method gets to be more appealing

1. Sensor nodes take out all irrelevant information and unwanted info before sending information to the CH which enhance the power consumption.
2. CHs just conserve the intra cluster path that enhances the scalability of the WSN effectively.

The rest of this paper is systematized as follows. In Sect. 2, the related work is introduced. System model for proposed GWO meta-heuristic methods are in Sect. 3. The comparative analysis and results are discussed in Sect. 4. Finally, the conclusion is given in Sect. 5.

2 Related Work

In this literature review, we show several routing methods recommended for cluster head (CH) selection then also highlighted described the limitations and a full awareness on the routing strategies following the different meta-heuristic such as PSO, GA, WOA and GWO algorithm are highlighted.

Sahoo et al. [12] proposed PSOECSM protocol directing heterogeneous WSN aimed at enhancing cluster head choice technique in addition to sink mobility remain resolved by means of PSO which reflects the reduced power CH along with closest CH from the moving sink in addition to dimensions of the bunch. PSO-UFC [13] are represent the PSO based strategy to gain the workout perform for inadequate clustering in addition to fault understanding mechanism to balance the intra and also inter cluster energy consumption. Latiff et al. [14] proposed PSOMSB targeting various characteristics specifically, network lifetime, information delivery as well as power use are believed to be to that the distance among the node as well as sink is recognized as. Hu et al. recommended immune orthogonal learning particle swarm optimization algorithm (IOLPSOA) [15]. Nevertheless, the long-haul transmission in this particular strategy eats a great deal of energy.

A different routing technique that are derived from GA are emphasized. Kuila et al. [16] exploited GA for controlling the load balancing of the system. The projected method did not contemplate the CH choice. Gupta and Jana [17] proposed GACR which involved the distance as well as standard deviation components in the objective function of its directing to obtain improved number of rounds prior to the very first node is dead. Bhola et al. [18] proposed genetic algorithm which shows efficiently consumption of energy in WSNs. This proposed algorithm is actually based on leach protocol. By using hierarchical leach protocol cluster heads are selected and to find out the optimal route GA is used.

Arora et al. [19] projected an ant colony optimization (ACO) which constructed self-organized method for WSNs. This is also introduced for energy consumption in efficient manner [20]. In this algorithm also we have to first choose Cluster heads by maximal residual energy and after that members joining process has been started under the cluster head. Gharaei et al. [9] proposed ICRPSO to Inter and also intra cluster-based routing using PSO. In ICRPSO, inter and intra cluster motion moving sink for grouping by using PSO discussed. ICRPSO deficiency as follows: (a) disregard traffic evidence for the spiral movement of the sink moves; and (b) the arbitrary movement of the sink in the group has a high energy usage.

Mirjalili and Lewis [21], proposed whale optimization algorithm (WOA) is recognized as among swam smart programme which is a novel nature-inspired meta-heuristic optimization algorithm, humpback whales swim around prey in a shrinking group and along a spiral shaped path at the same time to create distinct bubbles along a group or '9' shaped path.

Sahoo et al. [22] designed and introduced a hierarchical hybrid approach for distributed clustering using GA and PSO algorithm, but in broad level WSNs. This is dual levels of clustering where GA is utilized for the cluster that belongs at ground level and for higher level clustering PSO is utilized which provides better convergence. The results prove introduced approach became successful in reducing the energy consumption effectively which straight away point out towards the increment of network's lifespan.

Mirjalili et al. [8] grey wolf optimizer (GWO) a leader choice mechanism is recommended grounded on alpha, beta, as well delta wolves to upgrade as well as change the remedies in the archive as well as a power system mechanism continues to be incorporated to GWO to be able to enhance the non-dominated remedies in the archive.

Visu et al. [23] in their paper proposed "Bio inspired dual cluster heads optimized routing algorithm for wireless sensor networks". In this paper, they describe the clustering and routing strategies. From every cluster data or information is moved forward to their respective CH which assigns as an aggregator and mainstay of the routing system as well. Technically, CH in the cluster consumes the most energy as compare to the further nodes are in same cluster as the aggregated data is need to be transferred to the base station named, sink node through single or else multiple hops communication system. This causes imbalance in energy in the network to resolve this problem, and to optimize the energy consumption by the CHs, a double-cluster-based Krill Herd Optimization algorithm is introduced [24]. Routing protocols in wired and wireless network are absolutely different. The conventionally utilized WSN routing techniques are utilized because of its various advantages-like energetic routing, on-demand routing, hybrid routing, etc. This examination advances an improvement strategy to manage energy compelled climate and to derive the lopsided utilization of energy, for remote sensor network directing. As indicated by this proposed enhanced steering, at first the hubs which sent in the networks are grouped utilizing the underlying centroid calculation. Then, the sensed data got aggregated by the primary centros wids. The secondary centroids help by providing detail of the route trust value aimed at individually route. Kill Herd Optimization helps to figure out the minimize the decision path. Then, the data which are aggregated are moved forward through that established optimized path. The introduced trust-based Krill optimized aggregated data through the given path which square in shape [25]. In wireless sensor networks (WSNs), the technique, of gathering the sensor nodes (X), is one of the crucial ways to make the system long last. Gathering of nodes for cluster formation, and the election to elect a node as cluster heads (CH) among them for each cluster is performed.

The proposed method in this particular paper utilizes GWO for the choice of CH. The purpose for picking GWO more than other meta-heuristic strategies would be that the GWO highlights a quicker convergence rate. Additionally, GWO brings about the ceaseless decrease of search space notwithstanding assurance factors are less. Additionally, it stays away from local optima.

3 The Proposed Grey Wolf Optimization Algorithm

After observing at clustering as a major problem, the only aim is to gain the finest trade-offs among the energy consumption and efficient data transmission. The problem is still on the increment of the number of clusters. As such, we propose a multi-objective clustering using grey wolf algorithm to get rid of hierarchy issue. The proposed method will be compared and investigated with existing meta-heuristic algorithm and the well-known benchmarking multi-objective approaches.

This algorithm is influenced by the social hierarchy and hunting mechanism of grey wolf. According to the algorithm, it mimics the social hierarchy and hunting strategy of grey wolves. There are four wolves or layers of the social hierarchy which described in Fig. 2.

The alpha (α) wolf: Indicates the best objective valued solution
The beta (β) wolf: Indicates the second best objective valued solution
The delta (δ) wolf: Indicates the third best objective valued solution
The omega (ω) wolf: Indicates rest of the solutions

In the above model, the alpha is viewed as the strongest participant of the package. Alpha (α) is in the roof of the hierarchy, and also considered as the strongest prospect of the package. Alpha is generally male wolf but might be female too. Alpha wolves gave the order, and that is adhered to by all of the additional wolfs in the package. Beta wolves are generally accountable to apply the orders of alpha. Alpha wolf additionally looks for the sleeping spot for the pack.

Subsequently, beta grey wolves play an important job in the hierarchy. These are the next most crucial wolves in the package. Alpha wolves take the choices by using beta wolves. The beta wolves additionally coordinate in the responses purpose. Subsequently, delta wolves will come and also categorize as guards, predators, spies and caretaker. Then will be the role of omega wolves. These wolves are believed to be as babysitters and are permitted to consume within the last.

Hence, the primary 3 wolves that denoted as α, β, and δ take responsibility to guide the looking mechanism of the protocol. The remainder of the wolves (ω) are

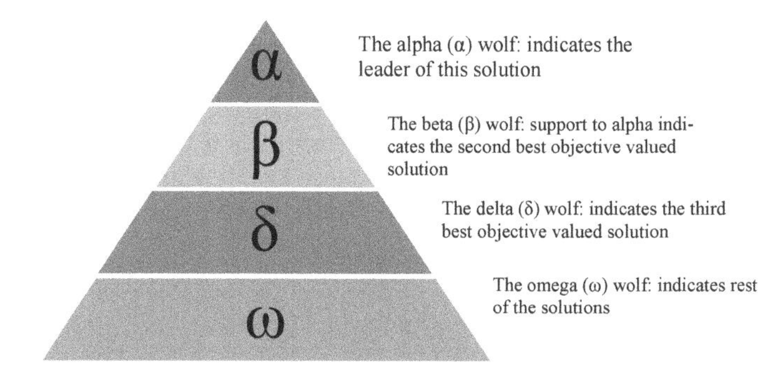
α
β
δ
ω
The alpha (α) wolf: indicates the leader of this solution
The beta (β) wolf: support to alpha indicates the second best objective valued solution
The delta (δ) wolf: indicates the third best objective valued solution
The omega (ω) wolf: indicates rest of the solutions

Fig. 2 Hierarchical model of GWO

thought-about and ordered to follow them. Throughout the hunting, grey wolves follow a collection of well-ordered phases: encircling, hunting then also attacking.

Along with the social leadership, the subsequent formulas have been suggested to be able to mimic the encircling behaviour of grey wolves through hunt [8].

$$\vec{D} = \left| \vec{C} . \vec{X}_p(t) - \vec{X}(t) \right| \tag{1}$$

$$\vec{X}(t+1) = \vec{X}_p(t) - \vec{A} . \vec{D} \tag{2}$$

$$\vec{D}_\alpha = \left| \vec{C}_1 . \vec{X}_\alpha - \vec{X} \right| \tag{3}$$

$$\vec{D}_\beta = \left| \vec{C}_2 . \vec{X}_\beta - \vec{X} \right| \tag{4}$$

$$\vec{D}_\delta = \left| \vec{C}_3 . \vec{X}_\delta - \vec{X} \right| \tag{5}$$

$$\vec{X}_1 = \vec{X}_\alpha - \vec{A}_1 . \vec{D}_\alpha \tag{6}$$

$$\vec{X}_2 = \vec{X}_\beta - \vec{A}_2 . \vec{D}_\beta \tag{7}$$

$$\vec{X}_3 = \vec{X}_\delta - \vec{A}_3 . \vec{D}_\delta \tag{8}$$

$$\vec{X}(t+1) = \frac{\vec{X}_1 + \vec{X}_2 + \vec{X}_3}{3} \tag{9}$$

In above Eqs. 1 and 2, $\vec{X}_p$ denoted the position vector of the grey and t denoted the current iteration. Where, grey wolf position vector is denoted by $\vec{X}$ and coefficient vectors are $\vec{A}$ and $\vec{C}$.

The coefficient vectors $\vec{A}$ and $\vec{C}$ are evaluated as

$$\vec{A} = 2. \vec{a} . \vec{r_1} - \vec{a} \tag{10}$$

$$\vec{C} = 2. \vec{r_2} \tag{11}$$

In Eqs. 10 and 11, $\vec{r_1}$ and $\vec{r_2}$ are the casual vectors lies between 0 and 1 and the element $\vec{a}$ is reduced linearly after 2 to 0 throughout the iterations.

The last phase of the system of chasing will be the attack. The procedure for assaulting might be mathematically identified utilizing the operators mentioned above. This is accomplished by lessening the valuation of $\vec{D}$ as well as letting down the assortment of contrast of $\vec{A}$ in the number of [−2a, 2a], whereas actually decreased by two to zero till the iterations. In case the standards of $\vec{A}$ invention of [−1, 1], the role of the exploration representative determination likely be in among the present place as well as the prey's situation. If $|A| < 1$, the wolves outbreak the target. Consequently, it could remain observed that based on the GWO method, the exploration managers upgrade their places based on the roles of alpha, beta and delta participants. The hunt for target starts once the wolves deviate from one another to locate the target. This particular research is determined by the roles of the alpha, delta members and beta. The circumstances for exploration or maybe hit are determined through the ethics of $\vec{A}$ as follows:

As per the diverge and search, $|A| > 1$.
As per the converge and attack, $|A| < 1$.

3.1 *Objective Function*

The objective function is the combination of various performance parameters collective to surround a phrase that is usually both maximized or perhaps minimized. Objective characteristic refers to two different settings that choose the ability of the individual. The drive parameters used in the objective function that spoke as follows.

- Objective 1 (Average energy of nodes):

$$F_1 = \frac{1}{N}\sum_{i=1}^{N} E_{(i)} \tag{12}$$

where $E_{(i)}$ symbolize the energy of the ith node along with N stand for the whole nodes of the network in Eq. (12).

- Objective 2 (Residual Energy):

$$F_2 = 1/\sum_{i=1}^{N}\left(\frac{E_{R(i)}}{E_T}\right) \tag{13}$$

where $E_{R(i)}$ is denoted the summation fraction of remaining energy of ith node and total energy is denoted by E_T in Eq. (13).

The objective function is the integration of objectives as in single expression is follows in Eq. (14)

$$F = 1/\left[(\gamma * F_1) + (\theta * F_2)\right] \quad (14)$$

where γ and θ are the constant values and $\gamma + \theta = 1$.

3.2 Radio Model

In this energy radio model, the quantity of energy use is determined by the distance among nodes. The power usage for moving the z bit statistics information with inside the distance 'd' is represented by $E_{tx}(z, d)$ besides additionally supplied as follow

$$E_{tx}(z, d) = z \times E_{\text{elec}} + z \times E_{\text{efs}} \times d^2 \quad (15)$$

The energy required whilst receiving z-bit of data as follows in equation

$$E_{rx}(z) = z \times E_{\text{elec}} \quad (16)$$

where E_{elec} and E_{efs} are the required energy intended for transceiver circuitry and energy of free space. $E_{rx}(z)$ is denotes the energy required whilst receiving the z-bit data.

4 Simulation and Result Analysis

The simulation options determine the simulation setting wherein the suggested GWO method is designed to use. The MATLAB application model 2019 is placed on a structure through system of 8 GB RAM, 1 TB HD, Intel processor i5 by CPU consecutively on 3.07 GHz in addition Window 10. The exact same sensor system produced arbitrarily is utilized in all of the simulations. It is assumed that nodes spread with the place of $100 \times 100\ \text{m}^2$ (Table 1).

4.1 Comparative Evaluation

The effectiveness of the proposed GWO-P method is analyzed with existing state of art algorithms; so as to test the recommended work is better one. For evaluation purpose, we are now analyzing current procedure as WOTC [11], GA-LEACH [10] and ICRPSO [9] methods and the results are evaluated in different metrics like, stability period, network remaining energy and throughput.

Table 1 Simulation parameters

GWO parameters	Values
Size of networks	100×100 m^2
Total nodes (N)	100
Sink node	1
Node energy (in Joules) (E_o)	0.5
E_{elec}	50 nJ/bit
Threshold distance (d_o)	87 m
(E_{efs})	10 pJ/bit/m^2
Inertia weight	0.7
Size of data packets	2000 bits
Number of total particles	30
C_1	2
C_2	2
Simulation run	20
Maximum iteration	150

4.1.1 Stability Period

Results well-known show that during GWO-P, the primary node is lifeless after 1278 rounds whereas for WOTC, GA-LEACH and ICRPSO, the standards of stability period remain individually 755, 814 and 1080 rounds as shown in Fig. 3. The network lifetime of GWO-P is 9801 rounds, whereas the WOTC, GA-LEACH and ICRPSO covers 8935, 7754 and 8359 rounds, respectively, computed.

4.1.2 Comparison of Network Lifetime of Simulated Methods with Stability Period

Table 2 represents the number of rounds taken for FND, HND and LND along with stability period for simulation methods. Here, the improvement of 65.2%, 57% and 18.3% of stability period of GWO-P against the existing methods WOTC, GA-LEACH and ICRPSO, respectively, and the improvement of 9.6%, 26.3% and 17.2% of network lifetime of GWO-P against the existing methods WOTC, GA-LEACH and ICRPSO, respectively.

4.1.3 Network Remaining Energy

In this metric, when information communication is assumed, the system's vitality begins decreasing. This is very basic to watch the conduct of network's outstanding energy through increment in number of rounds. GWO-P accomplishes better when contrasted with WOTC, GA-LEACH and ICRPSO calculations, individually such

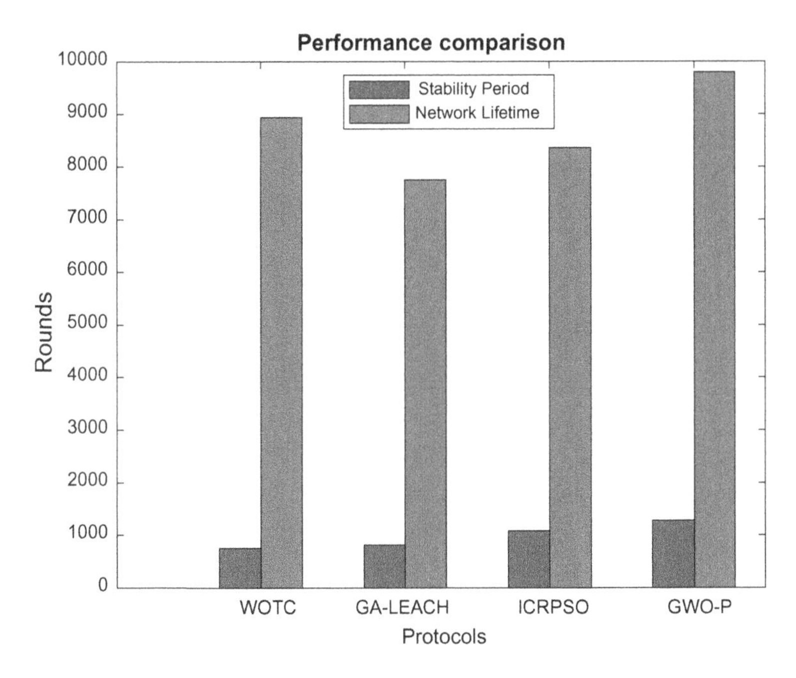

Fig. 3 Comparison of network lifetime with stability period of GWO-P with existing methods

Table 2 Comparison of network lifetime of simulated methods with stability period

Protocols	FND	HND	LND	Stability period
WOTC	755	3498	8935	755
GA-LEACH	814	3354	7754	814
ICRPSO	1080	4021	8359	1080
GWO-P	1278	4802	9801	1278

that it covers a more prominent numeral of rounds whereas the information communication is in improvement as appeared in Fig. 4. The vitality of a hub remains saved on an individually basis round because of the base energy utilization came about because of the vitality effective correspondence.

4.1.4 Throughput

The throughput of GWO-P is seen to be increased as the effective transmission of 108,487 information packets were finished as shown in Fig. 5. In any case, WOTC, GA-LEACH and ICRPSO sent 66,965, 69,714 and 77,835 packets of information, separately.

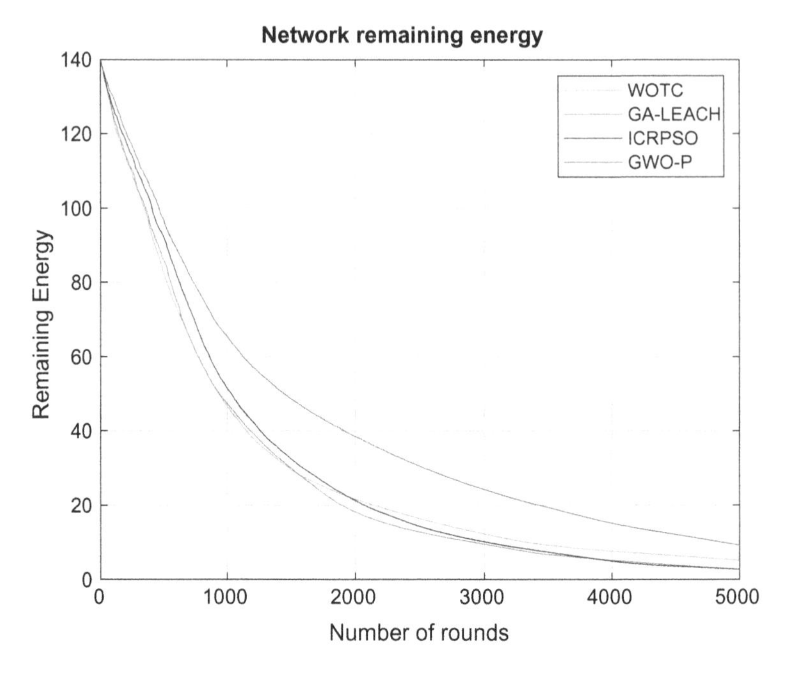

Fig. 4 Comparative analysis of network's remaining energy GWO-P with existing algorithms

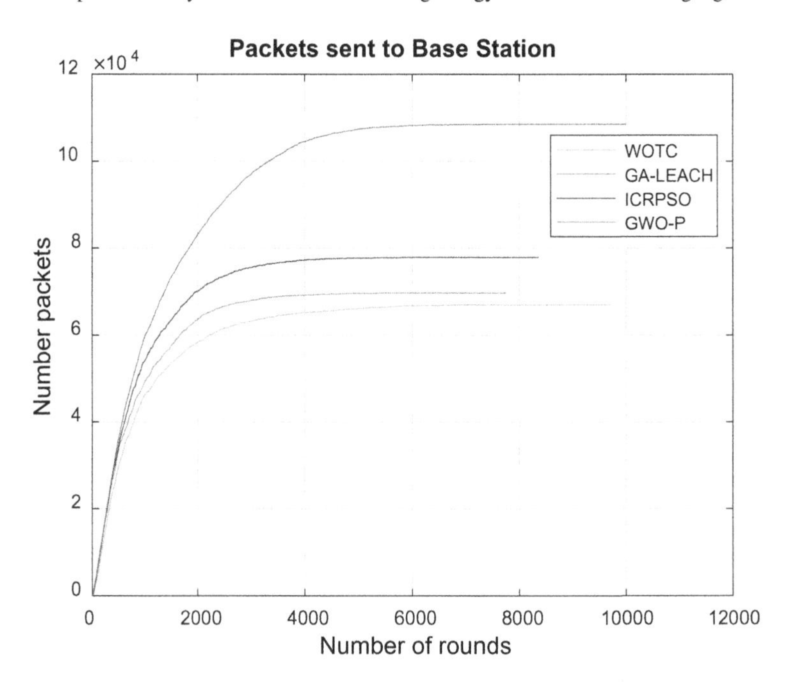

Fig. 5 Comparative analysis of throughput GWO-P with existing algorithms

5 Conclusion

The different meta-heuristic directing calculations have been reported in the accessible writing so far so as to acquire network stability and life span. The proposed grey wolf optimization for wireless sensor networks is roused from the everyday schedule of grey wolves that utilization four unique positions, portrayed by α, β, δ and ω to attack the prey. These various wolves are utilized to play out the activity of investigation and misuse in the search space. In the proposed strategy, upgraded number of bunches is taken by the assembly of the estimation of α wolf, as α wolfs spans to their best worth. So as to evaluate its presentation, the proposed GWO-P calculation is broadly introduced and reproduced in MATLAB. There is the improvement of 65.2%, 57% and 18.3% of stability period of GWO-P against the existing methods WOTC, GA-LEACH and ICRPSO respectively and the improvement of 9.6%, 26.3% and 17.2% of network lifetime of GWO-P against the existing methods WOTC, GA-LEACH and ICRPSO, respectively. It is likewise discovered that in different other execution measurements GWO-P beats the contender calculations for length of stability, network lifetime, expectancy and so on. Still, there is always an opportunity to further enhancement of proposed work by extending by numerous clustering methods to cumulative mobile sink nodes efficiently.

References

1. Akyildiz IF, Su W, Sankarasubramaniam Y, Cayirci E (2002) Wireless sensor networks: a survey. Comput Netw 38(4):393–422
2. Arampatzis T, Lygeros J, Manesis S (2005) A survey of applications of wireless sensors and wireless sensor networks. In: Proceedings of the 2005 IEEE international symposium on intelligent control. Mediterranean conference on control and automation. IEEE, pp 719–724. Retrieved from http://ieeexplore.ieee.org/abstract/document/1467103/
3. Abbasi AA, Younis M (2007) A survey on clustering algorithms for wireless sensor networks. Comput Commun 30(14):2826–2841
4. Sahoo RR, Singh M, Sahoo BM, Majumder K, Ray S, Sarkar SK (2013) A light weight trust based secure and energy efficient clustering in wireless sensor network: honey bee mating intelligence approach. Procedia Technol 10:515–523
5. Sahoo BM, Gupta AD, Yadav SA, Gupta S (2019, April) ESRA: enhanced stable routing algorithm for heterogeneous wireless sensor networks. In: 2019 International conference on automation, computational and technology management (ICACTM). IEEE, pp 148–152
6. Akkaya K, Younis M (2005) A survey on routing protocols for wireless sensor networks. Ad Hoc Netw 3(3):325–349
7. Mirjalili S, Saremi S, Mirjalili SM, Coelho LDS (2016) Multi-objective grey wolf optimizer: a novel algorithm for multi-criterion optimization. Expert Syst Appl 47:106–119
8. Mirjalili S, Mirjalili SM, Lewis A (2014) Grey wolf optimizer. Adv Eng Softw 69:46–61
9. Gharaei N, Bakar KA, Hashim SZM, Pourasl AH (2019) Inter-and intra-cluster movement of mobile sink algorithms for cluster-based networks to enhance the network lifetime. Ad Hoc Netw 85:60–70
10. Nayak P, Vathasavai B (2017, January) Genetic algorithm-based clustering approach for wireless sensor network to optimize routing techniques. In: 2017 7th International conference on cloud computing, data science & engineering-confluence. IEEE, pp 373–380

11. Ahmed MM, Houssein EH, Hassanien AE, Taha A, Hassanien E (2017, September) Maximizing lifetime of wireless sensor networks based on whale optimization algorithm. In: International conference on advanced intelligent systems and informatics. Springer, Cham, pp 724–733
12. Sahoo BM, Amgoth T, Pandey HM (2020) Particle swarm optimization based energy efficient clustering and sink mobility in heterogeneous wireless sensor network. Ad Hoc Netw 102237
13. Kaur T, Kumar D (2018) Particle swarm optimization-based unequal and fault tolerant clustering protocol for wireless sensor networks. IEEE Sens J 18(11):4614–4622
14. Latiff NAA, Latiff NMAA, Ahmad RB (2011, April) Prolonging lifetime of wireless sensor networks with mobile base station using particle swarm optimization. In: 2011 Fourth international conference on modeling, simulation and applied optimization. IEEE, pp 1–6
15. Hu Y, Ding Y, Hao K, Ren L, Han H (2014) An immune orthogonal learning particle swarm optimisation algorithm for routing recovery of wireless sensor networks with mobile sink. Int J Syst Sci 45(3):337–350
16. Kuila P, Gupta SK, Jana PK (2013) A novel evolutionary approach for load balanced clustering problem for wireless sensor networks. Swarm Evol Comput 12:48–56
17. Gupta SK, Jana PK (2015) Energy efficient clustering and routing algorithms for wireless sensor networks: GA based approach. Wireless Pers Commun 83:2403–2423
18. Bhola J, Soni S, Cheema GK (2020) Genetic algorithm based optimized leach protocol for energy efficient wireless sensor networks. J Ambient Intell Humaniz Comput 11(3):1281–1288
19. Arora VK, Sharma V, Sachdeva M (2019) ACO optimized self-organized tree-based energy balance algorithm for wireless sensor network. J Ambient Intell Humaniz Comput 10(12):4963–4975
20. Sahoo BM, Rout RK, Umer S, Pandey HM (2020, January) ANT colony optimization based optimal path selection and data gathering in WSN. In: 2020 International conference on computation, automation and knowledge management (ICCAKM). IEEE, pp 113–119
21. Mirjalili S, Lewis A (2016) The whale optimization algorithm. Adv Eng Softw 95:51–67
22. Sahoo BM, Pandey HM, Amgoth T (2020) GAPSO-H: a hybrid approach towards optimizing the cluster-based routing in wireless sensor network. Swarm Evol Comput 100772
23. Visu P, Praba TS, Sivakumar N, Srinivasan R, Sethukarasi T (2020) Bio-inspired dual cluster heads optimized routing algorithm for wireless sensor networks. J Ambient Intell Humaniz Comput 1–9
24. Senniappan V, Subramanian J (2018) Biogeography-based Krill Herd algorithm for energy efficient clustering in wireless sensor networks for structural health monitoring application. J Ambient Intell Smart Environ 10(1):83–93
25. Saravanan D, Janakiraman S, Kalaipriyan T, Naresh MV (2018) An efficient routing model using Krill Herd optimization algorithm—a survey. Int J Pure Appl Math 119(14):377–383

Deep Learning-Based Computer Aided Customization of Speech Therapy

Sarthak Agarwal, Vaibhav Saxena, Vaibhav Singal, and Swati Aggarwal

Abstract Video frame interpolation is a computer vision technique used to synthesize intermediate frames between two subsequent frames. This technique has been extensively used for the purpose of video upsampling, video compression and video rendering. We present here an unexplored application of frame interpolation, by using it to join different phoneme videos in order to generate speech videos. Such videos can be used for the purpose of speech entrainment, as well as help to create lip reading video exercises. We propose an end-to-end convolutional neural network employing a U-net architecture that learns optical flows and generates intermediate frames between two different phoneme videos. The quality of the model is evaluated against qualitative measures like the Structural Similarity Index (SSIM) and the peak signal-to-noise ratio (PSNR), and performs favorably well, with an SSIM score of 0.870, and a PSNR score of 33.844.

Keywords Video frame interpolation · Speech videos · Speech entrainment · Phoneme · Structural Similarity Index · Peak signal-to-noise ratio

1 Introduction

Speech is a vocalized form of communication, which is used to convey thoughts and information through the use of sound and corresponding mouth movements. Every word is made up of a combination of phonemes, which is an indivisible unit of sound in a particular language. While speaking each of these phonemes, the lips, the teeth and the tongue move in different ways. A combination of phonemes, joined by the appropriate movements of the mouth, is all that are required to construct a word in a particular language.

With the onset of innovative deep learning techniques, it is possible to learn these mouth movements and generate them synthetically in the form of a video. Video Frame interpolation is one such technique, which can be used to join different phonemes by synthesizing intermediate frames with proper mouth movements. Video

S. Agarwal (✉) · V. Saxena · V. Singal · S. Aggarwal
Netaji Subhas University of Technology, Dwarka, New Delhi 110078, India

A. Choudhary et al. (eds.), *Applications of Artificial Intelligence and Machine Learning*, Lecture Notes in Electrical Engineering 778,
https://doi.org/10.1007/978-981-16-3067-5_36

frame interpolation is a technique used to synthesize plausible intermediate frames between two subsequent frames of a video. This technique has widely been used for upsampling a video to a higher frame rate. It is also used for the purpose of video compression, by allowing storing only relevant frames, and creates new frames when decompressed. In recent years, new techniques have emerged which help to tackle this problem of video frame interpolation. We will discuss these techniques in Sect. 2.

In this paper, we present an unexplored application of this technique. We propose an end-to-end convolutional neural network (CNN) adopting a U-net architecture, to learn optical flows of pixels in the frames, and generate intermediate frames involving proper mouth movements, by joining different phonemes to create Hindi language speech videos. Since we use a CNN to generate **Sp**eech Videos using a frame **in**terpolation technique, the model is called **'SpinCNN'**. Generation of speech videos using frame interpolation can prove to have widespread applications, especially in the health care sector. Patients suffering from speech disorders like Broca's Aphasia can produce fluent speech while mimicking an audio visual speech model. This process is called speech entrainment [1]. Speech entrainment videos used nowadays consist of fixed speech exercises, which cannot be changed according to the requirements of a particular patient. To allow patients to create custom and individualized speech entrainment exercises, we present an approach to synthetically create speech videos.

In the following sections, the related work is discussed in Sect. 2, which is followed by an in-depth explanation of the proposed approach in Sect. 3. Finally, the results produced by the model are analyzed in Sect. 4. Possible application areas are discussed in Sect. 5, followed by the future work being discussed in Sect. 6.

2 Related Work

Several techniques have been used to tackle the problem of Video Frame Interpolation, the most naive method being Linear Frame Interpolation, or weighted frame averaging. In this technique, the linear combination of the intensities of the same pixels in subsequent frames is taken to generate intermediate frames. Such a technique produces intermediate frames of poor quality, consisting of blurriness and ghosting.

Numerous techniques have attempted to address frame interpolation by estimating optical flow, i.e., the pattern of apparent motion of objects in an image. An optical flow-based technique, called Motion-Compensated Frame Interpolation (MCFI), is the current state of the art in video frame interpolation. MCFI techniques can be of two types: pixel-based [2] or block-based [3]. The former works pixel by pixel, whereas the latter works on a block of pixels at a time, which reduces the computation time. Both of these categories consist of two parts: motion estimation and motion compensation. Motion estimation, as the name suggests, is the process of calculating the motion, or velocity of different pixels, or blocks, between two subsequent frames by estimation of optical flow. Motion compensation uses these calculated motion estimates to move these pixels, or blocks, in the desired direction.

Simple MCFI techniques fail to calculate motion vectors accurately, leading to the appearance of visual artifacts in the generated results. Guo and Lu [4] tried to improve the MCFI technique and tackle the generation of ghosting and visual artifacts. They tried to improve the motion estimates, and correct the incorrectly calculated motion vectors used during motion compensation. While MCFI-based techniques work well and produce good results, they are computationally expensive, making them difficult to apply to real-time applications. In recent years, the convolutional neural networks have become increasingly popular in solving the problem of video frame interpolation [5–7]. Niklaus et al. [8] used a CNN model to learn spatially adaptive convolutional kernels for every pixel. The process described by them is computationally expensive since a kernel has to be learned for every pixel in the frame. Long et al. [9] used a CNN to learn optical flow but did not use it for the purpose of frame interpolation. Our proposed model uses a U-net architecture, inspired from the research of Ronneberger et al. for the purpose of biomedical image segmentation [10]. Yahia [11] uses a similar kind of a model for the purpose of video frame interpolation for animated videos.

While many approaches have been developed to tackle video frame interpolation, the possible applications of this technique still remain largely unexplored. In this work, we present a novel application of this technique by proposing a model that synthesizes Hindi language speech videos by joining phonemes with the generated smooth intermediate frames.

3 Proposed Approach

In our proposed approach, two kinds of dataset are used. They are discussed in Sect. 3.1. Following that, the word construction process is described. Finally, an in-depth explanation of our proposed model, SpinCNN, along with the training details are discussed.

3.1 Dataset

The core idea of the approach is the synthesis of intermediate frames involving proper mouth movement to join different phonemes. So, the dataset must possess all kinds of phonemes, and the model should be able to learn the different mouth movement patterns in order to create plausible intermediate frames. The Hindi language consists of 11 *swar*, or vowels, 35 *vyanjan*, or consonants and 4 *sanyukt vyanjan*, or compound consonants. Different combinations of a consonant (*vyanjan*) or a compound consonants (*sanyukt vyanjan*), with a vowel (*swar*), can produce all phonemes in the Hindi language. This work uses two kinds of dataset. The first dataset, called the phoneme dataset, consists of short videos of all different phonemes in the Hindi language. For the purpose of this work, we use 334 useful and valid phoneme videos, and all of them created and stored as a part of the phoneme dataset.

The mouth, while speaking, can move in various manners. The lips can move from a closed position to a partially or a completely open position, or a rounded position to a spread position, and vice versa. The teeth and the tongue also play a major role while speaking. Different teeth and tongue movements result in different types of sounds being produced. The second type of dataset, called the training dataset, consists of 425 videos containing all these kinds of mouth movements possible while speaking the Hindi language.

The training dataset is used for training the proposed model, and allows the model to learn all these different kinds of lip, teeth and tongue movement patterns. The videos in the phoneme dataset are joined using the videos generated by the proposed model, using the method explained in Sect. 3.3. The videos of all phonemes created in both the datasets were preprocessed, so as to contain only the mouth portion, and the unneeded portion of the face is removed. Each frame in the video has a height and width of 256 pixels each. Both datasets were shot using a NIKON D5100, under proper lighting conditions with minimal movements and proper enunciation.

3.2 Word Construction in Hindi Language

In order to construct a particular word, its constituent phonemes are extracted. For instance, the Hindi language word ' 'नाम'', which means 'name' in English, can be divided into two constituent phonemes, ' 'ना'' and ' 'म''. For each extracted phoneme, a corresponding video is collected. In order to join two phonemes coherently, the transition between the first phoneme video to the second one must be as smooth and accurate as possible. Hence, the intermediate frames to be synthesized depend only on the position of the mouth in the last frame of the first video, and the first frame of the second video. These two frames are passed on to the proposed SpinCNN model, which in turn, produces a fixed number of intermediate frames (β). For instance, intermediate frames for the word ' 'नाम'' will be generated between the last frame of the phoneme video of ' 'ना'', and the first frame of the phoneme video of ' 'म''. The entire process of word construction is illustrated in the form of a flowchart in Fig. 1.

3.3 Proposed Model for Intermediate Frame Synthesis: SpinCNN

An end-to-end convolutional neural network adopting a U-net architecture is used for the process of intermediate frame synthesis. A U-net is a fully convolutional neural network, consisting of an encoder and a decoder, where equal-sized layers that are created during encoding are used again during decoding. Since we use a CNN to generate **sp**eech videos using video frame **in**terpolation, the proposed model is

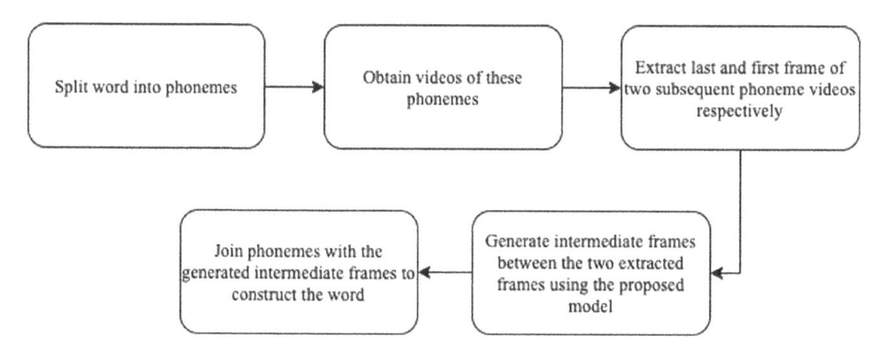

Fig. 1 Flowchart for word construction

called **'SpinCNN'**. The architecture of the proposed SpinCNN model (Fig. 2) here is largely inspired by the work of Long et al. [9].

Let,

1. C_x denotes a (conv-batchnorm-lrelu) × 2-maxpool block, where each convolution uses x kernels of size 3 × 3, and a stride of 1.
2. D_x denotes an upsampling-concatenation-(conv-batchnorm-lrelu) × 2 block, where each convolution uses x kernels of size 4 × 4, and a stride of 1.
3. P_x denotes a dropout layer with x being the dropout probability.
4. S_1 denotes a 1 × 1 convolution layer.

Our proposed SpinCNN model (Fig. 2) has the following architecture:

$$C_{32} - C_{64} - C_{64} - C_{128} - C_{256} - P_{0.5} - D_{128} - D_{64} - D_{64} - D_{32} - D_{32} - S_1$$

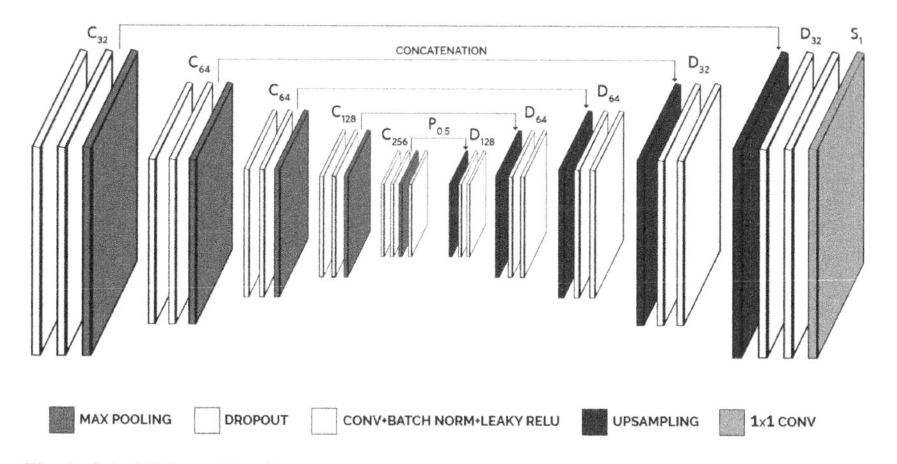

Fig. 2 SpinCNN model architecture

The pseudo-code of the proposed SpinCNN model has been discussed in Algorithm 1. Two subsequent image frames, each being a 3 channel RGB image, are concatenated and passed to the model (Line 5, Algorithm 1). We apply a convolution layer on the input several times, which gives us a latent representation of the input. This is done so that the model is able to learn the high level features of the image. In each convolution step, a 3×3 kernel is convolved with the input (Line 9, Algorithm 1). After each convolution, the result is saved for it to be used in future (Line 14, Algorithm 1), followed by maxpooling (Line 15, Algorithm 1). The latent representation obtained is then passed through a dropout layer in order to reduce the chances of overfitting (Line 17, Algorithm 1).

During the deconvolution process, the latent representation is bi-linearly upsampled to twice its size (Line 19, Algorithm 1). During maxpooling, a large amount of fine details are lost. In order to gain access to these fine details and regain structural information, a previously saved equal-sized layer from the convolution process is concatenated with the present layer (Line 20, Algorithm 1).

Each convolution is followed by a batch normalization layer (Line 10, 24, Algorithm 1), which normalizes the input to the layer. This is done to speed up the training process, and make weights easier to initialize. A Leaky ReLU layer with a slope of 0.2 is then applied as an activation layer (Line 11, 25, Algorithm 1). Leaky ReLU is used since it avoids the dying ReLU problem.

Finally, a 1×1 convolution layer is applied to the output layer (Line 29, Algorithm 1), which helps the model produce the desired fixed number of intermediate frames (β).

Algorithm 1: SpinCNN Model

```
initialize and populate conv_filters[]// number of filters in
different conv layers;
initialize and populate deconv_filters[]// number of filters in
different
deconv layers;
initialize save_layer[] as empty list;
initialize β;
next_input ← concatenated input frames;
foreach filterValue in conv_filter do
set count to 0;
while count < 2 do
next_input ← convolution_layer(next_input, filters=filterValue,
filter_size=3);
next_input ← batch_normalisation(next_input);
next_input ← LeakyReLu(next_input);
increment count by 1;
end
append next_input to save_layer;
next_input ← maxPooling (next_input);
end
next_input ← dropout(next_input);
foreach filterValue in deconv_filters do
next_input ← upSampling(next_input);
```

```
concatenate corresponding layer from save_layer with next_input;
set count to 0;
while count < 2 do
next_input ← convolution_layer(next_input, filters=filterValue,
filter_size=4);
next_input ← batch_normalisation(next_input);
next_input ← LeakyReLU(next_input);
increment count by 1;
end
end
final_output ← convolution_layer(next_input, filters = 3*β,
filter_size = 1);
```

3.4 Training

The aim of SpinCNN is to generate a fixed number of intermediate frames (β), between two subsequent frames. This is done by carefully constructing 'image bundles', consisting of $\beta + 2$ image frames, and passing it to the model for training. A single image bundle consists of 2 input frames, along with β target intermediate frames that are used to train the CNN model. For the scope of this entire work, we have chosen $\beta = 10$, which has been chosen by careful subjective analysis of the mouth movement patterns and the speed of transition from one mouth position to another.

Most of the time, the adjoining frames in a video are almost identical to each other, and may not help the model learn mouth movements. In order to capture substantial mouth movements, the frames in an image bundle should not always be subsequent. The sequential distance between each frame in a particular image bundle is determined by the value of the intermediate frame distance, or α. For clarity, the construction of an image bundle is illustrated in Figs. 3 and 4. When $\alpha = 1$ and $\beta = 10$ (Fig. 3), frame number (1) and (12) are concatenated and sent as an input to the model. The model, in turn, produces 10 intermediate frames, i.e., 30 channels, since each image contains 3 channels (RGB). These outputs are then compared with

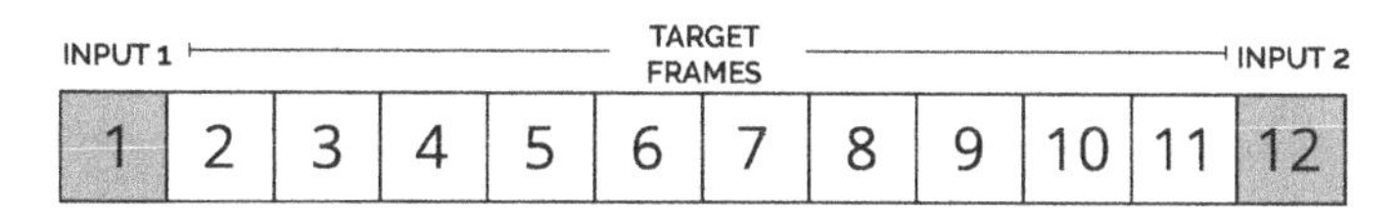

Fig. 3 Image bundle construction ($\alpha = 1$, $\beta = 10$)

INPUT 1
TARGET FRAMES
INPUT 2
1 3 5 7 9 11 13 15 17 19 21 23

Fig. 4 Image bundle construction ($\alpha = 2$, $\beta = 10$)

the target frames, i.e., frame numbers (2–11). Similarly, when $\alpha = 2$ and $\beta = 10$ (Fig. 4), frames (1) and (23) are the inputs, while frames (3–21) are the target frames. An α value of 2 allows the construction of image bundles that consist of proper and substantial mouth movements. A total of close to 1500 image bundles are constructed, and then shuffled. This dataset is then divided into a training set and a testing set, each containing 75% and 25% of the total image bundles, respectively.

The model uses Adagrad [12] as an optimizer, and it is batch-trained for 500 epochs with a batch size of 15, by minimizing the L_1 loss [13] on each pixel of the frame. L_1 loss is calculated by taking the sum of all absolute differences between each pixel across all channels in the generated as well as the target frames, divided by the number of pixels across all channels (N), as shown in Eq. 1.

$$L_1 = \frac{1}{N}\sum_{i=1}^{N}\left|p_i^{\text{Gen}} - p_i^{\text{Target}}\right| \quad (1)$$

4 Testing and Results

The dataset created was split into training and testing sets, in a ratio of 75–25%. The image bundles in the testing set were compared with the generated frames in order to evaluate the quality of the generated frames. In addition to evaluating the quality of the intermediate frames produced in a subjective manner, two evaluation metrics are used to compare the outputs as well.

4.1 Evaluation Metrics

The generated frames are compared with the target frames using two evaluation metrics, namely Structural Similarity Index (SSIM) [14] and peak signal-to-noise ratio (PSNR) [15].

SSIM attempts to measure the perceived change in structural information of a frame, and quantify the similarity of frames created. Higher the SSIM value, greater the similarity between two images. An SSIM value of 1.0 indicates complete similarity between two images.

PSNR is a log scale ratio between the maximum possible intensity of a pixel and the noise or dissimilarities that affect the quality of the image. It is based on the mean square error in the pixel intensities, and attempts to penalize the noise introduced in the generated frames.

Table 1 Testing set evaluation results

SSIM	0.870
PSNR	33.844

4.2 Evaluation of Synthesized Intermediate Frames

The image bundles in the testing set, which consists of 25% of the total image bundles, are evaluated against the two evaluation metrics. The mean SSIM and mean PSNR are found and reported in Table 1.

SSIM attempts to measure the change in structural similarity between two images. A mean SSIM value of 0.870 shows that the model is able to learn the changing structure of the frames well, and produce structurally correct intermediate frames. PSNR tries to penalize ghosting and occurrence of visual artifacts.

The higher the value, the clearer and better the results are. A PSNR value of 33.844 indicates that the results are not perfect, and possess some blurriness. After observing other works involving video frame interpolation [6, 7, 16, 17], it was found that the PSNR value obtained here is comparable to the values reported by these works.

For a particular pair of input frames, the synthesized interpolated frames are shown in Fig. 5. Two input frames were passed as input to the proposed SpinCNN model. SpinCNN, in turn, produces 10 intermediate frames. These generated interpolated frames are compared with the ground truth, and a mean SSIM score of 0.892 and a mean PSNR score of 34.326, are reported.

4.3 Evaluation of Constructed Words

A combination of phonemes is joined by interpolated frames involving correct mouth movements to construct a word. The quality and smoothness of the word produced are measured through a subjective appeal.

As an example, a Hindi language word ना, which literally means 'Name' in English, is constructed and tested. The word is constructed by joining the phonemes म and 'नाम' with the generated interpolated frames. The process of word construction is illustrated in Fig. 6. The constructed word, when subjectively evaluated, was found to be pleasingly smooth and coherent.

5 Proposed Application Areas

Speech videos are being used extensively in the health care industry for the purpose of speech therapy. However, the limitation of such videos is that they consist of fixed exercises, and are not customizable. Being able to synthetically create such videos allows flexibility and customization as per user requirements.

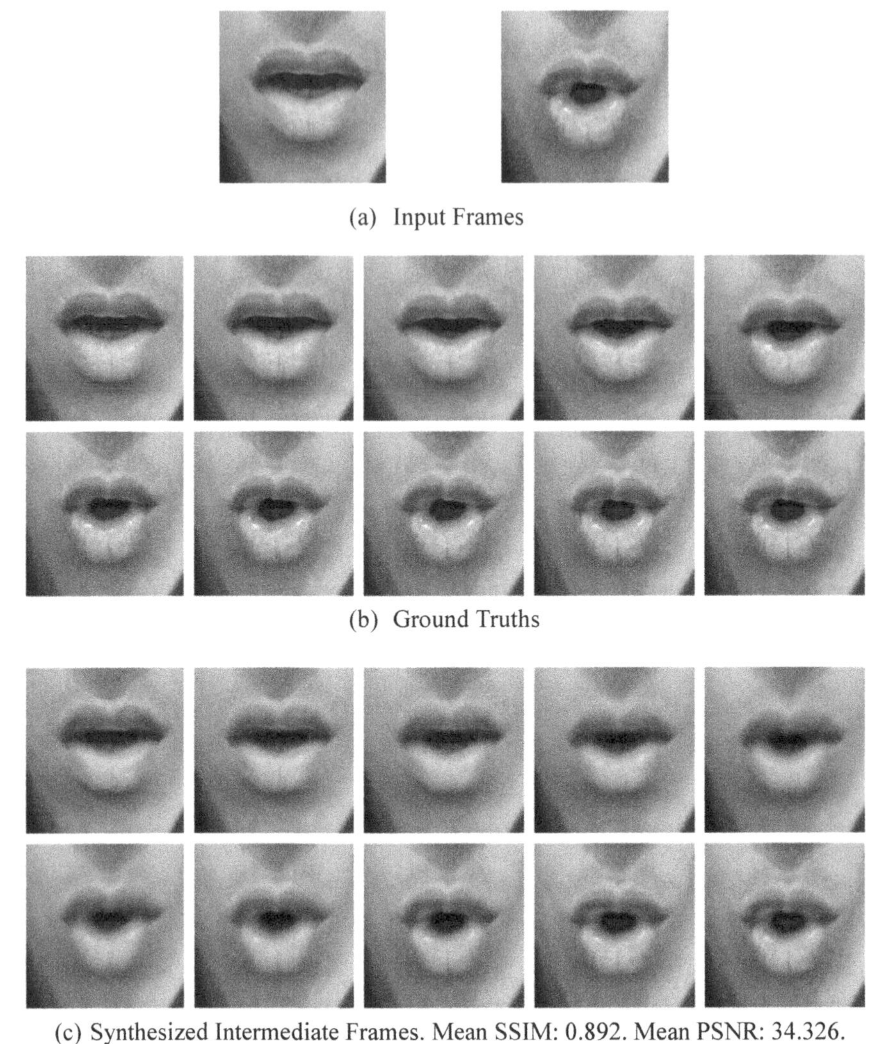

(a) Input Frames

(b) Ground Truths

(c) Synthesized Intermediate Frames. Mean SSIM: 0.892. Mean PSNR: 34.326.

Fig. 5 Evaluation of synthesized intermediate frames for a particular pair of input frames

Broca's Aphasia is a speech disorder in which the patient suffers partial loss of ability to produce language. Speech entrainment [1] is a process that helps such patients produce fluent speech while mimicking an audiovisual speech model. Further work by Fridriksson [18] suggests that the process of speech entrainment can even be used for the rehabilitation of such patients. As mentioned before, speech entrainment videos used today provide no customization to the user. Synthetically generated speech videos allow the patients to create speech entrainment exercises as per their needs.

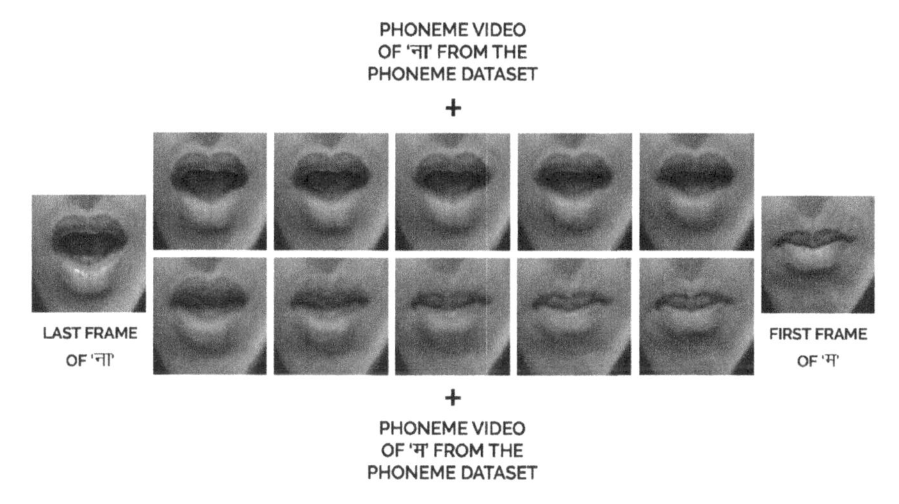

Fig. 6 Word construction process for the Hindi word 'नाम'

Complete or partial loss of hearing is a common disorder in today's world. Lip reading is a process which allows such patients to comprehend others' speech. According to Guilliams and Segui [19], speech videos consisting of proper lip movement can help people with hearing impairment, practice lip reading. Synthetically generated speech videos can act as a practice tool for such patients, by allowing them to create individualized lip reading exercises.

6 Conclusion and Future Work

In this paper, we present an application of video frame interpolation for the purpose of generating Hindi language speech videos by joining a combination of phonemes. Our proposed end-to-end CNN model, SpinCNN, synthesizes plausible intermediate frames by learning optical flows when trained on a dataset of images capturing different kinds of mouth movements.

The model was able to generate visually pleasing Hindi language speech videos by constructing words through the process of joining phonemes with interpolated frames in between. Experimental results report a mean SSIM score of 0.870 and a mean PSNR score of 33.844, which indicate that our proposed model was able to produce structurally correct interpolated frames, but with slight blurriness.

In future, this work can be extended to other languages apart from the Hindi language. Furthermore, different hyperparameters and layered architectures can also be explored to increase the sharpness of the frames generated.

References

1. Fridriksson J, Hubbard HI, Hudspeth SG, Holland AL, Bonilha L, Fromm D, Rorden C (2012) Speech entrainment enables patients with Broca's Aphasia to produce fluent speech. Brain J Neurol
2. Zhai J, Yu K, Li J, Li S (2005) A low complexity motion compensated frame interpolation method. In: Proceedings of IEEE international symposium on circuits and systems 2005
3. Ha T, Lee S, Kim J (2004) Motion compensated frame interpolation by new block-based motion estimation algorithm. IEEE Trans Consum Electron
4. Guo D, Lu Z (2016) The grid: motion-compensated frame interpolation with weighted motion estimation and hierarchical vector refinement. Neurocomputing
5. Liu Z, Yeh R, Tang X, Liu Y, Agarwala A (2017) Video frame synthesis using deep voxel flow. In: ECCV 2017
6. Sharma A, Menda K, Koren M (2017) Convolutional neural networks for video frame interpolation. Neurocomputing
7. Jiang H, Sun D, Jampani V, Yang MH, Miller EL, Kautz J (2017) Super SloMo: high quality estimation of multiple intermediate frames for video. Interpolation. arXiv:1712.00080v1 [cs.CV] 2017
8. Niklaus S, Mai L, Liu F (2017) Video frame interpolation via adaptive convolution. In: CVPR 2017
9. Long G, Kneip L, Alvarez JM, Li H, Zhang X, Yu Q (2016) Learning image matching by simply watching video. In: ECCV 2016
10. Ronneberger O, Fischer P, Brox T (2015) U-Net: convolutional networks for biomedical image segmentation. arXiv:1505.04597v1 [cs.CV] 2015
11. Yahia HB, Frame interpolation using convolutional neural networks on 2D animation. MA thesis, University of Amsterdam, Amsterdam, The Netherland
12. Duchi J, Hazan E, Singer Y (2011) Adaptive subgradient methods for online learning and stochastic optimization. J Mach Learn Res
13. Zhao H, Gallo O, Frosio I, Kautz J (2016) Loss functions for neural networks for image processing. IEEE Trans Comput Imaging
14. Wang Z, Bovik AC, Sheikh HR, Simoncelli EP (2004) Image quality assessment: from error visibility to structural similarity. IEEE Trans Image Process
15. Horé A, Ziou D (2010) Image quality metrics: PSNR vs. SSIM. In: International conference on pattern recognition (ICPR) 2010
16. Sharma A, Menda K, Koren M (2017) Frame interpolation using generative adversarial networks. Neurocomputing
17. Amersfoort J, Shi W, Acosta A, Massa F, Totz J, Wang Z, Caballero J (2017) Frame interpolation with multi-scale deep loss functions and generative adversarial networks. arXiv:1711.06045v1 [cs.CV] 2017
18. Fridriksson J, Speech entrainment treatment for Broca's Aphasia. University of South Carolina at Columbia, Columbia, SC, USA. http://grantome.com/grant/NIH/R21-DC014170-01A1
19. Guilliams I, Segui A (1988) Interactive videodisc for teaching and evaluating lipreading. Eng Med Biol Soc

Face Mask Detection Using Deep Learning

Sandip Maity, Prasanta Das, Krishna Kumar Jha, and Himadri Sekhar Dutta

Abstract With the spread of coronavirus disease 2019 (COVID-19) pandemic throughout the world, social distancing and using a face mask have become crucial to prevent the spreading of this disease. Our goal is to develop a better way to detect face masks. In this paper, we propose a comparison between all available networks, which is an efficient one-stage face mask detector. The detection scheme follows preprocessing, feature extraction, and classification. The mask detector has been built using deep learning, specifically ResNetV2, as the base pre-trained model upon which we have our own CNN. We use OpenCV's ImageNet to extract faces from video frames and our trained model to classify if the person is wearing a mask or not. We also propose an object removal algorithm to reject prediction below absolute confidence and accept only predictions above it. For the training purpose, we are using the face mask dataset, which consists of 680 images with mask and 686 images without mask. The results show mask detector has an accuracy of 99.9%. We have also used other pre-trained networks like MobileNetV2 as our base network and compared our results. ResNet50 gives us the state-of-the-art performance of face mask detection, which is higher than other face detectors.

Keywords CNN · COVID-19 · Deep learning · TensorFlow · Machine learning

1 Introduction

World Health Organization (WHO) announced in the circumstance report 96 [1] that coronavirus disease 2019 (COVID-19) has affected over 2.7 million people all over the world, which caused deaths of over 180,000. The severe acute respiratory syndrome (SARS) [2] and the middle east respiratory syndrome (MERS) are also similar big-scale respiratory diseases that happened within the past few years [2, 3]. Liu et al. [4] reported that COVID-19 spreads higher than SARS in terms of spreading.

S. Maity (✉) · P. Das · K. K. Jha
Calcutta Institute of Technology, Uluberia, Howrah, West Bengal 711316, India

H. S. Dutta
Kalyani Government Engineering College, Kalyani, West Bengal, India

A. Choudhary et al. (eds.), *Applications of Artificial Intelligence and Machine Learning*, Lecture Notes in Electrical Engineering 778,
https://doi.org/10.1007/978-981-16-3067-5_37

Therefore, public health is considered of the highest importance for governments [5], day by day, the number of people who are conscious about their health is increasing. According to Leung et al. [6], the spread of coronavirus can be reduced by using surgical face masks. Currently, WHO recommends people with respiratory symptoms to wear face masks or look after the people with manifestation. Many public sectors have mandatorily required people to use face masks [5]. Therefore, it has become a crucial task to help our society with the help of computer vision by creating a face mask detection system, but there have been a few types of research for facemask detection. The massive amount of variations such as indoor and outdoor conditions, image qualities, occlusions expressions, poses, and skin colors makes the face mask detection a complicated task for outdoor video clips. Due to considerable variation in situations, we need to make our model good enough to work correctly in every situation.

In recent years, convolutional neural networks (CNNs) have faced quick development in multiple areas. Face detection has various applications in many areas, such as Protect Law Enforcement, autonomous driving [7], education, firefighting [8], surveillance, and so on [9]. In this work, we present an effective face detector based on ResNet50 as the base network, which can satisfactorily address the occlusion and false positive issue. More specifically, following a similar setting as RetinaNet [10], we utilize a feature pyramid network and different layers from the network to detect the faces with different scales. Neural networks are implemented with nature-based optimization for better accuracy [11]. More acceptability can be achieved through the implementation of deep learning in nature-based optimization [12].

Since the face mask dataset [13] is small as there are only a few sources available, so here feature extraction is challenging. Here we have used transfer learning to use the pre-trained networks, which were trained on massive datasets. The proposed comparison is tested on a face mask dataset [13]. In this paper, we have proposed a model for detecting faces using single-shot detection an image by classifying each pixel and using a pre-trained ResNet50 network to detect masks on the faces which act as a binary classifier.

1.1 The Major Contribution of This Paper

In this situation, many researchers going developed the facemask detection model where the detection model helps the environment to stop spread the COVID-19. Our study of research is also developed as an efficient detection model and provides cooperation in the environment. The proposed network can detect absolute faces and multiple faces and make predictions from a single image. At present, general research in various use cases of deep learning is showing rapid growth. Although some works were carried out previously on face detection in various situations, no comprehensive work was dedicated to facemask detection. Therefore, it is worth it to use the full potential of deep learning, especially recurrent neural networks and computer vision, to solve real-world problems like detecting face masks, which is crucial for public health. The detection is based on our own CNN model and deep learning-based ResNet50. The organization of this paper is in Sect. 1—the introduction of this paper. The motivation of the work and literature reviews of the existing method is explained in Sect. 2. Section 3 depicted the proposed detection method. The experimental results and discussion are depicted in Sect. 4, and the conclusion is described in Sect. 5.

2 Motivation

Multiple models are available for facemask detection based on machine learning and deep learning. Existing methods are worked fine, but some of the models failed to recognize the facemask due to the situation of the environment and the different activities of the people. Our motivation is based on these existing methods, where we developed an efficient model to detect facemask of various situations.

2.1 Literature Survey

Marco Grassi et al. had developed the fastest image preprocessing by the linearize-shaded elliptical mask centered over the faces. For features extraction used association with discrete cosine transform, for classification, radial basis function, and MPL NN, it approved an increase of performances without altering the global weight and also a learning time diminution for multilayer perception NN [14]. Lei Li et al. had introduced a new detection model based on image analysis for 3D image mask data evaluation attack by reflectance contrast. In this method, the face data is processed early with intrinsic decomposition to determine its reflectance data. They extracted the features of the histogram from orthogonal planes to denote several reflectance image data between the original face data versus 3D face mask data. A convolutional neural network is used to get the information to depict several objects or surfaces react severely to changes in perception [15]. Mingjie Jiang et al. had proposed RetinaFaceMask; they get high accuracy, which is a productive mask detector. They introduced a novel technique that is not only efficient, but also capable of getting high accuracy in different situations. This paper is probably contributed to health care. The system model of RetinaFaceMask performs with ResNet. They extracted the different features which improved their model performance as well as accuracy. They worked with a large dataset trained by the neural network. Their proposed algorithm performs effectively, which found the 2.3%, 1.5% high rate and then compares to other methods, respectively. The precision and recall compared the other method, which is the baseline public dataset. They examine the prospect of implementing RetinaFaceMask by the NN MobileNet for embedded systems [16]. Shiming Ge et al. had worked on a dataset named MAskedFaces(MAFA), with 35,806 masked face data and 30,811 Internet image data. Faces in the dataset have different orientations, while at least one part of each face is covered by a mask. Based on the MAFA dataset, they had proposed LLE-CNN for the detection process, which is based on some major modules. Their module first worked with two pre-trained convolution neural networks to extract facial regions from the image and present them with very high-dimensional descriptors. They used a locally linear embedding algorithm for finding the similarity on the face images and trained the normal face with a large dataset. They had used unified CNN for performed regression and classification tasks. They had successfully built the module, which is able to recognize the face regions. By using the MAFA database, they achieved 15.6%, which is performed by several state of the arts [17]. Kaihan Lin et al. describe their research work on ResNet-101 is used to extract features, RPN performs to find ROIs, and they had used through the RoIAlign reliably securing the exact locations to find binary mask by fully convolution Nnetwork. Further, for the bounding box, GIoU is used and improved the accuracy by using the loss function. They had compared with mask regional convolutional neural network, faster regional convolutional neural network, and multitask cascade convolutional neural network, and they had proposed the G-Mask technique has got results on face detection dataset and benchmark, and WIDER FACE, AFW benchmarks [18]. Toshanlal Meenpal et al. had introduced a method to explore the mask segmentation from any object size of the input image. They had used a fully convolutional network for segmentation and gradient descent to train the model while minimizing the loss function by using binomial cross-entropy. They had also used FCN to avoid unwanted noise data and remove the chance to wrong prediction. Their model performs with high accuracy, and the basic advantage of this model can predict more than one face at a time in a single frame. They achieve the experimental accuracy which is 93.88% [19]. N. Ozkaya et al. had designed and developed an intelligent system with a novel proposal depending on ANN for generating face mask with face objects from fingerprints with absolute errors which are 0.75–3.60. Their experimental results had demonstrated that it is the prospect to find out the face mask from fingerprints without knowing one single piece of information about the face. Last, it is shown that fingerprints and faces are related to each other very closely. The experimental results are very supportive, multilayer perception consists of four layers, and ANN is a stronger and more reliable module [20]. Face detection has been extensively studied as the fundamental problem of computer vision. Before the convolutional neural network (CNN) renaissance, numerous machine learning algorithms were applied to face detection. With the looks of the primary real-time face detection method called Viola-Jones [21] in 2004, face detection has begun to be applied. The well-known Viola-Jones can perform real-time detection using Haar features and cascaded structure. Still, it also has some drawbacks, such as large feature size and low recognition rate for tricky situations. Despite optimizing the features of Haar [4], the Viola-Jones detector failed to handle real-world problems and was affected by many factors like brightness and orientation of faces. Viola-Jones was able to perform well to detect only frontal faces in well-lit conditions. It failed to perform well in less light conditions and with absolute images. Multiple new handcraft features like HOG [22], SIFT [23], SUFT [24], and LBP [25] are proposed.

In recent years, we have seen significant growth in the development of deep learning techniques in the field of semantic image segmentation, and such models achieve the best state-of-the-art results. The first goal for deep learning is the FCN by Long et al. (2015). FCN exudes the fully convolutional layers in well-known deep architectures, such as VGG-16 [26], AlexNet [11], and GoogleLeNet [12], ResNet [27], to convolutional layers to make spatial heat maps as the output of the model.

3 Proposed Method of Face Mask Detection

The propounded model of face mask detection is worked with our own CNN model and deep learning-based ResNet50. The detection process is performed with a different parameter. The ResNet50 is compared with others model and achieved the effective outcome where we used the small number of features of the face mask data. The model performed in several phases like preprocessing, face detection and cropping, feature extraction, and classification (Fig. 1).

3.1 Preprocessing

We propose to detect whether the person is wearing mask from the video containing single or multiple faces in different situations. We process every frame of the video through our facemask detector network. Irrespective of the input image size, it is resized to $224 \times 224 \times 3$ and fed to the CNN to extract features and make predictions. The output from this stage is then sent for post-processing.

3.2 Face Detection and Cropping

We grab the dimensions of the frame and then construct a blob from it. The blob is passed through the network to obtain the face detections. We feed the faces to the pre-trained networks for making future predictions.

3.3 Feature Extraction and Prediction

Primarily, the pixel values of the face and background are processed for global thresholding. After that, it is passed through a median filter to get rid of the high-frequency noise. Next, the gaps in the segmented area are closed. For faster processing, we

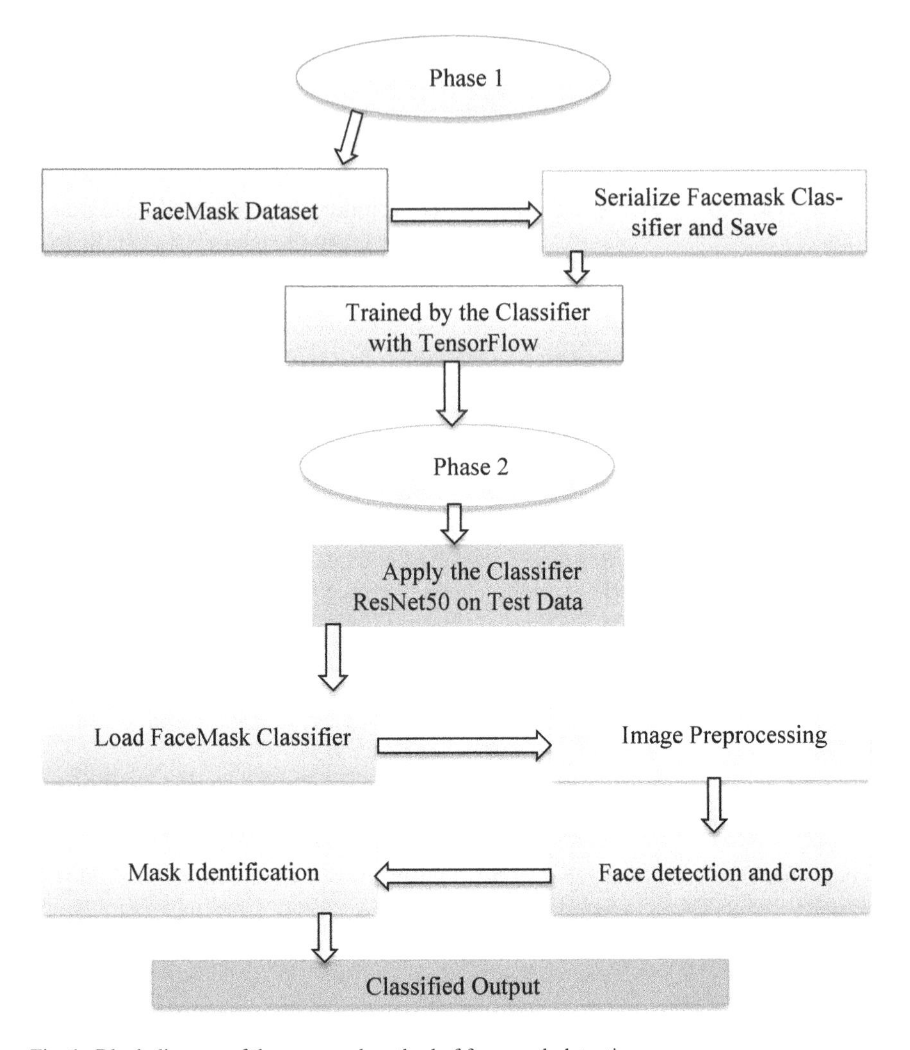

Fig. 1 Block diagram of the proposed method of face mask detection

will make batch predictions on all detected faces at the same time rather than single predictions. Then the segmented area is covered with a box with respective predictions. Filter out weak detections by omitting the confidence, which is lower than the minimum confidence. For the thresholding, we had performed the global thresholding where it separates the object from the background and calculates the threshold value with image value. The image is defined as $p(k, s)$

$$p(k, s) = \begin{cases} 1 & \text{if}(k, s) > R \\ 0 & \text{if}(k, s) \leqslant R \end{cases} \tag{1}$$

where R is the threshold value.

3.4 Architecture

The predefined ResNet50 network architecture has been used for feature extraction and making predictions. Our proposed model consists of 48 convolution layers with one max pool and one average pool layer. The initial image is preprocessed to the standard size of 224 × 224 × 3 pixels and fed to the network as input and passed through the proceeding layers for the feature. The images are passed through multiple convolutional layers and also some max pooling layers.

The input passed through a convolutional layer, in which convolution image with another window at the same time max pooling is applied, reduces each layer to a 50% reduction in the number of parameters. This is an important part of feature extraction. Lower-level features can be obtained by initial layers, whereas the subsequent layer can generate mid-level and higher-level features. The partial information obtained by segmentation is kept in the pixel-wise classification. Segmentation is obtained by converting VGG layers to convolutional layers. As it is the binary classification, so it creates two different channels for face and background.

4 Results and Discussion

The designed face mask detector semantically segments out the face locations with the respective label. Alongside this, the results of the proposed model are also excellent in recognizing side faces. The model has been trained on a human face mask dataset, which contains almost 900 images. Out of the images, 80% of images were used for training and validation, and the remaining data was used for testing the performance of the model. We have also shown the refined predicted mask after it is subjected to post-processing. Alongside, our model can also detect multiple faces from a single frame using single-shot detection. The post-processing has improved the pixel-level accuracy pretty well. The average accuracy for detecting facial masks is 99.9% (Figs. 2, 3, 4 and 5).

4.1 Performance Analysis

While we are training the model, accuracy and loss for validation data could be variating with different cases. Usually, with every epoch, loss should be going lower, and accuracy should be going higher. If the training accuracy is going high rapidly after a certain number of epochs, our model will start overfitting, which we can solve with certain methods. In our experiment, almost every model except the VGG-19 performs very well and reaches very good accuracy. Training loss also decreases with the number of increasing epochs (Figs. 6, 7, 8 and 9).

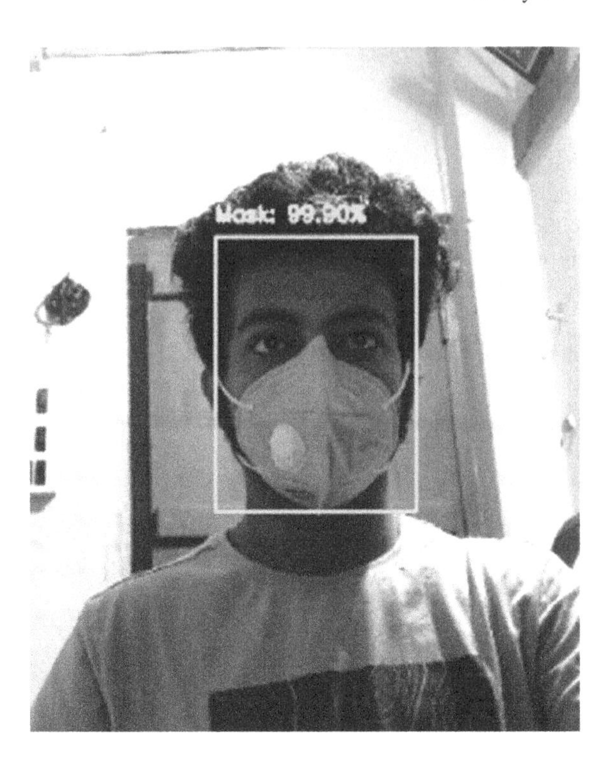

Fig. 2 Front face with mask

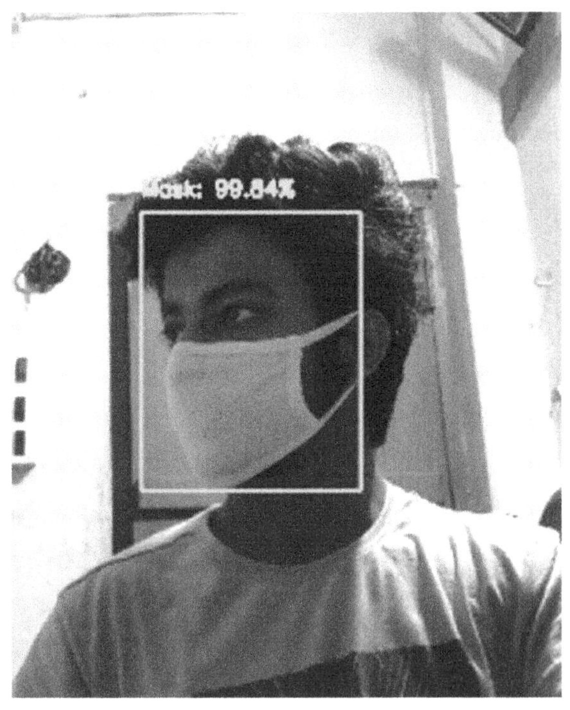

Fig. 3 Side face with mask

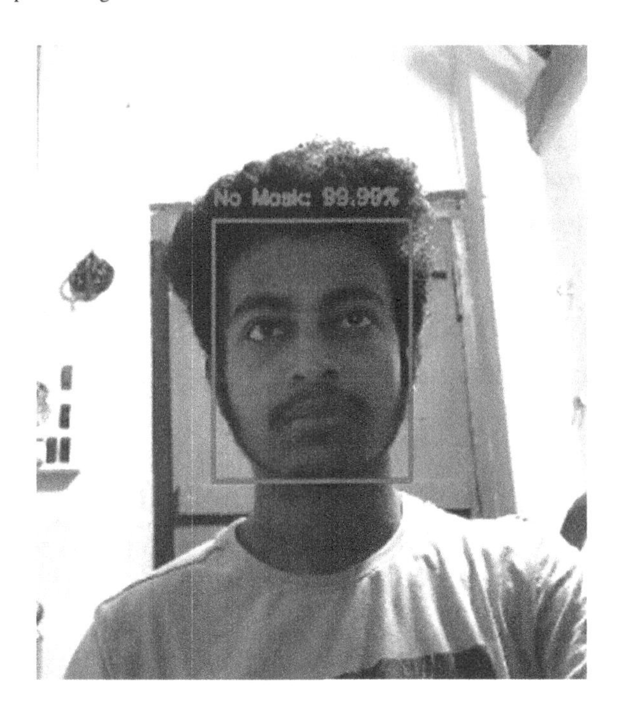

Fig. 4 Front face without mask

Fig. 5 Side face without mask

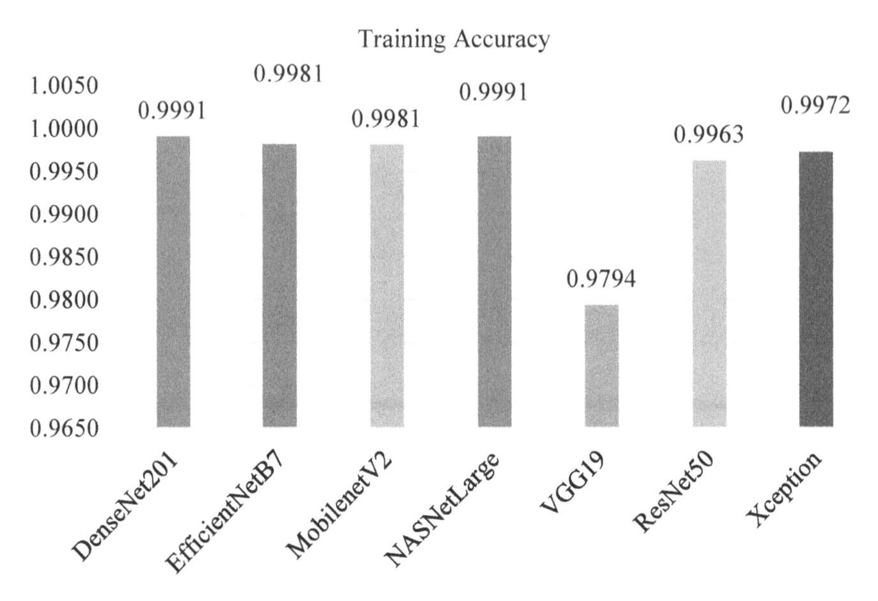

Fig. 6 Training accuracy of models

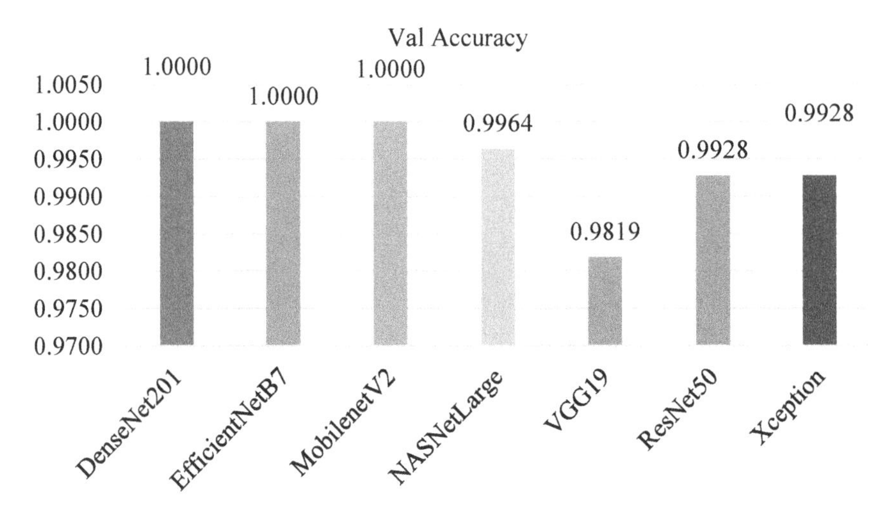

Fig. 7 Val accuracy of models

4.2 Comparative Analysis

In this work, we have compared the effective networks for face detection, and the results have been listed below. Most of them were able to predict with utmost accuracy with minor differences, and we have chosen the best performing network. The deeper

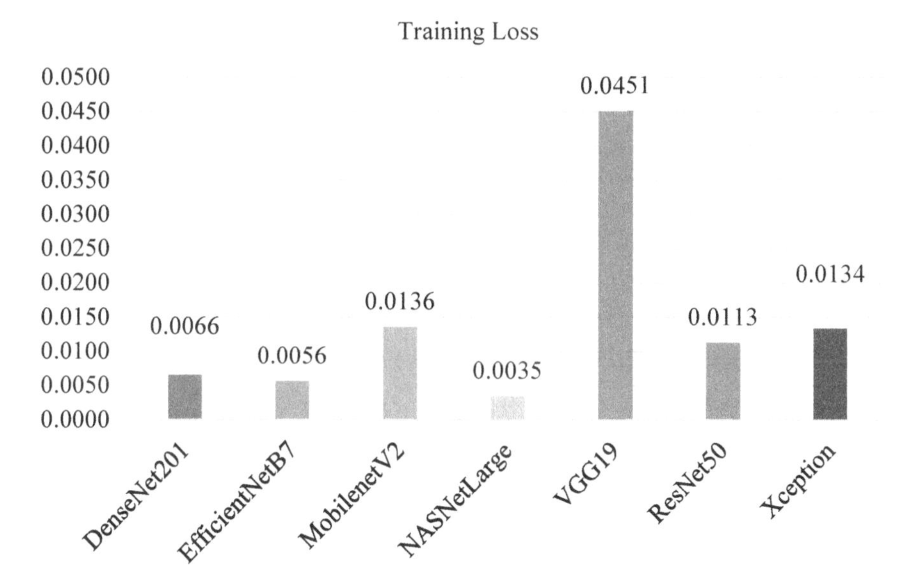

Fig. 8 Validation accuracy of models

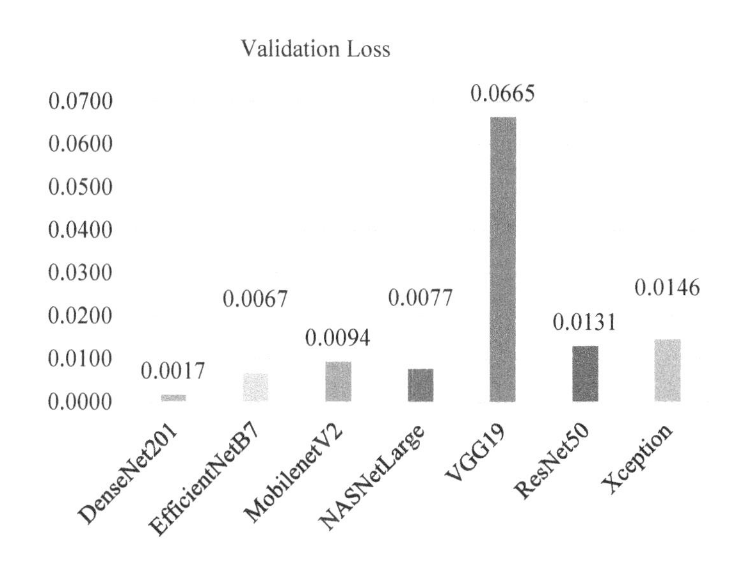

Fig. 9 Validation loss of models

networks seem to predict a bit better than shallow networks. In Table 1, we have listed the comparative analysis of such models (Table 2).

Table 1 Architecture of ResNet50 network

Layer name	Output size	18-layer	34-layer	50-layer	101-layer	152-layer
conv1	112×112	7×7, 64, stride 2				
conv2_x		3×3 max pool, stride 2				
	56×56	$\begin{bmatrix} 3 \times 3, 64 \\ 3 \times 3, 64 \end{bmatrix} . \times 2$	$\begin{bmatrix} 3 \times 3, 64 \\ 3 \times 3, 64 \end{bmatrix} . \times 3$	$\begin{bmatrix} 1 \times 1, & 64 \\ 3 \times 3, & 64 \\ 1 \times 1, & 256 \end{bmatrix} \times 3$	$\begin{bmatrix} 1 \times 1, & 64 \\ 3 \times 3, & 64 \\ 1 \times 1, & 256 \end{bmatrix} \times 3$	$\begin{bmatrix} 1 \times 1, & 64 \\ 3 \times 3, & 64 \\ 1 \times 1, & 256 \end{bmatrix} \times 3$
conv3_x	28×28	$\begin{bmatrix} 3 \times 3, 128 \\ 3 \times 3, 128 \end{bmatrix} . \times 2$	$\begin{bmatrix} 3 \times 3, 128 \\ 3 \times 3, 128 \end{bmatrix} . \times 4$			
conv4_x	14×14	$\begin{bmatrix} 3 \times 3, 256 \\ 3 \times 3, 256 \end{bmatrix} \times 2$	$\begin{bmatrix} 3 \times 3, 256 \\ 3 \times 3, 256 \end{bmatrix} \times 6$		$\begin{bmatrix} 1 \times 1, & 512 \\ 3 \times 3, & 512 \\ 1 \times 1, & 1024 \end{bmatrix} \times 23$	$\begin{bmatrix} 1 \times 1, & 512 \\ 3 \times 3, & 512 \\ 1 \times 1, & 1024 \end{bmatrix} \times 36$
conv5_x	7×7	$\begin{bmatrix} 3 \times 3, 512 \\ 3 \times 3, 512 \end{bmatrix} \times 2$	$\begin{bmatrix} 3 \times 3, 512 \\ 3 \times 3, 512 \end{bmatrix} \times 3$	$\begin{bmatrix} 1 \times 1, & 1024 \\ 3 \times 3, & 1024 \\ 1 \times 1, & 2048 \end{bmatrix} \times 3$	$\begin{bmatrix} 1 \times 1, & 1024 \\ 3 \times 3, & 1024 \\ 1 \times 1, & 2048 \end{bmatrix} \times 3$	$\begin{bmatrix} 1 \times 1, & 1024 \\ 3 \times 3, & 1024 \\ 1 \times 1, & 2048 \end{bmatrix} \times 3$
	1×1	Average pool, 1000-d fc, softmax				
FLOPs		1.8×10^9	3.6×10^9	3.8×10^9	7.6×10^9	11.3×10^9

Table 2 Comparative analysis of models

Models	Performance matrices			
	Training accuracy	Training loss	Validation accuracy	Validation loss
DenseNet201 [28]	0.9991	0.0066	1.0000	0.0017
EfficientNetB7 [29]	0.9981	0.0056	1.0000	0.0067
MobileNetV2 [30]	0.9981	0.0136	1.0000	0.0094
NASNetLarge [31]	0.9991	0.0035	0.9964	0.0077
VGG-19 [26]	0.9794	0.0451	0.9819	0.0665
Xception [32]	0.99719	0.01341	0.99275	0.01464
ResNet50 [27]	0.99625	0.01134	0.99275	0.01307

5 Conclusion

We were able to predict face mask on human faces from RGB images containing direct or acute faces. We have shown our results on the human faces dataset, which contains a decent amount of data. We considered a small number of features for predictions and got excellent results. The incorrect prediction problem has been solved, and an appropriate bounding box with the prediction confidence has been drawn around the segmented area. Here the resulting level reached up to 99.625% training accuracy with 1.13% training loss. By the inclusion of more adaptive learning, training loss may be reduced. This modified ResNe50 can provide better performance compared to other pre-trained models. The method has its applications in several tasks, like detecting the facial part. This model can be utilized in different object identification, viz. seat belt detection, helmet detection for two wheelers, and multiple people maintaining social distances or not. In the future, many real-time problems can be solved by the same algorithm, which had been used to detect a face mask. A device may be developed for implementation purposes, which will cater to solutions to several problems. Future work may be drawn in, viz. low-intensity light, huge crowd, and the different posture of partial face recognition.

References

1. World Health Organization et al (2020) Coronavirus disease 2019 (covid-19): situation report, vol 96
2. Rota PA, Steven Oberste M, Monroe SS, Allan Nix W, Campagnoli R, Icenogle JP, Penaranda S et al (2003) Characterization of a novel coronavirus associated with severe acute respiratory syndrome. Science 300(5624):1394–1399
3. Memish ZA, Zumla AI, Al-Hakeem RF, Al-Rabeeah AA, Stephens GM (2013) Family cluster of Middle East respiratory syndrome coronavirus infections. N Engl J Med 368(26):2487–2494
4. Liu Y, Gayle AA, Wilder-Smith A, Rocklöv J (2020) The reproductive number of COVID-19 is higher compared to SARS coronavirus. J Travel Med
5. Fang Y, Nie Y, Penny M (2020) Transmission dynamics of the COVID-19 outbreak and effectiveness of government interventions: a data-driven analysis. J Med Virol 92(6):645–659

6. Leung NHL, Chu DKW, Shiu EYC, Chan K-H, McDevitt JJ, Hau BJP, Yen H-L et al (2020) Respiratory virus shedding in exhaled breath and efficacy of face masks. Nature Med 26(5):676–680
7. Stefaniga S-A, Gaianu M (2018) Face detection and recognition methods using deep learning in autonomous driving. In: 2018 20th international symposium on symbolic and numeric algorithms for scientific computing (SYNASC). IEEE, pp 347–354
8. Bhattarai M, MartíNez-Ramón M (2020) A deep learning framework for detection of targets in thermal images to improve firefighting. IEEE Access 8:88308–88321
9. Almasi M (2019) An investigation on face detection applications. Int J Comput Appl 975:8887
10. Zhang H, Chang H, Ma B, Shan S, Chen X (2019) Cascade retinanet: maintaining consistency for single-stage object detection. arXiv:1907.06881
11. Jha KK, Dutta HS (2020) Nucleus and cytoplasm-based segmentation and actor-critic neural network for acute lymphocytic leukaemia detection in single cell blood smear images. Med Biol Eng Comput 58(1):171
12. Jha KK, Dutta HS (2019) Mutual Information based hybrid model and deep learning for acute lymphocytic Leukemia detection in single cell blood smear images. Comput Methods Progr Biomed 179:104987
13. https://github.com/prajnasb/observations/tree/master/experiements/data (Aug 2020)
14. Grassi M, Faundez-Zanuy M (2007) Face recognition with facial mask application and neural networks 4507:709–716. https://doi.org/10.1007/978-3-540-73007-1_85
15. Li L, Xia Z, Jiang X, Ma Y, Roli F, Feng X (2020) 3D Face mask presentation attack detection based on intrinsic image analysis. IET Biometr 9.https://doi.org/10.1049/iet-bmt.2019.0155
16. Jiang M, Fan X, Yan H (2020) RetinaMask: a face mask detector. V2
17. Ge S, Li J, Ye Q, Luo Z (2017) Detecting masked faces in the wild with LLE-CNNs:426–434. https://doi.org/10.1109/CVPR.2017.53
18. Lin K, Zhao H, Lv J, Li C, Liu X, Chen R, Zhao R (2020) Face detection and segmentation based on improved mask R-CNN. Discr Dyn Nature Sochttps://doi.org/10.1155/2020/9242917
19. Meenpal T, Balakrishnan A, Verma A (2019) Facial mask detection using semantic segmentation, pp 1–5. https://doi.org/10.1109/CCCS.2019.8888092
20. Ozkaya N, Sagiroglu S (2008) Intelligent face mask prediction system:3166–3173. https://doi.org/10.1109/IJCNN.2008.4634246
21. Viola P, Jones MJ (2004) Robust real-time face detection. Int J Comput Vis 57(2):137–154
22. Dalal N, Triggs B (2005) Histograms of oriented gradients for human detection. IEEE Comput Soc Conf Comput Vis Pattern Recogn CVPR 2005 1:886–893
23. Lowe DG (2004) Distinctive image features from scale-invariant keypoints. Int J Comput Vis 60(2):91–110
24. Bay H, Ess A, Tuytelaars T, Van Gool L (2008) Speeded-up robust features (SURF). Comput Vis Image Understanding 110(3):346–359
25. Ahonen T, Hadid A, Pietikainen M (2006) Face description with local binary patterns: application to face recognition. IEEE Trans Pattern Anal Mach Intell 28(12):2037–2041
26. Mateen M, Wen J, Song S, Huang Z (2019) Fundus image classification using VGG-19 architecture with PCA and SVD. Symmetry 11(1):1
27. Reddy ASB, Sujitha Juliet D (2019) Transfer learning with ResNet-50 for Malaria cell-image classification. In: 2019 international conference on communication and signal processing (ICCSP). IEEE, pp 0945–0949
28. Jaiswal A, Gianchandani N, Singh D, Kumar V, Kaur M (2020) Classification of the COVID-19 infected patients using DenseNet201 based deep transfer learning. J Biomol Struct Dyn:1–8
29. Marques G, Agarwal D, de la Torre Díez I (2020) Automated medical diagnosis of COVID-19 through EfficientNet convolutional neural network. Appl Soft Comput:106691
30. Saxen F, Werner P, Handrich S, Othman E, Dinges L, Al-Hamadi A (2019) Face attribute detection with MobileNetV2 and NasNet-Mobile. In: 2019 11th international symposium on image and signal processing and analysis (ISPA). IEEE, pp 176–180
31. Saxen F et al (2019) Face attribute detection with MobileNetV2 and NasNet-Mobile. In: 2019 11th international symposium on image and signal processing and analysis (ISPA). IEEE

32. Chollet F (2017) Xception: deep learning with depthwise separable convolutions. In: Proceedings of the IEEE conference on computer vision and pattern recognition

Sandip Maity is currently pursuing B. Tech. in Computer Science and Engineering from Calcutta Institute of Technology, Howrah, West Bengal. His research areas include big data, deep learning, and image processing.

Prasanta Das received his BCA degree from Prabhat Kumar College, Contai, under the Vidyasagar University, Medinipur. He has completed an MCA degree from Calcutta Institute of Technology under the Maulana Abul Kalam Azad University of Technology, Kolkata, India. His research interests include medical image processing, machine learning, and deep learning.

Krishna Kumar Jha received Master of Technology from the University of Calcutta, India, and pursuing Ph.D. in Technology from Maulana Abul Kalam Azad University of Technology, Kolkata, India. He is presently working as Assistant Professor at the Department of Computer Science and Engineering at Calcutta Institute of Technology, Howrah, West Bengal, India. His research areas include medical image processing and big data. He has published more than 08 research papers in various international journals and conferences and reviewed papers for reputed journals.

Himadri Sekhar Dutta received his M.Tech. degree in Optics and Opto-Electronics from the University of Calcutta, Kolkata, India, and Ph.D. in Technology from the Institute of Radio Physics and Electronics, Kolkata, India, respectively. He is presently working as Assistant Professor at ECE Department of Kalyani Government Engineering College, Kalyani. He was Chairperson of the IEEE Young Professional, Kolkata Section, for two consecutive years (in 2016 and 2017) and actively participated in different activities conducted by IEEE. His research areas include medical image processing, embedded systems, and opto-electronic devices. He has published more than 70 research papers in various international journals and conferences and reviewed papers for reputed journals and international conferences.

Deep Learning-Based Non-invasive Fetal Cardiac Arrhythmia Detection

Kamakshi Sharma and Sarfaraz Masood

Abstract Non-invasive fetal electrocardiography (NI-FECG) has the possibility to offer some added clinical information to assist in detecting fetal distress, and thus it offers novel diagnostic possibilities for prenatal treatment to arrhythmic fetus. The core aim of this work is to explore whether reliable classification of arrhythmic (ARR) fetus and normal rhythm (NR) fetus can be achieved from multi-channel NI-FECG signals without canceling maternal ECG (MECG) signals. A state-of-the-art deep learning method has been proposed for this task. The open-access NI-FECG dataset that has been taken from the PhysioNet.org for the present work. Each recording in the NI-FECG dataset used for the study has one maternal ECG signal and 4–5 abdominal channels. The raw NI-FECG signals are preprocessed to remove any disruptive noise from the NI-FECG recordings without considerably altering either the fetal or maternal ECG components. Secondly, in the proposed method, the time–frequency images, such as spectrogram, are computed to train the model instead of raw NI-FECG signals, which are standardized before they are fed to a CNN classifier to perform fetal arrhythmia classification. Various performance evaluation metrics including precision, recall, F-measure, accuracy, and ROC curve have been used to assess the model performance. The proposed CNN-based deep learning model achieves a high precision (96.17%), recall (96.21%), $F1$-score (96.18%), and accuracy (96.31%). In addition, the influence of varying batch size on model performance was also evaluated, whose results show that batch size of 32 outperforms the batch size of 64 and 128 on this particular task.

Keywords Non-invasive fetal ECG · Arrhythmia detection · Convolutional neural network · Deep learning

K. Sharma
Department of Applied Sciences and Humanities, Jamia Millia Islamia, New Delhi 110025, India

S. Masood (✉)
Department of Computer Engineering, Jamia Millia Islamia, New Delhi 110025, India
e-mail: smasood@jmi.ac.in

A. Choudhary et al. (eds.), *Applications of Artificial Intelligence and Machine Learning*, Lecture Notes in Electrical Engineering 778,
https://doi.org/10.1007/978-981-16-3067-5_38

1 Introduction

Fetal cardiac arrhythmia basically refers to the abnormality in fetal heart rate (FHR) and/or cardiac rhythm, i.e., the condition in which fetal heart rhythm is either too fast or too slow. A healthy fetus has a heartbeat of 120–160 beats/min, beating at a regular rhythm, which is significantly higher than the heart rate of an adult (50–70 beats/min). Fetal arrhythmias are diagnosed in 1–3% pregnancies [1], out of which about 10% are considered as probable sources of morbidity. Benign fetal arrhythmias, such as PACs with less than 11 bpm and sinus tachycardia, usually do not require a treatment before or after birth. The hemodynamic fluctuations-related postnatal fetal arrhythmias involve interventions, as in few cases these may lead to preterm deliveries [2]. Sustained fetal arrhythmias prompt to the possibility of hydrops fetalis (a serious condition in which fluid builds up in two or more areas of the baby's body, causing severe swelling), cardiac dysfunction, or even fetal demise [3]. Thus, if cardiac arrhythmia is not diagnosed or left untreated, it can pose a risk to mother as well as fetus, including congestive heart failure. Earlier mild cases of arrhythmias were thought as benign but now prenatal cardiologists assert that any kind of irregular heartbeat should be identified and monitored closely to prevent any fatal fetal distress.

NI-FECG is an encouraging non-invasive alternative to fetal diagnosis and monitoring, which is performed by placing surface electrodes on a pregnant woman's abdomen to obtain a FECG signal. NI-FECG promises to assist fetal arrhythmia diagnosis by the means of uninterrupted analysis of the fetal heart rate (FHR) for the beat-to-beat variations. It can also assist in morphological analysis of the PQRST complex [4, 5] and thus present a number of advantages over the existing invasive modalities: reduced cost, analysis at local level (pregnant women not going over long distances for analysis), motion estimation, information on ventricular and atrial activity, and opportunity of long-term continuous remote monitoring. The capacity of the NI-FECG to deliver a precise estimation of the fetal heart rate has been shown by several recent studies. However, till today, the clinical usability of NI-FECG has rarely been explored. The NI-FECG extraction is also a challenge due to the temporal and frequency overlap between the fetal and the maternal electrocardiograms as they need modern signal processing methods [6, 7].

Although there are various methods of automatic classification methods of adult ECGs that have been proposed, little or no work has been done to analyze the NI-FECG signals, as they usually exhibit as a combination of substantial noise, fetal activity and a greater amplitude of maternal activity. This makes the precise extraction and further analysis of the FECG waveform a perplexing task to perform.

Deep learning (DL) is the mainstream of machine learning, which provides a structure where tasks like extraction of features and classification are executed together. With the development of artificial intelligence (AI), deep learning methods, such as feedforward artificial neural network (ANNs) and the recurrent neural networks (RNNs), long short-term memory (LSTM) and gate recurrent unit (GRU) are widely applied to the medical data [8]. The current study aims to automate the classification process of NI-FECG signals into arrhythmic (ARR) and normal rhythm (NR) using the state-of-the-art deep learning-based convolutional neural network.

2 Literature Review

This section takes two different sets of literature into account: (i) the non-invasive FECG analysis and (ii) to understand how machine learning and deep learning have been employed in detecting cardiac arrhythmia.

2.1 Literature Overview of the Non-invasive FECG Analysis

Clifford et al. [5] and Behar et al. [4] analyzed the clinical attributes extraction from the NI-FECG morphology. The fetuses included in these studies did not consist of any reported cardiac condition. This limited the research conclusions of whether the estimation of these physiological attributes was precise enough to offer actionable medical information.

The study [9] demonstrated the viability of the non-invasive FECG as a supplementary technique to identify the fetal atrioventricular block and hence could support clinical decisions.

In the research work [10], a systematic review is carried out to highlight normal fetal CTIs using NI-FECG and all the outcomes including fetal CTIs (P wave duration, PR interval, QRS duration and QT interval) were assembled as early pre-term ($\leq$ 32 weeks), moderate to late preterm (32–37 weeks) and term (37–41 weeks), concluding that NI-FECG establishes efficacy to quantify CTIs in the fetus, mainly at advanced gestations.

The study has established that NI-FECG assists in the identification of fetal arrhythmias. The diagnosis based on the extracted NI-FECG recordings was compared with the reference fetal echocardiography diagnosis to establish that NI-FECG and fetal echocardiography established the existence of an arrhythmia or not. This research work shows, for the first time, that NI-FECG allows to recognize fetal arrhythmias and also offers added evidence on the rhythm disturbance than echocardiography.

2.2 Literature Overview Cardiac Arrhythmia Detection (Adult and Fetus)

Bengio Y. in his work [11] advocated the popularity of deep learning methods, stating that deep learning architectures learn features at multiple levels of abstraction (i.e., layers) which allows in mapping the input to the output without being provided with hand-engineered features.

Karpagachevi et al. [12] classified the ECG signals taken from the PhysioNet arrhythmia database into five types of abnormal waveforms and normal beats, using extreme learning machine (ELM) and support vector machine (SVM). Experiment results show that ELM offers enhanced accuracy for all the classes, thus strongly recommended the use of the ELM-based method for classifying ECG. Though, the study gives good results on the classification but is only applied to adult ECG.

In the study, [13], a deep learning-based neural network with six hidden layers was proposed to recognize premature ventricular contraction (PVC) beats from the ECG recordings. The network was trained with six features which were extracted from ECGs for the purpose of classification. Although the researchers used a deep learning technique, they still used the hand-engineered features from the ECG data.

In another study by Pourbabaee et al. [14], a deep CNN was trained to extract features from the raw signals and to classify the paroxysmal atrial fibrillation (PAF) and the normal beats.

Andreotti et al. [15] compared the state-of-the-art feature-based classifier with a deep learning-based CNN. The short segments of the ECG were classified into four classes (AF, normal, other rhythms, or noise) and thus establishing that deep learning algorithms are proficient of categorizing short ECG recordings. It is also established in the study that deep learning models are aided from the augmented dataset while feature-based classifiers did not benefit from dataset augmentation.

In the study [16], Alin Isin and Selin Ozdalili proposed a deep learning-based structure previously trained on a general dataset which was used to carry out automatic ECG arrhythmia diagnostics on the MIT-BIH arrhythmia database. In the experiment, AlexNet, a transferred deep CNN, was used as the feature extractor. The study concluded that ECG arrhythmia detection approaches based on non-deep learning methods were outperformed by the transferred deep learning feature extraction approach.

The research work of Gao et al. [17] proposed a long short-term memory (LSTM) recurrent neural network with focal loss (FL) for detecting arrhythmia on a heavily imbalanced dataset using the MIT-BIH arrhythmia dataset. The supremacy of using LSTM with FL was recognized by analyzing with the cross-entropy loss function-based LSTM.

All the abovementioned works demonstrated the application of many widely used machine learning and deep learning network structures in detecting arrhythmia from adult ECG, but deep learning techniques for NI-FECG signals are yet to be explored. The recent study conducted by Zhong et al. [18] proposed a deep learning approach to detect fetal QRS. This study is the first of its kind where a deep learning technique is applied on non-invasive FECG signals. In the proposed work, fetal QRS complex was identified from single-channel raw NI-FECG signals without canceling the maternal ECG signals using a deep learning-based CNN model. A precision of 75.33, recall of 80.54%, and F-1 score of 77.85% were achieved by the proposed CNN.

Another study conducted by Lee et al. [19] proposed a much deep CNN architecture to detect fetal QRS using the NI-FECG signals, without the channel selection presented the positive predictive value of 92.77% with a mean sensitivity of 89.06%.

3 Experimental Design

The CNN classifier developed in the work ran on the deep learning toolbox and signal processing toolbox of MATLAB2018. The computer system had a 64 bit Microsoft Windows 10 operating system, configured with an 8 GB RAM and Intel®Core ™ i5-9300H processor. The epochs were set to 30 with five iterations per epoch. Model approximately took 11.53 s per epoch for training, though the respective epoch setting is not guaranteed to be the best configuration for the CNN network.

3.1 Data Source and Description

The training data used in this study is non-invasive fetal ECG arrythmia database (NIFEA DB) (February 19, 2019) taken from physionet.org. Dataset has been provided with an open access to the users. The dataset contains 26 samples including 12 arrthymic and 14 normal rhythmic fetal samples, obtained from 24 pregnant women, out of which two had normal rhythmic twins. This dataset contains 500 NI-FECG recordings, which were recorded constantly for varying periods ranging from a minimum of 7 min to a maximum of 32 min. Each of these records has a sample frequency either of 500 or 1000 Hz and is indicated in the header of each file. NI-FECG records contain one chest lead and four to five abdominal leads (recorded by placing five–six abdominal electrodes on maternal abdomen and two chest electrodes). A sample of normal rhythm (NR) and arrhythmic (ARR) signal is shown (see Fig. 1).

3.2 Signal Preprocessing

The raw NI-FECG signals are preprocessed to remove any disruptive noise from the NI-FECG recordings without considerably distorting the fetal or the maternal ECG components. For this, the mean removal technique is applied in which the mean of the signal is subtracted from every sample point, resulting in the removal of the unwanted DC component (noise) in the NI-FECG signals. Then, a ten-point moving average filter is applied to remove high-frequency noise. The low-frequency noise components are removed with the help of a high-pass filter after the removal of high-frequency noise. Each of these steps is applied to all the records collected from the chest, and abdominal channels and filtered signals are acquired for the next step.

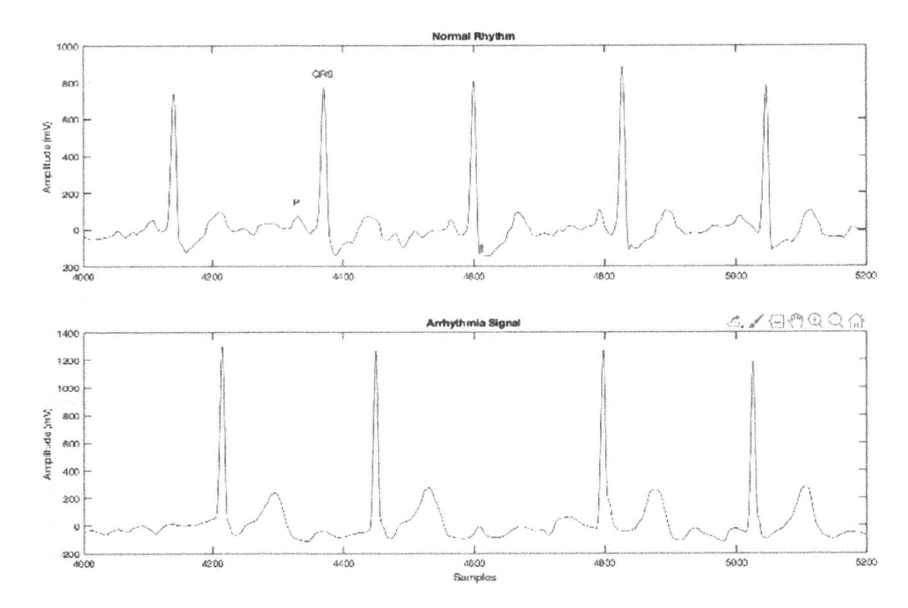

Fig. 1 Sample of normal rhythm (NR) signal and arrhythmia (ARR) signal

3.3 Feature Extraction

Extracting features from the data can assist in improving the train as well as test accuracies of the classifiers. Here, these time–frequency images, such as spectrogram, are computed, and then they are used to train the models. Time–frequency moments extract information from the spectrogram (see Fig. 2).

For the CNN network, each 1D feature is converted into a 2D feature to be as an input. Two time–frequency moments in time domains are spectral entropy (see Fig. 3) and instantaneous frequency (see Fig. 4).

Spectral entropy and instantaneous frequency vary by nearly one order or magnitude. Instantaneous frequency mean for the CNN can be a bit on the higher side for an effective learning. Since large inputs may slow down the network convergence rate, thus the mean and the standard deviation, the train set was used to standardize the train and test data. Standardization is a powerful method to improve the network performance during training. The 500 NI-FECG recordings were distributed into train and test sets with a ratio of 85:15. Since CNNs need a fixed window size, so the frame size is set to 100 ms.

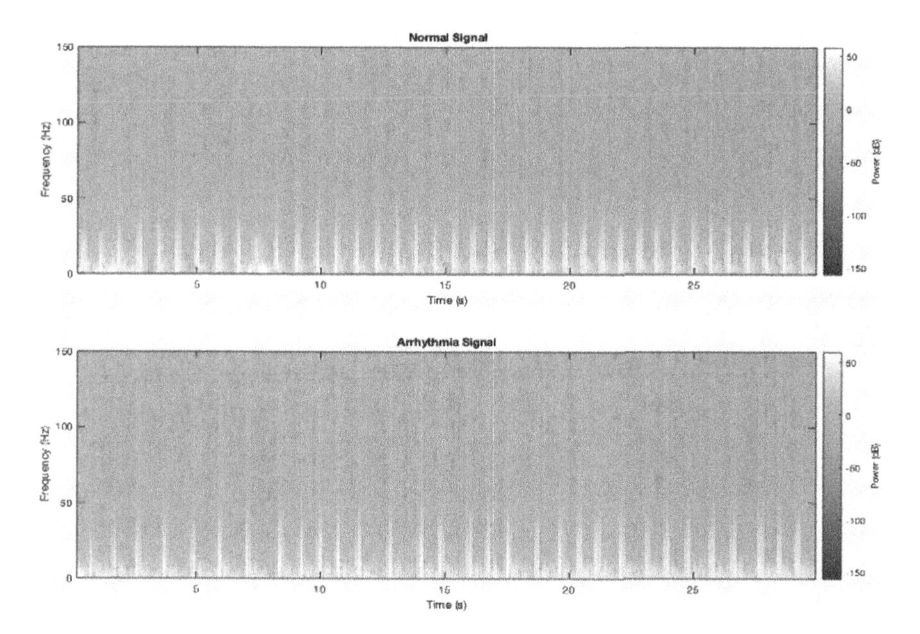

Fig. 2 Spectrogram of normal and arrhythmia signals

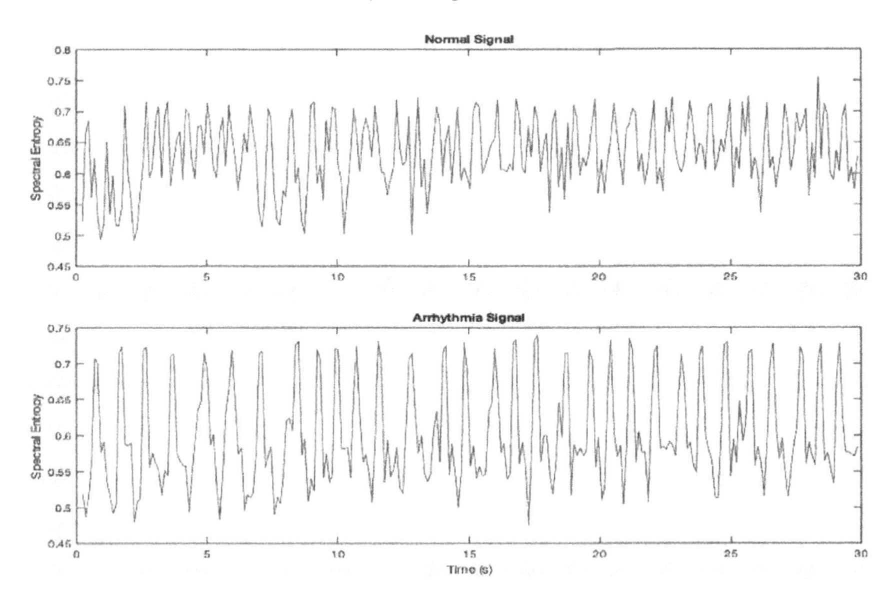

Fig. 3 Spectral entropy of normal signal and arrhythmia signal

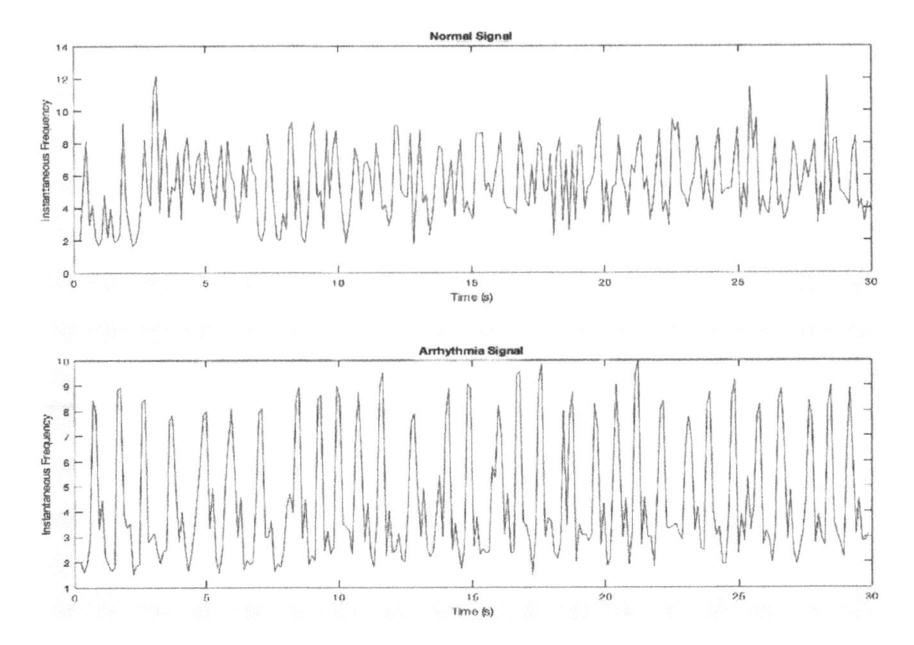

Fig. 4 Instantaneous frequency of normal signal and arrhythmia signal

3.4 CNN Architecture

An NI-FECG signal is fed as an input the CNN model, which is employed for the learning task. A sequence of labels (ARR and NR) are the outputs of the model. The schematic of the CNN architecture being used is shown (see Fig. 5). It contains seven convolutional layers, which are followed by two fully connected layers, a softmax layer, and a final classification layer.

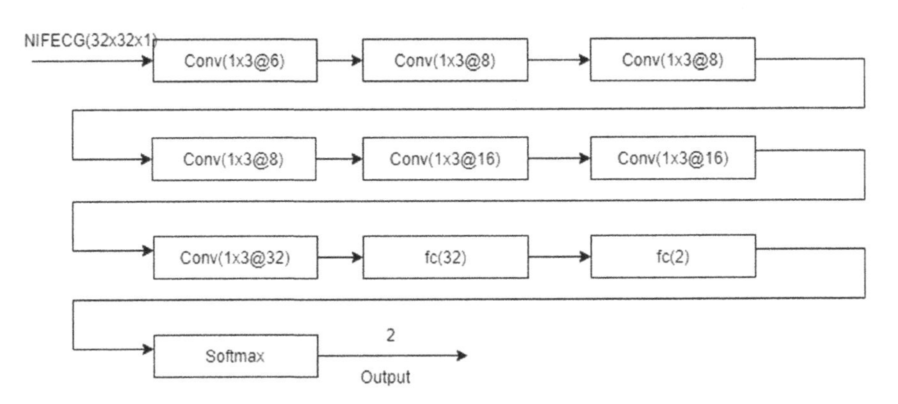

Fig. 5 Proposed CNN architecture

The first convolutional block contains a convolutional 2D layer (having six filters with a kernel size of 3), a batch normalization layer, and an activation function layer (ReLu) and a max pooling 2D layer. The second, third, and fourth convolutional blocks contain a convolutional 2D layer (having six filters with a kernel size of 3), a batch normalization layer and an activation function layer (ReLu), and a max pooling 2D layer. The fifth and sixth convolutional blocks contain a convolutional 2D layer (having 16 filters with a kernel size of 3), a batch normalization layer, and an activation function layer (ReLu). The seventh convolutional block contains a convolutional 2D layer (having 32 filters with a kernel size of 3), a batch normalization layer, and an activation function layer (ReLu). All the max pooling layers had a pool size of 2. The distribution over the two selected classes, i.e., ARR and NR, was produced by the concluding classification layer and softmax.

Batch normalization layer is used between the convolutional layers and the ReLu layers to speed up the training of convolutional neural networks. ReLu layer performs a threshold operation to each element of the input, but it does not change the size of its input. Max pooling layer divides the input into rectangular pooling regions and then computes the maximum of each region. Fully connected layer combines all the features learned by the previous layers and thus classifies the input. It acts independently on each time step in the case of sequence data, as is the case in this study. A softmax layer, following the fully connected layers, applies a softmax function to the input. Following the softmax layer is the classification layer, which computes the cross-entropy loss for the classification problems with mutually exclusive classes.

4 Result and Discussion

In this study, we tried to investigate whether reliable fetal arrhythmia classification could be attained using the deep learning approach without canceling the MECG signals from the NI-FECG signals. Deep supervised learning technique and convolutional neural networks (CNN) structure are used to achieve the goal of classification of fetal arrythmia, using the NI-FECG signals.

4.1 Performance Metrics

Performance measures, including accuracy, precision, recall, and $F1$-score, which were evaluated using a confusion matrix, were used for each class (NR and ARR). They were defined as follows:

$$\text{Accuracy} = \frac{\text{TP} + \text{TN}}{\text{TP} + \text{TN} + \text{FP} + \text{FN}}$$

Accuracy gives the ratio of no. of accurate predictions to the total no. of input samples, which reflects the consistency among the real and the test results.

$$\text{Precision} = \frac{\text{TP}}{\text{TP} + \text{FP}}$$

Precision determines how precise or accurate the model is in predicting that from the total predicted positives, how many actually belonged to the positive class.

$$\text{Recall} = \frac{\text{TP}}{\text{TP} + \text{FN}}$$

Recall calculates how many of the actual positives our model has captured labeling it as true positive.

$$F1 = \frac{2 \times (\text{Precision} \times \text{Recall})}{\text{Precision} + \text{Recall}}$$

$F1$-score is the harmonic mean of recall and the precision and is helpful in seeking a balance between precision and recall.

4.2 Result

The performance indices of Zhong et al. [18] and Lee et al. [19] and the proposed model at 85:15 train–test ratios have been shown in Table 1.

The model performance has been evaluated considering three different batch sizes, as shown in Table 1. It shows the classification accuracy on the testing set, when the train–test ratio is set to 85:15 and learning rate 0.01 and Adam optimizer is used. The precision, recall, and $F1$-score are shown in Table 2.

Table 1 Performance metrics of all compared models

	Precision	Recall	$F1$-score	Accuracy
Zhong et al. [18]	89.03	91.57	90.28	91.33
Lee et al. [19]	92.89	90.27	91.56	93.27
Proposed model	*96.17*	*96.21*	*96.18*	*96.31*

Italics signifies that it belongs to the proposed model

Table 2 Classification accuracy of different batch sizes

Parameter	Value 1	Value 2	Value 3
Batch size	32	64	128
Accuracy on test set (%)	**96.31**	93.23	91.47

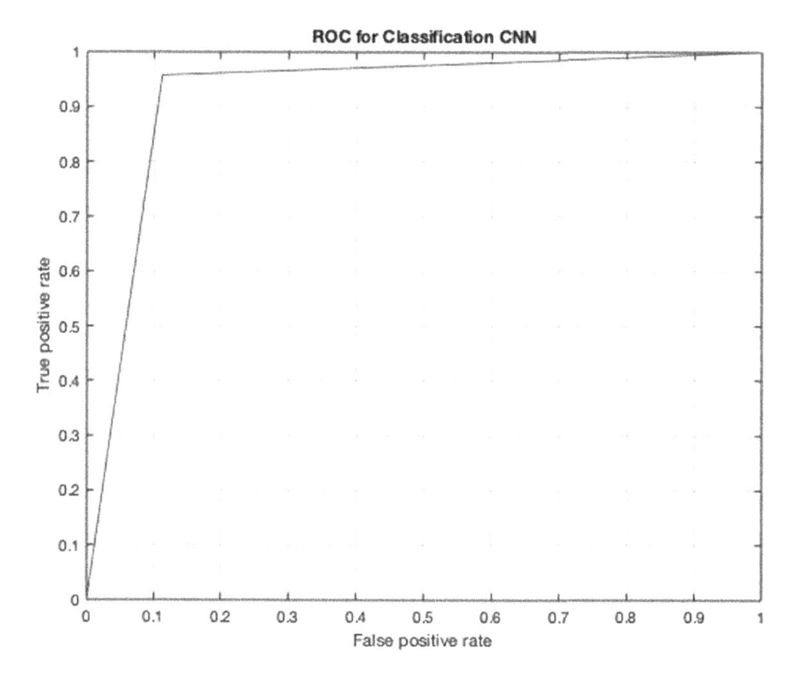

Fig. 6 ROC curve for the arrhythmic class

For the dataset and network structure used in the current research work, the optimal batch size is 32. To more intuitively compare the effectiveness of the proposed method, we analyzed the results using the ROC curve (see Fig. 6).

4.3 Discussion

The model was also analyzed by varying dropouts for the proposed network with an increasing dropout proportion and on comparing the results. It was found that the performance of the network did not improve. To confirm the effectiveness of the proposed model, the comparison is carried out with the algorithm proposed by Zhong et al. and that by Lee et al., respectively, and the results are graphically presented (see Fig. 7).

5 Conclusion

This work proposes a deep learning-based CNN model for classifying the fetal arrhythmia based only on NI-FECG signals without canceling the MECG signals.

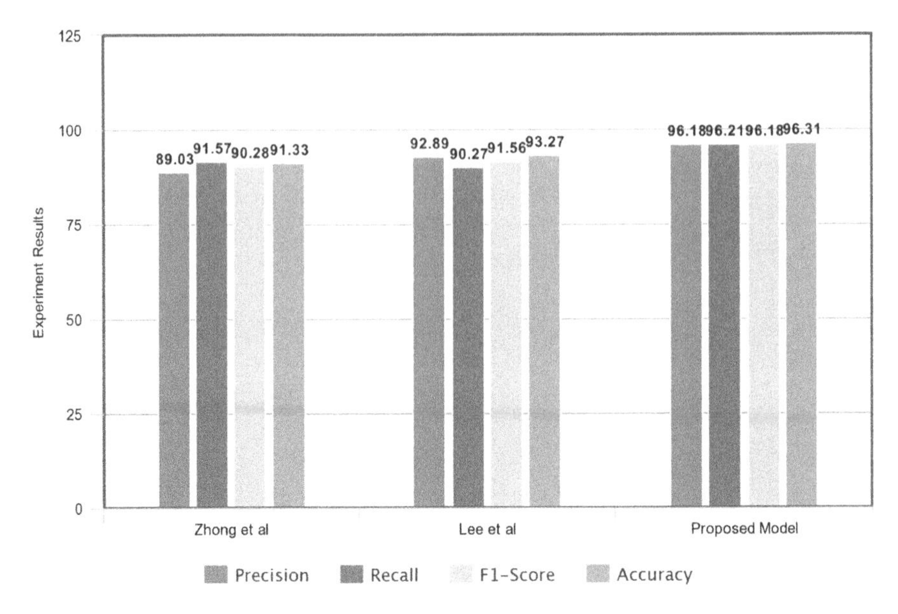

Fig. 7 Comparison of proposed model with Zhong et al. [18] and Lee et al. [19] models

The fetal arrhythmia can be classified with 96.31% accuracy, which is a good performance achieved by the proposed model. Our result shows that deep learning algorithms are capable of classifying the NI-FECG recordings. The DL method has an advantage that they do not require hand-engineered features, over other traditional methods. Moreover, if such pre-trained models are available, then it facilitates the imitation of those approaches, which can be fine-tuned to work for other databases or scenarios. The proposed method can efficiently assist the prenatal cardiologists to diagnose, analyze, and classify the NI-FECG signals in a more accurate way.

The present research work was conducted only on two NI-FECG signal types. So, in order to generalize the results, various NI-FECG beat types should be incorporated in the future works. Larger database with more patients and longer recordings is needed in the near future to provide the researchers with the opportunity to identify if their proposed algorithms are efficient in extracting features without any clinical distortion and classify the fetal cardiac arrhythmias.

References

1. Aggarwal S, Czaplicki S, Chintala K (2009) Hemodynamic effect of fetal supraventricular tachycardia on the unaffected twin. Prenat Diagn 29:292–293
2. Capuruço CA, Mota CC, Rezende GD, Santos R (2016) P06.03: Fetal tachyarrhythmia: diagnosis, treatment and outcome. Ultrasound Obstetrics Gynecol 48:182–182
3. Yaksh A, van der Does LJ, Lanters EA, de Groot NM (2016) Pharmacological therapy of tachyarrhythmias during pregnancy. Arrhythmia Electrophysiol Rev 5(1):41–44

4. Behar J, Zhu T, Oster J, Niksch A, Mah DY, Chun T, Greenberg J, Tanner C, Harrop J, Sameni R, Ward J (2016) Evaluation of the fetal QT interval using non-invasive fetal ECG technology. Physiol Meas 37(9):1392
5. Clifford G, Sameni R, Ward J, Robinson J, Wolfberg AJ (2011) Clinically accurate fetal ECG parameters acquired from maternal abdominal sensors. Am J Obstet Gynecol 205(1):47
6. Behar J, Andreotti F, Zaunseder S, Oster J, Clifford GD (2016) A practical guide to non-invasive foetal electrocardiogram extraction and analysis. Physiol Meas 37(5):R1
7. Clifford GD, Silva I, Behar J, Moody GB (2014) Non-invasive fetal ECG analysis. Physiol Measur 35(8):1521
8. Esteva A, Kuprel B, Novoa RA, Ko J, Swetter SM, Blau HM, Thrun S (2017) Correction: corrigendum: dermatologist-level classification of skin cancer with deep neural networks. Nature 546(7660):686–686
9. Lakhno I, Behar JA, Oster J, Shulgin V, Ostras O, Andreotti F (2017) The use of non-invasive fetal electrocardiography in diagnosing second-degree fetal atrioventricular block. Maternal Health Neonatol Perinatol 3(1):14
10. Smith V, Arunthavanathan S, Nair A, Ansermet D, da Silva Costa F, Wallace EM (2018) A systematic review of cardiac time intervals utilising non-invasive fetal electrocardiogram in normal fetuses. BMC Pregnancy Childbirth 18(1):370
11. Bengio Y Learning deep architectures for AI. Now Publishers Inc. Foundations Trends Mach Learn 2(1):1–127
12. Karpagachelvi S, Arthanari M, Sivakumar M (2012) Classification of electrocardiogram signals with support vector machines and extreme learning machine. Neural Comput Appl 21:1331–1339
13. Jun TJ, Park HJ, Minh NH, Kim D, Kim YH (2016) Premature ventricular contraction beat detection with deep neural networks. In: 15th IEEE international conference on machine learning and applications (ICMLA). IEEE, pp 859–864
14. Pourbabaee B, Roshtkhari MJ, Khorasani K (2016) Feature learning with deep convolutional neural networks for screening patients with Paraoxysmal atrial fibrillation. In: 2016 IEEE international joint conference on neural networks. IEEE, pp5057–5064
15. Andreotti F, Carr O, Pimentel MA, Mahdi A, De Vos M (2017) Comparing feature-based classifiers and convolutional neural networks to detect arrhythmia from short segments of ECG. In: 2017 computing in cardiology (CinC). IEEE, pp 1–4
16. Isin A, Ozdalili S (2017) Cardiac arrhythmia detection using deep learning. Proc Comput Sci 120:268–275
17. Gao J, Zhang H, Lu P, Wang Z (2019) An effective LSTM recurrent network to detect arrhythmia on imbalanced ECG dataset. J Healthcare Eng 2019
18. Zhong W, Liao L, Guo X, Wang G (2018) A deep learning approach for fetal QRS complex detection. Physiol Measur 39(4):045004
19. Lee JS, Seo M, Kim SW, Choi M (2018) Fetal QRS detection based on convolutional neural networks in noninvasive fetal electrocardiogram. In: 4th international conference on frontiers of signal processing (ICFSP). IEEE, pp 75–78

Security and Privacy Challenges and Data Analytics

Minimizing Energy Consumption for Intrusion Detection Model in Wireless Sensor Network

Gauri Kalnoor and S. Gowrishankar

Abstract The security is one of the major concerns in today's existing technology. Wireless Sensor Network (WSN) can be deployed in critical areas and network can be compromised by the malicious attack. Due to its unattended deployment strategy in remote places, security plays a major role and thus the primary line of defense is Intrusion detection system (IDS). The existing IDS cannot perform efficiently due to the mechanisms applied. Thus, a novel approach is designed and modeled to obtain high performance of WSN. In our proposed work, the probabilistic model which provides the direct way to visualize the model using joint probability, referred as Bayesian Network is combined with the stochastic process model called as Hidden Markov Model. This combined novel approach is a graphical model represented with nodes and edges. The evaluated results when obtained by applying the novel approach is observed and high detection rate is obtained when compared with the existing algorithms like weighted support vector machine (WSVM), K-means classifier and knowledge-based IDS (KBIDS). Maximum throughput and less transmission delay are obtained. The experiments are carried out for different attacks with various trained and test data. Thus, the novel approach gives overall high performance in WSN.

Keywords Bayes network · WSN · IDS · Markov model · Training · Knowledge-based intrusion detection

1 Introduction

The application fields which are largely used in WSN are monitoring data in health care application, military reconnaissance and surveillance and also smart home applications [1–3]. Some of the crucial fields has made security the most important and in demand for WSN. It is most unrealistic in the sensor network that requires continual power supply in some of the traditional techniques of encryption/decryption such that, frequent access control and key management can be obtained. This is due to its limitations of power supply, communication range and computation capability

G. Kalnoor (✉) · S. Gowrishankar
BMS College of Engineering, Bangalore, India

A. Choudhary et al. (eds.), *Applications of Artificial Intelligence and Machine Learning*, Lecture Notes in Electrical Engineering 778,
https://doi.org/10.1007/978-981-16-3067-5_39

[4]. Thus, the design of IDS is required to meet the requirements of security and to overcome these limitations of WSN.

The monitoring of network traffic, notifying network system or the administrator and checking the suspicious activities are the major role of Intrusion Detection System (IDS). But, in some of the occurrences, the IDS may also respond to the anomalous traffic and malicious nodes, which further takes action like holding the user or the IP address of source from permitting access of the system [5]. Many different IDS types are obtainable and stop the mission of traffic that is uncovering shady. Different types of IDS exist like host-based intrusion system (HIDS), network intrusion detection system (NIDS) and hybrid-based intrusion detection system. The IDS are either signature-based or statistical anomaly-based for intrusion detection. The IDS can be imperfect which is impossible as traffic at network is too high. The wireless ado and wired networks are susceptible to most of the different forms of threats in case of security. The threats are caused due to the unreliable and open communication channel, the dynamic structure in topology and also lack in central coordination. The attacks made by the intrusion can be singular or multiple. Thus, the main objective of IDS is to maximize the rate of true positive and minimize the false positive rate.

The work proposed is discussed in detail in the following sessions. The related survey carried out is discussed in Sect. 2. Probabilistic Model known as Bayesian Network is explained in Sect. 3 and also a stochastic model is framed and discussed, followed by Results and Analysis in Sect. 4. The work is concluded in Sect. 5.

2 Related Work

The two directions are possible to improve the accuracy in detection, i.e., minimize rate of false alarm and increase the rate of detection. The article [6] explained about implementing the DC (Dendritic Cell) Algorithm which is derived from the techniques called immune-inspired danger techniques where different inputs signals can be classified by the DC Algorithm. Thus, the input signals causing damage were considered to be anomalous and other signals were considered as normal signals. The experimental results show that the DC Algorithm has high rate of detection but low rate of false alarm was not achieved.

The authors in the article [7], explains the methods with two different types of agents, for example, T-cell agents and the dendritic cells agents, which work in partnership with each other. These agents count the value that indicates danger, and detect the attacks those are malicious. The scheme attained low rate of false alarm but lacks in required detection rate. Also, a weighted support vector machine (SVM) has been used to increases the boundaries among the anomaly and normal clusters so that classification errors can be minimized, thus enhancing the accuracy in detection effectively.

The factors based on which the IDS can be classified is the behavior of detection system and nature of attacks, i.e., signature-based IDS and anomaly-based IDS. The

IDS based on signature, recognizes the intrusion attacks those are well known and results in the best accuracy. But such IDS fail to detect new types of intrusion attacks, whose signatures are not defined or stored in the repository of attacks. Based on the identifications of features from observed network traffics or the node's utilization of resources, of the intrusion attacks, the anomaly-IDS can be designed. Thus, studies have shown that, many proposed models using optimization and machine learning methods were applied for better performance of WSN. Some were designed using K-nearest neighbor (KNN) [8], random forest (RF) [9], decision tree (DT), particle swarm optimization (PSO), genetic algorithm (GA), and support vector machine (SVM) [10], extreme gradient boosting (XGBoost). Other studies were proposed using combination of SVM and GA, GA and deep belief network (DBM), GA with FL (fuzzy Logic).

Yang et al. [11] has proposed the new algorithm for intrusion detection based on the technique of normalized and cut spectral clustering for sensor network. The main objective was to reduce the degree of imbalance among the classes in intrusion detection system (IDS). At first, the authors have discussed about the design of a normalized cut spectral among the datasets and then in the second stage, the classifier is trained for network intrusion detection using the new set of data. Extensive experiments have been conducted and the results have been analyzed in detail. Thus, the simulation results show that the degree of imbalance is reduced among the classes, thus reserving the distribution of data on one hand improving the performance of detection effectively on the other hand.

Alqahtani et al. [12] proposed the new model to detect the malicious intruder based on classifier known as extreme gradient boosting (XGBoost) and genetic algorithms named as GXGBoost model. This model was designed to improve the performance of traditional models for detecting the minority classes of malicious nodes and attacks in the most highly imbalanced traffic of data of wireless sensor networks. Using the techniques such as tenfold cross-validation and holdout, the set of experiments were carried out based on WSN-detection system known as WSN-DS dataset. Thus, the validation test results of tenfold cross showed that there was high performance based on the proposed approaches with other learning classifiers ensemble. The results also showed high rates of detection with 92.9, 99.5, 98.2 and 98.9% for scheduling, blackhole, flooding, and gray hole attacks, respectively, along with the normal traffic, rate of detection was 90.9%.

3 Model-Based IDS

The methodology using Novel Bayes Hidden (NBH) approach designed is discussed and the framework is shown in Fig. 1. It is performed at the node level where intermediate nodes are connected to the base station.

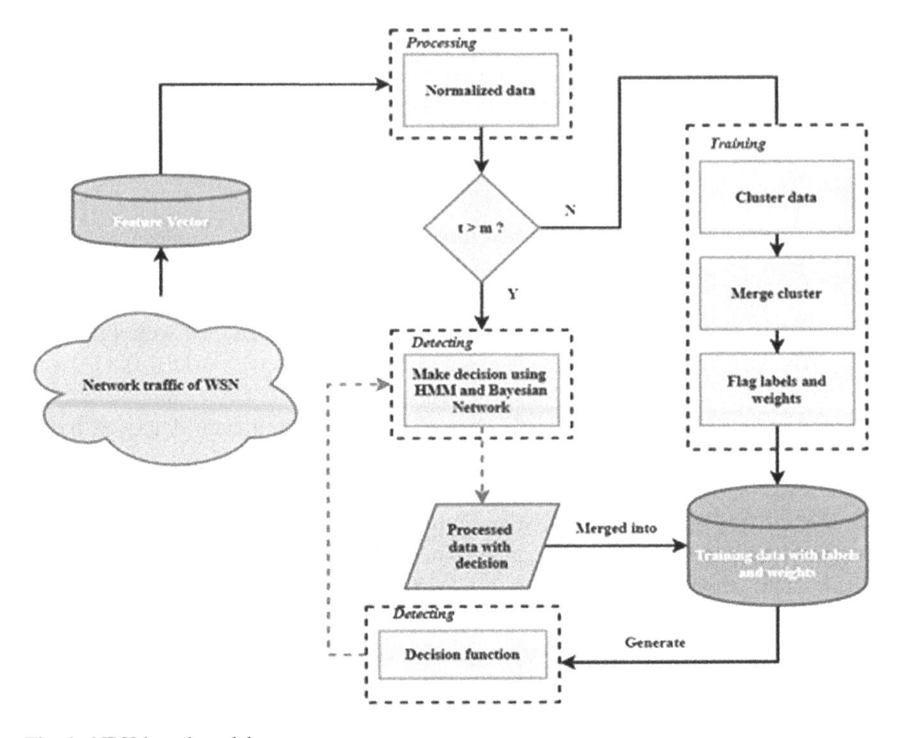

Fig. 1 NBH-based model

3.1 Bayes Belief Network Model

A Bayesian network is also acknowledged as the belief network model. It is a model which represents acyclic and directed graph denoted by G (V, E), "V" represents the set of vertices and "E" represents the set of edges directed and joining the vertices. In the Bayes Net, the loops of any length are not acceptable. Every Vertex V comprises random variable names and tables of probability distribution. This indicates that the ways, the probability of variables values are dependent on the probable combinations of its parental values.

3.2 Hidden Markov Model Architecture

HMM consists of finite set of states, where every state is associated with the probability distribution. The transitions possible among the states are concerned with the set of probabilities called as state transition probabilities. Based on the associated distribution probability, an observation or outcome is generated, at a specific state. The outcome of the state is visible to an observer, but the states are hidden. Thus, it is called as Hidden Markov Model.

The five elements discussed below are used, to define the HMM more precisely:

1. N: the number of states in the model
2. The number of symbol of observations within the alphabet, say M. M is said to be infinite, if the symbol observations are "continuous".
3. $A = \{a_{ij}\}$, a set of transition probabilities of a state.

$$a_{ij} = p\{q_{t+1} = j | q_t = i\}, 1 \leq i, j \leq N, \tag{1}$$

where q_t represents the present state.
The transition probabilities that should satisfy the constraints of normal stochastic,

$$a_{ij} \geq 0, 1 \leq i, j \leq N \text{ and } \sum_{j=1}^{N} a_{ij} = 1, 1 \leq i \leq N$$

4. In every state, the probability distribution,

$$B = \{b_j(k)\}$$
$$b_j(k) = p\{o_t = v_k | q_t = j\}, 1 \leq j \leq N, 1 \leq k \leq M \tag{2}$$

where v_k represents the kth symbol of observation in the alphabet and o_t, the present parameter vector.
The stochastic constraints should be satisfied:

$$b_j(k) \geq 0, 1 \leq j \leq N, 1 \leq k \leq M \text{ and}$$
$$\sum_{i=1}^{M} b_j(k) = 1, 1 \leq j \leq N \tag{3}$$

The state distribution initially $\pi = \{\pi_i\}$, where $\pi_i = p\{q_1 = \text{i}\}, 1 \leq i \leq N$.
Thus, the complete notation $\lambda = (A, B, \pi)$ denotes the HMM as a discrete probability distribution.

The general design of an initiated model of Hidden Markov is shown in Fig. 2. In the architecture, the random state variables are represented by oval shape. These variables adapt to any set of values. At time t, the hidden state of random variable is denoted by $Z(t)$. Here, $Z(t)$ Ɛ $\{Z_1, Z_2, Zm\}$. At time t, the observation is the random variable denoted by $X(t)$, where $X(t)$ Ɛ $\{X_1, X_2, XN\}$. The conditional dependencies are denoted by the arrows, and the diagrams are often called as trellis diagram.

This property is named as Markov Property. Also, the significance of $X(t)$, which is called the observed variable, is dependent only on the significance of $Z(t)$ the hidden variable. The Hidden Markov Model considered in our work, includes the state space of $Z(t)$ as discrete value. But the observations of variables themselves can be a discrete or continuous value.

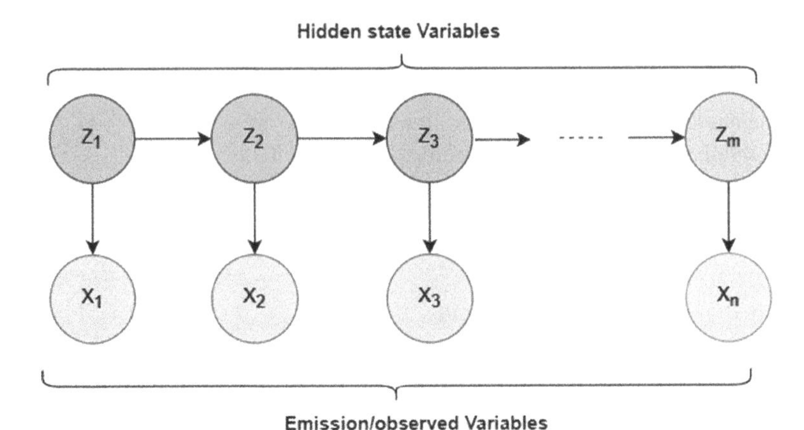

Fig. 2 Architectural model of HMM

Methodology for Intrusion Detection

The IDS framework is designed using the model which is comprised of different levels of processing.

Reading of Training Data

The model based on IDS is trained with the dataset and then tested. Dataset are extracted from many standards for building IDS. Once the dataset is chosen, it is pre-processed.

Pre-Processing of read data from dataset

The standard set of data is in the text format and consists of 41 dimensions with 4,900,000 records of which is considered as 10% of the total size. It has categorical, continuous and binary types of data type attributes. The text data format is converted to CSV (i.e., Comma Separated Values), which can be easily read and analyzed. The samples of about 35,000 records with 5 attributes have been considered. The attributes selected and the samples of data connected are tabulated. The size of data is reduced when the no. of attributes is reduced from 41 to 5 and no. of records in the dataset is reduced to 35,000 from 4,900,000. The symbols replace with the discrete values.

The state transition model describes for both normal and attack records as discussed below.

- No. of state hidden variables $N = 5$, which is equal to no. of variables chosen. Thus, the dimension of "state transition matrix" is 5×5.
- No. of emission symbols those are distinct $M = 18$. So, the size of matrix for emission transition probability is 5×18.
- The probability distribution at initial stage
- $\pi = \{0.000581, 0.261902, 0.08983, 0.375828, 0.271858\}$.

In general, the way to adjust the HMM parameters can be obtained by the learning problem, such that, the given training set called as the observations, and represents the best model for intended application. Thus Baum-Welch Algorithm is used to train the model of IDS. This Baum-Welch Algorithm is also known as "*forward–backward algorithm*".

All the above combined explanation is summarized in the pseudo code mentioned in the modified-Bayes algorithm below:

INPUT:

$X_{\text{All}} = \{x_1, x_2, \ldots, x_t\}$ feature vector
ΔT time of updating

OUTPUT:

1 flag of normal
− 1 flag of anomaly

1. Normalize each feature vector $x_t \in X_{All}$, $t = 1, 2, \ldots$. by (1).
2. Shift x_1 in feature space constructed by normal data $X_{\text{nor}} = \{x_1, x_2, \ldots x_{nr}\}$ by (2) unit the shift distance m_h falls below a certain threshold ε.
3. Record the tracks that x_1 has traveled as a similar set $s_1 = \{(p_1, d_1), (p_1, d_1) \ldots, (p_k, d_k)\}$ $k = 1, 2, \ldots$. And a cluster center c_1.
4. Cluster the training data $X_{\text{Train}} = \{x_1, x_2, \ldots x_{\text{tr}}\}$:

   For each feature vector $x_t \in X_{\text{Train}}$.

      If the distance $d_{x_t p_k}$ between x_t and point p_k in similar set $S_t \subseteq S$ less than d_k.
      Add $\left(x_t, d_{x_t p_k}\right)$ into s_i.
      Else
      Generate a new cluster center c_i and a similar set s_t by

   MSCA

      if c_t is equal to $c_j \in C$
      $s_j = s_t \cup s_j$
      Else
      Add c_t into C and s_t into S.
      End if
      End if

   End for

5. Merge cluster into two clusters and assign label for $x_t \in X_{\text{Train}}$
6. Assign weights for $x_t \in X_{\text{Train}}$ by its relative distance to cluster center (10)
7. Create the decision function by HMM and Bayes Model
8. Flag each subsequent feature vector x_t in $X_{\text{Test}} = \{x_{\text{tr}+1}, x_{\text{tr}+2}, \ldots x_{\text{te}}\}$ as 1 or − 1 by decision function. Combine the feature vector x_t and its label into training data.

If time $t = m + k\Delta T, k = 1, 2 \ldots$
Update the cluster midpoint by MSC Algorithm
Allocated weights for each feature vector x_t of the training data again.
Update decision function by HMM.

9. End If

4 Results and Analysis

To assess the performance of the proposed methodology, the experiments have been conducted. This experiment scenario was simulated by using NS2 (as shown in the Table 1).

Thus, it is showed that our proposed method has advantages over the mainstream methods, mainly in terms of detection rate and FAR. However, our method has achieved the stable performance in all the experiment type scenarios, mainly in different network structures (Figs. 3 and 4).

The time slice length Δt is one of the important factors and the key parameter that can affect the efficiency of the network. Thus, whenever there are variations in the time slice Δt, the accuracy of detection and the amount of energy consumed is calculated, by updating the interval to 300 ms. As Δt increases, the average DR and energy consumption is decreased gradually during data transmission. However, the average FAR rises steadily in such cases. Thus, the updating of time slices $\Delta t = 10$ s, the trade-off is achieved between the DR accuracy and energy consumption.

Running time (ART) is known as cost of energy, which is nothing but the cost of updating the interval and the average accuracy of detection. When $\Delta t = 10$ s, as lowering the Δt (20 s), the high accuracy in detection and low FAR can be achieved. But, when considering the cost, it's much higher. It is noted that $\Delta t = 10$ s, ARP = 62.17 s and when $\Delta t = 100$ s, ART = 1.35 s. Whereas the average FAR and detection rate decreased and increased, respectively, with increase in Δt. When $\Delta t = 500$ s, the stability was achieved. Finally, it is concluded that, when $\Delta t = 300$ s, there was

Table 1 Simulation scenarios of different attacks

Experiment	Attack type	Network structure
Experiment 1	Black hole	FN
Experiment 2	Flooding	FN
Experiment 3	Rushing	FN
Experiment 4	Multiple-attacks	FN
Experiment 5	Black hole	HN
Experiment 6	Flooding	HN
Experiment 7	Rushing	HN
Experiment 8	Multiple-attacks	HN

*FN flat network, HN hierarchical network

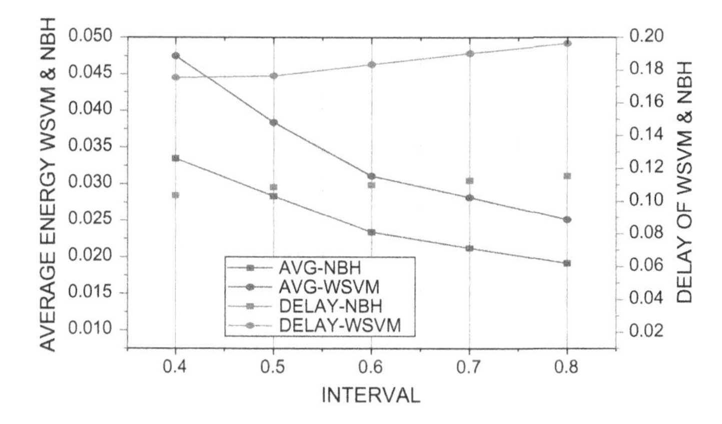

Fig. 3 Interval versus average energy and delay using proposed work NBH compared with WSVM

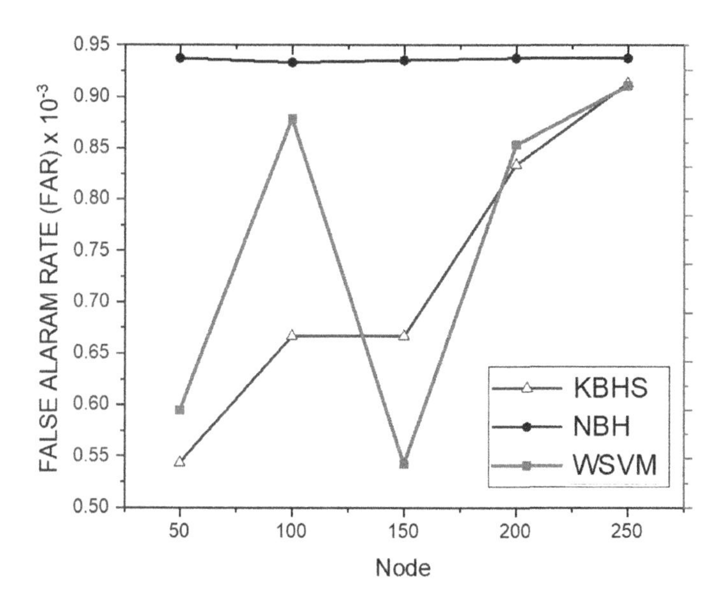

Fig. 4 Node versus FAR based on proposed methodology KBHS and NBH compared with WSVM

a balance between energy cost and detection accuracy, with ART = 3.36 s, much lesser when compared to that at $\Delta t = 10$ s.

5 Conclusion

The proposed work is modeled based on the probabilistic model known as Bayesian belief network model which is combined with the stochastic model known as Hidden Markov Model. The model is analyzed and designed using the novel approach and applying decision function which is updated periodically. Experiments are performed for different types of attacks and intrusion detection is performed by applying the novel algorithm considering the network structure for WSN. The training and testing of dataset is performed using the HMM as applicator. The simulation results is compared with the existing systems using WSVM and KBHS and better performance is obtained using our proposed model NBH. The parameter estimation and data training are done by applying powerful approach of HMM and Bayesian model in the decision function and IDS is created to find whether the traffic of the network is normal or intrusions have been detected. The model is run at the base station to detect the attack. Thus, high throughput and minimum delay are obtained with the proposed model and also detection rate, accuracy and other parameters are maximized when compared with the mainstream IDS techniques deployed in WSN.

References

1. He D, Kumar N, Chen J, Lee CC, Chilamkurti N (2015) Robust anonymous authentication protocol for health-care applications using wireless medical sensor networks. Multimedia Syst 21:49–60
2. Li M, Lin HJ (2015) Design and implementation of smart home control systems based on wireless sensor networks and power line communications. IEEE Trans Industr Electron 62:4430–4442
3. Ball MG, Qela B, Wesolkowski S (2015) A review of the use of computational intelligence in the design of military surveillance networks. Recent Adv Comput Intell Defense Secur 621:663–693
4. Qu HC, Jian S, Tang XM, Wang P (2015) Hybrid computational intelligent methods incorporating into network intrusion detection. J Comput Theor Nanosci 12:5492–5496
5. Huang JY, Liao IE, Chung YF, Chen KT (2011) Shielding wireless sensor network using Markovian intrusion detection system with attack pattern mining. Inf Sci 231:32–44
6. Patil S, Chaudhari S (2016) DoS attack prevention technique in wireless sensor networks. Proc Comput Sci 78:715–721
7. Su MN, Cho TH (2014) A security method for multiple attacks in sensor networks: against false-report injection, false-vote injection and wormhole attacks. Open Trans Wirel Sensor Netw 2:13–29
8. Sarigiannidis P, Karapistoli E, Economides AA (2015) Detecting Sybil attacks in wireless sensor networks using UWB ranging-based information. Expert Syst Appl 42:7560–7572
9. Limamd S, Huie L (2015) Hop-by-Hop cooperative detection of selective forwarding attacks in energy harvesting wireless sensor networks. In: 2015 international conference on computing, networking and communications (ICNC), pp 315–319, Mar 2015
10. Tan X et al (2019) Wireless sensor networks intrusion detection based on SMOTE and the random forest algorithm. Sensors 19(1):203

11. Yang G, Yu X, Xu L, Xin Y, Fang X (2019) An intrusion detection algorithm for sensor network based on normalized cut spectral clustering. PLoS ONE 14(10):e0221920. https://doi.org/10.1371/journal.pone.0221920
12. Alqahtani M et al (2019) A genetic-based extreme gradient boosting model for detecting intrusions in wireless sensor networks. Sensors 19(20):4383

A Blockchain Framework for Counterfeit Medicines Detection

Tejaswini Sirisha Mangu and Barnali Gupta Banik

Abstract The emergence of counterfeit medicines has given rise to a significant setback globally. These drugs may be contaminated, contain the wrong ingredient, or have no active ingredient at all, endangering approximately a million lives per year. The World Health Organization (WHO) estimates that 73 billion euros (79.26 billion USD) worth counterfeit medicines are traded annually. The imperfect supply chain is one of the primary causes of this issue. The present system does not have a proper record of the drug or vaccine that has been manufactured, produced, distributed, and finally reached to the consumer. There is no record of change in ownership of the drugs from manufacturer to consumers. Due to the lack of transparency of the system, the data are not shared between the systems. These loopholes play a significant role in producing and distributing counterfeit medicines. Blockchain is an emerging technology that can help in solving this issue of counterfeit medicines. Using Blockchain, tracking the drugs from its manufacturing until its delivery to the consumer can be possible. Trusted people can use an authorized Blockchain to store transactions and allow trusted parties to join the system and push all the records to Blockchain.

Keywords Counterfeit drugs · Medicines · Blockchain technology · Supply chain

1 Introduction

Medicine is a kind of drug used to detect, cure, or predict diseases. Pharmaceutical companies discover, develop, produce, and market drugs to cure patients suffering from various diseases or illnesses. Drug discovery is a very time-consuming process in which multiple medications and vaccinations are found and analyzed to see which

T. S. Mangu · B. G. Banik (✉)
Department of Computer Science and Engineering, Koneru Lakshmaiah Education Foundation, Deemed to be University, Hyderabad, Telangana 500075, India
e-mail: barnali.guptabanik@ieee.org

T. S. Mangu
e-mail: tejaswinisirisha.m@klh.edu.in

A. Choudhary et al. (eds.), *Applications of Artificial Intelligence and Machine Learning*, Lecture Notes in Electrical Engineering 778,
https://doi.org/10.1007/978-981-16-3067-5_40

diseases can be cured and how. Then, the dosage of the components is formulated to ensure the safety of the patient. They undergo numerous tests to verify if the drug is safe. After a few more procedures, the drug is manufactured and produced. Due to the illegal activities and the production of counterfeit drugs, the medications taken by the patients are rather degrading their health and also causing deaths.

1.1 Background

Counterfeit drugs are now leading us to a dangerous situation. The WHO estimates that the counterfeit drug revenue is about 200 Billion USD and is growing by 20% every year. Around 10–15% of the global pharmaceutical trading involves these counterfeit drugs. Counterfeit products are vandalizing economies, industries, and consumers' health and safety and have become a global concern. The companies are suffering from loss in revenue and reputation also. The productions of these drugs are continuously increasing worldwide, especially in developing countries. The majority of counterfeit drugs are from China, India, and the USA. India is not only the third-largest producer of generic drugs and vaccines and exports to almost 200 other countries.

Nonetheless, it is also a global nub for counterfeit drugs. India accounts for about 10% of the total production of drugs. According to WHO, about 35% of the total counterfeit medication worldwide originated from India. This kind of offense drags the lives of the patients endangered. Also, it causes dents to the profit-making pharmaceutical companies, which results in cutting down the research expenses and increasing the cost of medicines. To tackle this problem, we will need a solution where counterfeit medicines are no longer able to enter the markets. If they enter, they must be identified instantly [1].

1.2 Outline

The key objective of the healthcare industry is ensuring that the patient gets the right medication. However, in various countries, mainly undeveloped or developing countries, it is still a challenge. In the supply chain of health care, an intrusion of counterfeit drug chain is happening at various access points. This is high time to acknowledge that this is not just one supply chain. Every ingredient may include counterfeit drugs. Manufacturers are the primary link in the chain of supply and the main point from where the counterfeit drugs can enter a market. The ingredients could be wrong, dosages could be different, or a few components might be missing. Sometimes, the mismatch of the drugs and their bottles could also take place. Often, the counterfeit manufacturers pack the fake drugs to look like the real ones, almost indistinguishable. Quality control needs to be done here. While genuine manufacturers do the quality check, the fake ones lie about their credentials [2].

The next link is the distributors. In most of the cases, the manufacturer does not sell. The distributor companies sell medicines. To sell counterfeit drugs, the distribution networks need to be set up, ensuring the replacement of fake goods with real ones. While shipping the drugs to pharmacy or from manufacturers to distributors, counterfeit drugs are mixed. The main nub of this problem is the pharmacies. Going to the pharmacies and buying real branded medicines might be fine, but nowadays, everything is available online. This increases the chances of getting counterfeit medicines. The prices of medicines online differ relatively lower than the original ones. Also, the medicines that come from different countries might have been changed. Even when the cover might look the same, the composition of the drugs would be different. Online pharmacies, especially ones with loose regulations, would prefer selling the counterfeit medicines as they cost less than the original ones, and the profit they would make out is immense. Despite the shutting down of these online pharmacies by the government, this problem keeps arising as there is a worldwide demand for cheaper drugs [3]. The generic model of the supply chain of medicine movement is demonstrated in Fig. 1.

Blockchain technology can be used to track the journey of the drugs and the ownership of it at each level. Traceability of these medicines through Blockchain will help to detect and eradicate the counterfeit drugs.

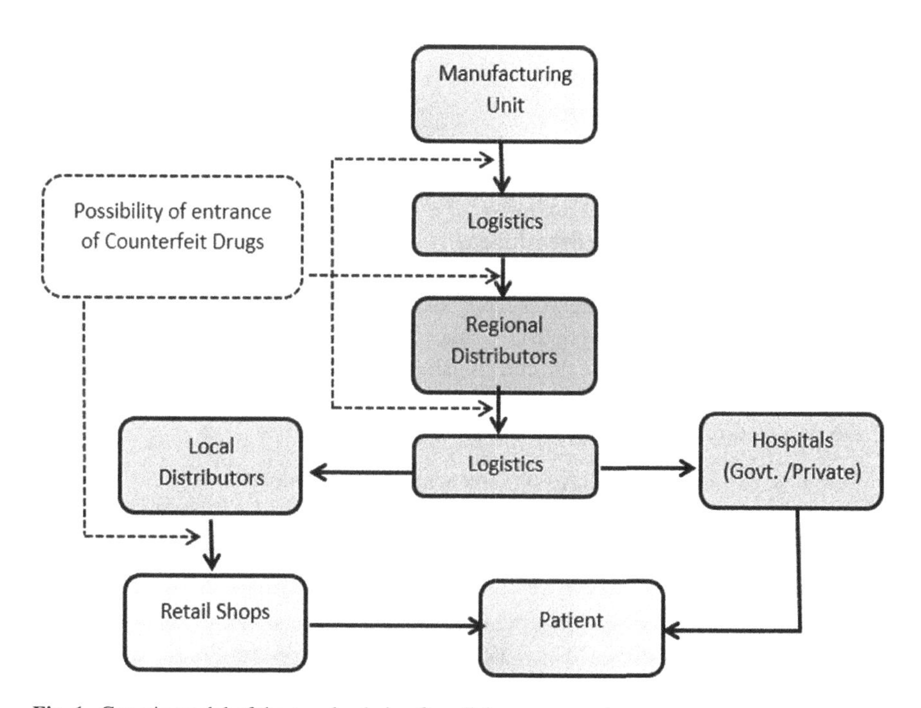

Fig. 1 Generic model of the supply chain of medicine movement

2 Related Knowledge—Blockchain Technology

Blockchain can be comprehended as a shared copy of an immutable ledger, facilitating the process of recording transactions and tracking assets in a network. Any transaction can be tracked and traced on a Blockchain network. It reduces risk, cost, and time spent compared to the traditional ways. In simple words, Blockchain is a data structure that creates a digital ledger of data and can share it among a network of independent parties. It cryptographically chains the blocks chronologically [4].

Blockchain technology creates a possibility of a decentralized world and allows general to create value, trust, transparency, and efficiency so that the users can be empowered without being bound to the centralized or third-party power. This is possible through the smart use of distributed ledgers, cryptography, and computation techniques. The information can be distributed without being copied. This technology not only cuts off the middlemen but also drops the need for match-making platforms [5].

The key characteristics of Blockchain can be are as follows [6]:

- Decentralization, i.e., it is not a possession of a single entity;
- Transparency so, there is no use of tacking the data.
- Immutability so that the data cannot be tampered.

2.1 Decentralization

The decentralized systems distribute power and control more uniformly throughout a system, away from a central authority. The database and Internet structures used from past so far have typically been driven by Internet service providers (ISP). In that system, one needs to log in to the intranets to access databases, so the intermediaries' house information, i.e., it is a centralized system. However in decentralized systems, the individual user will be in control of their data, assets, reputation, and will have peer-to-peer transactions. With the current Blockchain-enabled system, one can send cryptographic tokens peer-to-peer without any intermediary.

Decentralization allows Blockchains to be:

- Less likely to fail as they rely on many separate components.
- Not easy to hack as the networks are spread across millions of computers and be democratic.
- Tougher to tamper with the Blockchain and harm other users.

Even if 100 nodes stop working, the Blockchain persists, assuming that there is at least one node up and running. The Blockchain does not stop functioning even when the power of an entire country is cut off. This makes the Blockchain very resilient to attacks. Blockchain has a set of linearly connected blocks comprising information that is secured with cryptography. The immutable ledger entry serves as a watermark, a notarization, or an agreement of the transaction, thus automating the audit phase.

Implementing Blockchain for counterfeit drugs can help in increasing the efficiency of the quality check and ease auditing and tracking process [7].

2.2 Components and Working Principle

The device, client, or individual user on the network is referred to as a "node." Nodes are the computers that run the Blockchain software which takes transactions, broadcast it, and try putting them together into a block, in a race against other nodes. So, every node in the network is required to authenticate every transaction. Participants can be a person, device, or an entity. When a new node attempts to join the network, the entire transaction records are downloaded onto its system and will join other nodes in updating the ledger at the time of validation of new transactions. There is no chance of hacking into this system. Even if someone does try to hack, there is no use as the flow of ledger is unidirectional in Blockchain [4].

The consensus is vital in Blockchain, which synchronizes the transaction to the various distributed ledgers. The people linked to these mechanisms approve the transactions on the chain, which in turn helps in ensuring the validity and authenticity of the Blockchains transaction.

Blockchain technology uses cryptographic techniques. Public key cryptography is used to create verifiable historical records of transactional data. It helps us to build systems that can be truly decentralized. There are three primary types of Blockchains—public, private, and hybrid. The public and private Blockchains are both immutable distributed networks secured by consensus protocols and cryptography. The public and private Blockchains differ in terms of who is allowed to join in the network, carry out the consensus protocol, and preserve the shared ledger. The bitcoin Blockchain is an excellent example of a public Blockchain. In this, anyone can read and write from anywhere in the world. In a private Blockchain, only trustworthy and recognized participants are allowed to read or write on it. A hybrid Blockchain combines both private Blockchain and public Blockchain. This gives flexibility to the business associates to choose what data they want to keep confidential and what data make public and transparent [8].

An example of a spreadsheet can be taken whose copy is duplicated hundreds of times across a network. Now, visualize that this network updates this spreadsheet regularly. This is the basic principle of Blockchain. The details or information about transactions, etc., is appended continuously into the database. This is updated to all the servers/computers in the Blockchain network instantly. As the records are now available to all, it can be concluded that the information is public and verifiable. As there is no central party handling the information, and the information is simultaneously existing in millions of places, it is difficult to hack, thus providing the public with a transparent and safe system [8].

3 Literature Review

Counterfeit medicines are fake medicines which might be contaminated or contain wrong compositions and quantities of ingredients. These illegal drugs are harmful to one's health. The Food and Drug Administration (FDA), along with other industries, is vigorously monitoring the reported cases earnestly to combat counterfeit drugs. The amount of deaths caused by an overdose of chemicals in drugs, especially deadly amounts of fentanyl, is rapidly increasing in the USA. According to a report by the Drug Enforcement Administration (DEA), more than 700 deaths caused by fentanyl were reported in the USA between 2013 and 2014. According to the Centers for Disease Control and Prevention (CDC) report, that during this period, deaths caused by synthetic opioids had shot up 79%. A large portion of these deaths appears to be caused due to fentanyl. According to the report, few drugs seemed to be oxycodone, but the pills were not containing oxycodone. They were having the chemical U-47700. U-47700 is an illegal synthetic opioid which is not for human. It led to at least 17 overdoses and quite a lot of deaths in the USA alone. There are also cases reported about fentanyl poisoning due to unintentional intake through counterfeit narcotic tablets. While the patients purchase Norco from the street, for acetaminophen-hydrocone tablets for back pain, they receive a dose of fentanyl and U-47700, which is harmful to health. One such case is reported in "The American College of Emergency Physicians' Annals of Emergency Medicine" [9].

In 2015, the FDA had found a counterfeit version of Botox in the USA. It was allegedly sold to doctors' offices and medical clinics nationwide. An unauthorized, unlicensed supplier had sold this drug. Both the injections looked identical. The counterfeit package was missing entries of LOT: MFG: EXP on the outer pack, and the active ingredient on the vial was written as "Botulinum Toxin Type A" in place of "OnabotulinumtoxinA" [10].

According to estimations, one out of ten medical products distributed in underdeveloped and developing countries is either of low quality or fake, according to the research from WHO. As per Dr. Tedros Adhanom Ghebreyesus, WHO Director-General, substandard medicine affects vulnerable communities mostly. If a mother who gives up food or other basic needs to pay for her child's treatment, unaware that the medicines are substandard or falsified, and then that treatment causes her child to die. This is unacceptable. Countries have agreed on measures at the global level—it is time to translate them into tangible action [11].

As there is a need for proper supervision so that counterfeit drugs are detected as soon as they enter the market, researchers have found Blockchain technology to be an excellent solution. Satoshi Nakamoto introduced Blockchain in 2008 through Bitcoin [12]. Blockchain uses a distributed ledger system, which has shown its potential and adaptability in recent years. Most of the focus was on the financial industry in the beginning. Recently, the use of Blockchain in various industries like legal, healthcare, energy, etc., has come into focus.

The primary purpose of this technology is to increase transparency and trust. It helps in verification and auditing the goods or transactions. Traceability, privacy, and security also make this technology worth employing in various industries. Blockchain can store vital information like temperature, IoT devices, and location attached to the packages, which prevents tampering. By using this, the glitches in a system can be detected. This will be very useful as we can make the system stronger by rectifying the shortcomings [13].

According to The FDA, the USA has started using e-pedigree software to track drug shipments by providing a history of the drug. It uses RFID tagging for all the products with a description of the drug history. With the growth in such technologies, the development would be done at a great pace and entirely collapse the global threats like counterfeit drugs, saving the lives of countless people [14].

4 Proposed Framework

To eradicate fake medicines, a technology-enabled system is required to keep track of the flow of goods along the supply chain. This can be done through digital ledgers. Blockchain technology is a perfect solution to this issue. Utilizing this technology, the whereabouts of the drugs and vaccines can be available.

Block is an individual entity in a Blockchain that stores information. The blocks are connected in a unidirectional way due to which it is challenging to hack the Blockchain. In the healthcare industry, blockchains can be implemented to track and trace medicines. The supply chain starts from the manufacturer. First, drugs or vaccines are produced. Then, all the information about it is to be fed into the Blockchain. The QR code can be used to check the data at any point in time. The essential information like ingredients, manufacturing date, expiry date, temperature to be stored at, etc., would be entered into the Blockchain [15].

Each block also stores a timestamp, hash function, and transaction details. The information is stored safely using cryptographic methods. Encryption of data is done, and the blocks are connected through hash functions to be done [16]. A digital signature identifies the person who has done the transaction and serves the purpose of authentication, data integrity, and non-repudiation. After each transaction, the digital signature is verified. Next is, when the distributor/wholesaler receives the goods, they can verify the credibility through physical checking and also by scanning the QR code. The transaction details between the wholesaler and the manufacturer is stored into the next block. When these medicines arrive at the pharmacies, the following transaction between pharmacies and distributors is added into the Blockchain.

Similarly, the transaction between patients and pharmacies will be added to the chain. The patient can check the authenticity of the drug by scanning the QR code. As the Blockchain updates all the nodes simultaneously when the transaction occurs, the others involved in the supply chain can also verify if the drug has reached the patient [17].

In this process, every party, from the manufacturer to the patient, can easily find out if the drug is safe to use or not. Also, any investigation department who wants to check on the medicines can easily do so. The advantage of this system is that if any

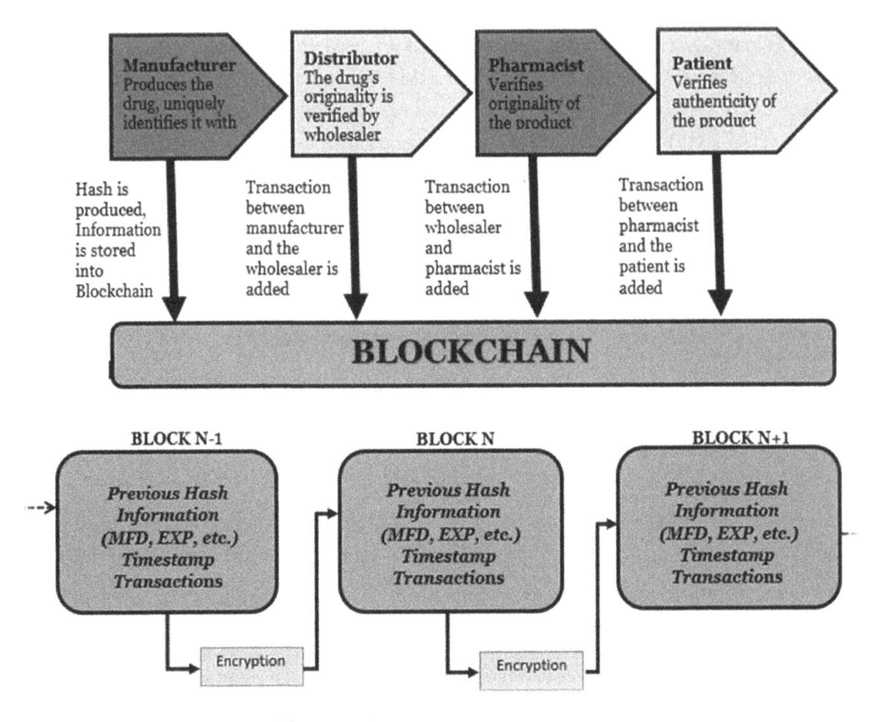

Fig. 2 Model of the proposed framework

medicine is found out to be a counterfeit product, the company or people involved in it can easily be caught and punished accordingly. Also, deaths and health issues due to intake of expired or wrong dosages or ingredients can be reduced on a large scale. The time delays, human errors, extra cost, etc., can be reduced radically [18]. Collaboration between pharmaceutical companies, government, and other industries can help in implementing this idea on a large scale. With the application of this technique, we can have a transparent, accountable, and secure system which would revive the trust in the healthcare industry for the coming future [19]. The model of the proposed framework has been depicted in Fig. 2.

5 Security in Blockchain

Blockchain is a trust less, immutable, and network consensus-based system, and the blockchain systems are very secure. Various cryptographic properties like digital signatures and hashing the data stored are secured well. As it works like an immutable ledger, once entered, the data cannot be changed or tampered. If done so, due to the network consensus, it can be easily be recognized, and the attempt would be shut down. As all the nodes in the network have a copy of each transaction, any change can easily be identified and instantly not execute the unacceptable activity.

6 Validation

Validation in Blockchain is done using consensus algorithms like proof of work (PoW) or proof of stake (PoS). Proof of work uses an enormous amount of electricity to secure the network as mining new coins take a lot of computing power. This method was used even in the bitcoin Blockchain. This mechanism was used to reach consensus between many nodes on a Blockchain network and provide a way to secure the Blockchain in terms of accuracy and robustness.

7 Future Scope

The implementation of Blockchain in the healthcare industry has started recently. The Blockchain-based networks like Medileger and TraceRx are already being used. Some additional features of the proposed framework can be added. For example, the chain can be started when the patient orders the medicine by uploading their prescription. Through this, it can be noticed that whether the patient is indeed advised to use the medication. It can also use another Blockchain to check the origin of raw materials used for making drugs or vaccines. It can make a ledger of people getting admitted, discharged, etc. As there is no maximum number of blocks defined for Blockchain, the data can be added easily. Using Blockchain to store the patient's admission, discharge, consults, etc., will make it easy to access later [20]. As anyone cannot tamper this, the information stored is safe and secure. The Blockchain can be applied in various other ways also, for example, collecting consent forms and time-stamping them, permission access to patient data, health information exchange, etc. With developing technology, the various issues can be addressed and solved to close all the loopholes and provide better and safe access to goods and information worldwide.

8 Conclusion

In this paper, the effects and harmfulness of counterfeit drugs have been discussed. The shortcomings in the supply chain due to which counterfeit medicines are flooding the markets have been studied. Also, the framework of Blockchain technology to detect the entrance of counterfeit drugs has been proposed and studied. The supply chain can be tracked when the drug is manufactured until it reaches the patient. This promotes transparency and credibility of the healthcare industry by proving trust. With the help of government and industries, the counterfeit medicines can be detected as soon as they enter the market and be punished by law accordingly.

References

1. Shah N, Sattigeri B, Patel N, Desai H (2017) Counterfeit drugs in India: significance and impact on pharmacovigilance. Int J Res Med Sci 3(9):2156–2160. https://doi.org/10.18203/2320-6012.ijrms20150596
2. Hall C (2012) Technology for combating counterfeit medicine. Pathogens Glob Health 106(2):73–76. https://doi.org/10.1179/204777312X13419245939485
3. Wertheimer AI, Chaney NM, Santella T (2003) Counterfeit pharmaceuticals: current status and future projections. J Am Pharm Assoc 43(6). ISSN 1544-3191. https://doi.org/10.1331/154434503322642642
4. Xu M, Chen X, Kou G (2019) A systematic review of Blockchain. Financ Innov 5:27. https://doi.org/10.1186/s40854-019-0147-z
5. Blackstone EA, Fuhr JP, Pociask S (2014) The health and economic effects of counterfeit drugs. Am Health Drug Benefits 7(4):216–224
6. Gordon WJ, Catalini C (2018) Blockchain technology for healthcare: facilitating the transition to patient-driven interoperability. Comput Struct Biotechnol J 16. ISSN 2001-0370. https://doi.org/10.1016/j.csbj.2018.06.003
7. Yoon HJ (2019) Blockchain technology and healthcare. Healthcare Inform Res 25(2):59–60. https://doi.org/10.4258/hir.2019.25.2.59
8. Agbo Cc, Mahmoud Q, Eklund J (2019) Blockchain technology in healthcare: a systematic review. Healthcare 7:56. https://doi.org/10.3390/healthcare7020056
9. Prekupec MP, Mansky PA, Baumann MH (2017) Misuse of novel synthetic opioids: a deadly new trend. J Addic Med 11(4):256–265. https://doi.org/10.1097/ADM.0000000000000324
10. Le Nguyen T (2018) Blockchain in healthcare: a new technology benefit for both patients and doctors. In: 2018 portland international conference on management of engineering and technology (PICMET), Honolulu, HI, pp 1–6. https://doi.org/10.23919/PICMET.2018.8481969
11. Pandey P, Litoriya R (2020) Securing E-health networks from counterfeit medicine penetration using blockchain. Wireless Pers Commun.https://doi.org/10.1007/s11277-020-07041-7
12. Stephan L, Steffen S, Moritz S, Bela G (2019) A review on blockchain technology and blockchain projects fostering open science. Front Blockchain 2. ISSN:2624-7852. https://doi.org/10.3389/fbloc.2019.00016
13. Hasselgren A, Kralevska K, Gligoroski D, Pedersen SA, Faxvaag A (2020) Blockchain in healthcare and health sciences—a scoping review. Int J Med Inform 134:104040. ISSN 1386-5056.https://doi.org/10.1016/j.ijmedinf.2019.104040
14. Kelesidis T, Falagas ME (2015) Substandard/counterfeit antimicrobial drugs. Clin Microbiol Rev 28(2):443–464. https://doi.org/10.1128/CMR.00072-14
15. Choi JB, Rogers J, Jones EC (1889) The impact of a shared pharmaceutical supply chain model on counterfeit drugs, diverted drugs, and drug shortages. In: 2015 Portland international conference on management of engineering and technology (PICMET), Portland, OR, pp 1879–1889. https://doi.org/10.1109/PICMET.2015.7273165.
16. Hossein KM, Esmaeili ME, Dargahi T, Khonsari A (2019) Blockchain-based privacy-preserving healthcare architecture. In: 2019 IEEE Canadian conference of electrical and computer engineering (CCECE), Edmonton, AB, Canada, pp 1–4.https://doi.org/10.1109/CCECE.2019.8861857
17. Kumar R, Tripathi R (2019) Traceability of counterfeit medicine supply chain through Blockchain, pp 568–570. https://doi.org/10.1109/COMSNETS.2019.8711418

18. Acri K, Lybecker N (2018) Pharmaceutical counterfeiting: endangering public health, society and the economy. Fraser Institute, pp 24–30. https://doi.org/10.2307/resrep23987.7
19. Sylim P, Liu F, Marcelo A, Fontelo PA (2018) Blockchain technology for detecting falsified and substandard drugs in distribution: pharmaceutical supply chain intervention. JMIR Res Protocols 7
20. Haq I, Muselemu O (2018) Blockchain technology in pharmaceutical industry to prevent counterfeit drugs. Int J Comput Appl 180:8–12. https://doi.org/10.5120/ijca2018916579

Static and Dynamic Learning-Based PDF Malware Detection classifiers—A Comparative Study

N. S. Vishnu, Sripada Manasa Lakshmi, and Awadhesh Kumar Shukla

Abstract The malicious software are still accounting up as a substantial threat to the cyber world. The most widely used vectors to infect different systems using malware are the document files. In this, the attacker tries to blend the malevolent code with the benign document files to carry out the attack. Portable document format (PDF) is the most commonly used document format to share the documents due to its portability and light weight. In this modern era, the attackers are implementing highly advance techniques to obfuscate the malware inside the document file. So, it becomes difficult for the malware detection classifiers to classify the document efficiently. These classifiers can be of two main type, namely, static and dynamic. In this paper, we surveyed various static and dynamic learning-based PDF malware classifiers to understand their architecture and working procedures. We also have presented the structure of the PDF files to understand the sections of PDF document where the malevolent code can be implanted. At the end, we performed a comparative study on the different surveyed classifiers by observing their true Positive percentages and $F1$ score.

Keywords Malware · Classifier · Machine learning · Obfuscation · Adversary · Portable document format (PDF) · Feature extraction · Processing · Malware · Parser

1 Introduction

Modern-day attackers are advancing their fabrication techniques for designing malware in robust ways through obfuscation methods to eliminate the detection and elevating their destruction capabilities. As a result, malware is still a significant threat

N. S. Vishnu (✉) · S. M. Lakshmi · A. K. Shukla
Lovely Professional University, Phagwara, Punjab, India

S. M. Lakshmi
e-mail: manasa.23777@lpu.co.in

A. K. Shukla
e-mail: awadhesh.23866@lpu.co.in

A. Choudhary et al. (eds.), *Applications of Artificial Intelligence and Machine Learning*, Lecture Notes in Electrical Engineering 778,
https://doi.org/10.1007/978-981-16-3067-5_41

to the security of the digital devices. Most of the malware attacks are carried out by embedding the malevolent code with the routinely used files such as images, PDFs, excel, and other document files. For the purpose of blending the dangerous code with benign files, it is very important for an attacker to have an in-depth knowledge about the structure of those benign files. There are three main benefits of embedding the malware with these routinely used documents. The first benefit which can achieved through this technique is that these files are used in abundance for diverse applications by majority of users [1]. So, these can be used in carrying out the social engineering attacks [2]. The second benefit can be that majority of the users operate on outdated software and applications which make them vulnerable to these kinds of attacks [1]. The vulnerabilities existing in the applications can be manipulated to design effective malware which indeed can trick the detectors from recognizing them. The longer the victim continue to use the outdated and vulnerable applications, the longer will be the lifespan of the malware infection residing in their systems [3]. Many researchers have found that self-learning mechanisms can be deployed on to the detection systems to strengthen their detection skills and efficiency. Presently, most of the malware detection systems operate based on the deep-learning strategies [1]. By assisting this technology, the system will be capable of making decision on its own and can be beneficial in detecting the hidden malware codes within the infection vectors [3]. But these systems are also prone to two widely known attacks, namely poisoning attacks and evasion attacks [2]. The poisoning attacks aim to exploit the system at the training phase, while the evasion attacks aim to manipulate the system at the testing phase [4]. For example, consider the system to be a child and the data which is imparted to the system as the chapters taught to the child at school. The main aim of the attacker would to destroy the system, while in our example case, to fail the child in the assessment. So, to achieve this, the attacker may either try to feed the system with illegitimate data or can give carefully designed input to confound the system. In the case of example, the lecturer may try to teach the child with inappropriate data or can even give complex questions at the time of evaluation to fail the child. The feeding of inappropriate data can be considered as a poisoning attack, and the process of rendering tricky question during the assessment can be considered as an evasion attack. Majority of research administered on the self-learning detection techniques have manifested that these methodologies are capable of precisely recognizing and classifying the files blended with hidden malware [1]. The research work conducted on carrying out evasion attacks on this kind of detection system have staged that a diligently fabricated input to these systems can be able to conquer them [1]. For example, adding perturbations to a clear image was able to misclassify it by a machine learning image recognition system.

In this paper, we will be discussing about the different self-learning based malware PDF classifiers which were proposed by various authors till now. The portable document format (PDF) files are widely used by the majority of users due to its lightweight and portability feature. It is a widely accepted format for sharing and viewing the data, which makes it an attacker's choice of infection vector. Here, we will be discussing the structure of the PDF files and diverse strategies through which the malware PDF detection systems can be evaded by fabricating the adversarial samples. These PDF detectors are mainly of two different types, static and dynamic detectors. We have surveyed most of the PDF detector proposed till present by different researchers to understand the mechanisms used by these detectors for the classification process.

2 Structure of PDF File

The method of interpreting a PDF file can be depended upon the type of parser used to view or manipulate it [1, 3]. There are some elements in every PDF file which are essential for the representation of PDF files [1]. These elements are basically classified in to two main categories, the general structure and file content [1]. The first category includes the information regarding the storage of content within the file [1]. However, the second category describes the way the content is displayed to the user accessing the file [3]. The general structure of a PDF file can be considered as a network of objects in which each of these objects will have distinct functions to perform [1]. There are four main components in the PDF structure, and they are:

2.1 Header

This part of the PDF consists of one-line text code [5]. This text is initiated with a % sign and signifies the version of the PDF file [6].

2.2 Body

This portion consists of an order of objects which denotes the activities carried out by the PDF file. Within these objects, there are chances of involving different kinds of blended file types like images and video files [1]. Even executable malevolent codes can be inserted in to PDF files by introducing them in one or more objects of the body component [1]. Every object declared in the body part of PDF file contains a distinct number which denotes its reference number [1]. The referencing number can be beneficial in calling a particular object. Whenever an object needs to be called, the reference numbers can be utilized. Every distinct object which is declared in the body ends with a keyword marker "endobj" [3].

2.3 Cross-Reference Table

By the name itself, we can interpret that this part of the PDF contains table. Let us understand the functionality and role of this table [1]. This table generally contains the information about the location of various objects inside the PDF file. It also informs the parsers from where to begin the parsing process [3]. It is represented in the file using "xref" marker and under that we can see various rows where each row indicates the references to different objects mentioned in the body field. Only those objects which are referenced in the cross-reference table will be parsed at the time of parsing the PDF file [1].

2.4 Trailer

This field describes some of the essential elements that are necessary for the file interpretation [1]. The adress of the first object comes under one of them. It also contains the references to the metadata associated with the references [3]. The end of the trailer field is denoted by using a marker "%%EOF," which also denotes the end of the file and signals the parser to stop its parsing activity [1] (Fig. 1).

3 Parsing Procedure

When a particular PDF file is selected for parsing by a parser, then it performs this activity in the following manner: At the beginning, the parser goes to the trailer field and determines the first object which needs to be interpreted [1]. Then, sequence of objects is parsed with the help of references mentioned in the cross-reference table. Then, finally the parsing procedure is terminated when it detects the "End of File marker" [5]. One of the significant features of using PDF files is that they eliminate the requirement of recreating files from scratch when some new objects are introduced into the priorly existing files [1]. Instead, they create new field structures which are dedicated for storing the newly introduced contents [1, 7].

4 Learning Based Malicious PDF Detection Process

"Machine Learning" technology is widely being utilized almost in every field [2]. This technology helps in making the system more intelligent and enable them for decision making [2, 8]. Due to its wide range of applications, it is also deployed in determining malevolent content hidden in files [9]. Here, we are going to discuss how this technology can be assisted in effectively detecting the malignant code in a PDF file under inspection [10]. Several self-learning systems were proposed by researchers for inspecting the PDF files in the past decade [11, 12]. In this section, we will be discussing about the architecture of the self-learning based malicious PDF detectors and the steps involved in classifying a given PDF file (Fig. 2).

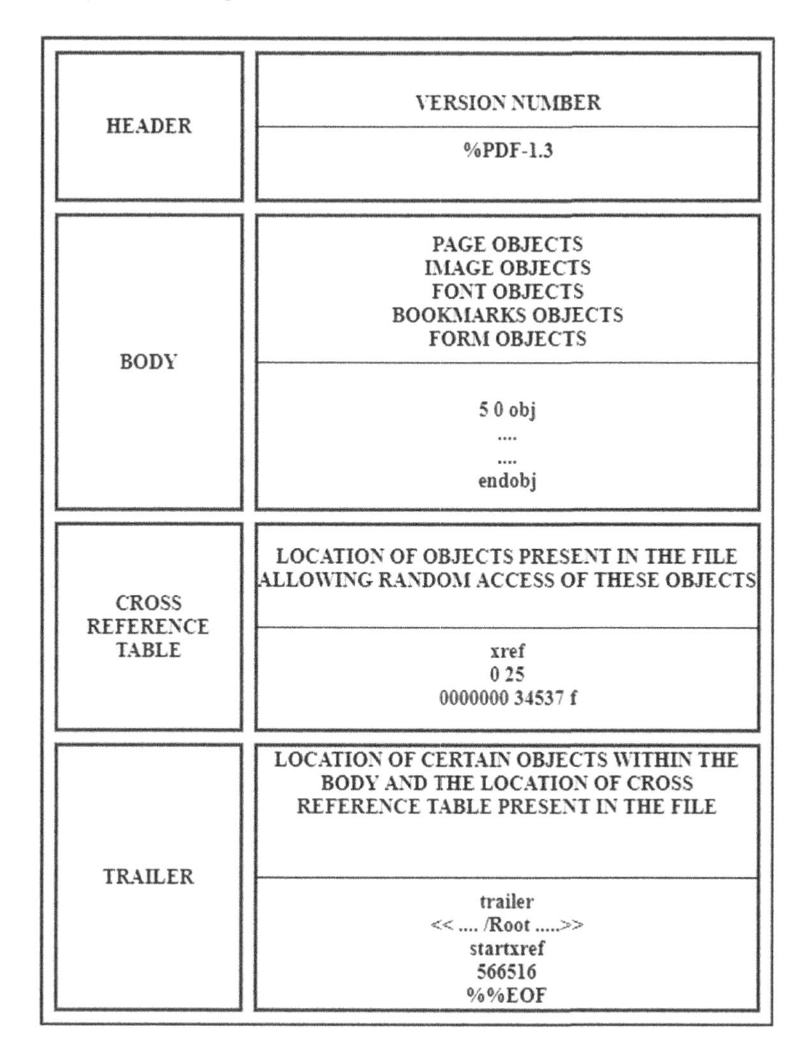

Fig. 1 Structure of PDF file [1]

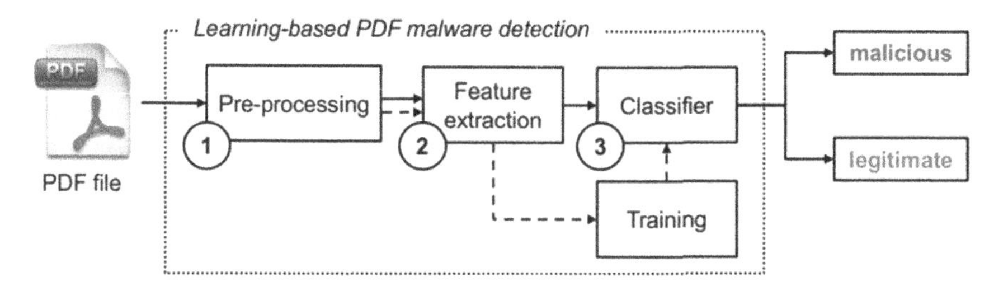

Fig. 2 Working of PDF malware classifier tool [1]

The main objective of these systems is to efficiently classify whether a specific file comes under the benign category or the malicious one [13]. For determining that, the system needs to examine and process the patterns and structures of the file's internal components [13, 14]. Even though, there are many proposed detectors, all of them are observed to follow these three basic procedures to analyze the PDF files [8], which are.

4.1 Preprocessing

This is the first process carried out on the PDF file which is chosen for the inspection purpose. In this procedure, the internal code which is critical such as JavaScript or ActionScript code is executed in a supportive environment under isolation condition to understand its nature [1]. It is processed under isolation conditions, so that its execution does not affect the system [15]. The other essential data elements such as metadata of internal objects and keywords are also extricated from the PDF files [10].

4.2 Feature Extraction

This is the next activity carried out by the detectors after the "Preprocessing" activity is completed. This procedure utilizes the proffered knowledge gained from its preceding step to accomplish its inspection activity [3]. In this step, the data extracted from the previous stage is transformed in to the form of vector numbers which will be able to convey the existence of certain components such as keyword and "API calls" in the PDF file [16].

4.3 Classifier

This is the stage where the algorithms are deployed in analyzing the extracted information from the previous stages [17]. The classifiers are first trained with sample sets to increase its knowledge about determining the differences between a benign and malignant PDF file [7]. A system can render quality efficiency in classifying the PDF files, if they are trained efficiently [14]. The efficacy of the system is not solely depended upon the algorithm, but also on the training set. The process of training is conducted before the system is actually deployed for the real-time application [4]. The main objective of the classifier is to precisely classify a given file to be malicious or not [1].

5 Types of Classifiers

The malware detectors built for inspecting the PDF files by assisting self-learning techniques can be classified in to two main categories, namely static detectors and dynamic detectors [10]. The static detectors perform static analysis of the file without carrying out the execution of the code within the file [10, 17], while the dynamic detectors performs their analysis by executing the actual code in supportive environmental conditions [10, 16]. We will discuss more about these two detection types in Fig. 3.

5.1 Static Classifiers

Most of the malware attacks implemented by utilizing the PDF files make use of JavaScript or ActionScript code [10]. So, these types of detectors mainly focus on the sections of the file where these kinds of codes are detected. But we cannot conclude that a particular PDF file as a malevolent one just by looking at the presence of the JavaScript code. Even benign files may also have such codes in them [10]. So, the static detectors inspect for the patterns of usage of specific variables, keywords, functions, and API calls for determining the nature of the file [16]. PJScan is a static PDF detector which decides the maliciousness of the PDF files based on the frequency of usage of variables, operators, and functions [18]. Some of the other

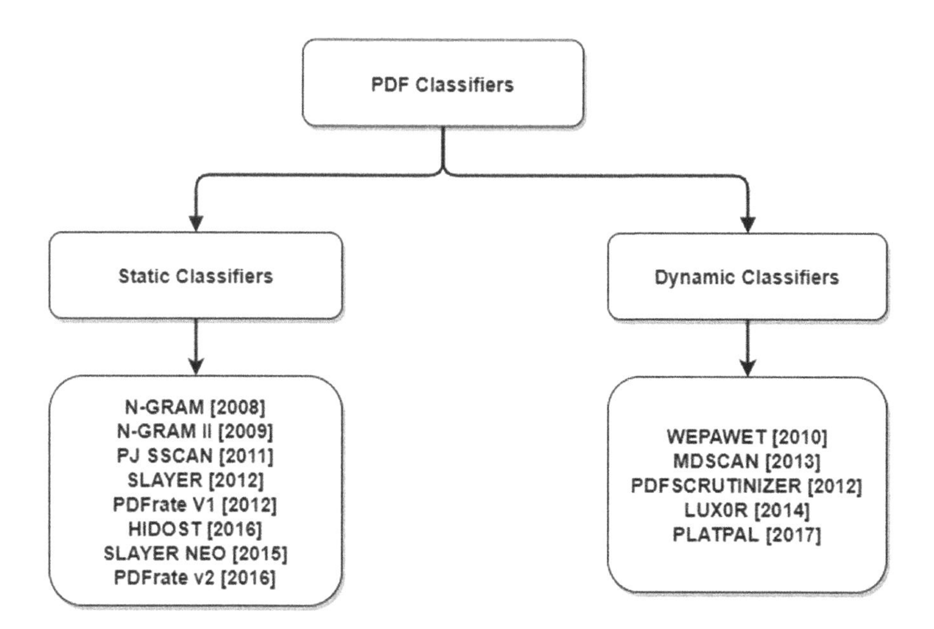

Fig. 3 Types of PDF malware classifiers [1]

detectors were based on tokenization and matching methods [16]. But in cases where these JavaScript contents are heavily obscured by the attacker inside the PDF files, the static detectors may fail by misclassifying the file [2]. The attacker can make the malicious content very much hidden, so that it cannot be detected by the parser at the time of parsing process and this code can be dynamically called at runtime to carry out the malignant actions. So, the main flaw existing in these class of detectors are that they cannot be capable of accurately classifying a deliberately obfuscated samples [1].

5.2 Dynamic Classifiers

Dynamic detectors are also mainly focused on the JavaScript code of the file as presence of them rendered to the elevated suspicion [12]. But this class of detectors does not look for the patterns or the frequencies of the keywords and other elements [4]. Instead, it will directly extract these JavaScript code and try to execute them in supportive isolated conditions and examine their actions [18]. The confrontation which was being faced in the static detectors due to practice of hiding techniques can be overcome by deploying these kinds of detectors [16]. The dynamic detectors unlike static does not examine the file just by carrying out an inspection on their internal structures [16]. They examine the nature of the file through extraction of flash and scripting codes and execute them to understand its actual nature [19]. Even though these kinds of detectors are capable to determine the malevolent code efficiently, they may fail in cases of unsupportive environments [10]. Some of the JavaScript code may need some other additional feature support for carrying out its actions [17]. For that, we need to emulate the execution process of the file contents by proffering it with its necessitated requirements [10]. Another, flaw can be that the dynamic detectors may fail when the malwares are inserted into files without the usage of JavaScript code [10].

6 Static Classifiers

6.1 N-Gram

Shafiq et al. [20] presented an approach to detect the presence of embedded malware codes in the benevolent files. The authors stated that at those times the marketed malware security software were not able to determine the previously known malware signatures present inside the files at the inspection process. So, they have fabricated a system which was capable of determining the hidden "Malcode" within the files called *N*-Gram [20]. They proclaimed from their analysis that the majority of the non-malicious file "Byte Sequencing" demonstrated a first-order dependence structure.

They also made use of "Entropy Rates" for examining the differences of distribution values of a specific file to detect the residence of malignant codes within the file [20]. On evaluating above-proposed model by the authors, it had exhibited efficient performance in dealing with blended malicious files.

6.2 N-Gram II

Tabish et al. [21] propounded a unique strategy for the identification of the malware which worked by examining the bytes of the file. In this approach, the content which are examined are not stored for any future purposes, it is a "Non-Signature"-based approach. Thus, the authors claimed that their system had the capability of identifying the unknown malware varieties. The presented system was evaluated against huge samples of files containing wide varieties of malware families and file formats [21]. It was observed that the system had staged to display an accuracy of 90% in classifying the files [21].

Figure 4 illustrates the architecture proposed by Tabish et al. [21] in 2009. The authors claim that their methodology utilized to build this system can eliminate the need to have any prior information regarding the type of the file chosen for the inspection purpose [21]. Therefore, this model could be able to detect the malware in cases where the attackers try to manipulate the header section of the PDF files [21]. In the architecture presented by the authors, the system is divided in to four different modules [21]. The first module is the "Block Generator Module" which divides the "Byte-level content" inside the file into equal sized smaller blocks [21]. The next module "Feature Extraction Module" performs the computational statistics on the blocks generated in the previous level [21]. The other two modules, namely "Classification and Correlational" module carry out various analysis on the results obtained from the previous model to detect the signs of malware [21].

6.3 PJscan

Laskov and Šrndic [18] proposed a static malware detection system which was able to identify the malignant JavaScript content contained within the file [18]. Authors claimed that their model was possessed to have lowered the processing time taken for analyzing the file than compared with the previously presented models, and it was also showing an impeccable efficacy in identifying known and unknown malevolent types [18]. This method was best suited for processing huge datasets due to its low inspection time and high efficiency [18].

The architecture of the PJscan model is shown in Fig. 5. When a particular PDF file is given to the system for analysis, the file is undergone through a feature extraction module which contains sub-categorial modules such as a JavaScript extractor to draw out the JavaScript code within the file contents [18]. Then, all the entries of these

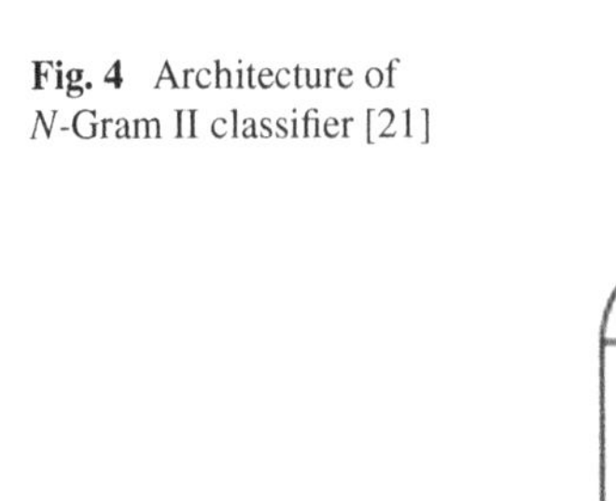

Fig. 4 Architecture of *N*-Gram II classifier [21]

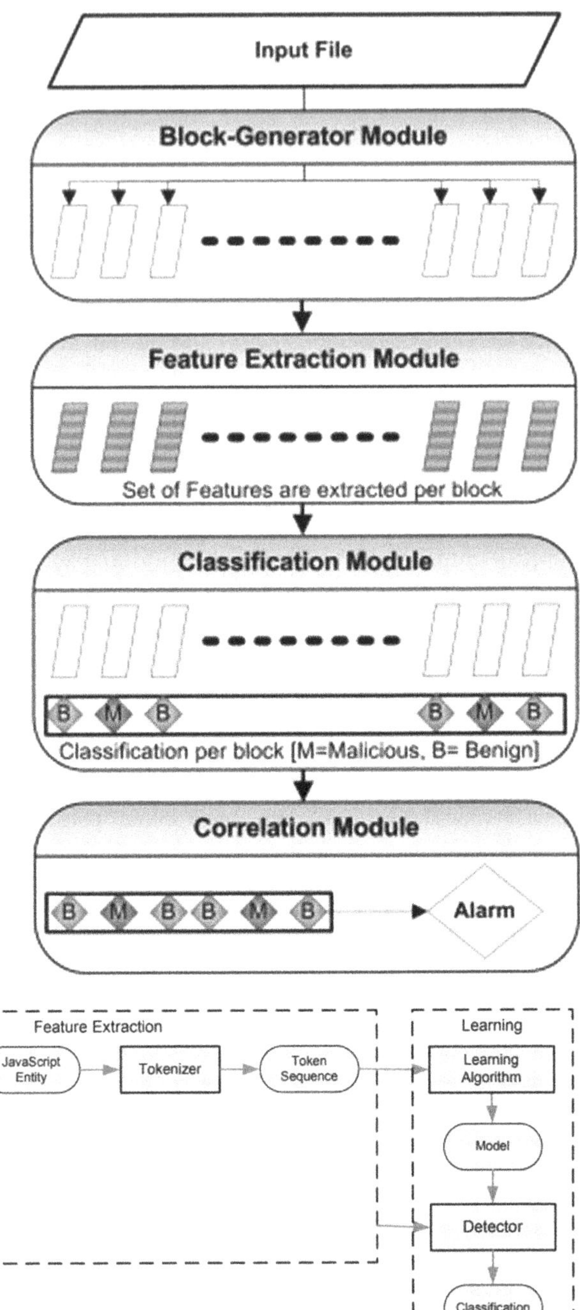

Fig. 5 Architecture of PJScan [18]

JavaScript code are then assigned with tokens [18]. Then, the algorithm is used for the classification purpose.

6.4 Slayer

Maiorca et al. [22] proposed a module to extricate the features within PDF file under the examination process. The authors have presented an effective classifier and feature extraction module which were combined to form an effective static detection model. This tool was said to proffer high flexibility [22]. One can utilize the tool either as a "Stand-alone tool" or as an additional "Plug-in Feature" to enhance the capability of existing detection systems [22].

6.5 PDFrate V1

Smutz and Stavrou [17] propounded an architecture to apprehend the malevolent traits of PDF files in a robust way by assisting self-learning approaches. This method was capable of arresting the information from the metadata of objects present in the PDF structure [17]. This detector functioned on basis of "Random Forest Classification" method, which is a classifier capable of recognizing the features from the various "Classification Trees" formed [17]. This classifier was well known to render good detection rates even in cases of unknown malware characteristics [17]. The "Random Forest Classification method" is still used in many of the recently proposed detection systems due to its efficient mechanisms [17].

Figure 6 shows the arrangement made by Smutz and Stavrou [17] for the efficient malware detection. This system undergoes through two levels of testing. The first

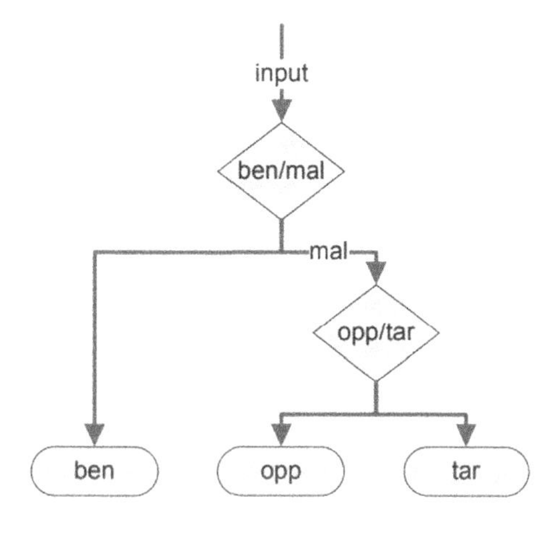

Fig. 6 Dual-level classification strategy carried out in PDFrate v1 [17]

testing is conducted to classify the file into either benign or malicious category [17]. If the file comes under the malicious category, then the file is made to go for another level of classification level to determine whether this malicious file is an opportunistic one or a targeted one [17].

6.6 Hidost

Šrndic and Laskov [4] presented a static detection system called Hidost for efficiently identifying the non-executable malware content present inside the PDF file structures. As the attackers began to append non-executable code into the benign files to eliminate the detection by static and dynamic classifiers, there came the need of constructing systems to identify these kinds of malware residing in the files [4]. This paper was an extension to the previously proposed approach by the same authors in the year of 2013 [4]. In that paper, the authors presented an approach of incorporating the content with their logical elements present in the file for achieving an elevated classification accuracy. Although, this model was fabricated to detect the malware in the PDF and flash files, the authors state that their approach can extended to detect malware in other formats also. On evaluating this system, it was found that this approach has dominated majority of the commercial VirusTotal detection software in accuracy [4].

Figure 7 displays the architecture of the Hidost system proposed by Šrndic and Laskov [4]. The authors claim that their system is successful in discriminating between the malevolent and the benign files. We cannot take only one vector as a common feature for classifying different files [4]. The processing and detection methods are separated in to different steps in this model, so that this system can be further be extended to efficiently detect other file malwares [4].

6.7 Slayer NEO

Maiorca et al. [8] proposed a model called Slayer NEO to detect the malevolent PDF documents by retrieving the information from the content and internal structure from the inspected PDF file. Authors asserted that this model had an efficient parsing mechanism indulged with it, which in turn will assist in determining the presence of non-JavaScript malicious content as well. With the deployment of this classification algorithm, the authors state that their model has outperformed various other static PDF malware detection systems [8].

Figure 8 shows the architecture presented for the Slayer NEO PDF malware detection model. At first, the PDF file is passed through a feature extraction module to draw out the presence of any embedded malwares inside it [8]. For that purpose, it uses parsers PeePDF and Origami [8]. After the feature extraction, the different

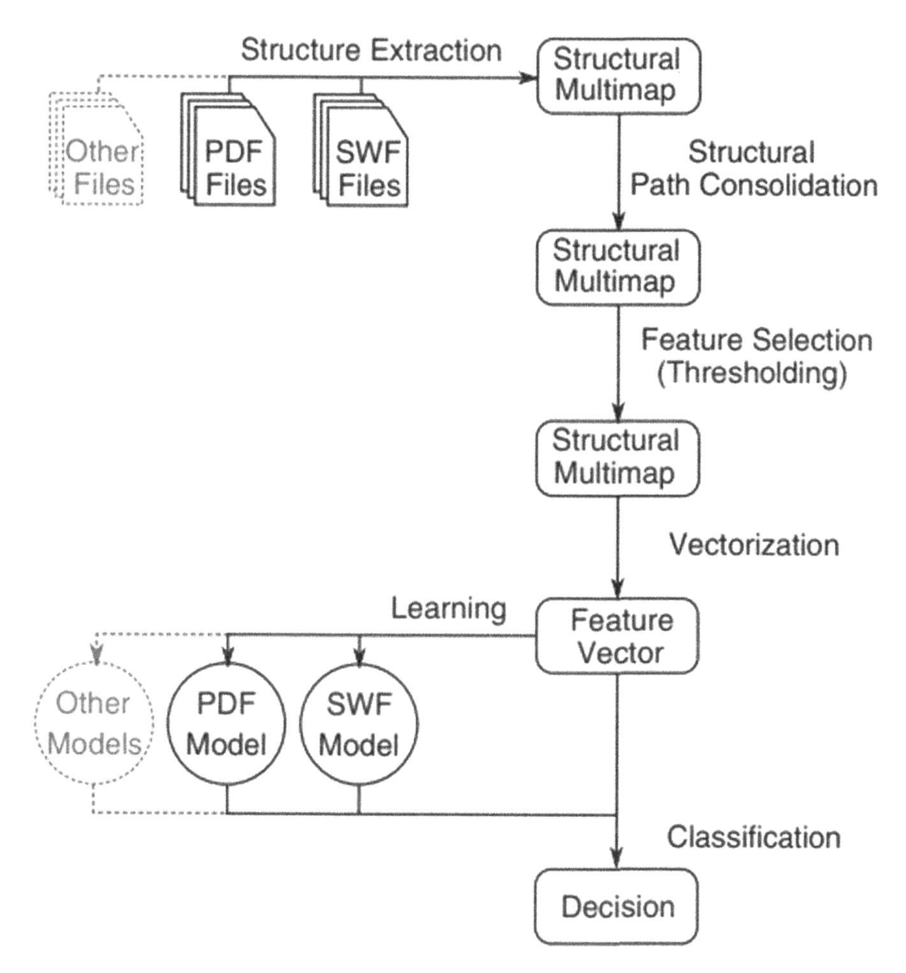

Fig. 7 Architecture of Hidost [4]

structures of the files are separated, and the feature vector is generated to classify the given file as a benign or a malevolent one [8].

6.8 *PDFrate V2*

Smutz and Stavrou [2] fabricated a unique and robust approach which was capable of recognizing adversarial samples. For building this system, the authors have experimented with huge number PDF samples against different classifiers to identify the specific range of samples for which the "Ensemble Classifier" was staging lower efficacies [2]. Then, based on the study of the characteristics of those samples, the authors proposed an ensemble classifier which operated on the basis of "Mutual Agreement

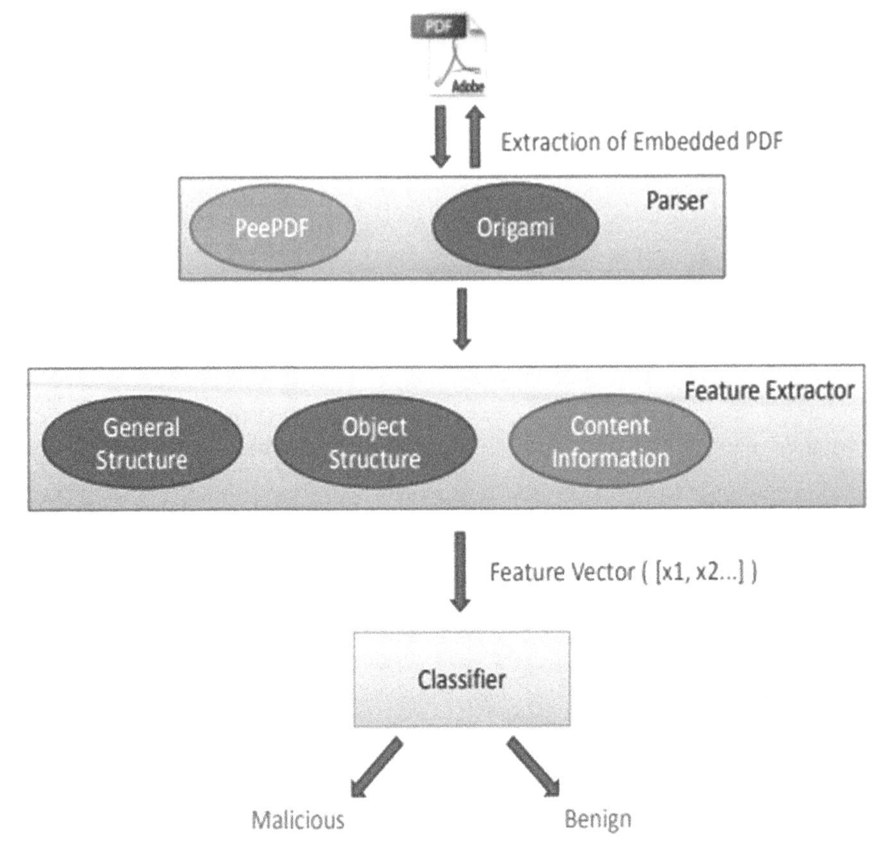

Fig. 8 System architecture proposed for slayer NEO [8]

Analysis" method [2]. They also suggested that their approach can be generalized to counteract various "Gradient Descent" and "Kernel Density Estimation attacks."

7 Dynamic Classifiers

7.1 Wepawet

Cova et al. [6] proposed an efficient strategy to carry out meticulous examination on JavaScript content of the PDF file. In this, the model built presented by the authors tend to blend the detection function with the emulating technology to be capable of recognizing the scripted code for performing the analysis [6]. It assisted the self-learning technology to recognize the malware content hidden within the JavaScript code through imitating procedures and comparing with the stored anomaly patterns

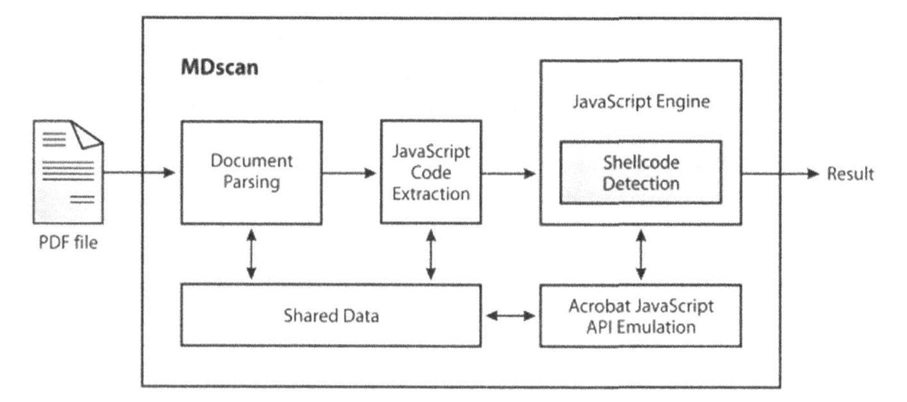

Fig. 9 Architecture of MDScan [14]

[6]. The authors stated that their model was also capable to add the newly detected malware signatures itself eliminating the need of explicit coding.

7.2 *MDScan*

Tzermias et al. [14] proffered a system which was capable analyzing the code statistically and additionally staged dynamic processing of the contained code within the document to recognize the malware residing in the PDF files. On evaluating the system against the datasets of PDF samples, the presented model was observed to be capable of detecting the heavily encoded malevolent PDF files [14].

Figure 9 shows the architecture of the MDScan proposed by Tzermias et al. [14]. In this model, the PDF document selected for inspection is parsed for detecting the JavaScript code within them. From both these modules, a dedicated module called shared data is connected. Then, the detected JavaScript code is passed through JavaScript Engine which has an inbuilt feature of detecting the shell code [14]. Then, it is run on an emulated environment to understand the nature of the file [14].

7.3 *PDF Scrutinizer*

Schmitt et al. [13] fabricated a dynamic PDF classifier named "PDF Scrutinizer" which was not only able to classify the PDF files efficiently but also proffered the reasoning behind its decision making. The authors stated that their model analyzed all the suspect files which are embedded with other .exe files separately from the normal files which are not embedded. This system, unlike the other classifiers, classified files under any of the three categories which are malicious, benign, and suspicious classes

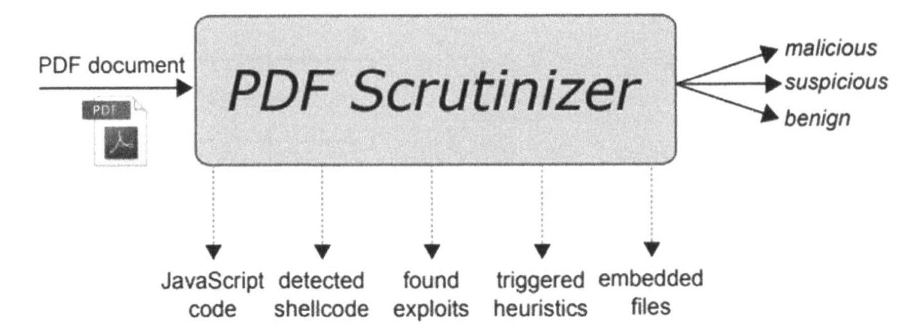

Fig. 10 Functionalities proffered by PDF Scrutinizer [13]

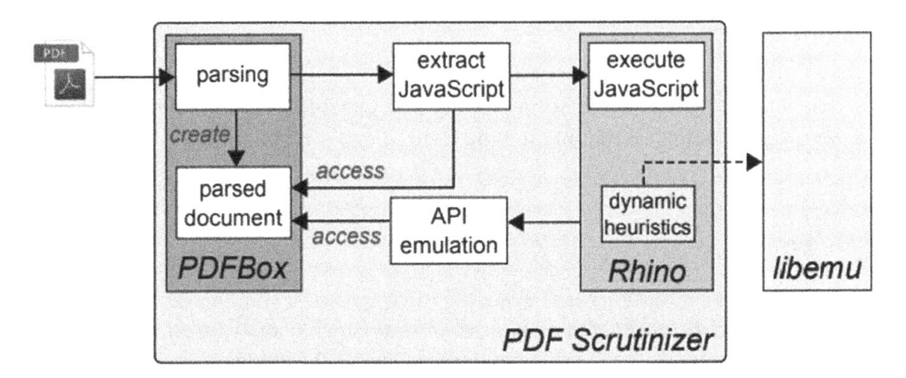

Fig. 11 Architecture of PDF Scrutinizer [13]

[13]. The PDF Scrutinizer utilized an interface called "PDF Box" for processing the PDF files during the evaluation [13].

Figures 10 and 11 display the functionalities and architecture of the PDF Scrutinizer model presented by Schmitt et al. [13] in 2012. The PDF Scrutinizer have different functionalities as we can see from Fig. 10. The PDF document which we provide to the system goes through various levels to check the possibilities of malicious content inside it [13]. The first one is to find the presence of JavaScript code inside the file under inspection [13] and then to observe the shell code existing in it. Based on the finding, a statistical analysis is conducted to classify the file as malevolent, benign, or suspicious file [13].

7.4 *Lux0R*

Corona et al. [16] fabricated a dynamic classifier called "Lux0R" which stands for "Lux 0n discriminant References." The model worked based on a "Lightweight"

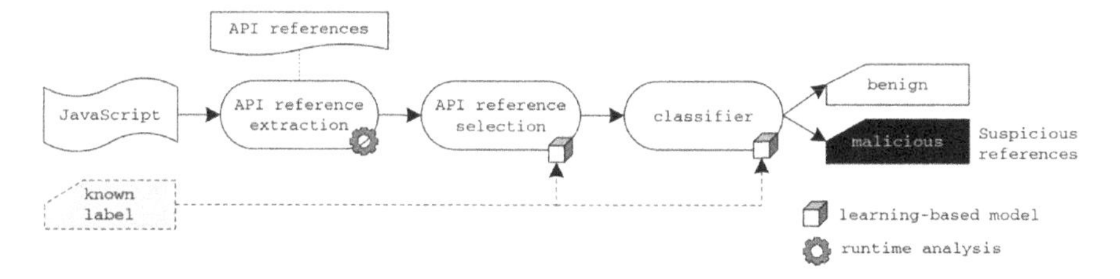

Fig. 12 Architecture of Lux0R [16]

method to analyze the nature of the JavaScript code by considering the "API references" which includes different functions, variables, and other elements declared in the file [16]. Lux0R selected the code present within the JavaScript section of the files and extracted its "API" elements to analyze them with the help of "Machine Learning" algorithms [16]. This method was considered to be an efficient strategy to counteract against the "Mimicry Attacks" [16].

Figure 12 shows the architecture of the dynamic learning-based PDF malware classifier Lux0R proposed by Corona et al. [16]. In this, instead of inspecting the whole PDF document, only the JavaScript code is inspected. Once the JavaScript code is extracted from the file, that code is sent for extracting the API references [16]. After that, based on API reference selection, the PDF is classified on the next stage which is Classification module [16].

7.5 PlatPal

Xu and Kim [10] prototyped a platform diversity tool "PlatPal" which appends to the "Adobe Reader" application to examine the inner mechanisms of the file under the parsing process, and also it executes the file code in sandboxes to determine the effect of the code to the host's machine [10]. The main principle used by the authors in building this system was that a benign file would be possessing similar traits on execution irrespective of the application used to execute it, while a malevolent one can stage diverse behaviors on processing by diverse applications [10]. This model is highly scalable and flexible solution to detect the embedded malware with reduced "False Alert" responses [10].

Figure 13 shows the architecture of the PlatPal proposed by Xu and Kim [10]. In this, the PDF file is sent to different virtual machines installed with various operating systems to check its behavioral properties at runtime. After running the file on these virtual machines, the results obtained from these machines are compared for the classification procedure. With this process, the impact of the file can be detected on various host platforms.

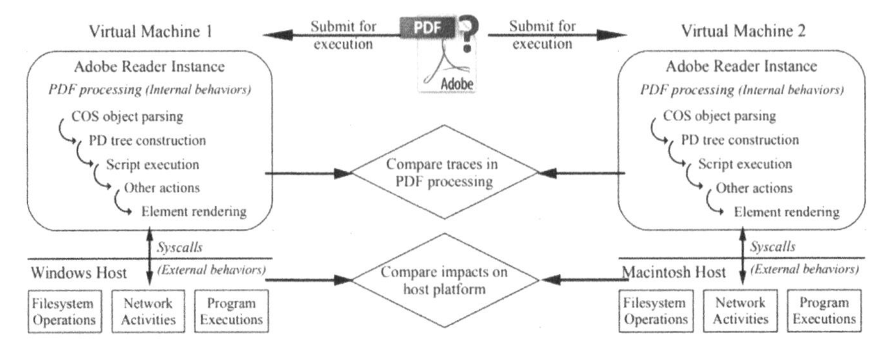

Fig. 13 Architecture of PlatPal [10]

8 Comparative Study Among Different Learning-Based PDF Malware Classifiers

Figure 14 displays the true positive percentages of different classifiers surveyed in this paper. The values plotted in the graph are taken from Table 2. We can see clearly that the TP percentage of the PJscan is low compared to that of the others whose TPs are plotted in the graph. The Hidost, Slayer, Slayer NEO, LuxOR, and PDFrate v2 are approximately having similar TP percentages, but these slight differences may be crucial and play a significant role while the evasion attacks are being implemented on the classifiers (Table 1).

The above line graph depicted in Fig. 15 shows the $F1$ score of the various classifiers which were reviewed in this paper. $F1$ score is generally used to calculate the accuracy measure of the classifiers while conducting a test and its recall. We can see that the values of the $F1$ score always lies between 0 and 1. In this, the Slayer, LuxOR, and Slayer NEO are having the top most $F1$ score as compared to the others, and also you can check Table 2 for observing the exact $F1$ score values of these classifiers.

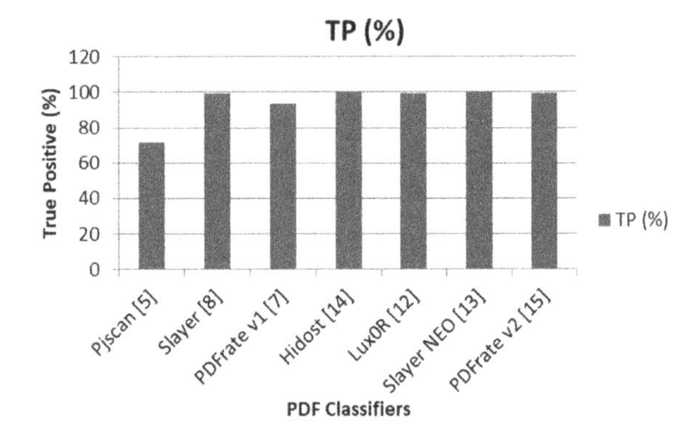

Fig. 14 Bar graph of TP percentage of the different classifiers

Table 1 Comparative analysis between different learning-based PDF malware classifiers [1]

Author(s)	Name	Year	Type	Features	Classifier
Shafiq et al. [20]	*N-Gram*	2008	Static	Raw bytes	Markov
Tabish et al. [21]	*N-Gram II*	2009	Static	Raw bytes	Decision trees
Laskov and Šrndic [18]	*PJscan*	2011	Static	JS-based	SVM
Maiorca et al. [22]	*Slayer*	2012	Static	Structural	Random forests
Smutz and Stavrou [17]	*PDFrate v1*	2012	Static	Structural	Random forests
Šrndic and Laskov [4]	*Hidost*	2016	Static	Structural	Random forests
Maiorca et al. [8]	*Slayer NEO*	2015	Static	Structural	Adaboost
Smutz and Stavrou [2]	*PDFrate v2*	2016	Static	Structural	Classifier ensemble
Cova et al. [6]	*WepaWet*	2010	Dynamic	JS-based	Bayesian
Tzermias et al. [14]	*MDScan*	2013	Dynamic	JS-based	–
Schmitt et al. [13]	*PDF Scrutinizer*	2012	Dynamic	JS-based	–
Corona et al. [16]	*LuxOR*	2014	Dynamic	JS-based	Random forests
Xu and Kim [10]	*PlatPal*	2017	Dynamic	–	–

Table 2 True positive (TP) percentage, false positive (FP) percentage, and $F1$ score of different classifiers [1]

Name	Year	Ben. samples	Mal. samples	Train. (%)	TP (%)	FP (%)	$F1$ score
Pjscan [18]	2011	960	15,279	50	71.94	16.35	0.832
Slayer [22]	2012	9989	11,157	57	99.5	0.02	0.989
PDFrate v1 [17]	2012	104,793	5297	9.1	93.27	0.02	0.801
Hidost [4]	2016	576,621	82.142	33.7	99.73	0.06	0.825
LuxOR [16]	2014	5234	12,548	70	99.27	0.05	0.986
Slayer NEO [8]	2015	9890	11,138	57	99.81	0.07	0.969
PDFrate v2 [2]	2016	104,793	5297	9.1	99.5	0.05	0.667

9 Conclusion and Future Scope

In this paper, we surveyed the different types of learning-based PDF malware classifiers. Basically, the PDF classifiers are of two types, namely static and dynamic classifiers. The static classifiers detects the signs of malwares by parsing through the whole document. However, the dynamic classifiers run the document in emulated environments to inspect the behaviors of the file during the execution. Most of the

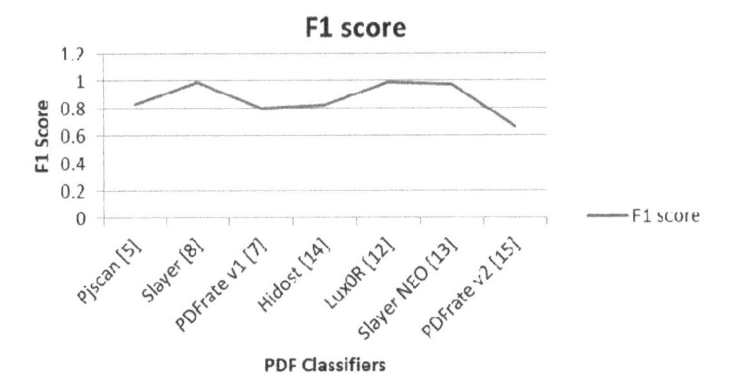

Fig. 15 Line graph for $F1$ score of different classifiers

dynamic classifiers mainly focus on the JavaScript code in the document as majority of the malicious samples have JavaScript content embedded in to it. At first, we have presented the structure of the PDF files and how they are parsed using a parsing software. Followed by that, we discussed the architectures of different static and dynamic classifiers and then conducted a comparative analysis on the surveyed models. The features, classification methodology, true positive (TP) percentage, false positive (FP) percentage, and F1 score were compared among the different classifiers. From the graphs, it was clearly observed that the PDF classifiers were showing a similar efficiency while compared to the others, but this variance can be crucial in detecting a critical sample set while evasion attacks being performed. The dynamic classifiers are more efficient in classifying the adversarial sample more efficient than compared to the static ones because of the usage of the sandboxed environments. This work can be further advanced by making hybrid classifiers which can be able to perform an initial static check and performing a dynamic check during its execution time, which can lead to the elimination of drawbacks faced in both types of pre-existing classifiers.

References

1. Maiorca D, Biggio B, Giacinto G (2018) Towards adversarial malware detection: lessons learned from PDF-based attacks 1(1). Retrieved from http://arxiv.org/abs/1811.00830
2. Smutz C, Stavrou A (2017) When a tree falls: using diversity in ensemble classifiers to identify evasion in malware detectors, pp 21–24 (Feb). https://doi.org/10.14722/ndss.2016.23078
3. Maiorca D, Biggio B (2019) Digital investigation of PDF files: unveiling traces of embedded malware. IEEE Secur Priv 17(1):63–71. https://doi.org/10.1109/MSEC.2018.2875879
4. Šrndić N, Laskov P (2016) Hidost: a static machine-learning-based detector of malicious files. EURASIP J Inf Secur 2016(1):1–20. https://doi.org/10.1186/s13635-016-0045-0
5. Tzermias Z, Sykiotakis G, Polychronakis M, Markatos EP (2011) Combining static and dynamic analysis for the detection of malicious documents. In: Proceedings of the 4th workshop on European workshop on system security, EUROSEC'11. https://doi.org/10.1145/1972551.1972555

6. Cova M, Kruegel C, Vigna G (2010). Detection and analysis of drive-by-download attacks and malicious JavaScript code. In: Proceedings of the 19th international conference on World Wide Web, WWW'10, pp 281–290. https://doi.org/10.1145/1772690.1772720
7. Dang H, Huang Y, Chang EC (2017) Evading classifiers by morphing in the dark. In: Proceedings of the ACM conference on computer and communications security, pp 119–133. https://doi.org/10.1145/3133956.3133978
8. Maiorca D, Ariu D, Corona I, Giacinto G (2015) A structural and content-based approach for a precise and robust detection of malicious PDF files. In: ICISSP 2015—1st international conference on information systems security and privacy, proceedings, pp 27–36. https://doi.org/10.5220/0005264400270036
9. Biggio B, Corona I, Maiorca D, Nelson B, Šrndić N, Laskov P, Giacinto G, Roli F (2013). Evasion attacks against machine learning at test time. Lecture notes in computer science (including subseries lecture notes in artificial intelligence and lecture notes in bioinformatics), vol 8190 LNAI(PART 3), pp 387–402. https://doi.org/10.1007/978-3-642-40994-3_25
10. Xu M, Kim T (2017) PlatPAL: detecting malicious documents with platform diversity. In: Proceedings of the 26th USENIX security symposium, pp 271–287
11. Kang AR, Jeong YS, Kim SL, Woo J (2019) Malicious PDF detection model against adversarial attack built from benign PDF containing javascript. Appl Sci (Switzerland) 9(22). https://doi.org/10.3390/app9224764
12. Castro RL, Schmitt C, Dreo G (2019) AIMED: evolving malware with genetic programming to evade detection. In: Proceedings—2019 18th IEEE international conference on trust, security and privacy in computing and communications/13th IEEE international conference on big data science and engineering, TrustCom/BigDataSE 2019, pp 240–247. https://doi.org/10.1109/TrustCom/BigDataSE.2019.00040
13. Schmitt F, Gassen J, Gerhards-Padilla E (2012) PDF Scrutinizer: detecting JavaScript-based attacks in PDF documents. In: 2012 10th annual international conference on privacy, security and trust, PST 2012, pp 104–111. https://doi.org/10.1109/PST.2012.6297926
14. Lu X, Zhuge J, Wang R, Cao Y, Chen Y (2013) De-obfuscation and detection of malicious PDF files with high accuracy. In: Proceedings of the annual Hawaii international conference on system sciences, pp 4890–4899. https://doi.org/10.1109/HICSS.2013.166
15. Nissim N, Cohen A, Moskovitch R, Shabtai A, Edri M, BarAd O, Elovici Y (2016) Keeping pace with the creation of new malicious PDF files using an active-learning based detection framework. Secur Inform 5(1):1–20. https://doi.org/10.1186/s13388-016-0026-3
16. Corona I, Maiorca D, Ariu D, Giacinto G (2014) Lux0R, pp 47–57. https://doi.org/10.1145/2666652.2666657
17. Smutz C, Stavrou A (2012) Malicious PDF detection using metadata and structural features. In: ACM international conference proceeding series, pp 239–248. https://doi.org/10.1145/2420950.2420987
18. Laskov P, Šrndić N (2011) Static detection of malicious JavaScript-bearing PDF documents. ACM international conference proceeding series, pp 373–382. https://doi.org/10.1145/2076732.2076785
19. Chen Y, Wang S, She D, Jana S (2019) On training robust PDF malware classifiers. Retrieved from http://arxiv.org/abs/1904.03542
20. Shafiq MZ, Khayam SA, Farooq M (2008) Embedded malware detection using Markov n-Grams. Lecture notes in computer science (including subseries lecture notes in artificial intelligence and lecture notes in bioinformatics), vol 5137 LNCS, pp 88–107. https://doi.org/10.1007/978-3-540-70542-0_5
21. Tabish SM, Shafiq MZ, Farooq M (2009) Malware detection using statistical analysis of byte-level file content categories and subject descriptors. Csi-Kdd, pp 23–31. https://doi.org/10.1145/1599272.1599278
22. Maiorca D, Giacinto G, Corona I (2012) A pattern recognition system for Malicious PDF files detection, pp 510–524. Retrieved from https://link.springer.com/content/pdf/10.1007%2F978-3-642-31537-4_40.pdf

MOLE: Multiparty Open Ledger Experiment, Concept and Simulation Using BlockChain Technology

Rahul Johari, Kanika Gupta, and Suyash Jai

Abstract Key sectors like medicine, finance, education, IoT use BlockChain based applications to derive many benefits, this technology has to offer. The foundation of Bitcoin that is blockchain technology has received extensive attraction in research recently. The BlockChain Technology provides benefits in collaboration, trustability, identification, credibility and transparency. This technology acts as an untampered ledger which gives end user permission for various operations to be managed in decentralized manner. Various areas are centered around this, covering domains-like financial services, hospitality sector, healthcare management, E-Governance and so on. However, there are several technical hindrances of this emerging technology such as security and scalability issues, that need to be taken care of. This paper presents an insight into concept of blockchain, highlights the initiatives undertaken by Government of India and a real time simulation showcasing MOLE: Multiparty open ledger experiment, based on BlockChain Technology using MHRD'S Virtual lab.

Keywords Blockchain technology · Virtual lab · Consensus algorithms · Decentralization · MOLE experiment

R. Johari (✉)
SWINGER (Security, Wireless IoT Network Group of Engineering and Research) Lab, USICT, GGSIP University, Sector-16C, Dwarka, Delhi, India
e-mail: rahul@ipu.ac.in

K. Gupta · S. Jai
ABES Engineering College, Ghaziabad, India
e-mail: kanika.gupta@abes.ac.in

S. Jai
e-mail: suyash.17bit1038@abes.ac.in

A. Choudhary et al. (eds.), *Applications of Artificial Intelligence and Machine Learning*, Lecture Notes in Electrical Engineering 778,
https://doi.org/10.1007/978-981-16-3067-5_42

1 Introduction

In recent days, cryptocurrency which is cryptographic in nature has become a center of attraction in both academics and industry. The most successful cryptocurrency has gathered a huge successful implementation with its market potential capping around a billion dollar [1]. Bitcoin network are specially designed that operates without any interference from any intermediating party and exploits the features and benefits offered by the blockchain technology, which was first came into light in 2009 [2]. Blockchain are usually kept in shape of blocks and it is a public ledger. Distributed consensus algorithms and asymmetric cryptography have been used for public ledger consistency and for users' security. This emerging technology, is blessed with key characteristics like decentralization, immutability and transparency (DIT). Blockchain can optimize the cost and efficiency with these traits. Furthermore, it allows various transactions to be completed without any intermediary or bank [4]. Blockchain can be used in a satisfied manner for various financial services such as remittance, online payment and digital assets [3]. However, it can also be used in various areas including public services [5], smart contracts [6], reputation systems [7], security services [8] and IoT [9]. Blockchain is used in these fields in multiple ways. Blockchain cannot be changed that is immutable. Transactions cannot be altered. Various fields that needs honesty and reliable use, could use this technology to their customers in potential market. As it is decentralized in nature, so it avoids the failure situation at single point. For the application of smart contracts, if the contract has been merged with the network of blockchain, it can be done automatically by miners of data.

BlockChain technology has great market for the building the future Internet systems, but it has faced various technical hinderances. The first one is scalability which is a huge concern. The cryptocurrency, that is, bitcoin block size is restricted to 2 MB, while a chunk is executed in block of each fifteen minutes. Consequently, the network of bitcoin is adhered to a capacity of 10 operations per second, which is not capable of dealing with trade of very high frequency. However, greater the block size, more will be the storage space which results in slower propagation and speed in the blockchain network. This results into centralization of transactions as few users would like to operate with big blockchain. Therefore, the combination of size of block and privacy has been a hard challenge to deal with. Secondly, through mining strategy, which is selfish in nature, miners of data could achieve revenues greater than their fair share [10]. They hide their blocks for greater revenue in the market. So, branching must take place very frequently, which hampers the development of blockchain. Hence some solutions are also proposed to fix these types of problems in blockchain. Some issues of privacy leakage can also be seen in the usage of this technology when users make operations with their public key and private key [11]. Various literature on blockchain technology from several sources are available like different types of research forums, posts, blogs, wikis, codes, journal articles, conference proceedings et al. and in [12] author(s) have showcased elaborate technical survey about decentralized digital currencies including Bitcoin. The paper focusses on the concept of blockchain technology instead of various digital currencies used in the market. Research Institute of Nomura made a detailed technical document about blockchain [13]. Proposed research work concentrates on the blockchain technology, various initiatives and its simulation on virtual lab.

2 Blockchain Architecture

Blockchain is a combination of various blocks, which holds a complete list of transaction records like conventional public ledger [14]. A block has only one parent block with a previous block hash contained in the header of block. It is worth to note down that blocks hashes [15] can also be stored in blockchain Ethereum. The foremost block of a blockchain which has no parent block is called as genesis block. For using digital signature, each user owns its pair of public key and private key. The private key is used to maintain confidentiality which are used in signing the transactions and then digitally signed transactions are rotated throughout the full network of blockchain. The elliptic curve digital signature algorithm [16] is typically used in blockchain technology. Following are the important traits of blockchain:

- Not centralized: In traditional centralized systems, every transaction needs to be checked through central trusted party (example: the central bank authority). There are various consensus algorithms used in blockchain with the aim of maintaining data privacy and consistency in distributed network [17].
- Persistency: Various invalid operations would not find place in blockchain networks by the honest miners and some transactions can be validated quickly. It is not possible to rollback or delete any transactions once they are into the blockchain network [18].
- Real Entity: Each user can interact with the blockchain network with its own address which hides the real entity of the user [19] (Table 1).

Table 1 Comparison among private, public and consortium blockchain

Property	Private blockchain	Public blockchain	Consortium blockchain
Read Permission	Couldn't be public	Public	Couldn't be public
Efficiency	Maximum	Minimum	Maximum
Centralized	Maximum	No	Partial
Consensus determination	One organization	All miners	Selected set of nodes
Consensus Process	Permissioned based	Permission less	Permissioned based

3 Various Initiatives of Government for Blockchain Technology

3.1 Initiatives

The report has been drafted by Ministry of Electronics and Information Technology (MeitY) which discussed about the non-centralized ledger technology and its need on different use cases which can take advantage of National Level Blockchain Framework. They also traced different areas where in, this emerging technology can be applied like banking, finance, cyber security, IoT et al. and it can prove to be one of research areas in the future internet market [20]. Various center of excellence have been built up to help the researchers to provide the framework in which they can conduct the research on different problem statement and its use cases. They also executed several projects in banking domain which helps the market in various ways.

3.2 Industry Interaction

The foremost initiative, of Government of India, was design and development of property registration system, primarily blockchain enabled in nature at Shamshabad District, Telangana State. There are various continued projects under the Government of India (GoI) which includes academic certificates authentication, management of hotel registry system and life cycle of vehicle. The Government of India (GoI) also helped IT companies as in February 2020, India's leading IT service provider, Tech Mahindra announced to collaborate with blockchain application is designed and developed by Netherland-based incubator Quantoz to provide secure digital framework for digital payments. It also helped another leading IT company by initiating consumer loyalty platform for Tata Consultancy Services on R3's enterprise blockchain Corda.

4 Properties of BlockChain Technology

As well known, the three pillars of Blockchain Technology are: Decentralization, Transparency and Immutability, which are briefly detailed as follows.

4.1 Decentralization

The real meaning of decentralization is not having a central unit. Now if we talk about this concept in blockchain, it means that blockchain does not have a central governing unit and it is autonomous in nature.

4.2 Transparency

Transparency means something with zero opacity. Now if we talk about this concept in blockchain, it means that blockchain transactions are public and can be viewed on the network by anyone. It is having zero privacy.

4.3 Immutability

Immutable means something that cannot be changed or altered. Once a transaction is packed or pushed into blockchain, it cannot be tampered with.

4.4 Functioning of Blockchain Technology: Role of Hashing

The three pillars of blockchain technology are decentralization, transparency, and immutability as mentioned above with their meanings. Cost and efficiency can be optimized using this approach. The request of software and their use that are built on blockchain architecture will only advance. A hash can be compared with a fingerprint (that is totally unique). Secure hash algorithm (256), a very popular cryptographic approach [21, 22] is used to calculate the hash value. Hash value is basically the combination of the alphabetical and numerical data. The understanding of hash is the cumulative approach to understand the working of blockchain. When a block is produced, a hash value is generated for the same and if any alterations have been done in the block [23], and then it will change the hash value too. The changes are easily identified. The hash value from the predecessor is the ultimate verdict within the block [24].

5 Experimental Set Up: Simulation Performed

Government of India has set up Virtual labs, which have been created with the certain objectives which are detailed as.

5.1 Aim and Objective of Virtual Lab

The virtual lab cater the students at all levels like at graduate levels, post graduate level, researchers and different academicians. It also provides access to various types of labs remotely in all disciplines of Engineering and Technology. With the help of this, students would be able to learn basic and advanced concepts with the aid of practical simulation which is taking place remotely and it also encourages students to conduct more experiments out of their curiosity.

The virtual labs provide full learning management system in which students can learn about different tools of simulation that includes video lectures, case studies, animated demonstrations, web-resources and they can self-evaluate by giving assessments. The virtual lab is meant to share all the resources and equipment's which are costly in nature and it is available to limited number of users due to restrictions of geographical boundaries and time. Using virtual lab, one can easily elaborate the concept and relation between blocks and chain. Apart from that with the help of open and distributed ledger, students would also be able to explain and apply the concepts of blockchain.

5.2 MOLE Algorithm

The MOLE Algorithm has been designed to showcase the concept of handling of financial transaction that is; credit and debit of money among multiple entities involved in transaction.

MOLE : BlockChain based Multiparty Open Ledger Experiment (MOLE)

NOTATION:

User	:	U_t	: U_i (value ranging from U_i for i=1 to n)
Credit Transaction	:	C_t	
Debit Transaction	:	D_t	
Amount	:	A	
Genesis Block	:	G_b	

Trigger: Transaction Starts

1. System credit amount : let's Say X to U_a
   $U_a \leftarrow$ {System}

2. $User_a$ credit amount : let's say Y to $User_b$
   If Amount with U_a is > 0
   $U_b \leftarrow A_y \mathbf{C_t} \{ U_a \}$
   $A_y \mathbf{D_t} \{U_a\}$

3. $User_b$ credit amount : let's say Z to $User_c$
   If Amount with U_b is > 0
   $U_c \leftarrow A_z \mathbf{C_t} \{ U_b \}$
   $A_y \mathbf{D_t} \{U_a\}$

4. The Credited Amount(C_a) is divided into chunks of equal sizes : C_0 .C_1…
5. A Genesis Block(G_b) is created to store Hash Value of C_0: $H(C_0)$
6. G_b now holds the value : $H(C_0)$ + TimeStamp(TS_0) + $H(G_b)$
7. More Blocks would be connected to Genesis Block, as the number of nodes or participating entities in transaction increases.

Transaction Ends

5.3 *Simulation Using Virtual Lab*

There are several steps of simulation in virtual lab [25] which need to performed sequentially to achieve desired result. Three experiments involving Node Connection (with and without loop). All these experiments have been simulated in FOSS (Free and Open Source) based GoI initiated MHRD'S-IIT virtual Lab. The snapshot of the same are depicted in Table 2.

Table 2 Simulation of NODE connection and MOLE experiment based on Blockchain technology using virtual lab

Experiment performed: Simulation snapshot	Description	Status
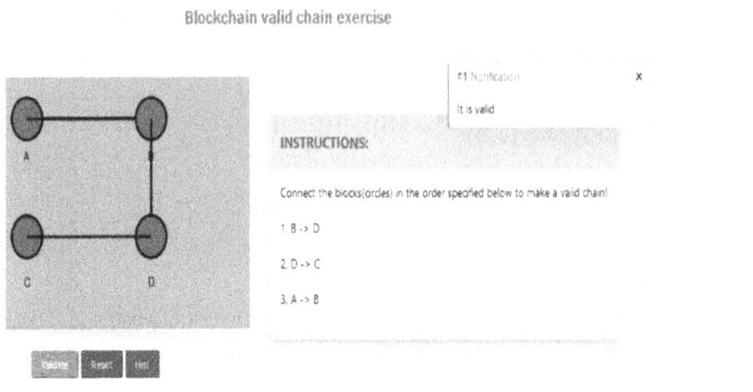 Blockchain valid chain exercise #1 Notification × It is valid A B C D INSTRUCTIONS: Connect the blocks(circles) in the order specified below to make a valid chain! 1. B -> D 2. D -> C 3. A -> B Validate Reset Hint **Experiment Number 1: Node Connection Experiment**	Four nodes A, B, C and D part of Blockchain technology group Connections in the form: A → B → C	VALID transaction
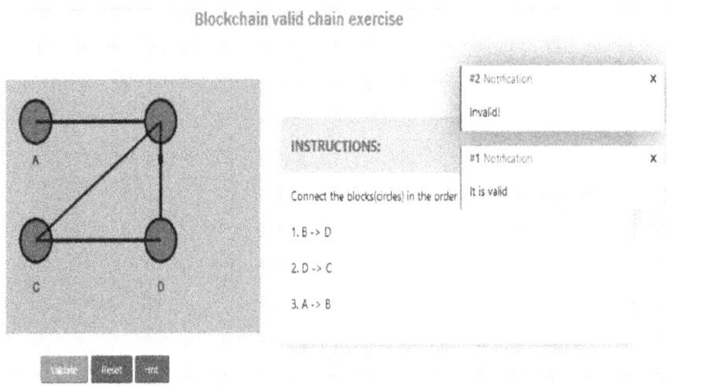 Blockchain valid chain exercise #2 Notification × Invalid! A B C D INSTRUCTIONS: #1 Notification × Connect the blocks(circles) in the order It is valid 1. B -> D 2. D -> C 3. A -> B Validate Reset Hint **Experiment Number 2: Node Connection Experiment (with loop)**	Four Nodes A, B, C and D part of Blockchain Technology Group Connections in the form: A → B → C → d	IN-VALID Transaction
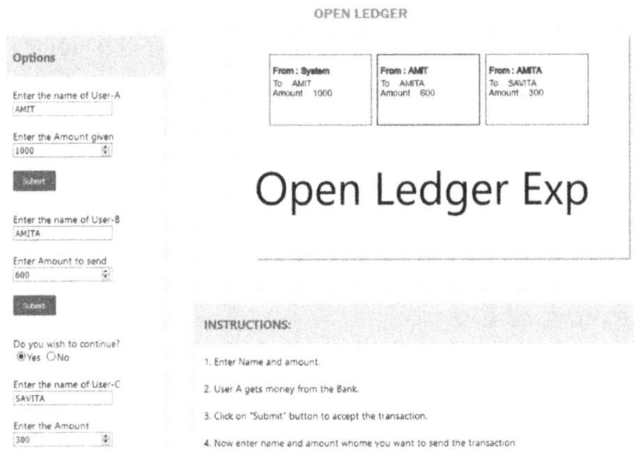 OPEN LEDGER Options From : System To AMIT Amount 1000 From : AMIT To AMITA Amount 600 From : AMITA To SAVITA Amount 300 Enter the name of User-A AMIT Enter the Amount given 1000 Submit Open Ledger Exp Enter the name of User-B AMITA Enter Amount to send 600 Submit INSTRUCTIONS: Do you wish to continue? Yes No 1. Enter Name and amount. 2. User A gets money from the Bank. Enter the name of User-C SAVITA 3. Click on "Submit" button to accept the transaction. Enter the Amount 300 4. Now enter name and amount whome you want to send the transaction **Experiment Number 3: Open Ledger Experiment**	System/bank transfers Salary 1000 USD to Amit. Amit transfers 600 USD to AMITA AMITA transfers 300 USD to SAVITA	VALID transaction

6 Conclusion

This paper describes the concept of blockchain and its three basic principles of decentralization, immutability and transparency. It also presented the tabular comparison between different types of blockchain and its various applications [26]. Various important initiatives taken by the Government of India to promote this technology has also been discussed. The concept of virtual lab and the simulation have been shown using MHRD's virtual lab simulation environment.

References

1. Antshares digital assets for everyone, 2016 [Online]. Available https://www.antshares.org
2. Vukolić M (2015) The quest for scalable blockchain fabric: proof-of-work versus BFT replication. In: International workshop on open problems in network security. Springer, Cham, pp 112–125
3. Decker C, Seidel J, Wattenhofer R (2016) Bitcoin meets strong consistency. In: Proceedings of the 17th international conference on distributed computing and networking, pp 1–10
4. Kraft D (2016) Difficulty control for blockchain-based consensus systems. Peer-to-Peer Netw Appl 9(2):397–413
5. Chepurnoy A, Larangeira M, Ojiganov A (2016) A prunable blockchain consensus protocol based on non-interactive proofs of past states retrievability. arXiv:1603.07926
6. Bruce JD (2014) The mini-blockchain scheme. White paper [Online]. Available http://cryptonite.info/files/mbc-scheme-rev3.pdf
7. van den Hooff J, Frans Kaashoek M, Zeldovich N (2014) Versum: verifiable computations over large public logs. In: Proceedings of the 2014 ACM SIGSAC conference on computer and communications security, pp 1304–1316
8. Eyal I, Gencer AE, Sirer EG, Van Renesse R (2016) Bitcoinng: a scalable blockchain protocol. In: 13th {USENIX} symposium on networked systems design and implementation ({NSDI} 16), pp 45–59
9. Meiklejohn S, Pomarole M, Jordan G, Levchenko K, McCoy D, Voelker GM, Savage S (2013) A fistful of bitcoins: characterizing payments among men with no names. In: proceedings of the 2013 conference on internet measurement, pp 127–140
10. Barcelo J (2014) User privacy in the public bitcoin blockchain
11. Moser M (2013) Anonymity of bitcoin transactions: an analysis of mixing services. In: Proceedings of Munster bitcoin conference, Munster, Germany, pp 17–18
12. Bonneau J, Narayanan A, Miller A, Clark J, Kroll JA, Felten EW (2014) Mixcoin: anonymity for bitcoin with accountable mixes. In: International conference on financial cryptography and data security. Springer, Berlin, Heidelberg, pp 486–504
13. Maxwell G (2013) CoinJoin: bitcoin privacy for the real world. In: Post on bitcoin forum
14. Ruffing T, Moreno-Sanchez P, Kate A (2014) Coinshuffle: practical decentralized coin mixing for bitcoin. In: European symposium on research in computer security. Springer, Cham, pp 345–364
15. Miers I, Garman C, Green M, Rubin AD (2013) Zerocoin: anonymous distributed e-cash from bitcoin. In: 2013 IEEE symposium on security and privacy. IEEE, pp 397–411
16. Sasson EB, Chiesa A, Garman C, Green M, Miers I, Tromer E, Virza M (2014) Zerocash: decentralized anonymous payments from bitcoin. In: 2014 IEEE symposium on security and privacy. IEEE, pp 459–474
17. Nayak K, Kumar S, Miller A, Shi E (2016) Stubborn mining: generalizing selfish mining and combining with an eclipse attack. In: 2016 IEEE European symposium on security and privacy (EuroS&P). IEEE, pp 305–320

18. Sapirshtein A, Sompolinsky Y, Zohar A (2016) Optimal selfish mining strategies in bitcoin. In: International conference on financial cryptography and data security. Springer, Berlin, Heidelberg, pp 515–532
19. Solat S, Potop-Butucaru M (2017) Brief announcement: Zeroblock: timestamp-free prevention of block-withholding attack in bitcoin. In: International symposium on stabilization, safety, and security of distributed systems. Springer, Cham, pp 356–360
20. Crypto-currency market capitalizations, 2017 [Online]. Available https://coinmarketcap.com
21. Johari R, Parihar AS (2019) BLAST: blockchain algorithm for secure transaction. Int J Secur Appl (IJSIA) 13(4):59–66
22. Szabo N (1997) The idea of smart contracts. Nick Szabo's Papers Concise Tutorials 6
23. Shukla S, Dhawan M, Sharma S, Venkatesan S (2019) Blockchain technology: cryptocurrency and applications. Oxford University Press
24. Thompson J (2017) Blockchain: the blockchain for beginnings, guild to blockchain technology and blockchain programming. Create Space Independent Publishing Platform
25. http://vlabs.iitb.ac.in/vlabs-dev/labs/blockchain/labs/blockers-intro-blockchain-psit/simulation.html
26. Johari R, Gupta K, Jha SK, Kumar V (2019) CBCT: cryptocurrency based blockchain technology. In: International conference on recent developments in science, engineering and technology. Springer, Singapore, pp 90–99

Intrusion Detection Based on Decision Tree Using Key Attributes of Network Traffic

Ritu Bala and Ritu Nagpal

Abstract As computer usage is increasing, network security is becoming a huge problem. As time passes, attacks are also increasing on the network, these attacks are nothing else, but it is like the intrusions that are causing **a lot of** much damage to our entire system. IDS is used to protect our data and network from these attacks and to secure our systems from intruders. The technology of data mining is used a lot in order to examine and analyze enormous network data. Data mining is an efficient method that is applied to IDS to detect a large amount of network data, and reduces the pressure of compilation done by humans. This paper compares the different techniques of data mining used to implement IDS. Information gain and rankers' algorithm are used for attribute selection and J48 and random forest classify the data for NIDS and the dataset used is KDDCup99. We have selected 9 attributes from the KDDCup dataset and the experiment is done on the WEKA tool. Moreover, the results show that the detection accuracy with only 9 attributes is almost the same as it was with all 41 attributes.

Keywords NIDS · WEKA · KDDCup99 · J48 · Random forest · Information gain · Rankers algorithm

1 Introduction

Due to the excessive use of computer networks, data's safety and security problem is increasing day by day. Attacks are sometimes called intrusion to cause much damage to the system and our data. IDS helps us in this work and keeps our system free from attacks. Earlier, the methods used were encryption, firewalls, virtual private networks, etc. Now-a-days these cannot be trusted entirely. These techniques do not suffice to protect our data, that is why today there is a need for such a technology that can investigate and analyze the wrong and illegal activities occurring in our systems and networks.

R. Bala (✉) · R. Nagpal
Guru Jambheshwar University of Science and Technology, Hisar, India

A. Choudhary et al. (eds.), *Applications of Artificial Intelligence and Machine Learning*, Lecture Notes in Electrical Engineering 778,
https://doi.org/10.1007/978-981-16-3067-5_43

Hence a dynamic approach has been designed for the security of our network which is called an intrusion detection system (IDS). It is of two types: Signature-based IDS and anomaly-based IDS [1]. It is further categorized as host-based and network based depending on the situation of the environment. Host-based invigilate the conduct of the host system and network-based invigilate the behavior of the network.

Lee et al. [1] in 1999 constructed the classification model with 41 attributes from the unprocessed traffic that was collected at MIT Lincoln Laboratory [2] called KDD Cup 1999 (The 1999 Knowledge Discovery and Data Mining Tool).

2 Data Mining in IDS

Data mining is a broad concept that is most commonly used in computer science. It is the process of extracting out new valid, meaningful and significant information from the large database. In the last few years, data mining techniques like classification, clustering and association rules have successfully found intruders. Business organizations and commercial accountants mainly use it, but now it is increasingly used in research to extract valuable information during experiments and observations [3]. Because of the excessive use of computer networks, there is a large volume of existing and newly arrived data on the network that need to be processed. That is why data mining-based IDS has gained attention in research. Data mining-based idea is used to examine the covered pattern of intrusion and the relationship hidden in the data [4]. It is used for the detection of variants, control false alarm and improve efficiency [5].

3 WEKA Tool

This tool is used to perform data mining and machine learning tasks. It is a group of many machine learning and data mining algorithms, especially classification, data preprocessing, regression, feature selection and visualization. Its programming is done in the JAVA language. It is used extensively in research. It has 49 data preprocessing tools, in addition to this has 76 classification algorithms, 15 attribute evaluators and 10 search algorithms for the selection of features. All the files must be in ARFF format to be run in it.

4 KDDCup 1999

KDDCup99 is the oldest and the most commonly used dataset which is used to access anomaly detection. It identifies good connections called normal and bad connections called intruders. Nine weeks of data is collected for training and test data. It contains 490,000 instances and every instance has 41 features labeled as either attack or normal [6]. These features are categorized into three groups as basic, traffic and content. There are 24 training attack types which are grouped as

- Denial of Service: In this attack, the attacker send so much request on the server and make the memory and other resources too busy that the request of genuine user to access the machine is denied, e.g., smurf attack, land attack, etc.
- R2L: In this type of attack, the attacker finds a way to get access to the machine through negotiating the security-like guessing the password, etc.
- U2R: In this attack, the attacker has the benefit of local access to the system and makes an effort to access to the administration, e.g., buffer overflow attack.
- Probe: In this attack, the attacker benefits by collecting the information about the victim machine, he gains this information by taking advantage of their weaknesses, e.g., Port scanning.

5 Related Work

The most important project in network security is to make an effective NIDS. Experts have done a lot of work in this field. The work is divided into 3 parts. First, find the important and relevant features from the attribute set, then upgrade learning algorithms and then assess the performance on any dataset. Li et al. [7] gave an ideal IDS in which they sorted the required features by applying the information gain method and chi-square method. The outcome of this work indicates that the accuracy of the detection is still maintained even by using some selected attributes. In [8], the author used PCA to select features and the features which had higher Eigenvalues were retained. In [9], the author explains that by using adequate training parameters and selecting the right features, high accuracy can be achieved. The performance of an IDS can also be improved. In [10] writers used function rating set of rules to lessen the function area through the usage of three rating set of rules primarily based totally on support vector machine (SVM), multivariate adaptive regression splines (MARS) and linear genetic programs (LGP). In [11], creators propose "Enhanced Support Vector Decision Function" for function selection that is primarily based totally on essential factors. First, the function's rank, and second, the correlation among the features were adopted. In [12], writers advise an automated function choice process primarily based totally on correlation—primarily based feature selection (CFS).

6 Algorithms Applied

6.1 J48

J48 is an open-source algorithm in WEKA used to build the decision tree and made known by Ross Quinlan. This algorithm works on supervised learning. The decision trees produced are used for classification. During decision tree construction, first of all find out the instances belonging to the same class. If found then, the tree is represented by a leaf and labeled by same class. Second, the information gain calculates every attribute and then results in the best attributes for branching the tree [13–15].

Entropy is the term used to calculate information gain [16]. The formula of entropy (E) is

$$E = -\sum_{i=1}^{n} P_i \log_2 P_i \tag{1}$$

Gain (S, X) of attribute X w.r.t. total sample (S) is

$$\text{Gain}(S, x) = E(S) - \sum_{j \in \text{values}(X)} \frac{\text{mod}(Sj)}{\text{mod}(S)} E(S_j) \tag{2}$$

Information Gain can be calculated as

$$\text{Splitinfo}(S, X) = -\sum_{k=1}^{c} \frac{\text{mod}(Sk)}{\text{mod}(S)} \log_2 \frac{\text{mod}(Sk)}{\text{mod}(S)} \tag{3}$$

$$\text{Gain Ratio} = \frac{\text{Gain}(S, X)}{\text{Splitinfo}(S, X)} \tag{4}$$

6.2 Random Forest

Supervised learning algorithm that is used in both classification and regression, but most commonly used in the classification. This algorithm builds the decision tree using each data sample and takes the prediction of each decision tree. Then, voting is performed on every predicted result and selects that prediction that gets maximum votes and gives the best results. This algorithm also reduces the problem of over fitting to some extent. This algorithm works efficiently with large amount of data. Its accuracy is very high even after missing a large proportion of data.

Table 1 Comparison of J48 and random forest

Algorithm used	Number of attributes	TP	FP	Precision	Time (s)
J48	42	0.996	0.004	0.996	2.54
	12	0.997	0.005	0.997	1.26
	10	0.997	0.006	0.995	0.33
	09	0.997	0.005	0.995	0.33
Random forest	42	0.999	0.003	0.999	9.31
	12	0.999	0.004	0.997	4.46
	10	0.998	0.004	0.997	3.4
	09	0.998	0.004	0.997	3.43

7 Experimental Result

7.1 Parameter

True positive (TP) gives the correct result means if there is an attack, identify it correctly.

False positive (FP): means there is actually an attack, but it predicts it as normal.

Precision: tell the correct percentage of positive prediction.

Time: Gives how much time it takes to build the model.

Accuracy = TP + TN/TP + TN + FP + FN * 100.

Precision = TP/TP + FP * 100.

7.2 Results

Table 1 shows the comparison of J48 and random forest with varying numbers of attributes. As shown in the table, if we reduce the attributes, there seems no significant difference between TP, FP, and precision results, but there is a big difference in the time it takes to build a model. If we get almost identical or we can say better results with a smaller number of attributes then it makes no sense that we have to use complete 42 attributes (Figs. 1, 2, 3 and 4).

8 Conclusion

In this paper, we have used information gain for the extraction of attributes for intrusion detection. To classify these extracted attributes, we have compared J48 and Random forest decision tree algorithms. The accuracy of the parameters has shown some progress by taking only 08 attributes compared to using 41 attributes.

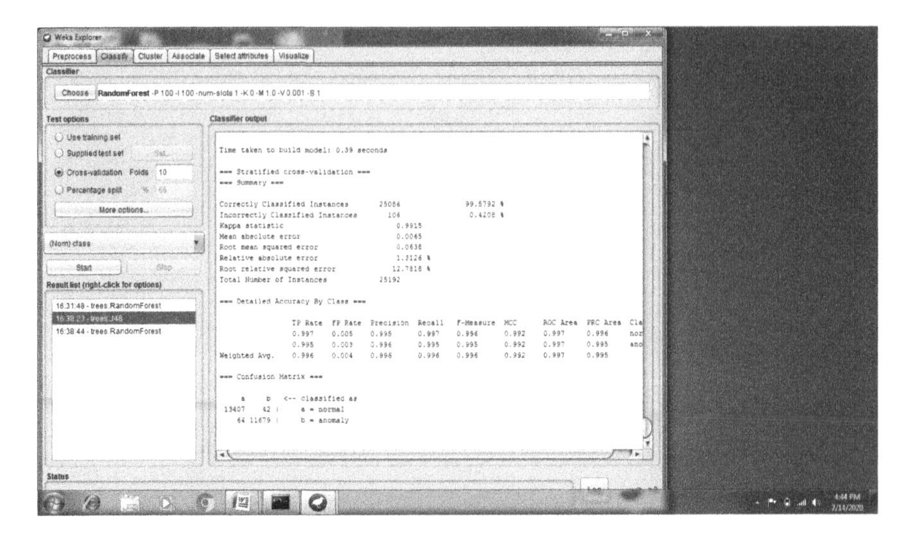

Fig. 1 J48 algorithm with 9 attributes

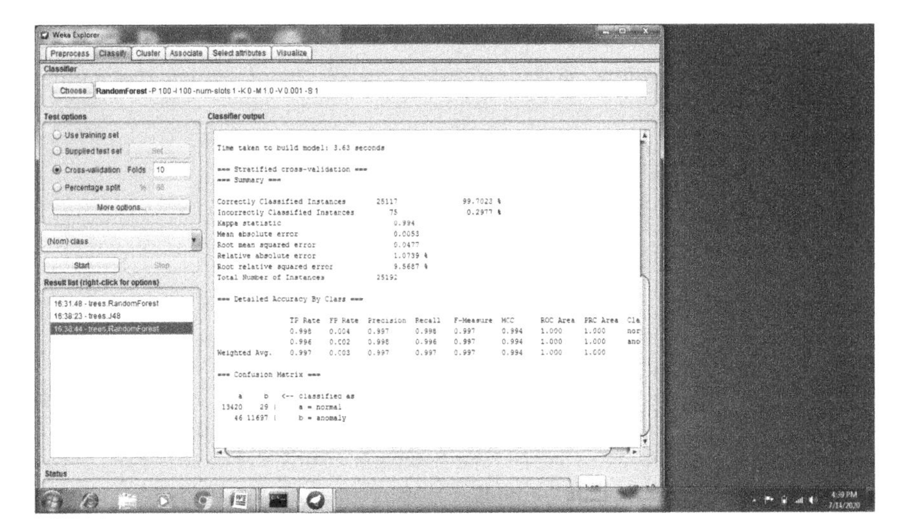

Fig. 2 Random forest algorithm with 9 attributes

The significant difference is in the time to build the model which shows remarkable improvement with fewer attributes. The experiment is performed on the WEKA tool using KDDcup99 Dataset. According to the results, J48 with 09 attributes gives better results. Still, some more work is required in this field. We propose to use other classification algorithms apart from J48 and Random Forest on varied datasets in the near future.

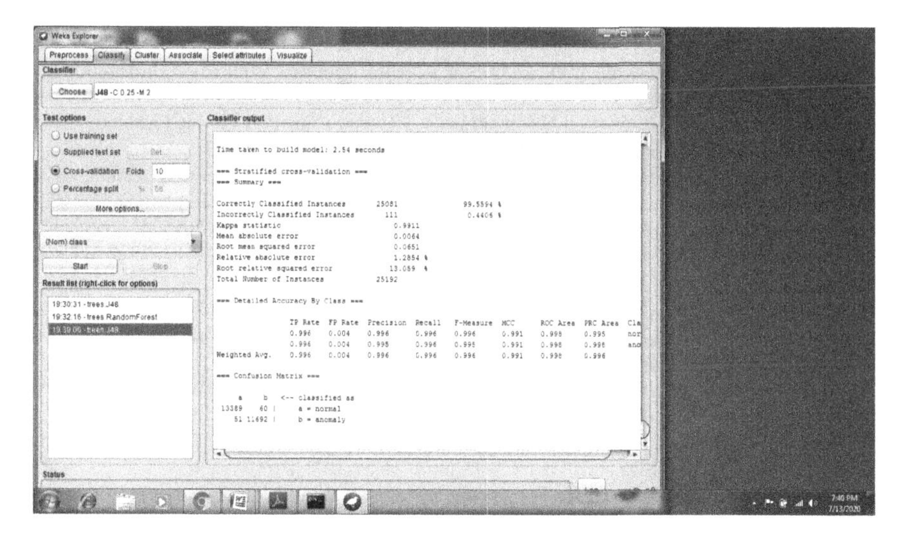

Fig. 3 J48 algorithm with 42 attributes

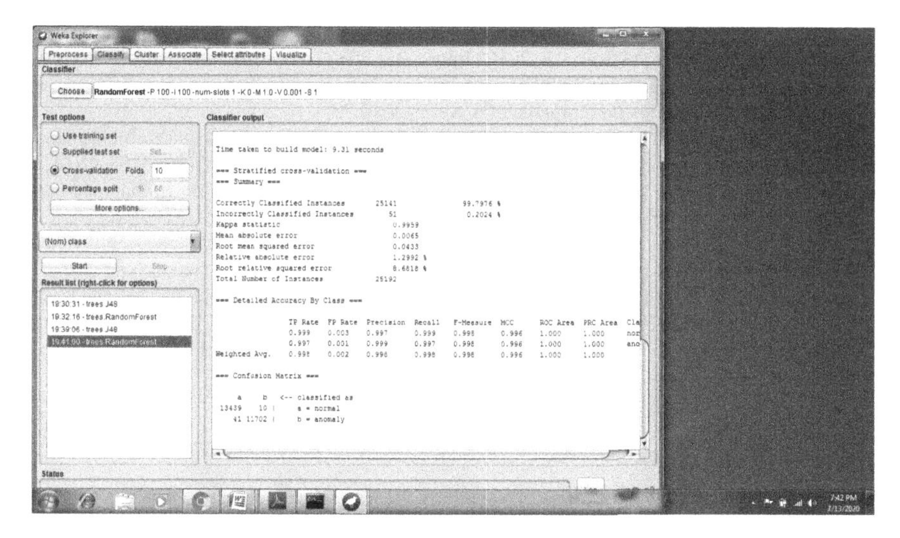

Fig. 4 Random forest with 42 attributes

References

1. Lee W, Stolfo S, Mok K (1999) A data mining framework for adaptive intrusion detection model. In: Proceeding of 1999 IEEE symposium on security and privacy. IEEE, pp 120–132
2. MIT Lincoln Laboratory-DARPA Intrusion Detection Evaluation http://www.ii.mit.edu/ist/ideval/docs/docs_index.html

3. Thuraisingham BM (2011) Data mining for malicious code detection and security applications. In: 2011 European intelligence and security informatics conference, pp 15–18. https://doi.org/10.1109/EISIC.2011.80
4. Dutt I, Borah S (2015) Some studies in intrusion detection using data mining techniques. Int J Innov Res Sci Eng Technol 4(7)
5. Lu T, Boedihardjo AP, Manalwar P (2005) Exploiting efficient data mining techniques to enhance intrusion detection systems. In: IRI-2005 IEEE international conference on infomation reuse and integration, pp 512–517. https://doi.org/10.1109/IRI-05.2005.1506525
6. Modi U, Jain A (2015) A survey of IDS classification using KDDCUP 99 dataset in WEKA. Int J Sci Eng Res 6(11):947–954
7. Li Y, Fang BX, Guo L (2006) A lightweight intrusion detection model based on feature selection and maximun entropy. In: 2006 international conference on digital object identifier, pp 1–4. https://doi.org/10.1109/ICCT.2006.341771
8. Ahmad I, Abdulah AB, Alghamdi AS, Alnfajan K, Hussain M (2011) Feature subset selection for network intrusion detection mechanism using genetic Eigen vectors. Proc CSIT 5
9. Abdulla SM, Najla B, Zakaria O (2010) Identify features and parameters. World Acad Sci Eng Technol 4(10):1553–1557
10. Sung AH, Mockamole S (2004) The feature selection and intrusion detection problems. In: Proceedings of the 9th Asian computing science conference 2004, lecture notes in computer science. Springer, pp 468–482
11. Zaman S, Karray F (2009) Features selection for intrusion detection systems based on support vector machines. In: Proceedings of the 2009 6th IEEE conference on consumer communications and networking conference. https://doi.org/10.1109/CCNC.2009.4784780
12. Nguyen H, Franke H, Petrovic S (2010) Improving effectiveness of intrusion detection by correlation feature selection. In: 2010 international conference on availability, reliability and security, pp 17–24. https://doi.org/10.1109/ARES.2010.70
13. Yu J, Kang H, Park D, Bang H, Kang DW (2013) An in-depth analysis on traffic flooding attacks detection and system using data mining techniques. J Syst Arch 59(10):1005–1012
14. Kim G, Lee S, Kim S (2014) A novel hybrid intrusion detection method integrating anomaly detection with misuse detection. Expert Syst Appl 41(4):1690–1700
15. Mukherjee S, Sharma N (2012) Intrusion detection using Naive Bayes classifier with feature reduction. C3IT-2012 Proc Technol 4:119–128
16. Meena G, Choudhary R (2017) A review paper on IDS classification using KDD99 and NSL-KDD dataset in WEKA. In: 2017 international conference on computer, communication and electronics (comptelix), pp 553–558. https://doi.org/10.1109/COMPTELIX.2017.8004032

An Extensive Review of Wireless Local Area Network Security Standards

Sudeshna Chakraborty, Maliha Khan, Amrita, Preeti Kaushik, and Zia Nasseri

Abstract During recent times, wireless local area network is very important and changed to a vital topic since the World Wide Web was introduced in 1995; we have many types of wireless network such as wireless local area network, wireless communication, wireless application protocol (WAP), wireless transaction and WMAN. Among all these types, WLAN gained much popularity and used widely in today's world in many areas for different purposes: office, airport, library, university, hospital, military campus and many more. In another hand, the security of all these functionalities has become a crucial topic in today's WLAN due to not existing any physical border around WLAN channel, so that, the security of information can be leaked by attackers very easy. For these purposes, IEEE introduced very famous security standards like WEP, WPA and WPA2 for securing communication between two endpoints. This survey gives a brief introduction to types of WLAN security, to study types of attack in wireless LAN, the vulnerabilities, and study on the existing security standards and will focus more on WPA2 which fix the problem of WEP and WPA and then explore the vulnerability of each standard; finally, this paper will end up with some useful mitigation and suggestion how to improve the wireless LAN security.

Keywords Access point · Wireless local area network · WEP · WPA · WPA2 · Standards · Comprehensive · Mitigation

1 Introduction

For the first time, 802.11i technology was ratified on June 24, 2004, which specifies the mechanism for WLAN, in which it is originally called wired equivalent privacy (WEP) to the equivalent of WLAN security algorithm by IEEE 802.11. Though, it is

S. Chakraborty (✉) · M. Khan · Amrita · P. Kaushik · Z. Nasseri
Computer Science and Engineering, School of Engineering and Technology, Sharda University, Greater Noida, India

P. Kaushik
e-mail: Preeti.kaushik@sharda.ac.in

A. Choudhary et al. (eds.), *Applications of Artificial Intelligence and Machine Learning*, Lecture Notes in Electrical Engineering 778,
https://doi.org/10.1007/978-981-16-3067-5_44

proved by many researchers that WEP cannot achieve the required security of data in three main security goals: confidentiality, integrity and authentication of data. So, it is also being approved by many researchers and considered as a broken way in its security design. As result, the usage of WEP was decreased day by day and superseded by another powerful security standard called WPA in 2012 [1]. This supersede by WPA solve a little the security challenges that occurred on WEP security protocol.

Wi-Fi protected access (WPA) uses the legacy hardware and simply can say that it is an intermediate solution to fulfill the security requirement that WEP has. WPA adopts temporal key integrity protocol (TKIP) for integrity and confidentiality purpose, in which it still uses Rivest Cipher 4 (RC4) for data encryption. A key mixing function is included in TKIP and as well included IV spaces to construct the unrelated and fresh key per packet. WPA uses an algorithm called Michael in order to improve the data integrity. Moreover, WPA implements a packet sequencing mechanism in order to bind a monotonically increasing sequence number for each packet, this functionality helps in replay packets detection [2, 3]. In a result, TKIP solves all known vulnerabilities in WEP, but it had still some limitation due to the uses of legacy hardware and relies on message integrity check (MIC) algorithm also known Michael algorithm in which it provides inadequate security [4]. For this vulnerably, it has been replaced with a more secure algorithm called counter mode/CBC-MAC protocol (CCMP). However, TKIP is being used by many people and should be discouraged [5]. Considering the use of new hardware, a new solution is proposed in order to enhance the security in the MAC layer. IEEE802.11i uses CCMP to provide authentication, confidentiality and replay protection; furthermore, it uses a new authentication method called 802.1X and key management due to provide a better mutual authentication by generating a fresh session key for transmitting data between two end points. In this survey paper, a theoretical comparison of different secretly standards has been provided and assessed based on the effectiveness of each standard on network performance using IEEE802.11n amendment (Fig. 1).

The organization of this paper is as follows: how Wi-Fi works in Sect. 2, wireless LAN security standards will be discussed in Sect. 3, attacks on wireless LAN in Sect. 4, the literature survey is discussed in Sect. 5, Sect. 6 will be basic solutions for mitigating security flaw in AP, countermeasure is discussed in Sect. 7, and finally end Sect. 8 with conclusion.

2 How Wi-Fi Works

Wi-Fi is the name of technology in the world of wireless that is using radio waves in order to provide wireless Internet and network connection with high speed. It does not have any physical wired connection for transmission of data between sender and receiver; here, AP plays an important role and do the primary job that is broadcasting the wireless signal.

A simple equipment like AP and client can provide the environment for wireless connection; here, the client for communicating to wireless network needs to go

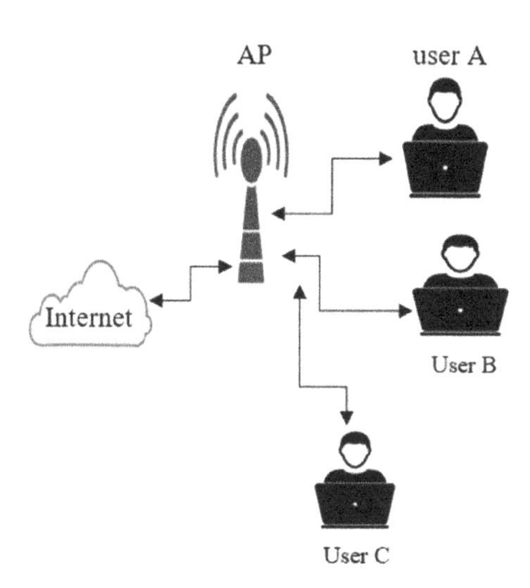

Fig. 1 WLAN

through five steps until to connect to wireless network: scanning the signal, joining, authenticating, association and finally re-association.

First step is starting with scan, scanning is the process of finding available Wi-Fi networks around us, in this step, client needs to identify the network by scanning, after the scan step is completed and found the appropriate network, then the client will go for two option either choosing basic service set (BSS) or extended service set (ESS), and BSS is a wireless network with only one AP but ESS is the wireless network with many APs which are interconnected to each other.

Scanning and joining cannot guarantee the network access; thereby, it is only the primary steps to connect to wireless LAN; in order to complete access to network, clients need to pass the next two steps authentication and association; for connecting to wireless network, client does not need to connect and access physically to the network; client just needs to be in boundary the wireless network signals.

In two ways, client is able to connect to BSS network:

1. Automatically.
2. Manually.

In manually, client chooses and joins to network manually, but in automatic, joining client has choice to pick an AP base on the power of signal strength, and in both methods, parameters are configured on AP device.

Authentication is a security method in wireless network in which the station is authenticated by authentication server; the purpose of authentication is that the station escapes from false AP and for the AP escape from false station. There are different types of authentication method, but here 802.1X uses EAP. The client can connect to access point successfully only when the authentication is successful, and the 802.1X authentication has many advantages.

1. Administrator allows all the access process to the network according to the appropriate standards.

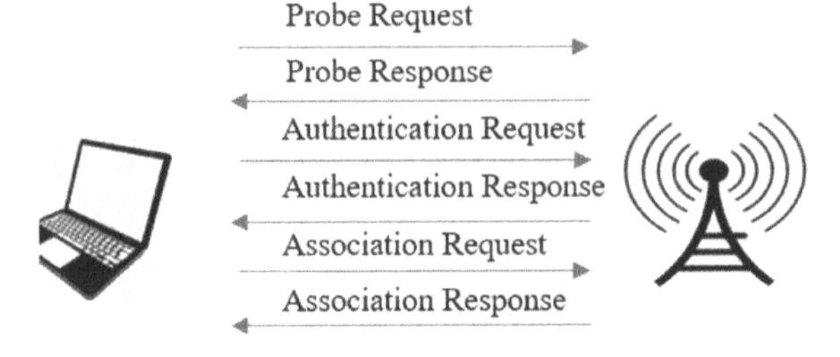

Fig. 2 Client authentication process

2. Administrator is responsible to define responsibilities for each user in network.
3. An authorized port cannot be compromised by a non 802.1X client (Fig. 2).

3 Wireless LAN Security

Wireless LAN is more susceptible to different attacks since the Internet usages grew up, and day by day usage of mobile devices such as mobile phone, laptop and other smart devices has been increased, so securing the credential data changed to a concern topic; thus, there is needed to apply security mechanism in order to protect the communication; for this purpose, there are three different security standards like EWP, WPA and WPA2 the most famous to enhance and raise the security of WLAN up and at least get close to the wired network security, we will go deeper on each security one by one [6, 7].

3.1 Wired Equivalent Privacy

WEP is the first cryptography algorithm which ratified in 1997 to enable the authentication and privacy; firstly, this standard uses RC4 for encryption which has pre-established shared secret key and liner hash function; second, it uses 24 bits for initialization vector (IV) and with a key length of 40 bits [8].

Weakness of WEP:

1. There is no key management system and replay detection.
2. The main drawback is to using static encryption.
3. Using XOR function for ciphering the pain text with key.
4. Using single IV and shared secret key makes the possibility of more vulnerability by using many tools like air snort and WEP crack.

3.2 Wi-Fi Protect Access

WPA is known as the second wireless LAN security standard which is the best supersedes to solve the WEP security drawback. This standard operates in two modes: One is enterprise mode, and second is personal mode which uses pre-shared key for authentication purpose but less secure than enterprise mode which uses IEEE802.1X or EAP for mutual authentication with key length of 128 bits key for encryption and 64 bits key for authentication and 48 bits for initialization vector, using message integrity code (MIC) which is also called Michael for data and header integrity; furthermore, it uses TKIP instead of RC4 algorithm [9].

Weakness of WPA:

1. WPA is based on RC4 for encryption which is weak and vulnerable against to denial of service attack [2].
2. Affects more on network performance due to pretty heavy cryptography computational in each packet, so it causes large amount of overheads.
3. TKIP is relying on MIC which provides inadequate security.
4. The main problem is not compatible with hardware and OS such as Windows 95.
5. EAPoL and MIC messages are sent in clear text so it can be hacked very easy.

1. In the first step, the image is taken as an input, and the global threshold value is being calculated.
2. In the second step, the input image was converted into black and white image.
3. In the third step, the distance transform of the image, i.e., the distance between the separations of the points in the image has been calculated.
4. In the fourth step, the values of the distance transform are selected without repetition and sorted in ascending order.
5. In the fifth step, the total area of the binary image was calculated, and the areas formed by the corresponding distance transform values are calculated. After calculation of the values of the areas, the area formed by each of the distance transform values is then divided with the value of the area of the whole binary image and is termed as feature values.
6. After getting all the feature values, the values are then divided into intervals such as 0.0000–0.0999, 0.1000–0.1999 and so on. After the division, the maximum value from each interval is used to draw a bar chart. The curve, obtained by joining all the top points of the bar chart, is observed and studied.
7. In this last step, datasets were prepared with a set of ten images from both text and non-text and were fed into the classifier for the training and the study [10]. Among all the classifiers, the classifier which produced the highest accuracy was sorted out, and then, the related confusion matrix of that classifier was studied for the performance of the proposed classification method (Fig. 3).

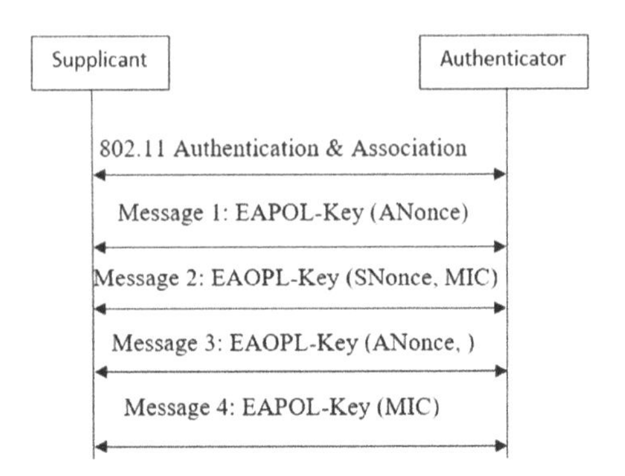

Fig. 3 Four-way handshaking

3.3 WPA2/IEEE802.11i

To overcome the security challenge of WEP and WPA, IEEE introduced a new standard named WPA2 in 2004 which is combined of AES encryption algorithm with CCMP as a mandatory which performs many complex round to make more complex the key than WEP which uses RC4 linear and differ from WPA because of message integrity and CCMP based on AES encryption algorithm rather than RC4 and TKIP [11, 12]. Counter mode with CBC-MAC protocol (CCMP) is a protocol for data integrity and confidentiality; this algorithm is consisting of two important protocol: One is AES encryption (CTR-AES) for encryption of data, and second is cipher block chaining-MAC (CBC-MAC) for data integrity and MIC for authentication [1].

The message between the authenticator and supplicant is as follows:

Message 1: First the authenticator sends and message by the name EAOPL-key frame which is contained ANonce for generating PTK and MIC, then supplicant uses the message in order to generate an ANonce and derives PTK.

Message 2: In second step, the supplicant sends an EAOPL-key frame which is contained SNonce, MIC and RSNE, then the supplicant derives a PTK; by default, the MIC is set to bit 1, and then, it will be confirmed by authenticator.

Message 3: Third step the authenticator sends a message 3 of EAPOL-key frame and derives the PTK.

Message 4: In fourth step, the supplicant replies with 4th and the last EAPOL-key frame to the authenticator in order to notify the authenticator if the temporal key is installed, and secure bit is set.

Advantage of WPA2

1. Protect of message privacy using CTR-AES.
2. Protect integrity of messages.
3. Protect the source and destination addresses from man–in-the-middle attack, Mac spoofing and message modification.
4. Provide strong encryption with mutual authentication.

Table 1 Review of WLAN security protocols

	WEP	WPA	WPA2
Key length	40–104 bits	128 bits encryption	128 bits or higher
Purpose	Provide basic security	Overcome the security flaw of WEP	Enhancement of WPA, by completely implemented od IEEE 802.11i
Data privacy	RC4	TKIP	CCMP using AES
Authentication	Open and shared	WPA-PSK enterprise	WPA2-personal and WPA2-enterprise
Data integrity	CRC-32	MIC	CBC-MAC
Key management	No key management	Robust key management generated by four-way handshake	Robust key management
Hardware compatibility	No need for new hardware	Only new NIC upgrade	Support only Wi-Fi device after 2006
Attacks vulnerability	DoS, PTW, FMS and chop-chop attacks	DoS, chop-chop, WPA-PSK attacks	DoS, MAC spoofing, dictionary attacks
Deployment	Easy setup	Complicated setup	Complicate setup

Weakness of WPA2

1. WPA2 using CCMP that will have effect on the performance of the systems.
2. In CCMP, control frame and management frame neither encrypted and nor authenticated, thus it is vulnerable against many attacks.
3. By predicting, initial number is still vulnerable against to dictionary attack.
4. New hardware is needed for more security.

Table 1 shows a brief information about each security protocols with all their functionality.

4 Attacks on Wireless LAN

There are different types of attack in wireless LAN which will be discussed by details, these are:

1. **Masquerading:** Masquerading attack happens when a network device mimics as a valid device. If this device deceives successfully, the target until it validates as a legitimate user. In this case, attacker will get all the permission and will not be detected. Here, the attacker tries to pretend as a particular user of that system to gain more access and increase the privilege escalation; this kind of attack may occur by use of stolen user ID and password or by finding security flaws in program or maybe bypassing the authentication.

2. **Snooping:** Snooping is the method of accessing personal information. This information can be used for different purposes, like access to more secure information that is useful and in the profit of attacker's company. This attack also known as information gathering or footprinting, when attacker starts to launching attack, this attack method is the first step that helps the attacker for further steps, as much as this information being precise will be more chance of successful attack on target. There are many powerful tools like maltego, nmap and google dark, in which the attacker uses to gain more information like DNS record, Whois and DNS to IP mapping, and finally, by this method, attackers are able to harvest email id, website, social profiles and more sensitive information.
3. **Modification:** After successful snooping attack, then attacker is able to read data and alter it, this modification of data is in a way even the sender and receiver cannot understand it, and for example, changing the number of electronic bank transfer and using tool like Burp Suite can intercept the wireless transmission and then alter the destination of the message in desire.
4. **Denial of service:** This attack is a cyber-attack in which the attacker tries to make the machine or network resource unavailable to legitimate users, and this DoS attack is usually accomplished by flooding the target machine with superfluous of request in order to overload the system to prevent the legitimate user access. DDoS is types of DoS attack instead of using one machine for flooding; here, the attacker uses many machines to flood the target or the flooding traffic originated from many sources instead of only one machine (Fig. 4).

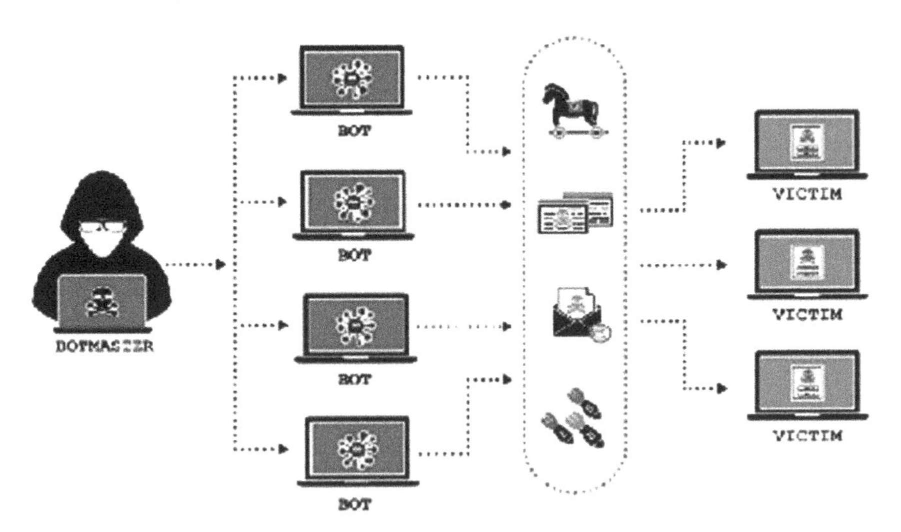

Fig. 4 DDoS attack

5 Basic Solutions for Mitigating Security Flaw in AP

It is obvious that every WLAN has vulnerability like DOS, DDOS and dictionary attack. AP is usually can named as the central control of each network and play an important rule, so that safety is the number one task to be taken care when setting up for the first time, unless the attacker can easily bypass the security of that network and steal the sensitive data from inside system, and below are few steps to be considered during setting up the AP.

Since the two standard WEP and WPA which are poor and outdate, also not able to provide on demand security, so it is mandatory to definitely configure WPA2 in order to encrypt our WLAN instead of using WEP and WPA which is outdate with security risks.

1. Authentication: Using authentication methods to authenticate all the connected devices, in authentication as much as the password chosen short would be more likely to be easily cracked by attacker.
2. Secure configuration: First step is to change all default configuration of AP like enable the MAC address filtering, MAC is the unique hardware address for each devices which are connected to network, then create list of MAC addresses and restrict which MAC addresses can connect and which MAC addresses are not allowed to connect, and third thing is to change the default password of admin to strong password with combination of special character, letter and numeric and disable the DHCP, instead use static IP address for each device.
3. So far, in WPA2 security standard, only password attacks like brute force and dictionary attack are known and popular in cybercrime nowadays. Here, this weakness refers to the ability of user, for example, as much as the complex password chosen by user, will be more secure, user must build their password as complex as possible, and it should be the combination of lowercase, uppercase and special symbols till to avoid password guessing and dictionary attack. We can also keep our password in a white paper instead storing it inside smart devices.
4. SSID displays the name of our network and will be available and accessible to all smart devices in the boundary of signal range; if we do not run a public hotspot, we should prevent from personal information that might from your side, your company and your location. If we hide the SSID of network, it could avoid some smart devices from seeing the AP's name; it cannot fully solve the security problem but at least makes it difficult for attacker to find the AP name.
5. Kept turn on the firmware update can also be added as a point of security rule in WLAN, as the attacker always get the advantage of security flaws in each software which are not up-to-date, then can easily take control over the admin rights and then let the malware infiltrate the networks. For this purpose, some AP has built-in functionality to update installed firmware; if this is not the case, then we should check it manually and regularly whether there is any update for the device to be downloading and install manually.

6. In case of any security flaw, there are many solutions to take action after the setup has completed, like setup firewall, as we know the majority part of WLAN are connected to the other network so that we should create our own firewall with appropriate rule and privacy to filter unwanted connection, and it is also helpful to use IDS and IPS system to detect the security flaw and attack as soon as possible.

6 Methodology of the Simulation

The steps of the methodology of the simulation are as follows (Table 2):

1 Setup a real laboratory of one AP, user and one attacker.
   (a) User: Windows 10 user or any smart device.
   (b) AP: Device which can provide Wi-Fi, it can be Alfa AP or smart device which provides Wi-Fi.
   (c) Attacker.
      - Attacker uses Kali Linux OS.
      - Configure radius server on Linux for user and AP authentication purpose.
      - Wireshark tools.

Table 2 Literature review

Name of author	Title	Technology and standards
Suroto et al., 2018 [1]	WLAN security: threats based on comparison and countermeasures	
Riddhi C. Somaiya, Dr. Atul M. Gonsai, Rashmin S. Tanna, 2016 [2]	WLAN security and efficiency issue based on encryption algorithm	RSA + AES
Mohmoud Khasawneh, Izadeen Kajman, Rashed Alkhudaidy, 2014 [13]	A survey on Wi-Fi protocols	WPA2
Ahmed M. Al Naamany, Ali Al Shidhani, Hadj Bourdoucen, 2016	IEEE 802.11 WLAN security	WEP-TKIP
Yi Ma, Hongyun Ning, May 2018 [6]	The improvement of wireless LAN security authentication mechanism based on Kerberos	Kerberos algorithm
Odhiambo E. Biermann	Integrated security model for WLAN	ISM, CCMP, EAP, RADUIS server
A. I. Androulakis, A. BHalkias	real-life paradigms of wireless network security attacks	WEP WPA

2. Penetration testing of WEP, WPA and WPA2 by using Aircrack toolkit like airmon-ng, airodump-ng, aireplay-ng and airbase-ng.
3. Setup a complete laboratory in NS2 simulator to compare the performance of each security standard.
4. Take action and suggest the solution according to the security flaw found in each standard.

Description:

1. User can be anyone who is authenticating to AP and send the first request to AP for authentication; different types of authentication are available in wireless communication: Open authentication, shared key authentication, EAP authentication and many more, in open authentication every smart device are allows first to authenticate and then try for communicating with AP, any device authenticate but only those device are able to communicate that has the same WEP key with AP's key, and only device can authenticate with AP which supports WEP protocol. But shared key authentication is different; during authentication, AP sends unencrypted text only to the devices that already request for communicating, when AP received that unencrypted text, then the client will encrypt and send back to AP, and the AP allows the client to be authenticate if the text string encrypted correctly. But this approach is vulnerable because here, the encrypted and unencrypted text string can be monitor, which is an open door for attacker; the attacker then can easily calculate the WEP key by comparing the unencrypted text with encrypted text, because of this drawback I cannot recommend using of shared authentication key. Likewise, shared authentication is not relied on RADIUS server. EAP authentication technology reduces the security flaws of the shared key authentication and provides the highest security in wireless communication by using EAP and with combination of RADIUS server. Here, the AP plays intermediate role to perform mutual authentication with RADIUS server which is connected as a wired LAN.
2. Device like Alfa AP which provides Wi-Fi or broadcasting the Wi-Fi signal in an environment and will give back reply to the request of client with ACK or deny message and asking for credentials.
3. Attacker is a person who is trying to lunch the attack by different methods: Brute forcing, making rogue access point, man-in-the-middle attack, DE authenticating the client and AP attacker uses Kali Linux OS and setting Wireshark tools on it to capture all the authentication and the traffics between client and AP. configuring RADUIS server on Linux OS in order to provide secure authentication between AP and client.

7 How to Reduce Chance of Attack

The main origination of threats in wireless communication comes from three basic origins: disruption, interception and alteration; below are some security solution for mitigating the security attacks in wireless communication.

(a) ***Encryption and Decryption Algorithm***

Encryption is the technique in which the clear texts are sent to destination in such a way which are not readable by intruder and will be decrypted in destination user, every device is already equipped with built-in encryption and decryption algorithm, and result messages are sent and received in a secure manner.

(b) ***Firewall Technology***

In a network, a firewall is a point in which the incoming and outgoing data packet are filters according to the rule which is assigned already in firewall, this device in network control and monitor the incoming and outgoing traffics and decide whether to reject or allow the traffic either inside the network or to outside of network.

(c) ***Avoid Accessing Public Wi-Fi***

Many places in which the people have more congestion there like hotels, airport, universities, hospitals, cafes and other public places that offer free Wi-Fi, using such a Wi-Fi makes chance of attack by attacker, attacker hack the system and gain access to sensitive data that are sent and received by us or even to the data stored data in our device and on complete system.

(d) ***Protecting Confidentiality***

Protecting the confidentiality is a security solution in which make it difficult for attacker to find the wireless signal and being intercepted by attacker, so that, it causes to reduce the risk of eavesdropping the wireless transmission.

(e) ***Reduce the Strength of Access Point***

By reducing the signal strength of wireless, it cannot guarantee the security because, it still provides the required coverage for users, but at least can minimize the chance for attacker to access wireless LAN.

(f) ***Steps to Reduce Risk of DoS Attack***

Daily routine audits on wireless network's performances and activities can easily identify the problems.

Doing wireless penetration testing using such tools like Aircrack-ng, Airmon-ng, Aireplay-ng and Airodump-ng.

Using demilitarized zone (DMZ) in our network in order to improve the network security by segregating devices on each side of firewall can be considered as a best option for administrator to take care in a network.

Inside the network, a DMZ plays a role that separates the network in two parts by keeping one or more devices inside the firewall and other outside of the firewall, placing DMZ in network between external firewall and routers cause to protect the internal devices from possible outside attack, and result the LAN is secure.

Using virtual local area network is another technique in a network to create desire security policies. The corresponding security policies decide whether the incoming frames are allowed to enter inside the corporate network or not, resulting the corporate data separated from the public data and services.

I strongly advise you to have IDS software and hardware inside the network in this purpose to detect types of malicious behavior that can threat the security and faith of computer systems.

8 Conclusion

Nowadays, wireless network is changed to the most famous technology over the world, from another hand the wireless network signal is not covered by any physical boundary, thus, network should be well protected from the dirty hands of hacker, and in this survey, first I give a brief overview of wireless network and then focus on three main security standards WEP, WPA and WPA2 with their security drawbacks and clarify the steps to be taken care during installing wireless access point.

References

1. Suroto S (2018) WLAN security: threats and countermeasures. Int J Inform Vis 2(4):232
2. Somaiya RC, Gonsai AM, Tanna RS (2016) WLAN security and efficiency issue based on encryption algorithm
3. Ma Y, Ning H (2018) The improvement of wireless LAN security authentication mechanism based on Kerberos. IEEE
4. Somaiya RC, WLAN security and efficiency issues based on encryption techniques. Int J Res Eng IT Social Sci 6(9):27–32
5. Saleh ME, Aly AA, Omara FA (2016) Data security using cryptography and steganography techniques. Int J Adv Comput Sci Appl 7(6):390–397
6. Khasawneh M, Kajman I, Alkhudaidy R (2014) A survey on wifi protocols. In: International conference on security in computer networks and distributed systems
7. Ning H, Ma Y (2018) The improvement of wireless LAN security authentication mechanism based on Kerberos. In: International conference on electronics technology, vol 06, no 04, pp 87–98
8. Chaitanya SBS, Kokil P (2013) Implementation of cryptographic algorithm for secure wireless communication. Int J Sci Res (IJSR) 5(6)
9. Dhiman D (2017) WLAN security issues and solutions. IOSR J Comput Eng (IOSR-JCE) 16(1):67–75
10. A. E., Wireless local area network security enhancement through penetration testing. Int J Comput Netw Commun Secur 4:114–129

11. Indrabi SJ, Saini N, Mohan M (2018) Secure data transmission based on combined effect of cryptography and steganography using vislible light spectrum. Int J Pure Appl Math 118(20):2851–2859
12. Islam MA (2017) Secure wireless text message transmission with the implementation of RSA cryptography algorithm. Int J Comput Netw Commun Secur 2
13. Bittau A, Lackey J, Handley M (2006) Security and privacy. IEEE Symposium 2006:15–40

Security Concerns at Various Network Phases Through Blockchain Technology

Anju Devi, Geetanjali Rathee, and Hemraj Saini

Abstract Among various recent techniques, IoT is considered as the most popular technique which connects various heterogeneous devices such as vehicles, smart phones, and tickets in order to automate their tasks according to the environment. In most of the IoT applications, the data management and entities control is done by a centralized authority where the security is considered to be a major issue. In order to secure the data from various centralized threats such as man-in-middle attack, single failure etc., it is a needed to propose a decentralized security scheme. Recently, Blockchain is considered as most secured decentralized network that ensures security and transparency while transmitting the data among entities. In this paper, we have discussed the need of blockchain security at different phases of network like data, users, and device by analyzing their different security metrics and key challenges.

Keywords Blockchain · Data security · IoT security · User security · Network security

1 Introduction

In order to reduce the human efforts, the technology changes the human lifestyle into more comfortable way by using various recent techniques such as augmenting reality, quantum computing, biometric, business intelligence, IoT, artificial intelligence, and deep learning, etc. Till now, all these technologies are authenticated by a centralized authority that can further damage the whole system. Among a variety of applications, nowadays, IoT is the most popular technology developed by Ashton in 1999 [1] with several applications such as healthcare, industrial, and smart city, however, the central framework that is authorized by a single person leads to various security and privacy concerns. Hence, security is needed in every phase such as as: device, data, and user. Blockchain is considered as best security technique to eliminate mentioned issues like security, privacy, transparency, traceability, and centralized, etc. The blockchain

A. Devi · G. Rathee (✉) · H. Saini
Department of Computer Science and Engineering, Jaypee University of Information Technology, Waknaghat, Solan, HP 173234, India

A. Choudhary et al. (eds.), *Applications of Artificial Intelligence and Machine Learning*, Lecture Notes in Electrical Engineering 778,
https://doi.org/10.1007/978-981-16-3067-5_45

technique is initially introduced by Satoshi Nakamoto in 2008 [2] that define the data as a block and broadcast it on to the network.

Each block of a blockchain is connected to the previous block through a hash value as shown in Fig. 1, where the first block is also known as the genesis block. As depicted in Fig. 1, every block consists of the data, hash value of the current block, nonce and hash value of the previous blocks. Each block is added into the existing blockchain after verified by the miners that will further store in chronological order. Once the blocks have been added into the blockchain, no one can alter any data without the permission of permanent ledger.

It is used in several applications like healthcare, smart contract, industrial, voting, transaction, and IoT, etc. It helps to reduce cost, improve security, transparency, efficiency and reduce the waiting time. There are two types of blockchain architectures as shown in Fig. 2: public and private. Private Blockchain is also known as a permission blockchain where the blocks are added after permission, however, a public blockchain is also known as permission less blockchain where without taking

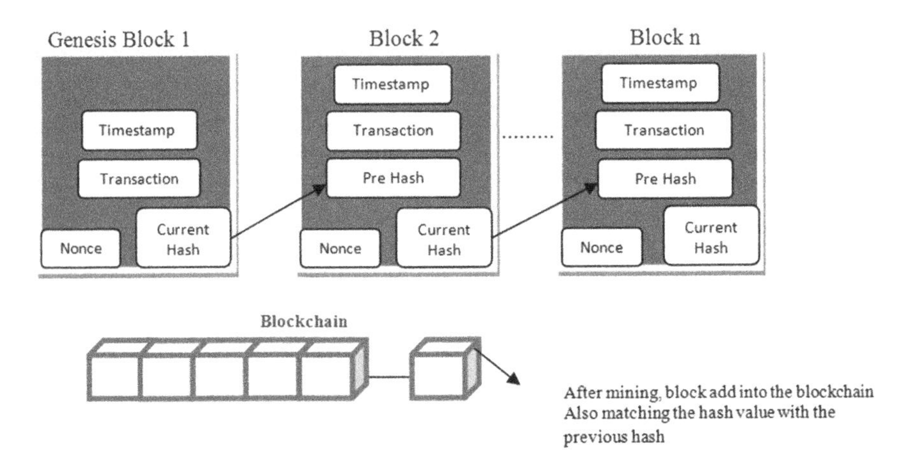

Fig. 1 Blockchain architecture

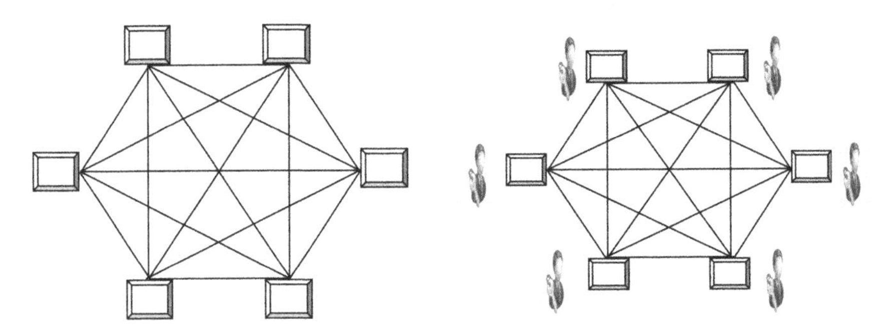

Fig. 2 Public and private blockchain

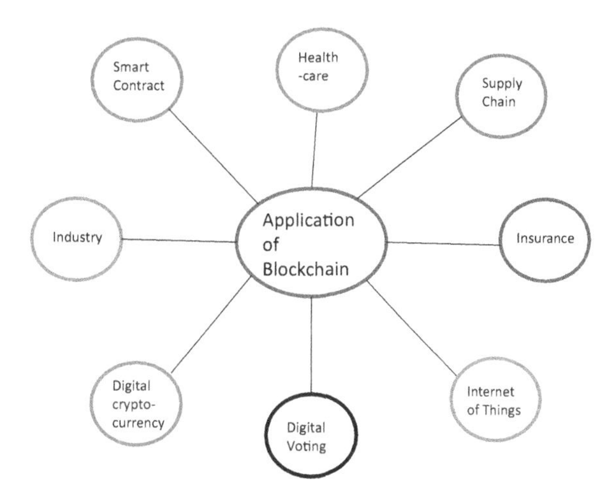

Fig. 3 Applications of blockchain

permission, one can join a network directly like Bitcoin, Ethereum, Lit coin, etc. The motivation of this paper is to determine the need of blockchain mechanism at different phases of network communication by analyzing its measuring parameters.

Recently, numbers of blockchain applications are becoming popular nowadays such as smart contracts, industrial, healthcare, as illustrated in Fig. 3 that increase the efficiency and security, enhanced traceability and transparency, and reduced the cost in the network.

1.1 Application of Blockchain

- **Smart Contracts**

A smart contract [3] is a computer code that can be built into the blockchain to facilitate and verify a contract agreement under a set of conditions that users agree to. When those conditions are met, the terms of the agreement are automatically carried out. However, in paper based contract, error possibility increases that requires the trust between involved parties.

- **Industrial applications**

Blockchain could be used as a marketplace to share or sell unused data. Since most enterprise data goes unused, blockchain could act as an intermediary to store and move this data to improve a host of industries [4, 5] and provides many industrial and business benefits.

- **Healthcare**

In healthcare, blockchain [6] is used to securely store their patients' medical record (signed) which provides patients with the confidence that the record can't be changed.

These personal health records could be encoded and stored on the blockchain with a private key that will be only accessible by authorized persons.

- **Digital crypto currency transfers**

Blockchain technique has the opportunity to transfer the funds [7] from one party to another more quickly and securely within some minutes and seconds. There is no need to wait for a longer period of time. In addition, the centre person (bank) is not involved within some miner charges. Further, within some minutes, funds may be easily transferred in one country to another, international or all worlds with no extra charges.

- **Digital voting**

The fraud voting chances always persist through disruptive persons etc. Voting with blockchain [8] carries the possibility to eliminate the election fraud. Each vote would be stored as a block on the blockchain by making them nearly not possible to alter. The blockchain protocol would also maintain transparency by reducing the personnel queries needed to conduct an election and provide officials with direct results.

- **Managing Internet of Things networks**

Blockchain-based applications monitor Internet of Things (IoT) [9, 10] networks and connected devices through wirelessly that can send and receive the data. Such an application could determine the trustworthiness of devices on a network such as smart cars or smart phones etc. The human life style is completely changed to automate their tasks within surrounding and reduces human efforts.

- **Insurance marketplace**

It can be used to support the insurance [11] marketplace transactions between different customers, policy holder and insurance companies. Blockchain can be used to buy, register insurance policies, support reinsurance activities etc. among insurance companies.

- **Supply chain**

Suppliers can use blockchain to record the origins of materials that they have purchased. This would allow companies to verify the authenticity of their products. Supplier send their product to customer with originality because of transparency, immutability, decentralized, traceability where the customer and the supplier trust will be increased. Nowadays, IBM, provenance, ambrosias, ever ledger and block verify are used for supply chain product approaches.

The organization of this paper divided into four sections. Section II elaborates the related work of security at different levels of a network through blockchain and also discussed the security issues on traditional systems and how to overcome to these issues through the blockchain. Section III briefly describes the blockchain parametric measures at the device, at data, and the application level. At the last, section IV deliberates the conclusion on different levels by using blockchain.

2 Related Work

Blockchain technique can be used in any phase of network communication by providing different levels of a security issue. This section illustrated the usage of blockchain technique in various phases such as at device phase, user phase and data phase.

Table 1 depicts a literature survey on blockchain in different applications on various platforms used by authors.

Banerjee et al. [3] have discussed various security issues in IoT technique and their resolution through blockchain technique by using reactive and proactive approach IoT datasets. These approaches are further categorized into **a** Collaborative approach and **b** Intrusion detection systems (IDS) and intrusion prevention systems (IPS) approach. Dongxiao et al. [4] have proposed a technique using Anonymous Reputation System atop a Proof-of-Stake approach (ARS-PS) for the application of the industrial internet of things (IIoT). In this paper blockchain mechanism is used in order to increase the transparency, privacy, and reliability of the system, however efficiency still remains an issue for the system. In addition, Hammi et al. [5] have discussed IoT, data exchanged through a centralized authority that connects to the environment. The authors have used blockchain as a decentralized system called bubbles of the trust that ensures the authentication of the devices, efficiency, low cost, identity and trust to each other. Danzi et al. [6] have proposed a lightweight mechanism within the blockchain to reduce the communication cost, delay and improve privacy, etc. They have provided some guideline to avoid unnecessary information of the storage by compressing only a portion of data. Further, Yao sun et al. [7] used optimal full function node within blockchain-enabled technology to increase the throughput and performance of the system. Angel et al. [8] have proposed zero-based knowledge protocol used with blockchain for IoT application. In this paper to eliminate the authentication and integrity of device, reduce data manipulation and counterfeit. There are still some issues present like system accuracy, robustness, and DoS attack. *The following* Table 2 *gives a summary of all the approaches proposed traditionally and through blockchain technique.*

Further, Wang et al. [9], have proposed em (expectation–maximization) algorithm and encryption method where if a server is not sensing any data and paying users then it may cheat for minimizing the cost by paying less even nothing and reduce the enthusiasm of users. *The depicted* Table 3 *gives a summary of all the approaches proposed traditionally and through blockchain technique.*

Once the server is captured by adversaries then the user's information can be disclosed and the system may collapse. User's authentication and privacy is a problem to prevent these limitations by writing a survey with a broad literature review on the blockchain 'users'. Minoli et al. [10] have deployed an IoT mechanism where large number of attacks occured due to security issues. Blockchain Mechanism (BCM) plays an important role in securing IoT applications and works at both layer applications as well as a lower layer of the communication model. Further, Huh et al. [11] have proposed blockchain integrated automated log-in-fingerprint verification

Table 1 Blockchain used at device phase

S. No.	Traditional approach	Blockchain approach
1	Centralized authority generate the PKG which can be compromised	Secure ICN to authenticate the user, eliminate centralized issue by using the HIBC approach
2	Centralized control and access devices, security not transparent and scalability limitations	Using the data centric approach for IoT devices to achieve trust, latency, scalability and data manage
3	Deices handle by third person can be changed the original data or product	Using block verifier and ever ledger approach for supply chain application tracking data and devices both
4	Not preserve the privacy of medical record, information is not secure	Using the MA-ABS approach based on blockchain to secure the records, immutability of data guarantee
5	Unable to detect malicious devices, higher communication cost, etc.	Using the malicious detection classification and algorithm to find out malicious nodes, lower cost, low time etc.
6	Privacy and traceability of data is still issue	Improves the anonymity, privacy, user's transaction and activity continuously traced
7	Security issue of information	Using anti-quantum blockchain technology integrity of data and digitally signed with private key
8	Manage and control of device by centralized, Latency and scalability still issue	Identity and Attribute management approach improve the robustness of user identity against fraud user's and securing millions of device
9	Device security issue, not able to detect the malicious nodes etc.	Avoid the collusion attack, improve the security of the system, lower resource consumption, higher operating efficiency etc.
10	Latency, Security, Stability of device problem	Reducing the block delivery time b/w the node in protecting the stability, low 51% attack success probability
11	Users privacy violate, security breach, can't protect vehicles from malicious etc.	Protect transaction information from malicious vehicles, privacy of the system as well as users
12	Malicious devices can damage all the system because key is managed by a single authority, Security and privacy issue present	Using blockchain based pseudonym management scheme to protect the system, vehicle's identity, location, privacy and security, high transaction speed etc.

(continued)

Table 1 (continued)

S. No.	Traditional approach	Blockchain approach
13		
14	Limitation of storage, quality, not guarantee of system security, can compromise device etc.	By using the Raspberry Pi and credit based pow mechanism to guarantee the system security, transaction and data access control in IIoT
15	Higher cost, not traceable and transparent, replace original product etc.	Using Ambrosus and Modum platform scheme for supply chain helps in transparency of products, reliable and secure system, increasing trust between consumer and producer, reducing, eliminating counterfeit products etc.
16	IoT devices not secure, immutability of information not guarantee etc.	In IoT environment using Light-weight authentication protocol scheme must be secure and lightweight, authenticate, integrity and non- repudiation function based on blockchain IoT

Table 2 Blockchain used at device phase

S. No.	Traditional approach	Blockchain approach
1	Lack secure of IoT datasets	Blockchain to ensure the integrity of IoT datasets and systems [3]
2	Single trusted marketplace, transparency of product and system limitations	To increase system transparency, reliability, privacy guarantee and also build the customers trust [4]
3	Process and exchange of data through a centralized authority	Decentralized system [5] called bubbles of trust, which ensures a robust identification and authentication of devices
4	Security, high communication cost, etc. are limitations	To reduce the communication cost, delay and improve privacy, etc. [6]
5	Denial of Service attack, throughput not high	To increase the throughput and performance of the system [7]
6	Data management, integrity, and authenticity of the device are not secure	To increase the authentication and integrity of the device, etc. [8]

platform to authenticate the users, tighten the security against tampering of user's information by the hackers. Pustisek et al. [12] have proposed a blockchain technique with an IoT application to improve the security and reduce the traffic through a viable approach. In Tang et al. [13] where IoT is growing rapidly to build the trust among platforms, the devices are controlled through a centralized approach. To eliminate these issues, blockchain technology is used to control the IoT environments by using IoT passport decentralized trusted platform. In addition, Rahulamathavan et al. [14] have proposed an IoT security by ensuring privacy, non-repudiation, integrity, and the confidentiality of data, users, and devices. In this paper to overcome these issues,

Table 3 Blockchain used at device phase

S. No.	Traditional approach	Blockchain approach
1	Disclosed the user's private information	Increase trust between users, minimize cost, and protect user information [9]
2	he deployment of the Internet of Things (IoT) results in an enlarged attack surf	IoT environments [10] BCMs play an important role, IoT Security solution and work at the lower layer of the communications models as well as at the application layer
3	Users authentication is a major issue in technology, easily open a lock by electric shock or stolen PIN, etc.	Tight security against tampering of user's information by hackers [11]
4	Low-power, low-bitrate mobile technologies	To reduce network traffic and enhance security [12]
5	Centralized approaches have been used to build federated trust among platforms and devices, but limit diversity and scalability	Propose a decentralized trust framework; IoT Passport is motivated by the familiar use of passports for international travel but with greater dynamism [13]
6	Integrity and non-repudiation to some extent, confidentiality, and privacy of the data or the devices are not preserved	IoT applications based on attribute-based encryption (ABE) techniques, blockchain is that the data generated by users or devices in the past are verified for correctness and cannot be tampered once it is updated on the blockchain [14]

the authors have used an attribute-based encryption (ABE) technique in which data generated by the users are verified that can't be tampered after updating it. Table 4 *gives a summary of all the approaches proposed traditionally and through blockchain technique.*

Further, Dwivedi et al. [15] have discussed security and privacy of medical data as a problem because of the delay in treatment and endangering the life of a patient. Then, the blockchain is used to secure and analyze the healthcare data by using a novel hybrid approach to accesses the control of medical data by reducing the modification of data. Moreover, Bhalaji et al. [16] have proposed IoT real-time modifications of information through intruders that may lead to a harmful impact on one's life. The solution to this problem is blockchain because a decentralized ledger is used to record the transaction and also prevents the modification of the data. In Chanson et al. [17] where the main issue was to ensure the security and privacy of data in the IoT technique. By using blockchain-based sensor data protection system (SDPS), the authors have solved the limitation to ensure the processing and exchanging of data preserved through privacy, efficient and scalable manner. Alphand et al. [18] have proposed a secure IoT resources single ACE authorization server to provide blockchain-based IoT Chain OSCAR and ACE architecture and to secure data access to provide a flexible and trusted way to handle data. Reyna et al. [19] proposed an IoT trust as a

Table 4 Blockchain used at device phase

S. No.	Traditional approach	Blockchain approach
1	Security and privacy problems of medical data	Blockchain is proposed to [15] provide secure, manage, and analysis of healthcare big data
2	Any modifications caused to this health-related data by the intruders may lead to a harmful impact on one's life	Blockchain [16] can be defined as a decentralized digital ledger that can be used to record transactions and at the same time prevents modifications to the recorded data
3	Issues of data security arising in the IoT	Blockchain-based [17] sensor data protection system (SDPS)
4	The single ACE authorization server	To provide an E2E solution for the secure authorized access to IoT resources, flexible and trustless way to handle authorization [18]
5	IoT trust a single authority in the security domain e.g., data reliability	Trust in distributed environments without the need for authorities [19]
6	A large amount of data is gathered from various IoT devices data integration and privacy is still a problem	A secure and reliable data sharing platform among multiple data providers, propose secureSVM, which is a privacy-preserving SVM training scheme over blockchain-based encrypted IoT data [20]
7	Authorization is a crucial security component of many distributed systems handling sensitive data or actions, including IoT systems	A novel mechanism for protecting the secrecy of resources on the public blockchain, without out-of-band channels or interaction between granters, provers or verifiers. WAVE is efficient enough to support city-scale federation with millions of participants and permission policies [21]

central authority where the security still leads to a challenge. To overcome the limitation of IoT as new technology, the blockchain technique is emerged and builds a trust in distributed environments without the involvement of the third party. Further, Shen et al. [20] have proposed a large amount of data that is transferred daily in real-life from various IoT devices but the integration of data and privacy is nowadays an issue. To make a reliable, secure and privacy-preserving information transfer among different data providers, the authors have proposed a secure Support Vector Machine scheme using blockchain. Moreover, Andersen et al. [21] have proposed a public blockchain for protecting the secrecy of the data without interaction between verifier. WAVE is a fully decentralized authorization system that is efficient to support city-scale with millions of participants.

3 Parametric Measures in Blockchain

This section discusses the number of parameters required to analyze the security through blockchain at different phases of network communication.

3.1 At Device Phase

Decentralized, time, throughput and scalable are the parameters used to determine the device metrics.

- **Decentralized**: Without depending upon the centralized authority, the decentralized mechanisms are used to control their devices and resources.
- **Performance**: The performance can be measured with the lightweight technique such as XML language and JSON lightweight technology for access control whereas JSON capability small overhead and network performance.
- **Times (sec) and no. of transaction**: When the number of transactions increases, the service time also increases. The Lightweight client service takes the minimum time to enhance the efficiency, minimize the delay; and to reduce the minimum execution time.
- **Throughput**: R is denoted as communication throughput that is calculated using CT transaction throughput, and L packet length as:

$$R \geq L(CT) \quad (1)$$

- **Scalable**: The Blend scheme defines a specific access authorization policy that is dynamic and heterogeneous.

3.2 At User Phase

Authentication and security metrics are the major parameters to analyze legitimacy of authentic user.

Authentication: Measurement of the authentication procedure lasts a time interval which is less than 50 ms on an average whereas testing time is considered negligible concerning the network times.

Security and Privacy: The attributes of multiple AA (attribute authorities) increase the security and slightly increases time complexity. Privacy guarantees during the verification as well as in the blockchain.

3.3 At Data Phase

- **Fast, inexpensive and reliable Transaction**: There are many different types of blockchain implementation and small networks that are often vulnerable to 51% attacks. Ethereum is used every hour where 10,000 transactions are sent which are difficult for an attacker to control over 51% of the network less expensive compared to Bitcoin-based Neuro Mesh.
- **Security and performance**: The data are stored on the cloud after encrypting by the sensor and when sharing the data then re-encrypt ensure to secure the transfer data to the users.

4 Conclusion

In this paper, a blockchain mechanism is discussed to overcome the centralized issues in different phases of IoT like data, devices, and applications. Further, the authors have mentioned on the tables how traditional IoT application problems are solved by the blockchain through various approaches. In different phases of devices, data and users, the measuring parameters through blockchain are discussed in this paper. The discussion of user's security concerns and its solution through blockchain will be communicated in future.

References

1. Ashton K (2009) That 'internet of things' thing. In: RFID
2. Nakamoto S (2008) Bitcoin: a peer-to-peer electronic cash system
3. Banerjee M, Lee J, Choo K-KR (2017) A blockchain future to Internet of Things security: a position paper. In: Digital communications and networks. https://doi.org/10.1016/j.dcan.2017.10.006
4. Liu D, Alahmadi A, Ni J, Lin X, Sherman Shen X (2019) Anonymous reputation system for IIoT—enabled retail marketing atop PoS Blockchain. IEEE Trans Indus Inform 15(6):3527–3537. https://ieeexplore.ieee.org/abstract/document/8640264.
5. Hammi MT, Hammi B, Bellot P, Serhrouchni A (2018) Bubbles of trust: a decentralized blockchain-based authentication system for IoT. Comput Secur. https://doi.org/10.1016/j.cose.2018.06.004
6. Danzi P, Kalør AE, Stefanovic C, Popovski P (2019) Delay and communication tradeoffs for blockchain systems with lightweight IoT clients. IEEE Internet Things J 6(2):2354–2365. https://ieeexplore.ieee.org/abstract/document/8671694/
7. Sun Y, Zhang L, Feng G, Yang B, Cao B, Imran MA (2019) Blockchain-enabled wireless internet of things: performance analysis and optimal communication node deployment. IEEE Internet Things J 6(3):5791–5802. https://ieeexplore.ieee.org/abstract/document/8668426/
8. Prada-Delgado MÁ, Baturone I, Dittmann G, Jelitto J, Kind A (2019) PUF-derived IoT identities in a zero-knowledge protocol for blockchain. https://doi.org/10.1016/j.iot.2019.100057
9. Wang J (2018) A blockchain based privacy-preserving incentive mechanism in crowdsensing applications. IEEE Access 6:17545–17556

10. Minoli D, Occhiogrosso B (2018) Blockchain mechanisms for IoT security. https://doi.org/10.1016/j.iot.05.002-2542-6605
11. Huh J, Seo K (2018) Blockchain-based mobile fingerprint verification and automatic log-in platform for future computing. https://doi.org/10.1007/s11227-018-2496-1
12. Pustišek M, Kos A (2018) Approaches to front-end IoT application development for the ethereum blockchain. In: International conference on identification, information and knowledge in the internet of things, Elsevier. https://doi.org/10.1016/j.procs.2018.03.017
13. Tang B, Kang H, Fan J, Li Q, Sandhu R (2019) IoT passport: a blockchain-based trust framework for collaborative internet-of-things. In: SACMAT '19, Toronto, ON, Canada ACM, June 3–6 2019. ISBN 978-1-4503-6753-0/19/06. https://doi.org/10.1145/3322431.3326327
14. Rahulamathavan Y, Phan RC-W, Rajarajan M, Misra S, Kondoz A (2016) Privacy-preserving blockchain based iot ecosystem using attribute-based encryption. In: UK-India Education Research Initiative (UKIERI)
15. Dwivedi AD, Srivastava G, Dhar S, Singh R (2019) A decentralized privacy-preserving healthcare blockchain for IoT. https://doi.org/10.3390/s19020326
16. Bhalaji N, Abilashkumar PC, Aboorva S (2020) A blockchain based approach for privacy preservation in healthcare IoT. pp 465–473. https://doi.org/10.1007/978-981-13-8461-5_52
17. Chanson M, Bogner A, Bilgeri D, Fleisch E, Wortmann F (2019) Blockchain for the IoT: privacy-preserving protection of sensor data. J Assoc Inform Syst 20(9):1271–1307. https://doi.org/10.17705/1jais.00567,ISSN1536-9323
18. Alphandy O, Amoretti M, Claeysy T, Asta SD, Duday A, Ferrari G, Rousseauy F, Tourancheauy B, Veltri L, Zanichelli F (2018) IoT chain: a blockchain security architecture for the internet of things. In: 2018 IEEE Wireless communications and networking conference (WCNC), 978-1-5386-1734-2/18
19. Reyna A, Martín C, Chen J, Soler E, Díaz M (2018) On blockchain and its integration with IoT challenges and opportunities, 0167-739X. https://doi.org/10.1016/j.future.2018.05.046
20. Shen M, Tang X, Zhu L, Du X, Guizani M (2019) Privacy-preserving support vector machine training over blockchain-based encrypted IoT data in smart cities. In: http://www.ieee.org/publications_standards/publications/rights/index.html. https://doi.org/10.1109/JIOT.2019.2901840
21. Andersen MP, Kolb J, Chen K, Fierro G, Culler DE, Popa RA (2017) WAVE: a decentralized authorization system for IoT via blockchain smart contracts. In: UCB/EECS-2017-234. http://www2.eecs.berkeley.edu/Pubs/TechRpts/2017/EECS-2017-234.html

Smart Infrastructure and Resource Development and Management Using Artificial Intelligence and Machine learning

Developing an Evaluation Model for Forecasting of Real Estate Prices

Ruchi Mittal, Praveen Kumar, Amit Mittal, and Varun Malik

Abstract Real estate prices are an important indicator of the economic health of a region. The real estate industry is also growing at a very fast pace and needs the confluence of technology to provide knowledge-enabled services. We tried to explore the drivers of real estate housing. The present investigation is being conducted using a sample of 414 unique UCI datasets on real estate pricing. The OLS multivariate is being performed with the help of control variables such as house cost, age, MRT station, number of accommodation stores, walking and geographic directions. The article provides evidence that almost all the control variables are perceived to be crucial in predicting house price. However, the article did not provide any evidence to support that the geographic coordinate (in longitude) influence sample houses' price. We argue the house cost is the most important determinate of real estate housing. Besides, ages, MRT station, the numbers of accommodation stores also help to improve the price of real estate housing. The finding of this study will provide investors and other stakeholders with important implications of real estate housing pricing in the best interests of capital appreciations.

Keywords Real estate · UCI · OLS multivariate · Regression model

R. Mittal · V. Malik
Chitkara University Institute of Engineering and Technology, Chitkara University, Rajpura, Punjab, India
e-mail: ruchi.mittal@chitkara.edu.in

V. Malik
e-mail: varun.malik@chitkara.edu.in

P. Kumar (✉) · A. Mittal
Chitkara Business School, Chitkara University, Rajpura, Punjab, India
e-mail: praveen.kumar@chitkara.edu.in

A. Mittal
e-mail: amit.mittal@chitkara.edu.in

A. Choudhary et al. (eds.), *Applications of Artificial Intelligence and Machine Learning*,
Lecture Notes in Electrical Engineering 778,
https://doi.org/10.1007/978-981-16-3067-5_46

1 Introduction

Real estate is the most lucrative backing in terms of its returns. Property valuation and determining the important attributes to evaluate real estate pricing play a key role to meet the ever-increasing demand and supply chain in this context [1]. With the use of technology in property valuation, it is now possible to keep track of the transactions related to buy, sell, rent, and loan and to analyze these transactions to design predictive models for the estimation for pricing and to appraise e-services [2–5].

The cities' management and sustainability become a major preoccupation at the global level. It often reflects the growth of the nation and economy. It is essential to estimate the price of a house based on certain critical parameters to make the existence of our globe certain. However, due to globalization and the emergence of new investment instruments (such as securities), real estate management has come to be disregarded. This situation is made worse due to the high fluctuation in real estate prices [6].

Housing typically represents a household's largest investment asset. Real estate prices also reflect the general health of the national and local economy [7]. Commercial, agricultural and residential properties represent a significant part of national assets [8]. As a result, the land business gets one of the significant enterprises around the world, whose running and advancement has an immediate relationship with the national economy and individual's life [9].

So with this specific background, we tried to get some fruitful insights into how real estate prices are determined. The present investigation is being conducted using a sample of 414unique UCI datasets on real estate pricing. We run an OLS multivariate relapse examinations with control factors, for example, house cost, age, MRT station, number of accommodation stores, walking and geographic directions. The article provides evidence that almost all the control variables are perceived to be crucial in predicting house price.

The article will proceed as follows: Related Work section covered the recent studies on the pricing of the house; the Research Methodology section will explain the procedure followed to test our hypotheses. Our empirical findings are discussed in the Data Analysis section, and the Conclusion section has contained a few concluding remarks.

2 Related Work

The real estate industry is also growing at a very fast pace and needs the confluence of technology to provide knowledge-enabled services. The data mining tools and techniques are widely used by the researchers and data analysts to analyze the data and predict the trends to enable fact-based decision making in every possible domain.

In this context, many researchers have adopted the use of technology and different techniques for data mining and machine learning for real estate valuation.

The authors in their study provide an understanding on the ways of estimating land price by addressing the issues associated with the price prediction of real estate and explains the benefits of ANN model over other models to overcome the difficulties of estimating the property prices [10]. Authors in [11] have presented a novel approach, validating better responses over the other approaches such as neural network or regression analysis for real estate valuation. The authors, in a separate study, aimed to predict the residential property price and concluded that a bagging ensemble model with genetic NN and fuzzy systems work better in comparison with other approaches [12].

Authors have presented a way to rank the mixed land based on a geographical function considering different types of properties that are close to each other, and their results indicated an improvement in the estimation of the real estate price [13]. Authors emphasized on the size of training data to have improved predictive accuracy in estimating the property price and used primary data to develop different predictive models for real estate valuation and performed statistical comparisons to check for the accuracy [14].

In a different study, authors used various data mining tools such as WEKA, RapidMiner, and KEEL for real estate price prediction and applied machine learning approaches based on regression, decision tree, and SVM to real data and compared these on various performance parameters. Authors revealed the fact that the machine learning algorithms gave almost similar results with different data mining tools; however, some significant dissimilarities were outlined while performing analysis with KEEL and RapidMiner [15]. Authors outlined the significance of applying data mining techniques in real estate valuation and to make predictions on the residential pricing. They have used primary data consisting of 506 records and analyzed it using artificial neural network giving an accuracy of 96% and validated the results using regression [16].

This study is based on the application of data mining techniques for property pricing on the real transactional database. The authors have used tree-based and network-based approaches for the analysis and modeling, and these models were checked for predictive accuracy and concluded that neural network makes better predictions as compared to the tree-based approach [17]. The authors in a different study have done a similar analysis based on various parameters and explained the potential of applying data mining functionalities for prediction and recommendation in the area of real estate. In their results, tree-based techniques have given better results with minimum factors as compared to neural network [18].

The authors have used the primary factors about real estate to predict residential pricing. They have made use of different regression techniques and presented the result as a combination of this exhibiting highest performance instead of depending upon just one best technique for the analysis. They have also developed a predictive model for residential pricing to benefit the customers of real estate [19]. They have identified the need for applying data mining techniques such as an artificial neural network for residential and commercial properties and exploiting its benefits to overcome the varied pricing in this sector. They have also emphasized on the accuracy of the results that can be achieved with the application of ANN in real estate price prediction [20].

Intending to estimate the real estate pricing in Taiwan, authors have used and evaluated various data mining functionalities such as ANN and SVM considering different factors associated with the area and presented the results indicating that the support vector regression outperforms the other technique. Authors have also outlined various important factors in predicting the property price in Taiwan efficiently [21].

Based on the above discussions, it is, therefore, hypothesized that:

H1: Age and house costs are adversely linked.

H2: Distance to the nearest MRT and house costs are adversely linked.

H3: The number of accommodation stores and house costs is positively linked.

H4: Geographic directions (in scope) and house costs are positively linked.

H5: Geographic directions (in longitude) and house costs are positively linked.

3 Research Methodology

3.1 Data Description

The present investigation utilized a sample of 414 unique UCI datasets on real estate pricing. We run an OLS multivariate relapse examinations with control factors, for example, house cost, age, MRT station, number of accommodation stores, walking and geographic directions.

$$\begin{aligned} \mathrm{HOUSPRC_t} = {} & \alpha 0\mathrm{INTERCEPTt} +_{\alpha 1} \mathrm{AGE_t} +_{\alpha 2} \mathrm{DISTANCE_t} +_{\alpha 7} \mathrm{CONVSTORE_t} \\ & +_{\alpha 8} \mathrm{GEOGLATITUDE_t} +_{\alpha 8} \mathrm{GEOGLONGITUDE_t} + \varepsilon_t \end{aligned} \tag{1}$$

3.2 Data Analysis Process

The process of analysis and knowledge discovery from data is explained in Fig. 1.

Initially, data is selected and extracted from the UCI repository, and finally, after following the intermediatory steps, regression analysis has been done to achieve the final result in the form of a real estate housing price predictive model.

Table 1 Description of variables understudy

Variables	Expected sign	Descriptions
$HOUSPRC_t$	Explanatory variables	House cost
AGE	−	Age of the property
DISTANCE	−	The distance to the nearest MRT station
CONVSTORE	+	Number of accommodation stores
GEOGLATITUDE	+	Walking directions, latitude
GEOGLONGITUDE	+	Geographic directions, longitude
ε_t	?	Error term

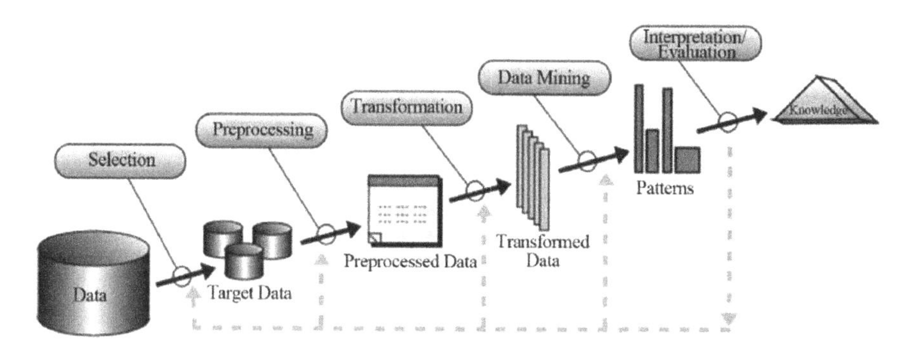

Fig. 1 KDD steps for knowledge extraction from data

4 Data Analysis

4.1 Descriptive Statistics

Descriptive measurements of the sample properties are depicted in Table 2. The mean of HOUSPRC is 37.98% which depicted that majority of sample houses are having a decent price in the market. The most extreme and least estimations of HOUSPRC are 117.50 and 07.60%. These insights depicted that sample comprises of certain houses with excellent market value and some with low market value.

Table 2 Descriptive insights

	N	Min	Max	Mean	SD
HOUSPRC	414	07.60	117.50	37.98	13.60
AGE	414	0.0000	43.80	17.71	11.39
DISTANCE	414	23.38	06.48	1083.89	1.26
CONVSTORE	414	0	10.00	04.09	2.94
GEOGLATITUDE	414	01.39	1.39	01.39	0.0000

The mean value of AGE is 17.71, which indicates that the sample houses are old properties. The minimum estimation of AGE is 0 which delineates that a few houses are new and built recently. Besides, the maximum value is 43.80 which portrayed that a few houses are very old and has a long corporate history. The descriptive measurements additionally accommodated the standard deviation of ESG is 11.39 which signifies the low scattering in sample properties.

The mean of DISTANCE is 1083.89 which shows that the majority of sample houses are far from Metro stations. The mean of the CONVSTORE depicted in Table 2 is 04.09, which is moderate and depicts that the sample houses have easy accessibility to the store.

4.2 Correlation Matrix

4.3 Real Estate Valuation Model

Further, before applying regression, the multicollinearity among independent variables must be checked [21]. A different parameter to handle multicollinearity is being suggested in previous literature. Moreover, the coefficients of connection underneath 0.9 may not cause genuine multicollinearity issues [22], while few researchers recommended the value underneath 0.8 shows no extreme multicollinearity [23]. Table 3 portrayed that the correlation among all variables are below this value.

Table 4 shows the results of real estate valuation model. It is clear from the table that the house's AGE is negatively and statistically significantly linked to the HOUSPRC at the 1% level and 5% level of significance (p-value < 0.01 and p-value < 0.05). These statistics confirmed our H1; it means that older the house the lower will

Table 3 Correlation grid

	1	2	3	4	4	6
HOUSPRC	1	−0.211	−0.674	0.571	0.546	0.523
		0.000	0.000	0.000	0.000	0.000
AGE		1	0.026	0.050	0.054	−0.049
			0.603	0.314	0.269	0.325
DISTANCE			1	−0.603	−0.591	−0.806
				0.000	0.000	0.000
CONVSTORE				1	0.444	0.449
					0.000	0.000
GEOGLATITUDE					1	0.413
						0.000
GEOGLONGITUDE						1

Table 4 The real estate valuation model

Model	Unstandardized coefficients		Standardized coefficients		
	B	Std. Error	Beta	t	Sig
INTERCEPT	−14,506.699	29,333.762		−0.495	0.621
AGE	−0.269	0.039	−0.225	−6.896	0.000
DISTANCE	−0.004	0.001	−0.395	−5.887	0.000
CONVSTORE	1.163	0.190	0.252	6.114	0.000
GEOGLATITUDE	13,668.470	2584.381	0.217	5.289	0.000
GEOGLONGITUDE	−2183.074	13,754.309	−0.009	−0.159	0.874
Adjusted R Square	0.566				

be the price. Moreover, the outcomes of OLS multivariate regression analyses show that the coefficient for the organizations' DISTANCE is negatively and statistically significantly connected with HOUSPRC at the 1% level and 5% level of significance (p-value < 0.01 and p-value < 0.05). These measurements indicate that a larger distance of a property from Metro station will reduce its value in the market (H2).

Further, inconsistent with our H3, we also found that easy availability to convenience stores and a higher number of convenience stores Metro station will increase the value of a property in the market (p-value < 0.01 and p-value < 0.05). In constant with our H4, we also found that the geographic coordinate (in latitude) also statistically significantly influences sample houses' price. However, contrary to our H5, this analysis didn't find any proof to support that the geographic coordinate (in longitude) influences sample houses' price. Moreover, the value of adjusted R Square is 0.566, which is very good and shows that input variables in the model are jointly able to explain 56.60% variation in the dependent variable housing price.

5 Conclusion

This paper examined the determinants of real estate housing. The present investigation is being conducted using a sample of 414 unique UCI datasets on real estate pricing. We run an OLS multivariate relapse examinations with control factors, for example, house cost, age, MRT station, number of accommodation stores, walking and geographic directions. The article provides evidence that almost all the control variables are perceived to be crucial in predicting house price. The prices of real estate are also fluctuating day by day; consequently, the results of this analysis showed that house pricing becomes a major preoccupation at a global level. So proper care must be taken while predicting house pricing. Finally, real estate prices also reflect the general health of the national and local economy.

6 Limitations and Future Research Directions

This research shall provide significant inputs to investors and shall help them determine real estate housing prices in a given area. Though the findings of this study are quite comprehensive, the historical use of book-based criteria to measure housing prices can inhibit the applicability of the results. Therefore, another study can be conducted using current and market-based criteria for housing prices and measuring housing as an investment. Moreover, the present analysis covered only UCI dataset on real estate pricing. Therefore, further research can acquire a more in-depth understanding of the issues brought up in this examination using another dataset.

References

1. Arnold AL, Miles ME, Wurtzebach CH (1980) Modern real estate. Warren, Gorham & Lamont
2. Bahia ISH (2013) A data mining model by using ANN for predicting real estate market: comparative study. Int J Intell Sci 3(04):162
3. Bansal G, Sinha AP, Zhao H (2008) Tuning data mining methods for cost-sensitive regression: a study in loan charge-off forecasting. J Manag Inf Syst 25(3):315–336
4. Buist H, Yang TT (2000) Housing finance in a stochastic economy: Contract pricing and choice. Real Estate Econ 28(1):117–139
5. Demetriou D (2017) A spatially based artificial neural network mass valuation model for land consolidation. Environ Plann B: Urb Anal City Sci 44(5):864–883
6. Fan GZ, Ong SE, Koh HC (2006) Determinants of house price: a decision tree approach. Urb Stud 43(12):2301–3231
7. Fu Y, Liu G, Papadimitriou S, Xiong H, Ge Y, Zhu H, Zhu C (2015, August) Real estate ranking via mixed land-use latent models. In: Proceedings of the 21th ACM SIGKDD International Conference on Knowledge Discovery and Data Mining, pp 299–308
8. Gao L, Guo Z, Zhang H, Xu X, Shen HT (2017) Video captioning with attention-based LSTM and semantic consistency. IEEE Trans Multimed 19(9):2045–2055
9. Gan V, Agarwal V, Kim B (2015) Data mining analysis and predictions of real estate prices. Iss Inform Syst 16(4)
10. Graczyk M, Lasota T, Trawiński B (2009, October) Comparative analysis of premises valuation models using KEEL, RapidMiner, and WEKA. In: International conference on computational collective intelligence, pp 800–812. Springer, Berlin
11. Gujarati DN, Porter D (2009) Basic econometrics. Mc Graw-Hill International Edition
12. Haila A (2000) Real estate in global cities: Singapore and Hong Kong as property states. Urb Stud 37(12):2241–2256
13. Hair JF, Black WC, Babin BJ, Anderson RE, Tatham RL (1998) Multivariate data analysis. Prentice Hall, Upper Saddle River, vol 5, no 3, pp 207–219
14. Jaen RD (2002, May) Data mining: an empirical application in real estate valuation. In: FLAIRS conference, pp 314–317
15. Ke XL, Diao FQ, Zhu KJ (2011) A real option model suitable for real estate project investment decision. In: Advanced materials research. Trans Tech Publications Ltd., vol 225, pp 234–238
16. Kempa O, Lasota T, Telec Z, Trawiński B (2011, April) Investigation of bagging ensembles of genetic neural networks and fuzzy systems for real estate appraisal. In: Asian conference on intelligent information and database systems. Springer, Berlin, pp 323–332
17. Kennedy P (2003) A guide to econometrics. MIT Press, Cambridge
18. Lin H, Chen K (2011, July) Predicting price of Taiwan real estates by neural networks and support vector regression. In: Proceedings of the 15th WSEAS international conference on system, pp 220–225

19. Peter NJ, Fateye OB, Oloke CO, Iyanda P (2018) Changing urban land use and neighbourhood quality: evidence from Federal Capital Territory (FCT), Abuja, Nigeria. Int J Civil Eng Technol 9(11):23–36
20. Peter NJ, Okagbue HI, Obasi EC, Akinola AO (2020) Review on the application of artificial neural networks in real estate valuation. Int J 9(3)
21. Trawiński B, Telec Z, Krasnoborski J, Piwowarczyk M, Talaga M, Lasota T, Sawiłow E (2017, July) Comparison of expert algorithms with machine learning models for real estate appraisal. In: 2017 IEEE international conference on innovations in intelligent systems and applications (INISTA), pp 51–54. IEEE
22. Varma A, Sarma A, Doshi S, Nair R (2018, April) House price prediction using machine learning and neural networks. In: 2018 second international conference on inventive communication and computational technologies (ICICCT), pp 1936–1939. IEEE
23. Yeh IC, Hsu TK (2018) Building real estate valuation models with comparative approach through case-based reasoning. Appl Soft Comput 65:260–271

Memetic Optimal Approach for Economic Load Dispatch Problem with Renewable Energy Source in Realistic Power System

Shivani Sehgal, Aman Ganesh, and Vikram Kumar Kamboj

Abstract Electric power industry is shifting from conventional energy sources to combined renewable sources and thus becoming the most challenging and difficult problems of electric power system. This necessitates the generation and dispatch of load at most economical cost. The main objective of economic load dispatch in power system operation, control and planning is to fulfil the energy load demand at the lowest price while fulfilling all the constraints (equality and inequality constraints). This paper presents the mathematical design of optimal load dispatch problem by considering the sources of energy generation from conventional power plants and renewable power plants (solar power plant), considering all the essential constraints of the realistic power system. In the proposed research the memetic optimizer developed by combining Slime Mould Algorithm with pattern search algorithm (SMA-PS) has been tested to find the solution of integrated thermal solar economic load dispatch problem and experimentally it has been observed that the proposed memetic optimizer is providing cost-effective solution to complex economic load dispatch problem of electric power system.

Keywords Economic load dispatch · Renewable energy sources · Memetic optimization

1 Introduction

Nowadays, the electric power trade is the most challenging and intricate problem due to the ever-increasing demand of electric energy and lack of availability of energy resources thus necessitates the economic load dispatch of power generation. With the development of electrification in transport industry the growth in energy demand

S. Sehgal (✉) · A. Ganesh · V. K. Kamboj
Lovely Professional University, Phagwara, Punjab, India

A. Ganesh
e-mail: aman.23332@lpu.co.in

V. K. Kamboj
e-mail: vikram.23687@lpu.co.in

A. Choudhary et al. (eds.), *Applications of Artificial Intelligence and Machine Learning*, Lecture Notes in Electrical Engineering 778,
https://doi.org/10.1007/978-981-16-3067-5_47

is also observed, thus forces the use of renewable energy sources along with the conventional energy sources [1]. In near future, use of electric vehicles is encouraged and so the energy demand. To produce, assess and utilize the generated power most effectively the economical load dispatch of the integrated system is required. The main goal of the economical load dispatch (ELD) problem of electric power generation system is to achieve the system load demand at lowest possible operating cost by scheduling the committed generating units, while satisfying the physical and operating systems (equality and inequality) constraints [2]. Increased dependency on renewable energy sources also helps to eradicate the greenhouse gas emissions and air pollutants with decreased dependency on fossil fuels in conventional plants. The continuously growing demand for energy has encouraged the researchers to consider the renewable sources of energy and more research is required to be done to reduce global warming and degradation of the ecosystem. Many uncertainties prevail while using the renewable energy sources (RES) due to their dependence on the weather condition and geographical positions. Many researches have been performed on economic load dispatch of RES integrated system. The research proposed in [3] proved that if weather conditions are good, the RES system can effectively boost the conventional thermal systems. Saxena and Ganguli [4] implemented the Firefly algorithm to reduce the cost of conventional power plant integrated with solar and wind generators. ElDesouky [5] proposed integrated RES system to reduce the operational cost and emissions from the combined plant. Suresh and Sreejith [6] suggested a dragonfly optimization algorithm to solve the load dispatch problem of combine solar thermal system and [7] used the dragonfly algorithm to solve the probabilistic load dispatch problems. The optimum dispatch problem of a dynamic system with integrated renewable sources is discussed in [8] taking the problem of emissions and cost with wind, solar and thermal systems.

2 ELD of Renewable Energy Integrated System

Economic Load Dispatch (ELD) problem may be stated as distribution of the loads on the various thermal generators existing in the power system and the entire operating cost of energy generation is reduced to minimum, subjected to the power balance and the generation capacity constraints also known as equality and inequality constraints, respectively[9]. The ELD problem is classified as convex and non-convex problem where linear constraints contribute to convex problem and non-linear constraints along the linear constraints develops the non-convex ELD problem. The generation capacity and power balance constraints are categorized as linear constraints and provide the simplified approximate results and a more specific and accurate problem is modelled by non-linear constraints like valve point loading (VPL), ramp rate limits (RRL) and prohibited operating zones (POZs) [10].

Earlier this power generation problem was dealing only with the conventional thermal power generators, which use non-renewable resources of energy. Nowadays, due to limited fuel resources and environmental concerns, alternate methods of energy generation like solar and wind have gained popularity other than conventional thermal power generation. These sources have gained a remarkable importance in the current scenario of research and development. The unlikely combination of coal and solar under suitable circumstances provides an elegant solution for large scale power generation with reduced emissions and pollutants. The capital investment on renewable sources like solar and wind are more but the operational cost is less. Moreover, these sources are weather dependent, and the out is intermittent and vary widely in a short span of time. So, these sources always require some backup when supply is less or unavailable. The complete mathematical formulation of economic load dispatch problem for combined thermal and renewable energy source [9, 10] has been formulated in the following sections and memetic optimizer SMA-PS has been implemented to different test systems.

3 Problem Formulation

The major aim of the ELD problem is to minimize the total operational fuel cost of the power generating units while satisfying the different equality and inequality constraints. The different objective functions are formulated for ELD problem considering the effect of renewable energy sources (RES) [2–11]. These are categorized and formulated with the following notations:

NG	Total number of generating units
${P^G}_n$	Power of the nth generating unit
n	Index for power generating units
a_n, b_n and c_n	Cost Coefficient of nth power generating units
${P^G}_{n(\min)}$, ${P^G}_{n(\max)}$	Minimum and Maximum power of the nth generating units
$F(P^G)$	Total Fuel cost of power generating units
P^{Loss}	Power transmission loss.

3.1 Economic Load Dispatch (ELD) Problem

The mathematical interpretation of classical ELD problem for an hour is characterized as:

$$F(P^G) = \sum_{n=1}^{NG} \left[a_n (P_n^G)^2 + b_n (P_n^G) + c_n\right] \tag{1}$$

Equation (1) of ELD for "H" number of hours is described as:

$$F(P^G) = \sum_{h=1}^{H} \left(\sum_{n=1}^{NG} \left[a_n (P_n^G)^2 + b_n (P_n^G) + c_n\right] \right) \tag{2}$$

Cubical ELD Problem. To establish the output power of online generating units the ELD problem intends to congregate the system load at least cost while fulfilling the system constraints. So, as to attain correct dispatch outcomes, a cubical function is used for modelling the unit cost.

$$F(P^G) = \sum_{n=1}^{NG} \left[a_n (P_n^G)^3 + b_n (P_n^G)^2 + c_n (P_n^G) + d_n\right] \tag{3a}$$

The Cubical ELD with Valve Point effect can be represented as:

$$F(P^G) = \sum_{n=1}^{NG} [a_n (P_n^G)^3 + b_n (P_n^G)^2 + c_n P_n^G + d_n)] + |\varphi_n \sin(\gamma_n (P_{n(\min)}^G - P_n^G))| \tag{3b}$$

Heat and Power ELD Problem. The heat and power ELD problem of a system aim to resolve the unit heat and power production of the generating units [12]. The mathematical formulation for heat and power ELD may be described as:

$$F_{\text{Power}}(P_n^G) = \sum_{n=1}^{NG} \left(c_n + b_n \times P_n^G + a_n \times (P_n^G)^2\right) \tag{4a}$$

$$F_{\text{Heat}}(P_n^G) = \sum_{n=1}^{NG} \left(g_n + h_n \times P_n^G + q_n \times (P_n^G)^2\right) \tag{4b}$$

$$F_{\text{Overall}}(P_n^G) = \sum_{n=1}^{NG} \left(c_n + b_n \times P_n^G + a_n \times (P_n^G)^2 + g_n \times P_n^G + h_n \times (P_n^G)^2 + q_n \times (P_n^G)^2\right) \tag{4c}$$

The objective function for heat and power ELD considering valve point loading effects can be reframed as:

$$F_{\text{Power}}(P_n^G) = \sum_{n=1}^{NG} \left(c_n + b_n \times P_n^G + a_n \times (P_n^G)^2 + \left|\varphi_n \sin\left(\gamma_n (P_{n(\min)}^G - P_n^G)\right)\right|\right) \tag{5a}$$

$$F_{\text{Heat}}(P_n^G) = \sum_{n=1}^{NG} \left(g_n + h_n \times P_n^G + q_n \times (P_n^G)^2 + \left|\varphi_n \sin\left(\gamma_n (P_{n(\min)}^G - P_n^G)\right)\right|\right) \tag{5b}$$

$$F_{\text{Overall}}(P_n^G) = \sum_{n=1}^{NG} \Big(c_n + b_n \times P_n^G + a_n \times (P_n^G)^2 + g_n \times P_n^G + h_n \times (P_n^G)^2 + q_n \times (P_n^G)^2 + \big|\varphi_n \sin\big(\gamma_n (P_{n(\min)}^G - P_n^G)\big)\big|\Big) \tag{6}$$

All the objective functions of ELD problem as mentioned above are subjected to equality and inequality constraints as follows:

Power Balance (Equality) Constraint. The total power generated by all the generating units in a power plant must be equal to the sum of load demand and real power loss. Mathematically it is represented as

$$\sum_{n=1}^{NG} P_n^G = P^{\text{Demand}} + P^{\text{Loss}} \tag{7}$$

$$\sum_{n=1}^{NG} P_n^G + P^{\text{Renewable}} = P^{\text{Demand}} + P^{\text{Loss}} \tag{8}$$

where

$$P^{\text{Loss}} = \sum_{n=1}^{NG} \sum_{m=1}^{NG} P_n^G B_{nm} P_m^G \tag{9}$$

if B_{i0} and B_{00} are loss coefficients, then the P^{Loss} equation can be modified as:

$$P^{\text{Loss}} = P_n^G B_{nm} P_m^G + \sum_{n=1}^{NG} P_n^G \times B_{i0} + B_{00} \tag{10}$$

The expanded version of the above equation may be represented as:

$$P^{\text{Loss}} = \begin{bmatrix} P_1 & P_2 & \cdots & P_{NG} \end{bmatrix} \begin{bmatrix} B_{11} & B_{12} & \ldots & B_{1n} \\ B_{21} & B_{22} & \ldots & B_{2n} \\ \vdots & \vdots & \ddots & \vdots \\ B_{n1} & B_{n2} & \ldots & B_{nn} \end{bmatrix} \begin{bmatrix} P_1 \\ P_2 \\ \vdots \\ P_{NG} \end{bmatrix} + \begin{bmatrix} P_1 & P_2 & \cdots & P_{NG} \end{bmatrix} \begin{bmatrix} B_{01} \\ B_{02} \\ \vdots \\ B_{0NG} \end{bmatrix} + B_{00} \tag{11}$$

Generator limit (Inequality) constraints. The power generation of each generating unit must lie within the specific generator limits (maximum and minimum values of operating power).

$$P_{n(\min)}^{G} \leq P_{n}^{G} \leq P_{n(\max)}^{G} \quad n = 1, 2, 3, \ldots, NG \tag{12}$$

Ramp rate limits. In order to maintain the rate of change of output power within a specified range as to keep the thermal shifts in the turbine within its safe boundaries and to increase the life, the operating range for online generating units is controlled by their ramp rate limits as follows. Mathematically

$$P_{n}^{G} - P_{0}^{G_o} \leq UR_n \quad n = 1, 2, 3, \ldots, NG \tag{13}$$

$$P_{n}^{G_o} - P_{n}^{G} \leq DR_n \quad n = 1, 2, 3, \ldots, NG \tag{14}$$

$$\max[P_{n(\max)}^{G}, (UR_n - P_{n}^{G})] \leq P_{n}^{G} \leq \min[P_{n(\max)}^{G}, (P_{n}^{G_o} - DR_n)]$$
$$n = 1, 2, 3, \ldots, NG \tag{15}$$

Prohibited Operating Zones. The thermal power generating units may lie in a certain specified range where operation of the generating unit is impossible due to some physical constraints like vibration in shaft, steam valve, other machine components, Such constrained or restricted regions are acknowledged as prohibited operating zones (POZ). Under such constraints, the operational region splits into isolated sub-regions, thus forming a non-convex problem.

$$\begin{cases} P_{n(\min)} \leq P_n \leq P_{n(\min),1}^{\text{POZ}} \\ P_{n(\max),m-1}^{\text{POZ}} \leq P_n \leq P_{\min,m}^{\text{POZ}}; \quad m = 2, 3, \ldots N_{\text{POZ}} \\ P_{n(\max),m}^{\text{POZ}} \leq n_i \leq P_{n(\max)}; \quad m = N_{\text{POZ}} \end{cases} \tag{16}$$

4 Proposed Methodology

In the proposed research, a memetic slime mould optimizer combined with pattern search optimizer has been used and named as SMA-PS. Slime Mould algorithm in its simple version simulates action and a transition in Physarum polycephalum slime mould in foraging, and its entire life cycle does not model. Simultaneously, the usage of SMA weights replicates positive and negative criticism of slime mould during foraging, shaping 3 distinct morphotypes. The venous configuration of the slime mould grows according to the contraction mode phase difference [13], here

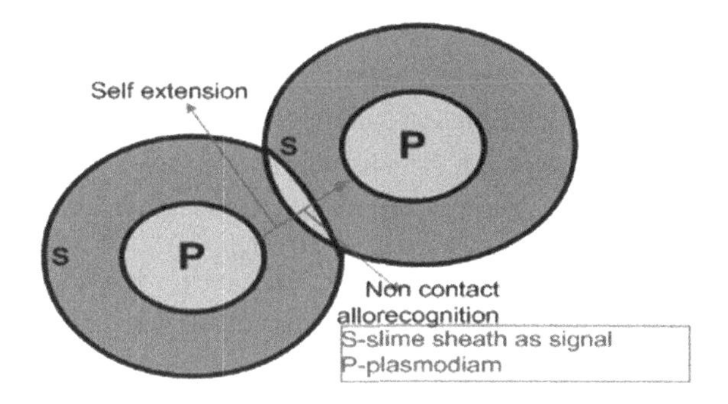

Fig. 1 Growing crops slime mould morphology

three parallels between the morphological updates in the venous structure and the level of contracture of the slime mould.

(a) Thick veins from across the radius as the contraction frequency varies from outward to inward.
(b) If the type of contraction becomes unstable, anisotropy starts to turn up.
(c) If the slime mould contraction series is no matter how long ordered in space and time the venous system will no longer be relevant.

Slime mould may also transform their search patterns as per the provenance nature of the food. If sources of food are of high quality, use region-limited search method for slime mould [14], thereby concentrating the check on sources of food discovered. If the density of the initial food provenance is less, the slime mould will stop leaving food sources to search other optional regional food sources [15]. This innovative research method can be expressed more often when various qualities of food blocks in an area are dispersed (Fig. 1).

4.1 Mathematical Modelling of SMA

This section illustrates mathematical modelling of SMA and describes the various methodologies for finding food. Which are namely approaching food, wrap of food, grabble food.

Approaching food. Depending upon the smell in the air, slime mould should address food. To describe its reaching nature in mathematics, the following equations are suggested to mimic the process of contraction.

$$\overrightarrow{S(i+1)} = \overrightarrow{S_b(i)} + \overrightarrow{vb}.\left(\overrightarrow{W} \cdot \overrightarrow{S_A(i)} - \overrightarrow{S_B(i)}\right), r < p \tag{17}$$

$$\overrightarrow{S(i+1)} = \overrightarrow{vc} \cdot \overrightarrow{S(i)}, r \geq p \tag{18}$$

In this case $\overrightarrow{vc}$, a parameter variable with a linear decrease from one to zero, $\overrightarrow{vb}$ a parameter variable with a range of $[-a, a]$, i reflect the current iteration, $\overrightarrow{S_b}$ individual position with the highest currently observed odour concentration, $\overrightarrow{S}$ individual location of slime mould, $\overrightarrow{S_A}$ individual and $\overrightarrow{S_B}$ individual are two of individuals randomly picked from swarm and slime mould individual weight as $\overrightarrow{W}$.

The equation p states the following:

$$p = \tanh|X(t) - bF| \tag{19}$$

Here, $t \in 1, 2, \ldots, n$, $X(t)$ reflects the fitness of $\overrightarrow{S}$, bF reflects the best fitness in all iterations. The equation $\overrightarrow{vb}$ states the following:

$$\overrightarrow{vb} = [-a, a] \tag{20}$$

$$a = \operatorname{arctanh}(-\left(\frac{i}{\max\_i}\right) + 1) \tag{21}$$

The equation $\overrightarrow{W}$ states the following:

$$\overrightarrow{W(\text{SmellIndex}(i))} = \begin{cases} 1 + r \cdot \log\left(\frac{DF - X(t)}{DF - wF} + 1\right), & \text{condition} \\ 1 - r \cdot \log\left(\frac{DF - X(t)}{DF - wF} + 1\right), & \text{others} \end{cases} \tag{22}$$

$$\text{Smell Index} = \text{sort}(X) \tag{23}$$

At which status suggests that X(t) actually graded in first-half of the population, r specifies the random number at the duration of [0,1]; DF symbolizes the optimum fitness achieved in the current iterative course, WF indicates a worst fitness value currently obtained in the iterative course, *smell index* replicates the order of sorted attributes of fitness (goes up in the question of minimal value). Equations (17) and (18) results are visualized in Fig. 2. The position of individual $\overrightarrow{S}$ quest can be improved to the best location $\overrightarrow{S_B}$ currently achieved, and tuning of $\overrightarrow{vb}$, $\overrightarrow{vc}$ and $\overrightarrow{W}$ parameters will adjust the position of the item.

Wrapping Food. This section mathematically simulates the mode of contraction of slime mould venous tissue structure. The higher the food concentration reached by the vein, the stronger will be the bio-oscillator-generated wave, the quicker the flow of cytoplasm and thicker the vein as in Eq. (22). In Eq. (22) the part r, is a simulation of venous contraction uncertainty. Condition simulates slime moulds to correct their patterns as per food quality. Larger weight near the region indicates higher food concentration and if food concentration decreases, the weight of the area also reduces, thereby forced to turn to explore other areas. Figure 2 shows the fitness values assessment process for slime mould.

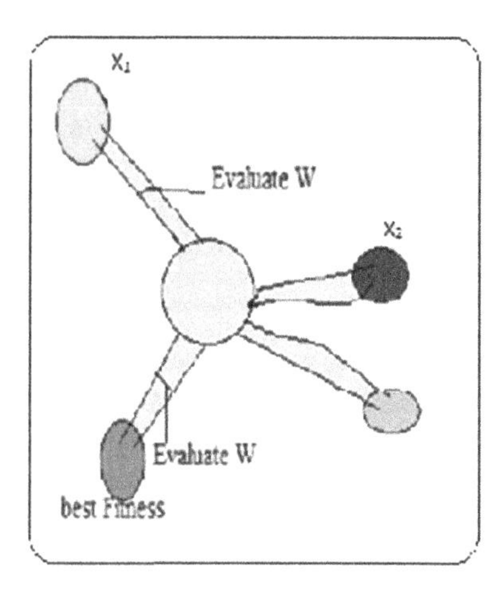

Fig. 2 Assessment of fitness

A mathematical model for upgrading its slime mould position shall be as follows:

$$\overrightarrow{S^*} = \begin{cases} \text{rand} \cdot \left(U_{\bigcup B} - U_{LB}\right) + U_{LB}, \text{rand} < z \\ \overrightarrow{S_b(i)} + \overrightarrow{vb} \cdot \left(W \cdot \overrightarrow{S(i)} - \overrightarrow{S_B(i)}\right), r < p \\ \overrightarrow{vc} \cdot \overrightarrow{S(i)}, r \geq p \end{cases} \tag{24}$$

Here, U_{LB} and $U_{\bigcup B}$ represent the quest range's lower and upper limits, rand and $r$signifies the random value between 0 and 1.

Food Grabble. Slime Mould obviously depends upon its propagation wave generated by the biological oscillator to alter the cytoplasmic flowing throughout the veins, and they appear in improved affordable food absorption role. To mimic differences in venous width of slime mould, $\overrightarrow{W}$, $\overrightarrow{vb}$ and $\overrightarrow{vc}$ are used to realize the varieties. $\overrightarrow{W}$ replicates arithmetically the oscillatory frequency of slime moulds next to the one at diverse food concentrations, but slime moulds can enter food quite effortlessly whenever they find food of high superiority, thus approaching food very slowly when the food level is lower, thereby increase the effectiveness of slime moulds in selecting the optimum food source. The value of $\overrightarrow{vb}$ vacillates between [−a, a] randomly and progressively goes to zero as iterations increases and the value of $\overrightarrow{vc}$ vacillates between [−1, 1] and eventually tends to be zero.

4.2 Pattern Search Algorithm (PS)

Pattern search (PS) optimizer is a black box mechanism and a derivative-free technique with local search ability and best for solution space at which optimization problem derivative is uncomfortable or uncertain[16, 17]. The study involves two movements when conducting its own operation: first one is exploratory search which is the local quest seeking to enhance the path to be pushed, and the other movement is the movement of pattern which is a greater quest for path improvement; in this movement, step performance can be increased unless the advancement has been modified. The movement of the pattern involves two points: one is the existing location while another is some arbitrary point with better value of the objective function that guides the course of the search; consideration of the new point is driven by Eq. (25).

$$x^{(\mathrm{iter}+1)} = x^{(\mathrm{int})} + \upsilon[x^{(\mathrm{iter})} - x^{(\mathrm{int})}] \tag{25}$$

Here, υ is a positive acceleration factor, used during multiplying the path length improvement's vector.

The position vector randomly generated was further changed using the pattern search (PS) method in the proposed hybrid SMA-PS algorithm, and the modified position vector $\overrightarrow{S}$ was applied to Slime Moulds to determine the location of the food source. The phase of exploration in SMA-PS is same as that of classical SMA and the phase of exploitation is improved using PS algorithm. The pseudocode for SMA-PS is as follows in Algorithm 2 and flowchart for the tendered SMA-PS is illustrated in flowchart in Fig. 3.

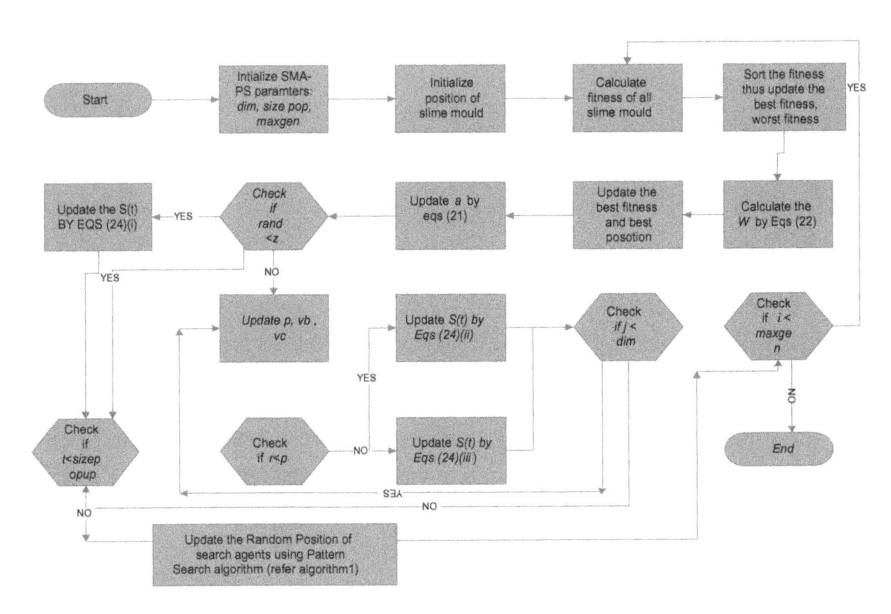

Fig. 3 Flow chart of hybrid SMA-PS algorithm

5 Results and Discussions

In this proposed research, an IEEE- 57 bus test system [18] is considered with seven generating units supplied from conventional energy source and two generators with renewable energy (solar) source are added as distributed generating units. The proposed SMA-PS memetic optimizer has been tested on Intel® Core ™ i7-5600 CPU @2.60 GHz for 500 iterations and 30-trial runs. The test data for IEEE-57 bus system has been taken from [18] and results for the proposed optimizer has been shown in Table 1 through Table 2. Comparison of results shows that the proposed optimizer has better efficiency as compared to many other optimizers.

Table 1 Test results for IEEE-57 Bus system without PV units

Unit	SMA-PS	DEIANT [18]	PSO [18]	ACO [18]
G_1 (MW)	444.4432	139.57	142.36	140.82
G_2 (MW)	100	92.84	94.69	93.67
G_3 (MW)	140	56.66	57.79	57.17
G_4 (MW)	100	73.94	75.42	74.61
G_5 (MW)	170.1131	461.27	470.5	465.43
G_6 (MW)	81.4812	97.33	99.28	98.21
G_7 (MW)	236.9625	353.58	360.66	356.77
Total cost ($/h)	15,581.9276	42,017.46	42,857.81	42,395.62

Table 2 Test results for IEEE-57 bus system with PV units

Unit	SMA-PS	DEIANT [18]	PSO [18]	ACO [18]
G_1 (MW)	437.3061	137.76	138.18	138.04
G_2 (MW)	100	90.97	91.24	91.15
G_3 (MW)	140	55.52	55.68	55.63
G_4 (MW)	100	72.45	72.67	72.60
G_5 (MW)	163.8646	451.00	452.36	451.91
G_6 (MW)	74.8227	95.23	95.51	95.42
G_7 (MW)	229.5467	344.48	345.51	345.17
PV_1(MW)	15.43	15.43	15.47	15.46
PV_2(MW)	12.03	12.03	12.07	12.06
Total Cost ($/hour)	15,220.2201	41,172.99	41,296.51	41,255.34

Algorithm 1: Pseudo-code of PS

Initialize the input parameters for pattern search algorithm i.e. (υ), (P^0) and (τ).

Initialize the current perturbation vector $P \leftarrow P^0$ and select the value for the starting point x^{int}.

Update x^{int} to x^{iter} using exploratory search around x^{int} to find an improved point x^{iter} that has a better value of objective function.

if $x^{iter} > x^{\text{int}}$

DO $P \leftarrow P/2$.

if $P_i < \tau$

DO $x^{final} = x^{\text{int}}$

else

update the solution vector using exploratory move.

else

DO $x^{final} = x^{iter}$, $P \leftarrow P^0$ and go to step-4.

end

end

Apply Pattern Move using following steps:

Obtain tentative x^{iter+1} by a pattern move from x^{int} through x^{iter}.

Obtain final x^{iter+1} by an exploratory search around tentative x^{iter+1}.

if $f(x^{iter+1}) > f(x^{iter})$

DO $x^{\text{int}} \leftarrow x^{iter}$ and go to step-3.

else

DO $x^{\text{int}} \leftarrow x^{iter}$, $x^{iter} \leftarrow x^{iter+1}$ and go to step-4

end

```
Algorithm 2: Pseudo-code of hSMA-PS
Initialize the parameters popsize,
Max_iteraition;
Initialize the positions of slime mould
S_t(t = 1,2,...,n);
While (i ≤ Max_iteraition)
Calculate the fitness of all slime mould;
    Update bestFitness, S_b
Calculate W by Eq. (22);
For each search portion
     Update p,vb,vc;
     Update positions by Eq.(24);
End For
Check S⃗ location using the pattern search meth-
od (refer to algorithm 1 pseudo code)
i = i + 1;
End While
  Return bestFitness, S_b;
```

The analysis of economic impact of RES integrated with conventional energy source is discussed by considering two cases of combined ELD problem. Case 1 is an ELD problem without considering the impact of RES and case 2 is an ELD problem with integration of RES. The results are compared with different existing algorithms [18]—DEIANT, PSO and ACO.

Case 1: ELD without PV Units. IEEE 57 bus system is considered with constant load demand of 1273 MW. The results presented in Table1 indicates the effectiveness of proposed SMA-PS algorithm as compared to DEIANT, ACO and PSO [18].

Case 2: ELD with PV Units. In case 2, two solar PV units are incorporated as DG units to study the impact of renewable sources. In this case, the load is economically apportioned among the conventional thermal generators and two PV generators. The results presented in Table 2 depicts the comparison of proposed SMA-PS algorithm with other existing algorithms. It is noticed that the operational cost in case 2 is considerably lower than in case 1. Therefore, the integrated solar thermal system reduces the operating cost of a power system.

6 Conclusion

In this paper the mathematical formulation of ELD problem with RES has been framed considering all the essential constraints, that is used to realize the economic impact of renewable energy sources in realistic power system. The proposed memetic optimizer has been tested to find the best possible solution for IEEE-57 bus system with two units of solar PV source and it has been observed that the proposed optimizer is providing better results as compared to other meta-heuristics optimizer algorithms. This work can further be expanded to solve the economic load dispatch problems with the impact of electric vehicles.

References

1. Zeynal H, Yap J, Azzopardi B, Eidiani M (2014) Flexible economic load dispatch integrating electric vehicles. In: Proceedings of the 2014 IEEE 8th international power engineering and optimization conference, PEOCO 2014
2. Kothari JS, Dhillon DP (2014) Power system optimization, 2nd edn. Phi Learning
3. Augustine N, Suresh S, Moghe P, Sheikh K (2012)Economic dispatch for a microgrid considering renewable energy cost functions. In: Innovative Smart Grid Technologies (ISGT), 2012 IEEE PES, pp 1–7
4. Saxena N, Ganguli S (2015) Solar and wind power estimation and economic load dispatch using firefly algorithm. Procedia Comput Sci 70:688–700
5. ElDesouky A (2013) Security and stochastic economic dispatch of a power system including wind and solar resources with environmental consideration. Int J Renew Energy Res 3:951–958
6. Suresh V, Sreejith S (2017) Generation dispatch of combined solar thermal systems using dragonfly algorithm. Computing 99:59–80
7. Das D, Bhattacharya A, Ray RN (2020) Dragonfly algorithm for solving probabilistic economic load dispatch problems. Neural Comput Applic 32:3029–3045
8. Anand H, Ramasubbu R (2020) Dynamic economic emission dispatch problem with renewable integration focusing on deficit scenario in India. Int J Bio-Inspired Comput 15(1):63–73
9. Kumar V, Bath KSK (2015) Solution of non-convex economic load dispatch problem using Grey Wolf Optimizer. Neural Comput Appl
10. Andervazh MR, Javadi S (2017) Emission-economic dispatch of thermal power generation units in the presence of hybrid electric vehicles and correlated wind power plants. IET Gener Transm Distrib 11(9):2232–2243
11. Afonaa-Mensah S, Wang Q, Bernard UB (2019) Dynamic economic dispatch of a solar-integrated power system: impact of solar-load correlation on solar power absorption and its effects on the total cost of generation. Energy Sources, Part A Recover Util Environ Eff 1–14
12. Jayakumar N, Subramanian S, Ganesan S, Elanchezhian EB (2016) Grey wolf optimization for combined heat and power dispatch with cogeneration systems. Int J Electr Power Energy Syst 74:252–264
13. Nakagaki T, Yamada H, Ueda T (2000) Interaction between cell shape and contraction pattern in the Physarum plasmodium. Biophys Chem 87:1–85
14. Kareiva P, Odell G (1987) Swarms of predators exhibit ‘preytaxis’ if individual predators use area-restricted search. Am Nat 130:233–270
15. Latty T, Beekman M (2009) Food quality affects search strategy in the acellular slime mould, Physarum polycephalum. Behav Ecol 20(6):1160–1167
16. Torczon V (1997) On the convergence of pattern search algorithms. SIAM J Optim 7(1):1–25
17. McCarthy JF (1989) Block-conjugate-gradient method. Phys Rev D 40(6):2149–2152
18. Rahmat NA, Aziz NFA, Mansor MH, Musirin I (2017) Optimizing economic load dispatch with renewable energy sources via differential evolution immunized ant colony optimization technique. Int J Adv Sci Eng Inform Technol 7(6):2012–2017

High-Throughput and Low-Latency Reconfigurable Routing Topology for Fast AI MPSoC Architecture

Paurush Bhulania, M. R.Tripathy, and Ayoub Khan

Abstract Multiprocessor system-on-chip (MPSoC) widely uses in various applications due to its capacity of delivering the aggressive performances in the low power cost. The System-on-Chips (SoCs) generally are used in consumer electronics with the integration of multiple functionalities. The main issue faced by the MPSoC architecture is higher delay in techniques of data transmission which causes because of network congestion. In order to overcome the same, an effective network technique or topology with an adequate routing requires to be developed for achieving a better performance. Also, if the introduction of artificial intelligence is done, then it will lead to improve in the performance of routing protocols. The mesh topology is integrated into torus topology to achieve an optimal routing for transmission of data packets from the source to the destination router of MPSoC. Also, XY-YX routing is accomplished over the MPSoC network topology for obtaining the shortest path through the MPSoC. The technique of predicting the response packet path is used to minimize the delay during the data transmission. Also, the performance of the proposed design is compared with three existing architectures MXY-SoC, K Means-MPSoC, and SDMPSoC. The future of the same can be incorporated in field of artificial intelligence by the various algorithms for transmission of data from one point to another. The results obtained using this technique will lead to high throughput and low latency level for fast AI MPSoC architecture.

Keywords Delay · XY-YX routing · Mesh topology · Multiprocessor system on chip · System-on-chip · Torus topology and artificial intelligence

---

P. Bhulania (✉) · M. R.Tripathy
Amity University, Noida, Uttar Pradesh, India
e-mail: pbhulania@amity.edu

A. Khan
Centre for Development of Advanced Computing, Noida, Uttar Pradesh, India

A. Choudhary et al. (eds.), *Applications of Artificial Intelligence and Machine Learning*, Lecture Notes in Electrical Engineering 778,
https://doi.org/10.1007/978-981-16-3067-5_48

1 Introduction

System-on-Chip (SoC) plays a significant role in various devices having compact size, high performance, and low-power requirement [1]. The structure of SoC design is mainly used in analyzing the low power-based system [2]. MPSoC combines the benefits of the parallel computation of multiprocessors with single-chip integration of SoCs [3]. The modern MPSoC presents a large number of processors to deliver the scalable computational power to the respective massive parallelism. On the other hand, the MPSoC requires an appropriate programming models and software development infrastructure [4].

The communication between the intra-chips creates the issues with the improvement of MPSoC. The issues faced by the design of NoC are power consumption, bandwidth, and latency constraints [5]. The growing demand of the embedded application leads to increase in the number of processing elements (PE) in the MPSoCs, as the current MPSoC has less amount of PEs. But the SoC increases almost linearly in the single chip to handle the future embedded applications [6]. Moreover, the MPSoC design has been developed in the various network topology structures such as hypercube-based topology [7, 8], 2-D mesh topology [9]. These network topologies use in transferring the data between the devices.

The major contributions of this research are given as follows:

- The mesh topology is integrated into the torus topology in order to achieve an effective data transmission having less delay. Also, the implementation of mesh-torus topology is easy in the network.
- The data transmission is carried out by using the XY-YX routing algorithm. The conventional XY-YX routing algorithm is improved as lightweight XY-YX routing algorithm for minimizing the overall hardware utilization.

Additionally, the power consumption is minimized by identifying the response packet paths, and delays are reduced by using this XY-YX routing.

2 Literature Survey

There are various researches already developed related to the MPSoC architecture. In this section, some of the recent researches with its limitations are surveyed.

Lu and Yao [10] introduced the dynamic traffic regulation for improving the performances of NoC-based MPSoC and the designs of the chip multi/many-core processor (CMP). In this work, two distinctive traffic regulations open-loop and closed-loop regulation were used for MPSoCs and CMPs, respectively. In the open-loop configuration, the network and IP blocks were treated as a black box. The output signals of IP cores were visible in black box for designing the regulation mechanism. In closed-loop configuration, the network and cores were treated as clear box. The network and internal states of traffic were monitored in the clear box, and the regulation was adapted based on the network and traffic status. Here, the routers were arranged in the mesh topology, and the XY routing algorithm was accomplished for transmitting the data from the source to the destination. This dynamic traffic regulation was used in the NoC-based MPSoC to minimize the packet delay and enhance the system throughput. However, the considerable area and power consumption of CMP are increased by using a big input buffer in the closed-loop scenario.

Liu et al. [11] presented the architecture of SMART NoC by using the single-cycle multi-hop bypass channels. The SMART NoC's communication efficiency was reduced by the communication contention which decreased the overall system's performance. The aforementioned problem was addressed by using the integer-linear programming (ILP) model that delivered better solutions by considering the inter-processor communication. The topology used in the SMART NoC architecture was mesh topology. Additionally, a heuristic algorithm, namely topo map, was used in the SMART architecture to perform the task mapping in the dark silicon many-core systems. This task mapping was used to solve the communication issue related to the polynomial time. The single-cycle multi-hop bypass channels were designed between the distant cores during runtime to minimize the communication latency. However, the SMART architecture used in the MPSoC was sensitive to the traffic conflict.

Guo et al. [12] presented the memory-reinforced tabu search (TS) algorithm to improve the MPSoC performances. This reinforced tabu search algorithm was developed with critical path awareness to accomplish the hardware/software (HW/SW) partitioning over MPSoCs. Initially, the amount of data was minimized to simplify the input task graph, and an entire data was preserved to enhance the processing efficiency. Next, the configuration of the crossbar was performed over the critical path algorithm based on the output task chain as well as the communication penalty of the task graph was minimized in the MPSoC. The search efficiency of the TS was improved by using the heuristic algorithm as the initial solution. The effectiveness and searching strength were improved by developing the hash technology and by integrating the dual memory tables. The task topology used in the MPSoC is comprised of in-tree, out-tree, fork-joint, mean-value analysis, and fast Fourier transform (FFT) task graphs. The runtime was minimized by using the system characteristics of MPSoCs, but the integration of HW and SW creates the run time problems in the MPSoCs.

Silveira et al. [13] used the preprocessing of fault scenarios which was mainly based on the forecasting fault tendencies. The fault tendencies were accomplished with a fault threshold circuit that was operated by the high-level software. In the case of fault detection, the preprocessing fault states were used for the rapidly reconfiguring the network with new deadlock-free routing. The variations in the asymmetrical topologies were identified by using the cross-correlation-based method. This work contained the fault-monitors which were located at each input port of the Phoenix NoC routers, and this Phoenix NoC was arranged based on 2D mesh topology. The defective links over the mesh topology were evaluated by using a fault monitor. The information related to the link state was sent to the Processing Element (PE) that was connected to the router. The efficiency of the routing algorithm degrades due to the reduction in the routing time interval.

As the increase in number of network topologies and their architecture [14], the complexity increases. To resolve the same, a centralized intelligence, known as "network mind," for supporting different network services can also be used for reducing the complexity. In this, a centralized AI control is deployed for connection-oriented tunneling-based routing protocols.

Also, the inclusion of artificial intelligence can help to improve the performance of routing protocols [15]. Nowadays, the application of artificial intelligence over routing protocols is only applied to real devices, i.e., in wireless sensor nodes. To implement, an intelligent routing protocol in a SDN topology is also an example of the same. This intelligent routing protocol is based on the reinforcement learning process that allows choosing the best data transmission paths according to the best criteria and based on the network status.

Powerline communications [16] are also a popular technology for providing infrastructure for applications related to IoT, smart grids, smart cities, in-home networking and have been experimentally considered for broadband access. In this, the implementation of the G3-PLC LOADng routing protocol is used in the nodes of a sensor/meter network, where the nodes share all the same medium.

3 Problem Statement

Current problems related to the MPSoC architecture are stated in this section, and it also explains how the proposed system overcomes the problems faced in the MPSoC design. The problems faced by the MPSoC architecture are mentioned as follows:

The utilization of big input buffer in the closed-loop scenario increases the power consumption and area utilization through the MPSoC architecture [10]. Moreover, the area of the MPSoC is increased due to the replication of the NoC [17] and increment in number of tiles [18]. The main reason behind higher area utilization is the usage of number of hardware components through the MPSoC architecture, and additionally, high-power consumption affects the performances of MPSoC. The throughput saturates at sometimes due to the highest request frequency [19]. The less throughput shows that the number of packets received by the desired node is less while transmitting the data packets. For an effective MPSoC, an effective reliable data transmission is required from the source node to the destination node. Also, the SMART architecture used in the MPSoC is sensitive to the traffic conflict [11], as the data transmission between the nodes is difficult during the network congestion.

Solution:

In this study, the mesh and torus are integrated together to achieve less delay while transmitting data. The lightweight XY-YX algorithm used in the MPSoC architecture obtains the shortest route for data transmission. The shortest path from the XY-YX algorithm helps to minimize the power consumption over the MPSoC architecture.

Moreover, the area utilization of the MPSoC architecture is reduced by processing only the node with higher priority during the data transmission. Here, the nodes with higher priority are identified by considering two different parameters packet arrival rate and congestion rate. Therefore, the processing of nodes only with higher priority leads to minimize the hardware components in the architecture.

4 Proposed Architecture

In this proposed framework, the hybrid work of mesh-torus topology is developed over the 4×4 MPSoC architecture. The XY-YX routing algorithm technique is developed for data transmission from the source to the destination. Since the transmission of data is carried out through the associated input/output ports. The prime objective of this architecture is topology reconfiguration technique during the congestion. The concern of integrating the mesh into torus topology is to accomplish an effective data transmission having less delay during the congestion. The overall MPSoC architecture is depicted in Fig. 1.

The main steps processed in this proposed MPSoC architecture are given as follows:

- Initially, the MPSoC is developed over the mesh-torus topology leading to reduction in the delay. Here, the MPSOC is designed in the range of 4×4 MPSoC architecture.
- Then XY/YX routing algorithm technique is developed for transmitting the data from the source to the destination.

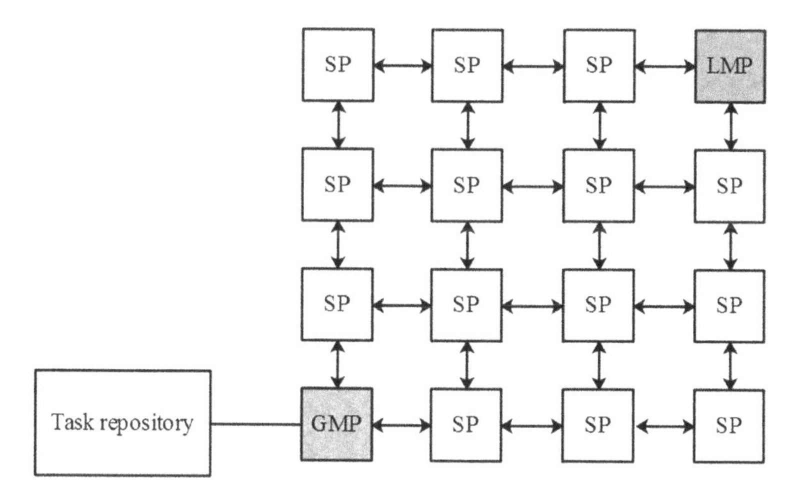

Fig. 1 Overall MPSoC architecture

- In this hybrid mesh-torus topology, the mesh topology is used when there is no congestion. Else, the torus topology is accomplished for transmitting the data packets.
- Finally, the priority of the nodes is considered in the lightweight XY/YX routing algorithm. The packet arrival rate and congestion rate are used to define the node's priority in the MPSoC.

4.1 Architecture of a System on Chip

The architecture of SoC used in the proposed MPSoC is depicted in Fig. 2. The modern embedded systems mainly depend on the system-on-chip platform, and this SoC has different components such as application-specific circuits, digital signal processors (DSP), one or more embedded microcontrollers, and read-only memory. The afore-mentioned components are integrated into the single package. These blocks are accessible from industries of intellectual property (IP) as hard cores or softcores. The hard core/hard IP block refers to the circuit where it is accessible in the lower abstraction level such as layout level. Additionally, the customization of hard IP is difficult with respect to the applications of the embedded system. The hard IP is modified for optimizing the cost functions. Similarly, the soft IP refers to the circuit that is accessible in a higher level of abstraction, such as register-transfer level. Moreover, the customization of the soft IP is easy to the desired application. Hence, the design of the embedded SoC improves the application-specific hardware, processors, and memories of creating the SoC.

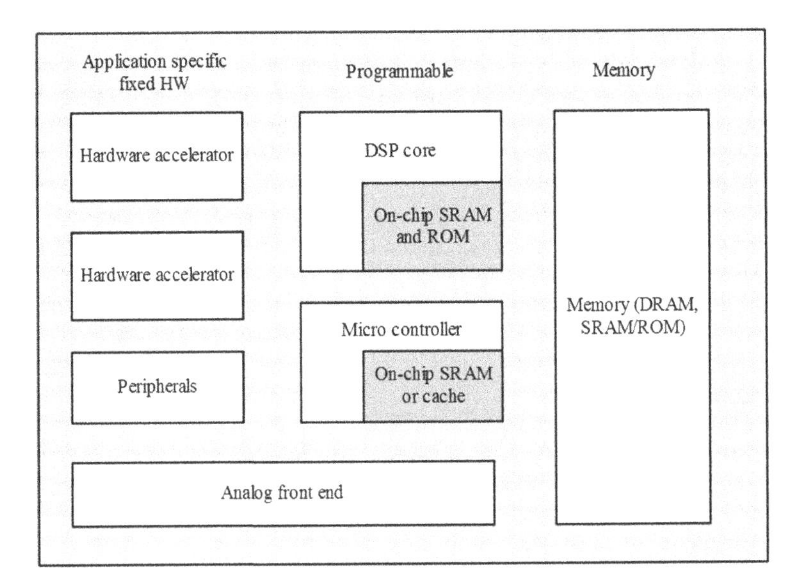

Fig. 2 Architecture of SoC

Figure 2 depicts the overall architecture of embedded SoC that has four main modules.

5 Experiment Results and Discussion

This section described the experimental results and discussion of the proposed MPSoC architecture. The design of the MPSoC architecture is developed by using the Verilog HDL in the Xilinx ISE 14.2 software. In developed MPSoC architecture, the network topology is created by combining both the mesh and torus topology. Additionally, the data transmission from the source router to the destination router is accomplished by using the XY-YX routing technique. In this study, a 4×4 mesh-torus topology is generated to analyze performance of the MPSoC.

The waveform and routing path for 4×4 MPSoC architecture without congestion are shown in Figs. 3 and 4, respectively. Similarly, the waveform and routing path for 4×4 MPSoC architecture with congestion are shown in Figs. 5 and 6, respectively, where baud_count represents the universal asynchronous receiver-transmitter

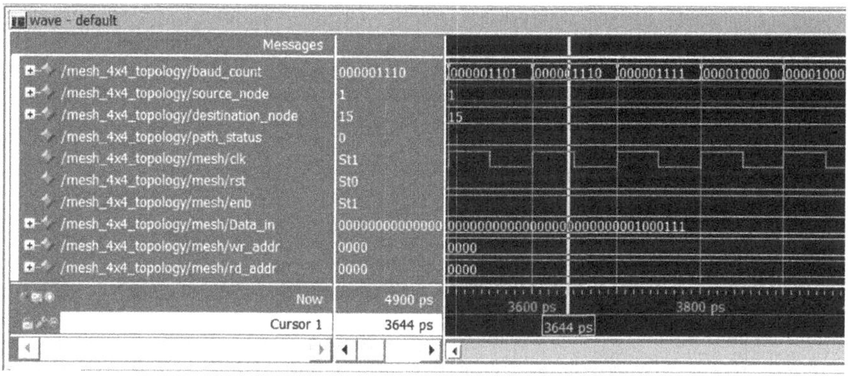

Fig. 3 Modelsim waveform for proposed MPSoC without congestion

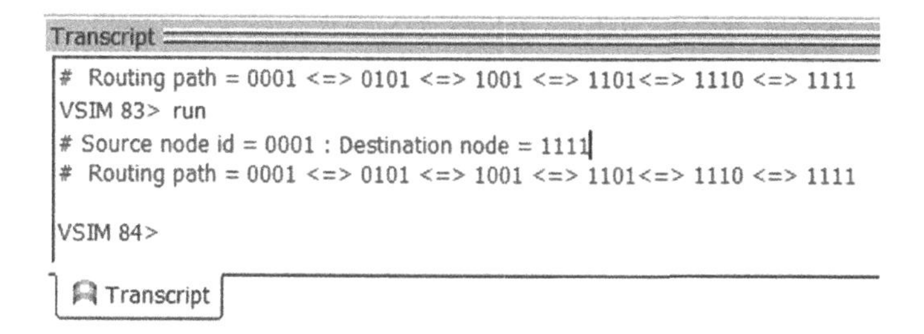
Transcript
Routing path = 0001 <=> 0101 <=> 1001 <=> 1101<=> 1110 <=> 1111
VSIM 83> run
Source node id = 0001 : Destination node = 1111
Routing path = 0001 <=> 0101 <=> 1001 <=> 1101<=> 1110 <=> 1111
VSIM 84>
Transcript

Fig. 4 Generation of routing path for proposed MPSoC without congestion

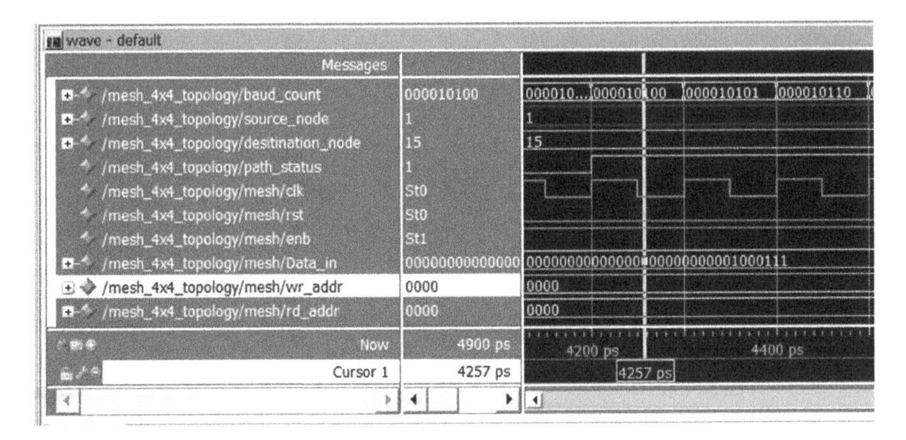

Fig. 5 Modelsim waveform for proposed MPSoC with congestion

```
Transcript
# Routing path = 0001 <=> 0010 <=> 0011 <=> 0111<=> 1011 <=> 1111
VSIM 91> run
# Source node id = 0001 : Destination node = 1111
# Routing path = 0001 <=> 0010 <=> 0011 <=> 0111<=> 1011 <=> 1111

VSIM 92>
Transcript
```

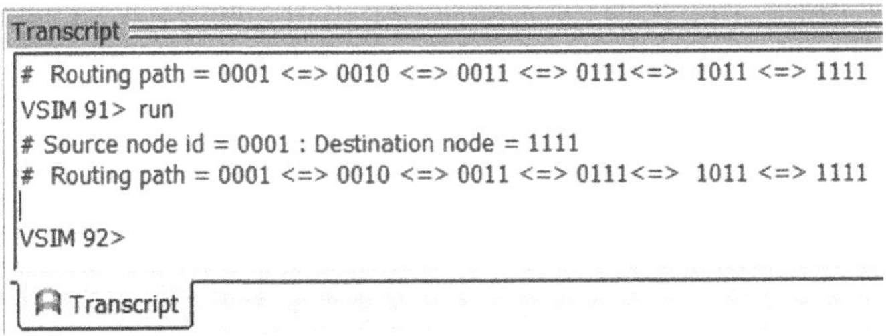

Fig. 6 Generation of routing path for proposed MPSoC with congestion

(UART) count, path_status defines the congestion, clk signal is used for synchronization, rst and enb are control signals, data_in is transmitted data, wr_addr represents the write address, and rd_addr represents read address. The source node and destination node are specified as 1 and 15 for analyzing the performances of the 4X4 MPSoC architecture. Here, path_status is equal to 0 defines the MPSoC architecture doesn't have congestion inside the network. Otherwise, path status is equal to 1 shows the congestion has occurred in the MPSoC architecture. In hybrid mesh-torus topology, the torus and mesh are enabled, when the MPSoC architecture is operated with and without congestion, respectively. From Figs. 3 and 6 know that the lightweight XY-YX routing effectively searches the optimal path under congestion constraint.

Table 1 Comparative analysis of proposed MPSoC with MXY-SoC

Virtex-5 family		
Parameters	MXY-SoC [17]	Proposed MPSoC
LUT	1471/28,800	1,326/28,800
Frequency (MHZ)	250	468.505
Delay (ns)	42	2.13
Power (W)	0.654	0.352

5.1 Performance Analysis

Comparative Analysis

The comparative analysis of the proposed MPSoC architecture with existing MPSoC architecture is described in this section. There are three existing MPSoC architecture used MXY-SoC [17], K Means-MPSoC [18], and SDMPSoC [20]. This comparison is made to evaluate the effectiveness of the proposed MPSoC architecture with the mesh-torus topology. Here, the MXY-SoC [17], K Means-MPSoC [18], and SDMPSoC [20] are developed in the Virtex-5, Virtex-6, and Zynq-7z020 family, respectively. The K Means-MPSoC design [18] and SDMPSoC [20] are implemented for evaluating the performance of the proposed MPSoC architecture.

Tables 1, 2, and 3 present the comparative analysis of the proposed MPSoC with MXY-SoC [17], K Means-MPSoC [18], and SDMPSoC [20], respectively. From the data present in Tables 1, 2, and 3, it is concluded that the performance of the proposed MPSoC gives better performance when compared with the MXY-SoC [17], K Means-MPSoC [18], and SDMPSoC [20]. For example, the LUT of the proposed MPSoC

Table 2 Comparative analysis of proposed MPSoC with MXY-SoC

Virtex-6 family		
Parameters	K Means-MPSoC [18]	Proposed MPSoC
LUT	9355/474249	688/46560
Flipflops	1755/5789	436/784
Frequency (MHZ)	165	559.550
Power (W)	1.759	1.293

Table 3 Comparative analysis of proposed MPSoC with MXY-SoC

Zynq–7z020-family		
Parameters	SDMPSoC [20]	Proposed MPSoC
LUT	1119	732
Flipflops	984	524
Frequency (MHZ)	667	680.672
Power (W)	0.428	0.100

is less compared to the MXY-SoC [17], K Means-MPSoC [18], and SDMPSoC [20] architectures. The reason behind the proposed MPSoC with lesser hardware utilization is the processing of the nodes which have higher priority. This kind of processing minimizes the area through the MPSoC architecture. Additionally, the delay of the proposed MPSOC is 2.13 ns; it is less compared to the MXY-SoC. The identification of the shortest path from the source to the destination using lightweight XY-YX routing algorithm is used to minimize the delay.

6 Conclusion

In this study, MPSoC architecture was enormously used to deliver the higher data packets while maintaining the power and complexity control. The architecture of 4×4 MPSoC was developed for analyzing the hardware utilization and delay performances for the transmission of data. The mesh technique was integrated with the torus one in order to reduce the delay occurred through the MPSoC. Also, the lightweight XY-YX routing was developed for achieving the committed data transmission from the source to destination. The hardware used in MPSoC was reduced by realizing the priority-based routing for transmission of data packets. The priority of the nodes was analyzed by evaluating two parameters congestion and packet arrival rate of the nodes. The lightweight XY-YX routing derived the shortest path which helped to reduce the power. Therefore, the proposed MPSoC architecture provided better performance when compared with the MXY-SoC, K Means-MPSoC, and SDMPSoC in terms of frequency obtained by the proposed MPSoC architecture was 559.550 MHz for Virtex-6 which is less when compared with the K Means-MPSoC. Furthermore, a fault-aware routing can be introduced in the MPSoC to protect the MPSoC from the temporary/permanent faults. The results obtained while simulation using the above technique and on comparing with other techniques will lead to high throughput and low latency level for fast AI MPSoC architecture.

References

1. Bansal R, Karmakar A (2017) Efficient integration of coprocessor in LEON3 processor pipeline for System-on-Chip design. Microprocess Microsyst 51:56–75
2. Durrani YA, Riesgo T (2016) Efficient power analysis approach and its application to system-on-chip design. Microprocess Microsyst 46:11–20
3. Shi F, Ji W, Gao Y, Wang Y, Liu C, Deng N, Li J (2010) Computationally efficient locality-aware interconnection topology for multi-processor system-on-chip (MP-SoC). Chin Sci Bull 55(29):3363–3371
4. Lombardi M, Milano M, Ruggiero M, Benini L (2010) Stochastic allocation and scheduling for conditional task graphs in multi-processor systems-on-chip. J Sched 13(4):315–345
5. Liu L, Yang Y (2010) Energy-aware routing in hybrid optical network-on-chip for future multi-processor system-on-chip. In: 2010 ACM/IEEE symposium on architectures for networking and communications systems (ANCS), pp 1–9. IEEE

6. Quan W, Pimentel AD (2016) A hierarchical run-time adaptive resource allocation framework for large-scale MPSoC systems. Des Autom Embed Syst 20(4):311–339
7. Pelissier F, Chenini H, Berry F, Landrault A, Derutin JP (2016) Embedded multi-processor system-on-programmable chip for smart camera pose estimation using nonlinear optimization methods. J Real-Time Image Proc 12(4):663–679
8. Damez L, Sieler L, Landrault A, Dérutin JP (2011) Embedding of a real time image stabilization algorithm on a parameterizable SoPC architecture a chip multi-processor approach. J Real-Time Image Proc 6(1):47–58
9. Singh AK, Kumar A, Wu J, Srikanthan T (2013) CADSE: communication aware design space exploration for efficient run-time MPSoC management. Front Comp Sci 7(3):416–430
10. Lu Z, Yao Y (2016) Dynamic traffic regulation in noc-based systems. IEEE Trans Very Large Scale Integr (VLSI) Syst 25(2):556–569
11. Liu W, Yang L, Jiang W, Feng L, Guan N, Zhang W, Dutt N (2018) Thermal-aware task mapping on dynamically reconfigurable network-on-chip based multiprocessor system-on-chip. IEEE Trans Comput 67(12):1818–1834
12. Guo Z, Zhang X, Zhao B (2019) A memory-reinforced tabu search algorithm with critical path awareness for HW/SW partitioning on reconfigurable MPSoCs. IEEE Access 7:112448–112458
13. Silveira J, Marcon C, Cortez P, Barroso G, Ferreira JM, Mota R (2016) Scenario preprocessing approach for the reconfiguration of fault-tolerant NoC-based MPSoCs. Microprocess Microsyst 40:137–153
14. Yao H, Mai T, Jiang C, Kuang L, Guo S (2019) AI routers and network mind: a hybrid machine learning paradigm for packet routing. IEEE Comput Intell Mag 14(4)
15. Sendra S, Rego A, Lloret J, Jimenez JM, Romero O (2017) Including artificial intelligence in a routing protocol using software defined network
16. Marcuzzi F, Tonello AM (2019) Artificial Intelligence based performance enhancement of the G3-PLC LOADng routing protocol for sensor networks. In: 2019 IEEE symposium on power line communications and its applications
17. Shahane P, Pisharoty N (2019) Modified X-Y routing for mesh topology based NoC router on field programmable gate array. IET Circuits Devices Syst 13(3):391–398
18. Khawaja SG, Akram MU, Khan SA, Shaukat A, Rehman S (2016) Network-on-Chip based MPSoC architecture for k-mean clustering algorithm. Microprocess Microsyst 46:1–10
19. Lin Z, Sinha S, Liang H, Feng L, Zhang W (2017) Scalable light-weight integration of FPGA based accelerators with chip multi-processors. IEEE Trans Multi-Scale Comput Syst 4(2):152–162
20. Rettkowski J, Göhringer D (2019) SDMPSoC: software-defined MPSoC for FPGAs. J Sig Process Syst 1–10

Comparison of Various Data Center Frameworks

Monalisa Kushwaha, Archana Singh, B. L. Raina, and Avinash Krishnan Raghunath

Abstract Our world is now becoming more attached to IT technologies. With the growing use of the Internet, there is a store and process exponential data. Data centers gained importance due to the emergence of Internet services. Data centers are responsible for the storage, management, and dissemination of data. Since the data is exponentially increasing, the data centers are more overloaded with data and their size is increasing. Since the data centers are more overloaded to fulfill the growing demands of the customers, this in turn is indirectly contributing to global pollution in terms of heavy consumption of power. Data centers consume a lot of power that is why many organizations want to design their data center their operations as green as possible. Thus, an efficient green data center framework is needed for the efficient design of data centers. To achieve this, the comparison of green data centers frameworks is being done which evaluates each of the data centers. The contrast inferred that the frameworks have the chances for enhancement in terms of components, attributes, energy-effective metrics, and implementation procedure. A new data center framework is proposed which is implemented considering the disadvantages of the existing data center frameworks.

Keywords Green data center · Green data center framework · Energy efficiency

1 Introduction

Data centers are the building blocks of the IT framework which provides the storage of data in a centralized manner, backups of data, and also networking and management of data in which the various electrical, computing, and mechanical systems

M. Kushwaha (✉) · B. L. Raina
Glocal University, Saharanpur, Uttar Pradesh, India

A. Singh
Amity University, Noida, Uttar Pradesh, India
e-mail: asingh27@amity.edu

A. K. Raghunath
Vellore Institute of Technology, Vellore, Tamil Nadu, India

A. Choudhary et al. (eds.), *Applications of Artificial Intelligence and Machine Learning*, Lecture Notes in Electrical Engineering 778,
https://doi.org/10.1007/978-981-16-3067-5_49

are designed to provide better efficiency without any impact on the environment [1]. Data centers are found everywhere useful and present in government organizations, IT companies, and educational institutions, and even in media [2]. They form the backbone of various IT services such as e-commerce, Web hosting, and social networking services [3]. Since the demand is increasing, the size of data centers also needs to increase which in turn will require more servers and another storage device for its construction. Servers are mainly responsible for the power consumption in the datacenters. Information technology (IT) accounts for 2% of carbon emissions which is almost as same as that of aviation industries.

According to the report of the Data Center Journal, the leading industry it was claimed that by 2017 the traffic on the data center will increase to 30% annually to 7.7 zettabytes [4]. Data centers consume up to 3.5% of global electricity such that it contributes to 250 metric tons of carbon emissions. The amount of data that is stored in a digital form is predicted to be rising shortly, and the amount of energy that is responsible for storing this data is also rising along with it. Harriet Parker, SRI Analyst, in his article "Efficient Energy Datacenters," quoted that it is expected that power use by the data centers is expected to be double every five years [5].

Further research was conducted by McKinsey and Company, "the carbon emissions from Datacenter are expected to quadruple by the end of 2020" [6]. It is a major challenge for us and mainly to the operators of the data centers and the company that is using it. Due to the running data centers, the excessive consumption of energy and carbon dioxide emissions has alarmed the companies concerning to reduce the carbon emissions. The rate by which the data is increasing and the associated carbon footprints from it is noteworthy. There is a need for an efficient framework that helps in the management and maintenance of green datacenters. No framework is perfect in its approach. So, there is a need to construct a framework that is close to perfection such that it can improve the building of data centers. A modular data center can be constructed on this framework.

2 Literature Survey

According to Nor and Selamat, "Review on green data center frameworks" [7], several data center frameworks were mentioned. In this paper, the comparisons of various frameworks have been done to get a look at green data centers which may serve the demanding needs shortly. The green initiatives of each framework are listed and analyzed independently. It focuses on the life cycle cost analysis to be also a metric that is taken to be in consideration while designing an efficient framework for designing a green data center. It is also known as total cost analysis (TCO). TCO analysis is performed for the following [8]:

1. Budgeting and planning.
2. IT equipment life cycle management.
3. To prioritize the proposals of capital acquisitions.

4. Selection of vendor.

By taking total cost analysis or life cycle analysis cost into consideration, the hidden costs can be uncovered. Total cost analysis is the total of the initial capital expenditures (CapEx) and the ongoing or long-term operational expenditures (OpEx). It is one of the important metrics which is to be considered while designing an efficient data center framework or any IT equipment. It is important to identify and weigh the values of the variables of TCO for the building and management of data centers. If by mistake its value is calculated wrongly, then it may cause loss to the company in millions of dollars [9]. It is estimated that 50% of electricity is utilized by data centers, 35% of the electricity to cool the heated IT equipments to keep it in operation, 10 percentage by uninterruptible power supply losses, and only 5 percentage by lighting.

The gain in the efficiency of energy in any of these areas has a major impact on TCO and the yearly operational expenses. It can be seen through a graph drawn below that only a 1% improvement of energy efficiency can lead to huge operational savings. The graph drawn shows the relation between TCO and the efficiency of data centers. Figure 1 shows that only by the improvement of 1% in the energy efficiency in the deployment of UPS at 20 megawatts of the data center. While capital expenditure is constant, the operational expenditure (OpEx) of a UPS over 5 years saves up to $1.3 million with the improvement in the energy efficiency of 1% only (92–93%), showcased in Table 1. The multi-mode UPS offers a 95% energy efficiency of $3 million. So, by considering the TCO model, the effect of only 1 percentage in energy efficiency gain is added up.

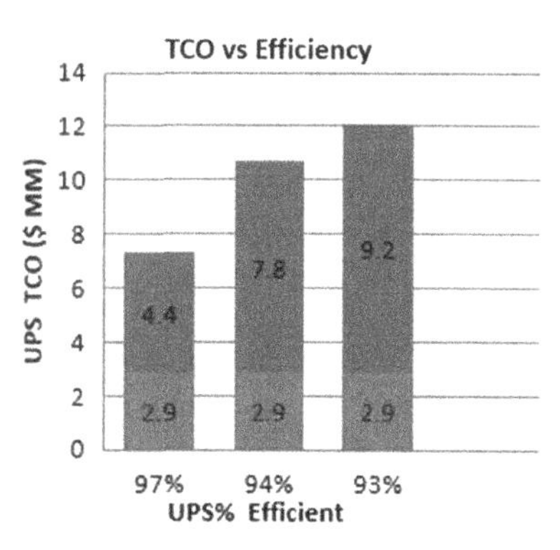

Fig. 1 TCO versus efficiency

Table 1 Tabular representation of TCO versus efficiency

UPS% efficient	97%	94%	93%
CapEx	$2.9	$2.9	$2.9
OpEx	$4.4	$7.8	$9.2
TCO total	$7.3	$10.7	$12.1

3 Comparison of Frameworks

Various frameworks of data centers are compared with each other. Each framework has its shortcomings and robustness. The frameworks which are compared are four pillar framework for energy-efficient high-performance computing data center, green IT framework using virtualization mechanism for data center, EU Code of Conduct by European Commission, efficient data center, data center of Google, data center of Facebook, data center by Energy Star, efficient data center framework by Green Grid.

3.1 Four Pillar Frameworks

The first framework evaluated in this research paper is the four pillar framework by Wilde et al. [10]. This framework is used for the "energy-efficient high-performance computing (HPC) systems." The four pillars are to build the infrastructure, high-performance computing (HPC) hardware, high-performance computing system (HPC) software, and high-performance computing (HPC) applications, as displayed in Fig. 2. The goal of the four pillars framework is to optimize the performance of the data center. Advanced power saving modes should be incorporated. One limitation of the four pillar framework is localized optimization.

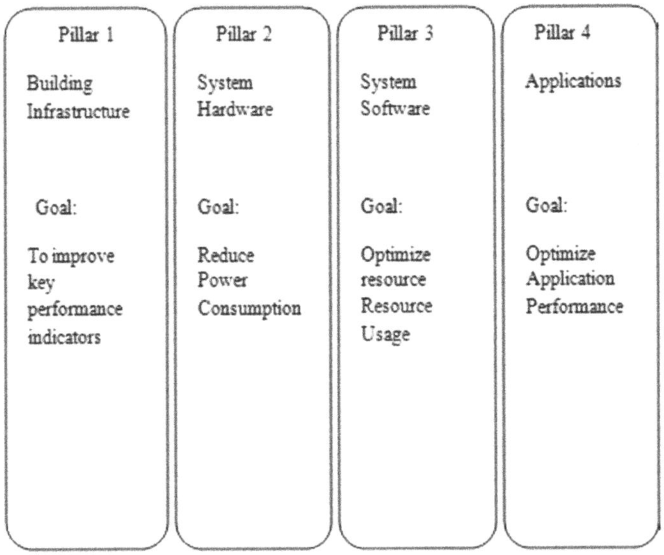

Pillar 1	Pillar 2	Pillar 3	Pillar 4
Building Infrastructure	System Hardware	System Software	Applications
Goal:	Goal:	Goal:	Goal:
To improve key performance indicators	Reduce Power Consumption	Optimize resource Resource Usage	Optimize Application Performance

Fig. 2 Four pillar framework

The first challenge which this framework faces is that based on some control parameters, the cooling infrastructure of the data center operates. It is not suitable to predict the future. It cannot differentiate a temporary power spike where it does not need adjustment due to the increase in the consumption of power by IT computing systems. The second challenge faced is in deciding what the cost optimum temperature of the cooling water should be concerning the surrounding environment. Since from the starting planning and designing phase to the maintenance phase, the human resource factor should be incorporated.

3.2 Energy Star Energy-Efficient Framework

Energy Star program was initiated in 1992 by the Environmental Protection Agency which is a part of the series of the voluntary program which aims at reducing the consumption of energy by taking step toward green initiative. Ways to decrease the consumption of energy by the data center explained by Energy Star Program are virtualization of the server (visualized in Fig. 3), organizing and improving stored data, adjusting the humidity and temperature of the data center, installing a waterside economizer, installing an air-side economizer, retrofitting air conditioners with

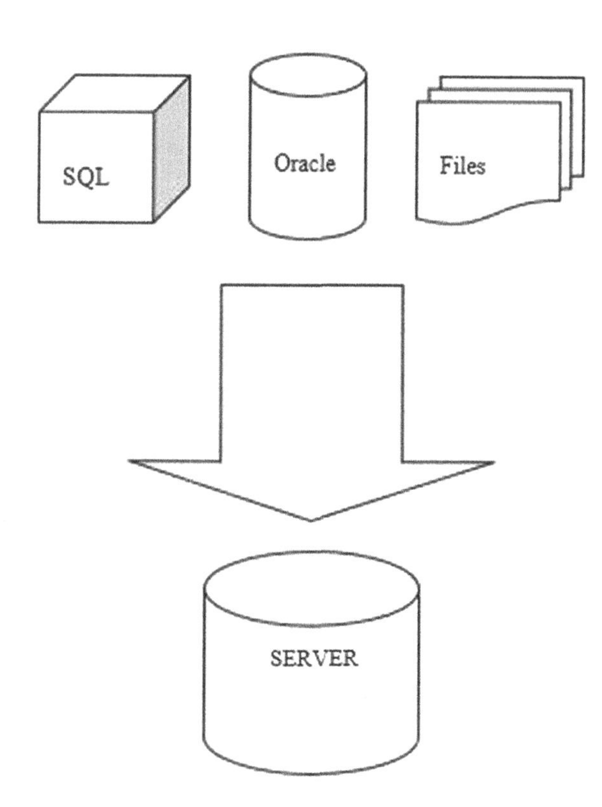

Fig. 3 Concept of virtualization

variable-speed fans, reviewing of the general airflow improvement tips, containing or enclosing server racks, taking advantages of hot/cold aisle layouts, and investing in technologies that are energy-efficient [11]. All these ways are categorized into three main components: HVAC adjustment, airflow management techniques, and IT opportunities. They have thoroughly defined ways to decrease energy consumption in detail.

3.3 Facebook Data Center

Facebook Data center is designed on maximizing the thermal, electrical, and mechanical efficiencies. The architecture of the Facebook data center comprises open racks, triplet columns, and battery cabinet which runs on 277 V AC which helps in permitting more energy into the data center in powering the servers [7]. Servers do not have their power supply rather depends on open rack in which the servers plug themselves to have a power supply at the back of the rack. The triplet columns are of five hundred pounds and can carry servers up to three thousand pounds. Instead of wasting heat produced in the server area of the data center, it is directed to the parts of the data center which require power in terms of heat for its processing. The cooling system of the data centers makes use of a hundred percent of air economization with evaporation.

3.4 Google Data Center

Annually, Google saves millions of dollars and helps in avoiding the thousand tons of carbon emissions due to its sustainable architecture of data center [12]. Google data center uses less than 50% of power than the other typical data center uses [13]. By switching to Google Apps, many companies have experienced fewer computing costs, power consumption, and carbon footprints by 65–90%. The United States General Service Administration has shifted to Google Apps in the year 2012 and shared their experience of how the power consumption was less and carbon footprints are reduced.

3.5 EU Code of Conduct on Data Centers

EU Code of Conduct on Data Centers was managed by the Joint Research Centre-The European Commission [14]. The initiative is taken to deal with the overconsumption of energy by the data center and in need to reduce the related environmental, economic, and energy supply security impacts [15]. The goal of this framework, summarized in Table 2, is to inform the owners and the operators of the data center

Table 2 EU Code of Conduct

Category	Description
Entire datacenter	Expected to apply to all existing IT, mechanical, and electrical equipment within the datacenter
New software	Expected during any upgradation or installation of software
New IT equipment	Expected for new or replacement of IT equipment
New build or retrofit	Expected for any data center built or undergoing a significant refit of the M&E equipment from 2010 onward
Optional practices	Practices without a background color

is to reduce the consumption of energy by the data center in such a way that the functions of the data centers are not hampered. EU Code of Conduct focuses on two areas—First is IT load that is the IT capacity available for the power consumed. Second is facilities load that is the systems that give support to the IT load like cooling systems, UPS, etc. [16]. EU Code of Conduct is a voluntary initiative and allows the stakeholders who are interested in this framework to come together and includes the coordination of other manufactures and vendors. It also lucidly defines the roles and responsibilities of vendors thereby helping in the improvement in building an efficient data center.

3.6 *Data Center Energy-Efficient Framework (DCEEF)*

It is a set of best practices which are developed by the New York State Energy Research and Development Authority (NYSERDA) [17] as an initiative to reduce the consumption of energy and provides the guidelines to the data center operators to evaluate and generate practical remediation road maps or both present and future energy savings. It involves all the best practices from five different domains making use of a holistic approach. The five different domains are facility designs and engineering, information technology, process, governance, and finance. Each domain consists of certain requirements that are tested against the three levels (Level 1, Level 2, and Level 3) of performance. Level 1 requirement is also known as "Low Hanging Fruits." It consists of the requirement which requires less investment in terms of capital and effort. Level 2 consists of a requirement that requires more capital and effort. Level 3 consists of is the last and the final stage in implementing energy efficiency which requires huge capitals and long-term impact on energy efficiency. It is a comprehensive approach that covers almost all aspects of the green data center. Since the requirements are tested against the various levels, it provides a broad outlook for the data center professionals to build the green data center.

3.7 Green Grid Energy-Efficient Data Centers Framework

Energy-efficient data center framework [18] outlines briefly the idea of green data center. Efficiency best practices are listed down along with the check-off box along with the date when it was executed. These best practices for the new data center as well as some can be immediately applied even to the traditional data center. In this framework, the best practices which are vertically represented in Fig. 4 by the Green Grid is in an unsorted manner. It is not prioritized according to its importance and its budget it will in building green data center. Managers of the data center may face difficulty in implementing the green data center by using the attributes represented by the Green Grid since the precedence in terms of its budget is not mentioned which will lead the problem of deciding the less budget green metrics (Table 3).

4 Proposed Green Data Center Framework

Considering the existing parameters, green data centers can be implemented along using the additional proposed parameters. Table 4 contains the additional parameters which are to be considered while building a new data center. Various data center frameworks are analyzed based on their strength and weakness in Table 3. Four pillar data center framework focuses on the optimization of the performance of the data centers. But it should also incorporate the advanced power saving modes. Since human resource plays a vital role in the organization, it should be included from the starting phase of the planning and designing phases until the operational and maintenance phases of the construction of data centers. It has components like facility design and engineering, information technology, process, and finance. Several improvements can be made like we can represent a process as IT management strategies and finance into asset management. Other modifications can be done by adding more components like building design and equipment, monitoring, and testing tools.

Energy Star energy-efficient framework focuses on the concept of server consolidation and aims at the reduction of consumption of energy by taking green initiative. It should also have the highly efficient thermal insulation and roofing and the proper energy-efficient logging and reporting.

Although Google and Facebook possess high gratitude for energy efficiency in the information technology world, they only focus on the important and novel initiatives they adopt. Facebook data centers make use of open racks, triplet columns, and battery cabinet that aid in allowing more energy in powering the servers of the data centers. But it should also work toward the continuous energy efficiency improvement program.

EU Code of Conduct Framework is the oldest approach which is based on best practices guideline. But it should also incorporate formal external auditing. In addition to it, it should include the certification standards which the International Standard Organization scheme includes. Energy Star Guidelines include almost all the

Fig. 4 Green Grid guideline [18]

Check-off Box	Efficiency Best Practice	Date Executed
	Itemized datacenter electric bill in hand	
	Optimization of datacenter design	
	Optimization of data equipment floor layout	
	Proper location of vented floor tiles	
	Rightsizing of UPS	
	Installation of "green" power equipment	
	Installation of a close-coupled cooling architecture	
	Deployment of server virtualization	
	Installation of energy-efficient lighting	
	Installation of blanking panels	
	Installation of efficient plumbing	
	Efficient server consolidation practices	
	Utilization of air conditioner economizer modes	
	Coordination of air conditioners	
	Proper configuration of server software	
	Proper alignment of datacenter staff	

parameters of the green data center. It uses the concept of virtualization which reduces approximately 50% of power consumption. Consolidating the server using the concept of virtualization helps in a 50% reduction of power consumption.

The data center energy efficiency framework makes use of total quality management, six sigma, and also lean six sigma methodologies. These methodologies are the quality standards that are used to improve the efficiency of the data center. Six

Table 3 Data center framework comparison

Sl. No.	Framework name	Advantages	Disadvantages
1	Four pillar framework	Optimizes the performance	No advanced power saving modes. Human resource should be from the start of the planning phase to the maintenance phases
2	Energy Star energy-efficient framework	Concept of server consolidation	Lack of proper energy-efficient logging and reporting
3	Facebook data center	Usage of open racks, triplet columns, and battery cabinet that aid in allowing more energy in powering the servers of the data centers	Lack of continuous improvement program
4	Google data center	Reduces tons of carbon emissions by its efficient architecture of its data centers	Server sleep modes should be enabled. Rainwater can be used to cool down the servers instead of freshwater
5	EU Code of Conduct framework	It outlines the roles of the stakeholders of the data centers	No formal external auditing and no certification standards
6	Data center energy-efficient framework	Use of total quality management and six sigma methodologies	Lack of data center maturity model
7	Green Grid energy-efficient data center framework	Best practices are vertically represented	Lack of prioritization of best practices according to its importance and its budget

sigma methodologies can be used which identifies the peak hours and permits the data center to put the servers at off-mode during non-peak hours. This helps to save unnecessary wastage of energy. It also can propose the use of the data center maturity model which helps in improving energy efficiency. It can help the operators of the data center to compare their performance with others and determine the level of maturity which helps in the improvement of energy efficiency today and the long run.

Green Grid energy-efficient data center framework makes use of the best practices. But they should be prioritized according to its importance and its budget it will in building green data center. Difficulty in implementing the green data center can be avoided by deciding the budget of the green metrics.

Table 4 Suggested parameters for proposed data center

Suggested parameters	Proposed DC	Existing DC
Human resource	✓	✓
Advanced power saving modes	✓	✗
Continuous improvement program	✓	✗
Contain server racks	✓	✓
High efficiency thermal insulation and roofing	✓	✗
Energy efficiency logging and reporting	✓	✗
Rain consumption for cooling	✓	✗
Enable server sleep modes	✓	✗
Server virtualization	✓	✓
External auditing	✓	✗
Data center maturity model	✓	✗
Certification standards	✓	✗
Prioritization of best practices	✓	✗

5 Conclusion

Based upon the existing frameworks, a new framework is designed considering the drawbacks of the existing frameworks. Working upon the disadvantages of the existing data centers, a new data center framework is proposed which makes use of the existing parameters along with the additional parameters. Certain parameters are suggested which should be implemented while building a new data center. A green data center is needed to curb the ever-increasing data rising exponentially and to reduce the carbon emissions and issues of global warming.

References

1. Mueen U, Azizah AR (2010) Server consolidation: an approach to make data centers energy efficient & green. Int J Sci Eng Res 1(1):1–7
2. Daim T, Justice J, Krampits M, Letts M, Subramanian G, Thirumalai M (2009) Data center metrics: an energy efficiency model for information technology managers. Manag Environ Qual Int J 20:712–731
3. Kant K (2009) Data center evolution: a tutorial on state of the art, issues, and challenges. Comput Netw 53(17):2939–2965
4. Shuja J, Madani SA, Bilal K, Hayat K, Khan SU, Sarwar S (2012) Energy-efficient data centers. Computing 94(12):973–994
5. Norhashimi BMN, Mohammed HBS (2015) Green data center frameworks and guidelines review. Int J Comput Inform Syst Ind Manag Appl 7:094–105
6. Kaplan JM, Forrest W, Kindler N (2008) Revolutionizing data center efficiency. McKinsey & Company, pp 1–13

7. Schmidt M (2020) Total cost of ownership TCO for assets and acquisitions—how to uncover the hidden costs of ownership. Business Case Analysis. [Online]. Available: https://www.business-case-analysis.com/total-cost-of-ownership.html
8. Handlin H (2013) Using a total cost ownership (TCO) four your data centers. DataCenter Knowledge. [Online]. Available: https://www.datacenterknowledge.com/archives/2013/10/01/using-a-total-cost-of-ownership-tco-model-for-your-data-center
9. Torsten W, Axel A, Hayk S (2014) The 4 pillar framework for energy efficient HPC data centers. Comput Sci Res Dev 29:241–251
10. Uddin M, Abdul Rahman A, Memon J (2012) Green information technology (IT) framework for energy efficient data centers using virtualization. Int J Phys Sci 7(13):2052–2065
11. Norhashimi BMN, Mohd HBS (2014) Review on green data center frameworks. In: 2014 4th world congress on information and communication technologies, pp 338–343
12. Google Inc. (2011) Google's green data centers: network POP case study. Google User Content. [Online]. Available: https://static.googleusercontent.com/media/www.google.com/en//corporate/datacenter/dc-best-practices-google.pdf
13. Google Inc. Efficiency. Google Inc. [Online]. Available: https://www.google.com/about/datacenters/efficiency/index.html
14. European Commission (2008) Code of conduct on data centres energy efficiency version 1.0. [Online]. Available: https://ec.europa.eu/information_society/activities/sustainable_growth/docs/datacenter_code-conduct.pdf
15. Ashwin S, Kaleem AU (2014) Understanding the maturity of EU code of conduct on data centres: a Mauritian case study. In: 2014 IST-Africa conference proceedings, pp 1–16
16. European Union (2016) Code of conduct for energy efficiency in data centres. EU Science Hub. [Online]. Available: https://ec.europa.eu/jrc/en/energy-efficiency/code-conduct/datacentres
17. New York State (2015) Data centers. New York State Energy Research and Development Authority. [Online]. Available: https://www.nyserda.ny.gov/Business-and-Industry/Data-Centers
18. Aljaberi MA, Khan SN, Muammar S (2016) Green computing implementation factors: UAE case study. In: 2016 5th international conference on electronic devices, systems and applications (ICEDSA), Ras Al Khaimah, pp 1–4. https://doi.org/10.1109/ICEDSA.2016.7818528

Soft Computing

A New Solution for Multi-objective Optimization Problem Using Extended Swarm-Based MVMO

Pragya Solanki and Himanshu Sahu

Abstract Multi-Objective optimization problems corresponds to those problems which are having more than one objective to be optimized together. For such problem rather than an optimal solution, a set of solutions exists which is trade-off among different objectives. There are several solution techniques exist including evolutionary algorithms. Evolutionary algorithm provides Pareto optimal solutions after evolving continuously through many generations of solutions. Mean-variance mapping optimization is a stochastic optimization technique which the swarm hybrid variant works well on single objective optimization problem. The paper aims at extending the swarm hybrid variant of mean variance mapping optimization to a multi-objective optimization technique by incorporating non-dominated sorting and an adaptive local learch strategy. The proposed solution is evaluated on standard benchmark such as DTLZ and ZDT. The evaluation results establish that the proposed solution generates Pareto fronts those are comparable to the true Pareto fronts.

Keywords Multi-objective optimization (MOO) · Evolutionary algorithm · Swarm intelligence · Mean variance mapping optimization-swarm hybrid (MVMO-SH) · Search problems · Heuristics optimization

1 Introduction

In real-world scenario, there are problem involving concurrent optimization of more than one competing and incommensurable problem, known as multi-objective optimization problem (MOP) [1, 2]. For single objective problems, optimization parameters are well defined, whereas for MOP, a set of alternative trade-offs exists. Set of solutions for such problem is known as Pareto optimal solutions [1], which is a

P. Solanki
Department of Computer Science and Engineering, MNNIT, Prayagraj 211004, Uttar Pradesh, India

H. Sahu (✉)
School of Computer Science, University of Petroleum and Energy Studies, Dehradun 248007, Uttarakhand, India
e-mail: hsahu@ddn.upes.ac.in

A. Choudhary et al. (eds.), *Applications of Artificial Intelligence and Machine Learning*, Lecture Notes in Electrical Engineering 778,
https://doi.org/10.1007/978-981-16-3067-5_50

tradeoff among different objectives. Looking at the broad picture, when all objectives are considered, these solutions are optimal as no other solutions with better fitness in that search space.

1.1 Multi-objective Optimization

Def: 1 (Generalized MOP): A Generalized MOP [1] is defined as

Find vector $\overrightarrow{v^*} = \left[v_1^*, v_2^*, v_3^* \ldots v_n^*\right]^{\mathrm{T}}$ which satisfies j inequalities $g(\vec{v}) \geq 0$ $i = 1, 2, 3, \ldots j$ and k equality constraints $h(\vec{v}) = 0$ $i = 1, 2, 3, \ldots k$ and will optimize the vector function

$$\mathcal{F}(\vec{v}) = [\mathcal{F}_1(v), \mathcal{F}_2(v), \ldots \mathcal{F}_k(v)] \tag{1}$$

for k objective to be minimized, a MOP can be described as follows:

$$\min y = \mathcal{F}_1(v), \mathcal{F}_2(v), \ldots \mathcal{F}_k(v) \tag{2}$$

where $\vec{v} = [v_1, v_2 \ldots v_n]^{\mathrm{T}}$ is the vector of V and V is the decision variable space. $y = y_1, y_2, \ldots y_n \in Y$ and Y is the objective space.

1.2 Search and Decision Making in MOP

Search and decision making are two major problem difficulties in solving an MOP. Search is optimization process where sampling is done to find solution set which are feasible for Pareto optimal solutions.

Decision making refers the problem of choosing a plausible compromising solution from the available Pareto optimal set. Based on how the decision process and optimization are combined, there are three following categories for multi-objective optimization methods.

Decision making prior to search: Target of the MOP is combined into one objective which includes all preferences set by the decision maker (DM).

Search before decision making: Without any preference information given, optimization is performed. The final choice of the solution is made by the DM from the outcome of the search which is a set of candidate solutions (preferably Pareto optimal).

Decision making along with search: During the interactive optimization process, the DM can clarify preferences. After every step of optimization, a number of alternative trade-offs is obtained, and on this basis, the DM decides and manages further objective preference and search.

1.3 Evolutionary Algorithms for MOP

Evolutionary algorithms (EA) are stochastic optimization techniques which work on two basic principles, first is "Selection" and the other is "Variation." In the selection process, candidate solutions having higher fitness score are selected to evolve. In the variation process, new candidate solutions are generated using mutation and recombination. These algorithms are trivial in nature yet presents itself as a very powerful technique for search process as it generates several optimal solutions in a single generation evolution. Performance metrics for MOP are generally described as [1, 2]:

- The generational distance between the generated solutions to the solution in the true Pareto front of the test cases.
- Uniform distribution of the generated solutions.

1.4 Swarm Based Optimization

Various algorithms already exist to solve MOO problems. A lot of work has already been done on those algorithms. A hybrid variant of mean variance mapping optimization algorithm is recently developed by incorporating local search and mapping techniques into the basic MVMO algorithm to extend it from a single particle MVMO to swarm-based MVMO [3]. The algorithm works well with the single objective optimization problems. The paper provides an extension of single objective MVMO-SH to a MOP algorithm. It incorporates the concepts of non-dominated sorting and adaptive local search strategy.

The remaining of the paper is organized as: Sect. 2 provides literature survey. Section 3 describes Proposed Methodology. Section 4 provides the implementational details and finally Sect. 5 gives the conclusion and future works.

2 Literature Survey

In the real-world scenario, many problems are posed as multi-objective optimization problems (MOPs). Such problem has a multiple objective to be balanced. Suppose a person needs to buy a car, number of conditions that should be fulfilled before selecting a car from available choices. The person must minimize the cost of the car, the car must have minimum maintenance cost, and the car should give maximum average in kmpl.

Several methods have been already devised for solving MOO problems. These methods are classified into different classes which are listed below.

1. **Scalarizing Methods**: MOO problems are scalarized into single objective problems. One way of doing this is by assigning weights to different decision variables and then combine those weighted decision variables to achieve a single decision variable.
2. **Apriori Methods**: While working with these methods, preference information must be communicated before making any decision using decision variables. Utility method is an example of Apriori method. In utility method, a utility function is defined that would define the preferences while working with decision variables.
3. **Aposteriori Methods**: These methods generate Pareto optimal solutions at every iteration of the algorithm. Evolutionary algorithms belong to this class of MOO problem solving methods.

Evolutionary algorithms are stochastic algorithms which works on natural evolution strategies. In evolutionary algorithms while employing evolutionary strategies, at each new generation, new solutions are evolved which are evaluated for their fitness to determine whether their fitness is better than their ancestors.

These methods work on an initial set of candidate solutions which is being modified at every generation with the new solutions that are developed in the process of evolution. The process of evolution includes selection and variation stages. In selection process, better solutions from the others in the candidate set are selected which are then modified and mutated in the variation stage of evolution.

Various evolutionary algorithms are already developed till now (Fig. 1).

- Schaffers Vector-Evaluated Genetic Algorithm [4]: Schaffer (1984, 1985) presented an MOEA called vector-evaluated genetic algorithm (VEGA). It represents selection by switching objectives. For each k objectives, selection is done separately and equal sized mating pool is filled which are later shuffled for crossover and mutation.
- Hajela and Lins Weighting-based Genetic Algorithm: It is an aggregation selection and parameter variation approach introduced by Hajela and Lin in 1992. It weighs the objectives, and the weighing factor is variable. It produces more than one solution in parallel using different weight combination for each individual.

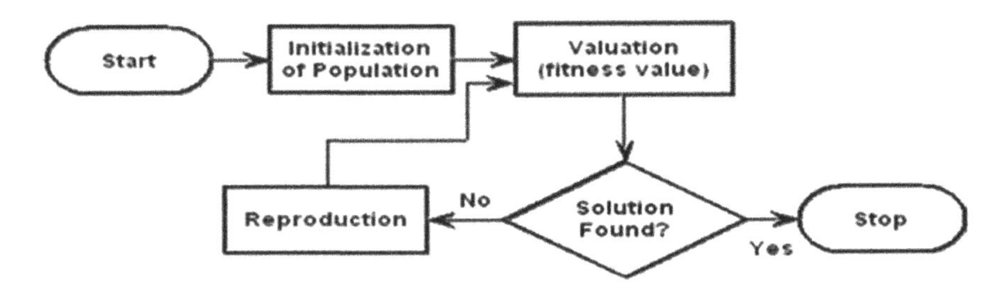

Fig. 1 Evolutionary algorithm iteration flowchart [2]

- Fonseca and Flemings Multi-objective Genetic Algorithm [5]: Fonseca and Fleming proposed a Pareto-based ranking procedure. The rank is given based on number of solutions dominating the solution evaluated in the current population.
- Horn, Nafpliotis, and Goldbergs Niched Pareto Genetic Algorithm [6]: The niched Pareto genetic algorithm (NPGA) is collection of tournament selection and Pareto dominance.
- Srinivas and Debs Non-dominated Sorting Genetic Algorithm-II (NSGA-II) [7]: The NSGA applies fast non-dominated sorting procedure to the combined population of parent set and offspring set generated. The NSGA is modified to include crowding distance algorithm to eliminate the weakness of NSGA.
- Strength Pareto Evolutionary Algorithm [8]: This algorithm uses the concept of Pareto dominance, non-dominated sorting procedure, and clustering techniques to evolve solutions in the next generation which are stored in an external repository.
- Multi-objective Particle Swarm Optimization [9]: It is proposed by Coello and Carlos in 2002. It uses an external repository, non-dominated sorting procedure to extend the traditional particle swarm optimization (PSO).

A recent addition to the heuristic approaches to optimization is the swarm hybrid of mean variance mapping optimization (MVMO-SH) [10]. The focus of the presented work is on extending it to solve MOPs.

2.1 Mean Variance Mapping Optimization

Mean variance mapping optimization (MVMO) is an addition to the heuristic approaches that uses statistical approach to solve a single objective problem. This method uses the values of mean and variance of the of the n-best solutions evolved till now in the process, for the mutation and mapping operation of the algorithm. The basic algorithm of MVMO is incorporated with the process of multi-parent crossover and a local search strategy to increase the search diversity of the algorithm. This variant of the basic MVMO is called the swarm hybrid variant of mean variance mapping optimization (MVMO-SH).

The algorithm begins with the initialization of random population within a range which is normalized in the range [0, 1]. This is followed by the update of archives, fitness evaluation, and local search and finally offspring generation using mutation and mapping. Interested readers can see the thorough description from [10].

2.2 Local Search Algorithm

Local search algorithm is used to generate a candidate solution set using line search strategy over the set of all non-dominated solutions found in the course of optimization till now which dominates the candidate solution as well as finding the neighbor solutions in a neighborhood radius which are better than the candidate solution and a centroid solution from the neighborhood set.

3 Proposed Methodology

The basic idea is to extend the swarm hybrid MVMO single objective optimization algorithm [3, 10, 11] into a multi-objective optimization algorithm by incorporating the concepts of Pareto optimality. This algorithm implements the MVMO-SH in conjunction with non-dominated sorting algorithm to rank the solutions in the population and an adaptive local search strategy to increase the convergence of the algorithm. The algorithm given below describes the proposed algorithm.

Algorithm 1: MOO using MVMO-SH

Input: *N* (population size), **max_count** (Maximum function evaluations)
ls (local search probability)
Output: H (Non-Dominated Set)

1. Generate initial parent population ***P*** within the range [0, 1]. Set $c = 0$.
2. Evaluate fitness for each particle in the population ***P***.
3. Sort population ***P*** using Non-Dominated Sorting and Crowding distance
4. Fill the external repository H with non-dominated solutions
5. Main loop for Multiobjective MVMO
for $c = 1\ to\ maxcount$ **do**
A. Classify the parent population into set of good solutions and bad solutions
B. **for i** $= 1\ to\ n$ **do**
1. Fill/Update individual archives
2. Crossover: Multi-parent for bad solutions otherwise single parent
3. Mapping selected solution's dimensions based on mean and variance
4. Evaluate fitness and store the generated solution,
end for
C. **for i** $= 1\ to\ n$ **do**
1. If random value > **ls** then call Adaptive local search [2]
2. Store all generated solutions L
end for
D. Create intermediate population $\boldsymbol{I_{pop}} = \boldsymbol{P} \cup \boldsymbol{C} \cup \boldsymbol{L}$
E. Sort intermediate population
F. Generate next parent population from sorted $\boldsymbol{I_{pop}}$
G. Store all rank 1 solution in external repository H
end for

4 Result and Analysis

The designed algorithm is implemented on MATLAB. To check the correctness of the algorithm, it has been evaluated over the known DTLZ and ZDT test problem suites. The proposed algorithm is bi-objective in nature, i.e., it takes two objectives to be optimized. All the graphs are generated using two objectives of the DTLZ and ZDT test suites, $f1$ is the first objective, and $f2$ is the second objective calculated as per the standard definitions of the DTLZ and ZDT test suites. The details for various test problems in the DTLZ and ZDT test suites on which our proposed algorithm works well are given in [12, 13].

The proposed algorithm when tested on DTLZ and ZDT test suites gives Pareto optimal fronts which are comparable to the true Pareto fronts for different DTLZ and ZDT problems. The figures (Figs. 2 and 3) shown, presents the graphs for Pareto optimal fronts where (Fig. 2) is for the DLTZ dataset and (Fig. 3) is for ZDT dataset. Every figure is divided in 2 subparts a and b, where (a) represents the Pareto fronts of different test problems generated by the proposed algorithm, and (b) represents the true Pareto fronts generated using Reference data available from [14].

5 Conclusion and Future Work

The proposed algorithm MVMO-SH to solve MOP is tested on well-known DTLZ (1, 2, 3, 4) and ZDT (1, 2) test problems. The Pareto fronts that are obtained using the proposed approach are significantly comparable with the true Pareto front for the above test cases.

The algorithm works fine but has a scope of analysis of different parametric settings as well as different local search strategies could be tried to provide a balance relation between exploration and exploitation. There is a scope of testing and enhancing the work further for non-separable problems which are not truly multimodal in nature.

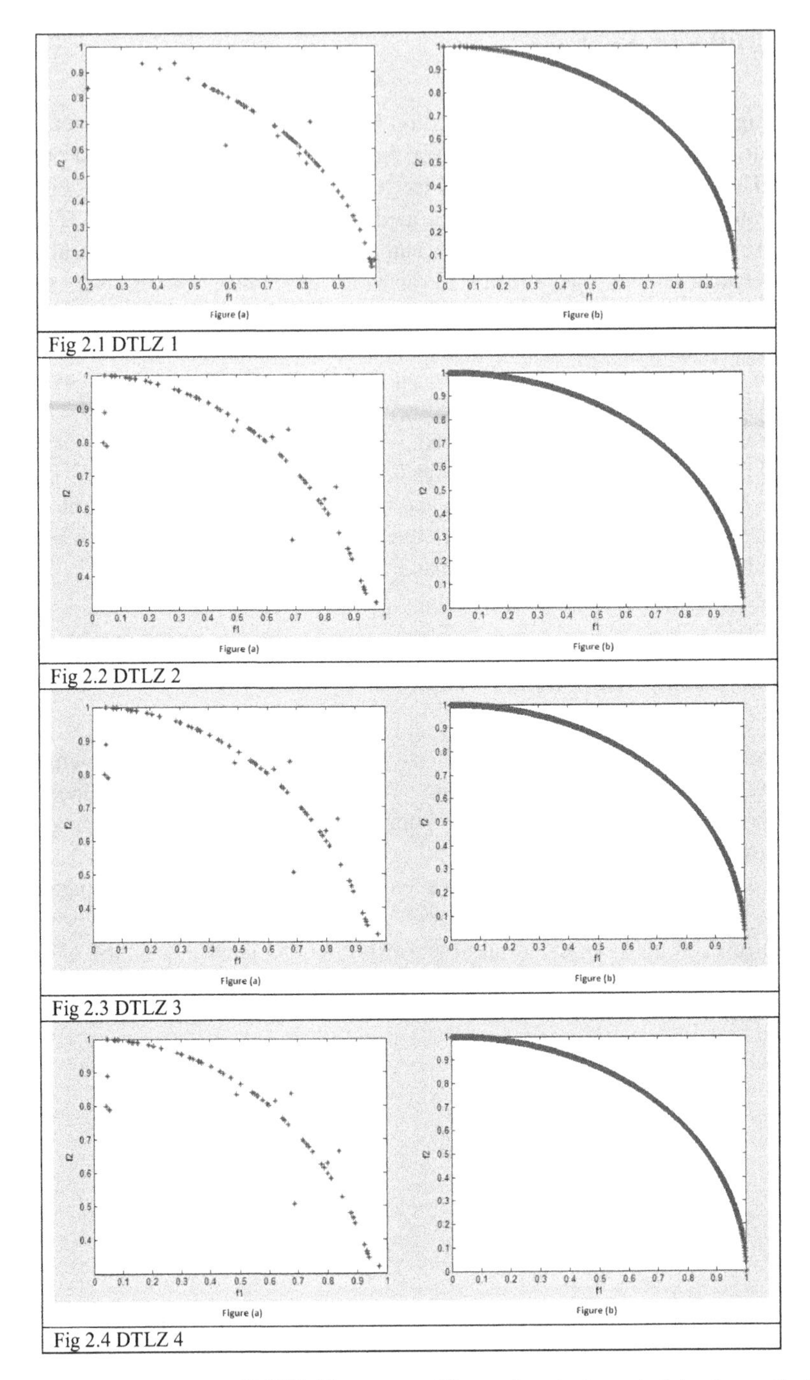

Fig. 2 Result for test cases of DTLZ. The generated Pareto front is shown in (**a**), whereas the true Pareto fronts is shown in (**b**)

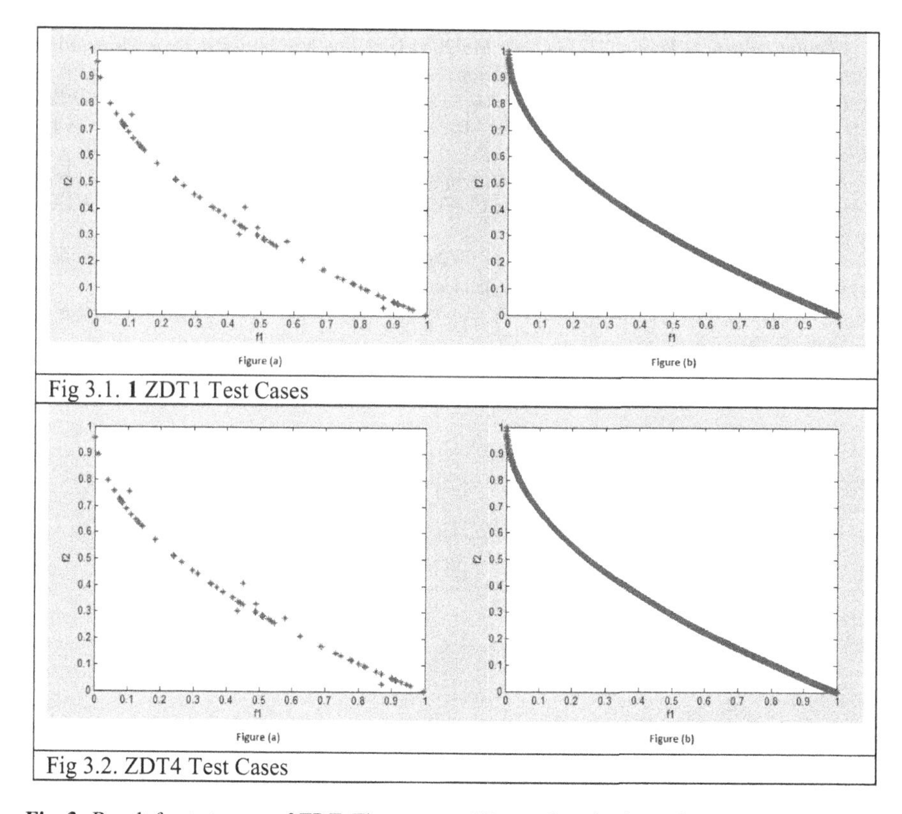

Fig. 3 Result for test cases of ZDT. The generated Pareto front is shown in (**a**), whereas the true Pareto fronts is shown in (**b**)

References

1. Deb K (2001) Multi-objective optimization using evolutionary algorithms
2. Zitzler E, Thiele L (1999) Multiobjective evolutionary algorithms: a comparative case study and the strength Pareto approach. IEEE Trans Evol Comput
3. Erlich I, Venayagamoorthy GK, Worawat N (2010) A mean-variance optimization algorithm. In: 2010 IEEE world congress on computational intelligence, WCCI 2010—2010 IEEE congress on evolutionary computation, CEC 2010
4. Schaffer JD (1985) Multiple objective optimization with vector evaluated genetic algorithms. In: 1st international conference on genetic algorithms
5. Fonseca CM, Fleming PJ (1993) Genetic algorithms for multiobjective optimization: formulation, discussion and generalization. In: ICGA
6. Horn J, Nafpliotis N, Goldberg DE (1994) Niched Pareto genetic algorithm for multiobjective optimization. In: IEEE conference on evolutionary computation—proceedings
7. Deb K, Agrawal S, Pratap A, Meyarivan T (2000) A fast elitist non-dominated sorting genetic algorithm for multi-objective optimization: NSGA-II. In: Lecture notes in computer science (including subseries lecture notes in artificial intelligence and lecture notes in bioinformatics)
8. Zitzler E, Laumanns M, Thiele L (2001) SPEA2: improving the strength Pareto evolutionary algorithm. In: Evolutionary methods for design, optimization and control with applications to industrial problems

9. Coello Coello CA, Pulido GT, Lechuga MS 2004 Handling multiple objectives with particle swarm optimization. IEEE Trans Evol Comput
10. Rueda JL, Erlich I (2013) Hybrid mean-variance mapping optimization for solving the IEEE-CEC 2013 competition problems. In: 2013 IEEE congress on evolutionary computation, CEC 2013
11. Rueda JL, Erlich I (2015) Testing MVMO on learning-based real-parameter single objective benchmark optimization problems. In: 2015 IEEE congress on evolutionary computation, CEC 2015—proceedings
12. Deb K, Thiele L, Laumanns M, Zitzler E (2005) Scalable test problems for evolutionary multiobjective optimization. In: Evolutionary multiobjective optimization
13. Mondanel L (1951) A benchmark study of multi-objective optimization methods. J Fr Med Chir Thorac
14. Benchmark dataset. http://jmetal.sourceforge.net/problems.html. Accessed 2020/08/09

Improving Software Maintainability Prediction Using Hyperparameter Tuning of Baseline Machine Learning Algorithms

Kirti Lakra and Anuradha Chug

Abstract Software maintainability is a prime trait of software, measured as the ease with which new code lines can be added, obsolete ones can be deleted, and those having errors can be corrected. The significance of software maintenance is increasing in today's digital era leading to the use of advanced machine learning (ML) algorithms for building efficient models to predict maintainability, although several baseline ML algorithms are already in use for software maintainability prediction (SMP). However, in the current study, an effort has been made to improve the existing baseline models using hyperparameter tuning. Hyperparameter tuning chooses the best set of hyperparameters for an algorithm, where a hyperparameter is that parameter that uses its value for controlling the training process. This study employs default hyperparameter tuning as well as the grid search-based hyperparameter tuning. Five regression-based ML algorithms, i.e., Random Forest, Ridge Regression, Support Vector Regression, Stochastic Gradient Descent, and Gaussian Process Regression, have been implemented using two commercial object-oriented datasets, namely QUES and UIMS for SMP. To evaluate the performance, a comparison has been made between the baseline models and the models developed after hyperparameter tuning based on the three accuracy measures, viz., R-Squared, Mean Absolute Error (MAE), and Root Mean Squared Logarithmic Error (RMSLE). The results depict that the performance of all the five baseline ML algorithms improved after applying hyperparameter tuning. This conclusion is supported by the improved R-squared, MAE, and RMSLE values obtained in this study. Best results are obtained when the grid search method is used for the tuning purpose. On average, the values of R-squared, MAE, and RMSLE measures improved by 20.24%, 12.26%, and 30.28%, respectively, for the QUES dataset. On the other hand, in the case of the UIMS dataset, an average improvement of 6.27%, 15.71%, and 16.39% has been achieved in terms of R-squared, MAE, and RMSLE, respectively.

K. Lakra (✉) · A. Chug
University School of Information, Communication, and Technology, GGSIP University, New Delhi 110078, India
e-mail: kirti.00516405319@ipu.ac.in

A. Chug
e-mail: anuradha@ipu.ac.in

A. Choudhary et al. (eds.), *Applications of Artificial Intelligence and Machine Learning*, Lecture Notes in Electrical Engineering 778,
https://doi.org/10.1007/978-981-16-3067-5_51

Keywords Software maintainability prediction · Machine Learning · Hyperparameter tuning · Random Forest · Ridge Regression · Support Vector Regression · Stochastic Gradient Descent · Gaussian Process Regression

1 Introduction

Software maintainability is the degree to which certain modifications such as the repair, addition, or omission can be done in the software code. There are two major approaches for measuring the maintainability of software. The first approach includes the extrinsic factors such as readability, modifiability, maintainability, modularity, etc. The second approach involves intrinsic factors like object-oriented (OO) quality metrics, including the size, coupling, and the cohesion metrics. Prediction of external factors is not only time-consuming but also very costly since it involves various additional factors, such as the surrounding environment, and the knowledge and viewpoint of the developers who were engaged in the programming of particular software. Thus, the approach based on the intrinsic factors is the most acceptable approach for software maintainability prediction (SMP). Software maintenance is a crucial factor in determining the total cost of the project. Maintenance can only be measured in later stages of the software development life cycle (SDLC) when the software is almost ready to be delivered, but, by then, it would be too late to streamline the quality of the software, thereby increasing the overall cost [19]. Hence, it becomes essential to build the prediction models for predicting maintainability well in advance, so that the effort and cost is reduced. Several maintainability prediction models employing different evolutionary, Machine Learning (ML), and hybrid techniques have been introduced in the past by various researchers [11, 12, 24]. However, there has always been a concern associated with the performance and precision of the SMP models. Therefore, in this study, hyperparameter tuning has been done to enhance the performance of the baseline ML algorithms [22]. In this technique, a set of best hyperparameters is selected for training purposes to build an optimal model for prediction. Hyperparameter tuning controls the overall behavior of the ML models. The output has further been optimized by using grid search method with hyperparameter tuning. In grid search, sorting of various combinations of parameters and hyperparameters is performed to obtain the best-suited result [4]. Five regression-based algorithms, namely Random Forest (RF), Ridge Regression (RR), Support Vector Regression (SVR), Stochastic Gradient Descent (SGD), and Gaussian Process Regression (GPR) have been used here for building the prediction models using two widely known datasets, viz., QUES and UIMS [14]. The performance is evaluated in conformity with the three accuracy measures, namely R-Squared value, Mean Absolute Error (MAE), and Root Mean Squared Logarithmic Error (RMSLE). The main purpose of the study is outlined as the Research Questions (RQs) below:

RQ1: Does the performance of the baseline algorithms improve with hyperparameter tuning?

RQ2: How much improvement is achieved in the performance of the baseline algorithms after employing hyperparameter tuning?

Further, the paper is organized in the following manner—Sect. 2 talks about the literature review performed for SMP. Section 3 provides the research methodology adopted in the current study; Sect. 4 highlights and discusses the results. Section 5 mentions threats to validity. Section 6 concludes the study and gives insights into future work.

2 Literature Review

There are several techniques and metrics in literature for SMP. Some of the techniques and relevant studies are mentioned in this section. Li and Henry [14], in 1993, studied the authentication of certain OO metrics with maintenance effort for the first time using QUES and UIMS datasets to determine the change per class. Kitchenham et al. [11], in 1993, gave distinct accuracy measures used in the prediction. Tang et al. [24], in 2003, gave six OO design metrics. In 2006, Koten and Gray [12] proposed that the accuracy predicted by Bayesian network model is more vital than, or no less than the regression techniques. Dubey et al. [2], in 2012, utilized multilayered perceptron for SMP. Dash et al. [1], in 2012, proposed the use of artificial neural networks in predicting maintainability of OO system. Kaur and Kaur [10], in 2013, analytically measured 27 distinct ML and regression techniques for SMP. In 2015, Malhotra and Chug [17] proposed the implementation for maintainability prediction using evolutionary models. Also, Malhotra and Chug [18], in 2016, made an effort to review the models, such as the statistical algorithms and ML algorithms to recognize several significant characteristics that can considerably affect the maintenance effort. In 2016, Jain et al. [8] investigated the evolutionary algorithms to check whether these algorithms perform better than ML classifiers. Mathur and Kaushik [21], in 2018, conducted the analysis of data using principal component analysis and determined the correlation matrix. Elmidaoui et al. [3], in 2019, outlined the influence of systematic mapping and summarized the published pragmatic SMP studies. In 2020, Gupta and Chug [5] proposed the cross-project method to predict software maintainability. Also, Gupta and Chug [6], in 2020, introduced an enhanced RF model using feature selection techniques for the purpose of SMP. It is observed that several models have already been employed for SMP. However, there is still a potential to improve the performance of these models. Hence, an attempt has been made in the current work to optimize the predictions of the already existing baseline ML models using hyperparameter tuning along with the grid search.

3 Research Methodology

The current study aims to find the effect of hyperparameter tuning in predicting maintainability. Initially, two OO datasets, namely QUES and UIMS, have been selected for conducting this study. An overall framework used here for SMP is shown in Fig. 1.

Initially, the two datasets have been divided into training and test sets. After that, five different ML algorithms have been applied on the training set for each dataset. Further, the performance has been evaluated in terms of R-squared, MAE, and RMSLE as the accuracy measures using the test set for both the datasets. Since the datasets used here are small-sized; therefore, while training, tenfold cross-validation has also been used. Subsequently, to improve the baseline performance of various

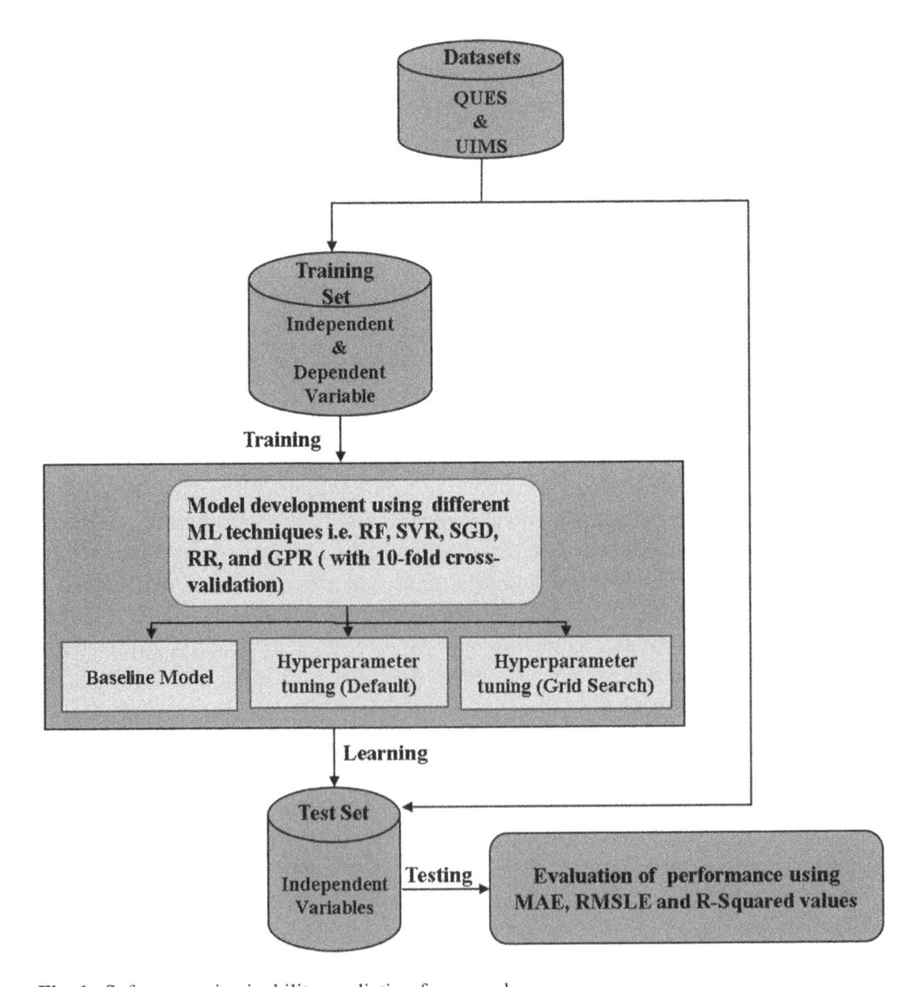

Fig. 1 Software maintainability prediction framework

ML algorithms, hyperparameter tuning has been done. This is followed by hyperparameter tuning using grid search method to further improvise the results. The SMP models developed using hyperparameter tuning have also been evaluated using the above three accuracy measures. The results obtained for all the ML models, with or without hyperparameter tuning have been compared and analyzed as described in Sect. 4.

This section is further subdivided into various subsections, giving a holistic overview of the research methodology adopted in the current investigation.

3.1 Metrics

The two OO datasets, QUES and UIMS utilized in the current study consist of ten independent variables. Out of these, five variables, i.e., DIT, WMC, NOC, LCOM, and RFC, belong to Chidambar and Kemerer suite [24]; the other four variables, i.e., DAC, MPC, SIZE2 and NOM, belong to Li and Henry suite [14]; and the remaining variable, i.e., SIZE1 is taken from traditional lines of code metric. The dependent variable in this study is CHANGE, which is number of changes in code lines with respect to each class.

3.2 Modeling Techniques

ML is a science of getting computers to learn from the experiences and act automatically without being specially programmed. The goal of ML is to make machines act like humans. In this study, five regression-based ML techniques have been used to predict maintainability as explained in Table 1. Although, a number of ML algorithms are available, only five particular algorithms have been selected to limit the scope of the current study. However, an attempt is made to analyze several different categories of ML algorithms by selecting various ML algorithms covering different categories, i.e., RR belongs to the category of generalized linear algorithms, RF is one of the ensemble algorithms, and SGD, GPR, and SVR belong to the class of artificial neural network-based algorithms. This further ensures the generalizability of the current study.

3.3 Hyperparameter Tuning

Hyperparameters are those parameters that define the architecture of the model. The process of searching an ideal architecture for a particular model being developed is known as hyperparameter tuning [15]. Several questions are addressed by hyperparameter tuning, such as, what degree of polynomial features must be used for a

Table 1 Regression-based ML techniques

Modeling techniques	Description
Random Forest (RF) [13]	RF is one of the ensemble methods used for solving regression, classification, and various other problems. RF is a strong modeling technique that can perform efficiently on large datasets having thousands of inputs without dropping the variables. It builds many Decision Trees (DT) during training and gives the right predictions. It tries to create an unrelated forest using techniques like bagging and feature randomness making accurate predictions
Support Vector Regression (SVR) [20]	Support Vector Machine holds up linear and nonlinear regression that is stated as SVR. It tries to categorize all the cases between the two classes, restricting the margin violation. Epsilon is the hyperparameter that controls the width of the street. SVR requires a training set, i.e., $T = \{x, y\}$ covering the domain of interest
Stochastic Gradient Descent (SGD) [7]	Stochastic means 'random.' In SGD, each iteration randomly picks a data point from the entire dataset to decrease the computations immensely, unlike gradient descent. SGD performs all the iterations with a single batch of data, thus, decreasing the computational time to a great extent
Ridge Regression (RR) [9]	Multiple regression-based data that is having multicollinearity is analyzed using RR. A certain degree of bias is added by this technique for estimation, thus reducing standard error. In RR, standardization of independent and dependent variables is done by subtracting their respective mean values and dividing by their standard deviation
Gaussian Process Regression (GPR) [16]	GPR is an extension of linear modeling. It is a nonparametric, kernel-based, probabilistic model used for regression purposes. GPR works well on a small dataset, and it addresses learning in both supervised and unsupervised frameworks

linear model; what is the maximum depth that should be allowed for a DT; what are the minimal samples needed in a DT for the end node; what is the learning rate of gradient descent, etc. Hyperparameters differ from model parameters as they are not trained directly from the data, unlike model parameters, which learn directly during the training phase.

Grid Search. It is a hyperparameter optimization technique, which is also known as exhaustive search [23]. Grid search is used for building models with all combinations of hyperparameter values. Each model having different values of hyperparameters is evaluated, and finally, the most appropriate architecture providing the best result is selected. This is done by fitting each model to the training set and then evaluating it on the validation data.

3.4 Cross-Validation Technique

Cross-Validation is a validation technique that foresees the prediction capability of the model on new and unrevealed data. It ensures that the algorithm is generalized, so that it is suitable for any individual dataset. k-fold cross-validation technique is used in this study and the value of $k = 10$. In this technique, each dataset is divided into ten subdivisions, out of which one is reserved for the testing purpose and the other nine portions are used for training the model till each of the ten partitions has been used for testing once.

3.5 Accuracy Measures

The accuracy measures used in the current study, i.e., R-squared, MAE, and RMSLE, are described as below.

1. **R-Squared**: It is the squared correlation between the real value and the anticipated value. Higher the R-squared value, better is the model. Its value ranges from 0 to 1.

$$\mathbf{R{-}Squared} = 1 - \frac{\sum_{j=1}^{j=m}\left((x_j - \widehat{x_j})\right)^2}{\sum_{j=1}^{j=m}\left((x_j - \overline{x})\right)^2} \tag{1}$$

Here, x_j and $\widehat{x_j}$ represent the real value and the predicted value, respectively, and $\overline{x}$ is the mean value.

2. **Mean Absolute Error (MAE)**: It is the ratio of the difference between the real and the anticipated value to the total number of classes.

$$\mathbf{MAE} = \frac{A_j - P_j}{A_j} \tag{2}$$

Here, A_j and P_j represent the real value and the anticipated value, respectively.

3. **Root Mean Squared Logarithmic Error (RMSLE)**: In RMSLE, a log of the predicted and the real value is taken. It is same as Root Mean Squared Error (RMSE) if real, and predicted values are small.

$$\mathbf{RMSLE} = \sqrt{\frac{1}{m}\sum_{j=1}^{m}\left(\log(P_j + 1) - \log(A_j + 1)\right)^2} \tag{3}$$

A_j is the real value, and P_j is the predicted value.

4 Results and Analysis

The current section highlights the results of the study conducted for SMP using various regression-based ML algorithms for Li and Henry datasets. The results obtained in the form of R-squared, MAE, and RMSLE values for all the ML algorithms with and without hyperparameter tuning are shown in Figs. 2 and 3 for QUES and UIMS datasets, respectively.

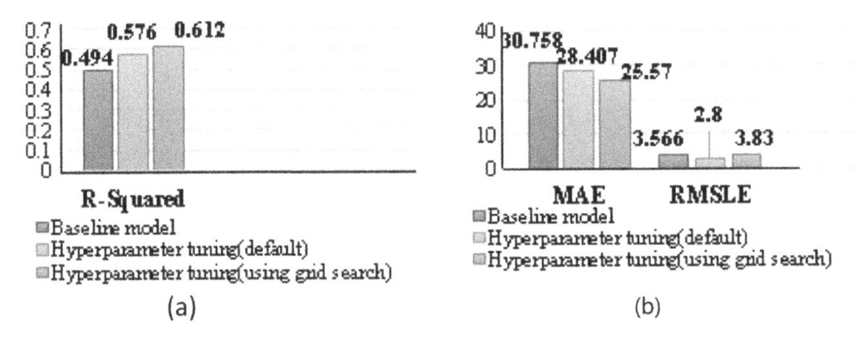

Fig. 2 Values of **a** R-squared, **b** MAE, and RMSLE for different ML algorithms, without hyperparameter tuning, with hyperparameter tuning (default) and with hyperparameter tuning (using grid search) for QUES dataset

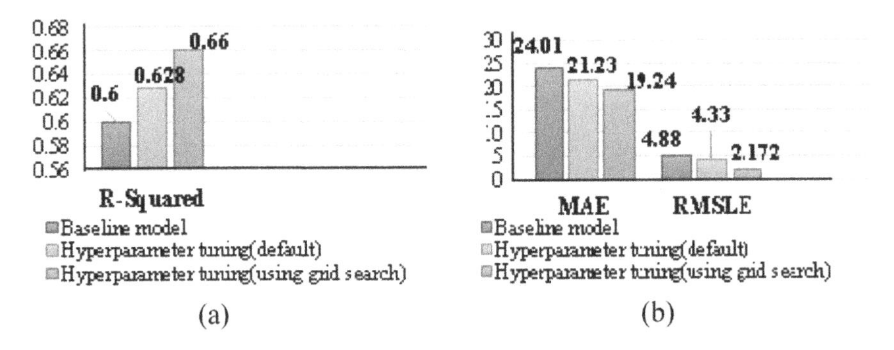

Fig. 3 Values of **a** R-squared, **b** MAE, and RMSLE for different ML algorithms, without hyperparameter tuning, with hyperparameter tuning (default) and with hyperparameter tuning (using grid search) for UIMS dataset

Further, the hyperparameters obtained for all the five algorithms using QUES and UIMS datasets have been represented in Tables 2 and 3, respectively. In both the tables, two sets of hyperparameters are given for each algorithm, i.e., hyperparameter tuning (default) and hyperparameter tuning (using grid search). These sets of hyperparameters are referred as the best parameters since they resulted in a more effective prediction as compared to the other hyperparameters, i.e., they have maximum contribution in the optimization of the baseline algorithms.

It is observed from Table 2 that in case of QUES dataset, the default value of 'min_sample_leaf' hyperparameter is '1' for RF algorithm, which changes to '3' on using grid search with hyperparameter tuning. Similarly, for RR algorithm, the default value of 'random_state' parameter changes from 'none' to '10' on using grid search with hyperparameter tuning. Further, other changes in the default values of the

Table 2 Hyperparameters for QUES dataset

Algorithms	With hyperparameter tuning (default)	With hyperparameter tuning (using grid search)
Random Forest	max_depth = None max_features = 'auto' max_samples = 88 min_samples_leaf' = 1 min_samples_split = 2 n_estimators = 10	max_depth = 10 max_features = 0.5 max_samples = 88 min_samples_leaf' = 3 min_samples_split = 4 n_estimators = 8
Ridge Regression	copy_X = True fit_intercept = True max_iter = None random_state = None solver = 'auto' tol = 0.001	copy_X = True fit_intercept = 2 max_iter = 88 random_state = 10 solver = 'auto' tol = 20
Support Vector Regression (SVR)	cache_size = 200 degree = 3 gamma = scale kernel = rbf max_iter = −1 verbose = False	cache_size = 170 degree = 5 gamma = scale kernel = rbf max_iter = −1 verbose = False
Stochastic Gradient Descent (SGD)	tol = 0.001 shuffle = True verbose = 0 fit_intercept = True random_state = None max_iter = 100	tol = 70 shuffle = True verbose = 0 fit_intercept = True random_state = None max_iter = 1000
Gaussian Process Regressor (GPR)	alpha = 1e−10 kernel = 1**2*RBF (length_scale = 10) random_state = None n_restarts_optimizer = 10 copy_X_train = True	alpha = 1e−10 kernel = 1**2*RBF (length_scale = 10) random_state = None n_restarts_optimizer = 50 copy_X_train = True

Table 3 Hyperparameters for UIMS dataset

Algorithms	With hyperparameter tuning (default)	With hyperparameter tuning (using grid search)
Random Forest	max_depth = None max_features = 'auto' max_samples = 31 min_samples_leaf' = 1 min_samples_split = 2 n_estimators = 100	max_depth = 3 max_features = 0.5 max_samples = 31 min_samples_leaf = 11 min_samples_split = 12 n_estimators = 70
Ridge Regression	copy_X = True fit_intercept = True max_iter = None random_state = None solver = 'auto' tol = 0.001	copy_X = True fit_intercept = 14 max_iter = 88 random_state = None solver = 'auto' tol = 80
Support Vector Regression (SVR)	cache_size = 200 degree = 3 gamma = scale kernel = rbf max_iter = −1 verbose = False	cache_size = 100 degree = 10 gamma = scale kernel = rbf max_iter = −1 verbose = False
Stochastic Gradient Descent (SGD)	tol = 0.001 shuffle = True verbose = 0 fit_intercept = True random_state = None max_iter = 1000	tol = 60 shuffle = True verbose = 0 fit_intercept = 10 random_state = None max_iter = 1000
Gaussian Process Regressor (GPR)	alpha = 1e−10 kernel = 1**2*RBF (length_scale = 10) random_state = None n_restarts_optimizer = 10 copy_X_train = True	alpha = 1e−10 kernel = 1**2*RBF (length_scale = 20) random_state = None n_restarts_optimizer = 3 copy_X_train = True

hyperparameters for various ML algorithms can be looked up in Table 2 for QUES dataset.

Afterward, it is clearly observable from Fig. 3, that the performance of the baseline algorithms has been successfully optimized using hyperparameter tuning for UIMS dataset.

Likewise, as given in Table 3, the default value of 'degree' parameter for UIMS dataset using SVR algorithm changes from '3' to '10'. Furthermore, the 'tol' parameter of SGD algorithm changes from '0.001' to '60' on using grid search with hyperparameter tuning. Other changes in the default parameters using different ML algorithms for UIMS dataset can be looked up in Table 3.

The two RQs raised in Sect. 1 have been answered as follows:

RQ1: Does the performance of the baseline algorithms improve with hyperparameter tuning?

As it is seen from Figs. 2 and 3, a positive change has been observed on using hyperparameter tuning with the baseline algorithms. The R-Squared values in Figs. 2a and 3a for both the datasets and for all the ML algorithms has increased, showing that the models have become more precise and accurate on using hyperparameter tuning. Further, the decreased values of MAE and RMSLE, as shown in Figs. 2b and 3b for all the datasets and algorithms, show an improvement in performance on applying hyperparameter tuning. This shows that the prediction error has decreased, i.e., the predicted value has come closer to the actual value. The accuracy of these models has improved further when hyperparameter tuning is accompanied by the grid search method.

RQ2: How much is the improvement in the performance of the baseline algorithms after employing hyperparameter tuning?

According to Figs. 2 and 3, the R-squared, MAE, and RMSLE values have improved significantly after hyperparameter tuning of the baseline algorithms, showing that the SMP models have become more efficient. The values of the above-mentioned accuracy measures have further increased on using the grid search method with hyperparameter tuning for both the datasets, i.e., QUES and UIMS, and all the ML algorithms. It is observed from Fig. 2a that an average improvement equal to 20.24% in R-squared values has been achieved for QUES datasets, considering all the five algorithms after hyperparameter tuning. However, the MAE and RMSLE values in the case of the QUES dataset have improved by 12.26% and 30.28%, respectively, as shown in Fig. 2b. Similarly, for UIMS dataset, an average improvement of 6.27% has been attained for R-squared values considering all the algorithms as depicted in Fig. 3a. Further, as seen from Fig. 3b, MAE and RMSLE values have improved by 15.71% and 16.39%, respectively, for UIMS dataset.

The results obtained above and the answers to both the RQs helped us conclude that the tuned results are better than the untuned baseline results because hyperparameter tuning has enhanced the learning capability of the baseline algorithms, making them super learners. These results further improved when hyperparameter tuning has been optimized using grid search method, which sorts through different combinations of parameters to give the best possible result.

On comparing the results of the current study with a few previous studies, it has been found that the proposed approach yields better results as compared to the prior studies. For example, the MAE values obtained by Mathur and Kaushik [21] for different algorithms using principal component analysis were higher than the MAE values obtained in the current study. The best MAE values as obtained in this study improved by 71.24% (from 89.22 to 25.57) and 53.92% (from 41.75 to 19.24), respectively, as compared to Mathur's study. Also, for UIMS dataset, improved results are achieved in the proposed study in terms of MAE, when compared to Dash et al.'s [1] study with an overall improvement of 39.71% (from 31.908 to 19.24). Further, the current study achieves improved R-squared values equal to 0.61 and 0.66 for

QUES and UIMS datasets, respectively, in comparison with R-squared values equal to 0.55 and 0.54 as obtained in one of the studies conducted by Dubey et al. [2] using principal component analysis.

5 Threats to Validity

The current section outlines various threats to validity. Main threat associated with this study is the usage of Li and Henry datasets, that are small-sized and have been developed using a specific programming language called Ada. So, the current study may not generalize for large datasets written in some other programming language. Also, research is required to generalize the results of the current study for other programming environments. Here, only one variation of hyperparameter tuning has been used, i.e., grid search method, which again becomes a limitation of this study. Further, only three accuracy measures have been used for comparing the performance, even though other accuracy measures are also available.

6 Conclusion and Future Scope

The primary purpose of the current study is to improve the performance of the baseline algorithms for SMP. Five different regression-based ML algorithms (RF, SVR, SGD, RR, and GPR) have been employed for the purpose of maintainability prediction on QUES and UIMS datasets. The primary goal of this study is to construct a suitable and optimal model using hyperparameter tuning of the above ML algorithms, for improving the baseline prediction. Default hyperparameter tuning and hyperparameter tuning with grid search method have been used apart from the baseline algorithms. R-squared, MAE, and RMSLE have been used for evaluating the performance. The timely and correct predictions will assist the technologists in utilizing the resources judiciously, increase efficiency, and minimize the corresponding maintenance costs. The current study resulted in an overall improvement of the baseline models after hyperparameter tuning. The results are summarized as follows:

- For QUES dataset, it is observed that an overall average improvement of 20.24%, 12.26%, and 30.28% has been attained in terms of R-squared, MAE, and RMSLE values, respectively, using all the five algorithms after employing hyperparameter tuning.
- Similarly, for UIMS dataset, hyperparameter tuning of the baseline algorithms resulted in an average improvement of 6.27%, 15.71%, and 16.39% with respect to R-squared, MAE, and RMSLE values, respectively, for all the ML algorithms taken together.

- The comparison between the results of the current study and the results of a few previous studies further indicates the supremacy of the proposed approach over several other approaches while predicting maintainability.

Hence, this study would allow the developers to investigate the capability of SMP models further and finally establish an efficient and universal model in the field of software maintenance for improving software quality. As a future extension to this work, authors plan to make hybrid models for SMP that would combine with other techniques of hyperparameter tuning using several other ML, evolutionary, or meta-heuristic techniques. This would result in building such software that would predict more accurately with the least precision errors.

References

1. Dash Y, Dubey SK, Rana A (2012) Maintainability prediction of object oriented software system by using artificial neural network approach. 9111204392:420–423
2. Dubey SK, Rana A, Dash Y (2012) Maintainability prediction of object-oriented software system by multilayer perceptron model. ACM SIGSOFT Softw Eng Notes 37:1–4. https://doi.org/10.1145/2347696.2347703
3. Elmidaoui S, Cheikhi L, Idri A, Abran A (2019) Empirical studies on software product maintainability prediction: a systematic mapping and review. E-Informatica Softw Eng J 13:141–202. https://doi.org/10.5277/e-Inf190105
4. Ghawi R, Pfeffer J (2019) Efficient hyperparameter tuning with grid search for text categorization using kNN approach with BM25 similarity. Open Comput Sci 9:160–180. https://doi.org/10.1515/comp-2019-0011
5. Gupta S, Chug A (2020) Assessing cross-project technique for software maintainability prediction. Procedia Comput Sci 167:656–665. https://doi.org/10.1016/j.procs.2020.03.332
6. Gupta S, Chug A (2020) Software maintainability prediction using an enhanced random forest algorithm. J Discret Math Sci Cryptogr 23:441–449. https://doi.org/10.1080/09720529.2020.1728898
7. Hajizadeh N, Keshtgari M, Ahmadzadeh M (2014) Assessment of classification techniques on predicting success or failure of Software reusability. CoRR abs/1409. 2:5–10
8. Jain A, Tarwani S, Chug A (2016) An empirical investigation of evolutionary algorithm for software maintainability prediction. In: 2016 IEEE students' conference on electrical, electronics and computer science SCEECS 2016, pp 1–6. https://doi.org/10.1109/SCEECS.2016.7509314
9. Jha S, Kumar R, Hoang Son L, Abdel-Basset M, Priyadarshini I, Sharma R, Viet Long H (2019) Deep learning approach for software maintainability metrics prediction. IEEE Access 7:61840–61855. https://doi.org/10.1109/ACCESS.2019.2913349
10. Kaur A, Kaur K (2013) Statistical comparison of modelling methods for software maintainability prediction
11. Kitchenham B, MacDonell SG, Pickard L, Shepperd M (1999) Assessing prediction systems. The information science discussion paper series
12. Koten C, Gray AR (2006) An application of Bayesian network for predicting object-oriented software maintainability. Inf Softw Technol 48:59–67. https://doi.org/10.1016/j.infsof.2005.03.002
13. Laradji IH, Alshayeb M, Ghouti L (2015) Software defect prediction using ensemble learning on selected features. Inf Softw Technol 58:388–402. https://doi.org/10.1016/j.infsof.2014.07.005
14. Li W, Henry S (1993) Object-oriented metrics that predict maintainability. J Syst Softw 23:111–122

15. Maher M, Sakr S (2019) SmartML: a meta learning-based framework for automated selection and hyperparameter tuning for machine learning algorithms. Adv Database Technol. EDBT 2019-March 554–557. https://doi.org/10.5441/002/edbt.2019.54
16. Malhotra R, Chug A (2012) Software maintainability prediction using machine learning algorithms. Softw Eng Int J 2:19–36
17. Malhotra R, Chug A (2015) Application of evolutionary algorithms for software maintainability prediction using object-oriented metrics. https://doi.org/10.4108/icst.bict.2014.258044
18. Malhotra R, Chug A (2016) Software maintainability: systematic literature review and current trends. Int J Softw Eng Knowl Eng 26:1221–1253. https://doi.org/10.1142/S0218194016500431
19. Mamone S (1994) The IEEE standard for software maintenance
20. Mantovani RG, Rossi ALD, Vanschoren J, Bischl B, De Carvalho ACPLF (2015) Effectiveness of random search in SVM hyper-parameter tuning. In: Proceedings of the international joint conference on neural networks, Sept 2015, pp 1–8. https://doi.org/10.1109/IJCNN.2015.7280664
21. Mathur B, Kaushik M (2018) Data analysis utilizing principal component analysis. Int J Eng Res Technol 11:333–348
22. Schratz P, Muenchow J, Iturritxa E, Richter J, Brenning A (2018) Performance evaluation and hyperparameter tuning of statistical and machine-learning models using spatial data. https://doi.org/10.1016/j.ecolmodel.2019.06.002
23. Shekar BH, Dagnew G (2019) Grid search-based hyperparameter tuning and classification of microarray cancer data. In: 2019 2nd international conference on advanced computational and communication paradigms ICACCP 2019. https://doi.org/10.1109/ICACCP.2019.8882943
24. Tang M-H, Kao M-H, Chen M-H (2003) An empirical study on object-oriented metrics, pp 242–249. https://doi.org/10.1109/metric.1999.809745

Karaoke Machine Execution Using Artificial Neural Network

R. Sripradha, Plauru Surya, Payreddy Supraja, and P. V. Manitha

Abstract Musicians and vocalists are facing the challenge of practicing their singing as instrumentalists are not available always or they are not very affordable. Karaoke can help to solve this problem and cater to the demands of these singers. When the vocals of a song have been removed and only the accompaniment or the instrumental background is left, the resulting music is called karaoke. This is used as a form of entertainment, wherein users sing along with the background music. This paper proposes a method to generate karaoke using the artificial neural network (ANN) tool in MATLAB. This is based on the out-of-phase stereo method. First, the training data is generated using the out-of-phase ssstereo method using audacity, to check its effectiveness; then, the same is implemented in MATLAB, and the generated data is used to train the artificial neural network. There is room of improvement in the proposed system, as it has been implemented with limited training data.

Keywords Karaoke · ANN · Neural network · Out-of-phase stereo method

1 Introduction

Karaoke originates from Japan, and it is a form of recreational activity in which an amateur singer sings along with the recorded music. This recorded music can be instrumental music performed by different artists or it can be the original music from which the singer's voice has been removed. The latter is performed by karaoke machines, and singers often prefer these as they feel closer to the original version of the song itself. There is often a video playing along with the music itself with the lyrics and cues as to when each line begins. It is often not possible to replicate the exact maneuvers performed by the instrumentalists in the original sound track. A reproduction of this kind is heavy on man-power, labor, and cost. Generating a

R. Sripradha (✉) · P. Surya · P. Supraja · P. V. Manitha
Department of Electrical and Electronics Engineering, Amrita School of Engineering, Amrita Vishwa Vidhyapeetham, Bengaluru, India

P. V. Manitha
e-mail: pv_manitha@blr.amrita.edu

A. Choudhary et al. (eds.), *Applications of Artificial Intelligence and Machine Learning*, Lecture Notes in Electrical Engineering 778,
https://doi.org/10.1007/978-981-16-3067-5_52

karaoke also means splitting the music or melody part and the vocals, hence it is also called audio-source separation. More generally, this is also the cocktail party problem—in a party or a crowded place, where there is a lot of noise and one person is talking, humans can naturally focus on this particular voice and not pay attention to the others. Computers do not have the capability to separate voices, in other words, have the ability to concentrate on only one, unlike the human auditory system.

It has many day-to-day applications that include but are not limited to improving the user's ability to sing through practice, family entertainment, singing contests, and studios. The separated vocals can also be used for content search and lyric generation. Extended applications include separating two or more voices, extracting only the voice of a person from a noise background, improving communication, and identification of a person based on voice which can prove invaluable to the law enforcers.

There are a number of ways of suppressing the vocals in a sound track, and these include using a low-pass filter to remove the voice [1]. This comes with the inherent disadvantage of the bass music also getting eliminated as the low-pass filter filters out all low frequencies including the voice [2].

The approximate fundamental frequency of the singer's voice was found, and this was used to eliminate the narrow bands containing this frequency and all its harmonics. In a case where there is no echo or where different synthetic techniques have not been used to modify the voice, it may have been possible. Out-of-phase stereo method does not suffer from these restrictions, and it can be used to a majority of the existing songs.

In this paper, the dataset has been generated using out-of-phase stereo method suggested in [1]. This dataset is used to train the artificial neural network to generate karaoke. The karaoke so generated is at par with the out-of-phase stereo method in terms of performance, because a neural network can only be as good as its training data. The novelty of the method lies in the fact that ANN is used for the generation of karaoke which has not been used anywhere else. This has been done using DFT in [3]. We believe that when better training data is obtained with human intervention—only background music or only voice—ANN can be used to automate this process.

The organization of the paper is as follows. Section 2 contains the literature survey. Section 3 contains a brief introduction to artificial intelligence (AI) and ANN, and this is followed by an explanation and block diagram of the proposed method. Sections 5 and 6 elucidate the observations, results, and the conclusion of the paper. The last section has references.

2 Literature Survey

The out-of-phase stereo method can be used on stereo music—music that has two different channels, one right and one left [2]. This method is simple and efficient and accurate to a certain degree. It is based on the assumption that the singer stands in the center and the musical instruments surround him/her. If the vocalist is spatially centered, the intensity of voice on each mic placed at different locations will be similar. In this method, the left signal is subtracted from the right signal, or vice-versa. The component common and equal in intensity to both the channels is the voice, and hence, it gets removed. The disadvantage posed in this method is within the assumption mentioned above, if it does not satisfy that criteria or is an instrument is centered with the voice, that instrument will also get removed along with the voice. FFT can be used to compare the powers of the left and right channel and can be used to separate them like the out-of-phase stereo method [2].

Ideal binary masks have been used with a probabilistic deep neural network (DNN) to separate vocal and real-time musical mixtures [4]. The music signals are converted into spectrograms in this method using short time Fourier transform (STFT). A binary mask is then used to remove the vocals from the accompaniment. The disadvantage of this method lies in its complexity.

Spatial audio object coding (SAOC) consists of an encoder, decoder, and renderer. The encoder produces a down-mix signal and spatial parameters from input audio and transmits it to the decoder. The SAOC uses the discrete Fourier transform (DFT) to do so. In this method, harmonics are extracted and eliminated [5]. The harmonic extraction is done from a clean vocals, and it involves pitch extraction, harmonic amplitude extraction, and MVF extraction. Harmonic elimination is then done using a filter that is designed based on the vocal spatial parameters and down-mix signal. Elimination of harmonics improves the karaoke.

Repeating pattern extraction technique (REPEAT) is a method in which the repeating background of songs is eliminated [6]. Songs normally contain a repeating background, and the vocals are superimposed over this. In the REPEAT method, a time–frequency mask is used to separate the repeating component. The size of window for detecting the repeating piece of accompaniment is varied for different components. This approach is simplistic and fast. It will not work if the accompaniment is non-repeating. While it can be used for karaoke generation in many contexts, it will not help the machine in solving the cocktail party problem.

Accompaniment separation from polyphonic music is executed based on automatic melody transcription [7]. First, the transcription takes place and then this transcription, consisting of a parametric representation of the lead vocals (MIDI note sequence with fundamental frequency trajectory for each note), is used to estimate, synthesize, and remove lead vocals using sinusoidal modeling, thus separating the lead and accompaniment which is similar to a karaoke.

Stereo type noise suppression is used for extracting the accompaniment signal or the karaoke part from the stereo music signal [8]. First, the vocals signals are extracted by finding the difference of the left and right spectrum of the signals in the frequency domain, which is first obtained through short Fourier transform (stft). After this is done, the stereo accompaniment signal is estimated by utilizing the canonical state space modeling and the Kalman filter theory.

It is assumed that the accompaniment is periodic and that the vocals are relatively aperiodic, like in the REPEAT method. The short time Fourier transform is used to convert the audio into an image, and periodically repeating patterns in this image is appears as peaks in the 2DFT spectrogram pattern recognition (image processing) is used to remove these repeating patterns. This separates the music from the voice [3].

Repeating nature of accompaniment is exploited to separate the vocals from the background music. Empirical wavelet transform (EWT), suitable for both non-stationary and non-linear signals, is first applied on the given signal to decompose it, and then, the repeating background is found and separated. EWT is applied for different frequencies to extract all repeating components [9].

Particle swarm optimization (PSO) can be used to find filter coefficients. These coefficients correspond to the different signals generated by instruments, as each note of a particular instrument has a repeating waveform. These filter coefficients are used to generate wavelets that have a very high correlation with these instruments and can therefore be used to analyze these signals [10].

Wavelets present in musical instruments, which are repeating notes, can be defined by scaling functions and wavelet functions [11]. Reconstruction of wavelets makes it possible to develop musical notes for instruments. This can be used for blind source separation of musical mixtures [12].

Out-of-phase stereo method is used for training and karaoke generation in the proposed method. If a stereo music is presented to this trained network, it generates karaoke, which is similar to the karaoke generated by the out-of-phase stereo method. ANN can thus be used for karaoke generation and better training algorithm can yield better results.

3 Artificial Neural Network

Artificial intelligence deals with the simulation of human intelligence in machines. In 1997, the first breakthrough in AI occurred, when IBM's deep blue became the first computer to beat a chess champion. Recent research in neurology has shown that the brain is an electrical network of neurons fire in all-or-nothing pulses. Digital signals also work in the form of 1's and 0's—all or nothing. Since any form of computation can be performed digitally and digital signals are quite similar to brain signals, it follows that it may be possible to construct an electronic brain.

Deep learning is a branch of machine learning that teaches computers to learn like humans, that is learn by example. ANN is one such deep learning model that is usually used for regression and classification. ANNs are connectionist systems have been designed to copy the biological neural networks that we possess. These systems learn to perform tasks by example, just as humans do. A very common example of a machine "learning by example" is image recognition and classification.

Regression is a statistical model where we try to find an outcome based on one or more variables called predictor variables. Regression can be classified based on linearity as linear and nonlinear and based on the number of predictor variables as simple regression (consists of one predictor variable) and multiple regression (consists of many predictor variables).

The training data considered for this paper has been generated using the out-of-phase stereo method. The relation between the input and output is linear, and the there is only one input. Classification of this model is simple linear regression, as explained above.

The nodes/neurons of ANN model the neuron in the brain, the connection between them model the synapses. The connection is real number, and the output of each neuron is modeled by nonlinear function. These connections have weights that affect the amount of or the strength of the signal to be considered. An initial weight is set, it then changes as the learning proceeds. Neurons send a signal only if the aggregate is greater than a certain value. Many neurons are connected together in a layer. Neurons of one layer are connected to the neurons of the immediately preceding and next layer. The layer that comes first and receives the input data is the input layer. The final layer that gives the result is called the output layer. These layers are of different types, e.g., fully connected, pooling, etc. Each layer performs some transformation or computation on the input signal. The network formed can be feed-forward networks or recurrent networks. The feed-forward networks do not have memory and never form a loop. Hyperparameters are some of the values that are set before the beginning of the learning process, and they do not change parameters like the number of hidden layers of neurons, learning rate, etc.

The neural network "learns" through examples by adjusting the weights to optimize the accuracy of the result, which is obviously carried out by decreasing the errors. The learning process is said to have been completed if additional examples do not decrease the errors or cause a change in the weights. A subtle shift in weights and errors continues to occur, and it never becomes equal to zero; it only tends to zero, so a threshold is provided. Learning rate states the size of change of the weights to decrease errors. If the learning rate is high, the training finishes quickly but the accuracy is usually much better with slower learning rates. Learning rate is often varied to achieve quicker convergence and prevent oscillations.

There are three major learning models—supervised learning, unsupervised learning, and reinforcement learning.

In the supervised learning model, the ANN is provided with both the input and output which are appropriately paired. Learning, in this case, would mean altering the input in such a way as to get the output specified. Usually, the mean-squared error is the cost function that is used. Common applications of supervised learning are classification, regression (function approximation) and sequential data.

In unsupervised learning, the input and the cost function are provided. This is often used when the data is unlabeled. It is used to determine structures or find correlations or features. Examples include clustering, dimensionality reduction, density estimation, feature learning, etc.

Reinforcement learning is a type of dynamic programming that trains an algorithm through experience using a system of reward and punishment. Common examples include a robot navigating to a particular location, a step in the right direction decreases the distance; in this context, it is equivalent to reward and a step in the wrong direction increases the distance, and this is equal to punishment. The aim is to decrease the distance, and no training dataset is provided to it.

The proposed method uses the supervised learning model.

4 Proposed Method

The block diagram of the proposed method is shown in Fig. 1. In this figure, the original song, consisting of two channels is split into left and right. The left channel is subtracted from the right one or vice-versa, and this is the process of the out-of-phase stereo method. The ANN tool in MATLAB, in its training phase, is fed the original song and this output. After training, the original song is directly fed to the ANN, and the neural network's output is the karaoke.

The out-of-phase stereo method has been used to generate the training data for ANN in MATLAB. The out-of-phase stereo method has been implemented using a free open-source software called audacity used across various platforms like windows, mac-OS, and Linux-like systems. This is a sound editing software used in professional studios.

Figure 2 shows a representation of the song when it is fed to Audacity.

The stereo track is split into two, one of them is selected and it is inverted using the effects menu. The stereo track is composed of left and right—tracks. When they

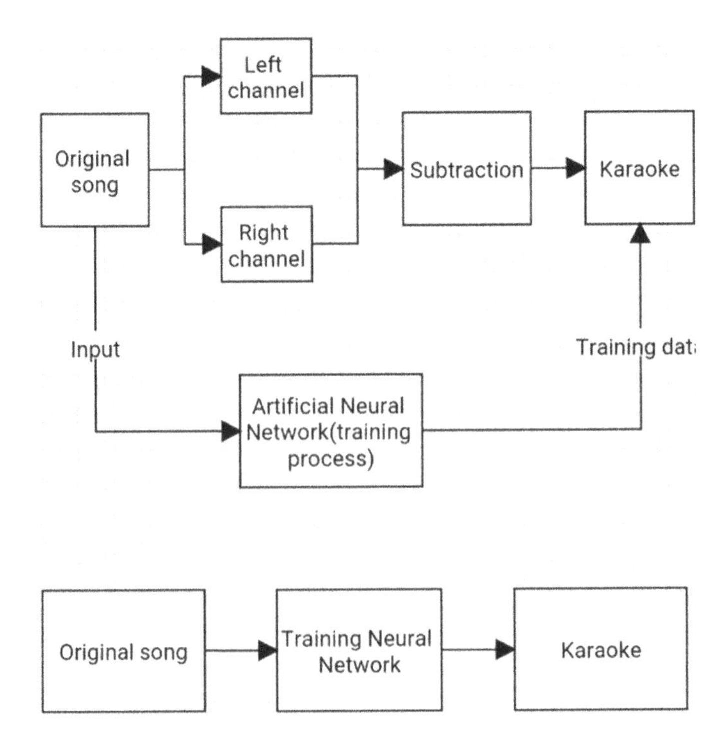

Fig. 1 Block diagram of the proposed method

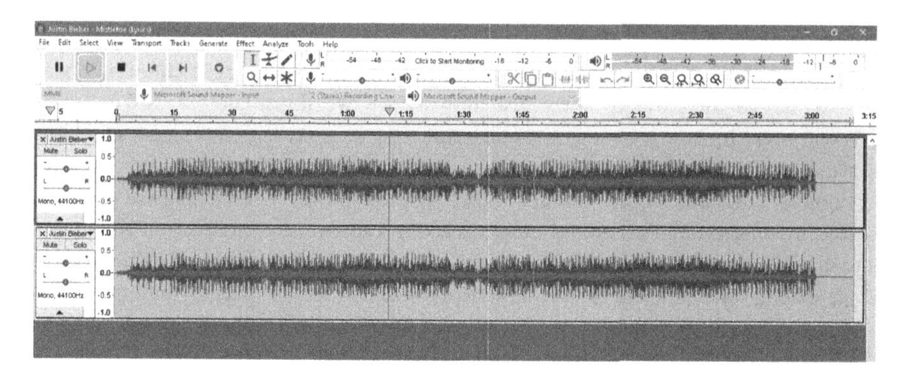

Fig. 2 Stereo music signal

are panned out, the audio is the same even after the inversion. But when the panning is brought down to zero percent and both the tracks are centered, the left and right track get subtracted from each other. The main vocals are removed after this process.

The dataset for this paper is the songs that are available to the public. Songs of different genres are used to provide a variety in the training of the neural network. It was noted that after training with 4–5 songs, the neural network stopped learning because the learning process was over. This happened because out-of-phase stereo method is a relatively simple algorithm and is easily learnt by the neural network.

Matrix laboratory (MATLAB) is a numerical computing environment and proprietary programming language developed by math works. MATLAB allows various operations that include matrix manipulations, plotting of functions and data, implementation of algorithms, creation of user interfaces, and interfacing with programs written in other languages. MATLAB has a variety of toolboxes each meant for a different applications. The artificial neural network toolbox is used in the proposed method.

After successful implementation of the out-of-phase stereo method in audacity, the same was implemented using basic coding in the MATLAB programming environment. The resulting variables saved to the workspace were used to train the ANN.

Songs that were not monaural, that is, stereo in nature was selected for the training purpose, majority of the recordings nowadays fall in this category. A MATLAB script was written to carry out the out-of-phase stereo method, and the workspace variables thus obtained were used for training the ANN.

5 Observations and Results

Supervised training method has been implemented where the input is the original song and the output, used only during training, is the karaoke.

The training data thus obtained was given to the neural network fitting tool in MATLAB. The Levenberg–Marquardtor the damped least square method has been used to perform the training. The neural network is a two-layer feed-forward network, has one input (the original song) and one output (the karaoke generated), as elucidated above.

Figures 3, 4, and 5 represent the ANN tool used in MATLAB. Songs of different genres are entered and reentered many times to improve the accuracy of the neural network. The training automatically completes when the performance reaches a good level, and the validation check is complete or the user stops the training manually. This is the time-consuming part of using neural networks. Once the network is trained, MATLAB has a feature where a function can be auto-generated the trained neural network. This function is called when karaoke is to be generated.

MSE of the karaoke generated using ANN has been measured with that generated using the out-of-phase stereo method for three different songs. The values obtained were 9.3e−6, 6.7e−5, and 7.4e−7. The PSNR values were also used for comparison to check the accuracy of the training, and the values of the three songs were 50.31, 41.75, and 61.29 dB. Figures 6 and 7 represent the spectrograms of the karaoke music signals for visual comparison of the signals. Figure 6 is obtained using the proposed method, and Fig. 7 is obtained using the out-of-phase stereo method. Upon visual comparison, it is obvious that the signals are very similar. The MSE and PSNR values are quantitative comparisons of similarity of the karaoke generated using the

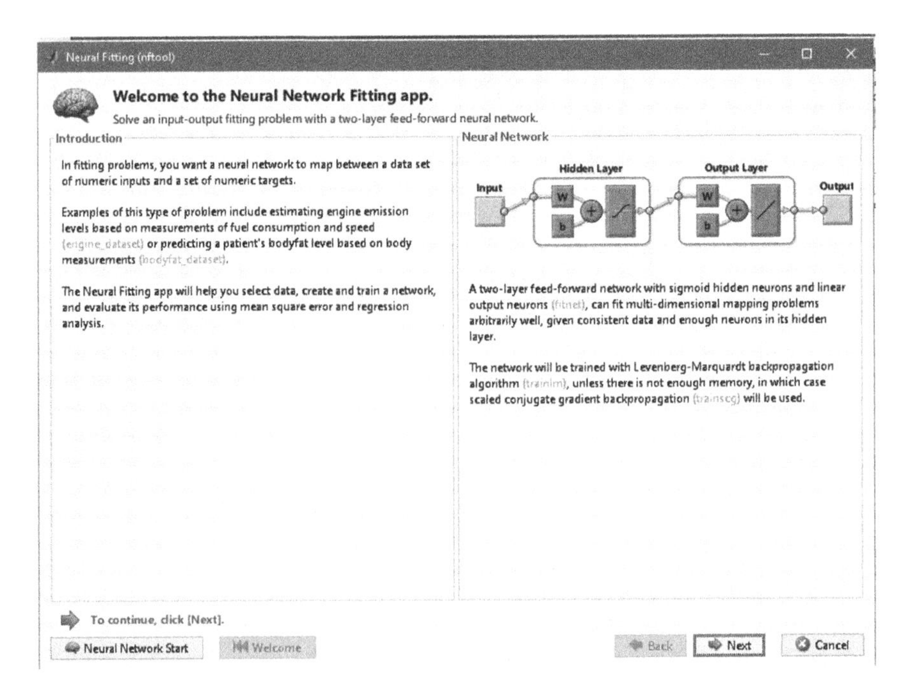

Fig. 3 ANN—neural network fitting tool, MATLAB

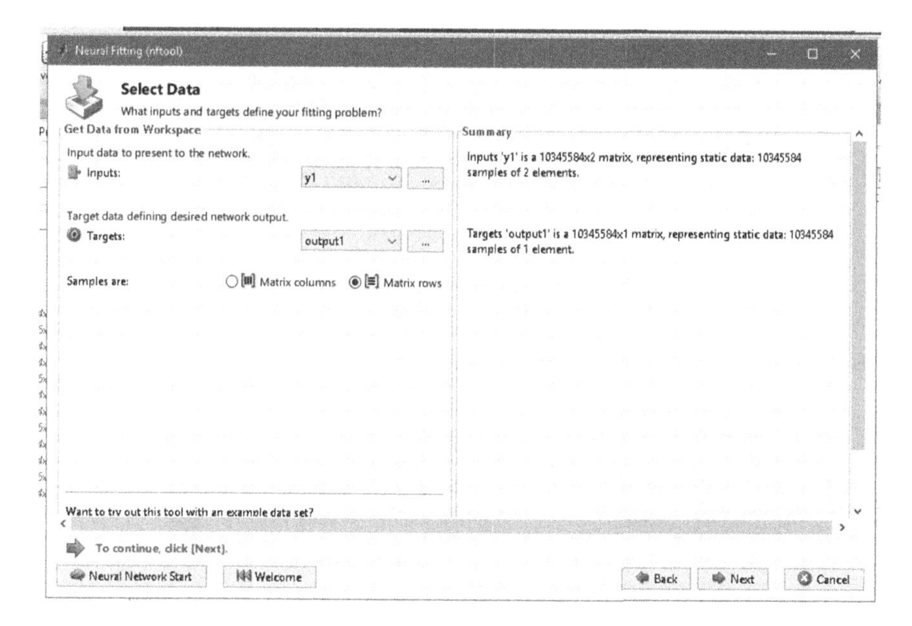

Fig. 4 Selection of training data

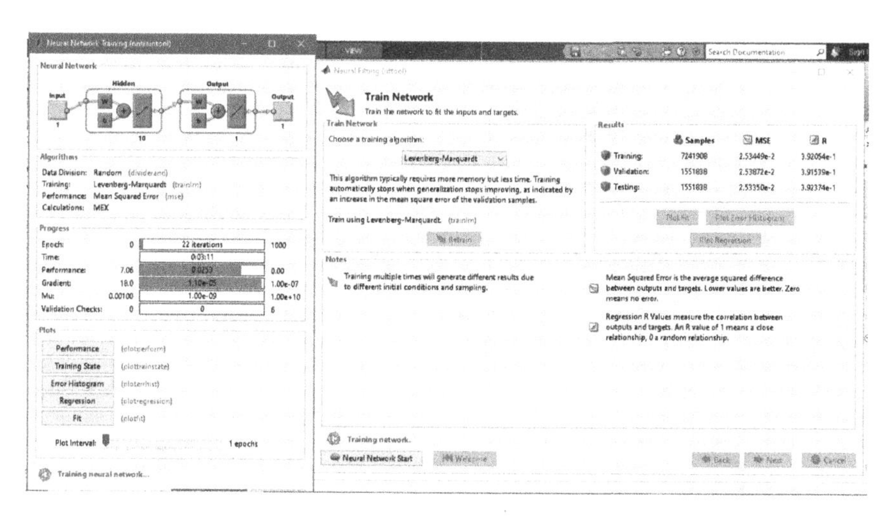

Fig. 5 Neural network training tool

two methods. These values (MSE and PSNR) the spectrograms indicate that the karaoke music signals are quite similar and that the neural network has been trained successfully to mimic this method of generating karaoke.

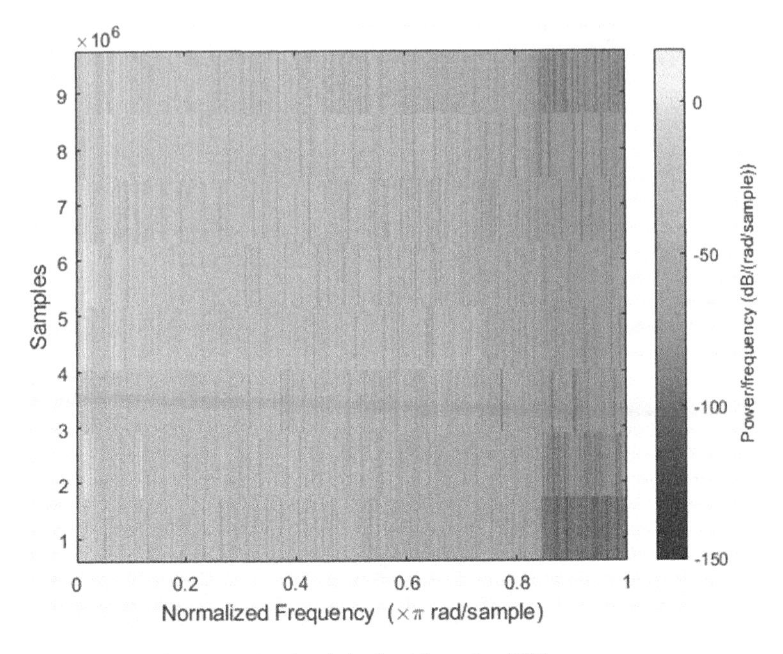

Fig. 6 Spectrogram of the karaoke signal obtained from the ANN

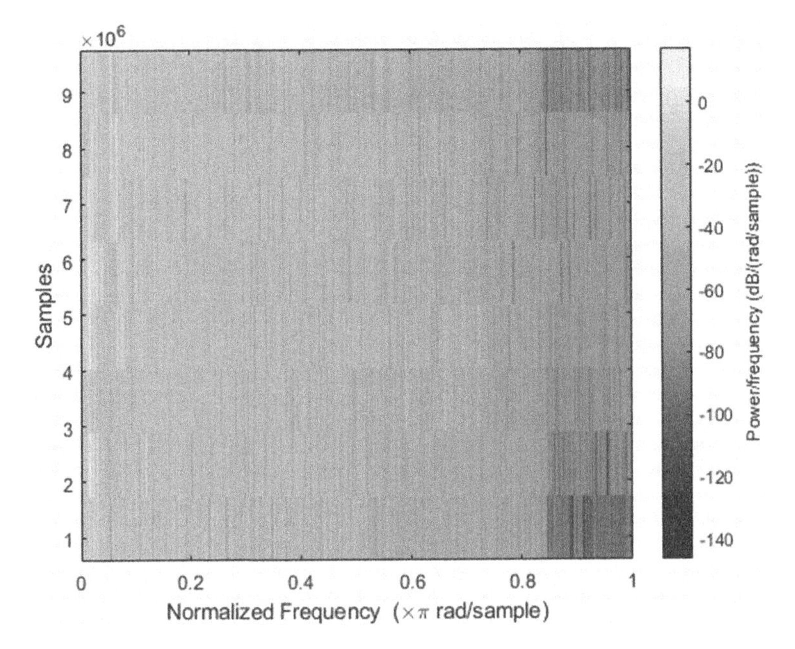

Fig. 7 Spectrogram of the karaoke signal obtained using the out-of-phase stereo method

6 Conclusion and Future Scope

The performance of any ANN is only as good as the training data. In this case, training data was generated using the out-of-phase stereo method. Songs were split into small interval for the training purpose. Songs from different genres were used. The trained ANN could mimic the output exactly as the MSE and the PSNR values proved.

The only major limitation of the proposed method is that it can be used only if the track is stereo, and it cannot be used in the case of monaural tracks, where there is only one microphone. This is a limitation because the out-of-phase stereo method is used for the generation of the dataset. If a different method were used, which did not depend on the stereo/monaural nature of the music, this would not be a limitation. This method's efficiency can be improved by bringing schemes to nullify the effect of echoes and some minor part of the vocals not being a part of both the channels. Also, since the data is stored in a variable (intensity per unit time), a generic karaoke of the same song cannot be used for the training process as the intensity values will differ greatly even if there is a millisecond difference between the tracks. A better set of training values can thus be used to get much better results. This is especially useful it the method of training data requires human intervention, as ANN can be used to automate the process once the neural network is trained to mimic the output generated.

References

1. Sharma AK et al (2014) An efficient approach using LPFT for the karaoke formation of the musical song. In: IEEE international advance computing conference (IACC). IEEE, pp 601–605
2. Bhalani NR, Singh J, Tiwari M (2012) Karaoke machine implementation and validation using out of phase stereo method. In: International conference on communication, information & computing technology (ICCICT), Mumbai, pp 1–3
3. Seetharaman P, Pishdadian F, Pardo B (2017) Music/voice separation using the 2D Fourier transform. In: IEEE workshop on applications of signal processing to audio and acoustics (WASPAA), New Paltz, NY, pp 36–40
4. Simpson AJR, Roma G, Plumbley MD (2015) Deep karaoke: extracting vocals from musical mixtures using a convolutional deep neural network. In: International conference on latent variable analysis and signal separation, pp 429–436
5. Park J, Kim K, Hahn M (2013) Vocal removal from multi-object audio using harmonic information for karaoke service. IEEE Trans Audio Speech Lang Process 21(4):798–805
6. Rafi Z, Pardo B (2013) Repeating pattern extraction technique (REPET): a simple method for music/voice separation. IEEE Trans Audio Speech Lang Process 21(1):73–84
7. Ryynanen M, Virtanen T, Paulus J, Klapuri A (2008) Accompaniment separation and karaoke application based on automatic melody transcription. In: IEEE international conference on multimedia and expo, Hannover, pp 1417–1420
8. Natori T, Tanabe N, Furukawa T (2016) Extraction method of accompaniment signal using stereo type noise suppression method. In: IEEE 12th international colloquium on signal processing & its applications (CSPA), Malacca City, pp 135–139

9. Kaur J, Gaikwad S (2017) Extraction of single channel from mixed audio sample using adaptive factorization. In: Third international conference on sensing, signal processing and security (ICSSS), Chennai, pp 45–47
10. Sinith MS, Tripathi S, Murthy KVV (2013) SSM wavelets for analysis of music signals using particle swarm optimization. In: International conference in signal processing and communication (ICSC), Noida, pp 247–251
11. Sinith MS, Nair MN, Nair NP, Parvathy S (2010) Identification of wavelets and filter bank coefficients in musical instruments. In: International conference on audio, language and image processing, Shanghai, pp 727–731
12. Sinith MS, Nair MN, Nair NP, Parvathy S (2011) Blind source separation of musical instrument signals by identification of wavelets and filter bank coefficients. In: IEEE recent advances in intelligent computational systems, Trivandrum, pp 129–133

A Review on Deep Learning Models for Short-Term Load Forecasting

Ksh. Nilakanta Singh and Kh. Robindro Singh

Abstract Short-term load forecasting (STLF) is a part of the smart grid (SG) system used in maintenance and management operations. Traditional machine learning (ML) techniques entail complicating and time-consuming processes of feature extraction and selection. Deep learning (DL) techniques of artificial neural network (ANN) have shown great potential in STLF. The modernization of SG and the availability of huge load data offer an opportunity for these DL techniques in STLF. Different techniques based on DL models have been proposed for STLF in the past few years. In this paper, a survey of DL model for STLF is presented. This literature survey includes papers published from 2016 to 2019. Common DL architectures such as stack auto-encoder (SAE), recurrent neural network (RNN), convolution neural network (CNN), and deep belief network (DBN) are frequently applied in combination with clustering methods. These DL architectures are briefly explained with a diagram before presenting a review of related papers. The strengths and limitations of the reviewed methods are discussed. Based on this review, the gaps in the existing research work on DL-based STLF are identified, and future directions are described. This paper is expected to serve as an initial guide for new researchers who are interested in the application of deep learning in STLF.

Keywords Short-term load forecasting · Deep learning · Deep neural network

1 Introduction

Electricity is the indispensable force that drives most of our everyday activities and considered a fundamental needs and backbone of our society. Its efficient generation and utilization will reduce carbon footprint and hence resulting in maintaining good environment and healthy economy. Load forecasting has been an extensively studied and researched field of utility grid as it aids in planning and maintaining the normal operation of power system. Based on temporal granularity or forecasting horizons,

Ksh. Nilakanta Singh · Kh. Robindro Singh (✉)
Manipur University, Canchipur, India
e-mail: rbkh@manipuruniv.ac.in
URL: http://www.manipuruniv.ac.in

A. Choudhary et al. (eds.), *Applications of Artificial Intelligence and Machine Learning*, Lecture Notes in Electrical Engineering 778,
https://doi.org/10.1007/978-981-16-3067-5_53

load forecasting can be of these four types—(1) very short-term load forecasting, (2) short-term load forecasting, (3) medium term load forecasting, and (4) long-term load forecasting [1]. STLF is an important part of grid system as it is used in power system dispatcher, assessing power system security, and generation scheduling, etc. It is a challenging task because electricity load can be affected by many factors such as season of the year, weather conditions, holidays, etc. The technique of forecasting has shifted from traditional ML model to ANN. Although traditional ML forecasting methods perform better than data-driven forecasting method, but they entail manual feature extraction and selection which are complicated and time-consuming. Also, they cannot deal with large amounts of data. DL is a type of ANN that has deeper inner hidden layers cascaded between input and output layers. It has complex structure of neurons and allow the network to identify hidden patterns in the dataset which has not been possible in shallow neural network. This type of network has been successfully applied to solve some of the most interesting problem in predictive analysis domain like image classification, pattern recognition, speech synthesis, natural language processing, etc. [2] Deep neural network (DNN) model has been tried in load forecasting and found satisfactory results [3]. Due to high dimension and gigantic size of data generated from current smart grid infrastructure, researchers have the opportunity to test DL model for forecasting electric load which has not been possible in the past.

Many review works have been done on load forecasting techniques [4–6], but few have a focus on DL techniques. Moreover, previous review work does not cover many of the papers. Recently there have been many researches works on load forecasting that use ANN and DNN. So, a comprehensive study of DL methods applied for STLF is presented to study and address these recent trends. This paper will also serve as a guiding tool to those researchers who want to study, improved DL models and efficiently applied on STLF. A more in-depth review is provided by including as many papers as possible, and including more recent works. The following part of this paper is organized as follows. Section 2 briefly explains the methodology used in this review work. Section 3 contains the review of different DL models. And Sect. 4 provides discussion and conclusion of this review work.

2 Methods or Approaches

The survey in this paper includes recent work from 2016 to 2019 from three prominent digital research library—ELSEVIER, IEEE, and WILEY. The keyword used for searching papers are "short-term load forecasting" and "deep learning."

Applying deep learning for electricity load forecasting is a relatively new technique although it has been used in other fields. Some research work shows a brief overview of the DL algorithms and methods that are used in electricity load forecasting [7]. Recent increase in the interest of DL-based STLF is catalyzed by three main factors, i.e., scalable on big data, unsupervised feature learning, and strong generalization capability. DNN is composed of multiple hidden layers, and each hidden layer can transform representation at one level into a representation at a higher level. Many of these layers combined and learn complex and nonlinear function from the input data. DL approaches have outperformed many of traditional ML approach and shallow ANN like weighted moving average, multi-linear regression, regression trees, support vector machine (SVM), and multilayer perceptron (MLP) in accuracy and mean absolute percentage error (MAPE) scores [8]. More hidden layers allow DL to learn complex features, though it increases computational complexity.

3 Deep Learning Forecasting Models

3.1 Stacked Auto-encoder

Auto-encoder is a network which reconstructs inputs through encoding and decoding of the original inputs. An auto-encoder has two part—encoder and decoder. Encoder is responsible for encoding input, x_i, to a hidden representation, $y_{i,}$ and decoder mapped back the hidden representation y_i to the reconstructed feature $\boldsymbol{x}'$. The network weight is train to optimize the approximation of $\boldsymbol{x}'$ to $\boldsymbol{x}$. These encoding and decoding function is given as follows.

$$y = s(Wx + b) \tag{1}$$

$$x' = s\left(W'y + b'\right) \tag{2}$$

where s is a sigmoid function, x is the inputs, y is the hidden code, x' is the reconstructed features, W is encoder weights, W' is the decoder weights, and b, b' are the bias.

SAE is the DL architecture with multiple layers of auto-encoder stacked together on top of each other. In stacked denoising auto-encoder (SDAE), noise is introduced on the input by masking some entries of it. Some random numbers of input are corrupted with an equal chance by masking them as zeros. Then, encoding is performed on this corrupted input to produce compressed hidden representation as in SAE. Similarly, decoding is performed to get feature similar to original inputs. This modification allows the SDAE to learn a robust representation of the original input rather than the simple identity and improves the robustness SDAE [9]. Figures 1 and 2 represent the network architectures of auto-encoder and denoising auto-encoder, respectively.

RS-SDA is another architecture of SDA that incorporate the RANSAC and SNE. RANSAC is an iterative method for eliminating the influences of outliers during the construction of RS-SDA models. A layer-wise implementation of the SNE is applied to automatically determine the number of units of hidden layers in the RS-SDA.

In [10], the application of the SDA and its extended version RS-SDA models is studied for forecasting both online and day-ahead hourly electricity prices. Their

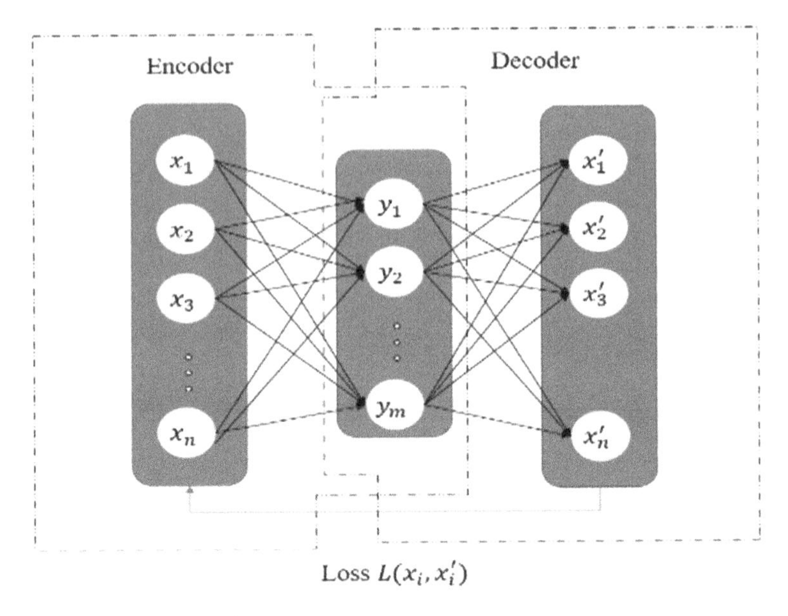

Fig. 1 Diagram of auto-encoder

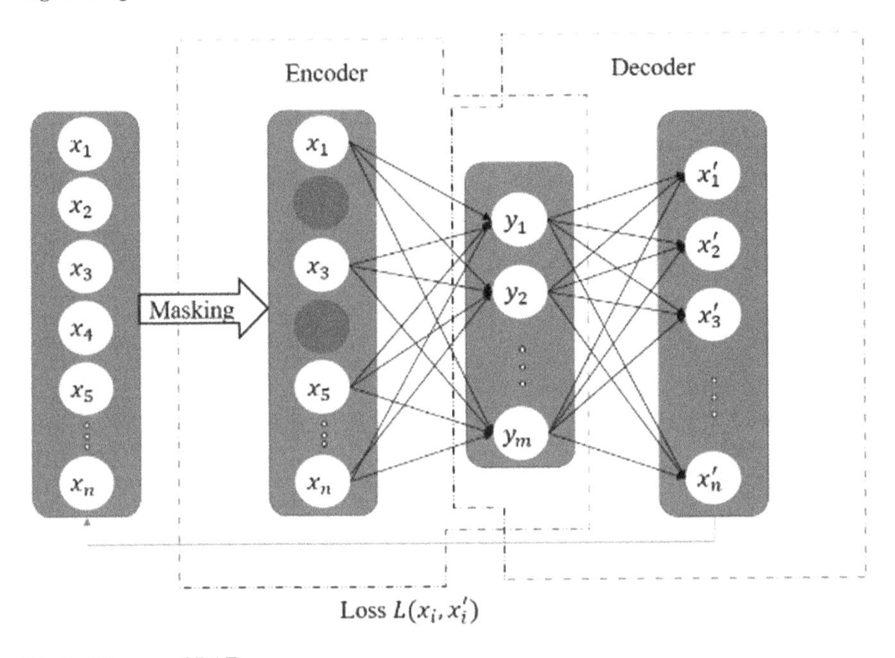

Fig. 2 Diagram of DAE

experiment used real data collected from Nebraska, Arkansas, Louisiana, Texas, and Indiana ISO hubs in the USA. In online forecasting, the SDA model is compared with other benchmark models. Then the effectiveness of the SDA model is validated in the day-ahead forecasting experiment. The proposed SDA model is compared with the recently reported method. Their results show that SDA models are capable of accurately forecast electricity prices, and the extended SDA model can further improve forecasting performance. In [11], a DL-based day-ahead load forecasting method is presented. Firstly, an SDA is used for feature extraction to refined more abstract from original load data. Then a support vector regression model is trained using the high-level feature data as input. Their comparison experiment shows satisfying results with lower MAPE values and higher efficiency. In [12], a combined method of SAE and backpropagation algorithms to forecast wind power is presented. This work consists of two part—a pre-training process where an SAE with three hidden layers is trained to extract important features from the historical data, and a tuning process where the backpropagation (BP) algorithm is used to find the weights of the whole network after adding an output layer to the SAE. Particle swarm optimization is used to find the best parameters of the network. The comparison experiment of the presented method with BP neural network and support vector machines shows a 12% improvement in accuracy when tested on a real wind farm dataset from EirGrid. In [13], a method of predicting residential peak load by applying the SAE algorithm is presented. Different auto-encoder architectures are examined such as Vanilla AE with latent structure 48–24–48, 5-layer SAE with latent structure 48–24–12–24–48, 7-layer SAE with latent structure 48–24–12–6–12–24–48, and 9-Layer with latent structure 48–24–12–6–3–6–12–24–48. The proposed method is compared with an ANN and extreme learning machine on CER Irish consumers' smart meter dataset. Their result shows the higher performance of SAE from other methods when tested using fivefold cross-validation with mean square error (MSE) and MAPE as performance matrices. The paper [14] presents a STLF model using SDA model. The SDA is used for learning features from noisy data. Four variables are used as input to the model—historical load, temperature, humidity, and daily average load. The presented method is compared with backpropagation neural network and AE on data from city in China, and the result shows good performance of SDA model in MAPE and MSE score.

3.2 Recurrent Neural Network

RNN is a special feed-forward neural network (FFNN) with extended feedback or recurrent connections. This connection supplies previous input information to the network and influences the current output. RNN is designed for processing sequential data and shows success in time-series forecasting problems. It maps input time-series data $X = \{x^1, x^2, x^3, \ldots, x^T\}$ to the corresponding output $Y = \{y^1, y^2, y^3, \ldots, y^T\}$. The training process minimizes errors in this mapping function. The learning mechanism of RNN is represented by the following equations:

$$a^t = b + W \cdot h^{t-1} + U \cdot x^t \tag{3}$$

$$h^t = \text{activation}\left(a^t\right) \tag{4}$$

$$y^t = c + V \cdot h^t \tag{5}$$

$$L = \text{loss\_function}\left(y^t, y^t_{\text{target}}\right) \tag{6}$$

The weight matrices are noted as U, V, W, and vectors b and c are bias. The activation function can be sigmoid function, hyperbolic function (tanh), or rectified linear unit (ReLU) [15].

The diagrams in Fig. 3 represent networks of a simple RNN and deep RNN where input x^i from series data is fit to the network one at a time. The hidden layer captures the current information to feed to the next cycle of RNN. There are more complex RNN and deep RNNs like bidirectional RNN, recursive neural networks, etc.

Long short-term memory (LSTM) is a variant of RNN where instead of simple recurrent connection, and each block has two parallel lines going in and out. As shown in Fig. 4, the top line in the LSTM block works as a cell, and the bottom line represents the hidden information. This LSTM structure has three inputs, two outputs and contains a memory cell c_t, an input gate i_t (determine what to write in the cell), a forget gate f_t (determine what to erase from the cell), and output get o_t (determine what to reveal from the cell). This structure of LSTM enables it to capture short/long-term dependencies across many time steps and solve the inherent problems of vanishing and exploding in RNN [16].

In [17], a novel RNN-based technique of short-term load forecasting is presented. This RNN architecture used a feedback layer made of neurons to provide the temporal

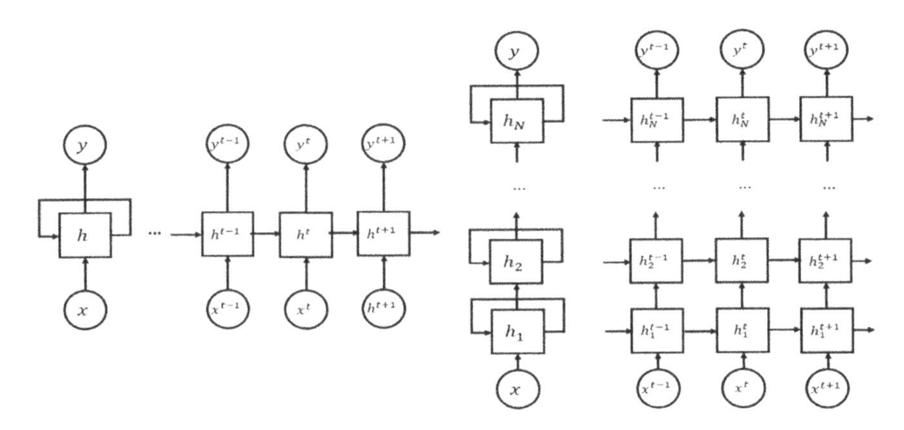

Fig. 3 Diagram of RNN and DRNN

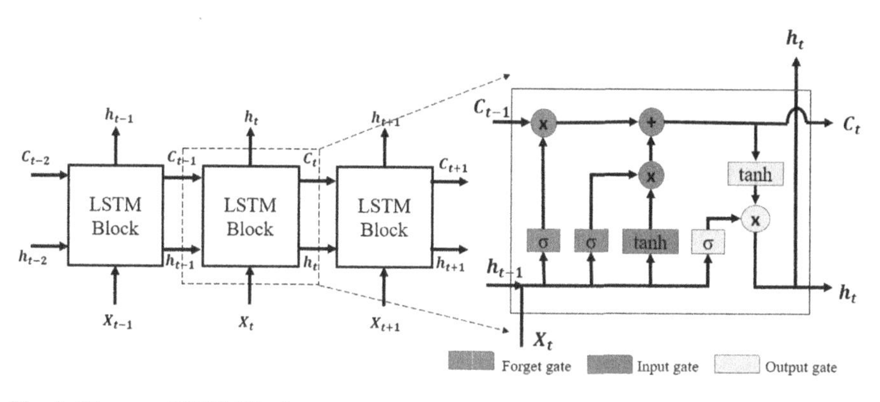

Fig. 4 Diagram of LSTM blocks

relations of the network. These arrangements enhance the nonlinear mapping capabilities and provide flexible feedback structure. The performance of the method was compared with several other methods like MLP, MLP trained by genetic algorithm, MLP trained by particle swarm optimization, and adaptive neuro-fuzzy inference system. The result shows the superiority of the proposed RNN in the MAPE score. The paper [18] presents a novel pooling-based DRNN for the resident load forecasting technique. The pooling in the sense that the load profile is batches based on the group of customers, and this allows and addresses the problem of overfitting when there is an increase in the number of the layer in DRNN. In [19], a novel approach of electricity load forecasting is presented where a single hidden layer FFNN is adapted to train a single hidden layer RNN. The result from the experiment shows that the proposed recurrent extreme learning machine achieved higher accuracy than the other ML methods. The main advantage of the proposed method is its simple and extremely fast training process and high generalization capacity which is a need in the time constraint nature of electricity load forecasting. In [20], a deep LSTM architecture is trained for short-term electrical load forecasting. This LSTM model is compared with ANN and ARIMA on an hourly interval time-series load data from Ontario, Canada. The result shows better performance of LSTM-based model from others in HRMSE value. In [21], two LSTM-based RNN architectures are investigated to forecast resident level electric load of one minute and one hours' time-steps datasets. Even though the simple and standard LSTM architecture can easily forecast on one hour's time-datasets, it fails to do the same in one minute time-steps datasets. It also fails to provide an accurate prediction further in the future timeline. However, the sequence to sequence-based LSTM architecture performed well in both resolutions of the datasets. Moreover, the S2S-based architecture provides flexibility in receiving an arbitrary length of input and predicting the arbitrary number of output time-steps values. Some works also try the quantile regression framework with LSTM where instead of point estimation a quantile value is predicted to represent future uncertainty. This paper [22] presents Quartile LSTM forecasting method which is based on the probabilistic forecasting strategy applied on LSTM. Comparison with other benchmark methods—FCNN with HED and Quantile-FCNN shows the superiority of their method in terms of more robust performance in modeling highly stochastic time series like the residential load and the network capture dependency of different inputs. In paper [23], a quantile loss guided LSTM for probabilistic load forecasting is present to handle the variability and uncertainty of future load profiles. In this work, traditional LSTM is extended to do probabilistic forecasting in the form of quantiles. The Pinball loss function is used to guide the training the network parameters. The experiment result shows superior performance of presented model from other benchmark methods. In [24], an LSTM model combined with the bat algorithm is proposed for STLF. The first step is the preprocessing that include correlation analysis, statistical analysis, and characteristic which are used to explore the best feature for selecting the training set of the network. The hyperparameter of the network is fine-tune using bat algorithm that is a bio-inspired metaheuristic algorithm. Their experiment yields promising results when compared to the other methods. In [25], an LSTM-based residential load forecasting framework is present. This framework learns resident behavior by using load measurements of both the whole household consumptions and selected appliances. They conclude that with the availability of more appliances load reading the result can be significantly increased. Their result is demonstrated by comparing the model with FFNN and KNN. In [26], a comparative experiment is performed where ARMA, SARIMA, ARMAX, and LSTM models are used to forecast electricity load with data having electricity load as well as temperature. Their result shows the superiority of LSTM from other models in terms of MAPE. In [27], a Bi-LSTM model based on the attention mechanism and rolling update is presented. The attention mechanism weighted the input to highlight the effective characteristic, and rolling update is used to update the data in real time. They show that these two mechanisms improved the performance of bidirectional LSTM and has higher prediction accuracy with lesser computation time and shows better generalization capability compared with other load forecasting models. In [28], a gated recurrent unit network (GRU) is used for STLF by considering the impact of electric price on load. In this work, historical load input is grouped based on features and then established the rule of classification tree to cluster the new data. Then GRU is trained by the feature of input and power load from the selected group. They show that the computation time of the GRU network is shorter than that of LSTM.

3.3 Convolution Neural Network

Convolution neural networks are developed for processing data with grid topology. The high local correlation data such as images, videos, and texts have a repeated pattern which CNN can easily capture. As image can be represented by 2D grids likewise time-series data can be viewed as 1D grids. Basically, CNN consists of three parts—(1) the convolutional layers performs convolution operation on input data with a special weight matrix called filter or kernel. (2) Pooling layers combine the output of convolution layers, and it is responsible for reducing the spatial size of

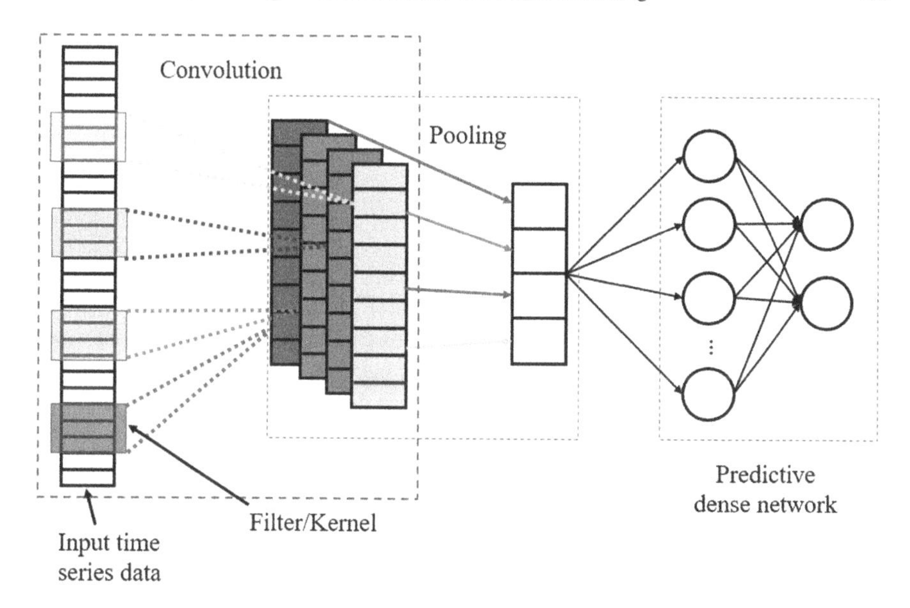

Fig. 5 Diagram of CNN

the convolved features and for extracting dominant features. (3) A fully connected layer combines all local features into global features and is used to calculate the final result [29]. Figure 5 shows a diagram of CNN with convolution module, pooling module, and a fully connected predictive neural network.

For time-series load data, 1D convolutional network is used. This network has many advantages such as lesser computation time, higher noise resistance, and ability to extract very informative and deep features. Recent work in load forecasting research has seen the application of CNN as a standalone method or combined with others. The effectiveness of using CNN on energy load forecasting has been investigated and highlights its potential by performing comparative studies [30, 31]. In [32], an STLF method is presented where load forecasting task is transformed into an image-processing problem. In this work, CNN model is used to extract features from input load data for clustering, and then a multilayer neural network is used to forecast future load variations. In [33], a method combining CNN and K-means cluster to forecast load on an electricity big data is presented. K-means clustering approach is used to partition the large dataset into small cluster. The prediction loads are generated by collecting prediction results from each CNN model trained with the clusters. In [34], a novel model of time coding multi-scale CNN is presented for short-term multi-step load forecasting. This network architect improves the CNN by extracting complex and significant feature of load sequences. Their comparative experiment result shows higher accuracy and excellent stability of their proposed model. More work in [35–38] shows frameworks that integrate CNN and LSTM for STLF. CNN modules are used to extract the local trends and repeated patterns appear in different regions, while the LSTM module helps in making accurate prediction by exploiting long-term dependencies.

3.4 Deep Belief Network

Deep belief network is a special type of DL network develop to address slow convergence and local optima trap problems in DNN architecture due to stochastic initialization of network parameters. A DBN architecture composed of multiple restricted Boltzmann machine (RBM) stacked on top of each other, and a regression layer stacked on the very top for prediction. RBM is a double layer network containing a visible layer and a hidden layer which form a bipartite graph [39, 40]. A network architecture of DBN is shown in Fig. 6. A DBN model training includes pre-training and backpropagation process. The pre-training process provides a good initial parameter for the network while the backpropagation process is used to fine-tune the network by adjusting the network weight by minimizing the error between predicted and actual output [41].

In [42], a deep belief network composed of two layers of RBM is used for electricity STLF. The model is compared with feed-forward multilayer perceptron on Macedonian electric consumption data, and the result shows the superior performance of the proposed DBN form other models in MAPE score. Combined model of DBN and copula model has been applied to forecast day-ahead and week-ahead power load [43–45]. The Copula model is selected to model the correlation between

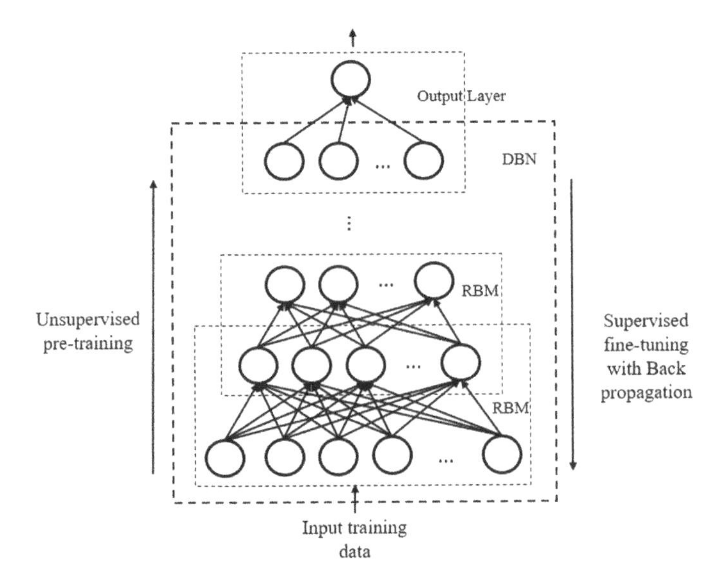

Fig. 6 Diagram of DBN

load data with electric price and other environmental variables. Then the DBN is trained for load predictions. Their comparative experiment result shows promising results. In [46], DBN architecture is combined with SVM for short-term load forecasting. In this work, SVM is used to classify the load type into normal load, peak load, and valley load. Then the sample points of the same load type are selected to train DBN to predict the future load. Their experimental comparison shows higher prediction accuracy than other baseline models (Table 1).

4 Conclusion

Smart grid system depends on STLF to predict future load and to keep up its daily operations. STLF has been around for many years, but only recently has seen a surge of that research that applied deep learning techniques. Traditional ML has successfully applied in STLF, but they are not scalable to large or big data. DL methods are able to process vast amounts of data and find more meaningful features using its deep layers. The highlights of this survey are as follows: (1) DL models-based STLF can achieve higher accuracy than other ML models. The most common found technique of STLF in literature is the hybrid model which combine DL model and other clustering methods. (2) RNN-LSTM is the most applied DL architectures in STLF. This is because they are inherently developed to process sequential data like time-series data. Electric load time-series data comes in different granular, i.e., 15 min, 30 min, and hourly interval are common types. LSTM model can process the daily and seasonal pattern in load data effectively and is one of the most common use models. (3) Though CNN is developed for image processing, its 1D variant model is suitable for load series forecasting. (4) DL models required a larger amount of data to train properly and to achieve acceptable performance, so more computational time and resources are required. (5) Special care needs to be made while initializing parameters of DL networks to improve the performance of STLF. Further research is required to develop new and better network parameter training algorithms. (6) MAPE and MSE are the commonly used performance measurement of the forecasting DL models. (7) A new direction of forecasting is needed to accommodate uncertainty in load forecasting. This can be addressed by developing the probabilistic-based DL model for STLF. Incorporating weather and seasonal information in the load data also help in achieving accurate prediction for forecasting methods. Distributed big data analysis of load data is also another direction of research.

Table 1 Reviewed paper summary

References	Learning algorithm	Temporal granularity	Dataset type	Data size	Performance metric
[10]	SDA	Hourly	Real	Hourly data from Jan 2012 to Nov 2014	MAPE, HR-hit ratio
[11]	SDA and SVR	Hourly	Real	Hourly data from 15th Aug 2015 to 15th Aug 2016th	MAPE
[12]	SAE and BP	15 min	Real	6075 samples (1st May 2014 to 21st June 2014)	RMSE, MAE, MAPE
[13]	SAE	30 min	Real	328 consumers × 325 days	RMSE, MAPE
[14]	SDA	15 min	Real	1st January 2013–31st October 2013	MAPE, MSE
[17]	RNN	Hourly data	Multivariable	10,920 consumer data from 1.1.2009 to 31.12.2010	MAPE
[18]	Deep RNN	Half-hourly electric consumption data, questionnaires, and customer profile	Multivariable		
[19]	Extreme learning with RNN	Daily	Real	Load data from 2011 to 2014	RMSE
[20]	LSTM	Hourly	Real	Load data from 2006 to 2016	Normalized RMSE
[21]	Deep LSTM	One minute resolution	Real	Power consumption from Dec 2006 to Nov 2010 (2,075,259 measurements)	RMSE
[22]	Quantile-LSTM	Half-hourly	Real	Load data from 1.7.2009 to 31.12.2010	Pinball loss
[23]	Probabilistic-LSTM	Half-hourly load, current day, current week	Multivariable	Load data from 1.7.2009 to 31.12.2010	Pinball loss

(continued)

Table 1 (continued)

References	Learning algorithm	Temporal granularity	Dataset type	Data size	Performance metric
[24]	LSTM with bat algorithm	15 min current day load	Multivariable	Quarter-hourly dataset from 1.5.2016 to 30.4.2017	RMSE, MAPE
[25]	LSTM-WA	Half-hourly	Real	Yearly Canadian household consumption data	MAPE
[26]	LSTM	Hourly load, temperature, humidity, wind speed	Multivariable	13 months data	RMSE, MAPE
[27]	Attention mechanism, rolling update and bidirectional LSTM	Half-hourly load, temperature, humidity, electric price	Multivariable	NSW data from 1.1.2009 to 1.6.2010, VIC data from 1.3.2015 to 1.7.2016	MAPE, RMSE
[28]	Gated RNN	Historical load, price, and temperature	Multivariable	Real world data from Jan 2006 to Dec 2010	MAPE
[32]	CNN	5 min	Real	Load data from Jan 2014 to June 2016	Relative error, Davies–Bouldin index
[33]	Combined CNN and K-means clustering	Hourly load, group, local zones, temperature, cloud cover, heat index, customer count	Multivariable	1.4 million load records from 2012 to 2014	MAPE, RMSE, NMAE and NRMSE
[34]	Multi-scale CNN with time-cognition	Hourly load and holiday	Multivariable	Ireland load dataset from 2014 to 2018	Average pinball score
[35]	LSTM and CNN	Hourly	Real	1.1.2015–31.12.2017 (26,304 samples)	MAE, MAPE, RMSE
[36]	Combined CNN and RNN	Hourly load, temperature, humidity	Multivariable	Load data from 10.2.2000 to 31.12.2012	MAPE, MAE

(continued)

Table 1 (continued)

References	Learning algorithm	Temporal granularity	Dataset type	Data size	Performance metric
[37]	Gated RNN and CNN	Hourly load, outdoor temperature, air pressure, humidity, wind speed	Multivariable	1 year dataset	RMSE
[38]	Time-dependency CNN and cycle-based LSTM	Daily load	Real	Hangzhou daily load from Jan 2014 to Mar 2017 and Toronto daily load from May 2002 to July 2016	Mean relative error (MRE)
[42]	Deep belief network (DBN)	Hourly	Real	Macedonian electric consumption data 2008–2014	MAPE
[43]	DBN and copula model	Hourly load, temperature, price, other parameters	Multivariable	One-year grid data from taxes 2013	MAPE, RMSE
[44, 45]	DBP with copula model	Hourly load, price, temp, humidity, pressure, wind speed, others parameters	Multivariable	One-year grid data from taxes 2016	RMSE, MAPE, HR
[46]	SVM and DBN	Hourly load	Real	Load data from 1.1.20117 to 31.12.2017 (8760 samples data)	Relative error

References

1. Eskandarnia EM, Kareem SA, Al-Ammal HM (2018) A review of smart meter load forecasting techniques: scale and horizon, p 15
2. Bansal S, Lodhi MRS, Nema DP (2018) State of art on short term load forecasting using artificial neural network, p 7
3. Hosein S, Hosein P (2017) Load forecasting using deep neural networks. In: 2017 IEEE power & energy society innovative smart grid technologies conference (ISGT), Washington, DC, USA, Apr 2017, pp 1–5. https://doi.org/10.1109/ISGT.2017.8085971
4. Baliyan A, Gaurav K, Mishra SK (2015) A review of short term load forecasting using artificial neural network models. Procedia Comput Sci 48:121–125. https://doi.org/10.1016/j.procs.2015.04.160
5. Raza MQ, Khosravi A (2015) A review on artificial intelligence based load demand forecasting techniques for smart grid and buildings. Renew Sustain Energy Rev 50:1352–1372. https://doi.org/10.1016/j.rser.2015.04.065
6. Deb C, Zhang F, Yang J, Lee SE, Shah KW (2017) A review on time series forecasting techniques for building energy consumption. Renew Sustain Energy Rev 74:902–924. https://doi.org/10.1016/j.rser.2017.02.085
7. A Almalaq, G Edwards (2017) A review of deep learning methods applied on load forecasting. In: 2017 16th IEEE international conference on machine learning and applications (ICMLA), Cancun, Mexico, Dec 2017, pp 511–516. https://doi.org/10.1109/ICMLA.2017.0-110
8. Wang H, Lei Z, Zhang X, Zhou B, Peng J (2019) A review of deep learning for renewable energy forecasting. Energy Convers Manage 198:111799. https://doi.org/10.1016/j.enconman.2019.111799
9. Vincent P, Larochelle H, Bengio Y, Manzagol P-A (2008) Extracting and composing robust features with denoising autoencoders. In: Proceedings of the 25th international conference on machine learning. ACM, pp 1096–1103
10. Wang L, Zhang Z, Chen J (2017) Short-term electricity price forecasting with stacked denoising autoencoders. IEEE Trans Power Syst 32(4):2673–2681. https://doi.org/10.1109/TPWRS.2016.2628873
11. Tong C, Li J, Lang C, Kong F, Niu J, Rodrigues JJPC (2018) An efficient deep model for day-ahead electricity load forecasting with stacked denoising auto-encoders. J Parallel Distrib Comput 117:267–273. https://doi.org/10.1016/j.jpdc.2017.06.007
12. Jiao R, Huang X, Ma X, Han L, Tian W (2018) A model combining stacked auto encoder and back propagation algorithm for short-term wind power forecasting. IEEE Access 6:17851–17858
13. Wang X, Wang J (2019) A stacked autoencoder application for residential load curve forecast and peak shaving, p 5
14. Liu P, Zheng P, Chen Z (2019) Deep learning with stacked denoising auto-encoder for short-term electric load forecasting. Energies 12(12):2445
15. Shi H, Xu M, Ma Q, Zhang C, Li R, Li F (2017) A whole system assessment of novel deep learning approach on short-term load forecasting. Energy Procedia 142:2791–2796. https://doi.org/10.1016/j.egypro.2017.12.423
16. Hossen T, Nair AS, Chinnathambi RA, Ranganathan P (2018) Residential load forecasting using deep neural networks (DNN). In: 2018 North American power symposium (NAPS), Fargo, ND, Sept 2018, pp 1–5. https://doi.org/10.1109/NAPS.2018.8600549
17. Mishra S, Patra SK (2008) Short term load forecasting using a novel recurrent neural network. In: TENCON 2008—2008 IEEE region 10 conference, Hyderabad, India, Nov 2008, pp 1–6. https://doi.org/10.1109/TENCON.2008.4766829
18. Shi H, Xu M, Li R (2018) Deep learning for household load forecasting—a novel pooling deep RNN. IEEE Trans Smart Grid 9(5):5271–5280. https://doi.org/10.1109/TSG.2017.2686012
19. Ertugrul ÖF (2016) Forecasting electricity load by a novel recurrent extreme learning machines approach. Int J Electr Power Energy Syst 78:429–435. https://doi.org/10.1016/j.ijepes.2015.12.006

20. Narayan A, Hipel KW (2017) Long short term memory networks for short-term electric load forecasting. In: 2017 IEEE international conference on systems, man, and cybernetics (SMC), Banff, AB, Oct 2017, pp 2573–2578. https://doi.org/10.1109/SMC.2017.8123012
21. Marino DL, Amarasinghe K, Manic M (2016) Building energy load forecasting using deep neural networks. In: IECON 2016—42nd annual conference of the IEEE industrial electronics society, Florence, Italy, Oct 2016, pp 7046–7051. https://doi.org/10.1109/IECON.2016.7793413
22. Gan D, Wang Y, Zhang N, Zhu W (2017) Enhancing short-term probabilistic residential load forecasting with quantile long–short-term memory. J Eng 2017(14):2622–2627. https://doi.org/10.1049/joe.2017.0833
23. Wang Y, Gan D, Sun M, Zhang N, Lu Z, Kang C (2019) Probabilistic individual load forecasting using pinball loss guided LSTM. Appl Energy 235:10–20. https://doi.org/10.1016/j.apenergy.2018.10.078
24. Bento P, Pombo J, Mariano S, Calado MdR (2018) Short-term load forecasting using optimized LSTM networks via improved bat algorithm. In: 2018 international conference on intelligent systems (IS). IEEE, 2018, pp 351–357. https://doi.org/10.1109/IS.2018.8710498
25. Kong W, Dong ZY, Hill DJ, Luo F, Xu Y (2018) Short-term residential load forecasting based on resident behaviour learning. IEEE Trans Power Syst 33(1):1087–1088. https://doi.org/10.1109/TPWRS.2017.2688178
26. Muzaffar S, Afshari A (2019) Short-term load forecasts using LSTM networks. Energy Procedia 158:2922–2927. https://doi.org/10.1016/j.egypro.2019.01.952
27. Wang S, Wang X, Wang S, Wang D (2019) Bi-directional long short-term memory method based on attention mechanism and rolling update for short-term load forecasting. Int J Electr Power Energy Syst 109:470–479. https://doi.org/10.1016/j.ijepes.2019.02.022
28. Wu W, Liao W, Miao J, Du G (2019) Using gated recurrent unit network to forecast short-term load considering impact of electricity price. Energy Procedia 158:3369–3374. https://doi.org/10.1016/j.egypro.2019.01.950
29. Krizhevsky A, Sutskever I, Hinton GE (2012) ImageNet classification with deep convolutional neural networks. In: Proceedings of the international conference on neural information processing systems (NIPS)
30. Amarasinghe K, Marino DL, Manic M (2017) Deep neural networks for energy load forecasting. In: 2017 IEEE 26th international symposium on industrial electronics (ISIE), Edinburgh, June 2017, pp 1483–1488. https://doi.org/10.1109/ISIE.2017.8001465
31. Koprinska I, Wu D, Wang Z (2018) Convolutional neural networks for energy time series forecasting. In: 2018 international joint conference on neural networks (IJCNN), Rio de Janeiro, July 2018, pp 1–8. https://doi.org/10.1109/IJCNN.2018.8489399
32. Li L, Ota K, Dong M (2017) Everything is image: CNN-based short-term electrical load forecasting for smart grid. In: 2017 14th international symposium on pervasive systems, algorithms and networks & 2017 11th international conference on frontier of computer science and technology & 2017 third international symposium of creative computing (ISPAN-FCST-ISCC), Exeter, June 2017, pp 344–351. https://doi.org/10.1109/ISPAN-FCST-ISCC.2017.78
33. Dong X, Qian L, Huang L (2017) Short-term load forecasting in smart grid: a combined CNN and K-means clustering approach. In: 2017 IEEE international conference on big data and smart computing (BigComp), Jeju Island, Feb 2017, pp. 19–125. https://doi.org/10.1109/BIGCOMP.2017.7881726
34. Deng Z, Wang B, Xu Y, Xu T, Liu C, Zhu Z (2019) Multi-scale convolutional neural network with time-cognition for multi-step short-term load forecasting. IEEE Access 7:88058–88071. https://doi.org/10.1109/ACCESS.2019.2926137
35. Tian C, Ma J, Zhang C, Zhan P (2018) A deep neural network model for short-term load forecast based on long short-term memory network and convolutional neural network. Energies 11(12):3493. https://doi.org/10.3390/en11123493
36. He W (2017) Load forecasting via deep neural networks. Procedia Comput Sci 122:308–314. https://doi.org/10.1016/j.procs.2017.11.374

37. Cai M, Pipattanasomporn M, Rahman S (2019) Day-ahead building-level load forecasts using deep learning vs. traditional time-series techniques. Appl Energy 236:1078–1088. https://doi.org/10.1016/j.apenergy.2018.12.042
38. Han L, Peng Y, Li Y, Yong B, Zhou Q, Shu L (2019) Enhanced deep networks for short-term and medium-term load forecasting. IEEE Access 7:4045–4055. https://doi.org/10.1109/ACCESS.2018.2888978
39. Kuremoto T, Kimura S, Kobayashi K, Obayashi M (2014) Time series forecasting using a deep belief network with restricted Boltzmann machines. Neurocomputing 137:47–56. https://doi.org/10.1016/j.neucom.2013.03.047
40. Lin Y, Liu H, Xie G, Zhang Y (2018) Time series forecasting by evolving deep belief network with negative correlation search. In: 2018 Chinese automation congress (CAC), Xi'an, Nov 2018, pp 3839–3843. https://doi.org/10.1109/CAC.2018.8623511
41. Ryu S, Noh J, Kim H (2016) Deep neural network based demand side short term load forecasting. Energies 10(1):3. https://doi.org/10.3390/en10010003
42. Dedinec A, Filiposka S, Dedinec A, Kocarev L (2016) Deep belief network based electricity load forecasting: an analysis of Macedonian case. Energy 115:1688–1700. https://doi.org/10.1016/j.energy.2016.07.090
43. He Y, Deng J, Li H (2017) Short-term power load forecasting with deep belief network and copula models. In: 2017 9th international conference on intelligent human-machine systems and cybernetics (IHMSC), Hangzhou, Aug 2017, pp 191–194. https://doi.org/10.1109/IHMSC.2017.50
44. Ouyang T, He Y, Li H, Sun Z, Baek S (2017) A deep learning framework for short-term power load forecasting, p 8
45. Ouyang T, He Y, Li H, Sun Z, Baek S (2019) Modeling and forecasting short-term power load with copula model and deep belief network. IEEE Trans Emerg Top Comput Intell 3(2):127–136. https://doi.org/10.1109/TETCI.2018.2880511
46. Yang J, Wang Q (2018) A deep learning load forecasting method based on load type recognition. In: 2018 international conference on machine learning and cybernetics (ICMLC), Chengdu, July 2018, pp 173–177. https://doi.org/10.1109/ICMLC.2018.8527022

An Evolutionary Approach to Combinatorial Gameplaying Using Extended Classifier Systems

Karmanya Oberoi, Sarthak Tandon, Abhishek Das, and Swati Aggarwal

Abstract Extended classifier system (XCS) is an extension of a popular online rule-based machine learning technique, learning classifier system (LCS), in which a classifier's fitness is based on its accuracy instead of the prediction itself, and a genetic algorithm (GA) and reinforcement learning (RL) component is utilized for exploratory and learning purposes, respectively. With the emergence of increasingly intricate rule-based learning techniques, there is a need to examine feasible methods of learning that can overcome the challenges posed by complex scenarios while supporting online performance. Checkers is a strategic, combinatorial game having a high branching factor and a complex state space that provides a promising avenue for scrutinizing novel approaches. This paper presents a preliminary investigation into feasibility of XCS in such complex avenues by taking 6×6 checkers as a specific case of study. The XCS agent was adapted to this problem, trained with random agent and was able to perform well against the alpha–beta pruning algorithm of various depths as well as human agents of different skill levels (beginner, intermediate and advanced).

Keywords Learning classifier systems · Extended classifier system · Gameplaying · Evolutionary learning system · Cognitive intelligence · Checkers

1 Introduction

Rule-based systems are present in many areas ranging from industry and automation and are widely used as support systems in decision-making processes to help other systems with error detection or simply to evaluate a situation. Complex systems use large amounts of online information and require rule-based systems to refine their knowledge base as they operate in real time.

One biologically inspired computational model for human-like cognition which uses a rule-base mechanism to model the search space and refines its rule to ascertain optimal actions for input states is the learning classifier system (LCS). It is one of

K. Oberoi · S. Tandon · A. Das · S. Aggarwal (✉)
Netaji Subhas Institute of Technology, New Delhi, India

A. Choudhary et al. (eds.), *Applications of Artificial Intelligence and Machine Learning*, Lecture Notes in Electrical Engineering 778,
https://doi.org/10.1007/978-981-16-3067-5_54

the classical computational intelligence approaches [1]. It was John Holland, who introduced the concept of classifier systems with adaptive capabilities [2–4]. The system proposed by him comprised of a rule-based representation of the problem environment coupled with the ability to improve itself from experience through breeding and generalization. Genetic algorithm [3] was used for adaptive evolution of new rules, and reinforcement learning was utilized to test the efficacy of existing rules. Rules are typically represented in the form of 'IF condition THEN action'. This makes classifier systems a highly interpretable model which can adequately describe the environment with the rule set.

Holland summarized [4] classifier systems as rule-based systems capable of processing rules in parallel, adaptively generating new rules and testing the efficacy of existing rules. Further, in 1995, Wilson introduced the XCS algorithm [5], which differed from traditional LCS in its use of the accuracy of a classifier's prediction to determine its fitness instead of the prediction itself, and used a GA to carry out exploration of the state space for improved performance.

Games in general offer an environment where it is easy to represent distinct situations and rank tasks or actions that are applicable at any moment. A robust sense of gain or loss, characterized by the outcome of application of a rule, along with a need for real-time strategy and critical decision-making, makes gameplaying an attractive avenue for evaluating the efficacy of rule-based systems. Hence, it has received a lot of attention from researchers working in the field of artificial intelligence (AI). Combinatorial games (a.k.a board games), like American checkers, Chess, Go, Othello and Backgammon, have been the subject of rigorous research in AI for a long time, wherein the focus lies on developing agents which can compete with human experts. Recent approaches to model agents are focused around deep reinforcement learning and sophisticated search techniques [6, 7]. Techniques like temporal difference (TD) learning [8–10] Sarsa and Q-learning [11, 12], minimax or alpha–beta pruning [13] have been researched widely using games as an avenue.

Recent studies [14, 15] and surveys [16] shed light on the prospect of using LCS as an effective online evolutionary rule-based learning technique and showcase its applications in different kinds of gameplaying. A survey found in IEEE Journal (2017) [16] highlights that there has been little application of LCS to combinatorial games. Examples of pre-existing work in this domain focus on a limited number of games, most of which are inherently simple and feature a tractable state space. Further testing of LCS and its variants on more convoluted scenarios (like games of greater complexity, offering high branching factor and an intractable state space) is a crucial step in scrutinizing the applicability of these algorithms, as they are becoming increasingly popular in the intersection of gameplaying and cognitive intelligence.

American checkers or English draughts is a combinatorial game played on an 8 × 8 checkered board. It involves two opponents altering the game state alternately while having complete information about the game state at each move. The vast and complex state space and high branching factor of checkers make it a good contender for building rule-based systems supporting real-time strategy and decision-making. Various AI techniques having different combinations of heuristic functions and optimization strategies have been used for building checkers playing agents [2]. Game search techniques like minimax combined with a large knowledge base developed from observing master players is the fundamental basis of Chinook, a checkers playing agent that was the first to win Man-Machine Checkers World Championship [17].

The remainder of this paper uses checkers as a case to study the XCS algorithm's performance on intricate combinatorial games and does not aim to advance the established state-of-the-art in checkers [18]. In this work, the 8×8 board has been reduced to the size of 6×6 for evaluation of the proposed agent. This has been done to expedite the time required to train the agent and aid this preliminary study.

Section 2 focuses on the motivation behind our specific case's analysis. Section 3 aims to provide the reader with background information relevant to the understanding of the remainder of the paper as it sheds light on the rules of checkers as well as the algorithmic structure of the XCS approach. Further, Sects. 4 and 5 explain the technicalities in the implementation of the environmental framework as well as the adaptation of the XCS agent to this problem. Section 6 analyses the results obtained by comparing the performance of the XCS agent with alpha–beta pruning algorithm and human agents. Finally, the conclusion and future work pertaining to the applicability of XCS are briefly explored in Sect. 7.

2 Motivation

Widespread and interdisciplinary interest in LCS is motivated by various factors which contribute to its increasing popularity in recent times. Following are some of the reasons that encourage the application of LCS in more and more domains every day:

1. LCS is rule-based, enabling easier integration with classic knowledge bases.
2. LCS is an online learning mechanism which allows it to learn and refine rules 'on the fly'.
3. Rule structure in LCS represents a cause and effect relationship. This can be utilized to understand the evolution of the optimal strategy and further derive strategic conclusions about the use-case.

However, the application of LCS to combinatorial scenarios is under-researched [16]. Some works like the application of XCS to Connect4, Othello [19, 20] and a probabilistic variant of LCS for battleship [21] inspire assessment of LCS algorithm in more involved scenarios. Combinatorial games involving large state spaces, having higher branching factor and requiring the agent to learn and generalize the behaviour of different kinds of pieces add non-uniformity to the learning process and remain an unexplored territory.

This paper is motivated by a need for thorough examination of the performance of XCS in such complex territories. Our use-case, i.e. checkers, offers this complex environment (with different pieces, such as Kings and Uncrowned, behaving differently coupled with a complex rule set, large state space and branching factor) is suitable for conducting preliminary investigations. Application of LCS to learn winning strategies in such scenarios is the step that needs to be taken in LCS research and hence motivates this work.

3 Background

In this section, the game of Checkers and its rules are explained. Further, a description of the XCS algorithm is given along with a summary of its working cycle.

3.1 Checkers

Checkers is a strategic board game for two players, involving diagonal movement of game pieces and mandatory capture moves by jumping over opponent pieces. There are many forms of checkers available with international checkers being played on 10×10 board, Canadian/Malaysian checkers on 12×12 board, and here, we will focus on American checkers which is played on an 8×8 board.

In Checkers, pieces always move diagonally. A piece making a non-capturing move may move only one square. A piece making a capturing move leaps over one of the opponent's pieces, landing in a straight diagonal line on the other side. If a player can make a capture, there is no option; the jump must be made as in Fig. 1. When a piece is captured, it is removed from the board. When a piece reaches the furthest row, it is crowned and becomes king. In practice, one of the pieces which had been captured is placed on top of this piece to mark it as a king. The king is limited to move diagonally, both in forward and backward direction. The objective

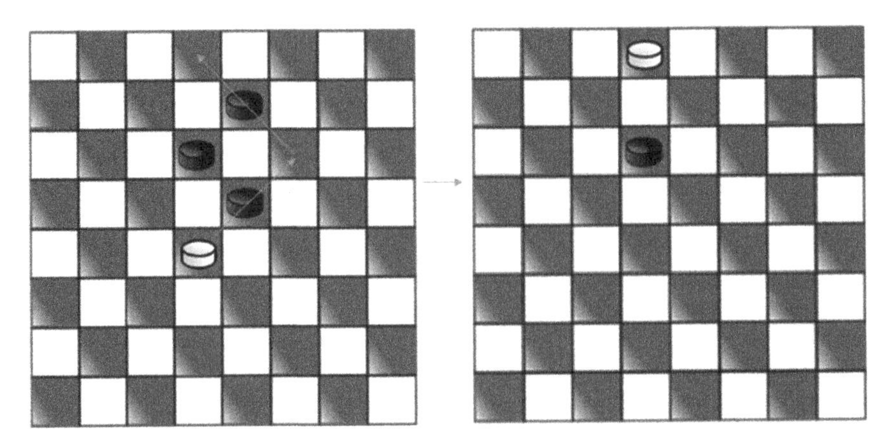

Fig. 1 Capture move in Checkers showing two black pieces being removed by a white piece using multiple hops

of the game is to capture as many discs of the opponent as possible, so the opponent has no valid moves left to make.

3.2 Description of XCS

Learning classifier system (LCS) [2–4] is a machine learning technique used to solve supervised, unsupervised and reinforcement problems [22]. It combines reinforcement learning, evolutionary computing and other heuristics to produce adaptive systems. Various types of LCS are present with extended classifier systems (XCS) [5] and zeroth-level classifier system (ZCS) being the most popular. ZCS was aimed at increasing the understandability and performance of LCS, while XCS was able to achieve optimal performance with maximally general classifiers. XCS consists of accuracy-based fitness allotment, application of GA to action set and an adoption of Q-learning credit assignment which distinguishes it from other classifier systems. Here, we provide a brief explanation of the working of XCS algorithm [18].

XCS consists of a population of classifiers each having an associated state, an action and expected pay-off [23]. The environment gives the current state of the system. This state is then matched against the conditions of the classifiers present in the population. Commonly used representation techniques of a classifier's condition include fixed length bit strings of ternary alphabets $(0, 1, \#)$ [22], real-valued alphabet [24], centre-based interval predicates [25] for real-valued inputs and other techniques [22]. Matching is done by comparing the current state with that of the condition and ignoring the don't cares (#). These matched classifiers form the match set $[M]$. If the number of distinct actions present in the match set is less than a predefined threshold value (θ_{MNA}) [23], then covering is performed. In covering, a new classifier is added into the population whose condition is a more generalized version of the current state (accomplished by introducing #'s with a probability $P_{\#}$). After $[M]$ is formed, an action is to be selected. For this, a prediction array [PA] is formed. Each element of [PA] represents a distinct action of $[M]$. [PA] tells about the pay-off which is expected if that action is taken.

Actions are generally selected in two ways: exploration (randomly selecting an action) and exploitation (selecting the action with optimal pay-off value). Random selection encourages exploration of the state space, whereas exploitation encourages taking the best possible action according to current knowledge. All those classifiers in $[M]$ whose actions match with the selected action constitute the action set $[A]$. The selected action is then applied to the environment, and the reward from the environment is given to all classifier in previous action set $[A_{-1}]$. The prediction value of the classifiers in $[A_{-1}]$ is inspired from the Q-learning mechanism as shown in Eq. 1.

$$Q(s_{t-1}, a_{t-1}) \leftarrow (1-\alpha)Q(s_{t-1}, a_{t-1}) + \alpha(r_t + \gamma(Q_{\max}(s_t, a))) \tag{1}$$

In XCS, $Q(s_{t-1}, a_{t-1})$ corresponds to the prediction value of the classifiers in $[A_{-1}]$, r_t is the reward signal received from the environment at time t, and $Q_{\max}(s_t, a)$ denotes maximum value in the prediction array formed at time t. Here, α and γ are the learning rate and discount factor, respectively [26].

Augmenting the reinforcement learning component of XCS is a GA process applied to the action set that introduces new classifiers. This is accomplished using genetic operators like mutation and crossover. Mutation, being a divergence operation, tends to alter the search space area being covered. Unlike crossover, it is applied on both the condition and action part. For condition, the bits are either interchanged (0 to 1 or 1 to 0) or a generality of a classifier is reduced (# to 1 or # to 0). For action, a random one is assigned. Crossover being a convergence operation is applied only on the condition part. A single point crossover divides the parent into two parts about a randomly selected point. The conditions to the right of this point are swapped producing new children. In two-point crossover, two cuts are made with the points being chosen randomly. The part between these cuts is swapped among parents, resulting in new children.

Roulette wheel selection and tournament selection are commonly used for selecting parent classifiers for breeding [27]. In the interest of maintaining generality, subsumption takes place from time to time reducing redundancy. A classifier subsumes another if it is more general and has the same condition and action. The subsuming classifier's numerosity is increased, and the subsumed classifier is removed from the population. Subsumption can be applied either in $[A]$ or with GA. However, for problems with smaller domain, it is usually discarded. Larger domains require subsumption to ensure that the size of the population does not explode. Figure 2 shows a flow chart of the XCS algorithm created in accordance with the above discussion.

4 Technical Framework

This section consists of necessary details regarding the implementation of a checkers playing program. It also includes a description of state and action representation employed during the remainder of this paper. Having described XCS, subsequent subsections deal with the intricacies that are specific to checkers.

4.1 Program Structure

The structure of a checkers game can be visualized as follows. Two players sit across from each other with a checkers board in between them. These are implemented as objects of a class 'Player' which employs function overriding to implement the individual strategies used by any instance. The game board is an object of another class called 'Board' which keeps track of the current state of the checkers board,

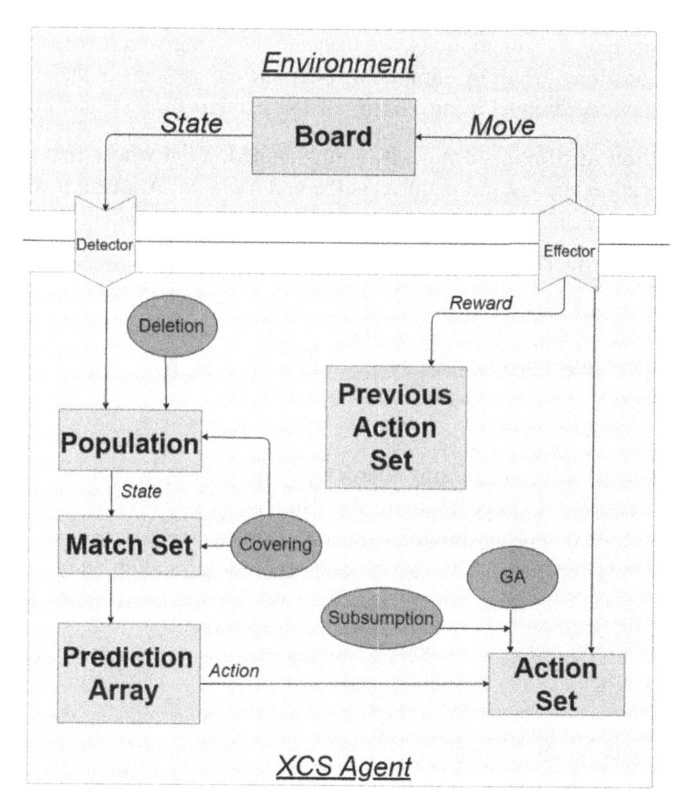

Fig. 2 Working cycle of the XCS algorithm showing the interaction between the XCS agent and the environment

possible moves for each player at any given time and provides methods to alter the state of board upon request by either of the agents.

4.2 *The Board*

To be able to manipulate the board upon interaction with an agent, the board must have a mechanism to send all available moves to the agent in question so that the agent may apply its control strategy and choose among one of these. Once this choice is made, there must be a method of altering the board using this choice. To accomplish this, we need to represent the current state of the checkers board as well as establish the format that a specific move on the board can take.

Representation of State. The board state is represented using a string of characters, where each character belongs to the set 0, 1, 2, 3, 4 where 0 is a blank cell, 1 and 3 are Agent 1's piece and king, respectively, and 2 and 4 are Agent 2's equivalent of the same. For the sake of simplicity, we truncate the string which would originally be $n2$ characters long to half by eliminating alternate empty spots of the board. Thus, our state is now represented using a string of 18 characters for a 6×6 board.

Representation of Moves. A move is, simply stated, a list whose first member is a starting location and second member is the end location. A location is also a list whose first member is a row number and second, a column number.

Interacting with the Board. Agents can call the board's *get_possible_next_moves()* function to return a list of possible moves and pass a move in the format specified above to a function *make_move(next)*. In this manner, we are ready to implement the game until one of the players loses the game (signalled by an empty list returned by get_possible_next_moves) or a predetermined maximum number of moves has been made.

Agents. Now that we have a working board, we come back to the problem of choosing a move from among a list of potential moves. This choice is the defining characteristic of an agent, and the way this choice is arrived at forms the description of a particular agent. Many types of agents are possible including random agent, Q-learning agents, minimax (alpha–beta) agents, etc. Next, we provide a brief description of the proposed 'XCS Agent'.

5 Proposed XCS Agent

The XCS agent proposed in this paper is a modification of the basic XCS first proposed in 1995 [5]. Many variants have since then been proposed which improve the robustness of the algorithm. Modifications to the algorithm are also necessary to combat some game-specific challenges as well as to attain improved performance. The structure of our algorithm builds upon the description of XCS given by Butz and Wilson [23] and applies it to the game of checkers. Here, we describe the algorithm in brief and later highlight the changes that have been made.

5.1 The Classifiers

A classifier can be visualized as an individual proposing a particular move to be made when our board is in a particular state. Thus, it is essentially a pair comprising a condition and a proposed action for this condition. Some of the other parameters associated with a classifier are its fitness, prediction error, numerosity and experience. The usage of these parameters is similar to the manner in which they have been used historically but will be clear in subsequent sections.

5.2 Parameter Settings

For setting parameter values, it is best to refer to current literature for optimum results. The commonly used parameter settings of the XCS agent were used as stated in [23] with a few changes. The discount factor (γ) was increased to 0.81 to increase the importance of previous moves in our multi-step game. Further, the maximum number of distinct actions was not limited to Θ_{MNA}, and covering was done till the match set comprised of all distinct valid actions in the current state. Initial prediction error and initial fitness were set to 0.01. The probability of generating hashes was set to 0.2. The crossover and mutation rates were set to 0.8 and 0.04, respectively. Finally, the size of the population was set to 1,000,000 as scaling of XCS to a non-Markov process such as checkers posed a significant challenge.

5.3 Reinforcement Component

The feedback given to an action is dependent on the implementation of the reinforcement component of the XCS agent and is therefore of prime importance. Very little information exists in the current literature regarding what comprises a good evaluation function for states in checkers. Here, we have used a simple and intuitive reward function that takes a weighted sum of the damage inflicted upon the opponent agent [28, 29]. This damage is aptly represented as a change in the number and location of opponent pieces before and after the move is performed. Apart from this immediate reward (Eq. 2) given to a selected action, we also maintained a set containing all action sets formed during the game. This set was targeted for the distribution of a delayed reward which acts as an ultimate indicator of the performance of these actions, judged based on the game's outcome. The value of this reward is a positive constant (or negative, in case of loss) solely determined using the outcome and independent of the specific state reached.

$$\begin{aligned}\text{immediateReward} \leftarrow\ & a(\Delta\,\text{normal pieces}) + b(\Delta\,\text{crowned pieces}) \\ & + c(\Delta\,\text{corner pieces}) + d(\Delta\,\text{threatened pieces}) \quad [a, b, c, d \epsilon R] \end{aligned} \tag{2}$$

5.4 Evolutionary Component

GA is performed at a certain rate determined by ΘGA. We firstly use a fitness proportionate selection method to establish parentage, and then, genetic operators such as dual-point crossover and mutation are applied with probabilities χ and μ, respectively. Instead of applying this component on the previous action set $[A_{-1}]$, we apply GA on a set of all winning classifiers till the current game. This modification is done to increase the probability of producing fitter children as their parents have contributed to the success of a previous game.

5.5 Modifications to XCS

Modifications and enhancements to the simple XCS algorithm explained in Sect. 4 were made to adapt it to the larger state space and branching factor of combinatorial games like checkers.

1. In the interest of exploring more actions for each state and tackling the large branching factor, Θ_{MNA} was removed to provide an evolutionary chance to all possible actions applicable at any state of the game.
2. A super action set ([SA]) has been proposed which is a chronological list of all action sets formed during the game. The outcome of the game is used to generate a delayed reward that is applied to this set in a discounted manner. We assume that the further an action set is from the end of the game, the less is its contribution to the outcome. This is accomplished by using γ with increasing powers.
3. Lastly, instead of applying GA to every action set after its formation (with a probability Θ_{GA}), GA is applied inside every individual action set in the super action set only if the game is won. This is done to enable reproduction within beneficial classifiers in the population.

```
modifiedXcsAlgorithm():
  [P] = recoverPopulation()
  [M] = []
  [M] = generateMatchSet([P], state)
  while valid actions available and not in [M] then
    generateCoveringClassifier([M])
  endwhile
  [PA] = generatePredictionArray([M])
  Action = selectAction([PA])
  [A] = generateActionSet(action, [M])
  [SA].append([A])
  applyAction(action)
  immediateReward = rewardFunction(action)
  updateActionSet(immediateReward, [A_-1])
  subsumption([A_-1])
  if isGameOver() then
    delayedReward = [-180,180]
    administerDelayedReward(delayedReward, [SA])
    if gameWon() then
        runGA([SA])
    endif
  endif
```

With the above changes in place, the working of the algorithm is presented as pseudocode above. The population is firstly recovered from the previous game, and the match set is initialized to empty. Next, it is populated using covering till there are no more valid actions left in the current state. Then, the prediction array is formed from the match set and is used to select an action. This action is used to form the action set which is appended to [SA]. The immediate reward obtained from the application of this action is then given to the previous action set([A_{-1}]) followed by subsumption. Lastly, after the game is over, a discounted delayed reward is administered to [SA], and if the game was won, GA is applied to each action set inside [SA].

6 Result Analysis

The development of the XCS agent was inspired by the technical framework [23] and written in Python 3 with modifications as discussed in Sect. 5.5. The size of the population was set to 1,000,000. The exploration probability was set to 0.5. The training was carried out for 110,000 games against a random agent on Linux platform and was run on a Microsoft Azure Virtual Machine (D4S_V3) with four CPUs.

As can be seen from Fig. 3, the winning percentage of the XCS agent increases from zero, reaching the 0.5 mark at the 569th game and then increases with a small slope. The winning percentage does not flatten out but continues to increase slowly as training progresses. The application of GA to winning action sets at the end of each game possibly contributes to the faster learning of the population.

The average fitness of the population increases as the number of games increases. The plot in Fig. 4, for the entire population, contains some noise which is attributed to the immense covering which takes place as the games proceed. This covering can be attributed to the algorithm exploring new search spaces. In the same figure, the

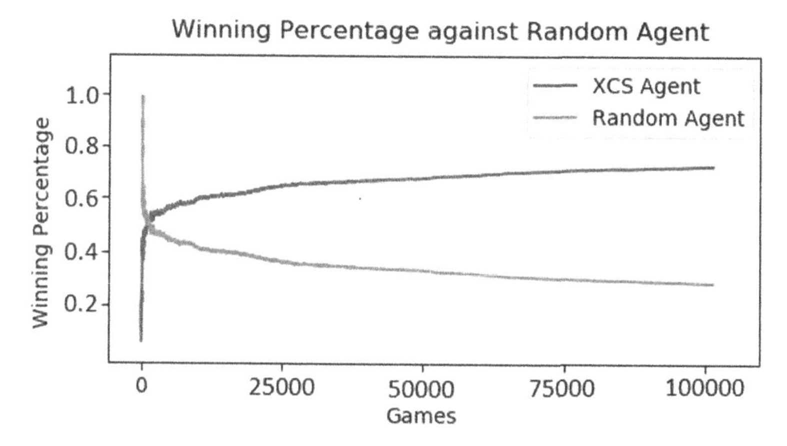

Fig. 3 Winning percentage of the XCS agent against random agent as a function of the number of games

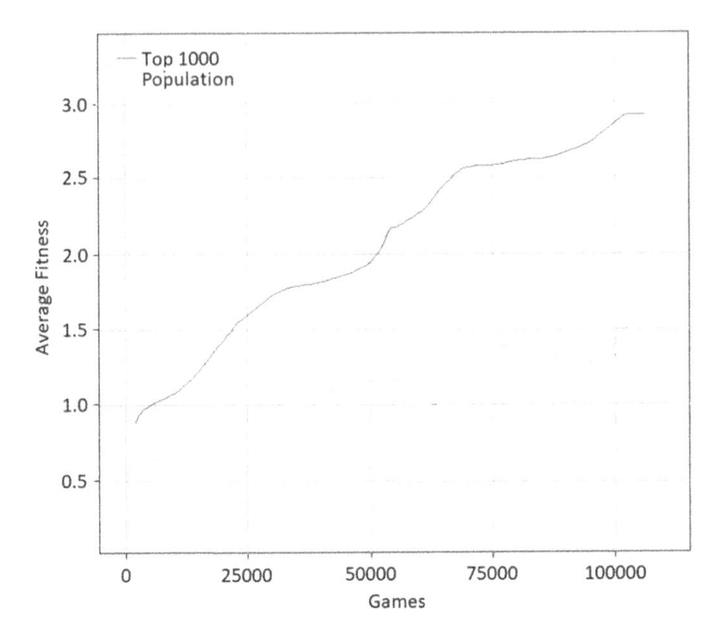

Fig. 4 Fitness of the population and the top 1000 classifiers as a function of the number of games

top 1000 fittest classifiers show a more stable trend with minute changes attributed to the change in the constitution of the top 1000 classifiers.

The average prediction error of the population firstly increases as depicted in Fig. 5. This is because the population is still growing. Hence, new classifiers are being created, thereby increasing the overall prediction error. As the number of games increases, there is a dip in the value. However, we see that the average prediction error of the entire population fails to follow a definitive trend [30]. This can be attributed to checkers being a multi-step problem, resulting in appearance of aliasing state problem [31]: the existence of more than one identical state, which requires the same action but returns with different rewards. The absence of a definitive trend may also represent the classifier's inclination towards winning the overall game (as accomplished using the addition of [SA] and delayed reward) rather than perform in a locally optimal manner and reducing the prediction error. Figure 5 also shows the average prediction error of the top 1000 fittest classifiers, where the drop in prediction error is more prominent.

The agent was tested against an alpha–beta agent having depths 2, 3, 4 and 5 for 100 games and performed better than the alpha–beta algorithm with depths 2, 3 and 4 as shown in Fig. 6. Although winning percentage observed against depth 5 was 42, it is impressive considering the short training period, large state space and branching factor. The changes that occurred in the population after every round of testing were retained for deeper alpha–beta search tests. Thus, being an online learning algorithm, the agent was able to adapt against its opponent in subsequent testing stages.

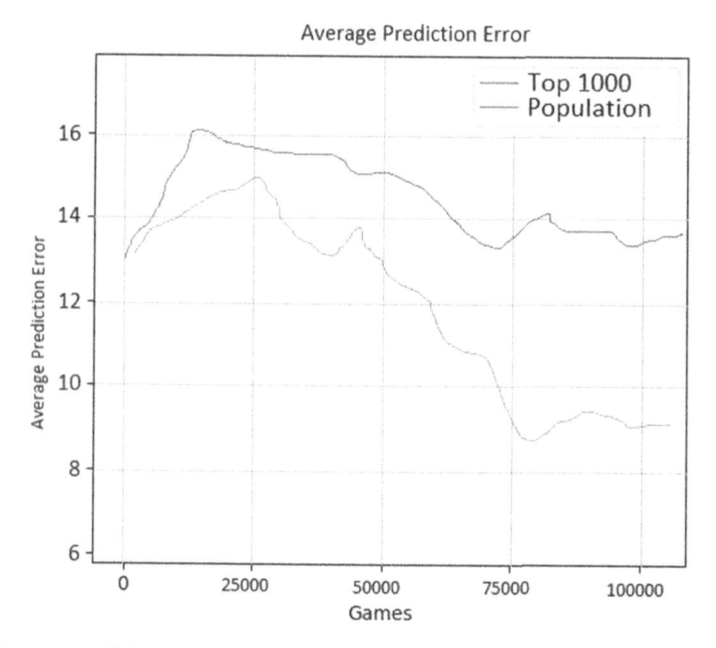

Fig. 5 Average prediction error of the population and the top 1000 classifiers as a function of the number of games

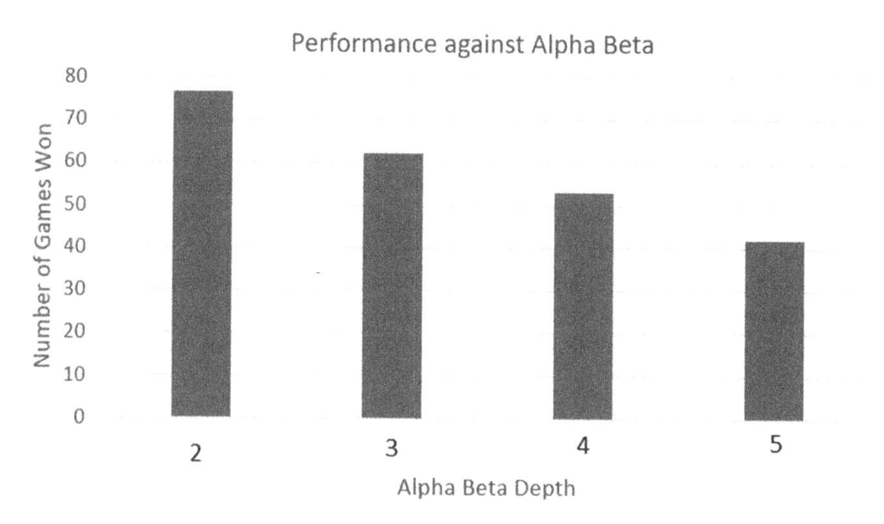

Fig. 6 Number of games won by the XCS agent (out of 100) against increasing depths of the alpha–beta algorithm

Table 1 Performance of the XCS agent against human agents of three different skill levels

Skill level	Won	Tied	Lost
Low	44	4	2
Average	39	7	4
High	24	16	10

The testing of the agent was also carried out against humans of three different skill levels. 50 games were played in each category. Low skill players were the players who were just introduced to checkers, whereas average players played checkers occasionally for recreational purposes. Highly skilled individuals were the ones who had notable experience in playing Checkers and were knowledgeable about winning strategies. The performance of the XCS agent was as depicted in Table 1.

7 Conclusion and Future Work

XCS uses combination of reinforcement learning and genetic algorithm to develop optimal rule sets. Through this paper, its usefulness in problems having high branching factor, high state space complexity and non-uniform behaviour has been investigated, by taking checkers as a case. The XCS agent was able to adapt to the game and gave promising results when put to the test against three agents: random, human and alpha–beta. The results of this preliminary work are encouraging and indicative of the applicability of a combination of GA and RL to complex combinatorial scenarios.

Possible avenues for future work stemming from this preliminary investigation include increasing the size of the board to 8×8, to tackle even larger state spaces. Since the 8×8 board of checkers has a large action space, therefore introducing generalization over the action space and replacing the prediction value parameter with a parameterized function could significantly improve the agent's performance and provide better results. This approach was used in XCSCA [32] and dramatically reduced the search space by focusing on the best action for each subproblem and thus finding the minimum set of rules necessary for solving the problem.

Training of the model for significantly longer periods of time with larger populations could help gain a more robust picture of the agent's performance and aid in the dismissing the possibility of local trends. An improvement in winning percentage can also be reasonably expected from a longer training session with random agent. Modifications such as addition of a neural network or probabilistic functions for reward administration, application of gradient-based updates [33] can be examined, and their results may suggest changes to XCS. The RL component of checkers is a relatively unexplored region of gameplaying, and usage of more complex and fruitful reward mechanisms is crucial to the agent's performance.

Additionally, XCS and other rule-based learning algorithms are applicable to a great variety of games which aim towards improving cognitive intelligence and reasoning in patients, especially infants, suffering from intractable disorders like autism [34]. Achieving better performance using such algorithms, which can infer reasoning from actions, can open new avenues of research into game-centric diagnostics and advance cognitive therapy.

Transferable XCS (tXCS) [35] has previously been used for both single step and multi-step benchmark problems which significantly increases the learning efficiency. The same can be used for the likes of checkers to reduce the training time and get early convergence.

References

1. Bull L (2004) Learning classifier systems: a brief introduction. In: Applications of learning classifier systems, pp 1–12
2. Snell F (1967) Progress in theoretical biology. Academic Press, New York
3. Holland J (2001) Adaptation in natural and artificial systems. MIT Press, Cambridge
4. Holland J, Holyoak K, Nisbett R, Thagard P, Smoliar S (1987) Induction: processes of inference, learning, and discovery. IEEE Expert 2:92–93
5. Wilson S (1995) Classifier fitness based on accuracy. Evol Comput 3:149–175
6. Silver D, Huang A, Maddison C, Guez A, Sifre L, van den Driessche G, Schrittwieser J, Antonoglou I, Panneershelvam V, Lanctot M, Dieleman S, Grewe D, Nham J, Kalchbrenner N, Sutskever I, Lillicrap T, Leach M, Kavukcuoglu K, Graepel T, Hassabis D (2016) Mastering the game of Go with deep neural networks and tree search. Nature 529:484–489
7. Silver D, Hubert T, Schrittwieser J, Antonoglou I, Lai M, Guez A, Lanctot M, Sifre L, Kumaran D, Graepel T, Lillicrap T, Simonyan K, Hassabis D (2018) A general reinforcement learning algorithm that masters chess, shogi, and Go through self-play. Science 362:1140–1144
8. Dubel ICL, Lefakis L (2006) Checkers reinforcement learning project: AI checkers player
9. Kwasnicka H, Spirydowicz A (2019) Checkers: TD (λ) learning applied for deterministic game
10. Tesauro G (1995) Temporal difference learning and TD-Gammon. Commun ACM 38:58–68
11. Eck N, Wezel M (2005) Reinforcement learning and its application to Othello
12. Van der Ree M, Wiering M (2013) Reinforcement learning in the game of Othello: learning against a fixed opponent and learning from self-play. In: 2013 IEEE symposium on adaptive dynamic programming and reinforcement learning (ADPRL)
13. Knuth D, Moore R (1975) An analysis of alpha-beta pruning. Artif Intell 6:293–326
14. Castillo C, Lurgi M, Martinez I (2003) Chimps: an evolutionary reinforcement learning approach for soccer agents. In: SMC'03 conference proceedings. 2003 IEEE international conference on systems, man and cybernetics. Conference theme—system security and assurance (Cat. No. 03CH37483)
15. Rudolph S, von Mammen S, Jungbluth J, Hähner J (2016) Design and evaluation of an extended learning classifier-based StarCraft micro AI. In: Applications of evolutionary computation, pp 669–681
16. Shafi K, Abbass H (2017) A survey of learning classifier systems in games [review article]. IEEE Comput Intell Mag 12:42–55
17. Schaeffer J, Culberson J, Treloar N, Knight B, Lu P, Szafron D (1992) A world championship caliber checkers program. Artif Intell 53:273–289
18. Schaeffer J, Burch N, Bjornsson Y, Kishimoto A, Muller M, Lake R, Lu P, Sutphen S (2007) Checkers is solved. Science 317:1518–1522
19. Browne W, Scott D (2005) An abstraction algorithm for genetics-based reinforcement learning. In: Proceedings of the 2005 conference on genetic and evolutionary computation—GECCO'05
20. Jain S, Verma S, Kumar S, Aggarwal S (2018) An evolutionary learning approach to play Othello using XCS. In: 2018 IEEE congress on evolutionary computation (CEC)

21. Clementis L (2013) Model driven classifier evaluation in rule-based system. In: Advances in intelligent systems and computing, pp 267–276
22. Urbanowicz R, Moore J (2009) Learning classifier systems: a complete introduction, review, and roadmap. J Artif Evol Appl 2009:1–25
23. Butz M, Wilson S (2001) An algorithmic description of XCS. In: Advances in learning classifier systems, pp 253–272
24. Wilson S (2000) Get real! XCS with continuous-valued inputs. In: Lecture notes in computer science, pp 209–219
25. Browne W (2004) The development of an industrial learning classifier system for data-mining in a steel hop strip mill. In: Applications of learning classifier systems, pp 223–259
26. Lanzi P (2002) Learning classifier systems from a reinforcement learning perspective. Soft Comput 6:162–170
27. Zhong J, Hu X, Zhang J, Gu M (2005) Comparison of performance between different selection strategies on simple genetic algorithms. In: International conference on computational intelligence for modelling, control and automation and international conference on intelligent agents, web technologies and internet commerce (CIMCA-IAWTIC'06)
28. Shuqin L, Weiming X, Xiaohua Y (2015) Study on the evaluation function parameters of the checkers game program on Weka platform. Int J New Technol Res 1(7)
29. Su Z, Li S, Jia Y, Zheng L, Fan S (2013) The realization of genetic algorithm in terms of checkers evaluation function. Appl Mech Mater 411–414:1979–1985
30. Tang K, Jarvis R (2005) Is XCS suitable for problems with temporal rewards? In: International conference on computational intelligence for modelling, control and automation and international conference on intelligent agents, web technologies and internet commerce (CIMCA-IAWTIC'06)
31. Barry A (2003) Limits in long path learning with XCS. In: Genetic and evolutionary computation—GECCO 2003, pp 1832–1843
32. Lanzi P, Loiacono D (2007) Classifier systems that compute action mappings. In: Proceedings of the 9th annual conference on genetic and evolutionary computation—GECCO'07
33. Butz M, Goldberg D, Lanzi P (2004) Gradient-based learning updates improve XCS performance in multistep problems. In: Genetic and evolutionary computation—GECCO 2004, pp 751–762
34. Hiniker A, Daniels J, Williamson H (2013) Go go games. In: Proceedings of the 12th international conference on interaction design and children—IDC'13
35. Li X, Yang G (2016) Transferable XCS. In: Proceedings of the 2016 on genetic and evolutionary computation conference—GECCO'16

Printed in the United States
by Baker & Taylor Publisher Services